D1739768

Hazards XVII

Process safety — fulfilling our responsibilities

Institution of Chemical Engineers, Rugby, UK

Hazards XVII
Process safety — fulfilling our responsibilities

Orders for this publication should be directed as follows:

Institution of Chemical Engineers,
Davis Building,
165–189 Railway Terrace, RUGBY,
Warwickshire CV21 3HQ, UK

Tel: +44 1788 578214
Fax: +44 1788 560833
Website: www.icheme.org/shop

Hazards XVII
Process safety — fulfilling our responsibilities

A three-day symposium organized by the Institution of Chemical Engineers (North West Branch) and held at UMIST, Manchester, UK, 25–27 March 2003.

This book contains the papers and posters presented at Hazards XVII. There is also an accompanying CD-ROM.

Contents listing for Hazards I–XVI and four other relevant IChemE conferences are available in a fully searchable form on the CD-ROM.

Organizing Committee

M.F. Pantony (Chairman)	Consultant
M.J. Adams	Symposium organiser/secretary
G.R. Astbury	Avecia Ltd
S.R. Beattie	Syngenta plc
D.C. Bull	Firebrand International Ltd
H.R. Cripps	Consultant
K. Dixon-Jackson	Ciba Speciality Chemicals plc
R.F. Evans	Health and Safety Executive
N. Gibson	Burgoyne Consultants Ltd
M. Hoyle	AstraZeneca
I. Kempsell	British Nuclear Fuels plc
T.A. Kletz	Consultant
G.A. Lunn	Health and Safety Laboratory
R.S. Mason	Consultant
M.C. McBride	Health and Safety Laboratory
I.F. McConvey	AstraZeneca
G.S. Melville	Burgoyne Consultants Ltd
N. Morton	Health and Safety Executive
M.L. Preston	Consultant
R.C. Santon	Health and Safety Executive
A.I. Thompson	Consultant

Corresponding Members of the Committee

R.L. Rogers	Inburex GmbH

INSTITUTION OF CHEMICAL ENGINEERS
SYMPOSIUM SERIES No. 149
ISBN 0 85295 459 X

Sponsors

This symposium is supported by the Health and Safety Executive
(HSE) and the Environment Agency and sponsored by:
ABB
ERM Risk

It is co-sponsored by:
Center for Chemical Process Safety (AIChE)
Society of Chemical Industry (SCI)
Chemical Industries Association (CIA)
The Royal Society of Chemistry (RSC)
European Process Safety Centre (EPSC)
Process Industries Safety Management (PRISM)
Society of Industrial Emergency Service Officers (SIESO)
Safety and Reliability Society
IChemE Subject Group for Safety and Loss Prevention (SLPSG)
IChemE Subject Group for Environmental Protection (EPSG)

Printed by Beshara Press, Tewkesbury, UK

Preface

Like as the waves make towards the pebbled shore
So do our minutes hasten to their end
William Shakespeare

'As an integral part of their competencies all engineers need to be equipped to appreciate, understand and implement the requirements of SHE management and practice to meet the working needs of industry and of their company (or other organization). Whilst the level of risk and degree of control is dependent on the industry sector concerned the basic principles do not change'[1].

Although the Engineering Institutions largely wrote the above guidance statement for use by academic undergraduate courses, the message conveyed could equally well be applied to graduates from any discipline of science, mathematics and technology. The message is equally applicable to employees at all levels in an organization – bottom to top. The new graduates of twenty-five years ago starting their careers in chemical engineering had relatively little principal statutory legislation to operate with. However, in the intervening years there has been a plethora of new principle legislation. Recently however, there have been moves both by politicians and regulators to consolidate this legislation. The advent of the European Union has also helped to harmonize SHE (Safety, Health and the Environment) legislation across countries. The influence of society has also been important in pressurizing the politicians and regulators into developing guidance that takes account of their concerns.

The plenary sessions are a microcosm of the overall conference which covers the wide range of SHE activities one has come to expect from a major hazards programme, for example: environmental protection, safety management, safe process design, storage, standards, risk factors and chemical reactions.

One major piece of legislation now impacting on industrial operations considers the impact of major accidents especially those previously considered to have relatively low risk factors. Therefore it is appropriate that the conference is

---

[1]*Incorporating Safety, Health and Environmental Risk Issues in Undergraduate Engineering Courses*, a document by the Inter-Institutional Group on Health and Safety of the Institutions of Chemical Engineers, Civil Engineers, Electrical Engineers, Mechanical Engineers and, the Hazards Forum

kicked off with two plenary papers covering the Control of Major Accident Hazards (COMAH).

Another plenary paper deals with societal concern head on giving practical guidance on how to semi-quantitatively measure societal impact using risk factors. There is a need to establish fundamentally what ALARP (As Low As Reasonably Practicable) statements mean and whether or not they are beneficial. Whilst amongst the professionals it may be possible to establish some consensus on the benefits of ALARP, if the consensus is not accepted by society then little will have been achieved. Perhaps societal acceptance is one of the last major and long lasting benefits to be achieved.

Two other very important contributions are on the effect of management failures and inherent safety. Good application of SHE principles and regulations require a top-down and bottom up approach by all the people and machinery concerned. The establishment of a Basis of Safety, Health and Environment (BoSHE) gives operators a unique opportunity to understand, simplify and represent information in a clear and useable form for their personnel and associated working partners. The issues associated with management failure (and the penalty of corporate manslaughter) are too broad to cover in this preface, however, it is incumbent on us to maintain good SHE life cycle chains where everyone involved in the chain plays their part. If the chain proves to be fragile at any section then this is where the focus needs to lie in the event of any major failure.

It is tempting to interpret the new legislation and claim that nothing has changed – the rules have simply been regularized. The main thrust of this temptation is that most professional organizations already largely meet the needs of the new legislation by conforming to the old legislation. To ensure beneficial progression with any new legislation, it is incumbent on SHE professionals to develop adaptive SHE systems and methods so that the hard lessons learned from corporate memory may be consolidated to help to avoid charges such as that of corporate manslaughter. If we are going to improve and *fulfil our responsibilities* then we need to monitor, adjust and improve our performance and that of our machines to the best standards available in the industry.

Dr Ian F McConvey
Chairman, North West Branch, IChemE

Contents

Keynote papers (plenary session 1)

Issues from COMAH (session 2)

Emergency planning and environmental protection (session 3)

Safety management (session 4)

Safe process design (session 5)

ix

Transport and storage (session 6)

Human factors and behaviours (session 7)

ATEX and other new standards (session 8)

Risk assessment and analysis (session 9)

Chemical reactions (session 10)

Issues from COMAH (continued) (session 11)

Risk assessment and analysis (continued) (session 12)

Safety management (continued) (session 13)

Issues from COMAH (plenary session 14)

Posters

xv

MAJOR ACCIDENT PREVENTION POLICY IN THE EUROPEAN UNION: THE MAJOR ACCIDENT HAZARDS BUREAU (MAHB) AND THE SEVESO II DIRECTIVE

Stuart Duffield
Major Accident Hazards Bureau,
European Commission, Joint Research Centre,
Institute for the Protection and Security of the Citizen,
Ispra, I-21020 (Va), Italy

The process industry is one of the major wealth producing activities of our modern day society; it accounts for 7% of global income and 9% of global trade. Its products are so diverse and widely used that our dependence on them is taken for granted and little consideration is given as to their origin. It is of paramount, strategic importance therefore, that the safety of this industry is assured. It is also of equal importance that the public has a rational perception of the risks posed by it to the environment and society at large. The awareness of this fact, together with the knowledge that the consequences of major accidents are no respecters of national boundaries, has resulted in a number of initiatives, and the formation of organisations aimed at maintaining and continually improving a "process safety culture". Primary among these initiatives has been the efforts of the European Commission in the formulation and implementation of the "Seveso Directives", (82/501/EEC and 96/82/EC). Closely coupled to this activity was the creation of the Major Accident Hazards Bureau (MAHB) located at the Commission's Joint Research Centre at Ispra in northern Italy. The bureau gives scientific and technical advice to the Commission and is responsible amongst other things for managing the technical working groups that have helped shape current and future legislation in the field of major accident prevention policy. This paper briefly describes the pertinent features of the Seveso II Directive, problems that have arisen in its implementation in the Member States and ongoing work to ensure its success. Then the principal elements of, and background to, the amendment to the Seveso II Directive are presented, and finally the functioning and achievements of the Major Accident Hazards Bureau are highlighted.

Process industry, Accident prevention policy, Seveso directives, Major accident hazards bureau

INTRODUCTION

The European Union is the world's leading producer of chemical products. The chemical industry supplies virtually all sectors of the economy and their products are so diverse and widely used that our dependence on them is taken for granted and little consideration is given as to their origin. It is also expected that the demand for chemicals will increase with the growth of the European economies including those of the Candidate countries.

The production and storage of chemical products is certainly not "risk free" and major accidents involving dangerous substances have occurred and will continue to occur world-wide in the process industry; the tragic recent accidents in Enschede and Toulouse are reminders of this fact. It is of paramount, strategic importance therefore, that the safety of

1

the process industry is assured. It is also of equal importance that the public has a rational perception of the risks posed to it and to the environment by this industry.

The awareness of this fact, together with the knowledge that the consequences of major accidents are no respecters of national boundaries, has resulted in a number of initiatives and the formation of organisations aimed at maintaining and continually improving a "process safety culture". In addition, the appreciation of the fact that a major accident in one sector of the industry gives no market advantage to a competitor if it leads to a general loss in confidence by the public in the industry, has recently led to a healthy openness and exchange of information regarding safety issues amongst the major industrial players.

Primary among these initiatives aimed at improving process safety has been the formulation by the European Commission of the "Seveso Directives", and closely coupled to this the setting up of the Major Accident Hazard Bureau (MAHB) located at the Commission's Joint Research Centre at Ispra in northern Italy. MAHB gives scientific and technical support to DG Environment, the directorate responsible for the legislation in this field, and operates the Commission's Major Accident Reporting System (MARS) database, the Seveso Plant Information Retrieval System (SPIRS), and the Community Document Centre on Industrial Risk (CDCIR). The Commission has also funded a considerable number of research activities, focused on industrial safety, in the Third, Fourth, Fifth and into the Sixth RTD Framework Programmes.

Other international organisations that are directly concerned with major accident hazards and emergency response include: the Council of Europe (COE), the International Civil Defence Organisation (ICDO), the International Labour Organisation (ILO), the International Programme on Chemical Safety (IPCS), the North Atlantic Treaty Organisation (NATO), the Organisation for Economic Co-operation and Development (OECD), the United Nation Economic Commission for Europe (UN/ECE), the United Nation Environment Programme (UNEP), and the World Health Organisation (WHO).

Major industrial initiatives have seen the creation of the European Process Safety Centre (EPSC), and include the work of the Loss Prevention Working Party of the European Federation of Chemical Engineering (EFCE), the European Chemical Industry Council (CEFIC), the American Institute of Chemical Engineer's Centre for Chemical Process Safety (CCPS), and the Design Institute for Emergency Relief Systems (DIERS) to name just a few. At the same time process safety has been included in the chemical engineering curriculum of many universities.

It is clear that accidents will continue to occur in the future, however there is the determination that through the diligent application of the Seveso II Directive their consequences can be minimised, and the risks posed to mankind and the environment reduced to a "tolerable" level.

HISTORICAL BACKGROUND

Over the last 30 years a number of major accidents has shaped European legislation in the field of prevention and control of major accidents occurring in the process industry.

A huge explosion in 1974 at the *Flixborough* plant in the United Kingdom resulted in 28 fatalities, personal injury both on and off-site, and the complete destruction of the industrial site. It also had a 'domino' effect on other industrial activity in the area, causing the loss of coolant at a nearby steel works which had the potential to cause a further serious accident.

2

In 1976 another explosion occurred at a chemical plant in *Seveso*, northern Italy where pesticides and herbicides were being manufactured. A dense vapour cloud containing tetrachlorodibenzoparadioxin (TCDD) was released from a chemical reactor, used for the production of trichlorofenol. Commonly known as dioxin, this was a poisonous and carcinogenic by-product of the uncontrolled exothermic reaction that was the cause of the accident. Although no immediate fatalities were reported, kilogram quantities of this substance, lethal to man even in microgram doses, were widely dispersed, resulting in the immediate contamination of some twenty five square kilometres of land and vegetation. More than 600 people were evacuated from their homes and as many as 2000 were treated for dioxin poisoning.

In 1984 the world's worst industrial accident occurred at the Union Carbide factory at *Bhopal*, India, where an erroneous introduction of water into a storage tank, containing 40 tonnes of methyl isocyanate, caused a runaway reaction and a subsequent release of the vessel contents into the atmosphere. The toxic cloud enveloped the near-by-populated areas and caused more than 2500 deaths and over 200,000 injuries. This disaster clearly identified the benefits of inherently safer approaches to chemical production, as the material released was a hazardous intermediate, the bulk storage of which was convenient but not essential. Similar arguments have been used to greatly reduce the amounts of chlorine stored on industrial sites. It is a sobering fact that a single accident such as that which occurred in Bhopal has resulted in the complete demise of the parent company, a once proud company which now no longer exists. Safety issues therefore can have a very important impact on corporate identity.

In 1986 an accident occurred at the Sandoz warehouse in *Basle*, Switzerland and highlighted the potential hazards caused to the environment by the process industry. Here fire-fighting water contaminated with mercury, organophosphate pesticides and other chemicals drained into the Rhine and caused massive pollution of the river through Germany, France and the Netherlands killing over half a million fish and contaminating drinking water.

In 2000 a tailing pond burst at a facility near the city of *Baia Mare*, Romania which was reprocessing old mining tailings and re-depositing the waste sludge into a new tailings pond. This led to about 100,000m^3 of waste water containing up to 120 tonnes of cyanide and heavy metals being released into the Lapus river, then travelling downstream into the Somes and Tisa rivers into Hungary before entering the Danube. It devastated large numbers of plant and wildlife species. Although nobody died or became seriously ill the impact might have been far more serious if the rivers had not been covered with ice for some 200 km downstream, or had the most severe floods for well over 100 years not occurred within weeks of the accident.

Also in 2000 a series of explosions at the company Fireworks S.E. that stored and assembled fireworks in the city of *Enschede* in the Netherlands caused the death of 22 person, 4 of which were fire-fighters, and injured almost 1,000 more. The incident inflicted extensive damage on a large area surrounding the factory, (within 200 m buildings were completed destroyed, within 750 m there was major structural damage). This area was mainly residential and close by was a large brewery with significant quantities of ammonia on site. Up until this point this accident was the worst that Europe had seen in terms of off-site consequences for over 50 years.

Then in 2001 the accident in *Toulouse* happened in which a huge explosion occurred in an ammonium nitrate production facility. The facility produced ammonium nitrate for

3

both the fertiliser and explosives industries, but the explosion occurred in a warehouse in which "off-spec" material was stored prior to its shipment for reprocessing. 30 people lost their lives, 8 outside the establishment including one pupil in a nearby school; 2500 people were injured some of whom very seriously; 10,000 homes were damaged, 600 completely destroyed; 2 school were wrecked and 70 were closed; one hospital was damaged and injuries and fatalities were sustained by the occupants of vehicles travelling on the highway running close to the factory. Close to the factory were two other facilities that had significant quantities of dangerous substances on site but fortunately no domino effect occurred.

The first piece of European legislation on the control of major accident hazards and the mitigation of their consequences was adopted in 1982[1] and was commonly known as the Seveso Directive after the accident that occurred in northern Italy. This directive was amended twice in 1987 and 1988 and provisions were laid down for a review of its scope following the experience gained with its implementation. The Member States, in accompanying resolutions concerning the fourth (1987) and the fifth (1993) Action Programmes on the Environment, had called for a review of the Directive in which there was a general desire to widen the scope of the Directive by including land-use planning policy, risk assessment and accident management. A resolution from the European Parliament also called for a review, and following these actions a new 'Seveso II Directive' was presented to Council and European Parliament by the Commission in 1994.

On 9 December 1996 the Seveso II Directive[2] was adopted by the Council, and following its publication in the Official Journal of the European Communities, entered into force on 3 February 1997. Member States then had up to two years to bring into force the national laws, regulations and administrative provisions to comply with the Directive. From 3 February 1999 the obligations of the Directive became mandatory for industry as well as for the public authorities responsible for the implementation and enforcement of the Directive. The fact that the original Seveso Directive was not amended but was replaced by a completely new Directive indicated that important changes had been made and new concepts had been introduced into the Seveso II Directive.

THE SEVESO II DIRECTIVE
The principal aim of the Directive is two-fold:

Firstly, the Directive aims at the *prevention* of major accident hazards involving hazardous substances.
Secondly, as accident will inevitably occur, the Directive aims at the *limitation of the consequences* of such accidents not only for mankind but also for the environment.

Both aims should be followed with a view to ensuring high levels of protection to mankind and the environment throughout the Member States in a consistent and effective manner.

The Directive is a new type of "goal orientated" legislation and places more emphasis on the socio-technical aspects of the control policy and attempts to bring more transparency and openness into the process by allowing for public consultation and by strengthening the role of MAHB as an information exchange system. For a comprehensive description of the background, contents and requirements of the Seveso II Directive the reader is referred to the excellent article by Wettig et al.[3], but the important new features appearing in the Directive are described below:

- The scope of the Directive is both broadened and simplified. There is no list of industrial installations, therefore there is no need to define the term *industrial activity*. In its place the concept of an industrial *establishment* is introduced, characterised by the presence of dangerous substances. There is a short list of named substances (Annex I, Part 1), and a more systematic list containing *generic categories* (Annex I, Part 2) such as toxic, explosive or flammable. Concerning the definition of these generic categories reference is made to other Directives relating to the classification, packaging and labelling of dangerous substances, preparations and pesticides. Depending on the quantities of dangerous substances present on site an establishment will be deemed either upper-tier or lower-tier. It is assumed that the risk of a major accident hazard arising increases with the quantities of substances present at the establishment, and consequently the Directive imposes more obligations on upper-tier than lower-tier establishments.
- The socio-technical aspects in an establishment are expected to be strongly affected by the obligation placed on the operator to provide a Major Accident Prevention Policy (MAPP), and for an upper-tier establishment, a Safety Report implemented by means of Safety Management Systems (SMS). These provisions are a major addition to the Directive and have been introduced after the discovery that most of the major accidents notified to the Commission over the years under the Major Accident Reporting System (MARS) had root causes in deficiencies in the management process[4,5,6].
- Similarly, the obligation of a land-use policy as set out in Article 12 will have important socio-technical consequences, especially for those countries where such an obligation was not part of national legislation prior to the Directive. In particular, planning policies are required to establish and maintain appropriate distances between establishments and residential and other areas, and when this is not possible additional technical measures need to be taken. The general public, which until now had the right to be informed on existing risks and on how to react in case of an accident, will have a much more active role in the overall process of risk management.
- The Competent Authorities are obliged to identify establishments, or groups of establishments, where the danger of an accident and its possible consequences may be increased because of the location and the proximity of the establishments and the dangerous substances present: the so called domino effect.
- The provisions for emergency planning and public information are reinforced, since the Safety Report becomes a public document and the public must be consulted in the preparation of emergency plans. The emergency plans also have now to be tested regularly.
- The Competent Authorities are obliged to organise a system of inspections under Article 18, comprising a systematic appraisal or one on-site visit every year: this is to be followed by a report.
- The Directive is concerned with dangers posed to the environment from hazardous installations following the inclusion of a generic category related to substances harmful to the aquatic environment.

- Finally, a concise and unequivocal definition of what constitutes a 'major accident', based on quantitative threshold criteria, is included in the Directive. It is expected that this will result in an overall reduction of the criteria for notification to MARS and lead to an increase in the homogeneity of data at the European level.

It can be seen that the Directive establishes a broader perspective as far as risk management of the storage and processing of hazardous substances is concerned. This is a perspective that should increase the awareness of the general public on risk control issues and help provide a rational basis on which the risks posed by the industry to the environment and society at large can be judged.

The Member States now have had over 3 years of experience with the Directive and the experiences of both the competent authorities and industry have been aired at international conferences[7,8,9] jointly organised by MAHB and the host Member State. Generally, concern is being expressed on the application of Article 12 on land-use planning; on the implications associated with the self-classification of chemicals by industry under the new European chemical policy and on public access to information in the aftermath of the terrorist act of September 12th 2002.

THE PROPOSED AMENDMENT TO THE SEVESO II DIRECTIVE
In the light of recent industrial accidents in Baia Mare/Romania, Enschede/Netherlands and Toulouse/France, and following studies on "single shot carcinogens[10] and substances dangerous for the environment[11] carried out by the Commission on request of the Council, it was necessary for the scope of the Directive to be broadened in order to fully achieve the goals of the Directive.

Since the Seveso II Directive was adopted by Council in December 1996, there has been a continual process of consultation with interested parties, concerning both the implementation of the Directive and possible improvements and amendments. This consultation has involved several international conferences and seminars, regular meetings with the Committee of Competent Authorities established under the Seveso II Directive, and the establishment and running of Technical Working Groups addressing various aspects of the Directive. An important aspect has also been the involvement of the general public and the first draft of the proposed amendment was published on the Internet by the Commission's Environment Directorate General in April 2001. This draft was also sent to Member States, EEA States, accession countries, environmental NGOs, European and national industrial federations and associations and some international organisations, with a request to distribute it further as appropriate.

Comments were invited and a public consultation meeting was held on 31 May 2001 in Brussels. Following the consultation meeting, written comments received were also published on the Internet with the permission of their authors. All comments received during the consultation process have been carefully evaluated and have been acted upon where the Commission felt it was appropriate. Discussions are still ongoing between the European Council and Parliament but the pertinent features of the amendment are given below:

- An amendment to unequivocally include mineral processing of ores and, in particular, tailings ponds or dams used in connection with such mineral processing of ores.

Industrial operators performing these activities will thus be obliged to put into effect Safety Management Systems, including a detailed risk assessment on the basis of possible accident scenarios. However, it is important to note that any mining activity would only be covered by the Directive if dangerous substances as defined in the Directive are involved and if they are present in quantities beyond the threshold levels set out in the Directive.

• Following the accident at SE Fireworks in Enschede and others involving fireworks, MAHB organised seminars first with the competent authorities at their meeting in Marseille, and then with a much larger representation at Ispra[12] to discuss what happened and to explore avenues in which the safety of this industry could be improved. The seminars identified the classification of fireworks as a key issue. The national regulations of most Member States for the storage of explosives are based on the classification system operated under the United Nations European Agreement Concerning the International Carriage of Dangerous Goods by Road (UN/ADR). The Seveso II Directive distinguishes between explosives on the basis of risk phrases according to the EC legislation on the classification, labelling and packaging of dangerous substances; the risk phrases refer only to the explosive sensitivity of substances (ease of ignition). The UN/ADR system on the other hand distinguishes between explosives on the basis of the hazard they represent – which may range from a mass explosion hazard for those explosives in Hazard Division 1.1 to a fire hazard for those in Hazard Division 1.4. This distinction is of particular relevance to pyrotechnics, and it is not reflected in the present Directive, which treats pyrotechnics as a single group. The amendment therefore proposes:

to alter the definitions for explosives in the Directive, making use of the UN/ADR classification scheme transposed into European law by Council Directive 94/55/EC, and to restrict the higher thresholds to explosives with HD 1.4 classification, principally consumer fireworks.

• During discussions on the Seveso II Directive in Council, questions were raised concerning the scientific and practical basis for the list of named carcinogens and the qualifying quantity assigned to them, and also concerning the qualifying quantities for substances dangerous for the environment. The Council, when adopting the Directive, therefore requested the Commission to carry out studies on these issues and to submit reports accompanied, if appropriate, by proposals for amending the Directive. In response the Commission established two Technical Working Groups that delivered their final reports in April 2000[10,11]. The reports propose extending the list of carcinogens with appropriate qualifying quantities, and significantly lowering the qualifying quantities assigned to substances dangerous for the environment.

Taking into consideration the conclusions of the Technical Working Groups and public consultation the proposed amendment in relation to "single shot carcinogens is to add to the list of 'carcinogens' already contained in Annex I, Part 1 the following substances:

1,2-Dibromo-3-chloropropane; 1,2-Dimethylhydrazine; Dimethyl sulphate; Diethyl sulphate; Benzotrichloride; Hydrazine; 1,2-Dibromoethane. Furthermore, the qualifying quantities for the whole group of 'carcinogens' should be increased from 1 kg to 0.5 tonnes for the application of Articles 6/7 and to 2 tonnes for the application of Article 9. Finally, a minimum concentration limit of 5% was introduced for all the carcinogens when in solution.

Regarding substances dangerous for the environment the amendment proposes:

to lower the qualifying quantities for substances dangerous for the environment as defined in Annex I, Part 2, item 9(i) (risk phrase R50, which should also be defined to include R50/53) from 200 to 100 tonnes for the application of Articles 6/7 and from 500 to 200 tonnes for the application of Article 9;

to lower the qualifying quantities for substances dangerous for the environment as defined in Annex I, Part 2, item 9(ii) (risk phrase R51/53) from 500 to 200 tonnes for the application of Articles 6/7 and from 2,000 to 500 tonnes for the application of Article 9; and

to amend the named substance *"automotive petrol and other petroleum spirits"* in Annex I, Part 1 in order to include gasolines, naphthas, kerosenes, and gasoils;while lowering the qualifying quantities from 5,000 to 2,500 tonnes for the application of Articles 6/7 and from 50,000 to 25,000 tonnes for the application of Article 9.

The proposed modifications to the qualifying quantities for substances dangerous for the environment would also serve to achieve consistency between the provisions of the Directive and those of the UN/ECE Convention on the Transboundary Effects of Industrial Accidents.

Furthermore, it is proposed that there should be a separation under the summation rule of toxic and eco-toxic hazards, reflecting the fact that these hazards are dissimilar, and concern that grouping toxic and eco-toxic hazards together in the summation rule would lead to an unreasonable increase in the number of establishments covered, particularly in the light of the ongoing process of classification of substances.

• In light of the major accident that occurred in Toulouse involving ammonium nitrate and taking into consideration the views of the European Parliament and the conclusions of an international seminar[13] organised by MAHB with invitees from Member State competent authorities, industry and academia it is proposed that the definition of the named substance "ammonium nitrate" in Annex I be changed to reflect more accurate the hazards associated with the different physical and chemical forms of the substance. Four entries on ammonium nitrate are proposed together with the qualifying quantities for Articles 6/7 and for Article 9, i.e.

Ammonium nitrate (5,000/10,000): fertilisers capable of self-sustaining decomposition;

Ammonium nitrate (1,250/5,000): fertiliser grade fulfilling the requirements of Directive 80/876/EEC (as amended and updated);

Ammonium nitrate (350/2,500): technical grade;
Ammonium nitrate (10/50): "off-specs" material and fertilisers not fulfilling the detonation test.

- The opportunity was taken at this time also to rectify some slight inaccuracies or ambiguities in the Directive. These were introduced as editorial amendments and do not change the scope or the application of the Directive.

THE MAJOR ACCIDENT HAZARDS BUREAU

The Major Accident Hazards Bureau (MAHB) was established with the specific remit to give independent scientific and technical support to the Commission and ensure the successful implementation and monitoring of EU policy on the control of major hazards and the prevention and mitigation of major accidents. Furthermore, in order to fulfil its information exchange obligations towards the Member States, the Commission established the Major Accident Reporting System (MARS), the Seveso Plant Information Retrieval System (SPIRS) and the Community Documentation Centre on Industrial Risks (CDCIR) which are managed and maintained by MAHB. The main customers for the services offered by MAHB, apart from the Commission, include all the actors in the legislative, regulatory and management activities concerned with process plant safety, (e.g., national and local authorities, industry, research organisations, safety consultants, trade unions). In order to facilitate an efficient and effective information exchange, MAHB has developed and maintains a dedicated web site (http://mahbsrv.jrc.it) from which information, guidance documents, scientific publications and software can be accessed and downloaded.

The principal tasks of the Bureau are briefly described below:
The maintenance and periodic updating of the Major Accident Reporting System database (MARS). This task involves the collection, in a consistent manner, of data on major industrial accidents involving dangerous substances from Member States; the analysis and processing of such data and the distribution of all non-confidential data and analysis results to the Member States. MARS is an up-to-date distributed information exchange and analysis tool, which is made up from two connected parts: one for each local unit (i.e. for each Member State Competent Authority) with which accident data is reported, and one central part for the Commission. Both parts can serve as data logging systems and, on different levels of complexity, as data analysis tools. The central database allows complex pattern analysis to be made, identifying and analysing the succession of disruptive factors leading to an accident. On this basis, 'lessons learnt' can be formulated for industry and the regulatory bodies to assist in further accident prevention. Examples of such analyses can be found in[14,15,16,17]. The trend in major accidents since reporting started is shown in figure 1.
The development and management of the Seveso Plants Information Retrieval System (SPIRS)[18]. This information system aims at containing all Seveso sites throughout the EU and Candidate Countries and figure 2 shows the current status. Also included in this system is a largely user-defined risk ranking tool so that comparative risk assessments can be performed.

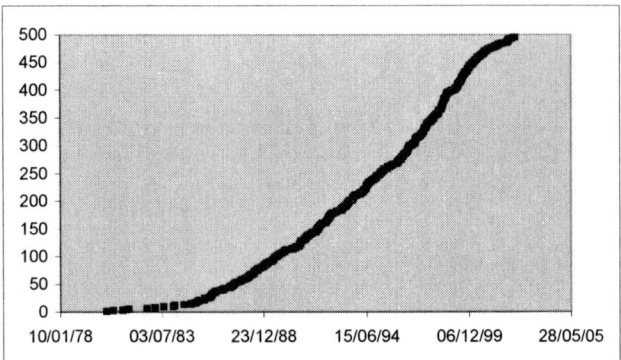

Figure 1. The trend in major accidents as reported to MAHB under the seveso directives

The management of the Community Documentation Centre on Industrial Risk (CDCIR). This task involves the acquisition, storage and assessment of relevant documents (guidelines, regulations, codes of good practice, accident case histories, risk studies, scientific literature etc.) related to major accident hazard control. It is perhaps unique in the fact that much of the contents are made up of 'grey literature' not readily available from alternative sources. In this context MAHB has developed an interactive feature on the MAHB web site containing summaries of all the material in the CDCIR, and provides an on-line search facility.

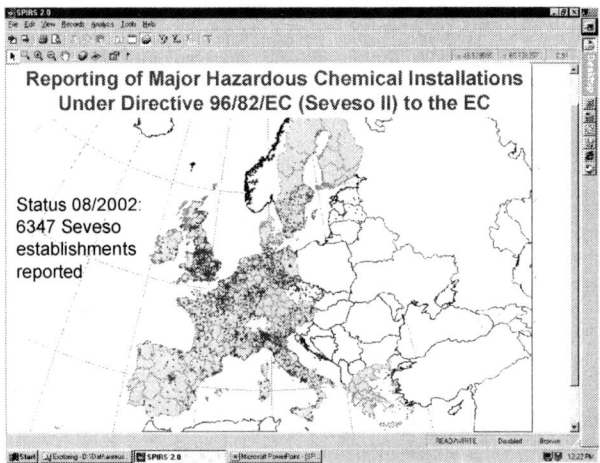

Figure 2. The current status of seveso establishment in the EU and candidate countries

The Directive has a substantial scientific and technical content, some of which is not fully defined within the legislation, and is a result of the fact that the state of scientific knowledge or of industrial practice is still evolving. This was recognised at an early stage by the Commission and the Member States and has led to various Technical Working Groups (TWGs) being established with the objective of producing non-prescriptive guidance on specific aspects of the implementation of the Directive. MAHB provides scientific, technical and administrative support to the functioning of the various Technical Working Groups, which are made up with experts drawn from Member State Authorities and representatives from industrial groupings; either those of the process industry in general or those specifically concerned with the safety or environmental issues. Guidance documents have been produced on *the Safety Report*[19], on *Information to the Public*[20], on *Safety Management Systems*[21], on *Land-use Planning*[22], on *Harmonised Criteria under Art. 9(6), i.e. derogations*[23], and on *Inspection Systems*[24], and can be viewed and downloaded from the MAHB web site: (http://mahbsrv.jrc.it).

MAHB also organises, on a regular basis, various technical meetings and international seminars covering topics connected with control of major hazards and the prevention and mitigation of major accidents; see for example[7,8,9,25,26,27,28].

To enhance the effectiveness of the support the Bureau provides, it is also involved in a number of research activities. These include: the development of a European harmonised risk assessment methodology to evaluate the risk level of industrial establishments, the assessment, through a benchmark exercise, of the uncertainties related to the different risk assessment methodologies commonly used in land-use planning activities, the development of standardised acute exposure levels of toxic substances for use in emergency planning and the development of novel techniques to detect the onset of thermal runaway events in batch-type reactors and the safe disposal of reaction products.

CONCLUSIONS

The safety of the European process industry is of strategic importance; similarly, it is equally important that the general public has a rational perception of the risks posed by it to the environment and society at large. It is our strong belief that the Seveso II Directive, an up-to-date piece of goal oriented legislation, provides the mechanism through which this can be assured by bringing transparency to the risk related decision making processes throughout the European Union. The Commission's Major Accident Hazards Bureau supports this initiative by operating and maintaining the Major Accident Reporting System, the Community Documentation Centre on Industrial Risk and by running the various Technical Working Groups set up to develop guidance for a coherent implementation of the Directive. The Bureau also fulfils a strategic role in providing an efficient information exchange system, for the authorities, industry, research community and the general public through the operation of its dedicated web-site.

The Directive has only recently been transposed into national law; the main challenge therefore will consist in ensuring that it is implemented in a consistent and effective manner throughout the Community.

11

ACKNOWLEDGEMENTS
The author is pleased to acknowledge the help and support of all his colleagues at the Major Accident Hazards Bureau and those in DG ENV, the Directorate General responsible for the Directive.

REFERENCES

1. Council Directive 82/501/EEC of 24th June 1982 on the Major Accident Hazards of certain industrial activities, Official Journal of the European Communities, 1982 *(OJ No: L 230)*.

2. Council Directive 96/82/EC of 9th December 1996 on the control of major-accident hazards involving dangerous substance, Official Journal of the European Communities, 1997 *(OJ No: L 10)*.

3. Wettig, J., Porter, S., Kirchsteiger, C., Major industrial accidents regulations in the European Union, *Journal of Loss Prevention in the Process Industries*, 12, (1999), 19–28.

4. Drogaris, G., Learning from Major Accidents Involving Dangerous Substances, *Safety Science* 16, 1993.

5. Rasmussen, K., The Experience with the Major Accident Reporting system from 1984 to 1993, EUR 16341 EN, 1996.

6. Kirchsteiger, C., Kawka, N., Characteristics of accidents notified to MARS, Proceedings of the EC-EPSC seminar on 'Lessons Learned from Accidents', Linz, EUR 17733 EN, 1998.

7. Papadakis, G.A., Seveso 2000, Risk Management in the European Union of 2000: The Challenge of Implementing Council Directive 96/82/EC, "Seveso II" European Conference, Athens, Nov 1999, EUR 19664 EN.

8. European Conference: Implementing Seveso II, London, Nov 2000 (see http://mahbsrv.jrc.it)

9. Wood, M., et al. Major Industrial Hazards in Land-use Planning, 2002 Seveso II Conference, Lille, March 2002, (see http://mahbsrv.jrc.it)

10. Christou, M.D., (ed), Carcinogens in the context of Council Directive 96/82/EC - Final Report of Technical Working Group 8, EUR 19650 EN, 2000.

11. Christou, M.D., (ed), Substances Dangerous for the Environment in the context of Council Directive 96/82/EC - Final Report of Technical Working Group 7, EUR 19651 EN, 2000.

12. Wood, M., et al. Pyrotechnic and Explosive Substances in the Context of the Seveso II Directive, Marseille Sept 2000, Ispra March 2001 (see http://mahbsrv.jrc.it)

13. Wood, M., et al. Proceedings of the Expert Workshop on Ammonium Nitrate, Ispra, Jan 2002 (see http://mahbsrv.jrc.it)

14. Papadakis, G.A., Amendola, A., Learning from Experience The Major Accident Reporting System (MARS) in the European Union, Proceedings of PSA 96 Conference, Crete, June 24–27 1996.

15. Drogaris, G., Major Accident Reporting System – Lessons Learned from Accidents Notified, EUR 15060 EN, 1993.

16. C. Kirchsteiger, C., Rushton, A., Kawka, N., Contribution of human errors to accidents notified to MARS, Proceedings of the EC-EPSC seminar on 'Lessons Learned from Accidents', Linz, EUR 17733 EN, 1998.
17. Porter, S., Kirchsteiger, C., The Challenge of Learning Lessons from Accidents, Proceedings of the EC-EPSC seminar on 'Lessons Learned from Accidents', Linz, EUR 17733 EN, 1998.
18. Kirchsteiger, C., Availability of Community Level Information on Industrial Risks in the EU. *Trans IChemE*, Vol 78, Part B, 2000.
19. G.A. Papadakis, G.A., Amendola, A., Guidance on the preparation of a safety report to meet the requirements of Council Directive 96/82/EC (Seveso II), EUR 17690 EN, 1997.
20. B. De Marchi, B., Funtowicz, S., General Guidelines for Content and Information to the Public (Directive 82/501/EEC-Annex VII) EUR 15946 EN, 1994.
21. N. Mitchison, N., Porter, S., Guidelines on a Major Accident Prevention Policy and Safety Management System, as required by Council Directive 96/82/EC (Seveso II), EUR 18123 EN, 1998.
22. Christou, M.D., Porter, S., Guidance on Land-Use Planning as required by Council Directive 96/82/EC, EUR 19695 EN, 1999.
23. Wettig, J., Mitchison, N., Explanations and Guidelines for the application of the Dispensation Rule of Article 9, paragraph 6 of Council Directive 96/82/EC on the control of major-accident hazards involving dangerous substances, EUR 18124 EN 1998.
24. Papadakis, G.A., Porter, S., Guidance on Inspections as required by Article 18 of Council Directive 96/82/EC (Seveso II), EUR 18692 EN, 1999.
25. Cacciabue, P.C., I. Gerbaulet, I., Mitchison, N., Proceedings of the EC seminar on Safety Management Systems in the Process Industry, Ravello, EUR 15743 EN, 1994.
26. Mitchison, N., Smeder, M., Proceedings of the EC seminar on Safety and Runaway Reactions, Frankfurt, EUR 17723 EN, 1997.
27. Kirchsteiger, C., Proceedings of the EC seminar on Lessons Learnt from Accidents, Linz, EUR 17733 EN, 1998.
28. Mitchison, N., Garcés de Marcilla Val, A., Smeder, B., Proceedings of the EC seminar on Accident Scenarios and Emergency Response, Toledo, EUR 18733 EN

GAUGING SOCIETAL CONCERNS

David Mansfield
AEA Technology

Risk managers are increasingly recognising that decisions regarding risks and their tolerability need to take into account both objective assessments of the risks, and society's expectations and concerns. Subjective factors such as choice, dread, mistrust and moral outrage play a part in determining what risks people are prepared to tolerate. The need to address societal concern is one of the key themes in the HSE's revised tolerability of risk publication 'Reducing Risks Protecting People'. This raises the challenge, 'how to measure societal concern'? This paper describes the development of a model to facilitate the analysis of societal concern for a wide range of familiar or emerging risk issues. The model provides a systematic, structured and transparent means to assess societal concern and provides both numerical and visual output to characterise the strength and nature of the concern. The model can provide valuable insight to support risk decisions, augmenting objective technical assessments of the risks.

KEYWORDS: Societal Concern, Societal Risks, Risk Perception, Risk Communication, Risk Decision Support, Modelling

OVERVIEW

Managing risks in today's complex socio-technical society requires an understanding of the risks presented and society's views, expectations, and concerns regarding the risks, their management and their tolerability. The recently published update of the HSE tolerability of risk publication 'Reducing Risk, Protecting People'[1] reflects this growing need to consider societal concern in risk decision-making. With this in mind, the HSE Risk Policy Unit commissioned researchers from AEA Technology and its subsidiary, Risk Solutions, to investigate if a predictive/analytical model of societal concern could be constructed. The result of the project was a semi-quantitative spreadsheet based model capable of characterising an issue in terms of its key societal concern factors and providing an overall indication of the potential level of concern, benchmarked against a number of established anchor points. The model allows risk managers to better understand a given issue in terms of its risk concern characteristics, to develop more effective risk communication and risk management strategies, to set priorities for further work, and to gain a view as to the overall potential level of societal concern compared with familiar risk issues.

The model is viewed as a key step forward in this area. Previous work in this field has concentrated on investigations in to the key psychological and sociological factors influencing concern and the technical evaluation of societal risk, with few attempts to relate the key factors or technical risks to the potential overall level of societal concern. The model developed in this project not only bridges this gap in the research by establishing relationships between the key factors and the overall level of concern, but also provides a practical tool which can be used to evaluate a wide variety of established or emerging health, safety and environmental risk issues. The model can be used to assess the concern associated with current or emerging issues through the use of focus groups. It can also be used to design surveys to elicit societal views on key issues. Furthermore, the insight it provides can be particularly useful in selecting appropriate risk management, control and communication strategies and prioritising research or risk reduction activities.

Full details of the model, its development and benchmarking are to be published in the HSE Contract Research Report series, available from the HSE.

SOCIETAL CONCERN – A DEFINITION
If an assessment is to be made of societal concern, the first step is to be clear as to what this is. Various terms are used in the risk literature eg individual risk, societal risk, risk perception and societal concern. Individual and societal risk primarily relate to technical risk estimates from an individual or group viewpoint. Risk perception is some measure of how an individual or body characterises a risk and the resulting estimate of the risk based on this viewpoint. Concern however arises from a more fundamental and emotive assessment of the characteristics of the risk. For this study, we have taken societal concern to be some collective subjective measure of individuals concern within society (as a whole or within some specified sub set as represented, e.g. by a focus group). The definition we have adopted is:

'Society's views, fears and expectations about a hazard or risk issue'

LITERATURE REVIEW
A review of work in the field of societal risk assessment and risk perception was conducted to identify suitable factors, methods, models and data to assist with the development of the model.

SOCIETAL RISKS
The literature search revealed a large body of work on the technical assessment and use of societal risk, including the use of F-N curves. An initial idea was to adapt the F-N curve approach to address societal concern. But it soon became apparent that this did not lend itself to some of the more subjective factors found to be important in generating concern. Instead, the model includes a number of parameters to characterise the societal risk profile.

CONCERN FACTORS
There has also been considerable research into the various factors that are important in public risk perception and concern. A number of factors appear time and time again in the studies.

The various factors highlighted by the studies were analysed to extract the key parameters for societal concern. The result was that 26 individual parameters were incorporated into the model. Where appropriate, some of these parameters have a further sub division to distinguish between people (ie health and safety, 'hs') risks and environmental ('e') risks. Not all the parameters have the same degree of influence on the model. The logic structure used to combine the parameters determines their overall influence. It is known that some of the factors incorporated do have a limited influence on the overall level of concern. The model also includes the facility to set individual weightings to each parameter if required. However, it was decided to keep these explicitly within the model to main a degree of completeness. A list of the parameters included in the model is given below.

Table 1. Typical concern parameters

Parameter / Reference	Understanding of hazard/uncertainty	Dread	Scale of Consequences	Vulnerability of potential victims eg children	Equity of distribution of risk and benefits	Preventability of harm in the future/controlability	Environmental impact and value of this	Global vs regional issues/factors	Accident history	Trust in those managing the risk and enforcing compliance	Natural vs man made hazard
HSC SASD 2001[2]	x	x	x	x	x						
Ives and Footitt 1996[3]	x	x	x	x	x	x	x	x			
Mendeloff and Kaplan 1989[4]				x			x				
Powell 1996[5]	x	x	x	x	x	x	x		x	x	
Chapman 1997[6]	x	x	x		x	x			x	x	x
Slovic et al 1980[7]	x	x	x		x	x		x			
The Royal Society 1992[8]	x	x	x	x	x	x					x
HSE 1989[9]	x	x	x		x	x	x		x	x	

1. Lack of reasonable choice over risk exposure
2. Lack of direct risk experience and knowledge
3. No source of information readily available to public/those at risk
4. (hs) Inequity - those at risk not those who benefit from activity - People health & safety aspects, 4 (e) As above - Environmental aspects
5. (hs) Vulnerability - those affected are from vulnerable group - People health & safety aspects, 5 (e) As above - Environmental aspects
6. Impact distribution - impacts will be concentrated in time and location, or many affected in any one event,
7. (hs) Perceived number exposed - proportion of population exposed or affected or feel/ viewed as affected - People health & safety aspects, 7 (e) As above - Environmental aspects i.e. environment or ecosystems exposed
8. Perceived harm potential - level of harm that could result from exposure
9. Perceived chance of harm from exposure - chance of harm occurring to an individual given anticipated level of exposure
10. Harm potential or chance of harm not known - uncertain
11. (hs) Nature of risks: 'spectacular dread' - risks of global catastrophe or threat to significant sections of society - People health & safety aspects, 11 (e) As above - Environmental aspects i.e. threat to global environment or significant ecosystems
12. (hs) Nature of risks: 'insidious dread' - threat to future generations/society - People health & safety aspects, 12 (e) As above - Environmental aspects i.e. insidious risk to species, ecosystems

13. Nature of harm/death evokes dread - phobia
14. Observability and delay - Effects of exposure difficult to observe - may be delayed or difficult to detect
15. Novelty - Hazard and risks relatively new compared to time in which effects may become apparent
16. Untreatable - harm not treatable, remediation limited if any
17. Not Reversible - source of harm cannot be removed by stopping activity, or long lag time for harm even if activity stopped
18. Scientific advice available to public on risks and their control is not clear or consistent
19. Scientific advice is subject of disagreement between experts/controversy
20. Past history of poor scientific advice from organisations involved or similar organisations/situations
21. Risks difficult, time consuming or expensive to control
22. Not in the interest of the organisation responsible for managing the risks to control the risks properly
23. Those responsible not strongly regulated with effective enforcement
24. Past history of specific or similar organisation/industry not managing risks properly
25. Public at risk or at large have no clear and effective means to interact with or apply pressure for improvement
26. Serious adverse reaction if high expectation of duty of care or trust placed in others not met

The model also includes an input switch to specify if 'management by others' is relevant or not. For example, some activities such as DIY or cycling are largely under the individual's control, and there is no element of the management of these activities by some third party. Initiating this switch deactivates those parts of the model representing the influence of poor hazard or risk management by, or a lack of trust in, external organisations responsible for managing a hazard or risk.

SOCIETAL RISK ESTIMATES AND CONCERN
One view may be that the level of public concern arises from the level of perceived risk. This hypothesis would explain differences between expert risk views and societal concerns about risk as simply a difference between the experts and public's assessment of the risk. This idea was tested by Slovic et al.[7]; they asked survey respondents to estimate the annual fatality rate for a range of activities and then compared these with expert assessments. The results showed that although the public's assessment of risk tended to underestimate high risk and over estimate low risks, the relative risk ranking of the activities were in-line with that derive from the expert assessments. However when Slovic et al. compared the estimates of fatality rate to the concern scores, they found there was only a low to moderate level of correlation. They concluded:

"Thus we can reject the idea that laypeople wanted to equate risk with annual fatalities, but were inaccurate in doing so. Apparently, laypeople incorporate other considerations besides annual fatalities into their concept of risk."

In their paper, Slovic et al. go on to explore some of the "other considerations" driving what we have defined as "societal concern", and their work has influenced our choice of concern parameters in the construction of our model.

Subjective societal concern factors appear to be based on a number of aspects of the risk characteristics, not just some assessment (however flawed) of the objective risk levels. As such, they cannot simply be dismissed, or 'corrected' by technical risk education. Instead it is essential to understand which parameters influence society's attitudes to risk and its tolerability, if risk criteria and decisions involving risk are to command widespread public support.

PERSONALITY TRAITS

The literature search also revealed a number of studies investigating how different personality traits, gender and age affected concerns about risk[10,11]. These concluded that although these factors do affect risk perception, the effects are secondary to factors such as choice, dread and trust. Some factors such as gender and age also exhibit different biases depending on the nature of the hazard. Given the relative simplicity of the model being developed, the complexity of incorporating personality traits coupled with the smaller effect of these, and the sparsity of data to anchor them, these secondary factors were not included within the current model. However, it would be possible to introduce general personality traits to the model at a later stage by adjusting the weightings within the model to reflect the different viewpoints. Incorporating variations of perception depending on the hazard type would be more complex and require an understanding of what hazard characteristics were influencing these biases for each personality type.

MODELS AND RELATIONSHIPS

Although the literature search unearthed numerous work on the various factors and parameters influencing societal concern, we found little data, correlations or frameworks that enabled the degree of influence of these parameters to be assessed, or which allowed an overall correlation or model for concern to be generated. The work by Slovic et al [7], provided the most comprehensive and qualitative assessment of risks and concerns. A decision was taken to try to build a model from first principles, which could then be benchmarked against findings reported in the literature, including the work by Slovic et al.

THE MODEL

The model has been developed using the key parameters highlighted by the literature as being important in generating societal concern. A structure has been developed to show how these various parameters relate to each other. This 'framework' has been based on the overall decision processes suggested in various studies augmented with logical reasoning. It adopts the following premise; that for societal concern to be generated:

- The hazard or risk issue must have some aspect of concern associated with it, eg due to dread, inequity, a lack of trust, or uncertainty

- Individuals in society must feel they do not have the ability to make an informed choice to avoid or control the risks, and
- Sufficient individuals in society must feel that they, their peers, or their values for society and the environment are at risk (ie the risks to society are significant)

This overall structure provides the 'top level' of the model (see Figure 1). This reasoning is then continued further to show how the 26 individual parameters relate to these three 'top level' conditions. The resulting structure is shown in Figure 2, and in many ways adopts a similar convention to that of a Fault Tree, using the equivalent of AND and OR gates to show how the various factors combine. However, it is important to recognise that this is not a Fault Tree. The factors are not events in a probability domain, and Boolean algebra does not apply.

Various options were considered to score/apply weightings to the various parameters feeding into the model. One option considered was to use pseudo-metrics such as newspaper column inches, risk comparison results, etc as a means to score the parameters. However it was difficult to find suitable metrics and data for many of the parameters. Also, some of the metrics would only be available for current issues and would not be suitable for assessing emerging issues. In the end a simple 1 to 7 semi-qualitative scoring scheme was devised since this enabled both current and emerging issues to be assessed by the model. The model includes detail guidance on the scoring of each parameter, using simple word descriptions to describe the parameter and indicate the type of characteristics relating to a score of 1 (low), 4 (median) and 7 (high). The scoring system is such that the higher the concern, the higher the score for all parameters.

Because the model is dealing with subjective scores, it was recognised that conventional algebra might not be applicable to the logic tree. A wide number of mathematical alternatives for propagating the scores were investigated for the different gate types. These included Boolean type approaches and methods based on both linear and logarithmic score propagation. The final selection was based on the method that provided the best fit to the desired characteristic of the particular logic gate. This method also maintains the 1-7 scoring system throughout the model, with the model output being an overall score based on 1 to 7 (with 7 being the highest level of concern). Separate scores are given for the 'health and safety - hs', 'environmental - e' and combined 'hs and e' elements of the model. A detailed discussion of the scoring methods and their evaluation can be found in the HSE Contract Research Report for the study.

The overall 'societal concern' score for a given issue, as estimated by the model, has limited application in its own right. The nature of the model and its subjective parameters means that delineation based on small differences between the scores for different issues or risks should not be attempted. However the model can provide a means to compare the scores of a given issue or issues of interest with those from some established benchmark issues. This allows issues to be broadly categorised in terms of their potential societal concern ie are they of low, moderate or high concern. The model also provides more detailed information on the nature of the concern, and this can provide very useful insight in to the issue and its management. This is discussed further below.

During the development of the model, it was recognised that some of the high level 'factors' in the model would be particularly useful in characterising the nature of the societal concern. Six concern factors were selected: lack of individual choice, the level of societal risk, dread, inequity, lack of trust in the information on the hazard/risk (lack of understanding) and a lack of trust in those responsible for managing the hazard/risks (where applicable). These factors are shown in Figures 2 and 3. The scores for these six factors calculated by the model provide a key part of the model output, providing an 'anatomy of concern' which can be used to gain insight into the why concern is being, or may be, generated and to help target effective risk management and communication activities. The model output includes a spider diagram to provide a simple but effective visual means to characterise the issue against these six factors. Examples of the model 'Spider diagram' output are shown in Figures 5 and 6.

Spider diagrams can have very different shapes for different issues. Compare Figures 5 and 6 – these are two spider diagrams that have underlying values which combine to give an almost identical top score.

It can be seen, for example, that Issue 1 is more driven by dread than Issue 2. Issue 1 also exhibits a high degree of trust in knowledge ("No trust in knowledge" is low) compared to Issue 2. Identifying these underlying characteristics can lead to a better understanding of the differences between issues and the key concern drivers.

MODEL TESTING AND BENCHMARKING

The model has been tested and benchmarked by a variety of techniques throughout its development. The lack of other models for societal concern, and the lack of hard data to calibrate the model, means that direct comparisons and validation cannot be used. Instead, the model has been benchmarked against some of the Slovic et al[7] study results, and has been tested using a range of scenarios (case studies) with the results benchmarked against concern rankings from various study groups.

The early tests were carried out internally by members of the project team, or in-conjunction with the project steering group members. These were then augmented by a series of study group tests carried out by members of Risk Solutions who had not been involved in the project. As the model neared completion, a one-day workshop was organised by the HSE where a number of case studies were run using the model using three groups of HSE personnel drawn from a wide variety of backgrounds and areas of expertise from major hazards to railways, nuclear and occupational health and safety. The findings from these various cases studies were then compared and contrasted against each other, and against an initial 'intuitive' ranking of the issues in terms of their potential for societal concern, to assess the models behaviour and results.

The following case studies were assessed at the HSE workshop.

Group 1	Group 2	Group 3
• Gas safety – pipelines	• Major LPG storage	• Railways – collisions
• Fairgrounds	• Nuclear power station	• Railways – overcrowding
• Stress	• Hospital acquired infection	• Contained use of GMOs
• Asbestos	• Adventure activities	

The overall feedback from the workshop and case studies was very positive. The general view of the participants was that the factors included in the model, and the high level factors drawn out by the spider diagram analyses, did seem to cover all the relevant aspects and provided appropriate information to inform the decision making process. Participants felt that the model provided a useful mechanism to analyse issues of potential societal concern. It was recognised that caution needs to be applied when making judgements based simply on the 'top score'. The real value in the model was the process of discussing, analysing and scoring the various factors feeding into the model and gaining an understanding of how these might influence the level of societal concern. The findings of the workshop were used to refine some of the model user guidance and score propagation methods.

Since the project finished in May 2002, the model has also been applied to workshop case studies sessions at a public policy seminar and at a presentation/workshop to the Health and Safety Commission. Topics have included some of those listed above and other topical issues such as mobile phone safety, genetically modified food, carbon monoxide poisoning and railway trespass/vandalism.

USING THE MODEL

The model provides a process by which an issue can be systematically discussed and evaluated in terms of its potential for generating societal concern. The model does not identify the concerns associated with the issue per se (in much the same way that the HAZOP method itself does not tell you what the hazards are – it is the study team who do this, prompted by the method), rather it provides a structured means to elicit the characteristics and conditions that could generate concern. The model is best used in a study team or focus group session. The group discuss each parameter in turn and then score this based on their overall view. If there is some divergence in the views for a particular parameter, then the model can be run using these different values to test the sensitivity of the results to the difference. The model then calculates the output scores and high level parameter scores from the input values provided by the study group. Depending on what is required, the group could be drawn from members or representatives of society at large or some specific interest group or sector. The model could also be used within an organisation, using an internal focus group, to look at emerging issues. In all these cases it is important not to bias the group, either by selecting those of a predisposed view, or by providing information or phrasing the issue in a way to influence the outcome. The information provided should normally be limited to factual information defining the issue and the effects or impact being considered.

The model provides a means to assess specific issues at some given point in time, and also to monitor how attitudes may be changing with time, for example as further information or events come to light. Potential applications include:

- As a complement to technical risk assessments in situations where societal concerns can influence policy and management actions
- Qualitative assessment of the strength and focus of societal concern for current issues, and the extent to which regulatory frameworks or management/operational controls address these

- Exploring new or emerging issues to inform strategic reviews and set priorities
- Focusing/tailoring risk communication
- Analysis to assist in the design of surveys and question sets to elicit views on risk issues
- Assessing the effects on societal concern of additional information or various communication and remedial action strategies
- A mechanism for visible, demonstrable engagement with society at large
- A means to build trust/confidence
- A means to raise awareness/level of debate
- A means to identify and consult with trusted independents
- Assessment of a range of views across interested parties and to assess the influence of experts or focussed information in focus groups

An indication of how the model output could be used to select appropriate management and communication strategies is shown in Figure 4.

The model is likely to be a useful tool for organisations responsible for aspects of public safety or environmental protection, and organisations whose commercial success relies heavily on the public's perceptions of its products, services, and socially responsible behaviour.

SOCIETAL CONCERN AND DECISION MAKING

Although the model provides a useful means to assess societal concern, it should be recognised that this is only one of many factors that need to be considered in any decision making process.

Some of the factors that may need to be considered in any overall decision are presented in Figure 7. Of particular note is the emphasis and importance placed on technical estimates of the risks vs. the level of societal concern. In some cases both the technical estimates of risks and the levels of concerns will be high (or low) and the results support each other. In other situations the results may be opposing. This may present an ethical dilemma when allocating resources, or devising strategies or action plans to minimise risk (ie reduce harm) **or** to address wider concerns that may be present in society (ie address views and perceptions, taking people seriously). In essence, is it better to try to reduce risks that are known to be significant but which do not generate societal concern (a familiar risk for example, road accidents) or to address issues of high societal concern which might actually present very small risk?

In cases where the risks are well understood, it may be better to act on the sound technical data and make efforts to communicate this such that the societal concerns are noted but to emphasise the facts and decision basis. Caution is needed if the cause for societal concern is a distrust of the technical risk estimates or a sense of inequity between those being put at risk compared to those taking the benefit. In these cases there is a need to understand and address any uncertainties in the risk estimates or inequity and act accordingly. Getting the balance right may be critical in terms of minimising risks and winning and maintaining public confidence and trust.

NEXT STEPS

A number of areas for possible future development and testing have been identified during the development of the model.

* Further testing and validation of the model.
  - The ideas currently being considered include a detailed peer review of the model by academic experts in the field of public risk perception and its measurement, and testing of the model on a wider range of benchmark issues.
* Improvements to the model scoring and guidance to users.
  - The current guidance to users is suitable for those familiar with hazard and risk concepts, but further clarification and rewording may be required if the model were to be used by lay people alone (ie without a facilitator familiar with the tool and parameter definitions).
  - Further investigation of group selection, training and elicitation techniques and how these can influence the model scoring and output, leading to improved guidance on this topic for model users is also being considered.
* Refinements to the model itself.
  Ideas currently identified for model development and refinement include:
  - Incorporating 'personality traits' within the model.
  - Improving the anchoring of weightings by carrying out specifically designed surveys of societal concerns.
  - More explicit modelling/ analysis of media amplification and the perceptions of pressure groups within the model, and investigation into the use of the model to indicate the presence of factors that could trigger media amplification or pressure group interest.
  - Investigation of improved weightings and mathematical mechanisms to combine the value of 'health and safety' and 'environmental' aspects of risk within the model.
* Applying the results of the model to risk management and communication.
  - Further research to provide advice on the extent to which societal concerns should be used as the basis for decision-making (vs technical risk estimates, etc).
  - Developing guidance on how the model results can be used to help decide on an appropriate/optimal risk communication / management strategy.

CONCLUSIONS

A multi-parameter model has been developed that can assess and analyse the degree of societal concern based on the characteristics of the issues concerned. The model draws on previous research to identify the key parameters for societal concern, but these have been rationalised to provide a more definitive set of parameters for modelling. The model also incorporates a structured representation of the relationships between the parameters, and uses this together with specially developed logic gates, to allow semi-quantitative scores to be propagated through the model. The inclusion of parameter relationships and logic represents a significant advance in the modelling of societal concern; previous work has been largely limited to developing statistical correlations between parameters.

It is considered that the model can provide a useful and practical tool to assist decision makers in assessing the potential for societal concern in a given situation or context. The model can provide both an overall indication of the level of potential societal concern and also provide an insight into the underlying factors that could generate that concern: helping decision makers to make better informed decisions and define better strategies for risk communication and management.

REFERENCE

1.   HSE, Reducing Risks, Protecting People, HSE 12/01, HSE Books, 2001, ISBN 0 7176 2151 0
2.   RAPU, Prioritisation of Regulatory Activity Directed at Issues of Societal Concern, A paper by the Risk Assessment Policy Unit (SASD), HSC, 2001
3.   Ives DP and Footitt AJ, Risk Ranking, UEA Norwich, HSE RSU Ref: 3444/R71.011 UEA, August, 1996
4.   Mendeloff JM and Kaplan RM, Are large differences in lifesaving costs justified? A psychometric study of the relative value placed on preventing deaths, Risk Analysis 9(3): 349-363, 1989
5.   Powell D An Introduction to Risk Communication and the Perception of Risk University of Guelph 1996
6.   Chapman S, Not in our back yard, University of Sydney Australian and New Zealand, Journal of Public Health, 21:614-620, 1997
7.   Slovic P, Fischoff B, Lichtenstein, S Facts and fears: understanding perceived risk. In: Societal risk assessment: how safe is safe enough (edited by RC Schwing and WA Albers), New York: Plenum Press, 1980
8.   BMA, Living with risk, The British Medical Association Guide, John Wiley & Sons, 1987, ISBN 0 471 91598 X
9.   The Royal Society, Risk Assessment: A Study Group Report, The Royal Society, London, 1983, ISBN 0 85403 208 8
10. Barnett J and Breakwell, GM Risk Perception and Experience: Hazard Personality Profiles and Individual Differences, Risk Analysis, Vol. 21, No. 1, 2001
11. Bouyer M et al, Personality Correlates of Risk Perception, Risk Analysis, Vol. 21, No. 3, 2001

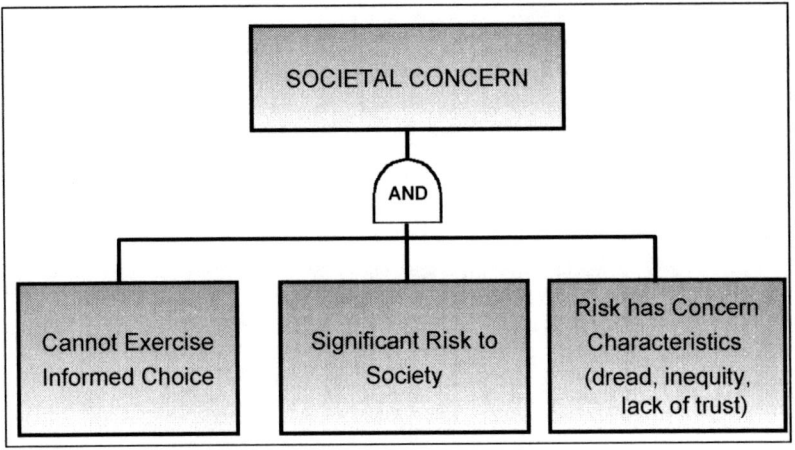

Figure 1. Model tree top level structure

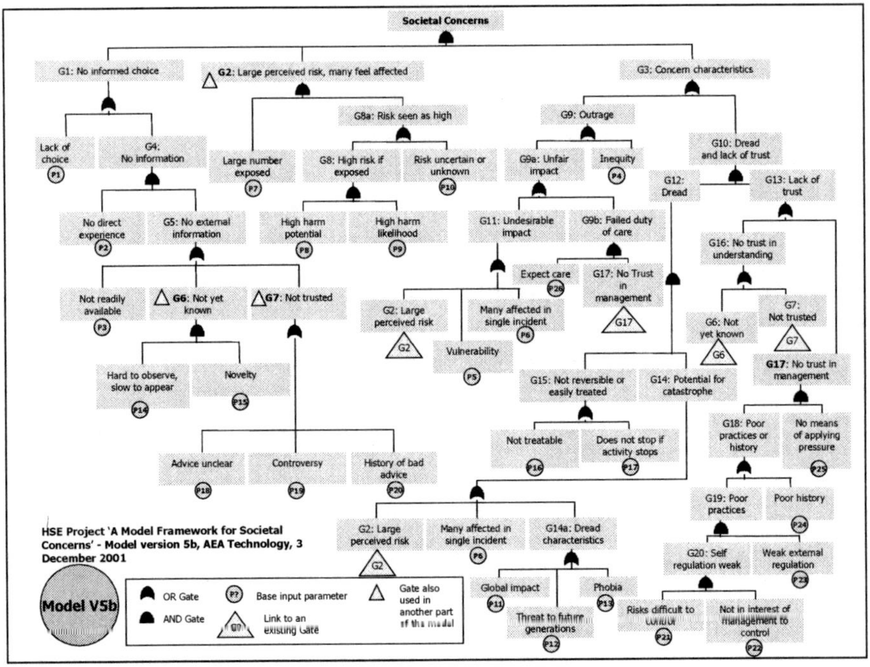

Figure 2. Societal concern model – overall structure

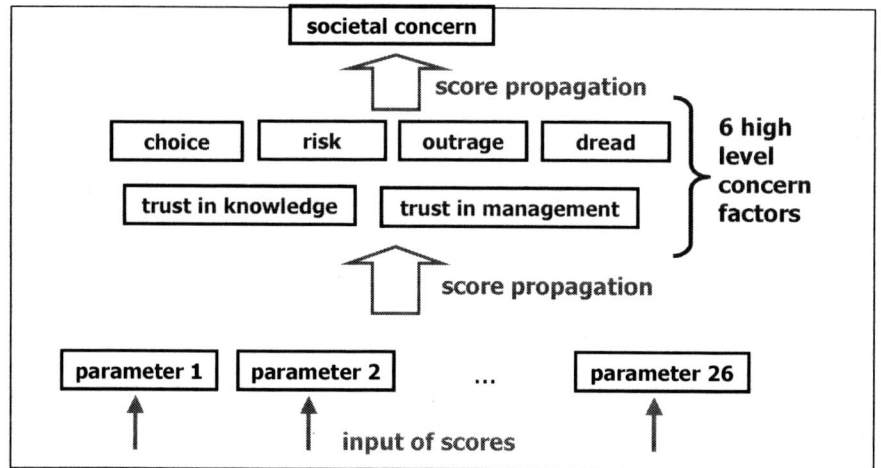

Figure 3. Model score propagation and 'high level' concern factors

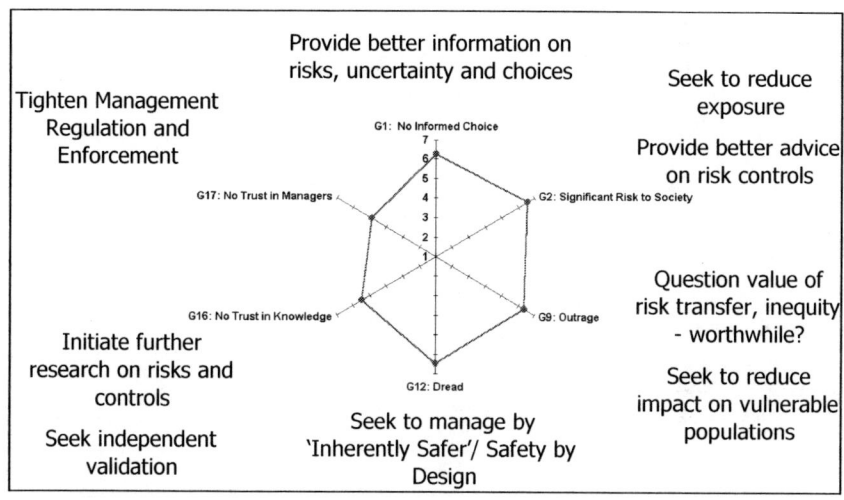

Figure 4. Selecting effective risk management and communication strategies

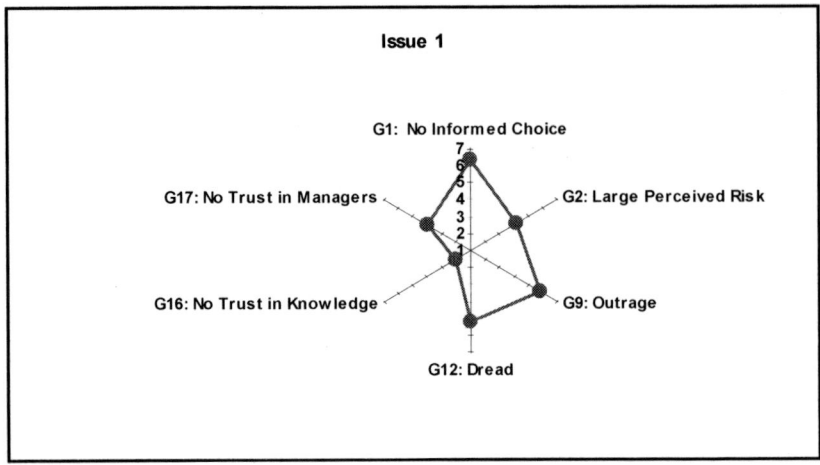

Figure 5. Model output spider diagram for an issue with a top score of 5.22

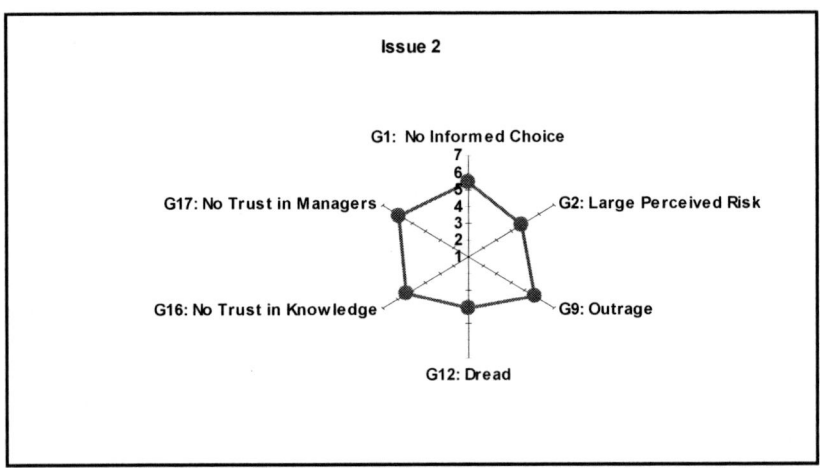

Figure 6. Model output spider diagram for an issue with a top score of 5.20

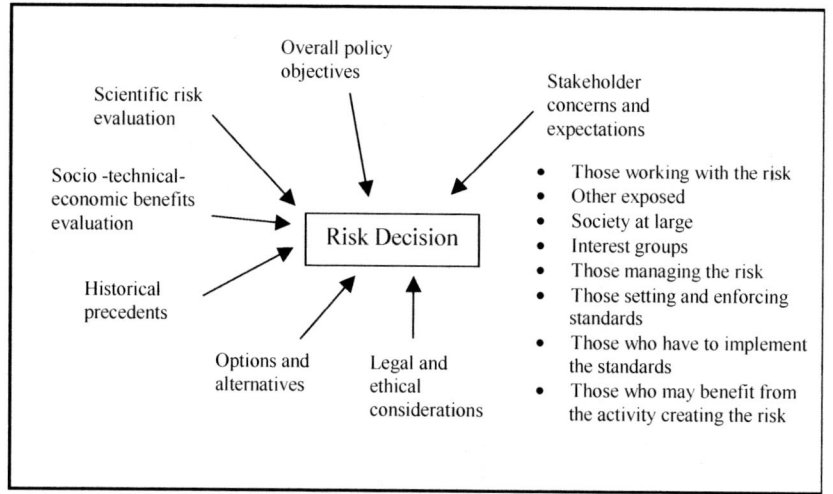

Figure 7. Overall decision context

CHEMICAL ENGINEERING – AN INHERENT SH&E IMPERATIVE

David W. Edwards
Visiting Fellow, Department of Chemical Engineering, Loughborough University, Loughborough, LE11 3TU, UK.

The principles of Inherent safety and inherently safer design are briefly described and the extension is made to health and the environment to yield the concept of Inherently Safer, Healthier and Environmentally Friendlier (ISHE) design and production. Six current problems and threats to the Process Industries (PI) that can be related to Safety, Health and Environmental (SHE) performance are proposed. Society's perception of the SHE performance of the PI often does not do them justice and the social construction of this perception is examined. Even though sustainable production is ill-defined, there is a push towards achieving it. Legislators are being driven to enact ever more stringent SHE regulations for the PI and the nature of the risk assessment methods to be used might change radically. Some production with problematic SHE issues is being transferred from the Developed Countries (DC) to the Less Developed Countries (LDC), where the problems may be amplified. A change to production by batch processes in the DC has been commented on and this might be deleterious for SHE performance; although no evidence for the trend has been found. Finally, the decline in numbers of students of chemical engineering in the UK is noted. Even though the SHE performance of DC companies is very good, it could be better and it is claimed that ISHE design and production, which is defined as avoiding or minimising SHE impacts, is the way to achieve this improvement. It is claimed that ISHE can help alleviate/counter these problems/threats. However, achieving ISHE design and production implies a move away from the traditional business aim of maximising return on investment (ROI) subject to acceptable SHE performance to a new aim of optimum SHE performance subject to maintaining an adequate ROI. This fundamental shift of emphasis is necessary for the long term sustainability (including profitability) of the PI in DCs.

Inherent SHE, public perception, sustainability, risk assessment

INTRODUCTION

It has become common practice in the process industries to group Safety, Health and Environmental matters together for the purposes of organisation, responsibility or reporting. The abbreviation SHE or permutations of it are often used.

Inherent SHE, or ISHE, is that safety, health and environmental performance that exists in a plant as an intrinsic attribute of it. A chemical plant with good ISHE is characterised by a lack of hazards, both in the short and long term, to human life and health and to the environment. Whereas, most chemical plants have acceptable SHE performance, because defensive measures are added to control identified hazards, it is better to build plant with inherently good SHE, because this cannot be compromised.

In this paper I examine a number of current problems and threats to the chemical industry and suggest that a move towards production plant that are optimised for ISHE subject to the constraint of making an adequate return on investment is an imperative counter measure. But first some definitions and nomenclature.

31

DEFINITIONS, CATEGORIES AND NOMENCLATURE

I shall follow the IChemE definition of hazard and risk as stated by Jones[1]:

Hazard – a physical situation with potential for human injury, damage to property, damage to the environment or some combination of these.

Risk – the likelihood of a specified undesired event occurring within a specified period or in specified circumstances. It may be either a *frequency* (the number of specified events occurring in unit time) or a *probability* (the probability of a specified event following a prior event), depending on the circumstances.

We can distinguish two broad categories of damage, that might result from a hazard:

Acute – a short-term event, which is possibly severe or even catastrophic;

Chronic – effects that persist over (long) time and could be severe.

Then, safety (S) implies freedom from acute danger posed by plant hazards, health (H) is about minimizing chronic effects upon anyone who interacts with the plant or product and Environment (E) refers to avoiding acute events and minimizing actual chronic impacts caused by the plant or product. Safety and health are essentially local considerations, whereas impacts upon the environment must be considered both locally and globally. In this paper SHE means anything to do with any of these three aspects of a chemical plant and product.

In order to develop my arguments it is necessary to categorise. These categories are broad, might give rise themselves to some discussion and will overlap. The main categories and abbreviations are as follows:

Process Industries (PI) = those employing chemical engineers on development and/or production of products.

Developed Countries (DC) = USA, Europe, Japan, Australia and some parts of Asia.

Less-Developed Countries (LDC) = those which are not DC.

Hence, DCPI = developed countries process industries, etc.

INHERENT SAFETY, HEALTH AND ENVIRONMENT (ISHE)

ISHE is the extension of inherent safety to cover health and environmental impact as well. Those readers who are in a hurry and are familiar with inherent safety may skip this section.

The traditional plant design philosophy and practice identifies hazards and then adds protective measures to control them. This method of secondary prevention[2] reduces the *probability* of accidents. However, there is an alternative philosophy of **Inherent Safety (IS)** or primary prevention, where the *possibility* of accidents is removed, by the use of safer chemicals and operations. In the practical approach to IS, called **Inherently Safer Design (ISD)**, hazards are identified early and then avoided or at least minimised, rather than controlled – so that accidents either cannot happen or their effects are minimal.

The principles of ISD were first enunciated by Kletz[3] after the Flixborough accident (28 killed) in 1974. At first interest was limited, but the appalling loss of life at Bhopal in 1984 (3000 immediate deaths and many thousands more thereafter) gave a greater impetus to discussion and a number of papers have appeared, which are referenced in Kletz's latest book[4]. Kletz's many papers and books give an exposition of the principles of ISD, which he has distilled into the qualitative application of six keywords:

- elimination – avoid hazards,
- intensification – use less of a hazardous material,
- attenuation – use less extreme processing conditions or a hazardous material in a less hazardous form,
- substitution – use a safer material or operation,
- limitation – minimise the effect of an incident,
- simplification – reduce the opportunities for error and malfunction.

A good example, used by Kletz, is the bungalow, which is inherently safer than a house, because stairs are the major cause of serious accidents in the home. Stairs are inherently unsafe, but they may be made 'safe' by lighting, fitting a handrail and child-gates, etc. It is important to distinguish between inherent safety and safety, because inherent safety is the more desirable quality. It is better to achieve safety inherently (live without stairs in a bungalow) rather than by modification (fitting a handrail, etc), because then unforeseen events (for example a rotten treadboard) cannot cause a problem.

Most chemical processes and the associated plant are safe – they have to be, but some are more inherently safe than others. For example, large inventories of toxic and/or flammable materials are inherently unsafe, while small inventories and/or non-toxic and non-flammable materials are inherently safe – what you don't have can't hurt anybody. Problems in an inherently unsafe plant may escalate catastrophically, while in an inherently safe plant they should not arise but, if they do, they are self-correcting or escalate harmlessly.

Therefore, it is almost self-evident that an inherently safe chemical plant is to be preferred over an inherently unsafe one, no matter how safe the latter is made by controlling the hazards. Also regulators, for example the Health and Safety Commission (HSC) and Health and Safety Executive (HSE) in the UK, are beginning to promote inherent safety and are pushing for inherently safe designs, for example with reduced inventories of hazardous materials.

ISHE can be considered on a number of levels. At the strategic level for example, if more than one product will satisfy a market need, we choose to make the one that has the best SHE performance and is made by the most ISHE process. At the most detailed level we may choose to use a particular type of gasket on our plant, in order to minimise emissions and leaks.

EVIDENCE OR LACK OF IT

This paper presents a personal, wide-ranging, subjective view of perceived problems facing the PI and the necessity for ISHE solutions. It is intended to be thought provoking and to throw up challenges. Such evidence as I have for the problems and solutions is referenced. However, there are many gaps and many of my arguments may be refutable by evidence to the contrary. If so, then please engage in the debate that needs to happen amongst the SHE and wider chemical engineering community.

THE PROBLEMS

PUBLIC PERCEPTION

Overall the safety and occupational health record of the DCPI is good. Environmental performance was poor in the past but has improved greatly in recent years. Legislation and inspection procedures have been in place for many years and the Industry has historically had a well-established "good neighbour" ethic, especially amongst the larger and more responsible companies – many of which have evolved hand in hand with their communities.

However, so far the PI have tended to adopt a 'we know what we are doing, let us get on with it' posture and, so far, the public has been happy to implicitly acquiesce – trading off employment and improved standard of life due to industrial production against this implicit permission to operate. Even so, it is debatable whether the public appreciates just how fundamentally their standard of life depends inextricably on the operations of the PI.

Recently, the public, who are becoming increasingly affluent but disconnected from industrial activity, encouraged by pressure groups and the media have become increasingly concerned about SHE issues. There is an emerging push towards sustainability, even though there is no consensus about what this means. However, it is undoubtedly true that the PI are large consumers of non-renewable resources and have a large impact upon the environment, whether by permit or unexpected incidents. The public are beginning to understand this and they also understand that many PI operations pose a risk of a large loss of life if the worst possible incidents were to be realised. Indeed a large onshore incident, with multiple loss of lives of people who are unconnected to the business of the offending facility, could sound the death-knell for the western process industries as we know them.

Familiarity, Control and Dread

It is helpful to consider other industries and social effects in trying to understand how public perception of the PI is constructed.

Railway safety, measured for example by the number of deaths per passenger mile travelled, has improved from 1967 to the present. Privatisation has made no difference to this trend[5]. This is not the perception of the public, which demands increased rail safety at enormous expense (£11 million per averted fatality for Automatic Train Protection) and to the detriment of road safety, where, overall, accidents continue to kill far more people and where simple preventative measures are comparatively cheap (£100k per averted fatality on local roads).

Alcohol and nicotine cause orders-of-magnitude more deaths, disease and social problems than illegal drugs, yet there is disproportionate attention upon the damage done by the latter. Illegal drugs are perceived as a much greater threat to one's offspring than alcohol and nicotine, which are more dangerous yet are legally promoted in the youth market.

Even though there is no sound evidence that the MMR vaccine causes autism, many members of the public decline its use and subject their children to the dangers of mumps, measles and rubella and increase the risk of a potentially devastating measles epidemic.

It seems that peoples' fear of possible incidents is constructed from the following three factors:

The size (dread) – the larger the potential loss-of-life, the greater the fear of the causing agent;

Unfamiliarity – agents that are not familiar or are not understood are feared more;

Lack of Control - risks posed by agents that are imposed upon the public or where the public is not in control or even involved will be feared and resisted more than those where the public has some control over or is involved in some way with the causing agent.

Hence, large loss-of-life incidents always attract more attention, even though they are very infrequent. Contrast railway accidents – always on the news, with car accidents, which only make the news, if there is a large loss of life. Some people will not drive on motorways because reported crashes involve large loss of life, whereas statistically travel on motorways is safer than on other roads. Autism is an unfamiliar dread disease, whereas measles, etc are not. A drugs overdose and possible death is worse than being drunk or having a cough.

Both railways and cars are familiar. However, whereas alcohol and cigarettes and individual vaccinations are familiar, illegal drugs and MMR are not familiar to most parents. Most adults drive cars and so the risk of accident is perceived to be under their control, travelling on a train relinquishes control to the driver and the people and systems operating the network. Alcohol and nicotine are legal and so controlled but other drugs are illegal and uncontrolled. Table 1 presents these results and shows the comparison, where the accepted 'alternative' has more ticks compared to the unacceptable. People are willing to accept a higher level of risk from sources that are familiar, over which they have a degree of control and where the size of possible incident is small. The problem for the process industries is that the public is unfamiliar with process plant hazards, they have no control over these hazards but they believe that the effects are potentially very large.

SUSTAINABILITY

This single word has recently generated intense debate and activity. It is still not clear what it means but the concept became established after the publication of the 'Brundtland Report[6]' in 1987, which defined sustainability as: "meeting the needs of the present without compromising the ability of future generations to meet their needs". There is General Agreement that sustainability must satisfy goals in the economic, environmental and social arenas.

I believe that there is also a general acceptance that much industrial production is unsustainable and that some products must be substituted and production methods changed - particularly as the larger LDCs ramp-up production.

Put simply, the industry cannot be sustainable when driven by maximising profit/minimising cost, because minimum cost production will inevitably consume non-renewable material and energy resources with non-optimum efficiency. However, it must be said that the root cause of unsustainability is not industry, which only makes its profit by providing goods and services for consumers. It is they, or rather you and I, who are the root cause by consuming too much.

Table 1. Presence or absence of features in everyday hazards influencing peoples' perception

Hazard	Familiar	In Control	Small Effect
Railway travel	✓	✗	✗
Car travel	✓	✓	✓
Illegal drugs	✗	✗	✗
Alcohol/nicotine	✓	✓	✓
MMR	✗	✗	✗
Individual vaccines	✓	✓	✓

RISK ASSESSMENT AND REGULATION
Simply put, a hazard is the potential for harm of a situation and risk is the likelihood of an effect originating from a hazard. Risk Assessment is taken to include: hazard identification, event scenario assessment, consequence assessment, risk assessment, risk comparison and decision-making. At present, there is immense interest in Risk Assessment across industry sectors and countries. Indeed the European Union has a major project underway at the Joint Research Centre, Ispra, which is attempting "technical harmonisation on risk-based decision-making"[7]. Much of this interest is driven by the awakening public concern (described above) about SHE, which extends across most human activities and is not limited to the PI.

Such developments, coupled with the public image problem discussed above pose the challenge that the DCPI might be hampered by regulation, drafted to allay public concern, which, although well-meaning, is miss-targeted, because of lack of understanding of the fundamental science and the impact of technology. There is also the danger that future EU legislation will lump the PI in with other sectors and insist on inappropriate methods and standards of risk assessment.

Finally, the public wish for the 'no risk society' could result in many process operations being moved to developing regions of the world which already offer significant economic benefits. If this occurs not only manufacturing employment but design skills, teaching and research will move outside of Europe in the long term. For example, India already offers design services at lower costs than Europe. Although this may be socially desirable it could make other objectives such as Sustainability, which benefits from the close coupling of producer and user, more difficult.

GLOBALISATION AND WORLD DEVELOPMENT
Many LDCs have burgeoning chemical industries, unfortunately this is being achieved at great cost to their people and their environment. This can only get worse as the pace of relocation of hazards from the developed to the developing world accelerates – this being the short-term response of the DCs to their publics' perception of unacceptable hazards. Most large multi-national companies claim that they insist upon the same good SHE standards wherever they operate but this does not alter the fact that loss of life and

environmental damage are much greater in LDCs than in the DCs. Indeed, if one travels to some LDCs you can feel the pollution – for example. If you look at accident reports and statistics for LDCs you will see that their safety and health performance is poor.

In the short-term the scale of human casualties and environmental degradation will increase in LDCs. You might say that sovereign states should look after their own people but we must accept that many environmental problems are global and affect us all.

DASH TO BATCH ?

Many commentators have remarked upon the changing emphasis of chemical production in DCs towards batch production. However, this flies in the face of better SHE performance. The key to ISHE production is to reduce inventories of hazardous materials, whereas batch reactors have much larger, often many orders of magnitude larger, inventories than a continuous reactor that has been designed to optimise mixing and heat and mass transfer. It seems that in the haste to bring new products to market as quickly as possible, good chemical engineering design is being ignored.

DECLINE OF ACADEMIC CHEMICAL ENGINEERING

The number of applications to and students taking up places on chemical engineering degree courses in the UK has been falling for the past 10 years. Figure 1 shows the most recent statistics. The commonly held view is that this is caused by the bad image of the industry amongst young people. There is certainly no shortage of well-paid jobs available for good students at the end of their courses. The situation with regard to post-graduate students is just as bad. It is very difficult to find UK PhD students.

So, even if we accept the arguments to follow that process plants must change radically, we will not have enough talented engineers to design this new generation of process plants.

THE SOLUTION – ISHE PLANT(S) ?

It is almost certainly true to say that current EU SHE practice is good and with the USA, probably defines the "current best practice" to which all others should aspire. The Industry is generally responsible in its behaviour and constantly strives to do better – most companies have well-established SHE programmes in place. Superficially it might seem that the industry is changing. Many large companies now produce SHE reports and the best of these are very good. If nothing else these demonstrate that large companies have high-minded visions and mission statements and that they are measuring and reporting performance. However, at best such documents report small incremental improvements in 'end-of-pipe' performance. I see no radical new plant, which is the only way to make significant, step-change improvements in SHE performance.

I believe that the problems identified above can be summed up in the phrase "the coming risk-tolerance crunch"[8]. This means that public tolerance of indirect risk is decreasing, while we are approaching a practical lower risk limit for engineered systems.The biggest challenge currently facing the PI is to ensure that it does not lose its permission to operate that is implicitly granted by stakeholders. This might happen if this 'crunch' comes when society perceives the SHE impact of the PI to be too negative, without manifest compensating benefits that they can understand. The only way to avoid the 'crunch' is by designing and operating ISHE plants.

ISHE DESIGN & PRODUCTION

This may be 'sloganised':

Neither you nor your environment can be hurt by anything that you do not have.

Risk is the likelihood of a hazard occurring and strictly speaking it is either a probability or a frequency, however there is currently a movement towards including the size of hazard in its definition. In either case, bearing in mind the comments about a practical lower limit for engineered systems, the only way to further reduce risk is to remove the hazards – or to reduce them, if the size is a factor. Therefore, we should develop products and achieve production plant that are: safe, have no long term adverse health effects and are not damaging to the environment, by not including agents (materials, operations) that could be detrimental in these areas or by minimising the inventories (amounts in plant) or instances of such agents. This is ISHE design.

This implies a fundamental and far-reaching change in the orientation of the process industries. Whilst it is important for the long term development of the industry that it retains its profitability (otherwise shareholders will disinvest), the industry must recognise that further improvements in safety, health and environmental performance together with 'Sustainable Development' are essential for the future. In the long-term the emphasis in the PI must evolve away from pursuing the current business objective of optimising economic performance subject to satisfying SHE constraints, towards a new objective of optimising SHE performance, subject to being able to generate or raise sufficient capital and making adequate returns on the capital employed.

In the long-term I believe that such a shift will guarantee economic survival. DCs currently have a great technological and consequent economic advantage but this will not last as the developing world catches up. We must provide the leadership to do things better in terms of SHE. This will provide the added value for the future and differentiate our products and production methods, so that we can still sell those products and the production technology in a world market that is increasingly aware of SHE issues.

Now I will address the problems identified above in turn.

PUBLIC PERCEPTION

ISHE production plant are the key to convincing society that there is no threat. Ideally, if there are no hazards then there is no risk. If the hazards have been minimised, then we can do no better, given that the product has a place in human activity.

In terms of the headings Familiarity, Control and Dread:

Familiarity can be engendered by better communication of what we do, how we do it and the benefits that society reaps from our activities. At a local level, is it not possible to engage ordinary people in design and development? This is done for consumer products, so why not for products of the PI?

An element of **Control** could also be given to society by the above engagement.

38

Addressing the **dread** factor is more problematic, but ISHE plant would minimise this response. There is a wealth of research in this area being done by social scientists – much of which is unknown to the engineering community but we must draw on this work. Some scientists and engineers claim that the public does not understand probability and so cannot be trusted to make rational decisions about major hazards, where the probability has been reduced to As Low As is Reasonably Practicable (ALARP). I believe that the public understands probability well and fully appreciates that an event with a miniscule probability can still happen today or tomorrow or in their lifetimes. The only way to demonstrate to an increasingly sceptical public that an activity has acceptable risk is to remove or reduce the size of hazard to an acceptable level.

SUSTAINABILITY

I have said that I do not believe that this concept has even been defined yet and may be it is not possible. You can only know if an activity is sustainable with hindsight and then it still cannot be projected into the future. However, if we take Sustainability to mean enduring, then clearly it must not stop. For production this implies not only continued economic viability and sufficient raw materials, which can only be assumed, but also avoiding events that might result in termination, such as high profile safety, health and environmental incidents – particularly those with deaths. The only reliable way to avoid such events, particularly over the long term, is to remove the hazards by designing ISHE plants. There is also the imperative to ensure that the public does not perceive the risk to be too high, even though we might not believe that it is, because, otherwise, they might shut us down anyway – as a precaution. This must be achieved by better communication and engagement as mentioned above.

Furthermore, such communication must be a dialogue, because it is the public as consumer that drives resource consumption. For example consumers buy bigger and bigger vehicles in the face of mounting concern about the environmental impact caused by cars. So, there are obligations on both sides and we must make the point that activities will only be sustainable if society does not demand products that are unsustainable, or consume too much of products that otherwise might be sustainable.

We must embrace sustainable production and lead the way – even at the expense of reduced profitability, because it is unreasonable to expect LDCs to embrace sustainable production methods if we do not. The dialogue aspect then is even more pressing, in order to educate the coming consumers of the LDCs. If they consume at our rate, then we are all doomed by a rapidly exhausted planet!

RISK ASSESSMENT AND REGULATION

In the future risk assessment for the PI will be driven to focus more on the human risk receptor, rather than just the inanimate risk source (the chemical plant), which has been the engineer's focus. Social scientists have many competing approaches to tackling the same problems, but a common theme is the focus on the human risk receptor. Furthermore, they seem more willing to tackle difficult, vague risks, that engineers would shy away from because there is no obvious method for quantification. They are often more concerned with understanding the drivers behind risk and perceptions of it than quantification.

The PI should engage with the social science community with the aim of moving towards a new or synthesised approach to risk assessment and management that can integrate 'soft', people-centred aspects with 'hard', numeric quantification. Such a new approach is vital for the chemical industry, where, I believe, the single biggest threat is from misinformed public opinion and misplaced regulation.

The Control of Major Accident Hazard Regulations (1999)[9] apply where a dangerous substance as listed in the Regulations is present in a quantity equal to or exceeding the entry for that substance. Therefore the obvious way to avoid these regulations is to ensure that any quantity of a hazardous substance is less than the prescribed threshold. This coincides with the key feature of ISHE design – avoid or reduce inventory of hazardous materials.

Regulation happens when it is deemed necessary. A good way to avoid excessive regulation is to anticipate future regulation, which as well as demanding new methods of people-focussed risk assessment is likely to encourage ISHE design. Indeed, inherent safety is already mentioned in the latest EU regulations and are a feature of the Contra Costa County Ordinance Code Chapter 450-8, Section 450-8.016(D)[10] (California, USA). Therefore, in order to secure a 'lighter touch' from regulators, the PI should engage in ISHE design now.

GLOBALISATION AND WORLD DEVELOPMENT

LDCs can only improve their SHE performance if they know how to do it. Therefore it behoves the DCs to share their good practice. In the long term this is good for business as well. DCs cannot compete with LDCs that have lower people and raw material costs and which also place a lower value upon human life and the environment. Their thresholds must be raised by encouraging and sharing good practice.

DASH TO BATCH

When I started work on this paper, this was to be a key theme. However, I can find no production data that either confirms or disproves a move from manufacture by continuous processes to batch production. If this is a trend, then we must assemble some data on production levels and also on reportable incidents in relation to production method.

DECLINE OF ACADEMIC CHEMICAL ENGINEERING

Having just left academia, I can say that this is happening and that it is deeply worrying. Although there is some recent evidence that admissions are stabilising at a new lower level, we must take action now to ensure that the previous downward trend is reversed to an upward one. Otherwise, we will see the closure of more Departments.

I believe that action has to be taken at the programme level to make chemical engineering degrees more relevant to young people and the present and future needs of industry. Programmes must emphasise SHE matters and particularly ISHE, but also more material related to the PI as a business and its interaction with society should be included. Chemical engineering curricula are already extremely crowded with 'hard' science and engineering and some of this must go, in order to make way for the new material. This would also have the benefit of making the subject less difficult, which I believe is a big disincentive to prospective students. Nowadays computer tools for technical calculations are all-pervasive – not all chemical engineers need to be able to do their sums with a pencil and calculator. Of course they need to understand the results and when they are at odds with reality but more important is the ability to be able to deploy these results in business and social contexts. For those engineers wanting to follow a career with the emphasis on technology, masters courses can be provided to enhance their more broader-based basic chemical engineering qualification.

CONCLUSIONS

It is now common practice in the PI to consider SHE issues together and this is the correct approach, because they are clearly related and interact – progress will only be made by considering SHE performance holistically. The concept of inherent safety and inherently safer design has been around for a long time and is widely acknowledged as the ideal for production plant, however its philosophy and practice must be extended to encompass the totality of SHE. There are now many imperatives for the PI to embrace ISHE and implement such design in its production facilities.

Study of public response to everyday hazards, leads to the scenario that the negative public perception of the hazards posed by the PI is constructed from three factors: unfamiliarity with the production methods and many products, lack of control over these hazards and fear of the potentially large consequences of serious SHE incidents. ISHE production plant are key to convincing the public that the PI have a place in our society – absence of hazard is the most convincing argument for permission to operate.

Sustainability is as yet not precisely defined and it may never be so. However, ISHE plants might endure for the long term because there can be no reason – large SHE incidents, particularly with loss of life – to shut them down. If they do not consume non-renewable resources and are economically viable in the long term, then their production is sustainable.

The PI must try to synthesise new methods for risk assessment from the traditional 'hard-science' based numeric quantification and presently unfamiliar 'soft' people-centred approaches employed by social scientists. This and the ISHE approach to production might then head off the danger of new, all-encompassing regulation placing infeasible constraints upon the industry.

The DCs must share their good ISHE practice with the LDCs and help to educate their people. In the long term this will safeguard the viability of the DCPI, because the DCs can only trade and compete with others who share the same attitudes about SHE.

A move towards batch production in DCs has been claimed but the reported statistics do not confirm this trend. If it is a trend we must consider the negative consequences for ISHE production.

The continuing decline in numbers of chemical engineering students must be reversed. Curricula must emphasise ISHE and the place of engineering and engineers in wider society. So that new products and processes are developed that take account of their impact on society.

I have stressed throughout the paper that the SHE performance of the DCPI is very good but it must attain new heights and this can only be achieved by ISHE plants. Moreover, I believe that plants must have optimal ISHE and that this implies that SHE performance becomes the objective function that drives our activities, subject to generating sufficient profit; rather than maximising profit subject to adequate SHE. The adoption of this new aim requires a leap of faith, with no guaranteed return – but, that is the scenario with most innovative projects.

REFERENCES

1. Jones, D.A. (ed.), 1992, *Nomenclature for Hazard and Risk Assessment in the Process Industries*, 2nd edn (IChemE, UK).
2. Zwetsloot, G.I.J.M. and N. Askounes-Ashford, 1999, Towards Inherently Safer Production: A Feasibility Study on Implementation of an Inherent Safety Opportunity Audit and Technology Options Analysis in European Firms, *TNO report R990341*, (Hoofddorp, Netherlands).
3. Kletz, T.A., 1976, Preventing Catastrophic Accidents, *Chemical Engineering* (US), 83, 8: 124-128.
4. Kletz, T.A., 1991, *Plant Design for Safety: a User Friendly Approach*, Hemisphere Publishing.
5. Evans, A.W., 2000, Risk on the Railways, *ESRC Risk and Human Behaviour programme Conference*, London, 11-12 September 2000.
6. World Commission on Environment and Development, 1987, *Our Common Future*, Oxford University Press.
7. Promotion of technical harmonisation on risk-based decision-making, Workshop 22-24 May 2000, Stresa, Italy, S.P.I.00.63.
8. Johnson, R.W., Unwin, S.D., McSweeney, T.I., 1996, Inherent Safety: How to Measure It and Why We Need It, *International Conference and Workshop on Process Safety Management and Inherently Safer Processes*, October 8-11 1996 Orlando, Florida.
9. http://www.hmso.gov.uk/si/si1999/19990743.htm, 1999, The Control of Major Accident Hazards Regulations, Statutory Instrument No. 743.
10. http://www.co.contra-costa.ca.us/

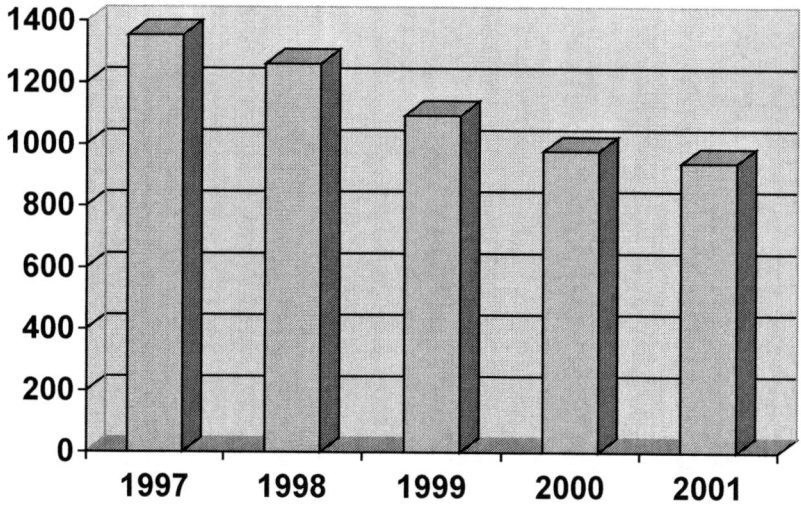

Figure 1. Number of chemical engineering students entering UK degree programmes

WHEN PROCESS SAFETY MANAGEMENT FAILS – THE RISKS TO CORPORATE EXECUTIVES

Michael Dore*
* Mr. Dore is a Partner with the law firm of Lowenstein Sandler PC, Roseland, New Jersey 07068. He has handled a number of cases involving process safety accidents in the chemical and pharmaceutical industries and has written extensively in this area. He is also an Adjunct Professor of Law at Rutgers Law School (Newark, New Jersey) and can be reached at mdore@lowenstein.com.

When process safety efforts fail to avoid accidents or mishaps, operating facilities are often called upon to participate with government agencies in investigations designed to determine the root cause and/or factors contributing to the adverse event. In the United States, numerous government agencies including the Environmental Protection Agency, National Transportation Safety Board, Occupational Safety and Health Administration ("OSHA"), Chemical Safety and Hazard Investigation Board, and others, respond to such process safety incidents. During any such governmental investigative response, chemical processing company executives and employees may be required to participate in efforts to unearth the cause of the incidents. These activities subject companies and individuals to potential regulatory, civil and even criminal exposures.

CHEMICAL PROCESSING ACCIDENTS

Accidents in the chemical processing industry often have catastrophic consequences. Massive property damage, environmental degradation, and sometimes even significant physical injuries and loss of life can result when process safety management efforts fail. Indeed, a Texas A&M University study identified more than 16,000 sudden chemical releases in 1998 that resulted in 61 deaths and 4,002 injuries. Moreover, in 1999, Dr. Paul L. Hill, chairman and chief executive officer of the Chemical Safety and Hazard Investigation Board, reported to Congress that "In 1996, chemical incidents claimed the lives of the equivalent of two fully loaded 737 passenger jets - 256 people perished. And an average of 256 people died the year before. And the year before that."[1]

Immediately following, and sometimes even coterminous with, incident response efforts, government investigations of such accidents will commence. In addition, facilities themselves may have Responsible Care, process safety management, or other independent obligations to conduct their own post incident investigations. Understanding the dynamic of such government and private party investigations can be crucial in protecting individual corporate executives from potential civil or criminal liability.

Critical incident response will often require a focus on the "responsibility" of particular individuals involved in those incidents. Responsibility for recent catastrophic events in the United States has been laid at the feet of terrorists such as Osama bin Laden and Timothy McVeigh. Other critical events ranging from Bhopal to the Exxon Valdez spill, however, have resulted in the imposition of civil and criminal liability upon corporate executives and employees. It can be anticipated that future catastrophic process safety failures will result in efforts to place the blame for those failures on the corporate executives and employees involved.

THE LIABILITY PARADIGM

It is clear that corporate entities involved in process safety incidents may be found liable for those accidents. Depending on the particular facts involved, civil or criminal liability based on negligence or strict liability may be imposed.

Absent special circumstances, individuals are not liable for the contractual obligations of their corporate employers. Individuals acting on behalf of the corporate entity, however, do not enjoy immunity from civil or criminal liability. In essence, the issue with respect to the imposition of such liability upon a corporate executive, is the level of participation that the executive had with respect to the liability creating activity.

As the New Jersey Supreme Court recently noted, a corporate officer can be held personally liable for a tort committed by the corporation when he or she is "sufficiently involved" in the commission of the tort. A predicate to liability is a finding that the corporation owed a duty of care to the victim, the duty was delegated to the officer and the officer breached the duty of care by his own conduct.[2] The breach of this duty of care, however, can involve either intentional or negligent conduct by the corporate executive.[3] Thus, personal involvement in or knowledge of activity which creates liability for corporate entities may well create that same liability for corporate executives themselves.

CRIMINAL LIABILITY

Inevitably, when corporate entities are exposed to criminal liabilities, corporate managers will also face personal liability risks. Initially, those exposures may flow from broad based state criminal provisions such as "reckless endangerment" statutes. In that regard, chemical facility operators should be aware that, in recognition of the fact that the chemical processing industry is a prime potential target for terrorist activities, states have strengthened the criminal statutes which have traditionally been used to respond to process safety incidents. Thus, by way of illustration, in March 2002 New Jersey enacted the "September 11, 2001 Anti-Terrorism Act". Along with numerous provisions dealing directly with the prevention and punishment of terrorist activities, this Act also modified the State's criminal statute prohibiting "Causing Or Risking Widespread Injury Or Death".[4] That statute has been used extensively to respond to chemical processing and other industrial mishaps and violations. It imposes criminal liability on a party who unlawfully causes an explosion or engages in the improper storage or release of harmful substances. The Anti-Terrorism Act expanded this statute to make it a crime punishable by up to five years imprisonment for a person to create a risk of widespread injury or damage by recklessly handling or storing hazardous materials. The Act also made it a crime punishable by up to ten years imprisonment if the handling or storing of hazardous materials violated any law, rule or regulation intended to protect the public health and safety.

In addition, media specific statutes such as the Clean Air Act or Clean Water Act or activity specific statutes such as the Occupational Safety and Health Act and the Resource Conservation and Recovery Act may be used to impose criminal or civil liabilities.

Moreover, in the area of health and safety regulations, courts have been willing to use the "responsible corporate officer" doctrine to impose civil and criminal liability upon executives whose conduct would not normally give rise to the imposition of that liability.

Under that doctrine, any corporate officer who had a "responsible share" in the violation of the relevant criminal statute could have individual liability imposed upon him.[5] Indeed, the criminal provisions of both the Clean Water Act and Clean Air Act incorporate the concept of criminal liability for "responsible" corporate executives. The Clean Water Act[30] was amended in 1987 to expand the definition of parties liable for criminal violations of the Act to include "any responsible corporate officer."[6] In addition, under The Clean Air Act the original definition of a "person" liable for criminal violations was expanded by §7413(c)(6), which provided that, with respect to criminal penalties, "the term person includes … any responsible corporate officer."[7]

The U.S. Senate Committee on Environment and Public Works, in discussing the amendments to the criminal penalty provisions of The Clean Air Act, reported:

[F]or the purpose of liability for criminal penalties, the term "person" is defined to include any responsible corporate officer. This is based on a similar definition in the enforcement section of the Federal Water Pollution Control Act. The Committee intends that criminal penalties be sought against those corporate officers under whose responsibility a violation has taken place, and not just those employees directly involved in the operation of the violating source.[8]

Such liability issues need to be considered at the earliest point in critical incident response efforts. Those initial efforts - which can take hours, days or weeks - will inevitably involve efforts to control and or contain the immediate adverse consequences of the incident. During this initial response period, liability considerations *per se* are seldom paramount.

Even at this initial point, however, the decisions made at this initial point may have consequences for later stages of the incident. By way of illustration, release reporting statutes often require immediate telephone notice of an incident to emergency response authorities. These statutes obligate parties to give very specific information. If the reporting employee later attempts to deny personal knowledge of particulars of the incident, the recording of the release reporting call may make such a denial of knowledge difficult to sustain. This is true, even if the "facts" conveyed in the release reporting call were provided to the reporting executive by third-parties. As such, "liability" considerations must be addressed at the earliest possible moment following a process safety failure.

PROTECTING THE CORPORATE EXECUTIVE

There are few places on earth more lonely than the space occupied by a corporate executive who is potentially responsible for a process safety mishap resulting in significant injuries to property, the environment, and individuals. Often support for such executives at that time can be seen by governmental authorities and the public as a failure of the corporation to take responsibility for the consequences of the catastrophic incident. On the other hand, abandoning such individuals in their time of need - particularly as they are subjected to regulatory and criminal investigations and the initiation of civil lawsuits - is seldom in the best interests of the corporation.

The best time to consider these issues is long before the adverse critical incident occurs. Most state laws require indemnification of employees charged with wrongdoing during the course of their employment - as long as those employees are ultimately exonerated. In addition, however, most states permit corporate indemnity provisions which, at a minimum, provide an ongoing defense to the corporate executive, regardless of whether that executive is ultimately found liable. Moreover, insurance policies are available to fund such D&O and other employee liabilities.

Deciding how broad to make these corporate indemnity protections and the classes of employees to which they should be extended, present significant issues for any corporation engaged in business activities with significant risks of catastrophic process safety failures. Whether to maintain flexibility in order to respond to the particulars of any individual incident or whether to decide in advance to "stand behind" (by either agreeing to defend and/or indemnify) all employees involved in such an incident is never an easy decision to make. Given the risks of such incidents to individual corporate executives and the all but inevitable fallout of those risks to their corporate employers, however, consideration of those issues before, rather than after, the critical incident is highly advisable.

GOVERNMENTAL INVESTIGATIONS
The risk of having criminal liability imposed upon individual corporation executives in the aftermath of process safety failures is increased by the fact that once the immediate process safety response activities are concluded, the "root cause" or other fault based investigations will begin.

In dealing with these investigations, corporate executives must know and appreciate the significance of what government agencies are involved in the investigative effort. All government agencies have different core constituencies, different statutory powers and different parties with direct influence over the essential character of their investigations.

In the United States chemical processing industry, Federal and State criminal authorities, EPA, OSHA and the Chemical Safety Board are the most likely to be involved in such investigations. In addition, depending on the facts of any particular incident, the National Transportation Safety Board, Food and Drug Administration and other State and Federal Regulatory Agencies may also be involved.

Often, multiple federal and state agencies are involved. In such circumstances, understanding the particulars of the relationship between these agencies can be vital. At times there are formal Memorandums of Understanding which identify the circumstances in which a particular agency will act as lead investigator, how investigative resources will be shared and other investigative protocol information.[9] Even in the absence of a formal Memorandum of Understanding, however, appreciating the dynamic of how different investigative agencies acquire and share information is crucial to protecting the interests of corporate executives.

To effectively represent the interests of those executives, the different constituencies and statutory focus of investigative agencies must be considered. When OSHA acts as the head of the investigative agencies, labor unions' influence on the investigation may be far different than when the National Transportation Safety Board is involved. In addition, the core competencies of the lead investigative agency can result in significant differences in substantive results. Thus, the agencies' access to technical expertise and familiarity with the nature of the operations being investigated, will often have a significant impact on the ultimate conclusions drawn by that agency.

Often of equal significance, the enforcement powers and protocols of different investigative bodies are vastly different. Some governmental authorities are vested with direct authority to pursue criminal charges -- while others need to involve other government bodies vested with the power to bring criminal claims. The authority of some agencies to conduct investigations is circumscribed by strict time limitations[10] while others are restricted only by lengthy criminal statutes of limitation. Some agencies can impose civil penalties subject to administrative appeal, while others may pursue such penalties only in highly structured judicial proceedings.

Of perhaps even greater significance than the technical expertise and enforcement powers of particular agencies, the investigative powers of those agencies vary widely. Only some can execute search warrants, issue subpoenas and compel witness testimony. Similarly, a limited number of agencies are authorized to propound written questions or to pursue testimony from witnesses outside of the regulated community.

Moreover, separate from the enforcement and investigative authorities of particular agencies, is the issue of the investigative preferences which the agencies use to exercise that authority. Thus, for example, some agencies require extensive headquarters authorizations with respect to their investigative activities while others do not.

With respect to the investigation itself, different agencies follow different procedures with respect to "ambush" or noticed interviews; simultaneous interviews of multiple witnesses; recording of witness interviews; preparation of interview reports or summaries; provisional warnings as to witnesses' constitutional rights; permission for participation by non-legal representatives; and sharing witness information with other government agencies. Moreover, the rules governing contact with witnesses represented by counsel and the confidentiality promises given to witnesses, vary from agency to agency.

Finally, the fallout from the investigation of different government agencies can be dramatically different. Different agencies follow different media disclosures policies. In addition, those agencies have different procedures with respect to the disclosure of tentative and final investigative conclusions; the peer review and internal vetting of those reports; protection of individuals and the release of investigative reports; the sharing of information across agency lines; and participation of the investigators in subsequent civil proceedings. Finally, those agencies have different abilities to impact the operations of target parties on the basis of failure to fully cooperate with the agencies' investigative effort.

All of these factors effect corporate executives' need to protect themselves in the aftermath of significant process safety incidents. An executive who is being investigated by an agency that may search his home or office without notice at any moment, or who may be confronted by investigators who will secretly record his responses to "ambush" interviews -- needs to be advised of the fact that such events are likely to occur. This is particularly true if the executive believes himself to be immune for personal liability because "he was only acting for the benefit of his corporate employer."

Moreover, warning employees of their constitutional rights and their ability to decline to participate in such governmental investigations, becomes far more important when a company is faced with a regulatory authority that conducts ambush interviews; executes search warrants; bans non legal advisors; and refuses to record the interview process than would be true in circumstances where a company was faced with a more orderly investigative process.

INTERNAL CORPORATE INVESTIGATIONS

Complicating the civil and criminal exposures of corporate executives can be pre-existing corporate policies requiring investigations (and sometimes disclosure) of the circumstances surrounding all process mishaps.

Such policies may parallel OSHA's process safety management regulations which require employers to investigate every incident that results in, or reasonably could have resulted in, a catastrophic release of highly hazardous chemicals in the workplace[11]. Similarly, any chemical company subscribing to "responsible care" standards is obligated to investigate significant process safety failures.

Though such policies are clearly well intentioned, they can seriously compromise post incident efforts to protect corporations and their executives from potential criminal and civil liability. Thus, at every stage of these internal investigations all parties must be aware that information uncovered during this process may be used to impose civil or criminal liabilities upon individual corporate executives.

THE ATTORNEY CLIENT ISSUE

With respect to both governmental and private investigations, the role of counsel for the corporation or for the individual executive can be crucial. When counsel conducts an investigation to provide legal advice to the Board of Directors and top management on an issue confronting the corporation, such as a pending lawsuit, there is substantial support for the proposition that the attorney/client privilege and/or work product doctrine protects the confidentiality of the investigation. Given the importance of the purpose of the investigation (i.e., to allow for the provision of legal advice), the documentation of those purposes through Board resolutions authorizing the investigation, designating the parties authorized to conduct the investigation, setting forth explicit confidentiality requirements and the obligation of corporate employees to cooperate with that investigation, is crucial.[12]

DETERMINING WHOSE COMMUNICATIONS ARE PRIVILEGED

The "scope" of corporate attorney/client privilege also needs to be considered in connection with the conduct of internal investigations. Clearly, all corporations consist of, and function through, individual directors, officers, employees, and other agents. Different states have different rules, however, as to the extent to which communication between counsel and particular individuals associated with the corporation are protected by the attorney/client privilege. Some states limit that privilege to communications with corporate "control group" members,[13] while other states expand the protections of the privilege to communications with all corporate employees involved in the "subject matter" of particular litigation events.[14] As such, counsel must be particularly careful to ensure that they know whether their communications with particular individuals will ultimately be covered by the attorney/client privilege under the applicable law.

FAIRNESS TO NON-CLIENTS

Of particular concern with respect to any internal corporate investigation is the obligation of counsel to ensure that those who are interviewed by, or at the direction of, counsel are fully aware of their client or non-client status. The American Bar Association's Model Rule of Professional Conduct 1.13(d) requires that "in dealing with an organization's directors, officers, employees, members, shareholders or other constituents, a lawyer shall explain the identity of the client when it is apparent that the organization's interests are adverse to those of the constituents with whom the lawyer is dealing." As such, notwithstanding in-house counsel's often close relationship with particular corporate employees, counsel must be careful to inform those employees that he or she does not represent them individually and that the disclosures of those employees may ultimately be revealed to third parties without their consent. This is a particularly important issue in circumstances where the investigation raises potential criminal concerns.

INCENTIVES FOR CORPORATE WAIVER OF THE PRIVILEGE

The potential for such disclosures of corporate employee communications increased dramatically in the Summer of 1999 when the Department of Justice issued new "Guidance On Prosecutions Of Corporations." That Guidance, in addressing the corporation's willingness to cooperate with the government's investigation as a factor in determining whether to charge a corporation with a crime, specifically states that:

one factor the prosecutor may weigh in assessing the adequacy of a corporation's cooperation is the completeness of its disclosure including, if necessary, a waiver of the attorney-client and work product protections, both with respect to its internal investigation and with respect to communications between specific officers, directors and employees and counsel. Such waivers permit the government to obtain statements of possible witnesses, subjects, and targets, without having to negotiate individual cooperation or immunity agreements. In addition, they are often critical in enabling the government to evaluate the completeness of a corporation's voluntary disclosure and cooperation. Prosecutors may, therefore, request a waiver in appropriate circumstances.

Thus, corporate clients have a strong incentive to waive their attorney/client privilege in circumstances involving potential criminal violations. As such, basic fairness requires individual employees be informed of their inability to prevent the ultimate disclosure of their confidential communications to in-house counsel.[15]

CONCLUSION

Dealing with the risks to corporate executives inherent in the aftermath of significant process safety failures requires an early and concentrated focus. That focus must include consideration of possible individual exposure before the issue is raised by the investigating authorities as well as consideration of appropriate executive corporate indemnification[16] and insurance[17] protections. It must also include immediate efforts to ensure that the interests of corporate executives are not damaged by the destruction or loss of crucial evidence relating to the critical incident. [18]

In addition, liability issues should be considered at the time that critical incidents are reported to regulatory authorities and at the time that spokesmen are selected to make disclosures to the public with respect to the incident.

Moreover, protection of the interests of both the corporation and its executives through the use of joint defense agreements should be considered. In addition, steps should be taken to maximize the protections afforded by the attorney client privilege throughout all governmental and internal investigations and executives should be prepared to respond to investigative techniques such as the execution of search warrants. [19]

Finally, every critical incident creates the possibility that key adversaries will surface who will attempt to increase the exposure of the corporation and/or its executives to regulatory, civil and criminal liabilities. Every step in the critical incident response effort presents the danger of creating such adversaries. For example, following a catastrophic explosion, promising the local community that the plant will never be reopened, may eliminate the danger that the local community will organize an effort to proceed criminally against an "offending" company - but that very announcement may ensure that union employees concerned about their future job prospects will organize to attempt to impose just such liability upon a corporation or its executives. Every critical incident will present different dangers. In all such circumstances, however, a recognition of who the company's key adversaries are, and a plan for dealing with those adversaries is essential.[20]

Finally, when process safety efforts fail and executives are exposed to the dangers outlined in this article, it will be all but impossible to get those executives to properly focus on these exposures. For that reason, consideration of these issues and the appropriate education and training of corporate executives must take place before - not after - the process safety failure has occurred.

REFERENCE

1. *See*, Guides to Chemical Risk Management, New Ways to Prevent Chemical Incidents, National Safety Council's Environmental Health Center (1999).
2. *Saltiel v. G.S.I. Consultants, Inc.*, 170 N.J. 297 (2002); *see generally*, Pachman, Does the Corporate Shield Protect the Corporate Officer from Personal Liability? Yes, No and Maybe, New Jersey Lawyer 16 (August 2002)
3. *See e.g., Wicks v. Milzoco Builders, Inc.*, 470 A. [24] 86 (Pa. 1983).
4. N.J.S.A. 2C:17-2
5. Commentators have questioned how the responsible corporate officer doctrine is, and should be applied. *See*, Dore and Ramsay, Limiting the Designated Felon Rule: the Proper Role of the Responsible Corporate Officer Doctrine In the Criminal Enforcement of New Jersey's Environmental laws, 53 Rutgers L. Rev. 181, 197 (2000) ("Thus, federal court decisions from *Dotterweich* to *Iverson* have disagreed as to whether the responsible corporate officer doctrine should be used to (1) determine who is a "person" subject to the criminal provisions of a substantive environmental statute; (2) impose vicarious liability on corporate officials on the basis of the conduct of others; (3) impose vicarious liability on corporate officials by virtue of the imputation of the mental state of others; (4) permit (or require) a fact finder to use the status and

authority of a defendant to determine that corporate officers are liable for the environmental violations of other corporate employees; (5) permit a fact finder to use the status and authority of a defendant to impose an evidentiary presumption (or a burden to disprove) that corporate officials are liable for the environmental violations of other corporate employees; (6) permit a fact finder to use the status and authority of a defendant as a factor in determining that corporate officials acted with the mental state necessary to impose liability for particular environmental offenses; (7) permit a fact finder to use the status and authority of a defendant as the exclusive basis for determining that certain corporate officials possessed the mental state necessary to give rise to liability; or (8) permit a fact finder to use the status and authority of a defendant to as the exclusive basis for determining that certain corporate officials engaged in the conduct necessary to give rise to liability.")

6. *See*, Note, Corporate Noncompliance with the Clean Water and Clean Air Acts: Theories to Hold a Director Personally Liable, 13 Va. Envtl. L.J. 99 (1993).

7. *See generally*, Barone, The 1990 Clean Air Act Amendments and Implications of the Self-Incrimination Clause: Are Environmental Managers Risking Criminal Liability? 5 Seton Hall Constitutional Law J. 967 (1995).

8. S. Rep. No. 94-717, 94th Cong., 2d Sess. 40 (1976), reprinted in 6A Legislature History of The Clean Air Act Amendments of 1977, No. 95-16, 4701, 4741 (1978).

9. *See*, e.g., Memorandum of Understanding between The United States Department of Labor Occupational Safety and Health Administration and The United States Chemical Safety And Hazard Investigation Board on Chemical Incident Investigations

10. *See*, 29 CFR 1903.14(a) (OSHA citation must issue within 6 months of accident)

11. 29 CFR 1910.119(m)(1)

12. For examples of forms which can be used for these purposes, *see*, Brewer, The Ethics Of Internal Investigations In Kentucky And Ohio, 27 N. Ky. L. Rev. 721 (2000); Martin, Conducting A Successful Internal Environmental Investigation, 6 Envtl. Law. 673 (2000).

13. *See e.g.*, Alaska R. Evid. 503(a)(2) (utilizing control group test); *see generally*, Hamilton, Conflict, Disparity, and Indecision: The Unsettled Corporate Attorney-Client Privilege, 1997 Ann. Sur. Am. L. 629; Note, Attorney-Client Privilege For Corporate Clients: The Control Group Test, 84 Harv. L. Rev. 424 (1970); *see generally*, Mr. Dore, Law of Toxic Torts: Litigation - Defense - Insurance §24 (West Group 2002)

14. Annot, Determination Of Whether A Communication Is From A Corporate Client For Purposes Of The Attorney-Client Privilege - - Modern Cases, 26 A.L.R. 5th 628 (2000); Gergacz, Attorney-Corporate Client Privilege: Cases Applying Upjohn, Waiver, Crime-Fraud Exception, And Related Issues, 38 Bus. Law. 1653 (1983); Waldman, Beyond Upjohn: The Attorney-Client Privilege In The Corporate Context, 28 Wm. & Mary L. Rev. 473 (1987)

15. Gallagher, Legal And Professional Responsibility Of Corporate Counsel To Employees During An Internal Investigation For Corporate Misconduct, 6 Corp. L. Rev. 3 (1983); Jonas, Who Is The Client?: The Corporate Lawyer's Dilemma, 39 Hastings L.J. 617 (1988); see, *United States v. Keplinger*, 776 F.2d 678 (7th Cir. 1985) (interviews with

corporate employees used in criminal proceedings against those employees). *But see, Dalrymple v. National Bank & Trust Co.*, 615 F. Supp. 979 (W.D. Mich. 1985); *Professional Serv. Indus., Inc. v. Kimbrell*, 758 F. Supp. 676 (D. Kan. 1991); *See generally*, Gallagher, Legal and Professional Responsibility of Corporate Counsel to Employees during an Internal Investigation for Corporate Misconduct, 6 Corp. L. Rev. 3, 23 (1983).

16. *See generally*, Monteleone and Conca, Directors and Officers Indemnification and Liability Insurance: An Overview of Legal and Practical Issues 51 Bus. Law. 573 (1996).

17. Robert A. Bregman, International Risk Management Institutes' Executive Liability Insurance Guide (IRMA 1998).

18. Palaez, Alfred S. 1995. *Caveat Spoliator.* Second Annual Conference on Defending Against and Mitigating Incidents Involving Injury or Property Damage in the Natural and LP Gas Industries, Oct. 16-18. Houston, TX: Institute of Gas Technology.

19. Kowal, When Unexpected Government Agents Drop In: Responding To Requests For Immediate Interviews, 54 Food Drug L.J.93 (1999); Davenport, Environmental Search Warrants: How To Prepare And How To Respond, Envtl. Counselor 15 (August 15, 1993);

20. In that regard, counsel for executives should be particularly alert to issues relating to the examination of computers. *See*, Federal Guidelines for Searching and Seizing Computers (U.S. Department of Justice Website).

21. *See, generally*, Michael Dore, *Be Ready to Respond To Critical Incidents*, 165 N.J.L.J. 852 (Sept. 3, 2001).

LESSONS LEARNED ABOUT PREPARING COMAH SAFETY REPORTS

Trevor Britton
Health & Safety Executive, Hazardous Installations Directorate, St Anne's House, University Road, Bootle, Merseyside, L20 3RA

Operators of COMAH scheduled premises are required to prepare a safety report making a number of demonstrations to show that they are able to prevent and limit the consequences of the major accidents identified from their processes. The regulator is required to come to conclusions about each operator's report and give their conclusions to the operator within a reasonable period of time. All the required reports have now been received. This paper outlines the key issues and lessons learned by the regulator about the preparation of such reports based on assessment experience so far. It summarises the key components of COMAH safety reports and what is good practice in writing them. The paper also takes a glimpse into the future by looking at some of the matters currently being considered involving COMAH and other permissioning regimes.

COMAH, Safety Reports

BACKGROUND

The Control of Major Accident Hazard Regulations (COMAH) apply to establishments that have, or can anticipate having, threshold quantities of dangerous substances. The regulations are goal setting and place a duty on operators of establishments to take all measures necessary to prevent and limit the consequences of a major accident. COMAH replaced the earlier Control of Industrial Major Accident Hazards Regulations 1984 (CIMAH).

COMAH is enforced in Great Britain by a new 'competent authority' (CA) which consists of the HSE and the Environment Agency in England and Wales and HSE and the Scottish Environment Protection Agency (SEPA) in Scotland. HSE and the Agencies work together jointly to implement the Regulations.

Central to COMAH is the requirement for operators of establishments, with higher thresholds of dangerous substances specified in the regulations (referred to as 'top-tier' establishments) to send a safety report to the CA. The safety report is a key element in identifying, preventing, controlling and mitigating major accident hazards. This is a continuation of a similar duty under CIMAH, but there are key differences between the two, which have led to deficiencies in COMAH reports submitted as follows:

- COMAH requires a report for an establishment as a whole, not just for individual hazardous installations containing dangerous substances
- The definition of 'dangerous substance' has been changed from a lengthy scheduled list of named substances to a very much shorter list. However it now includes 'generic' categories of substances, such as 'flammable', 'toxic' and 'dangerous for the

environment', based on their classification under the Chemicals (Hazard Information and Packaging for Supply) Regulations 1994, colloquially known as CHIP2.

- COMAH has more specific requirements in relation to safety management systems, including an additional requirement for a written Major Accident Prevention Policy to cover the overall aims and objectives with respect to control of major accident hazards.
- Increased emphasis on the prevention and limitation of the consequences of a major accident to the environment.
- Operators are now required to demonstrate how they are preventing and limiting the consequences of major accidents. Specifically, they are required to show the effectiveness of their management systems, that they have properly identified the major accident hazards, that the necessary prevention and limitation measures are in place, that the installations themselves are of adequate safety and reliability in design, construction and maintenance and that an on-site emergency plan have been drawn up.

IMPACT OF CHANGES
These changes have had a significant impact on the preparation of safety reports.

ESTABLISHMENTS AND DANGEROUS SUBSTANCES
Operators now have to consider the major accident potential across the whole of their sites, not just at the installations with the scheduled quantities and for a much broader range of substances, including preparations and mixtures, depending on their CHIP classification. As a result there are a further 100 plus COMAH top tier establishments contributing to a current total of over 350 such sites. The 'new entrant' sites include an increased percentage of businesses outside the traditional chemical industry, such as whisky warehouses, explosives sites and steelworks.

MANAGEMENT SYSTEMS
Furthermore, COMAH places emphasis on the importance of effective management arrangements for controlling major hazards and describes the framework for the systems that are required in Schedule 2 of the Regulations. This framework corresponds well to the model espoused in the HSE publication HS/G 65[1] *Successful Health & Safety Management* that should be familiar to all COMAH site operators. As a result, providing information in their safety reports about management arrangements has not caused UK operators too many problems, whereas in many EU countries this new development in Seveso II has been a major step change for operator and regulator alike.

ENVIRONMENT
A major challenge has been in showing that the major accident hazards to the environment have been systematically identified and as a result that the necessary measures to prevent a major accident to the environment (MATTE) are in place. Guidance on what is a MATTE[2] has been published but has proved difficult for operators to use during their risk predictive assessments. The Environment Agency has published guidance on undertaking an environmental risk assessment for COMAH[3].

The systematic identification of major hazards and assessment of their likelihood and severity is more complicated when considering the environment as a whole rather than just the people in it. If we look at the major accident potential in a simple way by considering the source of a release, the pathway to harm i.e. water, ground or air and the receptor, whether flora, fauna or the built environment, then the same model can be applied for both people and the environment as a whole. Clearly the release sources will be the same or similar and should be straightforward to identify. The pathway can be identified and the distribution modelled, however the response, in other words the harm, to the receptor from a quantity of release is a bigger problem given the diversity of receptors and the paucity of relevant dose-response data. This was a major topic that was highlighted in the COMAH CA Conference held in London (Nov. 2000).

The CA recognises the difficulty in providing appropriate data but expects operators to be thorough in the identification of sensitive receptors and to justify reasonable assumptions when describing the likely harm. Although these assumptions may introduce large uncertainties into the outcomes, these do not invalidate the importance of the risk assessment process and the consideration of whether existing control measures are adequate. The focus of the risk assessment should be on identifying release mechanisms and preventing them rather than whether the release affects an area of $5m^2$ or $7m^2$. However, some base line prediction (with uncertainties) or worst-case scenario has to be made to indicate the effectiveness of control, preventative or mitigation measures.

DEMONSTRATIONS

Even so, it has not been the above changes that have proved the most difficult challenge to operators in preparing safety reports and to the CA in assessing them. The requirement to demonstrate that certain controls are in place and the extent of the information required have proved the major stumbling blocks, so much so that up to 30% of the safety reports were initially returned for further work to be done. This did not mean that up to 30% of the UK COMAH sites were unsafe but it did mean operators had not made a case that their sites were safe and equally importantly that they had not shown they couldn't do more to prevent a major accident or limit its consequences.

CURRENT ISSUES RELATING TO DEMONSTRATIONS

The last tranche of safety reports due by 3 February 2002 has now been received. The majority was received by February 2001 so the CA has significant experience in assessing COMAH safety reports. There have been lessons learned by the CA as well as operators, resulting in revised procedures[4], which were introduced at the beginning of 2002 and are described elsewhere[5].

The key purpose of a COMAH safety report is to make the necessary demonstrations specified in Schedule 4 part 1. Assessors are finding that the following are the main reasons why operators have not been making these demonstrations as required:

- **Inadequate linking between major accident scenarios and the measures provided.** The links between the major accident scenarios and the necessary measures to prevent or limit the consequences of a major accident are not clearly made within the report. Too many operators are providing a description of what is done and asserting that they have all the measures that are necessary without making the links to scenarios, which are representative of the type of major accidents that could occur, to show this. The risk control systems that form part of the management arrangements, such as inspection and maintenance, change management, training etc. are important components of the full measures necessary when making these links.

- **Incomplete arguments as to why an operator believes that all necessary measures have been taken to prevent or limit the consequences of a major accident.** The guide to the COMAH Regulations[6] at paragraph 75 clearly states that prevention should be considered in a hierarchy based on the principles of reducing risk to **as low a level as reasonably practicable (ALARP)**. Operators are failing to show that there is a systematic risk assessment process, which is used to select, prioritise and schedule measures to reduce major accident risks ALARP.

- **Inadequate information.** The information provided about the measures in place and in what way they are relevant is often not enough or written in too general terms. Commonly operators are referring to company standards or practices without describing the relevance of the standards or guides they are referring to, what they cover, in what way they are relevant, how they operate, in what circumstances they are used and what limitations there are. Even so, the CA does not want to see the inclusion of copies of the company standards and guidance documents with a safety report, only an outline of how the standards are relevant.

MAKING THE DEMONSTRATIONS

ALARP PRINCIPLES
The demonstration that all measures necessary have been taken to prevent or limit the consequences of a major accident underpins the whole purpose of the safety report. This involves showing that the risks are ALARP, as explained above. The CA view of ALARP is discussed in HS/G 190[7] and in *Reducing Risks, Protecting People*[8] (R2P2). Essential considerations are:

- the scope for hazard elimination
- the scope for inherently safe design
- the extent to which relevant good practice is adopted

Where good practice has not been established, operators will have to show on a case specific basis that they have implemented risk-reduction measures to ALARP. **Good practice** is a much used and abused term. In this context, HSE use it as a generic term for those standards for controlling risk which have been judged and recognised by HSE as satisfying the law when applied to a particular case which is relevant and in an appropriate manner. This is discussed in more detail in R2P2[8] and supporting documents on HSE's website.

Where such codes or standards are relevant, the operator does not need to justify costs and technical feasibility against the acceptability of the risks involved, since these will have been considered when the codes or standards were prepared. However, the scope of the code or standard should be sufficiently clearly defined to know whether the specific circumstances addressed by the safety report are within scope or not.

The demonstration argument should show a clear bias towards safety, whether human or environmental, when justifying the measures in place are all that are necessary. Even in the cases where good practice, through following recognised standards, codes and guidance, is adopted throughout, operators should consider whether there are particular circumstances, because of the location of the establishment or because of serious consequences on site, that mean they should take additional measures to justify their arguments that risks are ALARP.

Where the report shows that a number of options for risk reduction exist, all options, or combination of options, that are reasonably practicable should have been implemented. The legal requirement to reduce risks ALARP means that the CA is likely to challenge a measure(s) provided if there are options that provide better protection but have not been chosen, for example on cost grounds.

ALARP arguments require an assessment of the risk that might be avoided. A comparison of net sacrifice and the benefits of risk reduction will give an overall view on the risks presented by a COMAH establishment. In making a demonstration in a COMAH safety report, this comparison is frequently made in monetary terms. If so, then case law indicates there has to be a considerable bias towards the cost of the sacrifice before an operator should accept that all measures necessary to prevent or limit a major accident have been taken. There is a requirement to show a gross disproportion between costs and benefits before deciding nothing further need be done. There are 2 simple questions, which should be answered by operators to take account of these issues: What more can I do? Why am I not doing it?

During assessment of COMAH safety reports, many operators have been asked to provide further information on extent and severity of their major accident hazards. Operators have been reluctant to provide information on the number of potential casualties, serious injuries and hospitalisations to be determined for each major accident hazard scenario given that it will be placed on the public register. Clearly this information is essential to enable a meaningful comparison between the sacrifice and benefits, when making the risk assessment and the consequent decisions as to whether further measures are necessary.

At the design stage, a life-cycle approach should be adopted taking account of the foreseeable risks throughout the lifetime of the installation and the measures required to prevent or limit the consequences of a major accident.

RISKS TO BE COVERED

Both individual risk and societal risk should be considered and both should include the risks to people off-site. HSE has published its approach to making ALARP decisions in R2P2[8]. It is the risk posed by reasonably foreseeable hazardous events from the duty holders' work activities both to employees and others not in their employ that must be addressed.

Societal or group risk is the risk of a number of multiple fatalities occurring in one event from a single major industrial activity and HSE has published criteria for addressing this risk. These criteria have been developed through the use of so-called FN-curves (obtained by plotting the frequency F at which such events might kill N or more people) to identify unacceptable, broadly acceptable and tolerable societal risks. These have been further developed for major hazards relating to dangerous substances and are available in an HSE document[9].

Even so, when considering the benefits at a particular site, if a measure results in a 'transfer' of risk to other people, the added risk to those people should be offset against the benefits the measures provide. For example, reducing the inventory of a hazardous substance by "just-in-time" delivery in road tankers rather than storage on site may be a transfer of risk. The added risk to those on the transportation route must be considered when making the ALARP decision but only to the extent over which the duty holder exercises control.

EXTENT OF INFORMATION PROVIDED

COMAH safety report requirements ask for demonstrations that the operator has reduced the risks to ALARP and that the establishment is being managed in such a way that this will continue to be the case. Some key points are:

- In showing that the risks are ALARP, the operator should **identify those events that dominate (safety critical events).** A necessary precursor to deciding the appropriate depth of information is a sufficient analysis of the hazards at the establishment. This is because the level of demonstration of safety and reliability of measures needed is proportionate to the severity of the hazard. The various major accident scenarios, their likelihood, the triggering events and their potential extent and severity must be analysed in the report.

- Operators can then **derive a set of major accident scenarios that are representative of the type of scenarios that are foreseeable.** The most serious events can be clearly identified and more attention paid in the report to demonstrating the measures in place for preventing these.

- Operators can say in their safety reports that they believe certain factors are not relevant to demonstrating that they are preventing or limiting the consequences of a major accident but should explain why. The safety report should **avoid over elaboration** in making justifications, particularly where this is disproportionate to the risks involved.

- The MAPP and SMS are key elements in ensuring continuing safety at establishments but it is relatively straightforward to present the demonstration if operators **follow the structure in Schedule 2 of COMAH.**

- In the preparation of safety reports, there has been a debate about the level of detail required in describing the prevention/control/mitigation measures in place. This information is primarily linked to the third demonstration in Schedule 4 – showing that there is adequate safety and reliability in the measures provided. The CA and the Chemical Industries Association (CIA) have prepared **guidance on the amount of detail required about the measures provided.** This guidance has been distributed to CIA members and published as Part 2 Chapter 8 of the Safety Report Assessment Manual[4], written for the guidance of CA assessors.

OUTLINE OF A COMAH SAFETY REPORT

The CA has said that it should not prescribe how operators present information and their demonstrations in their COMAH safety reports. It is clear however, in the light of experience, that some ways of presenting the information, required by Schedule 4 Part 2, are better than others in making the demonstrations, required by Schedule 4 Part 1. Some key points are listed here. The paragraph numbers in square brackets refer to the relevant paragraph of Schedule 4 Part 2.

1. SITE, ENVIRONMENT & MANAGEMENT INFORMATION

Factual information is required concerning

a. Description of the installations, processes and main activities [paras. 2(b)(c) 3(a)(b)]. The extent of detail will depend on how important they are as a source of or contributor to a major accident risk.
b. Description of dangerous substances on site. [Para. 3(c) provides a clear outline of the information required.]
c. Description of the environment around the site. This will include natural and built environment as well as populations, particularly vulnerable populations and where large numbers may gather. [para. 2(a)]
d. information on the management systems relating to major accident prevention. [para.1]. COMAH Schedule 2 outlines in some detail the type of information required.

This information is generally given in narrative text. HS/G190[7] discusses the factual information required, but those writing reports might find the list of criteria for assessors in Part 2 Chapter 1 of the SRAM[4] a useful checklist. The criteria in Chapter 2 cover general descriptions required and the criteria in Chapter 4 deal with management systems, supported by the type of evidence required. The descriptions of the management arrangements are key because these will show how the measures described in the safety report are to be delivered and maintained.

2. HAZARD IDENTIFICATION & MAJOR ACCIDENT SCENARIOS

The report should describe

a. What the arrangements are for hazard identification and risk assessment for the actual and anticipated substances and processes on site.
b. The major accident scenarios, which should include
   i. their probability or the conditions under which they can occur [para 4(b)]
   ii. an assessment of their extent & severity [para. 4(c)]

A table showing the links between the hazards and consequences identified and the preventive measures provided will help to demonstrate a systematic approach and could be used as a framework for developing more detailed demonstration arguments that all the measures necessary have been taken, later in the report.

This information will be reviewed at the initial stage of the CA's review and assessment of a report. The CA is looking for a systematic approach to risk assessment proportionate to the hazards and risks involved. It is important to remember that a major accident means an occurrence from an uncontrolled development in the course of operations leading to serious danger to people (whether workers on-site or other people off-site) or to the environment.

The 'extent and severity' information is essentially who might get hurt, how badly and how many it might be? To consider this, the hazard assessment of the identified major accidents must be carried out and then a prediction of the consequences made. In practice, this means providing information on potential casualties (both on site e.g. employees, contractors, etc. and people living and working nearby) for the representative set of major accident hazard scenarios, which form the operator's risk analysis. Information on extent and severity helps to determine what depth of demonstration is needed to show that prevention, control and mitigation measures adequately control major accident risks ALARP. The presentation of the extent and severity information can be in several forms and there is HSE guidance[10] expanding on what is required.

3.   MEASURES TO PREVENT OR LIMIT THE CONSEQUENCES OF A MAJOR ACCIDENT
The report should describe

a.   the technical parameters and equipment used for the safety of the installations [para. 4(c)]
b.   the equipment installed to limit the consequences of a major accident [para.5 (a)]

Where a number of hazards are comparable, information about the measures in place is sufficient so long as this provides evidence that the other similar hazards are also adequately controlled. A straightforward example of where this might be applicable is for similar storage vessels. The extent and type of information required can be more readily identified if a table linking hazards & consequences to the measures provided has been prepared.

Discussion on measures to limit the consequences of major accidents is as important as those of prevention, although the CA will review the latter first. In demonstrating that the measures provided have adequate safety and reliability [Schedule 4 Part 1 para.3], there is no need to include copies of recognised standards.

c.   The measures of protection and intervention to limit the consequences of a major accident including
   i.   the organisation of the alert [para. 5(b)] and
   ii.   the mobilisable resources [para. 5(c)]

Although the on-site emergency plan should be prepared together with the safety report, the CA does not expect a copy as part of the safety report. However the systematic analysis of the hazards and consequences, along with the risk assessment described in the report should enable the report to outline the emergency arrangements to limit these, confirm that an on-site emergency plan exists and explain the principles behind the plan, linked to the hazards and consequences.

4.   DEMONSTRATION ARGUMENTS
The report should show that the necessary measures have been taken to prevent and limit the consequences of a major accident

When showing that major accident risks are ALARP, the report should base its arguments on safety critical events. As discussed earlier, the depth of the risk assessment will depend on the extent and severity of the consequences and should be proportionate. Risk assessment techniques range from a simple qualitative approach to a detailed quantitative assessment. Fully quantified risk assessments, such as Quantitative Risk Assessment (QRA), are very costly and time-consuming exercises and because of this, some operators believe the effort is disproportionate to the benefits gained. One method for identifying the safety critical events, which may help to bridge the gap between qualitative and fully quantitative approaches, is to use a **risk matrix**. This type of approach has been widely used by many operators in their COMAH safety reports.

Risk is interpreted as the combination of consequence (severity) and likelihood (frequency). Both these are minimum requirements in Schedule 4 Part 2 of COMAH safety reports. A risk matrix enables this combination to be represented graphically. It is a reasonably quick and easy method to visualise the spread of risk and consequently is commonly used during (or after) hazard identification studies (such as a HAZOP), to screen hazards or to conduct a simple risk analysis. The main advantage of the matrix is its easy representation of different risk levels, and the avoidance of more time consuming quantitative analysis where this is not justified.

Consideration of costs and benefits require estimates. The basis for the risk estimate is usually qualitative, although it can be quantitative (for either the consequences or the frequencies or both). The matrix typically comprises a square divided into a number of boxes, with each box representing a different underlying risk level. Providing the risk analysis is based on cautious best estimates and the cost arguments appear realistic then the CA will use these to make a judgement on the risks involved and whether all the necessary measures have been provided.

For societal risks, if operators provide the required 'extent and severity' information' they can develop a considered argument on why they believe they have reduced risks ALARP. As a benchmark, HSE regards the risk of an accident causing the death of fifty people or more in a single event as intolerable, if the frequency is estimated to be more than one in five thousand per annum. FN curves have been drawn based on this figure and some operators have calculated their societal risks, relating to these, as part of their demonstration.

5.   IMPROVEMENT PLAN

Although the safety report should be able to demonstrate that an operator has taken all the necessary measures to prevent or limit a major accident, this is a an onerous test and it may well be that an operator will not be able to fully confirm this at the time the report is submitted. This is more likely for those operators who are required to prepare a safety report for the first time under COMAH. Relatively straightforward measures should be dealt with before submission of the report and the report can then confirm what is actually in place, however there may well be a number of measures which are more costly or time consuming to complete. Submission of the safety report should not be delayed because of these. As a result, the CA encourages the inclusion of an Improvement Plan, which includes the action an operator proposes to take and the timetable.

It is very easy to believe that the analysis in safety reports relating to hazard identification, consequence and risk assessment and linking the measures to prevent major accidents leads to Improvement Plans that are highly complex with sophisticated solutions. This shouldn't be the case. Much of the report should come down to systematic analysis of the hazards coupled with common sense solutions. The following is a recent example of an 'Improvement Plan'. This was for a steel works and concerned the handling of carbide.

Example
"The analysis we have done in conjunction with the preparation of the safety report has enabled us to find four shortcomings:

- the floor gullies in the work area will be sealed off and plugged in order to avoid the risk of explosion in the event of a very high emission of carbide
- the ground around the unloading bay and storage silo slopes so that there is risk of water accumulation. This will be rebuilt so that the slope will provide natural run-off
- delivery by road is a stand-by routine for rail delivery. When a road vehicle is unloaded, persons doing the work must stand outdoors. In wet or snowy weather, this involves greater risk than normal delivery by rail. So we are studying the possibility of rebuilding the unloading hall, so that lorries can also be unloaded indoors
- in winter, snow and ice are carried by railway wagons into the unloading hall. Puddles form in the unloading hall, with the associated risk of explosion. We are therefore investigating whether we can remove the snow and ice from the wagons before they are admitted into the unloading hall."

CONTINUOUS IMPROVEMENT
The inclusion of an Improvement Plan is a key point. It emphasises that the preparation of the safety report and the subsequent confirmation of the CA's conclusions is not the end of the process. In many ways it is a starting point because the report

- sets the baseline for preventing or limiting the consequences of a major accident at the establishment. Operators are required to think through the hazards and risks of their operations and assure themselves they have introduces appropriate control measures.
- forms a major part of the CA's inspection for the site. This will include matters to be followed up from the report and verification that the activities actually carried out on site are carried out in the way described in the report
- enables an informed dialogue between operators and the CA to reduce or remove risks of major accidents at the establishment.

The safety report is just one part of a strategy for regulatory oversight at top tier major hazard sites. The primary benefit of a safety report regime lies in the process of preparing the report. This requires operators to think through the hazards and risks of their operations and assure themselves that they have identified the hazards and risks in a systematic way and introduced appropriate control measures. The CA is anxious to avoid the preparation of safety reports being a 'one-off' paper exercise. The dangers are that the safety report becomes an expensive exercise in which all parties involved lose a sense of proportion as to the necessary measures to prevent major accidents and which has little if no relevance to the operation of the site on a day to day basis.

REVIEW & REVISION OF SAFETY REPORTS

COMAH does not require routine revisions to safety reports, but it does require operators to regularly review the content of safety reports

- as a result of changes to the safety management systems, establishment, installations, processes, nature and quantity of dangerous substance where these could have serious repercussions with respect to prevention of major accidents or the limitation of consequences of major accidents ('change reviews') and
- because of new facts or new technical knowledge, which in any case should be undertaken as a minimum at least every five years ('five yearly reviews'). An example of this is the guidance from HSE concerning ALARP and societal risk.

Policy is being developed on how revisions to COMAH safety reports should be prepared and handled. This will not only deal with the reviews required after 5 years but will also deal with revisions due to changes that have significant repercussions for major accidents and is likely to expect reviews and any consequent revisions of safety reports for whatever reason to be undertaken as early as possible, but taking into account overall resources and priorities.

The approach to preparing revisions to safety reports and their assessment must take account of major accident experience nationally and internationally as well as accident experience at a particular site over the period since the safety report was last reviewed. The review needs to be a robust process, which provides the public reassurance expected. The approach should also be flexible and although it needs to go over ground that has already been adequately covered in previous assessments, it need only do so to the extent of checking whether there are reasons to change the conclusions, for example because of new technology or new knowledge.

VALUE OF SAFETY REPORTS

Recent events, such as the railway accident at Ladbroke Grove, have encouraged HSE in particular to look at the value of regimes that require safety cases, safety reports or similar. These regimes are generally known as 'permissioning regimes'. As a result HSE has initiated a project to look at the value of COMAH safety reports for sites where a safety report was required by February 2002. This will be a long running project and will be reported[11] on as part of the Hazards XVII proceedings. HSE has also produced a discussion document[12] with a view to increasing transparency, stimulating discussion and seeking views on its approach to permissioning regimes and the fundamental principles that underpin HSE's approach.

It would be presumptuous to anticipate the conclusions of the work being done on this, however there were some telling comments by Lord Cullen in Part 2 of his report[13] on the Ladbroke Grove accident. He accepted the view that this type of regime (railway safety case) provided 'an appropriate means of managing safety' and 'an adequate assurance of safety for independent scrutiny'. He was also convinced that for goal setting legislation such as the Health & Safety at Work Act, there was a 'need for a framework required by legislation, within which the arrangements and procedures for the management of safety can be demonstrated and exercised in a consistent and effective manner' and that the 'framework within which management can exercise their *(sic)* responsibility for safety more effectively than under a highly prescriptive regime'.

CONCLUSION

This paper discusses some of the key problem areas that the CA has found in accepting COMAH safety reports. Some suggested ways forward in the form of an outline of a safety report have been given. This may be helpful for those operators who are currently dealing with the CA concerning their safety reports. It may also be useful for operators of sites that are new to COMAH top tier requirements. This applies to a number of sites from 30 July 2002, as a result of changes in classifications under the Chemicals (Hazard Information and Packaging for Supply) Regulations 2002, known as CHIP3 and as a result of the proposed amendments to Seveso II, which are likely to come into force in the UK in 2004.

A large amount of effort has been put into the preparation of safety reports by operators and by the CA in assessing them. This effort has been necessary to provide a good base line for the prevention of major accidents for future activities at COMAH top tier sites. The CA will expect operators to meet the levels of safety they describe in their reports and will develop this approach to ensure continuous improvement.

REFERENCES

1. Successful Health & Safety Management, HS/G 65, HSE Books 1997, ISBN 0-7176-1276-7
2. Guidance on the interpretation of a major accident for the purposes of COMAH, The Stationery Office 1999 ISBN 0-11-753501-X
3. Guidance on the Environmental Risk Assessment Aspects of COMAH Safety Reports, Environment Agency website, www.environment-agency.gov.uk
4. Safety Report Assessment Manual, Issue 3, HSE website, www.hse.gov.uk
5. Burns, J, 2002, What Everyone Wants? IBC's 4th Annual Conference on Safety Cases (IS1171)
6. A Guide to the Control of Major Accident Regulations 1999, L111, HSE Books PO Box 1999 Sudbury Suffolk CO10 2WA ISBN 0-7176-1604-5
7. Preparing safety reports: Control of Major Accident Regulations, HS/G 190, 1999 HSE Books PO Box 1999 Sudbury Suffolk CO10 2WA ISBN 0-7176-1687-8
8. Reducing Risks, Protecting People; HSE's decision–making process' HSE website, www.hse.gov.uk [also includes a number of documents referred to as the 'ALARP suite']
9. Guidance on ALARP decisions in COMAH, Semi-Permanent Circular (SPC)/ Permissioning /12, HSE website, www.hse.gov.uk

10. COMAH Safety Reports: Information about the extent and severity of the consequences of identified major accidents Semi-Permanent Circular (SPC)/ Permissioning/6, HSE website, www.hse.gov.uk
11. Thomas, R, 2003, COMAH Safety Report Regime-Evaluating the Impact on 'New Establishments', IChemE Hazards XVII proceedings
12. HSE Discussion Document DDE15 Regulating higher hazards: exploring the issues, HSE website, www.hse.gov.uk/disdocs
13. The Ladbroke Grove Inquiry Report Part 2, www.lgri.org.uk

COMAH COMPLIANCE FOR A FINE CHEMICALS & EXPLOSIVES FACILITY LOCATED WITHIN A SSSI

A Ennis[1] & RM Thomas[2]

[1]Haztech Consultants Ltd; Meridian House Business Centre, Winsford, Cheshire CW7 3QG
[2]Formerly with exchem organics; Great Oakley Works, Harwich Road, Great Oakley, Harwich, Essex, CO12 5JW. Now with Haztech Consultants Ltd.

SUMMARY

This paper describes experiences in the preparation of the COMAH report for the exchem organics Harwich site. exchem organics is a small fine chemicals site which was not covered under the CIMAH regulations but fell into the top tier of COMAH and hence had to submit a full report in February 2002. The site combines manufacturing of fuel additives and fine chemicals with explosives handling facilities and is located in the middle of a Site of Special Scientific Interest (SSSI).
The particular mix of activities on the site combined with the location presented a number of challenges in the preparation of the COMAH report. Constraints were also experienced due to the relatively small management team and it was decided to seek expert assistance from an external consultant. The report was prepared by exchem organics personnel with specialist assistance provided by Haztech Consultants Ltd.
Consequence modelling was required for a wide variety of identified hazard scenarios including gas dispersion, fire, condensed phase explosion and vapour cloud explosions. Results of the consequence modelling along with risk were considered in the compliance with the HSE "As Low As Reasonably Practicable" (ALARP) criteria[1,2] for the site.

INTRODUCTION

The SSSI contains one of only two populations of Sea Hogs Fennel in the UK and is the only UK breeding ground for Fisher's Estuarine Moth. In addition the site is surrounded by environmentally sensitive salt marsh with a colony of Common and Grey Seals inhabiting the area. The area is also designated a RAMSAR site under the Convention of Wetlands of International Importance for its importance as a wetland bird habitat. The site is home to a variety of wildlife other than those mentioned above including several species of owl, hawk and butterfly. There is also a wide variety of important wildlife located around the creeks and mudflats on the other side of the sea wall adjacent to the site. Although not part of the SSSI it is an environmentally sensitive area which is regularly monitored.

The site is a top tier COMAH site due to the amount of Nitroglycerine compounds held which are classified by Risk Phrase as Very Toxic. The site also holds significant quantities of Oleum and Nitric Acid along with a product classified as "R51 dangerous for the environment".

SITE OVERVIEW

The site consists of three main process plants and the explosives handling and storage activities. The manufacturing operations are concentrated into a relatively small area and are surrounded by the SSSI and farmland. Much of the farmland is also owned by the company and total site area is about 1200 acres. A dock provides the capability to accept and despatch materials by sea. As an explosives site, there is already a high degree of licensing and regulation in place for the explosives handling activities.

Due to the nature of the exchem organics chemicals manufacturing operations a very wide range of chemical products can be manufactured on the site. This created difficulty in fulfilling the requirements of the "Descriptive" section of the Safety Report Assessment Manual (SRAM). Thus, a typical range of generic chemistries was described. These were cross referenced to the detailed safety and operational procedures to demonstrate that the process was safe and of low environmental risk prior to production commencing. To describe all of the potential processes in detail in the COMAH report would have made for a very bulky document and been extremely time consuming and hence this generic approach was essential. In addition, the explosives operations encompass a vast range of materials. In this case, typical explosive materials were used as exemplars and their hazards covered in detail in the report. The assumption was then made that the effects from a Major Accident, in terms of blast overpressure and toxic effects, would be directly comparable for other materials with a similar risk phrase thus reducing the amount of work required to an acceptable level. It is not known at the time of writing whether this approach is acceptable to the Competent Authority (CA).

COMAH RELATED INFORMATION ALREADY AVAILABLE
exchem organics already had a considerable amount of information available that could be used in the COMAH report. In particular, the company had extensive information on the environment surrounding the site going back many years.

exchem organics carry out Hazard Studies and Risk Assessments on all new chemical processes and products likely to be manufactured on the site. These assessments not only include process safety but also allow a detailed environmental risk assessment to be in place before any significant steps towards production are taken. Risk assessments are carried out generally in accordance with the guidance in Reference 3 and Hazard Studies in accordance with References 4 and 5. These procedures assisted significantly in the demonstrations required by the COMAH regulations. The records are available for all of these in the company's record keeping system. There is also extensive information available within the company on chemical hazards and COSHH assessments.

The company also has extensive safe operating, maintenance and emergency procedures, which are administered by a quality system including full document control. A computer based planned preventive maintenance system was also in the process of implementation at the time of writing the COMAH report.

The site is regulated under the Explosives act of 1875 and 1923 and a licence is granted for certain controlled operations within the site, these being to manufacture store, test and destroy explosives within defined areas of the site. This ensures a highly regulated operation and an appropriate level of safety procedures and documentation. The company also has a Safety Management System (SMS) that is generally compliant with the guidance in HSG65[6].

Whilst this information is all relevant to preparation of the COMAH report, it was, however, noted that certain information was lacking which would be needed e.g. detailed consequence modelling and analysis. Additionally it was felt that a site-wide review of Major Accident Hazards (MAHs) would be beneficial.

CHALLENGES IN THE PREPARATION OF THE COMAH REPORT
The main challenges faced in preparation of the COMAH report can be summarised as follows:

RESOURCE
The exchem organics management team is relatively small and there is little spare capacity for additional work to be taken on without compromising the day to day operation and safety of the site. It was not considered acceptable that operational efficiency and safety should be compromised to write the COMAH report. It was identified at an early stage that significant additional resource would be needed from outside the company to complete the report within the required timescale.

SPECIALIST KNOWLEDGE
Although exchem organics has a considerable level of expertise within the company it was clear that additional specialist knowledge was required in areas such as Risk Assessment, Hazard Review and Consequence Modelling and knowledge of COMAH legislation and safety case preparation.

CONSEQUENCE MODELLING
Consequence modelling requires the use of specialist tools and the expertise to interpret the results. Neither the tools nor expertise were available within the company. Consequence modelling is discussed further in the sections below.

REPORT STRUCTURE
Structure of the report was seen as an issue at an early stage. Consideration was initially given to constructing the report following the structure of the guidance document HSG190[2] but this was dismissed as it would have led to an extremely repetitive, lengthy and difficult to read report. It was also apparent that the time taken to construct an applicable report structure could be better spent in other areas.

COST
The cost of preparing the report was an issue from the onset when set against budgetary constraints. The cost of using a consultant has to be balanced against the savings and benefits from freeing up exchem organics management for other tasks.

71

SELECTION OF CONSULTANT

For the reasons discussed above, a decision was taken to seek outside assistance from a specialist safety consultancy. It was essential that the consultant selected had the complete range of appropriate skills available in house. Selection was made based on a combination of experience, cost and also interpersonal aspects (since exchem organics and the consultants would have to work closely together for an extended period this was considered important).

A decision was taken to use Haztech Consultants Ltd. based on a combination of price, experience and ability and the availability of a CD-ROM based COMAH report structure. Attendance at a Haztech IChemE approved COMAH course was a contributory factor in the decision since it provided an opportunity for exchem organics management to meet the personnel who would be working on the report and assess the expertise available.

APPROACH TAKEN TO REPORT WRITING

It was considered essential that exchem organics retained ownership of the report and the information contained therein. Hence, the process was led by a senior member of the exchem organics management team (the Site Operations Manager). Other members of the exchem organics management team were used as appropriate with personnel from Haztech Consultants Ltd providing specialist knowledge and backup. The whole report was reviewed by a joint exchem organics/Haztech team and a gap analysis carried out prior to the final draft being produced.

This approach allowed the majority of the exchem organics management team to continue with normal duties whilst facilitating completion of the COMAH report in the required timescale. If key site management personnel are withdrawn from normal activities then there are risks that personnel become isolated from normal business roles and production and day-to-day safety become neglected.

Due to the size and complexity of the report it was necessary to employ a quality revision control system in order to prevent confusion and duplication of effort. A simple colour coding system was used in the text. Only the exchem organics Site Operations Manager was allowed to finalise any parts of the document. Files were identified both by date and revision letter as a double check.

There was also a single point of contact for communications between Haztech and exchem organics in order to ensure that all communications were properly recorded and directed. The master copy of the document was held by exchem organics and various people were assigned to work on specific sections. A programme of work was also set up so that progress could be monitored against milestones and deliverables. A target date for completion of 24 December 2001 was set to allow for any last minute adjustments and the final two weeks before submission date for printing and binding. This date was, in fact, met and the report submitted three days before the deadline.

REPORT STRUCTURE USED

As stated previously, HSG190[2] was judged unsuitable to use as a template. A prefabricated template was supplied by Haztech based on EU guidance[7] which is more clearly structured.

In order to provide flexibility for future modifications e.g. new plants and processes the report was constructed as a "core" and a number of installations. The core consists of three sections:

- Descriptive
- Hazard & Risk Assessment
- Safety Management

The core of the report contains features that are common across the whole site and the installation sections deal with information specific to the installation referring back to the core report where necessary. Demonstrations and supporting information are placed in the appendices. This significantly reduces the amount of repetition. The format used also allows for easy amendment of individual sections.

Since the site handles explosives, there is a certain amount of highly confidential information which it was considered imprudent to allow into the public domain for security reasons. The confidential information was, therefore, placed in a special appendix in a separate folder thus allowing it to be withheld more easily. Advice was taken from the Security Services as to which information was considered sensitive.

HAZARD IDENTIFICATION & RISK ASSESSMENT

Hazard Identification and Risk Assessment (HAZID & RA) are a key part of the COMAH report and site safety in general. Although a variety of HAZID and RA methods are already used on the site it was considered beneficial to carry out a site-wide review to ensure that all MAHs were identified. This review was carried out using a guide word approach and using a small team comprising a study leader provided by Haztech and relevant personnel from the exchem organics management and operations teams.

The outcome of the Hazard Review was a set of Hazard Summary Sheets. These sheets provide linkage between: Hazard, Risk, Consequences, Preventive measures, Protection systems, Mitigation measures and Emergency response. The methodology used is similar to Hazard Study 2 and consistent as described in the widely known Hazard Study methodologies from the IChemE[4] and CIA[5] and therefore acceptable to the Competent Authority. The process was aligned towards Major Accident Hazards and thus minor incidents were filtered out thus reducing the number of scenarios to be modelled down to a reasonable level. Risk assessment was carried out for the identified Major Accident Hazard scenarios in order to judge compliance with the ALARP criteria (see below).

The Hazard Summary Sheets form a key part of the COMAH report and potential Major Accident Hazard scenarios can clearly be identified from them. Outstanding actions are recorded in the same manner as a hazard study and may be carried forward into the Site Improvement Plan. Identified potential MAH scenarios are screened prior to consequence modelling in order to (a) eliminate and scenarios that are not MAHs and (b) identify any common scenarios that can be used in several cases.

CONSEQUENCE MODELLING

The mix of activities on the site necessitated consequence modelling of gas dispersion, pool fire and both TNT (condensed phase) and TNO Multi-Energy (vapour cloud) explosions. PHAST Version 6.0^8 has all of these models available.

Consequence modelling was carried out by Haztech personnel and interpretation of the results discussed and agreed with exchem organics. It should be noted that interpretation of consequence modelling results requires specialist knowledge and training. Several issues were raised from the consequence modelling which were then examined in further detail:

- It was possible to eliminate some Major Accident Hazard scenarios completely due to the low level of consequences
- Some events which were potential Major Accident Hazards were discovered not to be as large as originally estimated during the Hazard & Risk Assessment
- A few events were identified which had more severe consequences than originally thought

These events and their consequences were given consideration against the potential safety improvements that could be made and recommendations fed into the Site Improvement Plan.

GAS DISPERSION

The PHAST Unified Dispersion Model was considered appropriate for the gas dispersion cases being considered in the consequence modelling as it has been widely validated[9] and was known to be available within the HSE. Other gas dispersion models and modelling programs were considered[10] but on examination did not offer any advantage over using PHAST.

Some gas dispersion modelling had been done several years previously for a limited number of cases by another consultant. This was, however, judged to be not of use for the COMAH case as the events and source terms were not clearly defined and therefore could not be clearly linked to the identified Major Accident Hazard scenarios. A representative set of consequence modelling scenarios was, therefore, defined based on those in the Hazard Summary Sheets. Accurate source terms are essential for consequence modelling and hence considerable care was taken to ensure that the source terms used were appropriate for the particular scenario being modelled.

For gas dispersion modelling detailed wind and weather data were obtained from a local weather station and these were used to define two representative weather conditions to be used in the modelling, these being Pasquill-Gifford categories D5 (atmospheric stability class D with 5m/s wind speed) and F2 (stability F with 2m/s wind). D5 represents the most commonly found conditions on the site occurring >65% of the time and F2 a worst case for gas dispersion (stable conditions with low wind speed)[11]. There were no other cases that would benefit from being modelled since there was no warehouse fire Major Accident Hazard scenario which would have meant dispersion modelling for weather conditions D10 and D15. A degree of sensitivity modelling was used for each of the scenarios e.g. for a leak of Oleum, the model was run for 6mm, 12mm, 25mm and 50mm hole sizes as well as a catastrophic tank rupture. Both side views and cloud footprints were considered in order to obtain an estimate of the release consequences. Possible limitations of the dispersion model were also considered when assessing consequences.

For modelling of Oleum gas dispersion, the guidance in References 10 and 12 were taken into account. From this it was apparent that the release of Oleum in dry conditions did not pose a significant hazard but a release in wet conditions would result in the rapid evolution of H_2SO_4 fumes[18]. This is a particular problem for two reasons (a) H_2SO_4 is highly toxic and (b) the cloud disperses relatively slowly.

The Dangerous Toxic Load for H_2SO_4 is: DTL $= 1.30 \times 10^4$ ppm^2.min and hence a concentration as low as 114ppm for 1 minute may potentially be fatal. This compared to the DTL for SO_3 which is: DTL $= 4.655 \times 10^6$ ppm^2.min i.e. DTL for 1 minute exposure is 2158ppm. Thus it can be seen that any release of Oleum in wet conditions could cause a major incident and, since the Oleum tanks are located in a bunded area there was potential for the evolution of a sizeable cloud of H_2SO_4 with a relatively small depth of rainwater in the bund. For an instantaneous release e.g. catastrophic rupture, the cloud moves downwind in the form of a hemisphere. This can be seen on Graphs 1 and 2 showing the cloud centre line concentration after 1 minute and 2 minutes.

The size and concentration for this scenario caused potential problems both on- and off-site. The works restaurant and plant managers' offices were particularly vulnerable being located only 80m from the source with a potential total of 25 people present. The persistence of the cloud indicated that it could drift off-site under certain weather conditions and, although unlikely to cause fatality or serious injury, could potentially cause distress to members of the public. This was taken extremely seriously by the company and potential reductions in hazard and risk put in hand in the site Improvement Plan. For smaller hole sizes, H_2SO_4 is evolved over a period and hence can be modelled as a continuous release. At these lower release rates the cloud size is much smaller and the effects do not go off-site however there is still potential for injury or fatality to personnel on-site. A concentration versus distance diagram for this case can be seen in Graph 3.

The net result of this is that consideration of the process from the aspect of inherent safety has indicated a route by which the requirement to store Oleum on the site may potentially be eliminated completely and this is being pursued.

POOL FIRE

The site contains several bulk storage tanks containing principally Methanol or Methanolic liquors and also one or two other flammable liquids. These are contained in a bunded area located adjacent to one of the plants. The radiation due to a pool fire in the bunded area was calculated in order to assess the potential impact on the adjacent plant. The radiation versus distance curve for this case can be seen in Graph 4. It was concluded that this case did not pose any significant hazard to plant since all adjacent tanks were protected by safety relief valves and other plant was located a sufficient distance from the source so as not to be affected.

TNT EXPLOSION

Although consequence zones are set on the explosives license these are not sufficient for the purposes of COMAH and hence the explosion modelling had to be revisited for the explosives handling facilities.

The site has a number of magazines and explosives handling facilities including burning grounds and loading facilities. These are protected to a degree by their construction and external earth mounds with safety distances and maximum quantities set by the explosives license. This means that explosives handling facilities are located well away from other activities carried out on the site and consequences of any incident are therefore reduced. Nevertheless, there are several points at which an explosion could occur outside of the protected areas e.g. offloading from ship and road transport to magazines.

The TNT explosion model is well established in the literature[13,14,15] and although generally discredited for modelling of vapour cloud explosions[16,17], it can be seen that it is appropriate for modelling point source, condensed phase events. A number of explosives are, however, significantly more powerful than TNT (e.g. TNT/RDX mixtures, Nitroglycerine) and hence this must be taken account of in the model.

Explosion overpressure graphs were produced for a several quantities of explosives to provide sensitivity analysis. Typical overpressure radii can be seen in graph 5.

VAPOUR CLOUD EXPLOSIONS (VCE)

Vapour cloud explosions were modelled using the TNO Multi-energy model[21,22] which is the most widely used in Europe and is considered to be the most appropriate. The alternative Baker-Strehlow model[23,24] is mainly used in the USA and was not judged to offer any significant advantage over the TNO model. As stated above the TNT model has been discredited for VCE modelling. With all consequence modelling, it is essential that the source term should accurately reflect the potential hazard situation and here again considerable effort was expended in obtaining the correct parameters for use in the PHAST model.

The particular event identified was the loss of containment of a flammable solvent above its' flashpoint inside one of the process buildings. Since the process being carried out involves a period of operation at reflux it was considered that there was potential for a significant release of flammable vapour within the building.

The degree of uncertainty in the TNO model stems principally from the estimate of the volume of gas in the congested area and the degree of congestion. Thus, conservative estimates of both the volume and degree of congestion were used in order to ensure that the worst case scenario was used. Additionally, the TNO model assumes implicitly that the whole of the congested volume is at the stoichiometric concentration i.e. the worst case in terms of explosion violence and hence is again conservative since in a real world situation the mixture is extremely unlikely to be homogeneous within a compartment. A typical curve for overpressure versus distance is shown in Graph 6.

In order to provide some degree of sensitivity analysis two cases were run, one with 100% volume fill of the compartment (absolute worst case) and the other with a 10% fill (more credible case). As above, the explosion effects were assessed based on references 3, 11, 12 & 21. The results of the analysis indicated that other than the process plant itself, the site canteen and plant managers' offices were potentially at risk. A number of actions were raised for the Improvement Plan in order to reduce both the hazard and risk from this process.

COMMENTS ON CONSEQUENCE MODELLING
Interpretation of the effects of overpressure was made using the guidance in References 11, 18, 19. Damage criteria for various overpressures can also be found in these references. Potential for damage to buildings took into consideration the vulnerability of the various structures on the site and the guidance in Reference 20. There is a diverse mix of buildings on the site ranging from "blast-proof" structures specifically designed for explosives manufacture (one of the plant control rooms) to wood frame and prefabricated buildings (e.g. offices and amenities buildings) hence, vulnerabilities varied widely. The explosives license manages safety by setting separation distances and hence it was found that occupied buildings were generally at minimal hazard from condensed phase explosions.

The use of PHAST allowed for rapid modelling of the various potential Major Accident Hazard scenarios identified. Although much of the modelling could have been done by manual methods or other computer programs, the speed and output features of PHAST significantly reduced the time and cost of the exercise. The major disadvantages being (a) the first cost of PHAST and (b) the requirement of specialised knowledge to use to program effectively and interpret the results. As with many complex computer programs garbage in will inevitably result in garbage out.

The consequence modelling and analysis exercise proved very useful in the COMAH report to demonstrate that the consequences of the particular Major Accident Hazards were fully understood by the company and the results used in a positive manner to improve the safety of the site. It should be noted that consequence analysis was carried out against safety, health and environmental[25] criteria.

ENVIRONMENTAL ASPECTS
exchem manage all activities with due regard for the environmental sensitivities of the site and can demonstrate a year-on-year environmental improvement in the quality of the SSSI and surrounding area based on regular reports by English Nature and other environmental bodies. The company has also won a number of environmental improvement awards over a ten-year period. Whilst this information was useful in demonstrating the companys commitment to the environment, the potential for Major Accidents To The Environment (MATTEs) still had to be considered carefully against the criteria in Reference 25 using Source, Pathway, Receptor methodology. A considerable amount of time was spent on the environmental aspects of the COMAH report due to the sensitive nature of the site environs.

The potential to cause damage to the SSSI and surrounding areas was noted from the Hazard review process which identified a number of events that could potentially impact on areas of the SSSI. The potential hazard for a MATTE was, however, estimated as limited due to a combination of the location of hazardous material storage, and the relatively large area of the SSSI in relation to the damage potential of the identified Major Accident Hazards. Gas dispersion modelling was carried out to determine potential effects due to a release of Oleum and this indicated that both the limited plume size and duration would result in minimal impact on the environment in general. Difficulty in deciding appropriate harm criteria to the SSSI and other wildlife was experienced due to the lack of available, applicable scientific data. In particular there are no data relating to the effects of toxic gases on Fishers Estuarine Moth.

The low-lying nature of the site means that drainage is achieved by a number of dykes and ditches. In the event of a loss of containment of liquid, these can be isolated by means of a set of sluice gates to prevent contamination spreading by this pathway. There was also some debate on the effectiveness of some of the protective measures, in particular the risk of flooding and potential since the site is protected by a sea wall designed for a once in 200 year event. The sea wall is designed, constructed and maintained by the Environment Agency.

The Risk of a MATTE was estimated as being very low due to a combination of good engineering design of equipment, secondary containment and mitigation measures in place.

DEMONSTRATION OF ALARP

One of the major problems of interpretation in the COMAH guidance is the definition of the phrases "As Low As Reasonably Practical" and "All Measures Necessary" and their implementation in the COMAH report. It was originally stated that compliance with applicable standards or guidance would be sufficient to demonstrate that risks had been controlled to an ALARP level. In later versions of the HSE Safety Report Assessment Manual it became apparent that ALARP had been expanded to mean examination of all possible options, even if the options were not covered by current industry best practice or economically feasible.

Fortunately this did not have a great impact on the site since most major risks were already adequately controlled. Where risks were found that could potentially be reduced these measures were put into the site Improvement Plan for further consideration. Due consideration was made of ALARP criteria using the guidance in the HSE Safety Report Assessment Manual on the HSE web site. In particular a matrix approach was used to balance hazard and risk[26].

One example of this was the Oleum tanks which were located outside in bunded areas. Gas dispersion showed that there were potential off-site effects from larger releases where the release mixed with water (e.g. rainwater in the bund) as well as potentially severe effects for a number of occupied buildings on-site. This had not previously been thought to be a problem due to the size of the site and location of the tanks. Consideration against ALARP criteria indicated that this was potentially unacceptable to the company purely in terms of hazard even though the risk was relatively low and hence an action was raised in the site Improvement Plan to consider potential improvements. This particular item was given the highest priority as it was judged to be the worst Major Accident Hazard identified.

COST/BENEFIT

The cost of COMAH compliance has been considerably higher than originally estimated from HSE figures in Reference 27. This must be considered as both quantifiable, in terms of the direct costs of compliance (HSE charges, consultancy fees etc.) and non-quantifiable being the costs of removing staff from other, equally important safety work or direct manufacturing duties. Directly quantifiable benefits from the COMAH process have, so far, been negligible. Overall, the cost/benefit analysis of COMAH must be questioned given the high cost of compliance compared against the good safety record of the company and the UK chemical industry in general.

SITE IMPROVEMENT PLAN

A site improvement plan was developed based on the actions raised from the Hazard Summary Sheets and consequence modelling. All actions were rated based on safety, health and environmental criteria, consideration being given to reducing both Hazard and Risk. The potential benefit of the improvements was considered against the available capital budget in order to make the most effective use of the available money. All actions have also been given target completion dates based on projected available finances. The Improvement Plan is a live document covered by the site Quality System and will be reviewed and updated periodically to account for changes in priorities, plants and financial constraints.

One example of this was the Oleum tanks as described above, the action being to prevent water entering the area under the tanks, or otherwise prevent a release from mixing with water. Potential solutions include roofing over the area or double skinned tanks. The best solution, of course, being removal of the tanks from the site altogether and replacement with a less hazardous feedstock, this latter being the long-term objective.

CONCLUSIONS

It was identified at an early stage that given the financial cost of COMAH compliance it was essential that maximum value was obtained from the exercise. The benefits from COMAH can be summarised as:

-   Focus of attention specifically on the prevention of Major Accident Hazards as distinct from "slips, trips and falls" safety
-   Reclassification and prioritisation of some hazards on the site on the basis of
-   Improvement of Emergency Planning by clarification of the consequences of the identified MAHs and improved consequence modelling and assessment
-   Better communications with Local Authorities and the public from the consultation process and Emergency Planning exercise
-   Clarification of some aspects of the SMS and better understanding of how the it links together as a result of having to describe and write down the process
-   Potential long-term improvement in site safety from concentration on MAHs and rationalised priority list

As a direct result of writing the COMAH report a number of the hazards on the site were re-classified and a number of identified improvements already have been carried out. Others, however, remain long-term projects due to financial constraints.

Particular problems encountered during the report writing process were:

- A significant amount of time was spent by all of the exchem organics management team in preparation of the COMAH report thus taking them away from routine activities. It was apparent that there was an increase in pressure on certain key members of the staff during this period.

- It is difficult for a small company to have the level of expertise available to complete a COMAH report which will stand up to scrutiny by the experts within the CA, thus there is additional cost in the support needed from external consultants. There was also some difficulty in interpreting what proportionality actually meant for exchem organics.

- Uncertainty with regard to the CAs actual requirements due to regular changes in the SRAM and a lack of other public-domain guidance. Although the SRAM is intended for HSE assessor use, it is the only really useful source of information on the HSEs expectations in the COMAH report. Other guidance e.g. Safety Report Assessment Guides has appeared too late to be used in writing the COMAH report. Compliance with the key ALARP requirement was made difficult by changes in the guidance on the subject which were very difficult to track on the web site.

- Cost of COMAH report preparation both in financial and temporal terms may have a potentially adverse short-term effect on safety. The cost of preparing the COMAH report for this site is estimated at approximately £180,000 to date, taking into account consultancy costs, management costs etc. but not including CA fees which currently stand at £20,000. This represents a large portion of the companys annual profits. The cost of additional work required in order to fulfil the additional detail required after preliminary assessment has not yet been quantified.

SUMMARY

COMAH compliance produced a unique set of challenges for a small site with diverse business interests. It would not have been possible to produce the report without the assistance of outside consultants. The COMAH process has been financially extremely expensive for the site and also placed a lot of additional stress on the site management. Additional problems in compliance have been caused by the continuous changes in the HSE SRAM and difficulty in interpreting the HSEs exact requirements for COMAH compliance.

Consequence modelling was an important part of the COMAH process which proved valuable in identifying the main hazards on the site. Analysis of the output provided some surprising results for a number of identified Major Accident Hazard scenarios.

The actual value of COMAH in improving safety is yet to be confirmed but given the recent good record of the British chemical industry in respect of fatalities or injuries to members of the public it will probably take a considerable time for benefits to become apparent. A cost/benefit analysis on the cost of COMAH to the British chemical industry against the improvement in terms of the reduction in safety and environmental incidents would prove interesting as would a direct comparison with COMAH costs and implementation in other EU countries.

REFERENCES

1. Guide to COMAH regulations 1999; HSE; HSE Books; 1999
2. Preparing safety reports: Control of Major Accident Hazards Regulations 1999; HSG190; HSE Books; 1999
3. Pitblado R & Turney RD (Eds); Risk Assessment in the Process Industries 2nd Edition; IChemE, 1996
4. HAZOP: Guide to Best Practice; BJ Tyler, F Crawley & ML Preston; IChemE, Rugby, UK; 1999.
5. CIA; A guide to Hazard & Operability Studies; CIA 1987 reprint
6. HSE; Successful Health & Safety Management HSG65; HSE Books; 2000 Reprint
7. Guidance on the preparation of a Safety Report to meet the requirements of Council Directive 96/82/EC (Seveso II)
8. Process Hazard Assessment Software Tool "PHAST" Version 6.0; DNV Technica; London
9. Holt A, Witlox HWM; Validation Of The Unified Dispersion Model; Consequence Modelling Documentation; PHAST Documentation; March 2000, UDM6.0
10. AIChE CCPS; Guidelines for Use of Vapor Cloud dispersion Models 2nd Edition; AIChE CCPS; 1996
11. Lees FP; Loss Prevention in the Process Industries; 2nd Edition, 1996 Butterworth Heineman
12. Griffiths R (Ed); Sulphur Trioxide, Oleum & Sulphuric Acid Mist, IChemE Major Hazards Monograph; IChemE 1996
13. Gardner DJ, Hulme G, Hughes DJ, Evans RF & Brighton P; A survey of current predictive methods for explosion hazard assessments in the UK offshore industry; Hazards XII, European Advances in Process Safety; UMIST 19–21 April 1994, IChemE Symposium Series No.134
14. Lawrence WE & Johnson EE; Design for limiting explosion damage; Chem Eng, 7 Jan 1974, pp96–104
15. Strehlow RA & Baker WE; Characterization and evaluation of accidental explosions; NASA CR-134779 (Contract research report) June 1975
16. Van Den Berg AC, Van Wingerden CJM & The HG; Vapour Cloud Explosions: experimental investigation key parameters and blast modelling; Tr IChemE, Process Safety & Environmental Protection; August 1991
17. Marshall VC; TNT Equivalence - The Decatur Anomaly; TCE, Feb 1980, pp108–109
18. AIChE CCPS; Guidelines for Evaluating the Characteristics of Vapor Cloud Explosions; AIChE CCPS; 1994
19. IChemE; Explosions in the Process Industries 2nd Edition; Major Hazards Monograph, IChemE 1994
20. Chemical Industries Association; Guidance for the location of occupied buildings on chemical manufacturing sites; CIA, February 1998
21. Van Wingerden K, Hansen RO & Foisselon P; Predicting blast overpressures caused by vapor cloud explosions in the vicinity of control rooms; Process Safety Prog, Vol.18, No.1, Spring 99, pp17–25
22. Van Den Berg AC, Van Wingerden CJM & The HG; Vapour Cloud Explosions: experimental investigation key parameters and blast modelling; Tr IChemE, Process Safety & Environmental Protection; August 1991

23. Baker QA, Doolittle CM, Fitzgerald GA & Tang MJ; Recent developments in the Baker-Strehlow VCE analysis methodology; 31st AIChE Loss Prev Symp 9–13 March 1997; Paper42f pp1–16

24. Baker QA & Tang MJ; Vapor cloud explosion analysis; Process Safety Prog; Vol.15, Part 2, pp106–109, 1996

25. Guidance on the Interpretation of Major Accident to the Environment for the purposes of the COMAH regulations; DETR

26. Middleton M & Franks A; Risk Matrices; TCE, Issue 723, September 2001

27. HSC; Proposals for Regulations implementing the Directive on the Control of major accident hazards involving dangerous substances; Consultative Document; 1998

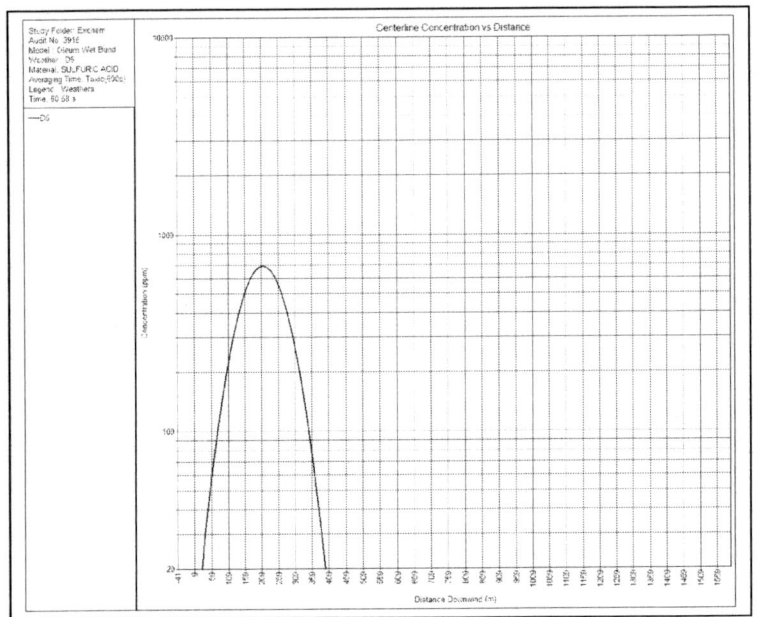

Graph 1. H$_2$SO$_4$ instantaneous release after 1 minute

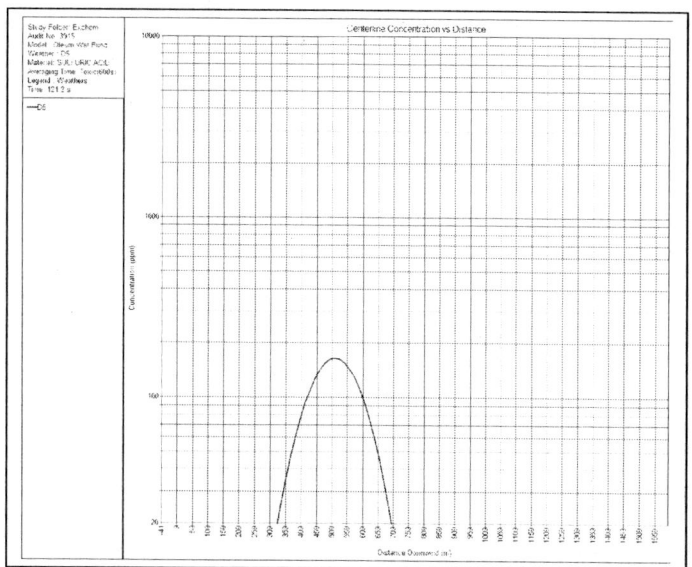

Graph 2.   H₂SO₄ instantaneous release after 2 minutes

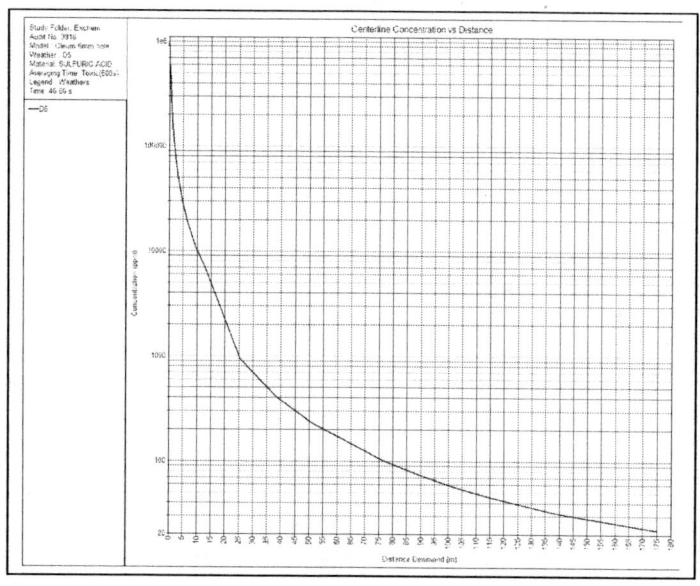

Graph 3.   Centreline concentration versus distance for H₂SO₄ continuous release

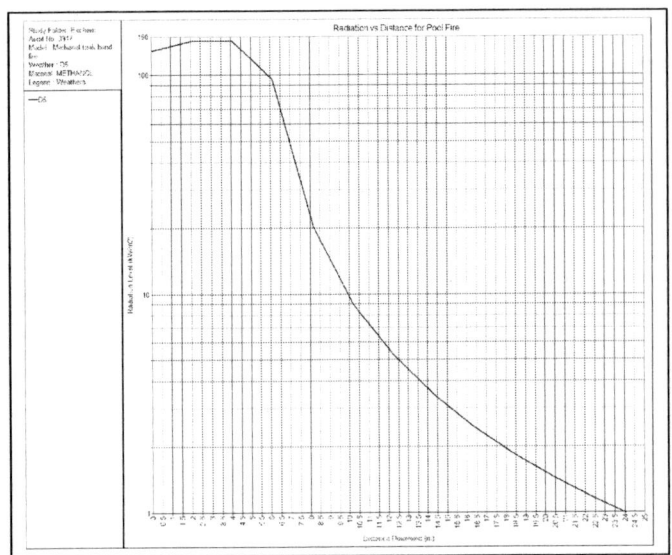

Graph 4. Thermal radiation vs distance curve

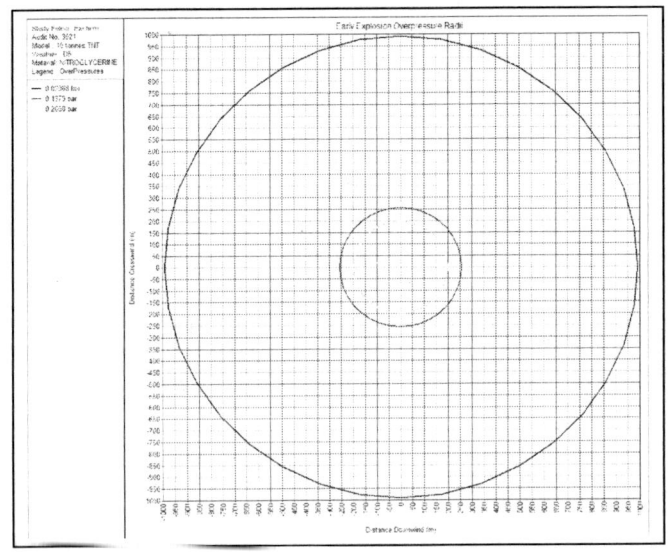

Graph 5. Overpressure radii for 10 tonnes of TNT

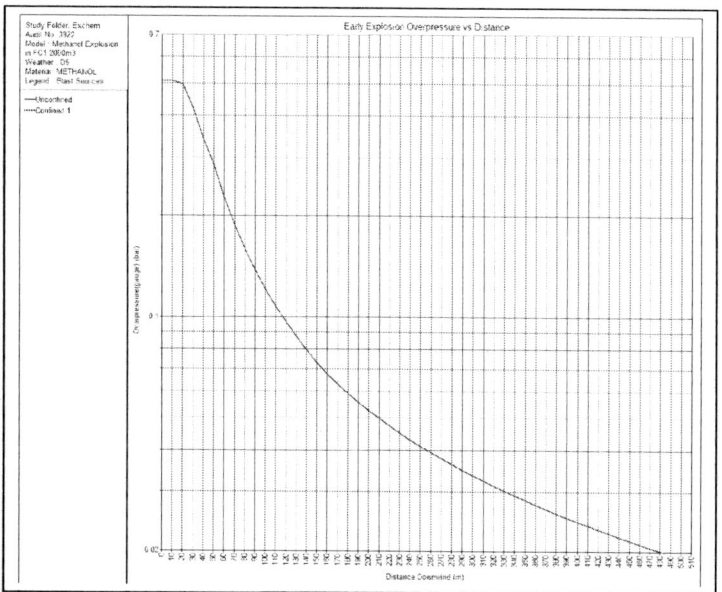

Graph 6. TNO explosion overpressure versus distance

SAFETY REPORT: MAINTAINING STANDARD AND REVIEWING

Sa'ari Mustapha* and Izani Mohd Zain**
*Department of Chemical and Environmental Engineering, Faculty of Engineering University Putra Malaysia, 43400 UPM Serdang, Selangor, Malaysia
**Department of Safety and Occupational Health (DOSH), Aras 2, 3 & 4 Block D3, Parcel D, Pusat Pentadbiran Persekutuan, 62506 Putra Jaya, Malaysia.

Synopsis

Malaysia promulgated Control of Industrial Major Accident Hazards (CIMAH) Regulations in February 1996. Since then there have been 159 major hazard installations (MHI) and 321 non-major hazard installations (NMHI). The classification of MHI and NMHI is based on threshold quantity (TQ) of hazardous substance(s). For MHI the quantity of the hazardous substance(s) in its vicinity is equal or exceeds the specified threshold quantity while NMHI poses the material(s) in between 10% TQ and TQ. According to the regulations, the MHI must submit a safety report and emergency response plan (ERP) to the Department of Occupational Safety and Health (DOSH) but the NMHI is only required to show its activities are safe and to submit of an emergency response plan (ERP). The importance of the safety report is to identify the nature and scale of use of the dangerous substances, describe the type, relative likelihood and consequences of major accidents and the arrangements for safe operation and control and mitigation of major accidents. In order to meet the requirement or standards of the safety report, the regulations require that the MHI(s) consult the registered Competent Person(s) for advice and approval (screening) of the report. It was anticipated that the use of the service of the competent persons for approval of the safety report will reduce the number of reports that are rejected, need modification or need correction. This paper discusses result of a review of the safety reports.

KEYWORDS: Safety report, standard, CIMAH, major accidents, review, errors/omissions

INTRODUCTION

According to The Occupational Safety and Health (The Control of Industrial Major Accident Hazards) Regulations, 1996, all Major Hazard Installations (MHIs) must submit Safety Reports to the Department of Occupational Safety and Health (DOSH). Its content must fulfill the legal requirements as stated in Schedule 6 (Subregulations 14(1) and 15(1)) i.e. *Information to be Included in Report of Industrial Activity"*. The preparation of the safety report must include:[1]

i) Information relating to every hazardous substance
ii) Information relating to the installation
iii) Information relating to the management system for controlling the industrial activity
iv) Information relating to the potential major accidents in the form of risk assessment

The safety report is a documentation of activities of an installation for the prevention, control, and mitigation of potential accidents. It is a management tool to manage risk on-site and a facilitating device for surveillance by the DOSH inspector. In preparation of safety report errors and omission must be avoided so as to comply with the legal requirement

© 2003 IChemE

(CIMAH). Furthermore, it should not be a one off preparation but rather a dynamic safety manual that requires be reviewing or updating continuously. The rationale is that the plants might undergo process modification, or change of work procedures, or surrounding development during the installation life cycle. Reviewing of the safety report must be carried out every three years or soon after any modification has been made.

THE SAFETY REPORT

In mid 1970's after several occurrences of major accidents in Europe, European Community governments passed regulations requiring chemical process facilities to demonstrate that their activities were managed safely. This major hazard directive is commonly referred to as the Seveso Directive. Britain has enforced the directive, which is known as the Control of Industrial Major Hazard (CIMAH) regulations in 1984. The government of Malaysia adopted the CIMAH regulations and promulgated the Regulations in February 1996. In the United Kingdom (U.K.), the enforcement of CIMAH regulations is under the Health Safety Executive (HSE) but in Malaysia it comes under the Department of Occupational Safety and Health (DOSH).

The objectives of CIMAH are:

i.   To prevent major accidents
ii.   To limit the consequences to people and environment

In order to meet the above objectives, a criteria was developed to recognize the installations that pose potential major accidents such as fire, explosion and release of toxic substances. In CIMAH it is based on hazardous substances (HS) and its threshold quantity (TQ). A hazardous substance is defined in the regulations as toxic, flammable, explosive and oxidizing substance. Above a certain amount or inventory (threshold quantity), is believed to be potentially a major accident that threaten the safety and health of workers on-site or to the public surrounding the installation. The Bhopal disaster is an example of a major accident where 2,500 people lost their lives and 200,000 were people injured due to an accident of releasing methyl isocyanate (MIC).

Lists of hazardous substances and the threshold quantity are presented in Schedule 1 and 2. Industrial activities are classified into Major Hazard Installation, Non Major Hazard installation and No-Threshold Quantity as depicted in figure 1:

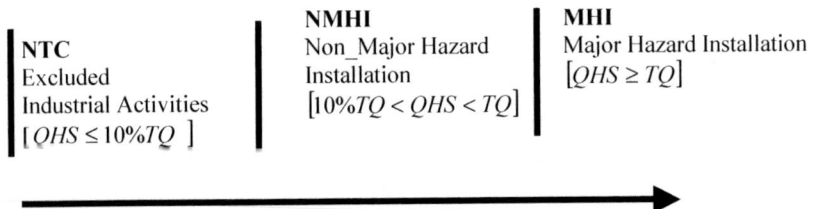

| NTC Excluded Industrial Activities $[QHS \leq 10\%TQ]$ | NMHI Non_Major Hazard Installation $[10\%TQ < QHS < TQ]$ | MHI Major Hazard Installation $[QHS \geq TQ]$ |

Figure 1.   Classification of industrial activity according schedule 1 and 2

Installations having hazardous substance equal or higher than the threshold quantity (TQ) are classified as Major Hazard Installation (MHI), whereas those that contain the material in between 10% threshold quantity and threshold quantity are categorized as Non-Major Hazard Installation (NMHI) and installations with less than or equal to ten percent of threshold quantity are called Non-Threshold Category (NTR).

From Notifications received by DOSH until 2001 there were 159 MHI and 321 NMHI. The state of Selangor registered the highest number of MHI, which is 37 installations whilst Johor had the highest number of NMHI i.e. 46.

As required by the CIMAH regulations, the MHI must submit a Safety Report and on site Emergency Response Plan (ERP) to The Director General of DOSH.

The objectives of the safety report are:

i.   To identify the nature and scale of the use of dangerous substance at the installation.
ii.   To give an account of the arrangements for safe operation of the installation, for control of serious deviations that could lead to a major accident and for emergency procedures at the site.
iii. To identify the type, relative likelihood and consequences of major accidents.
iv.   To demonstrate that the manufacturer has identified the major hazard potential of his activities and has provided appropriate controls.

The DOSH requires that the information of the safety report must include:[2]

i.   Hazardous substances: name/quantity, hazards, purity/impurity.
ii.   Installation: a map of the site: a scale plan of site showing locations and quantities of hazardous substances; process description and conditions; number of persons on site; nature of land use and population in the vicinity, and the nearest emergency services.
iii. Management System: staffing arrangements (person in charge, responsible for safety, to set emergency procedures); safe operation; design, construction, testing, operation, inspection and maintenance; staff/worker training programs.
iv.   Risk Assessment: description of potential sources, conditions and events; diagram to show features for accident control; description of measures taken to prevent, control and minimize the consequences; prevailing meteorological conditions; and consequences to the surroundings in the form of appropriate risk measures.

Other requirements imposed on manufacturers under MHI are to keep an up to date report and plan, to inform local authority and help them to prepare off-site ERP, inform the public and DOSH about any major accident.

The obligations of manufacturers under the NMHI category are notification of an industrial activity, preparation and submission of Demonstration Operation Document and Emergency Response Plan (ERP) to DOSH in case the Director General requests, for them and to keep an up to date report and plan and inform DOSH of any major accident.

In order to maintain the standard and quality of the safety report submitted to DOSH, its Major Hazard Division registers consultants as "Competent Persons" to approve the reports. Originally they were it called 'register persons' a term used by other government agencies for one who has been appointed by an authority to approve a report such as the report of Environmental Impact Assessment (EIA). The Criteria used to register a Competent person by DOSH is mainly based on academic qualifications and experience in the area. Such a person is required to possess a Bachelor of Science or Engineering and also meet other criteria such as adequate knowledge, skill expertise and experiences in area related to the content of safety report. So far, 69 competent persons have been registered by DOSH compared to 159 Major Hazard Installations (MHI) in Malaysia.

Although the competent person approves the safety report, it still requires to be reviewed by the DOSH. The manufacturer is also asked to present the report to the DOSH officers. Any unsatisfactory report due to the inadequacy or discrepancy of the information must be corrected before the manufacturer can re-submit the report to the DOSH. For a satisfactory report, the DOSH will acknowledge the acceptance by sending a letter to the manufacturer.

MAJOR HAZARD INSTALLATIONS

The government of Malaysia promulgated Control of Industrial Major Accident Hazards (CIMAH) Regulations in February 1996. From Notifications received by DOSH up to 2001 there were 159 MHI and 321 NMHI. The state of Selangor registered the highest number of MHI, which is 37 installations whilst Johor had the highest number of NMHI i.e. 46.

As required by the CIMAH regulations, the MHI must submit Safety Report and on site Emergency Response plan (ERP) to The Director General of DOSH. Other requirements imposed are to keep up to date report and plan, inform local authority, help local authority to prepare off-site ERP, inform the public and inform to DOSH about any major accident. The responsibility of manufacturers under the NMHI category are notification of an industrial activity, prepare and submit Demostration Operation Document and Emergency Response Plan (ERP) to DOSH if the Director General request, keep up to date report and plan and inform DOSH of any major accident.

Breakdown of MHI according to activity as tabulated in Table 1:

Table 1.   Breakdown MHI according to activity

Activity	Number
Chemical processing plant	25
Bulk storage of petroleum products	23
Petrochemical plant	19
Water treatment plant	22
Bulk storage of hazardous substance	12
Bulk storage and bottling of LPG	11
LPG cylinder storage	8
Air separation plant	5
Rubber glove manufacturing	5
Bulk storage and bottling of ammonia	3
Insecticide mixing	3
Textile manufacturing	3
Others	20
Total	159

METHODOLOGY

Thirty safety reports were reviewed in this study. For the purpose a checklist was prepared according to Schedule 6 (Subregulation 14(1) and 15(1)) i.e. "Information To be Included In The Report on Industrial Activity". It contained a total of four components and the list presented as follows:

(a) Information relating to every hazardous substance involved in the activity in relevant quantity as listed in Schedule 2.

    i) The name of the hazardous substance as given in Schedule 2 or for a hazardous substance included in either of those Schedules under a general designation, the name corresponding to the chemical formula of the hazardous substance;

    ii) A general description of the analytical methods available to the manufacturer for determining the presence of the hazardous substance, or references to such methods in the scientific literature;

    iii) A brief description of the hazards, which may be created by the hazardous substance;

    iv) The degree of purity of the hazardous substance, and the names of the main impurities and their percentages.

(b) Information relating to the installation, namely

    i) A map of the site and its surrounding areas to a scale large enough to show any features that may be significant in the assessment of the hazard of risk associated with the site;

    ii) Scale plan of the site showing the locations and quantities of all significant inventories of the hazardous substance;

    iii) A description of the processes or storage involving the hazardous substance and an indication of the conditions under which it is normally held;

    iv) The maximum number of persons likely to be present on site;

    v) Information about the nature of the land use and the size and distribution of the population in the vicinity of the industrial activity to which the report relates;

    vi) Information on the nearest emergency services (fire station, hospital, police station, community hall, etc.)

(c) Information relating to the system of management for controlling the industrial activity, namely

    i) The staffing arrangement for controlling the Industrial activity with the name of the person responsible for safety on the site and the names of those who are authorized to set emergency procedures in motion and to inform outside authorities

    ii) The arrangements made to ensure that the means provided for the safe operation of the industrial activity are properly designed, constructed tested, operated, inspected and maintained.

    iii) The arrangements for training of persons working on the site.

(d) Information relating to a potential major accident in the form of risk assessment, which contains the following:

    i) A description of the potential sources of a major accident and the conditions or events, which could be significant in bringing one about.

ii)   A diagram of the plant in which the industrial activity is carried on, sufficient to show the features which are significant as regards the potential for a major accident or its prevention or control.

iii) Information about prevailing meteorological conditions in the vicinity of the site;

iv)   A estimate of the number of people on site who may be exposed to the hazards considered in the report;

v)   The consequences to the surrounding areas in the form of risk assessment.

From the 30 safety reports (safety cases), which were reviewed, there were 599 errors/omissions found. The highest errors or omissions were observed under Part (d) (i.e. "Information relating to a potential major accident in the form of risk assessment") that contributed 286 errors, followed by Part (b) (i.e. Information relating to the installation) that gave a total of 182 of errors/omissions, 92 errors/omissions from Part (c) and the least errors/omissions from Part (a) which was only 39.

The highest errors/omissions from the itemized parameters which are under the sub-sections of the Parts (a), (b), (c) and (d) were investigated further. In this category the errors/emissions of "the quantity of hazardous substance" under the sub-section 'scale plan of the site showing the locations and quantities of all significant inventories of the hazardous substance' of Part (b) is the super highest (number one) among the highest that gave a total 23, followed by behind the errors/omissions of "justification of assumptions made" under the 'sequence study under risk assessment' that contributed 20 errors/omissions. The errors/omissions of the "constitution of site safety committee" under the sub-section of 'the staffing arrangement' of Part (c) and the "showing hazardous substance" that is a requirement under the sub-section 'scale plan of the site showing the locations and quantities of all significant inventories of the hazardous substance' had equal ranking i.e. were number three with number of errors/omissions of 14.

Other errors/omissions of the itemized parameters under the Part (a), (b), (c) and (d) can be obtained from the Table 2. The errors/emissions which are frequently observed in the result of the checklist are no information represented by the term 'no explanation' (unexplained of the itemized parameters) like personnel protection equipment (PPE) and training requirement, less judgment as represented by the term 'not considered' for example there is no information of spontaneous failure due to original defect, human error and failure due to external events and inadequacy of information noted as 'not inadequate' such as in the term "preventive and control measures of consequence for the accident scenario". Some kinds of the errors/omissions, which were also regularly, encountered in U.K. CIMAH safety reports (Lees, 1989).

RESULT AND DISCUSSION

Table 2. Summary of result of safety reports review (number of safety report is 30)

Items		No. of non-compliance	Comments
a) Information relating to every hazardous substance involved in the activity in relevant quantity listed in Schedule 2, namely:			
i. the name of the hazardous substance as given in Schedule 2 or for a hazardous substance included in either of those Schedules under a general designation, the name corresponding to the chemical formula of the hazardous substance;	Name as in Schedule 2> threshold quantity	0	
	Other hazardous substances < threshold quantity	6	Doesn't mention other hazardous substances with less than threshold quantity because of the misunderstanding of CIMAH requirements. This hazardous substance is also a potential to create major accidents (escalation effect).
ii. a general description of the analytical methods available to the manufacturer for determining the presence of the hazardous substance, or references to such methods in the scientific literature	Method of determining the existence of a substance	14	- Liquid form so can be detected visually. - Doesn't have hazardous detector at all. - No a specific best engineering practice regarding to the detector requirement
iii. a brief description of the hazards which may be created by the hazardous substance	Description of hazards for every substance	7	- Doesn't describe hazard
iv. the degree of purity of the hazardous substance, and the names of the main impurities and their percentages	purity/impurity of hazards for every substance	12	- not described that this information - do not realise that this information can be obtained from quality sheet.

Note: No. of companies, which do not mention or explained regarding to the item specified in Schedule 6 of CIMAH

(b) Information relating to the installation, namely

		No. of non-compliance	Comments
i. a map of the site and its surrounding areas to a scale large enough to show any features that may be significant in the assessment of the hazard of risk associated with the site;	map of site	5	- very hard to get map in Malaysia. - The map is out of date - The map is unclear
	with scale	15	- map without scale
	schools, hospitals, prisons, etc., if any	13	- map does not detail the items.
	industrial premises		
	Other hazardous installations		
	Roads, railways, airport, ports		
	Recreational areas		
	Reservoirs, rivers, lakes etc		
	Monuments, archaeological area, conservation areas		
ii. scale plan of the site showing the locations and quantities of all significant inventories of the hazardous substance	Attach a site plan	1	unclear
	A site plan with scale	9	- attach site-plan without scale. - Not up to date
	Showing hazardous substance	14	On site plan: - no indication of the hazardous substance
	Quantity of hazardous substance	23	- does not mention the quantity

iii. a description of the processes or storage involving the hazardous substance and an indication of the conditions under which it is normally held;	description of processes	4	- does not describe the process
	Simplified process flow diagram	5	- not attached process flow diagram
	Simplified piping and instrumentation diagram	5	- not provided P&ID
	Stated physical state, pressure and temperature	15	- process not mentioned parameter on the simplified P&ID
iv. the maximum number of persons likely to be present on site;	Maximum no. of workers day time	5	Not mentioned
	Maximum no. of workers night time	7	Not mentioned
	Located on site plan	21	- does not put the max. numbers of workers for day and night on the site plan. - No specific requirement on above.
v. information about the nature of the land use and the size and distribution of the population in the vicinity of the industrial activity to which the report relates;	land use	7	- does not describe surrounding area, neighbors or residential area
	population in the vicinity	11	- does not mentioned if any congested or populated area.
	population in the vicinity for day and night time	11	- the number of people surrounding not mentioned
	nearest emergency services	3	- the nearest emergency services such as bomba, hospital, ambulance, police etc. not mentioned
	Distance to the nearest emergency services	8	- does not mentioned the distance.

(C) Information relating to the system of management for controlling the industrial activity, namely-

i. The staffing arrangement for controlling the Industrial activity with the name of the person response for safety on the site and the names of those who are authorized to set emergency procedures in motion and to inform outside authorities

	No. of non-compliance	Comments
Company/plant organisation chart	4	- not attached at all the general org. chart attached, not specific to the plant
Any particular or unusual expertise required running plant safely.	3	- complex process requires special people to handle.
General description of attributes appropriate to plant management and supervisory staff	5	- no explanation on job and task at management and supervisory level.
Description of line management responsibilities and reporting relationships with safety and environmental implications.	4	- Job description with safety task.
General description of managerial attributes appropriate to the site including experience and qualifications of staff at different levels.	4	- qualification and experience of staff especially those who carry out critical task not mentioned.
Statement of general policy	9	- not attached policy - no requirement of policy if the number of staff is less than five.

	Item		Not mentioned
	Constitution of site safety committee.	14	Not mentioned: - committee members - structure
	Accident reporting and investigation	7	- no explanation
	Accident analysis	2	- no explanation
	System audit	6	- no explanation
	Personal protective equipments requirement	3	- no explanation
ii. The arrangements made to ensure that the means provided for the safe operation of the industrial activity are properly designed, constructed, tested, operated inspected and maintained.	Plant design	0	
	Plant modification	4	- no explanation on how to manage plant change
	Plant construction	0	
	Plant operation	2	- no explanation
	Plant procedures	7	- no list of plant operation procedures
	Process and operational control	0	
	Plant maintenance	3	- no explanation
	Permit to work	1	- no ptw system explained
	Test and inspection	4	- not mentioned
	Control of contractors	6	- contractor selection not mentioned - contractor evaluation not mentioned - contractor supervision not mentioned
iii. The arrangements for training of persons working on the site.	Training requirements	4	- no explanation

(d) Information relating to a potential major accident in the form of risk assessment which contains the following:

i. a description of the potential sources of a major accident and the conditions or events which could be significant in bringing one about.

	No. of non-compliance	Comments
Identify all hazardous substance	4	- does not identify all hazardous substance
Overall explanation on hazard identification technique	6	- no explanation
Spontaneous failure due to original defects	12	- not considered
Spontaneous failure due to those arising in the course of operation	1	- not considered
Failures in high risk activity (loading/unloading)	7	- not considered
Failures due to excursions from normal operating conditions	6	- not considered
Failures in maintenance, inspection and quality assurance	10	- not considered
Failures due to other on-site events	6	- not considered
Failures due to external events	13	- not considered
Human errors	13	- not considered
Overall explanation on consequence estimation/ calculation	5	- no explanation on effect of fire, explosion and toxic release

Item	Criterion	No.		Status
ii. A descriptions of the measures taken to prevent, control or minimise the consequences of any major accident.	Consequence of each event considered	8	-	not considered
	Consider escalation effect	16	-	not considered
	Adequate preventive measures for each event	9	-	inadequate
	Adequate control measures for each event	9	-	inadequate
	Adequate measures taken to minimise the consequences	8	-	inadequate
iii. A diagram of the plant in which the industrial activity is carried on sufficient to show the features which are significant as regards the potential for a major accident or its prevention or control.	Adequate piping and instrumentation diagram.	12	-	inadequate
	Indicate process parameter such as temperature, pressure, etc.	10	-	not indicated
iv. Information about prevailing meteorological conditions in the vicinity of the site;	Data from the nearest meteorology station.	5	-	no data at all
	Wind direction and distribution	5	-	data incomplete
	Wind stability	12	-	not mentioned
	Atmospheric temperature	12	-	not mentioned
	Relative humidity	13	-	not mentioned
v. an estimate of the number of people on site who may be exposed to the hazards considered in the report;	Estimate the number of on-site people affected	13	-	not mentioned
	Estimate the number of off-site people affected	15	-	not mentioned
vi. the consequences to the surrounding areas in the form of risk assessment.	Overall plant risk assessment methodology	12	-	no explanation

Justification on assumption made	20	- no explanation on assumption and limitation of methodology used.
Generation of likelihood on every significant events	12	- not conducted
Risk contours	12	- no risk contour.

CONCLUSION

The safety report is a tool or manual is which used for prevention, control and mitigation of potential accidents. The information contained in the report must be according to the standard as specified in the legal requirement (the CIMAH regulations) and satisfied by the authority i.e. the Department of Occupational Safety and Health (DOSH). Reviewing of safety reports approved by "a competent person" is a must since a significant number of errors/omissions were found in the checklist. However, it should be taken into consideration that the result obtained from the review is from the first group of safety reports after enforcement of the CIMAH regulations in Malaysia. Without the service of the competent persons, it is envisaged that DOSH officers might find more errors/omissions. However, there is a large space of improvement toward reducing the errors/omissions to a minimum such as through exchange of ideas, discussions or training of the people who are involved in the preparation of the safety reports. Symposiums, seminars or workshops are good platforms to achieve this objective.

REFERENCES

1. Occupational Safety and Health Act 1994(Act 514) and Regulations, 1996, ILBS.
2. Omar Mat Piah, "Background and Requirements of CIMAH From the Regulatory Perpective". Seminar on Control of industrial Major Accident Hazards Regulations, Kuala Lumpur. 27–28 May 1996.
3. Lees, F.P. and Ang, M.L, 1989, Safety Cases, Butterworth, p233.

THE ONGOING CHALLENGE OF DEMONSTRATING ALARP IN COMAH SAFETY REPORTS

Graeme Ellis, Senior HSE Consultant
ABB Eutech Ltd, Daresbury, Warrington, UK

The requirement in the 'Seveso II' directive to demonstrate that 'all measures necessary' have been taken to prevent and limit the effects of major accidents presents a considerable challenge to both the regulator in defining a suitable approach and to those charged with writing safety reports. Based on practical experience of COMAH risk assessments for several companies, this paper will present the approach developed by ABB Eutech. It covers the key issues and challenges that have arisen, how these have been addressed, the results of initial assessments and the direction of the regulator based on latest guidance. One of the key issues has been to demonstrate a proportionate approach whilst avoiding the excessive costs associated with extensive quantified risk assessments. The paper presents a screening methodology built around a semi-quantitative risk matrix that has been used successfully to make the necessary demonstrations. Recent guidance from the HSE is driving towards a greater level of quantification, especially for high consequence and high risk events, and these new requirements are discussed at the end of the paper.

KEYWORDS: COMAH, Predictive, ALARP, Risk assessment, Safety Critical Events

INTRODUCTION

The Control of Major Accident Hazards (COMAH) Regulations 1999 came into force in the UK in April 1999 implementing the European Union Directive commonly known as 'Seveso II'. The regulations require duty holders or site operators storing quantities of named or generic dangerous substances above prescribed threshold limits to prepare a safety report. A key requirement of the safety report is to demonstrate that 'all measures necessary' have been taken to prevent or mitigate hazardous events with the potential to cause serious harm to people or the environment. In keeping with UK Health and Safety law requiring risks to be reduced 'so far as is reasonably practicable', the HSE has interpreted 'all measures necessary' as reducing risks to 'as low as reasonably practical' (ALARP).

To achieve ALARP it is implicit that some form of risk assessment has been carried out to judge the current level of risk and then decide whether the costs to implement further risk reduction measures can be justified against the likely benefits. Whilst conceptually this requirement is clear the depth of the assessment could range from simple qualitative assessment at one extreme to full quantified risk assessment (QRA) backed by detailed cost benefit analysis at the other extreme. Faced with the requirement to assess a large set of representative scenarios, most operators have tried to avoid the high costs associated with extensive use of QRA.

This paper explores the development in risk assessment guidance from the HSE for COMAH including the most recently published material. The concept of proportionality was initially proposed to define the depth of the demonstrations required, but the guidance was non-prescriptive. To meet the deadlines for safety reports, operators were therefore forced to develop their own approaches to COMAH risk assessments. This paper describes the approach developed by ABB Eutech from well-proven process hazard analysis techniques that has been used with a number operators. This methodology is based on semi-quantitative risk assessment using a calibrated risk matrix. The judgements are supported by quantified consequence assessment where appropriate and in particular for high severity scenarios. Based on assessment feedback and recent HSE guidance, the requirements from the regulator have recently become more prescriptive. Experience in meeting these further requirements is described in this paper and issues to be faced during the ongoing assessment process are discussed.

To illustrate the practical aspects of the risk assessment methodology a case study is described in the paper. This relates to a liquid sulphur dioxide road tanker offloading and storage system, installed on the site of a COMAH 'top tier' speciality chemical manufacturer located a short distance from housing. The main hazard on such an installation is loss of containment releasing sulphur dioxide and creating a toxic plume. If the release is not quickly isolated, it is possible in certain wind and weather conditions for such an incident to result in serious injuries and fatalities.

INITIAL GUIDANCE ON DEMONSTRATING ALARP

The general duty under the 'Seveso II' Directive is that every operator shall take 'all measures necessary' to prevent major accidents and limit their consequences to persons and the environment. The fact that limitation measures are mentioned has been interpreted in the UK as recognising that risk cannot be completely eliminated and that some form of risk assessment is required to judge whether the measures to reduce risk are adequate. The HSE guidance on the COMAH regulations[1] and on preparing safety reports[2] introduces the concept of 'proportionality' in determining the depth of risk assessment required. This is related to the scale of hazard and the residual risk. A complex chlorine manufacturing site close to a centre of population therefore requires a more in-depth assessment than a simple chlorine storage system in a remote location.

Although not defined in the 'Seveso II' Directive, the HSE has interpreted the general duty as requiring operators to demonstrate that risks have been reduced to 'as low as reasonably practicable' (ALARP) to be consistent with UK Health and Safety legislation. A hierarchical approach is required for demonstrating ALARP, initially considering inherent safety principles to eliminate or reduce the hazard, then following Approved Codes of Practice, industry standards, company standards and good engineering practice. Any shortfalls in meeting these standards needs to be justified by the operator.

Initially the HSE avoided prescriptive guidance for COMAH risk assessments, but included guidance on their interpretation of the Directive. For all the dangerous substances held on site the assessment must include a hazard identification and consequence assessment. By considering the 'worst case scenarios', these may be found in some cases to be trivial, and it can be concluded that no major accident hazards exist. Where the consequences are non-trivial a risk analysis is required with information on frequency or likelihood. This requirement is fundamental to the demonstration that 'all measures necessary' have been taken.

Due to the complexity of hazards in the chemical industry it is essential that a systematic methodology is used to identify a representative set of foreseeable major accidents. Recommended processes for this exercise are hazard and operability (HAZOP) studies, reviews of past accidents and incidents, bespoke industry checklists or failure mode and effects analysis (FMEA). Where protection against major accidents is dependent on the action of an automatic shutdown system or human intervention, the risk assessment must consider whether the reliability of these measures ensures that risks remain ALARP. Where consequence assessment models have been used, these must be clearly referenced and a justification made for their use and for any key variables or assumptions made, such as wind speed and weather type in the case of toxic gas dispersion models.

INITIAL RISK ASSESSMENT METHODOLOGY
A methodology was developed by a group within ABB Eutech based on well-established process hazard analysis (PHA) techniques as practised by their former owner ICI. This aims to answer a number of key questions, namely; what can go wrong, how bad could the consequences be, how often could it happen, are the risks acceptable against established criteria and if not what further measures should be taken.

HAZARD IDENTIFICATION
Under the COMAH Regulations a 'suitable and sufficient' set of major accident scenarios must be identified for the site. These must represent the full range of potential major accidents including fires/explosions, toxic releases and damage to the environment. The ABB Eutech approach uses a team comprised of experienced and competent technical, operating and maintenance staff from the plant, lead by a process safety specialist. The ABB Eutech 'Process Hazard Review' (PHR) technique was used for the studies. This utilises a guide diagram with prompts to identify all the credible loss of containment events with the potential to lead to major accidents, including generic causes such as the failure of equipment and control systems or human error. The guidewords and examples of the prompts for causes are shown in table 1.

Table 1.   Main headings for hazard identification guide diagram

Type of event	Hazardous event	Prompts
Operated outside design limits	Internal explosion	e.g. static discharge
	Runaway reaction	e.g. double charging catalyst
	Physical overpressure	e.g. tube failure
	Temperature excursion	e.g. brittle failure
	Vessel overfill	e.g. operator error
Loss of containment under designed operating conditions	Long-term weakening	e.g. internal corrosion
	Seal failure	e.g. gasket blow-out
	Moving equipment failure	e.g. compressor rupture
	Maloperation of openings	e.g. drain valve
External events	Vehicle impact	e.g. road tanker collision
	Knock-on effect	e.g. explosion on nearby plant
	Loss of utilities	e.g. cooling water failure
	Fire	e.g. warehouse fire

For the sulphur dioxide storage system the guideword 'vehicle impact' prompted the team to consider the potential for the road tanker to move off due to driver error whilst still unloading causing the offloading hose to rupture.

For small to medium sites the above approach was used for all the operations involving dangerous substances where a loss of containment had the potential for a major accident. For larger sites this approach was felt to be too demanding, and it was more appropriate to consider a representative set of scenarios. This generally involved the study team considering the most severe or highest risk scenarios and checking that the protective measures were consistent for other less severe operations. For example a number of storage tanks containing flammable substances or a set of reactors with similar processes could be assessed generically with a considerable saving in time and effort.

CONSEQUENCE ASSESSMENT
Having identified a credible mechanism for loss of containment the study team next consider the consequences in terms of harm to people or the environment. An issue commonly encountered was in defining whether events were credible. For instance was it credible for a vessel to rupture catastrophically when pressure relief had been provided. As COMAH requires a focus on high severity events as well as more likely events, the general approach was to only take credit for passive protection measures such as bunds when determining the worst credible consequence.

The main aim of this stage is to understand the types of hazardous events that could occur based on the properties of the material released. These will generally include the effects due to thermal radiation from fires, overpressure from explosions, acute toxicity from vapour releases and impact on the environment. A qualitative judgement is initially taken by the team on the extent and severity of the consequences. For high consequence events, particularly those with the potential to cause on-site or off-site fatalities, quantification of the hazard range is generally carried out using specialised computer based programmes. Models for effects on the environment are not so well developed, and generally a qualitative judgement has been made.

Based on the results of the consequence assessment and an ABB Eutech guide diagram with word models, the severity is ranked on a scale of 1 to 5. A number of categories are used including safety impact to workers and the public, harm to the environment by airborne or liquid releases, acute health affects, media attention or action by the regulators. As an example for on-site harm to workers the effects are classified in table 2.

Consequence categories 3, 4 and 5 are defined as meeting the criteria for a major accident, the other levels were included on the assessment sheets to demonstrate the thoroughness of the hazard identification process and to take credit for associated protective measures.

For the example offloading hose rupture resulting in liquid sulphur dioxide being released, the consequences are a flash of vapour forming a toxic plume and further vapour release from the resultant liquid pool. Using published data of the fatal dose for sulphur dioxide and modelling the release in a number of weather conditions, it was found that the hazard range to a 1% probability of fatality extended to several hundred metres from the point of release. As this extends a considerable distance off-site the severity level was set as 'Catastrophic'.

Table 2.   Examples of consequence word models

Consequence category	Word model for on-site safety impact
5 – Catastrophic	Many fatalities
4 – Extremely serious	One or few fatalities. Many major injuries.
3 – Major	One or few major injuries. Many serious injuries, hospital for > 24 hr
2 – Serious	One or few serious injuries, hospital treatment. Many minor injuries.
1 – Minor	One or few minor injuries, medical attention.

FREQUENCY ASSESSMENT

Risk is a function of both the consequences of a hazardous event and the likelihood expressed as a frequency. To complete a risk assessment it is therefore necessary to make an estimate of the frequency. A qualitative judgement on a scale of low, medium and high may be suitable for simple hazards, but for events with the potential for major accidents it is judged that a semi-quantitative approach is required. The study team starts by identifying and listing the associated protection measures including prevention, control and limitation measures involving plant hardware, safe systems of work and human factors. The reliability of the protection measures throughout the life cycle of the plant are considered to determine how effective they are in providing protection. For safety critical systems it is likely that a specialist audit is required to provide the necessary demonstration for COMAH. As an example, the sizing of a pressure relief system for an overpressure event may not have suitable design documentation to ensure that it would provide effective relief.

For the tanker drive away hazard the prevention measure is a procedure for gates to be closed on either side of the road tanker when it has parked in the offloading bay. The gates are only opened by the operators when the offloading hose has been disconnected. Should the gate not be closed due to operator error and the drive away occurs causing the offloading hose to rupture, liquid sulphur dioxide will be released and the strong smell provides a warning. As a limitation measure the road tanker driver and operator have breathing apparatus available to allow isolation of the tanker outlet valve or operation of the emergency shut-down system at a remote location isolating the tanker outlet line. A local wind sock gives an indication of the cloud direction, and the on-site and off-site emergency plans can be brought into operation, including on-site personnel moving into toxic safe havens. For the initial safety report a semi-quantitative assessment of event frequency was carried out using the ABB Eutech word models as shown in table 3 for guidance.

Table 3.   Frequency word models

Frequency Category	Frequency range (per year)	Word model
A - Frequent	10^{-1} to 1	Has occurred during lifetime of plant
B - Probable	10^{-2} to 10^{-1}	Could occur during remaining lifetime of plant
C - Occasional	10^{-3} to 10^{-2}	Not expected to occur during remaining plant lifetime
D - Remote	10^{-4} to 10^{-3}	Incidents in industry on similar technology
E - Improbable	10^{-5} to 10^{-4}	Foreseeable but requires the failure of more than one layer of protection
F - Very Unlikely	10^{-6} to 10^{-5}	Credible but requires the failure of several layers of protection
G - Extremely Unlikely	10^{-7} to 10^{-6}	Very unlikely event in area with low occupancy

For high severity or high-risk events a quantitative approach may be required to refine the estimate of event frequency. This may use historical data from previous incidents or fault tree analysis for more complex events. Such methods are time consuming and need to be carried out by specialists. The costs of quantified frequency analysis are therefore high and its use needs to be proportionate to the scale of hazard and risk.

The likelihood of tanker drive away leading to loss of containment was judged to be remote based on known incidents in industry on similar technology. The risk is reduced by the various additional protective measures such as locked gates and emergency shut-down system and it was judged that the frequency was therefore reduced by an order of magnitude to improbable.

In general the semi-quantitative estimate of event frequency was found to be far more difficult for study teams than the assignment of a consequence category. This is due to the complexity of most processes and indicates the benefits of a fully quantified approach in providing clarity. It was found to be beneficial to develop some rules for carrying out the assessments to ensure consistency.

RISK ANALYSIS

With the consequence and frequency of the hazardous event categorised, a comparison can be made with suitable criteria to provide guidance on the overall level of risk and whether further improvements should be considered. For major accident hazards where the potential exists for serious injury or fatality to workers and members of the public, criteria have been set by the HSE in the document 'Reducing Risks, Protecting People' (often referred to as R2P2)[3]. Risk criteria are not normally expressed as a single value, above which risk is unacceptable and below which risk is acceptable. Instead there is a tolerability band within which risks must be reduced to ALARP. This brings in the need to carry out some form of cost-benefit analysis to determine if the costs of an improvement can be justified by the risk reduction achieved. Above the tolerable band the risks are said to be 'unacceptable' with improvements seen as essential in all but exceptional circumstances, and below the tolerable band the risks are judged as 'broadly acceptable', with no further working required to reduce risks.

R2P2[3] defines the upper boundary between intolerable and tolerable if ALARP levels of risk for members of the public as a fatality once in 10,000 years and for workers as a fatality once in 1,000 years, with the lower limit between tolerable and broadly acceptable levels of risk as 1 in a million years for all people. The above thresholds were used to produce the ABB Eutech calibrated risk matrix shown in figure 1 with an explanation of the risk levels in figure 2. Note that upper and lower ALARP regions have been introduced to distinguish between the level of gross disproportionality to be applied when considering the need for further measures.

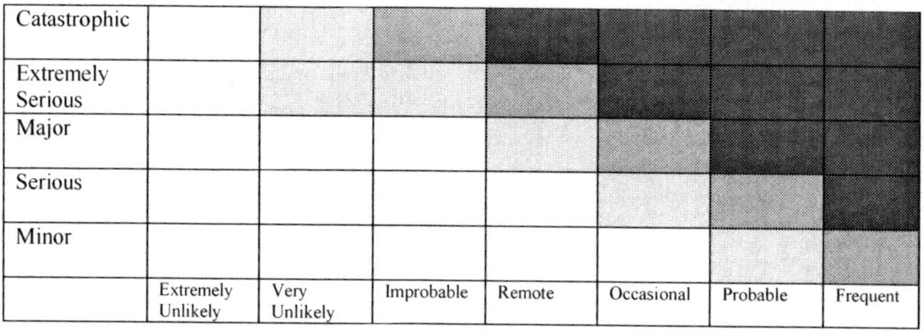

	Extremely Unlikely	Very Unlikely	Improbable	Remote	Occasional	Probable	Frequent
Catastrophic							
Extremely Serious							
Major							
Serious							
Minor							

Figure 1. Calibrated risk matrix

	Broadly acceptable	No need for detailed working to demonstrate ALARP
	Lower ALARP	Tolerable if cost of risk reduction would exceed the improvement gained
	Upper ALARP	Tolerable only if risk reduction is impracticable or if its cost is grossly disproportionate to the improvement gained
	Unacceptable	Risk cannot be justified except in extraordinary circumstances

Figure 2. Definition of levels of risk

For the tanker drive-away scenario the consequences of an offloading hose rupture were judged as catastrophic and the frequency of a tanker drive-away and failure to isolate the leak was estimated as improbable. The risk is therefore in the upper ALARP region meaning that further measures should be considered unless it can be shown that the cost of these is grossly disproportionate to the likely improvement. The improvements considered can either reduce the consequences of the event or reduce the frequency of the event, in either case it is necessary to make reference to current standards for the technology under review.

Whilst the risk matrix approach for COMAH safety reports has many advantages there are a number of issues as raised by Middleton[4]. Hazardous events can have several outcomes dependent on subsequent events such as ineffective pressure relief, ignition or delayed ignition of flammable substances or failure of secondary containment. The tendency is to consider the worst-case outcome to avoid an excessive number of points on the matrix but this can lead to an unrepresentative set of scenarios. The published criteria for individual risk refer to the cumulative risk from a medium sized operation, including all the hazardous events that could harm an individual. The risk matrix is being used to consider individual scenarios and it is therefore important to consider the number of events within each consequence category. It has been found effective to plot all the scenarios on the risk matrix to give a graphical representation of the spread of risk. For high severity events the risk matrix does not differentiate between events that could kill one or a few people off-site and those with the potential to cause tens of fatalities where criteria for societal risk need to be considered. Such events were classified in safety reports as 'catastrophic with the potential for multiple fatalities'. The author believes the limitation of risk matrices to individual risk needs to be accepted and that for high severity events a more detailed quantified risk assessment methodology is more appropriate.

ALARP DEMONSTRATION

Positioning events within the lower or upper ALARP bands does not necessarily mean that the risk is ALARP as further measures may be appropriate. A definition in more recent HSE guidance of this band is 'Tolerable if ALARP', with the responsibility on the operator to demonstrate that all reasonably practicable measures have been taken. Only when the assessment has been carried out and the necessary improvements implemented can the risk be classified as ALARP. The assessment is however subject to ongoing reviews based on further experience within the industry and advances in technology.

The CEFIC[5] code of practice for sulphur dioxide storage systems includes the following considerations for protection against tanker drive away:

- Automatic valves upstream and downstream of hose, operated locally and remotely,
- Trip on detection of low pressure in system,
- Wheel chocks or clamps to prevent movement,
- Fail-safe braking system on road tanker,
- Mechanical interlock to ensure barriers are dropped with hose connected.

None of the above protection measures were incorporated into the existing design and a qualitative cost benefit assessment was carried out by the study team. The first and third options are procedural and therefore less reliable than automatic systems. The second option will generally be more reliable but could be in a failed state. The forth option, whilst a reliable method, relies on the road tanker being fitted with the brake locking device and its' being maintained by the haulier. The new protective measure selected was an interlocked barrier system to ensure that the barriers are closed whenever the offloading hose is in use with periodic inspection of the system to ensure that it is still functioning correctly. The above measure was judged to give considerable risk reduction whilst only requiring modest cost, and a decision was taken to go ahead with this improvement.

SAFETY REPORT ASSESSMENT EXPERIENCE

The timetable for COMAH safety report submissions meant that a number of sites previously regulated under the CIMAH Regulations submitted reports in February 2001.

Early feedback indicated a rejection rate of around 35% with the main reason being a failure to link the major accident scenarios with the associated protective measures. As CIMAH did not require such a link to be made it is possible that these sites had failed to recognise the new requirements under COMAH. The ABB Eutech approach provides a major accident table format where each scenario has details of the causes, consequences, protection measures and event frequency. It is felt that this format provides the linkage required by the HSE and this view has been supported by successful assessment outcomes.

Sites new to the major accident regime have mostly submitted their safety reports in February 2002. Following earlier experience HSE has modified its' assessment strategy to initially consider only the predictive sections of the report. The objective is to ensure that a suitable and sufficient set of scenarios has been considered prior to a full assessment being started, thereby avoiding wasted effort for the HSE and cost for the operator. A number of common themes have emerged leading to safety reports either being rejected or the assessment process being put on hold pending further information.

The HSE are making a clear distinction between the extent and severity of major accidents for consequence assessments. The method described earlier provides an estimate of the extent or hazard range and from this the assessor determines if the harm will be very localised, extend across the site or extend into off-site industrial or public areas. The severity is defined as the number of people that might be harmed or for environmental effects the area that might be harmed. Operators are being asked to estimate the severity of their worst case scenario in terms of the number of fatalities to people on-site and off-site. A standard approach has been developed by ABB Eutech that initially calculates the areas under risk contours for lethal doses at the 1%, 10%, 50% and 90% levels. For toxic releases the plume direction towards the areas with the highest population density is used to find the worst case severity. Two weather conditions are considered with different probabilities for the proportion of people indoors and outdoors, the mitigating effect of being indoors is also calculated. The above approach is effective in giving a worst case figure for numbers of fatalities although the need for a conservative approach tends to over-estimate the severity. For this reason and a concern over how information on numbers of fatalities will be interpreted by local residents, operators are generally reluctant to release data in a public document.

Another key area for further information is the demonstration made in safety reports that risks have been reduced to ALARP. The approach described earlier results in a qualitative assessment of the need for further measures to be taken based on the judgement of the team against known standards and good engineering practice. The competence and knowledge of the team are critical and the leader needs to have a challenging approach to avoid acceptance of the status quo. It is argued that the degree of improvement actions arising from the risk assessment process provides a demonstration on the adequacy of the ALARP demonstration. For example an assessment that did not raise any improvements would no doubt raise suspicions that the team had not been sufficiently rigorous. The HSE are taking the requirement a step further and requesting that the assessment considers *all* the feasible additional measures that could be taken, providing a justification for those not implemented. The basis for this decision might be that the improvement is impracticable or that the costs outweigh the benefits.

The HSE requirement for ALARP demonstration would be very onerous if applied across the full range of major accident scenarios for a site. The concept of safety critical events (SCEs) has been introduced by Middleton[4]. These are defined as the events within each consequence category with the highest frequency and those with very high consequences. For each SCE a detailed ALARP demonstration is required to justify why further measures are not being taken. The HSE are currently preparing an extensive list of possible protection measures to be used as a checklist or guide diagram when carrying out such assessments.

To meet this new requirement the study team have searched for further protective measures against the hierarchy of inherent safety, prevention, control and limitation measures. No possibilities should be ignored at this stage. The company is likely to have discarded several options due to impracticability, the justification for these decisions needs to be recorded. For other options the costs may be considered too high compared to the risk reduction that can be achieved and a simple cost benefit assessment technique can be used to justify the decision not to implement any further measures. In practice it has been found that such assessments have resulted in new measures being agreed. With the sulphur dioxide tanker drive away hazard for instance, further measures included a procedure for removal of keys from the road tanker driver during offloading.

LATEST HSE GUIDANCE

The HSE has recently published a policy[6] and guidance[7] on making ALARP decisions in the context of COMAH. The policy document provides a basis for making decisions on gross disproportion when carrying out cost benefit analyses. A 'proportion factor' (PF) is defined as the total cost of implementing the measure divided by the value of the fatalities thereby prevented. Measures must be implemented if the PF is one or less when close to the Broadly Acceptable boundary or if the PF value is 10 or less when close to the Unacceptable boundary. Detailed quantified analysis is not essential as the policy advises the professional judgement of a team supported by a crude form of cost benefit analysis will be adequate. For new plants good practice contained in approved codes of practice, HSE guidance and industry standards should be considered as a minimum requirement, with confirmation of the relevance and currency to the specific hazard. It is recognised that for existing plants, applying good practice retrospectively will be subject to a test of reasonable practicability.

The policy document[6] states that ALARP decisions need to consider both the risk to individuals and societal risk. The latter requirement is new to COMAH as the HSE has previously stated that societal risk assessments were not required due to difficulties in interpreting the results. A distinction is made between societal risk and societal concern, the latter being the socio-political response to hazards with the potential to cause multiple fatalities or affect vulnerable groups despite a low level of risk. If societal concerns exist for a specific installation it may be necessary to implement further measures beyond any societal risk considerations.

The guidance document[7] is more specific to the needs of the chemical industry and COMAH in particular. It gives guidance on the depth of the risk assessment and ALARP demonstration required as interpreted on table 4.

Table 4.   Proportionality in risk assessments and ALARP demonstrations

Risk matrix position	Depth of risk assessment	Type of ALARP demonstration
Intolerable	Quantified Risk Assessment (QRA)	Risk reduction required regardless of cost
Upper ALARP	QRA if close to intolerable semi-quantitative	Apply relevant good practice plus Consider further risk reduction measures applying proportion factor of 10
Lower ALARP	Semi-quantitative Qualitative if close to broadly acceptable	Apply relevant good practice plus Consider further risk reduction measures applying proportion factor of 1
Broadly acceptable	Qualitative	Apply relevant good practice

For societal risk R2P2[3] gives the criteria of 50 fatalities once in 5000 years as the boundary between the intolerable and tolerable if ALARP regions. Other published 'anchor points' for societal risk have been reported by Ball[9] along with a number of FN curves to be used for societal risk assessments. To avoid the complexity and cost of a fully quantified risk assessment, HSE has developed a 'rough but rapid' technique to indicate the level of societal risk. An approximate risk integral (ARI) is calculated as a function of the maximum number of fatalities for the worst case event and the frequency of this event. HSE are providing a simple program for calculating the ARI from these values, further details on the development of the methodology and the equations involved can be found in a paper by Hirst and Carter[8]. The ARI approach now favoured by the HSE explains why they are requesting companies to provide an estimate of the number of fatalities for the worst case event as discussed earlier.

The HSE guidance document[7] gives the criteria for individual risk in agreement with R2P2[3] and the new criteria for societal risk as shown on table 5.

CONCLUSIONS

The COMAH Regulations place an onerous duty on operators of sites with large inventories of dangerous substances including the need to carry out detailed risk assessments. The procedures developed from established process hazard analysis techniques have proved effective for a number of COMAH risk assessments carried out by ABB Eutech. The structured approach demonstrates that a thorough evaluation of the risks posed by the operation has been undertaken, and the semi-quantitative use of a risk matrix avoids the use of time consuming and costly quantified risk assessments.

Table 5. HSE risk criteria

Risk matrix position	Individual risk frequency of fatality per year	Societal risk Approximate Risk Integral (ARI)
Boundary between intolerable and Tolerable if ALARP regions	1×10^{-3} for worker 1×10^{-4} for public	500,000
Boundary between Tolerable if ALARP and broadly acceptable regions	1×10^{-6} for all	2,000

The methodology provides the opportunity to review the existing protection measures against current industry standards, and to use a risk based analysis to determine if further risk reduction measures are required. For the illustrative example of a road tanker drive away and hose rupture the risk was judged to be at the upper end of the tolerable region. A recommendation has been made to install an interlocked barrier system to prevent the road tanker moving whilst the offloading hose is attached and further recent assessment has added a procedure for the driver's keys to be removed.

Many companies consider the COMAH Regulations to be a costly burden on their business and the strict application in the UK has lead to claims that a level playing field does not exist in Europe. The approach described in this report allows a focused and risk prioritised improvement plan to be developed. This will achieve genuine reductions in the risks of major accidents and in the long term will protect companies from the highly damaging effects of a serious accident. A second benefit from the involvement of site personnel in the risk assessment process is a greater awareness of the potential for high consequence, low frequency events. It is common to find that the primary focus for those involved in health and safety are regular low severity events such as slips, trips and falls. COMAH has helped to redress the balance in favour of the high severity events that, although rare, are an ever-present danger should attention to the critical safety measures lapse.

Recent guidance from the HSE has clarified a number of the earlier problems in determining a proportionate approach for COMAH risk assessments and ALARP demonstrations. The difficulty for operators and those writing safety reports is that the guidance is late and is being applied by HSE retrospectively. The requirement to include a measure of societal risk when making ALARP decisions is a new approach and it is likely that for operators of sites with high consequence events this will form part of ongoing assessments.

REFERENCES
1. HSE, 1999, A Guide to the Major Accident Hazards Regulations 1999, L111, HSE Books
2. HSE, 1999, Preparing safety reports: Control of Major Accident Hazards Regulations 1999, HSG190, HSE Books
3. HSE, 2001, Reducing Risks, Protecting People: HSE's decision making process, HSE Books

4. Middleton, M., 2001, Using risk matrices, *The Chemical Engineer*, 723: 34-37.
5. The European Council of Chemical Manufacturers Federations (CEFIC), 1990, CESAS recommendations for the safe storage and handling of Liquid Sulphur Dioxide.
6. HSE, 2002, HID's approach to 'as low as reasonably practicable' (ALARP) decisions, SPC/PERMISSIONING/09, available at www.hse.gov.uk/hid/spc
7. HSE, 2002, Guidance on 'as low as reasonably practicable' (ALARP) decisions in control of major accident hazards (COMAH), SPC/PERMISSIONING/12, available at www.hse.gov.uk/hid/spc
8. Hirst, I and Carter, D, 2002, A worst case methodology for obtaining a rough but rapid indication of the societal risk from major accident hazard installations, *Journal of Hazardous Materials*, 92:Issue 3:223–237.
9. Ball Professor D J and Floyd Dr P J, Societal Risks - Final Report, *School of Health, Biological and Environmental Sciences, Middlesex University*

DIRECTORS' AND ENGINEERS' RESPONSIBILITIES FOR SAFETY – A CAUTIONARY TALE

Brian R. Harris
Joint Managing Director, Nobels Explosives Co. Ltd.

IChemE North West Branch
Hazards XVII Symposium
24–27th March 2003

Directors' and Engineers' Responsibilities for Safety

A Cautionary Tale

The Cautionary Tale

Personal experience of the aftermath of a fatal
explosion at Cookes Works, Penrhyndeudraeth, Wales

14 June 1988

Joint Managing Director, Nobels Explosives Co. Ltd.

Background

•Nobels Explosives Co Ltd
wholly owned subsidiary of ICIplc
•Products
Commercial Explosives
Detonators
Propellants
Propellant Devices

•Headquarters
Ardeer, Stevenston, North Ayrshire

•Manufacturing Sites

Ardeer	2000 people	
Wigan	250 people	(Roburite Works)
Penrhyndeudraeth	100 people	(Cookes Works)

Organisation

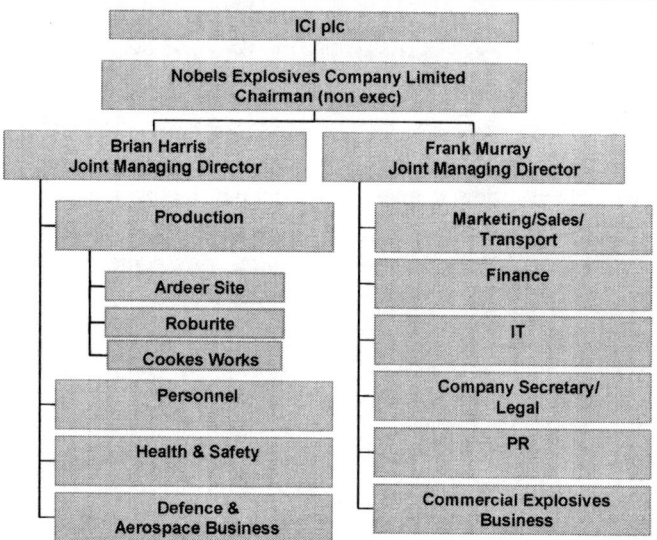

Cookes Works

- Manufacturing process
    - Nitration of glycerine to produce nitro-glycerine
    - Mixing to produce gelignite - a safe paste
    - Cartridging in paper to produce sticks of explosive
- Flat organisation

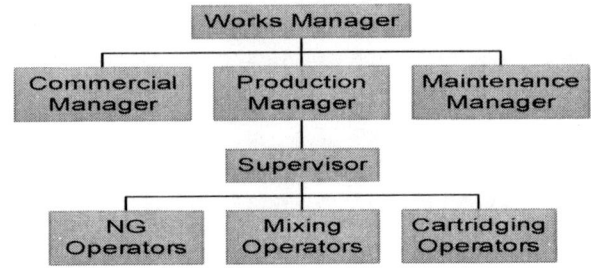

- A close knit family, Welsh speaking

My Background as at 1988

- 1966 - 1986

    ICI Petrochemicals
    - design, plant operation and senior management
    - exposure to management of high hazard processes
    - "schooled" by Trevor Kletz

- 1986 - 1987

    Nobels Explosives Co Ltd
    Production and Personnel Director

- 1988 -

    Nobels Explosives Co Ltd
    Joint Managing Director

Communication and Safety Culture

- Clear ICI Group Health and Safety Policy statement and arrangements

- Clear Nobels Explosives Co. Ltd Health and Safety Policy and arrangements

- Joint consultation
  Health and Safety committees

- Health & Safety first item at meetings Board, Executive, Management and consultation meetings

- Speaking to People

- Toolbox talks

Communication and Safety Culture

- All operations subject to hazard analysis and safety review

- Well defined audit programme
  - internal / external (ICI)
  - action follow up and close out
  - visibility and audit trail

- Regular review and audit by Explosives Inspectorate

Personal Involvement

- Personal site visits and audits
  - quarterly at Cookes Works

- Close inspection of randomly selected area

- Spoke with the plant operatives

- Informal discussion with representatives

- Priority given to spending to improve safety

- Reviewed with Works Manager the outcomes of audits and actions

Safety Arrangements Assessment

- From what I have said so far how would you judge the management arrangements with regard to safety back in 1988?

- Not untypical of a company that is striving to achieve a good safety performance and exercising care.

Monday, 14 June 1988

**Just another day
or so I thought!**

Personal Memories of 14 June

- It's Good to be Alive
- The call you never hope to get
- Preparation and travel
- An aerial view
- Arrival
- Assessment
- There are no bodies
- The families
- Press conference
- BBC TV interview
- 9 o'clock news
- The world has changed

Personal Memories of 14 June

• **It's Good to be Alive**

250 Kg of Nitroglycerine
+
250 Kg of finished explosive

• **The world has change**

15 - 18 June

- The alarm call

- The live radio interview

- Factory assessment

- Visiting the families

- HSE/Coroner

- Forensic investigation

- Making Safe/ Saving the Factory

The Following Week

- Investigation

- Plan for recovery

- Initial findings

- HSE attitude/personal exposure

The 3 P's

People
ignorance, attention or intent

Plant
inadequate design or poor maintenance

Procedures
inadequate or poor compliance

Failure of any one can lead to an unsafe event

The Causes/Evidence

• **Causes of the explosion:**
Foreign body entering mixer with raw materials

• **Causes of the deaths:**
Operators in the wrong place
- not following procedure
- not using the reinforced control room/bunker

• **Evidence:**
• Plant log book was filled in for batches yet to be started
• Well known that these two operators were regularly in the canteen at times inconsistent with the batch times
• Clearly not once off behaviour

The Causes/Evidence

• **Causes of the explosion:**
Foreign body entering mixer with raw materials

• **Causes of the deaths:**
Operators in the wrong place
- not following procedure
- not using the reinforced control room/bunker

• **Evidence:**
• Plant log book was filled in for batches yet to be started
• Well known that these two operators were regularly in the canteen at times inconsistent with the batch times
• Clearly not once off behaviour

Inspector's Questions

- How is Nobels Explosives Company organised to implement its safety policy?

- What is the relationship between this Company and your parent Company?

- How does the Board of the Company operate?

- What is <u>your</u> role in the Company and what does it entail?

- What are <u>your</u> qualifications for this role?

- Who appointed you?

More Inspector's Questions

- What training have <u>you</u> had for this role particularly with regard to your Health and Safety responsibilities?

- Explain to me <u>your</u> role in safety management in the Company.

- How do you discharge <u>your</u> safety accountability?

- What information do <u>you</u> receive?

- What do <u>you</u> do with it?

- How do <u>you</u> know that the information is valid?

Yet More Questions

- How do you know Company procedures are being followed?

- You have told me about audit processes and their findings. How do you know that agreed actions are properly closed out?

- What else would you like me to know?

    etc etc etc

- End of (interrogation) interview

- You receive a statement to sign and receive a copy

Uncertainty

- Personal introspection
- Other recent disasters
    - Kings Cross fire
    - Herald of Free Enterprise ferry sinking
- The increasing desire to prosecute individuals
- Would I be yet another test case?

- A year goes by!

Outcome

- **HSE decides to prosecute the Company**

- **29 March 1990 Mold Crown Court**
  Company prosecuted under section 2 of the Health & Safety at
  Work Act for failing to ensure the safety of employees by lack of
  supervision of compliance with operating instructions for the safe
  mixing of explosives.

- **Company pleaded guilty**
  - fined £100,000 + £30,000 costs
  - account taken of employees' own actions

- **Other costs**
  - personal "costs"
  - diverted management effort
  - commercial impact
  - increased insurance premium

Learning/Practice

- **Check that the Health & Safety Policy gives clear
  direction to the organisation**

- **Understand the scheme of organisation for Health &
  Safety and ensure it is understood throughout the
  organisation**

- **Clarify explicitly my accountability for Health & Safety**

- **Get any specific training I need**

- **Identify the information I need to discharge
  my accountability**

Learning/Practice

- Validate the assurance process and identify what I do to assure myself that what is supposed to happen does happen

- Know how I can answer the questions I was asked by the HSE

- Know how I demonstrate that I have not been negligent

Finally

As a Senior Executive, by knowing how you demonstrate that you are not negligent because of what _you_ do, not only are you assuring your own integrity and that of your company, more importantly you are increasing the probability that people in your organisation will not be injured and will return home each day to their families.

Sadly that was not the fate of two men on 14 June, 1988

MAJOR INCIDENTS AT WASTE TREATMENT SITES – CASE STUDIES AND LESSONS

Andrew Hitchings
COMAH Policy Advisor, Environment Agency of England and Wales, Block 1, Government Buildings, Burghill Road, Westbury-on-Trym, Bristol, BS10 6BF.
andrew.hitchings@environment-agency.gov.uk

Over the past few years there have been a number of incidents at hazardous or chemical waste treatment facilities that have resulted in unacceptable impacts on the environment and human health. These incidents have fostered a negative public opinion of this industry at a time when the forthcoming implementation of the Landfill Directive in the UK may require a substantial increase in capacity.

The Environment Agency (Agency) and Health and Safety Executive (HSE) are concerned that the management and technical standards across this sector keep pace with best practice and learn the lessons of these incidents. Accordingly the Agency and HSE undertook a number of joint site inspections during late 2002 to assess the current state of the industry.

This paper briefly examines a number of recent incidents, their causes and impacts. It then identifies the relevant key technical standards for the sector. The paper then summarises the outcomes of the Agency and HSE audit exercise. The detailed findings of the exercise will be published in a forthcoming report.

KEYWORDS: waste establishments, Environment Agency, Health and Safety Executive, chemical or hazardous waste treatment, accident prevention, technical standards, COMAH

BACKGROUND

Waste transfer and treatment facilities have been an integral part of the waste management industry for many years. They handle a wide range of waste materials (for example acids/alkalis, organic compounds, oxidising compounds, chlorinated solvents, pesticide residues) and undertake a range of activities (storage, treatment or transfer) that pose various threats to the environment and human health. Typical treatment operations include neutralisation, immobilisation, de-watering and blending to make combustible fuels (commonly known as secondary liquid fuel (SLF)).

The Environment Agency (Agency) issues waste management licences under Part II of the Environmental Protection Act 1990 (EPA) to facilities that keep, treat and dispose of waste. There are approximately 150 licensed hazardous or chemical waste treatment facilitates in England and Wales. Part II EPA regulates the deposit, keeping, treatment and disposal of controlled waste to ensure that it does not cause pollution of the environment or harm to human health.

Certain process activities such as recovery of organic solvents by distillation require authorisations under Part I of EPA. These are also issued by the Agency and require the Best Available Techniques Not Entailing Excessive Costs (BATNEEC) to prevent, minimise and render harmless releases to the environment and the selection of the Best Practicable Environmental Options (BPEO). There are approximately 40 authorised solvent distillation facilities.

The Pollution Prevention and Control Act 1999 (PPC) will replace waste management licensing under Part II EPA 90 and waste processing authorisation under Part I EPA 90 with an integrated licensing regime for chemical and hazardous waste facilities. New facilities or substantial modifications to existing facilities are licensed under PPC and existing sites will be transferred over to the PPC regime around 2005. PPC requires the use of the Best Available Techniques (BAT) to prevent and, where that is not practicable, generally to reduce emissions and the impact on the environment as a whole.

All waste treatment and transfer facilities are subject to the requirements of the Health and Safety at Work etc Act 1974 (HASAW). This includes a duty on employers to take all reasonably practicable measures to ensure the health, safety and welfare of employees, including the provision of safe systems of work, training, supervision etc. HASWA Act also includes a duty on employers to take all reasonably practicable measures to protect non-employees (contractors, local residents, other businesses etc). The Health and Safety Executive (HSE) enforce HASWA.

Certain waste transfer and treatment facilities are also subject to the requirements of the Control of Major Accident Hazard Regulations 1999 (COMAH). COMAH applies because of the presence or anticipated presence of threshold quantities of certain dangerous substances. COMAH requires operators to take all necessary measures to prevent major accidents and limit their consequences to persons and the environment. In England and Wales COMAH is enforced by a joint Competent Authority comprising the Agency and HSE. There are also additional planning consent requirements of sites subject to COMAH under the Planning (Control of Major Accident Hazards) Regulations 1999 (P(COMAH)). There are approximately 20 hazardous or chemical waste treatment facilities notified as under COMAH.

As you can see there is a raft of legislation whose purpose is to achieve high standards of operation within this industry sector. However despite this there has been a long history of incidents and a number of high profile incidents within the last 2–3 years.

RECENT INCIDENTS

Table 1 provides a summary of recent major incidents with off-site impacts at hazardous and chemical waste facilities. Perhaps the most significant being the incidents at the Cleansing Services Group (CSG) facility in Gloucestershire in October 2000 and the Distillex solvent recovery facility on Tyneside in April 2002.

FIRE AT CSG, SANDHURST, OCTOBER 2000

CSG operates a hazardous waste treatment facility and transfer station at Sandhurst, near Gloucester. At approximately 02.00 hours of 30 October 2000 a major fire started in a waste storage compound (Figure 1). The facility was unoccupied at the time and the fire service had difficulty accessing the facility because of fire blocking the only access route and a series of explosions of waste aerosol cans and larger drums. Approximately 60 people were evacuated by the emergency services and 13 persons mainly emergency services personnel, were taken to hospital as a precautionary measure but none were admitted. The fire service gained access to the site from upwind across fields and the fire was eventually extinguished at 18.00 hours.

Table 1. Summary details of recent incidents at hazardous and chemical waste treatment facilities

Operator, location and date	Incident description	Consequences and regulatory actions taken
Distillex Ltd, North Shields April 2002. IPC authorised (Part I EPA 90) distillation process (section 5.2(a)).	Fire and explosion caused by use of an angle grinder (see more detailed entry)	Site destroyed. Prosecution by HSE with £39k fine. Incident cost the company in excess of £1m.
P&R Disposal Services, St Helens. October 2001 Waste transfer station licensed under Part II EPA 90.	Fire and site extensively damaged. Cause unknown but malicious damage is a possibility. Fire associated with theft from site.	Contamination of local watercourses. Site destroyed. Warning letter issued by Agency.
Parke Environmental, Newport, Gwent. July 2001 Waste treatment facility licensed under Part II EPA 90 and COMAH lower tier (LT) establishment.	On-site fatality and off-site release of a cloud of hydrogen sulphide during a neutralisation reaction with in-complete characterisation of wastes.	HSE and Agency both served Prohibition Notices. Prosecution ongoing.
Cleansing Service Group Sandhurst. 30 October 2000 Waste treatment facility licensed under Part II EPA 90 and COMAH lower tier (LT) establishment.	Fire in a waste storage facility followed extensive flooding from the River Severn (see more detailed entry).	Agency served a Suspension Notice under Waste Management Licensing HSE served 2 Improvement Notices under HASAWA 74 Agency served Improvement Notice under RSR 1993 Prosecution pending.

Approximately 180 tonnes of mixed chemical wastes including some pesticides and chlorinated hydrocarbon solvents were consumed in the fire. The site is adjacent to the River Severn and flooded on 3 November. Emergency actions had to be taken to make the site safe and move fire damaged and other material beyond the reach of floodwaters. The site was surrounded by floodwater (Figure 2) and could only be accessed by boat. Serious flooding continued until 22 November and the site flooded again in December.

Detailed investigations were undertaken by the Agency and HSE. These investigations discovered the unforeseen presence of un-authorised wastes including 7 25 litre drums labelled "solvent contaminated with BSE" and 2 drums of radioactive waste.

Substantial and prolonged monitoring and modelling of the incident took place. This included:

- 17,500 tests on 500 environmental samples (air, water and land), none of which indicated any significant levels of contaminants off-site,
- At the time modelling of the accident by the HSE indicated that a "dangerous dose" of toxic materials would not have occurred at the site boundary,
- Blood tests on those exposed on the day of the fire were found to be negative for solvents and heavy metals,
- Radiological monitoring off-site concluded that there was no evidence for the presence of radioactive materials in the areas surveyed,
- The local Health Authority have undertaken multiple health surveys. These surveys offer evidence that the physical and/or physchological health of a significant number of Sandhurst residents involved in the surveys were affected following the fire, although these symptoms are generally though to be self-limiting. Health monitoring work continues.

At the time of the incident the site was subject to a waste management licence issued by the Agency under Part II EPA and was notified as a COMAH site. There was a high level of media and political interest in the incident and the HSE and Agency (acting as the COMAH Competent Authority) have submitted progress reports to the Deputy Prime Minister[1,2].

The cause of the accident has not been established. The most likely causes seem to be loss of containment of "laboratory smalls" or leakage of pyrophoric materials. Arson is also a possibility.

DISTILLEX EXPLOSION, APRIL 2002

Distillex Ltd is located ten miles to the east of Newcastle just north of the Fish Quay, North Shields. The facility refines organic solvents (including wastes) for toll recovery, for reuse by the originating company and for the recovery of reusable materials, which are subsequently sold on to other markets. The site is subject to an authorisation issued by the Agency under Part I EPA.

On the 12 April 2002 an intense fire devastated the Distillex facility (Figure 3). The office, the three drum storage areas, the warehouse, the boiler, the plant area and static road tanker were all destroyed. Commercial buildings adjacent to the site in line with the fire were also destroyed as a result of the extreme heat and embers from the fire, despite the actions of the fire service.

In the response to the incident the emergency services set up a ½ mile exclusion zone around the facility, evacuated 500 people, closed the Tyne tunnel road link and the metroline and diverted aircraft away from the air corridor. The incident attracted national media interest.

The cause of the incident is believed to be the use of a pneumatic angle grinder by an operator in an attempt to cut the steel frame off a 1 tonne Intermediate Bulk Container

(IBC), containing solid still bottoms. It is believed sparks given off from the grinder ignited vapour within the IBC. The fire spread rapidly to the skip and the drum storage area. At the time of the fire there were a number of containers of flammable and combustible materials stored outside of the bunded areas. This reduced the separation distances between flammable liquids and mean there was an absence of secondary containment both of which contributed to the spread of the fire.

Samples taken from soil and vegetation in areas affected by the smoke plume confirm that concentrations of combustion products were well below environmentally acceptable levels. Although the smoke plume was very dramatic, there is no evidence of lasting environmental damage as a result of the incident. However there was substantial distress caused to local residents and businesses. The local health authority reported 5 casualties as a result of the incident and the Police had 36 reported injuries on duty resulting from the incident.

BEST PRACTICE GUIDANCE

The Agency has recently published new best practice guidance for the waste treatment industry[3]. The Agency is encouraging operators to use it to review their current operations and in updating their working plans. The Agency will use the guidance in its reviews of licences. Facilities are expected to carry out improvement to reach the required standards at the earliest opportunity and to a programme agreed by the Agency.

The guidance identifies key issues for the sector including those relevant to accident prevention and mitigation.

ACCIDENT MANAGEMENT PLANS

Facilities should draw up accident management plans based upon a thorough assessment of the hazards and risks. These accident management plans should include the technical and managerial prevention and mitigation measures that are necessary based upon this assessment. Accident management arrangements should be regularly reviewed and tested.

WASTE CHARACTERISATION, SAMPLING AND CHECKING

Failure to adequately screen waste samples prior to acceptance and confirm the composition on arrival at the facility is key to safe management of waste facilities. You cannot manage the process risks or the treatment operations without adequate knowledge of the waste. A lack of knowledge has historically lead to subsequent problems that include: inappropriate storage and mixing of incompatible substances, accumulation of wastes and unexpected treatment characteristics (as in the Parke Environmental incident).

Pre-acceptance procedures should be sufficient to enable the facility to:

- screen out unsuitable wastes;
- confirm the details relating to composition i.e allow selection of verification parameters to be tested upon the arrival of the waste at site;
- identify any unexpected substances within the waste which may affect the treatment process or may react with other reagents;
- accurately define the hazards associated with the waste;

- ensure waste are stored on site in compliance with segregation, separation and engineering requirements;
- determine whether the waste is within the terms of possible onward permits and the cost of any onward disposal;
- meet legislative requirements.

Acceptance procedures should be sufficient to ensure:

- procedures for checking paperwork arriving with the load;
- procedures for safe unloading to allow inspection and sampling;
- visual load inspection;
- drum and package labelling procedures;
- timely sampling procedures for all incoming wastes including bulk wastes as well as drums, containers and laboratory smalls;
- written records of verification and compliance testing;
- written records of assessment of consistency with pre-acceptance information and documentation;
- policy for recording and dealing with non-conforming wastes;
- written records of rejection criteria applied to wastes;
- sample retention systems i.e period of retention, method of retention;
- written records of decision making re acceptance or rejection of wastes and decisions on future treatment or disposal options.

RECORD KEEPING
An internal tracking system and stock control procedure should be in place for all wastes. This system should record:

- what waste has arrived on site;
- what waste is currently held on site,
- where wastes are currently held;
- how long the waste has been on site;
- total quantity of wastes currently on site;
- compliance with licence conditions.

WASTE STORAGE AND SEGREGATION
Detailed procedures are necessary for ensuring the effective management for the storage of waste on site. These should be generated having conducted risk assessments to identify the correct manner in which licensed materials are to be stored and the standards of operation that are to be expected. This risk assessment should consider:

- location of storage areas;
- storage area infrastructure requirements and relevant standards to be met (materials of construction etc).
- conditions of tanks, drums, vessels and other containers;

- stock control systems
- segregated storage requirements

PREVENTION OF ACCUMULATIONS OF WASTE

Procedures and auditing systems should be in place to ensure waste does not accumulate. There should be a plan of disposal or treatment for each waste consignment accepted and if it is found that the plan is not being followed alternative arrangement need to be implemented in a timely manner.

Failure to ensure adequate removal of wastes has lead to large numbers of drums being stored on some sites. This causes increased accident risk as well as operational difficulties. Wastes involved are typically unchecked and drums are left to deteriorate. Such situations are often associated with large-scale site clearances and can be accompanied by competitive pressures and customer insistence to accept additional waste streams. Typically the wastes involved are difficult to handle and/or treat and may have been transferred between various facilities with a consequent unacceptable loss of information relating to original producer and composition.

TREATMENT

The key issues for control of the waste treatment operation include the following:

- ensuring the waste is suitable for the activity (pre-acceptance);
- adequately characterising the waste (acceptance procedures);
- appropriate and safe storage of wastes (storage);
- provision and maintenance and inspection of treatment equipment;
- operational control of the treatment process (key parameters and process monitoring equipment);
- appropriate disposal of effluents.

AGENCY AND HSE AUDIT EXERCISE

METHODOLOGY

The Agency and HSE undertook a joint audit exercise looking at standards within the hazardous and chemical waste treatment sector during late 2002. This exercise was prompted by growing concerns in the regulators about both the incident record of this sector and their impression that this industry as a whole operates below current best practice. An external report of this exercise will be published.

The audit exercise involved the joint inspection by both regulators of a cross section of 25 facilities selected so as to allow a picture of the "state of the industry" to be established. The exercise was launched at an industry seminar on 14 May 2002 jointly organised by both regulators and attended by the majority of industry.

An inspection template was drawn up that would allow for consistent inspections to enable conclusions to be drawn, and allow reporting on the greatest concerns.

The template follows the structure of the best practice guide, and was split into 6 selections:

- Management Arrangements
- Pre-acceptance Procedures to assess waste
- Acceptance Procedures
- Storage
- Treatment
- COMAH

Prior to the inspection the facility operator was informed of the visit and given a copy of the template to indicate the questions that would be asked. This was to enable the operator to prepare any response, have copies of any documents available and to plan for the inspection. If inspectors considered that by giving such notice the site conditions seen during the visit were not typical then a further unannounced follow up visit was undertaken.

FINDINGS
The inspection exercise was completed during June – October 2002 and all 25 facilities were jointly inspected by the Agency and the HSE.

Although the number of facilities visited was small and should not be used to draw statistical conclusions the regulators believe that the results are representative of this industry sector as a whole. The detailed findings and any subsequent conclusions drawn will be presented in the external report into the exercise to be published in early 2003. In summary the findings are: -

- On the whole industry was co-operative and some facilities had made recent steps to improve their standard of operation;
- On the whole management documentation was in-place, however in many cases it was of a poorer quality than expected;
- On almost half the sites tanks, vessels, pipework and values were not adequately identified and labelled;
- On a significant minority of sites there was concern over compliance with the COSHH Regulations;
- On a significant minority of sites there was no comprehensive preventative maintenance programme;
- On a significant minority of site there was not an adequate site storage plan;
- On a minority of sites waste reception areas/procedures were not considered adequate for the containment and segregation of wastes (this was a contributory factor in the Distillex incident);
- On a minority of sites labelling of wastes was not adequate to ensure that appropriate segregation took place;
- On a minority of sites layout and location of storage areas was not suitable with the standards contained in HSG51/71;
- On a number of the sites accumulations of waste were identified as a cause for concern;
- On two of the sites undertaking treatment operations concern was expressed over the standard of control instruments;

- One site was found to be holding COMAH qualifying inventories but was not notified under COMAH. A further 3 sites were considered to have operational flexibility which might bring them into COMAH.

The follow-up actions from the exercise have included:

- 6 sites where letters and/or site advice have been given on matters of concern;
- 1 Prohibition Notice served by the HSE on storage of laboratory "smalls" (although this was served just prior to the joint exercise visit);
- 3 Improvement Notices served by the HSE (although 2 of these were just prior to the joint exercise visit) on storage and management issues;
- 1 Enforcement Notice issued by the Agency to remove time expired waste accumulations.

JOINT WORKING

The Agency and HSE have both found the joint exercise very rewarding. The Agency and HSE already work closely under the COMAH regime and this exercise has shown that some of the benefits realised under COMAH can be extended to joint working under other regimes.

No evidence of conflicting interests or demands on operators has been exposed by the exercise, rather both organisations have been able to confirm the complementary nature of our requirements for this industry sector.

In the coming months the Agency and HSE will be considering whether joint working should be extended both within this sector and to other industry sectors where we have mutual interest.

FUTURE CHALLENGES FOR THE INDUSTRY

The way in which hazardous wastes are to be managed in the future is set to change significantly. This presents challenges and opportunities for both the industry and the regulators.

The Landfill Directive will result in a significant reduction in the number of landfill sites accepting hazardous wastes (current estimates[4] are 182 interim sites until July 2004 and 41 sites beyond this) and this may lead to serious shortfall in treatment and disposal capacity for a number of hazardous waste streams.

Unless there are significant reduction in hazardous waste generation it is likely that additional hazardous or waste treatment capacity (whether integrated at the producer end or as third part operations) will be necessary as part of the solution to the constriction of landfill capacity although there are still substantial areas of legislative, regulatory and market uncertainty.

In recognition of this problem the Department of Environment, Food and Rural Affairs (DEFRA) has recently announced the setting up of an Advisory Forum of Hazardous Waste in which the Agency will participate fully.

The Agency and HSE will be analysing the findings of this inspection exercise carefully over the coming months to ensure that accident prevention continues to be a major focus of compliance activity on these facilities with the aim of preventing further incidents of the nature described in this paper.

REFERENCES

1. Joint Report by the Health and Safety Executive and the Environment Agency, 2001, Report for the Deputy Prime Minister the Right Honourable John Prescott MP into the Major Fire on 30 October 2000 at Cleansing Services Group Ltd, Upper Parting Works, Sandhurst Lane, Sandhurst, Gloucester, GL12 9NQ, Environment Agency web-site, www.environment-agency.gov.uk/regions/midlands

2. Joint Report by the Health and Safety Executive and the Environment Agency, 2001, Progress Report for the Deputy Prime Minister the Right Honourable John Prescott MP into the Major Fire on 30 October 2000 at Cleansing Services Group Ltd, Upper Parting Works, Sandhurst Lane, Sandhurst, Gloucester, GL12 9NQ, Environment Agency web-site, www.environment-agency.gov.uk/regions/midlands

3. Environment Agency, Recovery and Disposal of Hazardous and Non Hazardous Waste (Other than by Incineration and Landfill), Waste Management Licensed Facilities, May 2002, version 1.1, Environment Agency web-site, www.environment-agency.gov.uk

4. Environment Agency, Hazardous Waste Management Market Pressures and Opportunities: Background Paper, Draft Report December 2002, Entec UK Limited

Figure 1. CSG Ltd, fire on 30 October 2000

Figure 2.   CSG Ltd, aerial view of site during flooding

Figure 3.   Distillex Ltd, fire on 12 April 2002

ASSESSING AND REDUCING FLOOD RISKS ON MAJOR HAZARDS SITES

Aidan Whitfield,
Environment Agency, Kingfisher House, Goldhay Way, Orton Goldhay, Peterborough.
PE2 5ZR. aidan.whitfield@environment-agency.gov.uk

During the 1980s and early 1990s, there were relatively few serious flooding events in the UK. Consequently it was widely believed that flood-defence measures have "tamed" the environment and that the risk of flooding was minimal. However, since the Easter 1998 floods in central England, there have been a number of serious flooding incidents across the UK and the public awareness of flood risks has increased significantly. Whilst media attention has focussed on the damage caused to domestic properties, flooding also occurred on a number of major hazards sites. This paper describes some of the incidents that have occurred in the process industries and the measures taken to reduce the risk and consequences of flooding.

The Environment Agency, working as part of the COMAH Competent Authority has set flood risk assessment on major hazard sites as an inspection priority. On top-tier sites, operators have been required to address flood risk as part of their COMAH Safety Report. On lower-tier sites the Agency has prioritised its inspection effort on those sites identified to be at risk of flooding. The Agency is also concerned about non-COMAH sites and will use its powers under the Pollution Prevention and Control (PPC) regulations to require operators to address flooding issues.

KEYWORDS: COMAH, Environment, Flooding, Risk assessment, Pollution Prevention and Control (PPC)

INTRODUCTION

In 1947 and 1953, there was major flooding in the east of England and the lessons learned from those events, shaped the UK flood defence strategy for the next 30–40 years.

Freshwater flooding affected the Fens of East Anglia in March 1947, following a six-week spell of severe winter weather. Heavy rain melted the accumulated snowfall producing the equivalent of 110 mm of rainfall over 24 hours and all of the major rivers leading into the Wash burst their banks. 250 square kilometres of the most productive farmland in the country was flooded, hundreds of families were made homeless, and thousands of livestock were swept away. Miraculously no-one was killed.

On 31st January 1953, a combination of spring high tides, a deep depression and northerly gales, combined to cause a tidal surge in the southern North Sea. It affected the coast from Lincolnshire to Kent and at its peak, the surge was 2½ metres above the high spring tide level. There was no flood warning system at the time and the surge struck at night, drowning many occupants of single storey beach chalets in their bedrooms. In all 307 people died, 24,500 houses were damaged and 40,000 people were evacuated. It was the worst peacetime disaster ever to strike Britain.

As a consequence of these floods, drainage work was carried out on many rivers, flood defence banks were built, a comprehensive flood-warning scheme was introduced and the Thames Barrier was completed in 1982. The effectiveness of these measures was demonstrated in 1978 when another tidal surge occurred on the north Norfolk coast. It was higher than the 1953 surge but caused significantly less damage. In Kings Lynn, for example, 15 people had drowned in 1953, whereas in 1978 there were no deaths despite extensive flooding in the town centre.

During the 1980s and early 1990s, there were relatively few serious flooding incidents in the UK due to a combination of successful flood defence measures and an absence of the relevant severe weather conditions. Unfortunately, this led to a sense of complacency in the minds of politicians and the general public, that the risk of flooding was not a serious issue in the UK. On numerous occasions, Local Authorities granted planning permission for developments in the flood plain against the advice of the Environment Agency. There was also a reduction in spending on flood defence works as Local Authorities diverted funds into other areas.

FLOODING AT BP OIL, NORTHAMPTON

The first major flooding of recent years occurred during Easter 1998. An active weather front remained stationary over central England and in many places more than 60 mm of rain fell in 48 hours. In total 5 people died, 4,500 properties were flooded and the cost of damages was estimated at £350m. The worst affected town was Northampton where two people died and 2,500 houses were flooded when the River Nene burst its banks. Most of the residents were unaware that they lived in a floodplain and were caught completely by surprise.

The flooding in Northampton also affected a number of commercial and industrial premises including a fuel distribution and storage terminal operated by BP Oil, which is a lower-tier establishment under the Control of Major Accident Hazards (COMAH) Regulations 1999. Most of the site was flooded to a depth of approximately 0.5 metres, though there was no loss of containment that might have caused pollution of the River Nene. Damage was restricted to the offices, some pumps and underground power cables. An HSE specialist inspector conducted a thorough inspection of the tanks to ensure that there had been no flotation or other damage and the site returned to full operation within 6 weeks. There are a number of reasons why the flooding caused so little damage, including:

- Northampton is a simple fuel storage and distribution terminal (no chemical processing).
- The entire inventory was contained in large storage tanks built on plinths that remained above the level of the floodwater.
- Prompt action by the staff prevented rainwater accumulating on the floating roof tanks which could have led to the collapse of the roofs.
- There were no materials stored in drums or bags which could have been washed away or affected by the floodwater.

- BP Oil was able to maintain the supply of fuel to customers by utilising its other storage and distribution terminals at Hemel Hempstead and Birmingham.
- BP Oil, being a major company, was able to pay for the cost of cleaning up the site and returning it to normal operation.

The flooding of the BP Oil terminal did not cause significant harm to human health or the environment and consequently it did not act as a "wake up call" to the COMAH Competent Authority regarding the risks of flooding on COMAH establishments. (The Competent Authority (CA) for the COMAH regulations in England and Wales comprises the Health and Safety Executive (HSE) working jointly with the Environment Agency (EA) (and in Scotland the HSE working jointly with the Scottish Environment Protection Agency (SEPA)).

LESSONS LEARNT FROM THE EASTER 1998 FLOODS
Following the Easter 1998 floods, the Environment Agency board commissioned an independent report, which concluded that whilst the flood warning system had worked properly, significant improvements could be made. The Agency carried out a wide-ranging review of its flood defence activities and identified the following improvements:

1. Publicity campaigns to increase public awareness of flooding and to encourage at-risk homeowners to develop a flood plan.
2. Publication of indicative flood plain maps on the internet.
3. Upgrading of river flow telemetry systems.
4. New computer models to be used for flood forecasting.
5. Better communications with the Met Office and the emergency services.
6. More understandable flood warnings codes.
7. The use of automated telephone phone messaging to disseminate flood warnings.

The Agency launched all these improvements during "Flood Awareness Week" in September 2000. This publicity campaign used television, posters and newspapers and was timed to coincide with the start of the winter flooding season.

FLOODING AT CSG, SANDHURST, GLOUCESTER
The autumn of 2000 was the wettest for 270 years and the prolonged heavy rainfall caused significant river flooding in many places, with North Yorkshire, the Severn Valley, and parts of Kent and Sussex particularly badly affected. In total 2 people died and 10,000 properties were flooded (though flood defences protected 280,000 properties). The total bill for damages was estimated at £1.0bn. The improved flood warning arrangements introduced by the Environment Agency following the Easter 1998 floods undoubtedly reduced the loss of life and assisted the work of the emergency services. The flooding also contributed to a major accident that occurred at a waste treatment and storage site operated by Cleansing Services Group (CSG) Ltd in Sandhurst near Gloucester. The incident started in the early hours of 30th October 2000 during a severe storm, when there was a major fire in a waste storage area. Approximately 180 tonnes of mixed chemical waste including flammables, toxics and chlorinated hydrocarbon solvents were consumed in the fire. Gloucester police

set up Gold Control to manage the fire as a major incident. 60 people were evacuated from their homes by the emergency services and 13 people, mainly emergency service personnel, were taken to hospital as a precautionary measure, though none were admitted. The site had notified the CA that it was a COMAH lower tier establishment and the fire was reported to the European Commission as a COMAH major accident because of the quantities of dangerous substances involved and the evacuation of local residents. Agency flood warnings indicated that the site, which is alongside the River Severn, would flood within days and actions had to be taken to make the site safe by moving fire-damaged and other material beyond the reach of flood waters. The flooding occurred on 3rd November, and local residents had to be evacuated for a second time. The CA issued a COMAH prohibition notice though this was withdrawn when the company notified the CA that the inventory had been reduced such that COMAH no longer applied. It was replaced by an improvement notice served under the Health & Safety at Work Act 1974. The Agency issued a notice of suspension for the Waste Management Licence and two years later, most waste management activities remain suspended, though CSG have been permitted to store empty vehicles and empty containers on site. The incident at CSG raised two particular areas of concern for the Competent Authority; the fire risks of storing mixed waste materials at waste transfer stations and the risk of flooding on major hazard establishments.

AGENCY ASSESSMENT OF FLOOD RISK
After the Autumn 2000 floods, the Agency again reviewed its flood defence procedures and published a report on the lessons learned. The CSG incident and a number of other flooding issues on major hazard sites (see Table 1) were reviewed by Agency staff from Process Industry Regulation (PIR), Waste Regulation and Flood Defence. Their conclusions were:

- Many major hazards sites are located on an indicative flood plain and are therefore susceptible to either fluvial or tidal flooding. (These locations were chosen because they provide level building land, access to good transport links, a supply of cooling water and a discharge route for liquid effluents).
- Many sites were built during the 1950s and 60s and the flood defences provided at the time might not be adequate to protect against the anticipated effects of sea-level rise and climate change.
- Many sites have never experienced flooding hence flood risk might not have been properly addressed as part of the on-site and off-site emergency plans.
- Flooding of major hazards sites could lead to the loss of containment of dangerous substances and have a significant effect upon the environment. Pollution could affect the water courses themselves, adjacent sensitive habitats and necessitate closing drinking water intakes with consequent disruption to public water supplies.
- Flooding could also have significant financial and operational implications for the site concerned. It could lead to some operators going into receivership, leaving the Agency and Local Authorities to deal with land contamination and clean-up issues. (Some 10% of domestic properties flooded in autumn 2000 were still uninhabitable a year later and the restoration costs averaged £40,000 per property).

146

Table 1. Flooding incidents and issues at other major hazard sites in the UK 1999–2002

Operator details	Flooding incidents and risk assessments	Flood defence improvements
Hickson & Welch Castleford, Yorks Organic chemicals manufacture COMAH top tier (TT)	This has been a chemicals manufacturing site since the early 1900s, built on both banks of the River Aire. In 1979 part of the site was flooded when river levels were high and water backed up through the surface water drains.	In 2000, work was completed on a "site kerb" and drainage improvements to protect the river from spillages and/or fire water run-off. In October 2000, when the River Aire reached its highest level for 200 years, this system protected the site from flooding.
Syngenta Ltd., Yalding, Kent. Pesticides and herbicides formulation, filling & packing. COMAH top tier (TT)	This has been an industrial site since the early 1900s, located alongside the River Medway. The lower part of the site, including some manufacturing buildings, was flooded in 1968, with further minor flooding of roadways occurring every few years. In October 2000 the lower parts of the site were flooded to a depth of 1 metre. Materials were removed from production areas and there was no loss of containment. Production was disrupted for two days whilst waiting for water levels to fall.	Following the 1968 flood, warehousing was relocated to higher ground and a site storm and fire water storage system was built to protect the river. The system also acts as a flood defence structure to prevent minor flooding. There is a flood emergency plan in place, which worked effectively in 2000.
Tessenderlo Fine Chemicals, Leek, Staffordshire. Organic chemicals manufacture. COMAH lower tier (LT)	This site is located on the banks of the River Churnet and was originally a 19th Century dye works. The lower part of the site flooded in October 1998 and November 2000. Flood	A new warehouse has been built on high ground above flood level and the design for a new water abstraction point has incorporated features to prevent flood damage. A new ETP has been designed and

Table 1.　Continued

Operator details	Flooding incidents and risk assessments	Flood defence improvements
	warnings enabled the toxic material to be moved from the warehouse on to higher ground. The drum storage area was flooded but no material was lost. The effluent treatment plant (ETP) was inundated with flood water, halting production for 1 day. This site is near the top of the river catchment hence flooding occurs rapidly following heavy rainfall, then recedes within hours.	will incorporate above ground tanks and a 1.2m high bund wall to keep out floodwater, adding approximately 2% to the total project cost. A staged emergency flood plan is in place.
Avonmouth Industrial Complex including; Astra Zeneca (organic chemicals manufacture, TT), BG Transco (gas storage, TT), BP Oil UK (fuel storage, LT), Britannia Zinc (metal production), Tera Nitrogen (Fertiliser manufacture, TT), Rhodia (Inorganic chemicals manufacture, TT)	These sites are built on flat ground alongside to the Severn estuary. There was no flooding in 1998–2000. COMAH safety reports have assessed that there is a risk of flooding up to 0.5 m deep occurring at a frequency of 1 in 100 years. This frequency may increase to 1 in 50 years due to sea level rise associated with climate change. There have been significant developments in the area since the last full topographical survey in 1976, particularly the M49 motorway, built across the area on embankments in 1996–98.	The Environment Agency is carrying out detailed modelling of the flood risks, including the use of aerial LIDAR (Light Detection And Ranging) surveys flown in January 2002. The work is due to be completed by the end of 2002 and it is anticipated that there will be a programme of works to improve the flood defences. Emergency planning arrangements will be reviewed to ensure there are designated site access routes above the level of any floodwater.

AGENCY POLICY ON MANAGEMENT OF FLOOD RISK

In May 2002, the Environment Agency introduced a new policy on the "Management of Flood Risks at Major Installations". The policy provides a structure for assessing flood risks on COMAH sites and those regulated under the Pollution Prevention and Control (PPC) Regulations 2000. In applying the policy, the Agency will distinguish between:

a) Planning application proposals for the developments of the new installations. These will be addressed by using Planning Policy Guidance Note PPG25 on Development and Flood Risk and by pre-application discussions on permitting.
b) Existing operating installations that are already permitted. These will be targeted on a risk and priority basis as part of a rolling programme:

- The Agency has conducted a top-level indicative screening of installation, using Geographical Information Systems (GIS) overlays of site-types located in the indicative flood plain (roughly equating to a 1 in 100 year return period for flooding without defences).
- The second screening will be carried by staff in Area offices, based on local knowledge of the sites. It will consider the extent of the potential hazard or consequences if flooding of the installation occurs. This may remove some sites from consideration.
- Prioritisation of the potential hazards will be in the following order:
    i.   COMAH top-tier establishments. The Agency will ensure that flood risk is addressed in safety reports submitted to the CA and this should be completed by mid-2003.
    ii. COMAH lower-tier sites. The Agency has listed those at risk of flooding and will inspect them on a rolling prioritised basis.
    iii. Other industrial sites subject to PPC, flood risk be addressed through the PPC application and permitting programme which is phased to last until 2007. (The PPC regulations require operators to ensure that the consequences of accidents are minimised. This is a new duty compared to the previous IPC regulations and the Agency is interpreting this as a requirement for the operators to address the issue of flood risk).

SITE SPECIFIC FLOOD RISK ASSESSMENT
Some of the issues that will need to be considered by site operators carrying out flood risk assessments include:

1   Before flooding occurs:
    a   How long in advance will flood warnings be issued (this may be only a few hours for sites at the top of river catchments or those subject to tidal flooding).
    b   How will flood warnings be received and who will act on them?
    c   Can critical parts of the site infrastructure be protected by raising them on plinths above flood level or surrounding them with flood walls?
    d   Can the process be shut down and dangerous substances secured in the time available before flooding occurs?
    e   Can staff be evacuated safely?
2   During a flooding incident:
    a   How deep will the site flood and for how long?

  b How will staff gain access to the site during the flooding incident? (All the available boats may be needed to evacuate local residents.)
  c Can containment of dangerous substances be maintained ? (Consider flotation of storage tanks and drums being swept off-site).
  d Can site services be maintained during the flooding event? eg electricity, compressed air, nitrogen, etc.
  e How will the concerns raised by press and media and local population regarding pollution risk be addressed?
3 After a flooding event:
  a Can sites services be restored rapidly?
  b How will the site clean-up be carried out? (this may involve the disposal of sludges contaminated with dangerous substances).
  d Will the stocks of spare parts be adequate?
  e How long are the lead times for replacement of critical items?
  f Is there a contingency plan to ensure continuity of supply to customers during interruption to business.

EUROPEAN FLOODING INCIDENTS

In August 2002, severe flooding in central Europe, affected several major hazard sites including:

1. The Fluorochemie factory in Dohna, Saxony that was successfully shut down before being completely inundated by the floodwaters of the river Elbe. Two railway tank wagons containing liquid hydrogen fluoride were abandoned in Schlottwitz station when the track was flooded and they had to be monitored by helicopter.
2. The Spolana chemical factory outside Prague had a chlorine leak when pipework around a storage tank was damaged by debris washed down by the floodwaters of the river Danube. The fire service plugged the leak but had to abandon the site when the flood water rose above the tank. Several days later when the floodwater had receded it was discovered that the entire tank inventory of 80 tonnes of liquid chlorine had been lost. Residents were required to shelter indoors, but there was no harm to people or the environment. There were also concerns that contaminated soil containing mercury and dioxins may have been washed away into the river during the flooding event.

CONCLUSIONS

During the last five years, there have been several serious flooding incidents across the United Kingdom. The focus of attention has been on the damage caused to residential property and only one major hazard site, a waste treatment and storage site operated by CSG Ltd in Gloucester, has been severely affected. It has been suggested that these flooding events are due to changes in the climate associated with global warming, and that further flooding is likely to occur in the future. A large number of process industry facilities are at risk because they are located in the indicative flood plain. That risk is being evaluated by the Environment Agency in order to ensure the highest standards of environmental protection.

REFERENCES
1. Bye and Horner 1998. Easter 1998 Floods - final assessment by the independent review team. Vol 1, ISBN 1-873-16066-6. Vol 2, ISBN 1-873-16067-4.
2. Environment Agency 2001. Lessons Learned – Autumn 2000 Floods. ISBN 1857055063.
3. DTLR 2001. Planning Policy Guidance Note 25 (PPG 25). Development and Flood Risk. ISBN 0-11-753611-3
4. Patterson & Poppleton 2001. Controlling and De-Contaminating Site Wastewater. IChemE Symposium Series No. 148. ISBN 0-85295-441-702.
5. Whitfield 2002. COMAH and the Environment, Lessons learned from Major Accidents 1999-2000. IChemE Transactions Volume 80 Part B.

Figure 1. BP Oil (UK) Ltd., Northampton terminal. Easter 1998.

Figure 2. CSG, Ltd., waste storage & treatment facility, Gloucester. November 2000.

Figure 3. Tessenderlo Fine Chemicals, Leek, Staffordshire. November 2000.

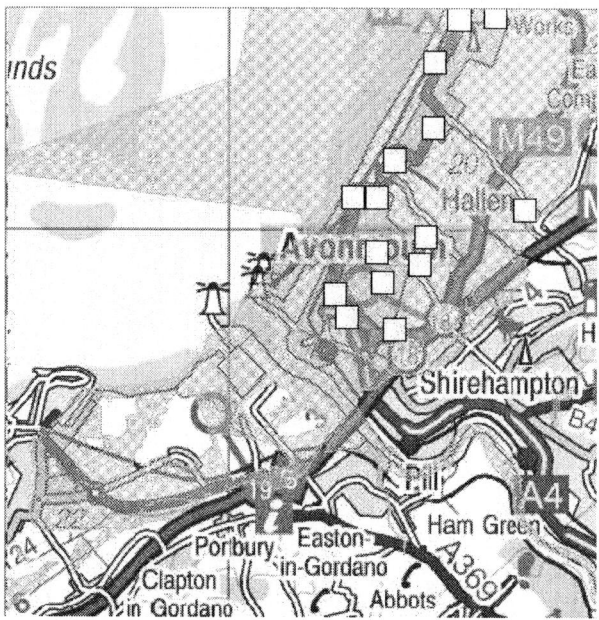

Figure 4. Avonmouth, agency website map – flood plain and process industry site

USE OF REAL-TIME MEASUREMENTS FOR ESTIMATING RELEASE RATE

Shahryar Khajeh Najafi*, Ernie Gilbert
SAFER Systems, L.L.C., Camarillo, CA

There are several key parameters that influence the action taken by the emergency responder on the scene of a chemical spill: the location of the source; the type of chemical is involved; and how much chemical is being released. The objective of this paper is to present a methodology for using meteorological data and information from chemical sensors to obtain the requested information. The meteorological data is used to locate the source. The sensors feedback is used to calculate the release rate and identify the chemical. To reconstruct the emission rate versus time, a dispersion model which is capable of incorporating the real-time data has been utilized.

The proposed methodology was tested against field trials. The maximum error between the simulated and measured rate ranged from 20% to 90% for the trials. It was concluded that of all the parameters affecting the performance of release rate estimation algorithm, the near source phenomena of aerosolization and dilution at the source, and the plume meander are the most important parameters. A two-tier approach is suggested for the rate estimation depending on the source of emission. For high momentum releases (e.g. tank/pipe), the goal would be to find the size of the rupture, and for low momentum release (e.g. pool), the goal would be to find the emission rate.

Backcalculation, sensor, dispersion

INTRODUCTION

A typical chemical release event may involve several derailed railroad cars leaking unknown amounts of chemical or a plant process area engulfed in a toxic gas. In events like these, it is very difficult to locate the leak and to determine the amount of chemical being released. Initial release rate estimates are very challenging and even an expert responder can only guess. Response agencies who intend to warn people in harm's way and evacuate would therefore have a great demand for new method for quick and accurate estimation of release rate.

For accident within a plant, the release rate can be calculated by a process engineer from a mass balance around the leaking vessel/pipe using the information available to him from process control unit. This method works best if the release rate is substantial. In the case of small leaks, the release rate estimation is very difficult. For mobile sources (e.g. train derailment or truck accident) release rate estimation is very difficult during an actual event.

The objective of this paper is to present a methodology of inverting downwind concentration measurements using an appropriate dispersion model to reconstruct emission rate versus time. We call this method backcalculation. Utilizing weather data and the position of the deployed sensors, the source location can be identified. The unique response of sensors to a specific chemical would also determine their identity.

Using real-time concentration measurement to estimate the rate has been studied previously by Lehning et al. (1994) and Piccot et al. (1994, 1996). Lehning studied the possibility of inverting the downwind concentration measurements to calculate the total emission rate and distribution of a non-uniform area source. Piccot used the open-path Fourier

Transformation Infrared (FTIR) spectrometer to measure the emission from area and volume sources. They used a Gaussian model to obtain an integrated average concentration along the spectrometer's path and compared the result with the measured values. They showed good performance of the model when an ad hoc method for stability calculation was used. These methods are mostly suitable for measuring emissions from fugitive steady sources. The present study proposes a methodology that can be utilized during an emergency event.

The chemical sensors can be categorized into two groups: point detection sensors and scanning devices sensors. The scanning device sensors include: Lidar (Light Detection and Ranging), Dial (Differential Absoption Lidar), Fourier Transform Infrared (FTIR) spectrometers, thermal imaging. Points sensors include Photo-Ionization Detector (PID), Flame-Ionization Detector (fiD), electro-chemical, and paper tape. The backcalculation algorithm works best with the point sensors. Due to cost constraints, portable sensors which can be dropped into the path of the cloud during an episodic release are more practical. The portable sensors transmit, via radio wave, their geographic location (latitude/longitude), gas concentration and the time of that measurement to the backcalculation algorithm. The activities of the sensors can be observed on a GIS based mapping system. The sensors can also be mounted on vehicle (with positive pressure inside) which can be driven through the cloud for sampling and transmitting the results to the base.

One of the advantages of portable point sensors is their flexibility in terms of rapid deployment. The open-path instruments like FTIR need some time to set up and it is very hard to move them around. This makes point sensors more suitable for emergency events.

With the recent terrorist attack and vulnerability of chemical plants to such threats, chemical companies can not afford to ignore the importance of a combined sensors/ backcalculation set-up as a tool for early warning and mitigation. The sensor technology has matured to a point that the sensors are durable, affordable, and able to measure concentration in real time. Most of them have a global positioning system and can transmit information via radio wave. Rapid detection of release enables prompt implementation of corrective action and a response procedure.

DESCRIPTION OF THE MODEL

An integrated Lagrangian puff model is used for this study. The model allows for the incorporation of real-time weather and sensors data for backcalculation. It uses the 16 sectors wind rose (N, NNE, NE, E, ...) where sectors are 22.5 degrees apart. The wind speed, direction, temperature, and solar radiation are measured. The polling frequency for these parameters is 3 seconds. The archived data which is used for backcalculation represent five minutes running average. The vertical and horizontal stabilities are calculated based on the 10 minutes running average of wind speed and wind direction. The polling frequency for the sensors is one second with an archived value of one minutes running average.

The dispersion model uses the Richardson Number criteria to invoke the Gaussian or dense gas sub models.

$$Ri = \frac{g\dfrac{(\rho_g - \rho_a)}{\rho_a}H}{U_*}$$

Where ρ_g, and ρ_a are the gas and air density, H is the height of element, and U_* is the friction velocity. Therefore, for a specific gas and weather data, the quantity of release which affects the height of the puff, would determine the type of the model to use.

OVERALL VIEW OF THE CONCEPT

Basically there are five major components in a backcalculation assessment:

1. Gas detection sensors
2. Meteorological measurement
3. Release location
4. Starting time of the release
5. Sophisticated dispersion model

To find the release location a reverse corridor is constructed from the position of each impacted sensor utilizing the weather data (Figure 1). The reverse corridor is a wedge drawn from the position of each sensor using the opposite wind direction measured by the meteorological tower. So if wind is blowing from the west the reverse corridor is drawn from the east. The angle of the corridor depends on the atmospheric stability.

To calculate the wedge angle, a downwind distance of one kilometre is selected (OA). The horizontal dispersion parameter (σ_y) is found from Pasquill-Gifford charts using appropriate stability curve. The wedge is then constructed by connecting the end of a line segment, which is perpendicular to the reverse mean wind direction, to the sensor location. The length of this line segment is 2.14 times the horizontal dispersion parameter. The factor of 2.14 sets the cloud width where the concentration falls to about one percent of the centreline value.

The intersection of those wedges would contain the most probable area for the source location. This technique is very useful in narrowing down the location of the source considering that most of the time the chemical presence is suggested by its smell and not its visibility.

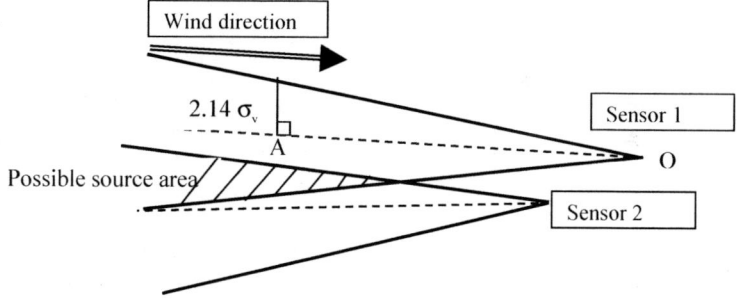

Figure 1. Construction of reverse corridor

The backcalculation uses a trial and error procedure (Figure 2) to estimate the rate. For each trial rate a concentration-time profile is predicted by the model at each sensor location (figure). Actual sensor information, namely, the time and measured concentration is compared against the model prediction. If there is a match within the convergence span in terms of time and concentration measurement for all the sensors, the predicted release rate is recorded. Otherwise, a new release rate is assumed. There are two loops for convergence; one loop converges on the time, the other converges on the concentration. A substantial change in measured concentration is an indication of a rate change. This process would render the rate vs. time profile for a transient source. The release rate is then fed to the dispersion model to render the final plume impact. This procedure is repeated as new information is received and updates are necessary.

The greatest accuracy would be obtained if the sample is taken close to the plume centreline and away from its edges. To take this fact into account, a weight factor is assigned to each sensor due to its position with respect to plume centreline. The weight factor would adjust the convergence criteria for each sensor.

The convergence criteria is defined as:

$(Cmeas - Cest)/Cmeas < Tol/Wn$

Where:

Cmeas = Measured concentration
Cest = Estimated concentration
Tol = Tolerance
Wn = Weight factor for each sensor $(0 < Wn < 1)$

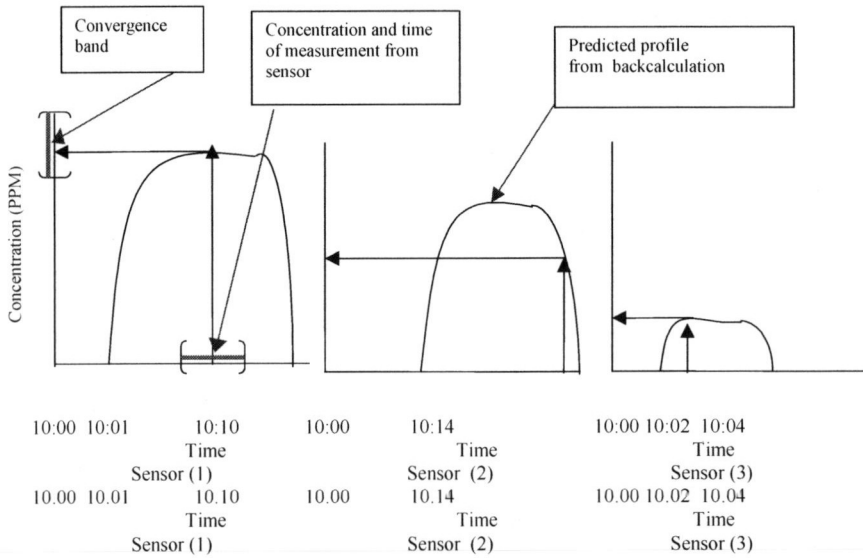

Figure 2.   Comparison of model predictions against the sensors measurements

A global tolerance level of 10-3 is assumed, but this value is adjusted (increased) as the sample points get farther away from the mean wind direction. Therefore, different convergence criteria is used depending on the location of the sample point to the prevailing wind direction.

In practice, the wind direction would not be maintained steady; therefore, it is difficult to sample close to the plume centreline (which coincides with wind direction). For this reason the weight factor was devised. With a handheld GPS and a knowledge about wind direction, an emergency responder can have a good sense of placing the portable sensors. The sample points must be clear from any obstacles (e.g. buildings, fences, and trees, etc).

DETAIL DESCRIPTION OF THE PROCESS

Consider the dispersion of an accidental release of a chemical (Figure 3). The impact of the plume on each sensor differs due to its travel time and/or wind shift with respect to the position of each sensor. Therefore, some sensors may not begin reading a concentration until later in an event while other sensors may be impacted early on. Some sensors may go out of commission if they become saturated or reach a maximum upper limit of their reading range.

Assume the following hypothetical situation for the first few moments of a release. The inner area represent the lowest level we will model, but not necessary the lowest level that can be measured. The outer area represents the area where some level of the cloud may be monitored at a range below the level of concern. Assume sensors 1, 2, 3, 10 are fixed and sensors 4, 5, 6, 7, 8, 9 are portable ones.

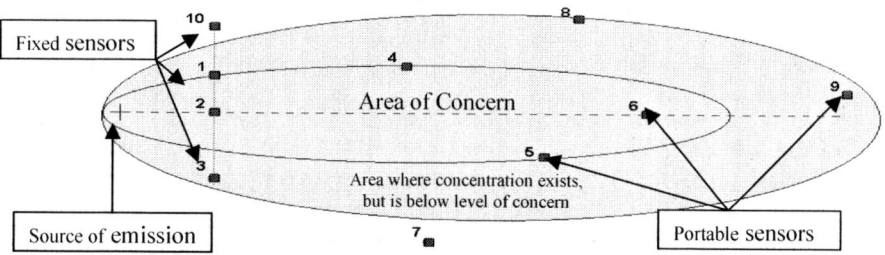

Figure 3.   Plume/sensor impact

Fixed sensor along with deployed portable sensors start transmitting the following information to the back calculation module.

Some gas monitoring sensors take 50 seconds to start detecting the chemical (sensor lag), this lag is factored in time convergence loop. The starting time of an event is a necessary piece of information for backcalculation algorithm. Backcalculation can accept either an starting time of event from the user or can the starting time based on the position of the closest impacted sensor to the source and the wind speed.

Table 1. Sensor data communication to backcalculation

Sensor ID	Location	Concentration (ppm)	Time	Saturation (ppm)
1	X1, Y1, Z1	C1	t1	C1limit
2	X2, Y2, Z2	C2	t2	C2limit
.	...	...	...	...
N	Xn, Yn, Zn	Cn	tn	Cnlimit

Suppose the event starts at time T. Sensors 1 and 2 pick up the gas concentration T4 minute after the release. After T7 minute into the release, sensor 2 reaches its maximum concentration range. At T9 minutes, plume impacts sensors 3 and 4.

At T7 minutes the operator starts the program and sites the release. Sensors 1 and 2 are available at this time. Sensor 2 has reached it maximum concentration. However, valid readings are available from sensor 1 to time T7 minutes and for sensor 2 up to T6.

The sensor poling frequency is 1hz, but a one minute running average is used for the backcalculation which is consistent with one minute interval puff releases for the dispersion model. The release rate calculated for the T4 interval is the assumed for the T1 through T4 where there was no actual measurement.

At T + 10 minutes the program runs again. This time, it has built an array of data similar to table 2. For time T through T7 seconds, the sensors 1 and 2 are used for rate calculation. For time T7 to T8, sensor 1 is used. For time T8 through T9, sensors 1 and 4 are used for rate calculations.

The process continues as more sensors with valid reading participate. The following table is a sample (based on three sensors) of how the release rate is calculated and sorted in the ascending order of time before it is passed to the dispersion modeling for the calculation of the plume impact.

For the case when the responder gets to the scene of accident and the event has been in progress for a while, the starting time of calculation would be different than the starting time of the event. As might be expected, the release rate history from the actual time of accident to the calculated starting time based on the deployed sensors is lost.

COMPARISON WITH FIELD DATA

The Desert Tortoise sensors data is used to test the accuracy of backcalculation model. A series of four large-scale (15–60 m^3) pressurized ammonia spill test were conducted by the Lawrence Livermore National Laboratory, Goldwlre et al. (1983). The tests, called the Desert Tortoise series, were conducted at Frenchman Flat in Nevada during August and September of 1983.

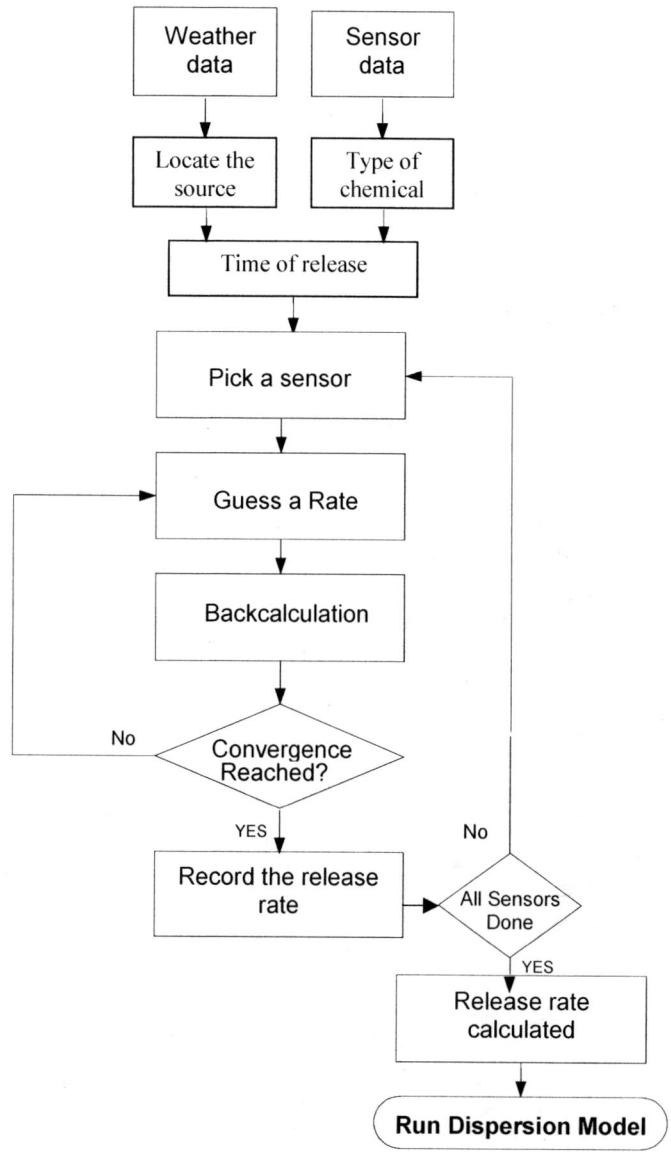

Figure 4. Algorithm for backcalculation

Table 2. Typical sensor/plume interaction behaviour

Time + (seconds)	Sensor 1	Sensor 2	Sensor 3	Sensor 4
T1	Fill in from 60	Fill in from 60	No reading	No reading
T2	Assumed Conc.	Assumed Conc.	No reading	No reading
T3	Assumed Conc.	Assumed Conc.	No reading	No reading
T4	First actual reading	First actual reading	No reading	No reading
T5	Valid reading	Valid reading	No reading	No reading
T6	Valid reading	Valid reading	No reading	No reading
T7	Valid reading	Maximum reading obtained	No reading	No reading
T8	Valid reading	Not used	No reading	Assumed Conc.
T9	Valid reading	Not used	Below range reading	Valid reading

Table 3. Release rate set up for dispersion calculation

Time (min)	Sensor # 1	2	3		Time (min)	Average release rate based on three sensors (lb/min)
T1	R11	R12	R13		T1	R1
T2	R21	R22	R23	⟹	T2	R2
T3	R31	R32	R33		T3	R3
T4	R41	R42	R43		T4	R4

Sensors were placed at 100 and 800 meters away from the source. At 100 meters, the sensors were placed 15 meters apart. At 800 meters, they were 100 meters apart. The seven sensors at 100 meters were numbered as G02 through G08, with G05 assumed to be on the centreline. There were five sensors at 800 meters (G20 through G24) at 800 meters, with G22 being on the centreline.

The following table shows the characteristics of the two tests chosen for this study:

Table 4. Field data for Desert Tortoise 1 and 4 trials

Test number	Desert Tortoise 1 (DT1)	Desert Tortoise 4 (DT4)
Chemical	Ammonia	Ammonia
Phase of release	Liquid	Liquid
Duration	126 (s)	381 (s)
Orifice diameter	3.19 (in)	3.72 (in)
Spill amount	10200 (kg)	41100 (kg)
Spill rate	4860 (kg/min)	6480 (kg/min)
Average wind speed @ 2 m	7.42 (m/s)	4.51 (m/s)
Average wind direction	223	229
Average directional variability	5.73	5.02
Relative humidity	13%	21%
Surface roughness	3 mm	3mm
Temperature	29 (c)	32 (c)
Stability class	D	E

The actual field sensor measurements were averaged over one minute intervals and the results were used to compare the model prediction against sensor readings. The one minute averaged concentration is used as input to the backcalculation. A sample of the process for DT4 for sensors at 100 and 800 meters is shown in Figures 13 and 14.

Two studies were performed:

a) Knowing the rate, compare the prediction of the model concentration-time profile with that measured by the sensors. This will show the goodness of the dispersion model utilized.

b) Knowing the measured concentration-time profile of each sensor, calculate the release rate. This will test the goodness of the backcalculation algorithm and the benefit of the weighing factors.

A) RELEASE RATE IS KNOWN

Liquid ammonia flashes to vapor, and aerosol upon release to the atmosphere. A fraction of the aerosol rains out to form a pool. The combined streams of vapor, aerosol, and emission from pool are dispersed by atmospheric flow. This process continues until the tank content is emptied or tank release and pool emission is mitigated.

The following graphs (Figures 5 & 6) show the calculated source for atmospheric dispersion. Comparing the total rate for dispersion with ammonia liquid discharge rate of 4860 kg/min for DT1 and 6480 kg/min for DT4 represent the effect of rainout. The initial cloud dilution due to jetting is not taken into consideration.

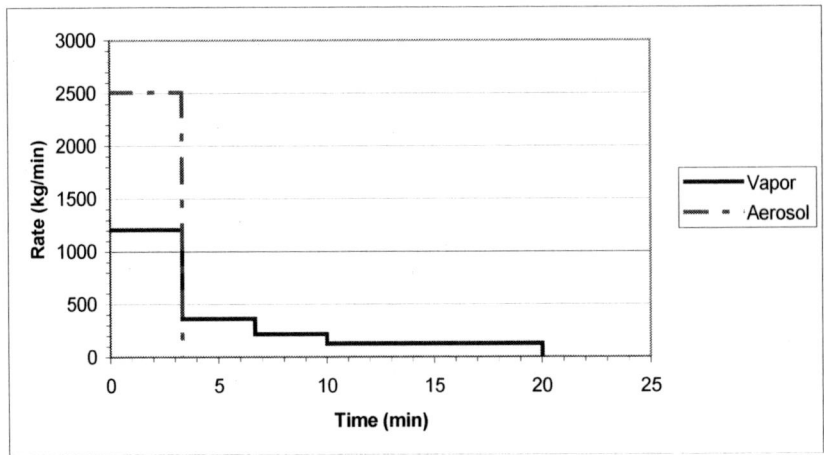

Figure 5.   Source term for dispersion-DT1 field trial

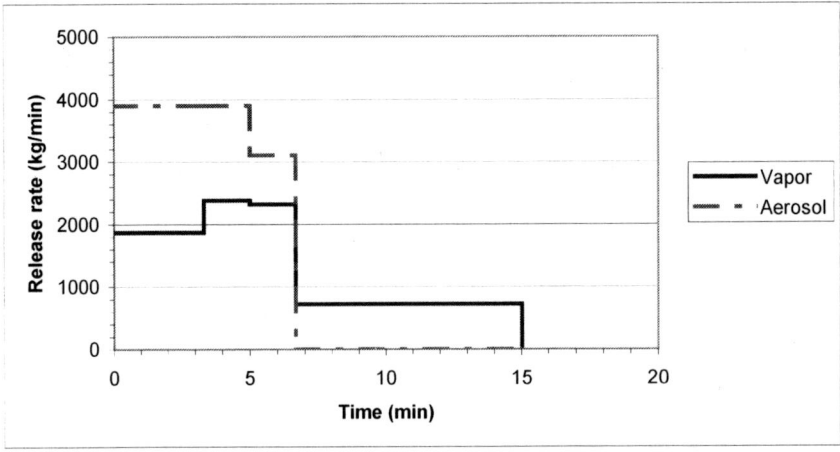

Figure 6.   Source term for dispersion-DT4 field trial

The result of the dispersion model is overlaid against the sensor data for DT1. Figure 7 & 8 show the comparison at 100 and 800 meters. The assessment is performed for the centreline values only. There is no resemblance between the two plots at 100 meters downwind. The duration and the time of the maximum concentration is different. The percent error between the peak sensor concentration and that of the model is 34%.

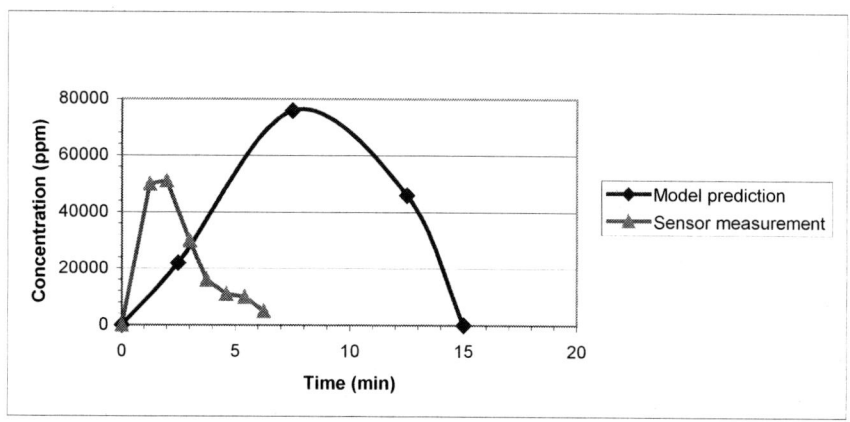

Figure 7.   Predicted vs. measured centreline concentration profile at 100 meters-DT1

The comparison of the model with sensor measurement looks very good at the distance of 800 meters. The concentration profiles seem reasonably matching in their trend and occurrence of the maximum concentration. There is a 30% error between the model predicted peak concentration and the sensor measured values.

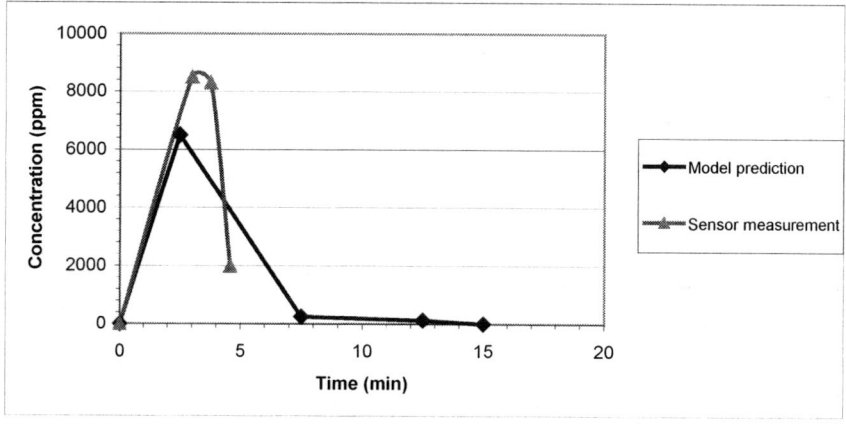

Figure 8.   Predicted vs. measured centreline concentration profile at 800 meters-DT1

Comparison of the concentration profile predicted by the model against the measured values at 100 meters for DT4 is interesting. The slope of the concentration plots after the cloud arrival is identical, and the peak concentrations occur at the same time. However, there is 65% error between the measured and model peak concentration.

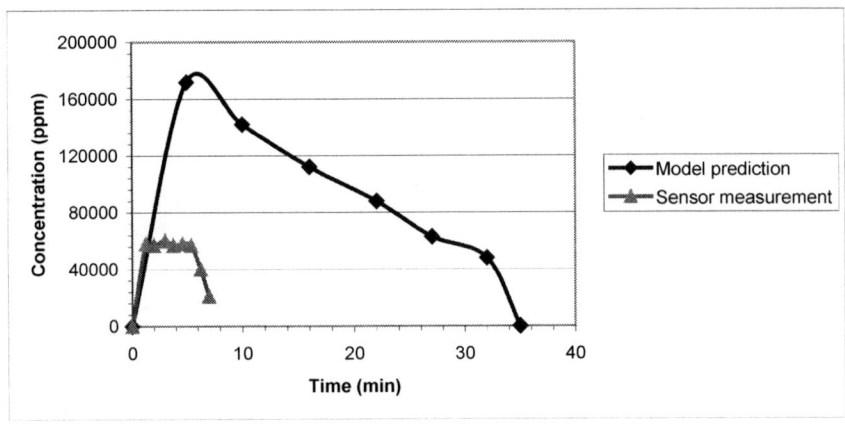

Figure 9.   Predicted vs. measured centreline concentration profile at 100 meters-DT4

The concentration profiles seem reasonably matching in their trend and occurrence of the maximum concentration at 800 meters for DT4 experiment. The maximum concentration occurs at almost the same time and the percent error between the field data and model simulation is roughly 58%.

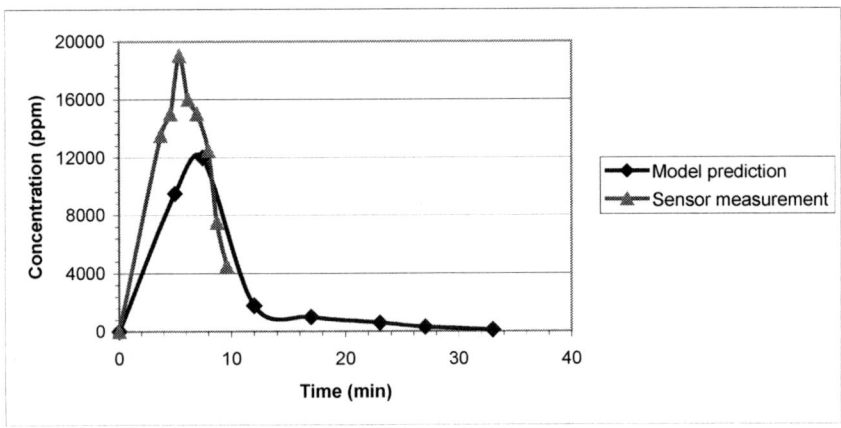

Figure 10.   Predicted vs. measured centreline concentration profile at 800 meters-DT4

The model prediction is higher than the measured concentration at 100 meters for both DT1 and DT4 and this trend it reversed at 800 meters. This can be due to the jetting effect which causes the trial cloud to dilute faster than the model up to the point that jetting effect is dominant. At the end of the jetting zone the modelled cloud dilute faster due to its higher density than the trial cloud. The persistence of the jetting effect up to the 100 meters downwind was observed by LLNL personnel during the field trials.

B) RELEASE RATE IS UNKNOWN

The two field trials DT1 and DT4 were selected to gauge the behavior of the backcalculation model. The one minute average concentrations for all impacted sensors (Figures 13–14) along with their time of measurements were input to the backcalculation algorithm. The one minute averaged sensor data produced 16 data points for DT1 and 56 data points for DT4. These data were input to the backcalculation module for release rate estimation.

Duration of liquid release from the tank was 2 and 6 minutes for DT1 and DT4 respectively. The emission from the pool continued for several more minutes after the valve was closed. The total rate for dispersion was 3700 kg/min for DT1 and 5800 kg/min for DT4 trials.

To simulate the field trials with backcalculation algorithm the molecular weight of ammonia was modified. This adjustment was necessary to obtain identical initial density between the model and field trials. Figures 11–12, represent release rate estimates for both field trials. The symbols represent the estimated values, and the horizontal line is the plot of actual rate.

Two distinct regions are observed for the release estimation curves. An almost steady rate during the discharge of liquid ammonia from tank and a sharp drop in rate after the valve is closed and emission is solely from the shrinking pool. According to sensors data the pool content was depleted in just a few minutes. The maximum percent error between the predicted and measured values for the period where the tank is contributing is 20% for DT1 and 90% for DT4.

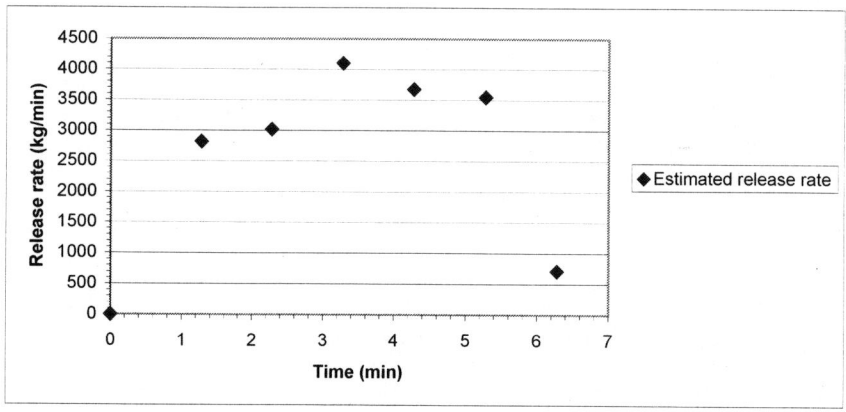

Figure 11. Model prediction against the actual release rate-DT1

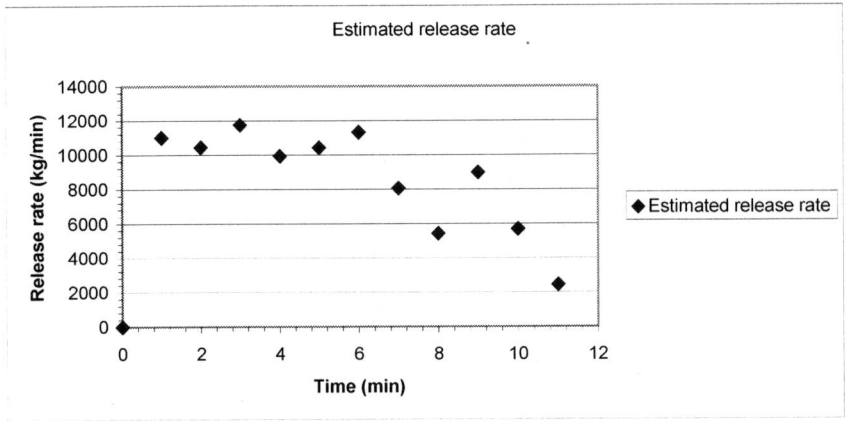

Figure 12. Model prediction against the actual release rate-DT4

One possible explanation for the magnitude of error is ignoring of the initial dilution at the source by the backcalculation algorithm. The liquid ammonia jetting out of vessel flashes to vapor and aerosol and is diluted by air entrainment. The effect of jetting is more pronounce in DT4 than DT1 due to higher release rate and temperature. The backcalculation algorithm assumes pure vapor and does not take into consideration the initial dilution of the cloud. This leads to overestimation of rate for both cases with DT4 having the poorest performance due to its higher dilution effect. This observation is an important piece of information for further enhancement to the backcalculation module.

The break up of error does definitely contain the plume meandering effect as well, although this is very hard to quantify in this study due to lack of information. However a test of sensitivity of the model to wind meander was performed. The sensors position was fixed and the mean wind direction was changed within fifteen degrees. The analysis of the simulation results indicated a model sensitivity of between 5% to 10% for a fifteen-degree plume meander.

CONCLUSION

This study was set out to accomplish two goals: (1) to develop a concept to obtain certain key parameters needed during an emergency response, and (2) using the field data to validate the concept. The novel idea of locating the source using weather data and impacted sensors location was introduced. The construction of several reverse corridors, which are based on the wind direction and stability, narrows the possible location of the chemical spill.

A backcalculation algorithm was presented which takes into account the spill location, time of release, and measured concentration to estimate the release rate. Two field data from Desert Tortoise (DT) series of tests, DT1 and DT4, were selected for validation purposes. It is recognized that there are many uncertainties involved in the rate estimation process: (1) averaging time of sensors, meteorological input, and model; (2) position of sensors with respect to the mean wind direction (cloud meandering); (3) near source phenomena (e.g. aerosolization) and initial dilution due to jetting effect.

Sensitivity analysis of the result indicates that 2, and 3 plays the most important part on the overall performance of the model. To reduce or dampen the effect of cloud meandering, several sensors should be deployed at any downwind distance where they are placed across the wind.

The near source phenomena can be included in the rate estimation model if two different parameters are used for trial and error calculations: (1) for high momentum releases from a tank and/or pipe the parameter of interest would be the size of the hole; (2) for other sources (e.g. pool only) the parameter of interest would be the emission rate. Following the above guidelines and using a good dispersion model, which can take into account the real-time measurements, can provide a good tool for rate estimation of chemical spills.

REFERENCES

M. Lehning, D. P. Y. Chang, D. R. Shonnard, and R. L. Bell, "An Inversion Algorithm for Determining Area-Source Emissions from Downwind Concentration Measurements", J. Air & Waste Manage. Assoc, 44: 1204–1213 (1994)

S. D. Piccot, S. S. Masemore, E. S. Ringler, S. Srinivasan, D. A. Kirchgessner, W. F. Herget, "Validation of a Method for Estimating Pollution Emission Rates From Area Sources Using Open-Path FTIR Spectroscopy and dispersion modeling techniques", J. Air & Waste Manage. Assoc, 44, 271, 279 (1994)

S. D. Piccot, S. S. Masemore, W.L. Bevan, E. S. Ringler, D. Bruce Harris "Field Assessment of a New Method for Estimating Emission Rates from Volume Sources Using Open-Path FTRI Spectroscopy", J. Air & Waste Manage. Assoc, 46, 159–171, (1996)

H. C. Goldwire, G. McRae, G. W.Johnson, D. L. Hipple, R. P. Koopman, J. W. McClure, L. K. Morris, R. T.Cederwall, "Desert Tortoise Seris Data Report, Pressurized Ammonia Spills", Lawrence Livermore National Laboratory, December 1985.

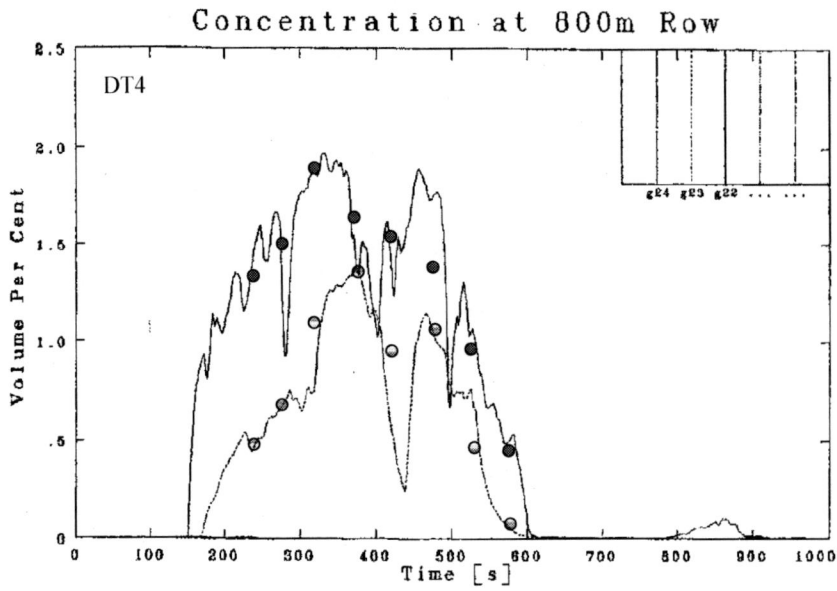

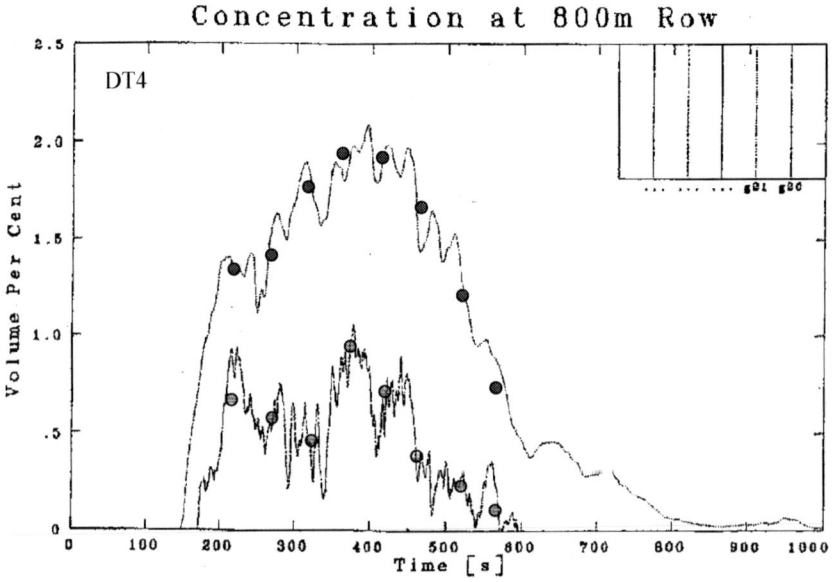

Figure 13. Sensor measurement at 800 meters-DT4 trial

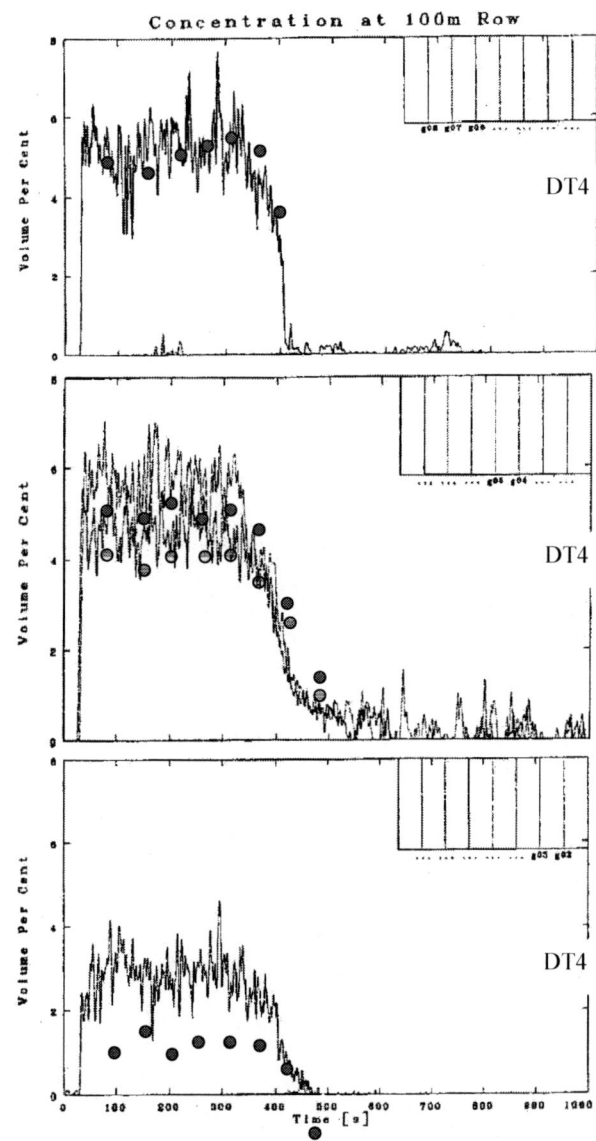

Figure 14. Sensor measurement at 800 meters-DT4 trial

171

USING BEHAVIOUR-BASED METHODS TO IMPROVE ORGANISATIONAL EFFECTIVENESS

Gordon Sellers[#] and Chris Marsh[*]
[#]Behavioural Science Technology, Bracknell RG12 1JB, UK, tel 01344 455090, GordonSellers@onetel.net.uk
[*]ConocoPhillips Limited, Humber Refinery, South Killingholme DN40 3DW, UK, Tel 01469 555956, Chris.Marsh@conocophillips.com

Companies are learning that behaviour-based methodology is broadly applicable to performance improvement. The successful use of this method rests on some familiar principles: behaviour analysis, employee engagement, implementation design, and some new developments: excellence in organisational functioning and the optimal use of information technology. When combined, these tools become a powerful engine to drive performance improvement in safety, quality, error reduction and customer service. This paper provides an overview of combined methods. It ends with a case study from ConocoPhillips' Humber Refinery, which launched behaviour-based safety towards the end of 2001. This was built around the refinery's Safety Emphasis Team which had a newly-appointed full time leader and 12 further members who were a cross section of the refinery workforce, with the remit *"to develop, recommend and implement a series of changes which will have credibility and benefits that help us all achieve an accident free career"*.

Behaviour, Performance Improvement, Safety, Quality, Error Reduction, Customer Service, Refinery

CONNECTING PEOPLE TO SYSTEMS

The behaviour-based approach to performance improvement is focused on a set of related problems:

- The need to increase the likelihood that the right behaviour will occur
- The need to engage personnel in the improvement effort
- How to assure further upstream diagnosis of system deficiencies
- The fact that significant performance variation is going undetected
- How to get behaviour-based performance data into the hands of workgroups and move the responsibility for improvement to them
- How to avoid over-reliance on the classroom training model
- The need for something other than disciplinary action as a tool for improvement
- The fact that positive feedback is an under-utilised but extremely powerful tool for improvement in human performance

We note that companies have used a variety of approaches to improve performance in areas such as occupational safety, reliability, patient safety, customer service, productivity, and error prevention. Those approaches include drafting policies and regulations, posting slogans, organising contests and awards, developing and delivering training, forming committees and councils, establishing best practices or quality circles, and issuing reprimands. Some of these approaches work well; some don't. When these approaches work well, the indication of success or effectiveness is that the workgroup produces fewer non-conforming behaviours, thus reducing the rate at which errors or incidents occur.

An increasing number of organisations have successfully introduced behaviour-based safety processes and now recognise that the behaviour-based methodology has potential value in other areas of performance improvement – such as quality improvement, error reduction and customer service. So they are now expanding their behaviour-based safety processes to include other areas which are important to their businesses. Some other organisations have gone straight into non-safety performance improvement, but until now the majority have used behaviour-based safety as their starting point.

WHY LEAD WITH SAFETY?
There are three good reasons for leading with safety: ease of buy-in; visible outcome measures; and sustainability.

Ease of Buy-in
It is relatively easy to get buy-in for safety at all levels of the organisation. Paul O'Neill (the former CEO of International Paper and of Alcoa, and until recently the US Secretary of the Treasury) is quoted as saying, "A truly great organization must be aligned around values that bind the organization together", "Safety is a way to show that human beings really matter", and "Leadership uses safety to make human connections across the organization…."

Improved safety is a clear benefit for all levels of an organisation:

- For the employees, who have a reduced risk of injury;
  For the managers, who spend less time investigating accidents and incidents; and
- For the organisation, which improves its public image as well as reducing the cost of injuries and damage.

This is not always the case with other performance improvement initiatives where the organisation may be perceived (often mistakenly) as the only winner.

Visible Outcome Measures
Outcome measures are very visible for safety. Most organisations have accident data stretching over many decades, whether expressed as the RIDDOR or OSHA rates (the accident rates that are legally defined as reportable/recordable in UK and USA), total injury rates, lost workday rates, etc. In our experience, there are rarely such well understood and well recorded outcome measures for other performance areas.

Sustainability
Behaviour-based safety has an impressive record of sustainability. Figure 1 shows the percentage of all implementations of the Behavioural Accident Prevention Process® led by BST across the globe, that started in a given year and are still functioning today. The majority of these sites have experienced major reorganisations, changes in site leadership, changes in ownership, downsizing or other disruptive events. Even with these changes, their BAPP® initiatives survive and their organisations continue to reap the benefits.

CRITERIA FOR DEFINING BEHAVIOUR-BASED SAFETY

Before an organisation decides to use its behaviour-based safety initiative as its starting point for behaviour-based improvement in another performance area, it is essential to confirm the effectiveness of the behaviour-based safety process. We use several criteria to help us judge that.

1. Is there an appropriate focus on the system rather than the worker?

   - Observation data should be used to analyse behaviours and never used against an employee. The first time that an employee is disciplined, formally or informally, as the result of an observation by a peer or by a supervisor, then that is a sure sign that employee buy-in to the process will immediately be replaced by hostility.

   - Does the process avoid any direct or indirect implication of 'blame the worker'? Focusing on routine workgroup behaviour does not mean focusing on blame or fault. Behaviour only becomes routine or common across entire workgroups when that behaviour is reinforced by system consequences. In this context, workgroup behaviour that is out of tolerance or out of compliance may be responsible for defects or errors and not be blameworthy. Instead, those non-conforming behaviours are somehow "baked" into the current system. What is called for is fact finding, not fault finding.

   - Is the distinction recognised between enabled behaviour and non-enabled behaviour? An enabled behaviour is under the control of the employee and takes little or no additional time and effort to perform it safely than at risk. Conversely a non-enabled behaviour is outside the control of the employee, due to the physical work environment or to the management systems; even if the employee wishes to do so, he or she cannot perform the task without being at risk. Clearly, different strategies are needed to resolve barriers to enabled and non-enabled safe behaviours.

2. Is observation data used to improve facilities?

   - Is there a systematic mechanism to use observation data for safety improvements? If not, employees will soon become disillusioned with the process: "Why waste our time doing observations? We report problems but nothing ever gets done to fix them!".

   - Is root cause analysis done? ABC Analysis (Antecedent Behaviour Consequence Analysis) is a powerful root cause analysis tool that uncovers why people behave the way they do. Antecedents set the stage for behaviour, and consequences encourage or discourage the repetition of the behaviour. Behavioural science teaches us that consequences have a much stronger influence on behaviour than do antecedents. Yet most organisations spend their time and other resources disproportionately on antecedents (slogans, signs, training, policies and so on), when they need to be focusing their efforts on delivering consequences for the identified behaviours that amount to excellence.

3. Does the effort contribute to a positive culture?

   - Where incentives are used, they need to reward the desired behaviours. Unfortunately in many cases incentives are based on the accident rate, which is out of the direct

175

control of any employee or manager; furthermore this typically drives accident reporting underground.

- Is attention diverted away from equipment and management system issues? This can occur if the effort is focussed solely on behaviours.

4. Are appropriate roles given to employees?

- Are front line employees involved in the process? A typical behavioural safety team, comprised mostly of front line employees, will be empowered to manage the observation process, analyse data, and address the issues that are under their control – those affecting enabled behaviours and localised difficult behaviours. On the other hand, management will be responsible for key appointments, organisational issues, and addressing issues that are outside the control of the behavioural safety team – those affecting widespread difficult behaviours and non-enabled behaviours.

- In the excitement and enthusiasm of empowering front-line employees to manage significant part of a behaviour-based safety process, it is all too easy to overlook other professionals who have an important role to play. Thus a safety professional would typically be involved in hazard identification, root cause analysis and remediation. An engineer would typically be involved in designing any equipment changes needed to remove non-enabled barriers.

5. Is the role of behaviour-based safety set in its proper context?

- Again, in the excitement and enthusiasm of implementing behaviour-based safety, it is easy to exaggerate its role. Wise organisations recognise that BBS is but one component of a comprehensive safety system.

CROSSING THE BRIDGE FROM SAFETY TO PERFORMANCE IMPROVEMENT

Before seeking to improve performance, you should ask a series of questions about the functioning of the organisation. These include:

- How effective is your organisation at reaching its objectives?
- How easily does your organisation respond to the necessity of change?
- How efficient are you at making things happen?
- Do you have high trust, good communication and high levels of teamwork – or the opposites?
- Do you have alignment for improving performance?

Following extensive research, BST has developed an Organisational Functioning Survey (Figure 2) which provides some answers to the above questions by first assessing the CAUSES under three main headings:

- Organisational Factors
  1. Procedural justice
  2. Leader-member exchange
  3. Management credibility
  4. Perceived organisational support

- Team Factors
  5. Work group relations
  6. Teamwork
- Safety Specific Factors
  7. Organisational value for safety
  8. Upward communication
  9. Approaching others

  These CAUSES feed through into EFFECTS:

- Organisational commitment
- Openness to change
- Job satisfaction
- Mutual trust and respect
- Organisational citizenship behaviour
- Excellent communication

  In turn, the EFFECTS result in OUTCOMES such as:

- Injury rate
- Level of safe behaviour
- Quality
- Customer service

USING BEHAVIOUR-BASED METHODS TO DRIVE CROSS-FUNCTIONAL PERFORMANCE IMPROVEMENT
We recommend using a staged approach, building confidence at each step before moving further forward:

1. Define Phase 1 performance objectives, establishing a continuous improvement mechanism for safety performance while setting the stage for expansion into other performance areas.
2. Complete assessment of organisational functioning.
3. Implement Phase 1 methods: primary behaviour-based safety.
4. Demonstrate success based on predetermined outcome criteria.
5. Set Phase 2 performance targets e.g. quality, decision making, medical errors, customer service, environmental events, student life, internal communication, management & supervisory alignment.

Two brief examples outline how this was done in practice:

QUALITY
This often naturally follows safety, but it is unwise to assume that all that is needed is to add some 'critical quality behaviours' to the existing 'critical safety behaviours'. The outcome measures are often less well defined than with safety; the inventory of critical quality behaviours is often much more specific to each task rather than being generic; the observer

team may have a different makeup; and the observation strategy is often different. Having appreciated and accommodated those differences, company which processes mineral clays has achieved significant reductions in the proportion of material that has to be reworked i.e. much more material is 'right first time'.

DECISION MAKING

When management at a petrochemical company analysed its environmental and safety incidents, they found that human error was a significant contributor. They focussed on the issue of response to atypical operational situations, where effective decision-making could mean the difference between a minor process upset and a catastrophic incident. Our assessment showed communications issues, where management's production decisions were not always clearly communicated to unit operators; also lessons learned during an operational upset were not always effectively communicated across shifts. We assisted site personnel to assess who made which kinds of decisions and how they made them, in particular were decisions made by groups or by an individual. We then used applied behaviour analysis to identify the antecedents (triggers) and consequences driving the existing decision-making process, before developing a model for decision-making in a team-based unit. Site personnel also developed a measurement tool based on five to seven decision points for each of three decision-making scenarios, allowing the organisation to collect data on the rate at which relevant site personnel use the identified excellent decision-making behaviours.

GETTING THE BASICS IN PLACE – A CASE STUDY FROM THE CONOCOPHILLIPS HUMBER REFINERY

The ConocoPhillips Humber refinery is the most advanced refinery in Europe. Construction began in 1966 and was completed in 1969, since when there has been very significant investment to enhance efficiency, safety and environmental protection. The site has about 740 ConocoPhillips employees, about 310 alliance/core contractors and a variable number of non-core contractors, typically 700. This total workforce is known as Team Humber.

Safety has always been a high priority at Humber Refinery, with significant resources being invested in engineering controls and safety management systems. As a result, accident rates have fallen significantly since the early 1990s. However by 2001 it was widely agreed that the safety performance had plateaued and Humber Refinery recognised that 'more of the same' traditional safety measures would not achieve continuous improvement.

As part of the refinery's safety improvement efforts, a Safety Emphasis Team was formed in July 2001 with a full time leader and 12 further members who are a cross section of the refinery workforce. The team has the remit "to develop, recommend and implement a series of changes which will have credibility and benefits that help us all achieve an accident free career". The Refinery General Manager is the Safety Emphasis Team Champion. One of the team goals for 2001 was to "recommend, implement and manage a Refinery Behavioural Safety Process that embraces the Conoco and Contract workforce (Team Humber)". Accordingly the team leader and other team members researched behavioural processes, which has included attending public seminars and visiting two of ConocoPhillips' US refineries which have implemented BST's Behavioural Accident Prevention Process® technology (or BAPP® technology). They were impressed by the achievements that they saw in the US refineries and, after detailed discussions, they engaged BST to assist them to implement a behaviour-based safety process – which the employees have named PUMA (Personal Undertaking to Minimise Accidents).

KEY DECISIONS
Three important decisions were made from the outset:

Integrated Implementation
It would have been impractical to implement behaviour-based safety simultaneously across the whole refinery, however to avoid a fragmented approach the decision was taken to use the Safety Emphasis Team as the core for behaviour-based safety, under which the process would be rolled out to different departments on an agreed but aggressive timetable of under two years.

Company Employees and Contractors
Another key decision was that the process should involve both ConocoPhillips employees and contractors, who were already represented on the Safety Emphasis Team. So the initial launch was with Operations staff in one of the four main process areas ('Division C'), plus the 'Heavy Trades' contractor group covering civil trades, mechanical trades and scaffolders.

Involving Supervisors and Managers
The main emphasis in behaviour-based safety is with the people who are most at risk of injury – and these are generally the front-line employees. However first-line supervisor or middle manager positions are among the most difficult jobs in an organisation. They are at the end of a chain of command, exposed to contrasting pressures and demands. Negotiating this tug-of-war successfully requires the right skills; a supervisor needs to be a good coach to front-line employees, while still achieving the objectives of the organisation. The skills that the supervisor demonstrates can make the difference between worker cynicism and worker support – and ultimately determine the level of organisation success in many aspects, not least in behaviour-based safety. As well as the normal briefing for managers and supervisors about PUMA, two further steps have been for managers to develop and commit to an inventory of critical leadership behaviours which complement the inventory of critical behaviours for front-line employees; and to provide supervisors with skills training to enhance their effectiveness in safety.

IMPLEMENTATION STEPS
The implementation steps were:

- Planning Meeting with the Safety Emphasis Team and key representatives of management, in particular to select the departments for the initial implementation. An important factor in this selection was the enthusiasm of the managers from Process Division C and the Heavy Trades Contractors.
- Implementation Design Workshop, consisting of ½-day focus groups each with up to 20 representatives of the selected departments, to identify the issues which needed to be addressed to ensure success of the implementation and to start developing action plans for the top priority issues, followed by a Design Meeting which tailored the implementation in the light of these issues.

- Briefing Sessions for all members of the selected workgroups, which took away the mystery about what was about to happen.
- Training of Steering Team (i.e. Safety Emphasis Team plus additional personnel representing the workgroups in the implementation). This included:
  - ➤ initial training during which the Team developed a draft Inventory of Critical Behaviours® - and then went out and conducted buy-in meetings with all members of the selected workgroups
  - ➤ second training focussing on observation & feedback;
  - ➤ Third training on observation quality – there is little point in gathering masses of unreliable data;
  - ➤ behavioural action planning, to remove barriers to safe behaviours. Two important parts of this training were to focus on a very small number of at-risk behaviours where the effort expended would yield the greatest results, and to integrate the action planning with other performance improvement processes in the refinery;
  - ➤ observer training, to ensure that the refinery has the capability of training future generations of observers.
- Management Training, which was key to ensuring that they were kept in the loop and understood their roles.
- Meanwhile PUMA has been launched with other process divisions and contractor groups.

Still to come will be:

- A *Datalink* analysis of the site's BAPPTrack® database, to identify trends that would not be immediately obvious from a visual inspection; and
- A *Sustainability Review* to evaluate the strengths and further improvement opportunities in the process. It uses BST's proprietary Sustainability Index, which is based on implementation experience at hundreds of sites and identifies issues that should be addressed to improve process long-term sustainability.

Throughout the implementation, BST provided a host of **off-site services** including: phone consultation with the Facilitator, Management Sponsor and others; review of faxed and mailed materials, including draft of the ICB checklist, committee meetings minutes, etc.; preparation and customisation of training materials; and writing of follow-up letters and recommendations. The site is also eligible for **BST User Network Benefits**, including attending Users Conferences – indeed ConocoPhillips Humber set a new precedent when they invited three of their US colleagues to visit the refinery before joining seven ConocoPhillips Humber representatives at BST's 2002 UK Users Conference.

If we were to start again, what would we do differently? When this question was asked of key members of the Safety Excellence Team and of the BST consultants involved, they all replied, "Very little".

SUCCESSES
Even at this early stage, several important successes can be recognised, including:

- Engagement from all employees throughout the organisation;

- The inventory of critical leadership behaviours is developing commitment and support among managers;
- By giving responsibility to front-line employees and contractors, there is real ownership and accountability at the sharp end;
- Reliable upstream data are being collected through observations in the plants, workshops, offices and even transportation, the latter largely through self-observations;
- Accident investigations are now identifying the critical behaviour chain, rather than persecuting the injured party;
- Action planning is being done by problem-solving teams close to the problem, using natural work groups and co-opting in engineering resource when needed to remove specific barriers;
- The accident data review group has a new focus;
- There has been a very subtle and gradual change in the refinery culture – "The way we do things around here"; and
- The 'Team Humber' concept embodied in PUMA is helping to break down the traditional barriers between ConocoPhillips employees and contractor employees.

CHALLENGES

Like any other initiative, PUMA has its share of challenges – and the Safety Emphasis Team and management are working together to overcome them:

- Failure of past change efforts;
- Leaders "too busy for PUMA";
- Teams working in off site locations;
- People with learning difficulties;
- Communication across the site to all work groups whilst keeping the information relevant to the individuals; and
- Maintaining the initial interest.

CONCLUSION

ConocoPhillips Humber refinery has made an excellent start in implementing an effective behaviour-based safety process. Its challenges in the next few years are to maintain the initial enthusiasm – and to expand the process to other important performance areas beyond safety.

Sustainability of the BAPP® Technology

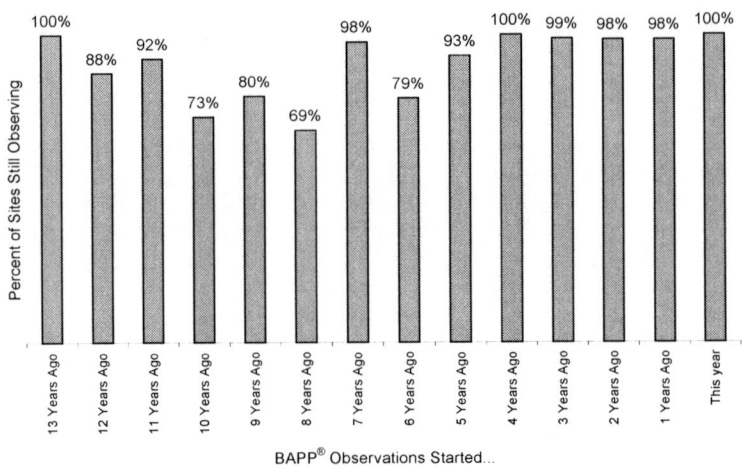

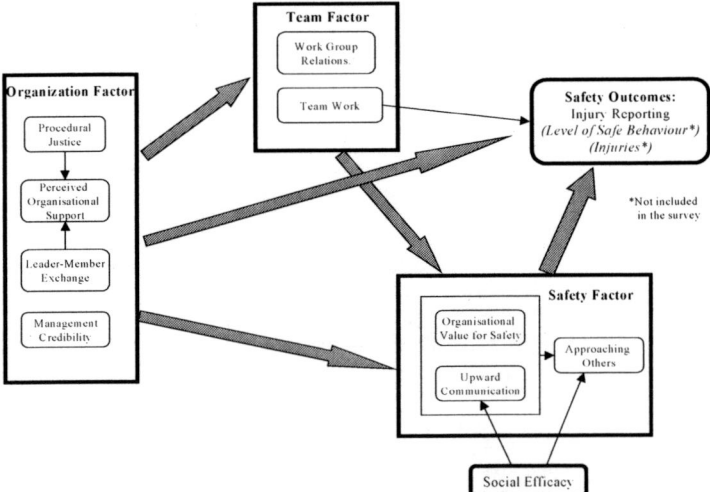

Figure 1. Sustainability of the behavioural accident prevention process®

© 2002 Behavioural Science Technology International and ConocoPhillips Limited

Figure 2. Organisational effectiveness – causes, effects and outcomes

182

THE MANAGEMENT OF ORGANISATIONAL CHANGE

Trevor A. Kletz
Department of Chemical Engineering, Loughborough University

The explosion at Flixborough in 1974 drew the attention of the process industries to the need to control modifications to plant, in order to identify and forestall possible adverse consequences. Changes to processes were soon treated in a similar way. Several decades elapsed, however, before it was realised that changes in organisation can also have unforeseen effects and that they should be systematically studied before they take place. This paper describes a number of incidents in which the immediate causes were technical but the underlying or root causes were changes in organisation, including changes in relationships with contractors. Some of the points to be considered in the management of organisational change are discussed.

KEYWORDS: Modification, change, management, accident investigation, Longford, railways

I feel so strongly about [knowledge] because of the horror I have for wrong medicine; because of the horror I have for changes which waste painfully learnt knowledge; because of the horror I have for plans which acknowledge the need for money capital but underplay the need for knowledge capital. – George Muir[1]

The explosion at Flixborough, UK in 1974 and other incidents drew the attention of the oil and chemical industries to the need to control changes to plants and processes. Many publications, including references 2, 3 and 4, have described accidents that have occurred because no one foresaw the results of such changes and have suggested ways of preventing such accidents in the future. Only in recent years, however, we have realized that changes in organization can also have unforeseen effects and should also be scrutinized systematically before they are made. In the UK this is now a legal requirement in high hazard industries[5] though it seems to be enforced rigorously only in he nuclear industry.

A common organisational change is to eliminate a job and distribute the jobholder's tasks amongst other workers. Although the jobholder is asked to list all his or her duties, sometimes one or two are missed, especially those carried out by custom and practice and not listed in any job description. For example, someone may have built up a reputation as a "gatekeeper", someone who knows how to get things done, where scarce spares may be squirreled away and so on. Another person may be the only fitter who really understands the peculiarities of a certain machine. Only after they have left are their distinctive contributions really recognised. Such changes are not "like for like".

The following are some examples of the unforeseen effects of major changes, one temporary but the others intended to be permanent.

ADMINSTRATIVE CONVENIENCE VERSUS GOOD SCIENCE

My first example, from James Lovelock's autobiography, *Homage to Gaia*[6], shows what happened when "administrative convenience ruled and good science and common sense came second", though the results were a decline in effectiveness rather than safety. He was working in a government-funded research centre that employed chemists and biologists. It was amalgamated with another similar institution some distance away. To the administrators it seemed sensible to move all the chemists to one site and all the biologists to the other as this would avoid the need to duplicate the services that each group required. The administrators did not realise, and did not listen to those who did, that research benefits from informal day-to-day contact between people from different disciplines. Both institutions declined. As we shall see, in the section on Longford, a similar loss of communication occurred when the professional engineers were moved from a plant to the company's head office.

AN INCIDENT ON AN ETHYLENE PLANT

The plant was starting up after a turnaround. At 2 am on the day of the incident the shift team started the flow of cold liquid to the demethaniser column. A level should have appeared in the base of the column two hours later. It did not, but problems elsewhere distracted the shift team and they did not notice this until 7 am. By this time the temperature at the top of the column was –82°C instead if the usual –20°C and the level in the reflux drum rose from zero to full scale in 10 minutes. This should have told the shift team that the column had flooded, had overflowed into the reflux drum and would now be filling the flare knock-out drum (see Figure 1). However, neither of the two high level indicator/alarms on this drum, set at 8% and 22% of capacity, showed any response.

It was 12 noon before anyone had a thorough look at the column. They found that the wires leading from the column level indicator were disconnected and that the valves between the knock-out drum and its level indicators were closed. Both vessels were shrouded with scaffolding and the state of the wires and connections was not easily seen. Liquid was now entering the flare stack. It failed as the result of low temperature embrittlement and was followed by a fire.

The immediate causes of the incident were the failures to restore the isolations of the level instruments before start-up and the slowness of the shift teams to realise what was happening. The underlying causes were far deeper and were due to both short-term and long-term changes in organization.

SHORT-TERM CHANGES

It was the practice on the plant to work 12-hour shifts instead of the usual 8-hour shifts during start-ups so that there were more people present than during normal operation. On this occasion the operators refused to do so (though they were willing to work overtime if necessary; this would give them more pay than working 12 hour shifts). However, the foremen and shift managers worked 12-hour shifts. They changed shift at 7 am and 7 pm while the operators changed at 6 am, 2 pm and 10 pm. This pattern of work destroyed the cohesion that had been built up over the years within each shift and lowered the competence of the team as a whole.

A report in the local newspaper said that, "A major influence over the behaviour of the operating teams was their tiredness and frustration". A trade union leader was quoted as saying that the management team members were more tired than the operators as they were working 12-hour shifts.

In addition to the usual shift personnel, two professional engineers were also present on each shift but their duties were unclear. Were they there to advise the shift manager or, being more senior in rank, could they give him instructions? Should they try to stand back and take an overview or should they "muck in"? On the day of the incident they did the latter and got involved in the detail of the problems that distracted everyone from the demethaniser.

LONG-TERM CHANGES

So far I have followed the published report on the incident[7] but there had also been more fundamental changes. The incident shook the company. It had a high reputation for safety and efficiency and the ethylene plant was considered one of its flagships – one of the least likely places where such a display of incompetence could occur, so what went wrong?

About seven years earlier there had been a major recession in the industry. As in many other chemical companies drastic reductions were made in the number of employees, at all levels, and many experienced people left the company or retired early. This had several interconnected results.

- Operating divisions were merged and senior people from other parts of the company, with little experience of the technology, became responsible for the ultimate control of some production units.
- There was pressure to complete the turnaround and get back on line within three weeks. This pressure came partly from above but also from within the team, as the members were keen to show what they could do. They should have aborted the shutdown to deal with the problems that had distracted everyone during the night but were reluctant to do so.
- There were fewer "old hands" on the plant who knew the importance, when there were problems, of having a look round and not just relying on the information available in the control room. A look round would have shown ice on the demethaniser column.
- Delayering had produced a large gap in seniority between the manager responsible for the ethylene plant and the person above him. This made it more difficult for the ethylene manager to resist the pressure to get back on line as soon as possible. Previously an intermediate manager had acted as a buffer and prevented commercial people and more senior managers speaking directly to the start-up team. Also, he would probably have aborted the start-up. Senior officers, not footsoldiers, order a retreat.

The company had an outstanding reputation for openness but was reticent about this incident and no report appeared in the open literature, apart from the local newspaper, until about twelve years later, after the company had sold the plant.

THE LONGFORD EXPLOSION

On 25 September 1998 a heat exchanger in the Esso gas plant in Longford, Victoria, Australia fractured, releasing hydrocarbon vapours and liquids. Explosions and a fire followed, killing two employees and injuring eight. Supplies of natural gas were interrupted throughout the State of Victoria and were not fully restored until 14 October. There was no alternative supply of gas and many industrial and domestic users were without fuel for all or part of the time that the plant was shut down[8,9,10,11].

The purpose of the unit in which the explosion occurred was to remove ethane, propane, butane and higher hydrocarbons from natural gas by absorbing them in "lean oil". The oil, now containing light hydrocarbons and some methane and now known as "rich oil", was then distilled to release these hydrocarbons and the oil, now lean again, was recycled. The heat exchanger that ruptured was the reboiler for the distillation column. The cold rich oil was in the tubes and was heated by warm lean oil in the shell.

As the result of a plant upset the lean oil pump stopped. There was now no flow of warm lean oil through the heat exchanger and its temperature fell to that of the rich oil, –48°C (–54°F). The official report describes in great detail the circumstances that led to the pump stopping. However, all pumps are liable to stop from time to time and the precise reason why this pump stopped on this occasion is of secondary importance. Next time it will stop for a different reason. In this case one of the reasons was the complexity of the plant. It had been designed to recover as much heat as possible and this resulted in complex interactions, difficult to foresee, between different sections. (Complex plants have complex problems.)

Ice formed on the outside of the heat exchanger when the flow of warm oil stopped but no one realised that the low temperature was hazardous. Despite long service the operators had no idea that the heat exchanger could not withstand low temperature and thermal shocks and that restarting the flow of warm lean oil could cause brittle failure. More seriously, some of the supervisors and even the plant manager, who was away at the time, did not know this. It was not made clear in the instructions.

The ignorance of the operators does not surprise me. When I was working in production, before I became involved full-time in safety, some operators' understanding of the process was limited. Trouble-shooting depended on the chargehands (later called assistant foremen) and foremen, assisted by those operators who were capable of becoming chargehands or foremen in the future. In recent years I have heard many speakers at conferences describe the demanning and empowerment their companies have carried out and wondered whether the operators of today are really better than those I knew in my youth. At Longford they were not.

Esso claimed that the operators had been properly trained and that there was no excuse for their errors. But the training emphasised the knowledge that the operators needed to do their job rather than the understanding they needed to deal with unforeseen problems. They were tested after training but only for knowledge, not for understanding. One operator was asked why a certain valve had to be closed when a temperature fell below a certain value. He replied that it was to prevent "thermal damage" and received a tick for the correct answer. At the Inquiry[9] he was asked what he meant by "thermal damage" and replied that he "had no concept of what that meant". When pressed, he said that it was "some form of pipework deformity" or "ice hitting something and damaging pipework". He had no idea that cold metal becomes brittle and may fracture if suddenly warmed.

Now we come to the crucial changes in organisation. Two major changes were made during the early 1990s. In the first, all the engineers, except for the plant manager, the senior man on site, were moved to Melbourne. The engineers were responsible for design and optimisation projects and for monitoring rather than operations. They did, of course, visit Longford from time to time and were available when required but someone had to recognise the need to involve them.

In the second change the operators assumed greater responsibility for plant operations and the supervisors (the equivalent of foremen) became fewer in number and less involved. Their duties were now largely administrative.

Both these changes were part of a company-wide initiative and were the fashion of the time. There was much talk of empowerment and reduced manning. The Report concluded that, "The change in supervisor responsibilities... may have contributed by leaving operators without properly structured supervision". It added, "Monthly visits to Longford by senior management failed to detect these shortcomings and were therefore no substitute for essential on-site supervision."

On the withdrawal of the engineers the Report said that it "appears to have had a lasting impact on operational practices at the Longford plant. The physical isolation of engineers from the plant deprived operations personnel of engineering expertise and knowledge, which previously they gained through interaction and involvement with engineers on site. Moreover, the engineers themselves no longer gained an intimate knowledge of plant activities. The ability to telephone engineers if necessary, or to speak with them during site visits, did not provide the same opportunities for informal exchanges between the two groups, often the means for transfer of vital information". None of this was recognised beforehand. Chats in the control room and elsewhere allow operators to admit ignorance and discuss problems in an informal way that is not possible when a formal approach has to be made to engineers at the company HQ. Empowerment can become a euphemism for withdrawal of support. There is a similarity with the changes in the research organisations described earlier.

On one occasion when I was a safety adviser with ICI Petrochemicals Division I was asked to move my small department to a converted house just across the road from the main office block. I objected as I felt that this would make contact with my colleagues a little bit harder as they would a little less likely to drop into our offices.

At Longford there were also errors in design. The heat exchanger that failed could have been made from a grade of steel that could withstand low temperatures or a trip could have isolated the flow of cold liquid if the temperatures of the heat exchanged fell too far. These features were less common when the plant was built than they became 30 years later but they could have been added to the plant. The designs of old plants should be reviewed from time to time. We cannot bring all old plants up to all modern standards – inconsistency is the price of progress – but we should review old designs and decide how far to go. Esso intended to Hazop the plant but the study was repeatedly postponed and ultimately forgotten. Another design weakness was the excessive heat recovery system already mentioned.

Exxon has a high reputation for its commitment to safety and for the ability of its staff. Was Longford a small plant in a distant country that fell below the company's usual standards or did it indicate a fall in standards in the company as a whole? Perhaps a bit of both. Exxon did not require Esso Australia to follow Exxon standards and the Longford plant fell far below them. Exxon were fully aware of the hazards of brittle failure but their audit of Esso did not discover the ignorance of this hazard at Longford. On the other hand, the removal of the engineers to Melbourne and the reductions in manning and supervision were company-wide changes. It also seems that in the company as a whole the outstandingly low lost-time accident rate was taken as evidence that safety was under control. Unfortunately, the lost-time rate is not a measure of process safety.

RAILTRACK
When British Railways was privatised and split into a hundred companies, Railtrack owned the track, signalling and stations, but other companies owned the trains, another group of companies operated them and British Railway's former maintenance groups became independent contractors. This left Railtrack without any engineering expertise or the ability to monitor its contractors.

This produced many interface problems between the various companies who had different objectives. For example, Railtrack had to compensate the train operating companies whenever trains were delayed and this made Railtrack reluctant to increase the time available for maintenance. There was also a literal interface problem, as responsibility for the rails and the wheels lay with different organisations. "Both sides of the wheel/rail interface may be operating within their respective safety based Standards, but the combined effect of barely acceptable wheel on barely acceptable rails is unacceptable".[12] This led to the rolling contact fatigue of the track (also called gauge corner cracking), the Hatfield train crash in October 2000 and the consequent upheaval while hundreds of miles of rail were replaced.

The engineering principle involved is hardly new. In 1880 Chaplin showed that a chain can fail if its strength is at its lower limit and the load is at its upper limit.[13] The Hatfield crash did not occur because engineers had forgotten this but because there were no engineers in the company's senior management.

OUTSOURCING
A marketing manager in a company that manufactured ethylene oxide foresaw a market for a derivative. The company operated mainly large continuous plants while the production of the derivative required a batch plant. The derivative was wanted soon and the company did not want to spend capital on a speculative venture. The manager therefore looked for a toll manufacturer. He found one able to undertake the task and signed a contract with them without consulting any of his technical colleagues. The toll manufacturer was very competent but was located in a built-up area. When it was realised that ethylene oxide was being handled there, this gave rise to some concern even though the stock on site was moderate.

A few years later the buildings were demolished as part of a slum clearance scheme. The HSE then refused permission for new ones to be built in their place. Before the local authority could develop the site they had to pay the toll company to move to a new location.

This incident occurred some years ago, before the CIMAH Regulations came into force. It probably could not happen today, but is a warning that outsourcing, of products or services, is a change that should be systematically considered before it takes place.

MULTISKILLING

Multiskilling presents specific problems, illustrated by the Flixborough explosion. The site was without a mechanical engineer for several months, as the only one – the works engineer –had left and his successor had not arrived. Arrangement had been made for a senior engineer from one of the owning companies to be available when needed but the men who designed and built the temporary pipe that failed did not realise that these tasks were beyond their competence and did not see the need to consult him[14]. Similarly, in many plants there is now no longer an electrical engineer but the control engineer is responsible for electrical matters. An electrical engineer is available for consultation somewhere in the organisation but will the control engineer know when to consult him? Will he know what he doesn't know?

The same applies at lower levels. Will the process operator who now carries out simple craft jobs be able to spot faults that would be obvious to a trained craftsman?

One of the underlying causes of the collapse of a mine tip at Aberfan in South Wales, which killed 144 people, most of them children, was similar. Responsibility for the siting, management and inspection of tips was given to mechanical rather than civil engineers. The mechanical engineers were unaware that tips on sloping ground above streams can slide and have often done so[14].

THE CONTROL OF MANAGERIAL MODIFICATIONS

As with changes to plants and processes, changes to organisation should be subjected to control by a system, which covers the following points:

- Approval by competent people. Changes to plants and processes are normally authorised by professionally qualified staff. The level at which management changes are authorised should also be defined.
- A guide sheet or check list. Hazard and operability studies are widely used for examining proposed modifications to plants and processes before they are carried out. For minor modifications several simpler systems are available[2]. So far as I am aware, only one similar system has been described for the examination of modifications to organisations[15]. Some questions that might be asked by those who have to authorise them are suggested below.
- Each modification should be followed up to see if it has achieved the desired end and that there are no unforeseen problems or failures to maintain standards. Look out for near misses and for failures of operators to respond before trips operate. Many people do not realise that the reliability of trips is decided on the assumption that most deviations will be spotted by operators before trips operate. We would need more reliable trips if this were not the case.
- Employees at all levels must be convinced that the system is necessary or it will be ignored or carried out in a perfunctory manner. The best way of doing this is to describe or, better, discuss, incidents such as those described above, which occurred because there was no systematic examination of changes.

SOME POINTS TO BE COVERED IN A GUIDE SHEET
Define what is meant by a change: Exclude minor re-allocations of tasks between people but do not exclude outsourcing, major re-organisations following mergers or downsizing or high level changes such as the transfer of responsibility for safety from the operations or engineering director to the human resources director. Accidents may be triggered by people but are best prevented by better engineering[16].

A recent survey in the US showed that nearly half the companies that replied to a questionnaire on the management of change said that they included organisational change[17]. However, they may not include the full range of such changes.

Some questions to ask are:

- How will we assess the effectiveness of the change in the short and long-term?
- What will happen if the proposed change does not have the expected effect?
- Will informal contacts be affected (as at Longford)?
- What extra training will be needed and how will its effectiveness be assessed?
- Following the change, will the number, knowledge and experience of people be sufficient to handle abnormal situations? Consider a number of past incidents in this way.
- If multiskilling is involved, will people who undertake additional tasks know when experts should be consulted? See the Section on Multiskilling.

Except for minor changes, these questions should be discussed by a group, as in a hazard and operability study, rather than answered by someone on their own. "None" or "not a problem" should not be accepted as an answer unless backed up by the reasons for this answer.

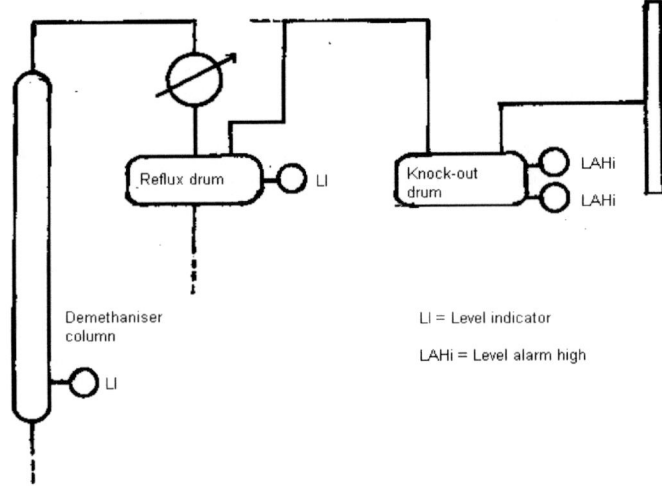

Figure 1. The level indicator on the column and the level alarms on the knock-put drum were out of order. The column filled with liquid which over-flowed into the to drum and then into the stack

REFERENCES

1. Muir, G., Dec, 2001, A better railway, *Modern Railways*, 58 (629): S1–S6.
2. Lees, F.P., 2001, *Loss Prevention in the Process Industries*, 2nd edition, Butterworth-Heinemann, Oxford, UK and Woburn, MA, Vol. 2, Chapter 21.
3. Kletz, T.A., 1988, *What Went Wrong? - Case Histories of Process Plant Disasters*, 4th edition, Gulf, Houston, Texas, 1998, Chapter 2.
4. Sanders, R.E., 1999, *Chemical Process Safety – Learning from Case Histories*, Butterworth-Heinemann, Oxford, UK and Woburn, MA.
5. Health and Safety Executive, 1999, *Preparing Safety Reports – Control of Major Accident Hazard Regulations 1999*, Report No HSG 190, HSE Books, Sudbury, UK, 1999.
6. Loveock, J., 2000, *Homage to Gaia*, Oxford University Press, Oxford, UK, p. 301.
7. Anon, 2000, A major incident during start-up, *Loss Prevention Bulletin*, No. 156: 3–6.
8. Dawson, D.M. and Brooks, J.B., 1999, *The Esso Longford Gas Plant Explosion*, State of Victoria, Australia.
9. Hopkins, A., 2000, *Lessons from Longford*, CCH Australia, Sydney, Australia.
10. Kletz, T.A., 2001, *Learning from Accidents*, 3rd edition, Butterworth-Heinemann, Oxford, UK and Woburn, MA, Chapter 24.
11. Boult, D.M., Pitblado, R.M. and Kenney, G.D., 2001, Lessons learned from the explosion and fire at the Esso gas processing plant at Longford, Australia, *Proceedings of the AIChE 35th Annual Loss Prevention Symposium, 23–25 April.*
12. Ford, R., Jan. 2002, Gauge corner cracking – privatisation indicted, *Modern Railways*, 59(640): 19–20.
13. Pugsley, A.G., 1966, *The Safety of Structures*, Arnold, London (quoted by Tait, N.R.S., 1987, *Endeavour*, 11(4): 192).
14. Kletz, T.A., 2001, *Learning from Accidents*, 3rd edition, Butterworth-Heinemann, Oxford and Woburn, MA, Chapters 8 and 13.
15. Conlin, H., 2002, Assessing the safety of process operation staffing arrangements, *Hazards XVI – Analysing the Past, Planning the Future*, Symposium Series No. 148, Institution of Chemical engineers, Rugby, UK, pp. 421–437.
16. Philley, J., 2002, Potential impacts to process safety management from mergers, downsizing, and re-engineering, *Process Safety Progress*, 21(2): 151–160.
17. Keren, N., West, H. H. and Mannan, M.S., 2002, Benchmarking MOC practices in the process industries, *Process Safety Progress*, 21(2): 103–112.

THE HAZARD OF MANAGEMENT

R. Ward
PhD, MBA, BE, ASTC.
F.I.E.Aust. M.A.I.M. CPEng.
Visiting Fellow, Faculty of Engineering, University of New South Wales, Sydney.

When engineers get onto the topics of hazards, risk, safety, reliability and related matters the thinking generally revolves around the physical equipment which engineers design and build, and operate and maintain, and manage to some extent. Any discussion may, then, progress to processes and from that to procedures and policies, into the areas where more general management is involved. But consideration of those matters rarely extends as far as management itself, to whether the management system and the people in it are sufficiently hazard-free and reliable to minimise risk.

Why should we look at management as a possible hazard? There is evidence suggesting some serious incidents are related to management, which raises another, following-on, question: why should management be, or present, hazards? The answer is simple, much of management activity is performed by using judgement, which is not a truly reliable practice and provides opportunity for error, and a research project has shown that there are several often-occurring, quite "normal", conditions which readily cause managers to make mistakes.

The paper based on this abstract will discuss how management judgement affects management reliability, and how that might be improved. It will conclude by presenting a very recent real and tragic event caused by application of apparently-sound judgement which results showed was incorrect.

WHAT DO MANAGERS DO?

One of the amusing memories of being a manager is having a young (and inquisitive) child ask: "What do you do at work, Daddy?" An honest answer, from one who was a manager in a chemical manufacturing firm, is something like: "I walked around quite a lot, and I talked to people, and I did some writing, and, oh, yes, I had lunch, then I went back to my office and did a lot of phone calls, more writing, more talking to people. Does that give you an idea of what my day was like?" The child's answer first displayed serious understanding, then produced a shattering question: "Do you mean you get paid for doing just that?"

There's more to management that, of course, and if we go to any of the management texts we find that managers plan, organise, lead, and control, with varying details added by some, such as "staffing" on organise and "motivating" on lead (eg Stoner *et al*, 1985). Whatever, however they may be stated, those are accepted as the classic functions of management. And behind, within, and surrounding those there's decision-making, which really is the core or central function of management.

Making decisions is something mysterious which, it seems, only humans do. The tasks related to it, problem-solving and problem-*finding*, usually "just happen", particularly for engineers: problems seem to find us, rather than the opposite. This combination, problem-finding-and-solving, can be the step leading up to decision-making. The reason for putting that in such a way: "the step leading up to decision-making", is that one can't make a decision until one knows what the problem is, and what are the alternative solutions - - -

without a problem and alternative solutions, there's no need for a decision. Both, a problem and alternative answers, are necessary as a lead-up to making a decision.

If a problem exists and a solution is demanded, and there's only one answer, no decision is needed. Or possible.

There is, actually, a very close link between problem-solving and decision-making; the two are very hard to separate from one another, and there's a very strong impression that some authors refer to one when they mean the other. Harrison (1987) stressed this difference by pointing out that decisions can often be made and implemented successfully in the absence of problems, and problems can be identified and solved without decisions. However, there is a one-way connection, in that solving a problem (by identifying possible solutions) may lead to a decision having to be made, but the reverse does not apply.

Other animals, certainly, can solve problems, but it does seem that only humans do this other act. There is a big difference between the two, problem solving is *finding an answer* (dogs and cats and, of course, monkeys have been seen to do that), but decision making is *choosing between alternatives* (and that seems to be the *homo sapiens* thing).

WHAT IS DECISION-MAKING?

Definitions abound. Going way back into the past we find many examples of what writers believe decisions, and decision-making, to be. For example, Cooper (1961) gave us:

Decision - is a commitment to action. Examples of action are the giving of an order to a subordinate, the signing of a letter to a customer to inform him of the terms on which the company will accept a contract, etc.

A decade more recently, another, somewhat different, was given by Massie (1971):

- a course of action consciously chosen from available alternatives for the purpose of achieving a desired result.

That spells out an important factor. Without alternatives, no decision is needed or can be made; there must be a reason for making the decision, which is to make a selection from among the available alternatives so that the desired result can be reached.

Also, the alternative selected must be "consciously chosen". Very often, even in large organisations, decisions are made "by default", that is, procrastination is allowed to set in and time is allowed to slip by so that finally events either make the decision or force the choice. Such is *not* decision-making. The "default choice" was around for a long time, well before the computer was invented.

Massie (1971) also pointed out there are three important ideas associated with decisions:

1. A decision involves a choice between alternatives; if there are no alternatives, then no decisions are required or can be made.
2. A decision involves mental processes at the conscious level, processes which contain important logical aspects and these suggest that decisions should be "rational". However, there are also emotional and sub-conscious factors which may not be consciously expressed and which may make a decision appear to be non-rational to an observer.

3. A decision is purposive, to attain some objective; that is, there is a reason for making it.

In the first numbered paragraph above there's reference again to the need for alternatives. Finding a reasonable range of alternatives from which to choose is the sort of activity that trained engineers do well, because that's really part of problem-solving, and our technical education system teaches that, but taking the next step, selecting which alternative to use, is the decision-step.

TYPES OF DECISIONS

The above suggested a uniformity of decisions, but that's not so, there's very real diversity of types, so we need to consider those different types of decisions which can be (and are) made.

Using another voice from the past, Drucker (1963) pointed to four characteristics which have a bearing on the nature of a business (or management) decision:

1. The degree of futurity of the decision. (In a short article titled "The Aphorisms of Peter Drucker" he stated: "Long range planning does not deal with future decisions, but with the futurity of present decisions.")
2. The impact of the decision on other functions, or departments, divisions, etc. in the organisation. (He advised: don't shake others up, well, not *too* much.)
3. The number of qualitative factors in it, such as principles of conduct, ethical values, social and political beliefs, etc. (This is referring to values as distinct from facts.)
4. Whether the decision, as a type, is periodically recurrent, or rare, or even unique.

Radford (1975) classified decisions under a number of headings which overlap, in some cases, those already mentioned above but are worth reviewing. He also pointed out that his classifications are not necessarily mutually exclusive, and a particular decision may have characteristics from more than one class. These characteristics of a decision may be:

(a) routine, well understood, and documented in all of its aspects, or it may be such as to require human involvement and judgement at one or more stages of the process,
(b) involve well-defined quantitative parameters, such as money, or it may refer to matters for which numerical values or outcomes are difficult to define (such as health and welfare),
(c) undertaken against a background of certainty in terms of the outcome of any course of action chosen, or in the face of uncertainty with regard to future events or the actions of others,
(d) such that well-developed mathematical techniques can be applied, or it may be that no well-defined structure has been developed for a problem of its type,
(e) a problem existing and requiring a solution at one point in time, or it may consist of a sequence of inter-related decisions,
(f) such that it can be resolved by an individual acting alone, or it may be the responsibility of a group of individuals who may not have identical views and approaches with regard to the problem,

(g)   arising in a small community of individuals, or in a large organisation consisting of a number of sub-groups with divergent interests.

Ansoff (1968) divided decisions into three general classes:

Strategic - centralised, made in partial ignorance, non-repetitive, very future-oriented.
Administrative - largely concerned with resource-allocation and the resolution of conflict between differing objectives.
Operating - decentralised, repetitive, large volume, need to sub-optimise forced by complexity.

As a broad classification, many writers (Stoner *et al*, 1985, for example) classify decisions as programmed and unprogrammed. The Radford Type (a) is clearly the "programmed" decision, as is Ansoff's reference to some decisions being repetitive. Drucker's fourth characteristic above deals with the same distinction, which he expanded to show the difference between programmed and unprogrammed decisions:

(i)   programmed decisions are routine or repetitive, to the extent that a definite procedure has been worked out so that this can be applied every time the specific situation arises,
(ii) unprogrammed decisions are unique, novel, and hence unstructured, with no cut-and-dried method for handling the particular situation because it hasn't arisen before, or because its precise nature and structure are elusive and complex, or because it is so important that it deserves a custom-tailored treatment.

In actual fact, in practice, very, very few decisions are either one type or the other; most have an element of each in them.

On this point, Radford drew attention to how the "experienced manager" "cheats": when an organisation decision is repeated, even with some minor variations, the manager facing it has an opportunity to vary and correct his methods and approach so that the result becomes closer and closer to being "right". That is, he has appeared to perform an unprogrammed decision, but behind the scenes he has really made the decision on a programmed basis by using a "partial precedent" (incomplete as a whole but applicable in part) as a guide.

Drucker pointed out that programmed decisions usually have only one right answer, because the problem being addressed involves restoring or maintaining the status quo - - - the operation at its present level. We understand such decisions - - - the programmed ones - - - the routines, pre-set and "by the book" - - - precisely *because they are not decisions after the first time*.

The first time these turned up they were unprogrammed, and they therefore required a decision, or a set of decisions, to cover the immediate situation. If that decision were proved to be correct, then recorded, when the problem was first experienced, on all those future occasions no decision would be necessary because some standard, precedent, rule, regulation, law, or "standard operating procedure" (quite commonly called "S.O.P.") takes over. So this class of unprogrammed decisions can become the programmed type which attends to a standardised situation.

However, there still remains a group of first-time, not-previously-experienced, answers-as-yet-unknown, and unique unprogrammed decisions which managers face from time to time.

Then there is the group of *really* unprogrammed decisions. Because these are "first time" decisions there will be an unknown and unquantifiable risk that the action decided will be wrong. All we can know about these "first time" decisions is there's an infinite number of ways to be wrong, and a limited set of right answers (often, or usually, only one?).

Drucker then subdivided this "first time" group into two classes:

(a) Those involving optimising, choosing the "best" or optimal answer. Here logical, rational, analysis succeeds. These respond to "management science", and are generally not business survival-or-death decisions because there is usually a range of acceptable answers, all close to "best".

(b) Entrepreneurial decisions, with not only no right answer but also usually no optimal answer, with a tendency towards high, or even full, uncertainty.

Those in (b) are a different type of decision from those in (a). The (b) type are likely to cause a major impact on the manager making them as well as on the organisation for which they are made.

Drucker's point about type (b) is that the aim in such a case cannot be to eliminate or reduce risk caused by uncertainty, or even to minimise risk, but to make the enterprise capable of taking bigger risks - - - but the right ones.

(As is often the case with Drucker's writing, this is an interesting and refreshing way of looking at decisions and risk.)

THE IDEA OF DECISION-RATIONALITY

Now for some ideas of how decisions are made, and there's a general assumption that the process is rational (Harrison, 1987). The idea of rational thinking goes back a long way, even to the ancient Greek philosophers. The father of the more modern developments of this appears to be Descartes, probably better known among engineers for his work on co-ordinate geometry. His proposition on rationality was that if we start with one basic premise it is possible to go step by step from that to an appropriate conclusion, with every step verifiable and repeatable under the same conditions. If someone else were to repeat the same process from the same premise, he'd get the same result.

So this is the idea behind rationality in decision-making, that if we start from the same premise, then each step in the reasoning can be followed and understood by another person operating via the same logical processes. A truly "rational", or logical, decision would therefore be one based on purely objective information and reasoning, on hard facts on the one hand and two-plus-two-equals-four reasoning on the other.

This is, of course, the sort of decision-making which originates from computer-analysis of problems, and it's absolutely faultless as far as it goes.

It's not the same as the situation involving two fishwives who were seen arguing abusively across a narrow street in Aberdeen, in Scotland.

Two academics were passing by, and one said to the other: "They'll never agree. They're arguing from different premises."

For another to try to suggest whether a decision was rational or non-rational can be extremely difficult. One person's believed totally rational and logical behaviour may be another person's perception of total, certifiable, insanity. Also, there's a fine distinction between the blundering idiot who follows protocol but leaves the swamp overflowing and the smart guy who won't follow protocol but applies initiative to drain the swamp while dodging the alligators.

Thinking further about rationality, it is obvious that unless full objective information is available then the necessary conclusion to satisfy the pre-determined objectives will probably not be stated, and the nominated end-point will not necessarily be reached. Also, the result can be, and often is, quite wrong, and the obvious reason for that wrong result is full and objective information is really never available.

However, even if full objective information were available, the "best" result may not be reached. The reason for that is simply that a totally rational approach would have to omit the subjective, gut-feeling, items of information, which would of necessity be unsupported by any of the objective evidence demanded by the rational process. Very often the gut-feeling information, which contains the emotional and inexpressible data, can be important.

The terms "objective" and "subjective" have been used above to identify the difference between a decision based on external, quantitative, "hard", information, as opposed to one based on internal, personal gut-feelings and beliefs, all "soft" evidence, quite unquantifiable but of which one must nevertheless take notice. Many writers (for example, Massie (1971), as well as many other, later, writers) have referred to these and have distinguished between them.

Referring further to the matter of the two sorts of information, there is no conceivable reason, in this author's opinion, why one "sort" of information "must" or "should" be more accurate than the other at all times, but of course many, especially engineers and accountants, prefer objective to subjective evidence, or information. Indeed, many people will ignore subjective assessments of a situation and refuse to recognise that such a condition exists. But it does.

Who remembers Harrison Ford, as Han Solo in "Star Wars", saying repeatedly: "*I've got a bad feeling about this!*"

He was right every time, too. Does that suggest anything about his ability to assess probability? Or were the situations stochastic, for some reason?

Nevertheless, it's often right for those people at least to attempt to suppress the internal (to a large extent conditioned-by-training) desire for firm, objective data, and to work on feelings, to some extent. Peter Drucker criticised this innate desire, gently but somewhat sardonically, in the movie **The Effective Executive** by stating what he called the only thoroughly proved law of statistics: "Tell me what facts you want and I'll tell you where to find them."

Equally, one may be sure other decisions would be improved if those who "normally" make them on emotions-bases were able to employ some objective, rational, techniques.

It becomes evident that programmed decisions, dealt with correctly by use of established routine, will have a high probability of success because they have been tested in previous applications, hence they have a small risk of the desired result not eventuating.

But unprogrammed decisions, whether they be of the optimising or entrepreneurial type, lack established routines developed from precedent, and will therefore have a lower probability of the "best" answer being chosen, and of satisfying. They therefore hold more risk of failing to satisfy, of *not* satisfying.

SO WHY IS MANAGEMENT HAZARDOUS?

It's tempting to argue from the above that decision-making is a smooth application of education and training. Well, education and training probably helps, but that's not the whole story behind successful decision-making.

Management is unreliable, hence a hazard to those who work as managers, to those who work under them, and to the firms employing them, because managers must make decisions, often unprogrammed, and managers can, and will, make mistakes when making decisions, particularly those unprogrammed. A manager's mistake can ruin his career, can injure or even kill a subordinate, and can send a company into liquidation.

Having made that statement, that managers are unreliable because they make mistakes, this author is bound to offer, first, proof of the proposition, and second, some analysis of why managers make mistakes.

Here's an illustration. In Australia we have recently been through an insurance crisis, with one major firm going into liquidation after buying out other firms, paying consultants, and rewarding executives, spending relatively enormous sums on all three. The chief executive has been in court, has been questioned about what happened, why did they do what they did, and the answer seems to be: "It seemed to be a good idea at the time." What seems to be coming out, dragged out by counsel's questions, is that the senior management made some very serious mistakes. The usual excuse for making mistakes is that the person was only human, rather than divine, but in our insurance debacle it appears the human quality greed has been involved as well as a sheer blunder.

And another, older, and relevant to this Conference. Engineers and managers in the process industries in your isles will remember the incident at Flixborough was caused by management not controlling a plant modification, by omitting a checking-action, and thus allowing the work to proceed without full investigation. We had our equivalent, a few years ago, at Longford, also traceable to management.

The making of a particular type of management mistake was explored by giving senior students in mechanical engineering, most of whom were already in supervising positions in industry, a series of management cases, each case containing three 'levels' of decision-making problem.

The first was a technical-engineering problem, of relatively trivial value in the general context.

The second level was an 'obvious', 'local' management problem, which the students are required to solve, or at least resolve, as it related to the week's lecture topic.

The third level was some other management-related feature of the factory (such as the conflict occurring between two characters), or of the company as a whole (for example, the entrenched hierarchical structure and its politics).

In addition there was a background problem, indicated by an accident which happened every week, but not mentioned in the terms of the assignment.

The mistake-causing elements were identified as follows:

strangeness - the cases were set in the chemical industry, in a continuous process system, 'strange' to most of the students, uncertainty - the students did not know what to expect would happen next week (although the author, obviously, did know),

precise tasks - there was a clearly indicated problem (the 'second level' management problem),

immediacy - the assignments were on a one-week cycle, to be returned one week after being handed out (mirroring real-world urgency),

time pressure - the immediacy was increased by the presence, hence pressure, of work in other subjects.

The "mistake" which a large number of students committed was classified as the "error of omission" (Ward, 1993). All solved the engineering problem, most answered the local management problem, many at least commented on the company problem (some with disgust), but only a very small number, two out of thirty, took any notice of the continuing accident-events - - - which ultimately led to a fatality.

WHY DID THE STUDENTS MAKE THAT MISTAKE?

The majority fell into error because they concentrated on what was directly before them, the section of the subject impressed on them by the week's lecture, and they ignored the accidents which were going on because that was the unprogrammed part.

Fortunately, most management decisions are programmed, they have been worked out in the past and can be found in policies and procedures (S.O.P.'s), what the lawyers call precedents. It's just as well there's a lot like that, because without all that awful documentation a lot of decision-making couldn't be delegated.

But managers do meet unprogrammed decisions and a manager must act with imperfect and incomplete knowledge of what's behind the need to decide, and without a methodology to deal with them. As already noted, the way comes from what we call "experience", an experienced manager can "cheat" by applying his knowledge of what worked in a similar situation last month, or last year, and make adjustments to suit a situation which may be similar, but not quite the same as, the one now being faced. That works in some cases, but not when the situation is something quite novel.

When a truly novel, unprogrammed, decision arises the opportunity for serious error increases because the manager must use judgement. That is the true background reason for the students' error. They were left (deliberately) to their own devices to choose what parts of the assignments they would answer, and the majority judgement was faulty.

So now for a look at a quality which is even more mysterious than simple decision-making (which is mysterious enough): judgement.

JUDGEMENT

Some decisions cannot be made entirely by use of objective techniques, and require use of the mysterious element we term "judgement". Here, this is not what is pronounced in a court of law by a judge, based on evidence presented by lawyers and witnesses, some expert and some lay, the balance of probabilities, beyond all possible shadow of doubt, it's what a manager uses when facing a truly unprogrammed decision. He uses a "judgement technique" which may, or may not, appear to be rational to others.

The type of decision-making in which an individual's judgement is required has been given the title "judgement call", a phrase originally, I believe, from the USA military, but the term has by now been taken up by business (Mowen, 1993). Expressing the need for such a type of decision is often indicated by another person saying to the decision-maker: "It's your call."

How may we define a judgement-call-decision?

It is a choice, in a high stakes environment, between two or more poorly-identified options, and the choice must be based on ambiguous information while facing conflicting goals, often with a close time-horizon.

What are the characteristics or parameters of judgement calls?

First, here are some typical situations, which identify decision elements to be weighed up (Mowen, 1993):

1. To shoot or not to shoot (clearly a military example), or to go ahead or to not go ahead.
2. To stay or to quit.
3. To retain present security or to seek future possible gain.
4. To accept risk or to retain security (related to 3. above).
5. To indulge in chance or to maintain control (combining 3. and 4. above).

Re the fourth above: bear in mind that risk is expressed pseudo-mathematically as the product of consequences (damage or injuries caused by an undesirable event following a decision) and the probability or uncertainty (of the event occurring). Some low-probability (very unlikely to occur) events have serious consequences (and may occur, though improbable).

How may we find help in deciding how to answer a judgement call? By using these understanding elements which will assist decision action:

1. Find the cause of the problem (which may be difficult if time is short).
2. Choose a frame of reference (how does the cause relate to the situation?).
3. Use reason rather than emotion (but recognise feelings may be helpful).

We should now review the steps which should help in any decision-making situation:

1. Be sure of the desired goals or objectives.
2. Observe a need, or a deficiency, in the path towards those goals or objectives.
3. Question: does that present a problem?
4. If so, identify and express the problem clearly.
5. Generate options to solve the problem.
6. Assess and evaluate the options.
7. Choose an option (strategy) and decide what to do (tactics).
8. Implement and act.
9. Monitor progress and results.

Now we can see why a judgement call decision can be difficult! It's because in an unprogrammed decision, a judgement situation, Steps 1 to 6 involve uncertain, possibly conflicting, goals, lack of accurate information, and time constraints, with a background awareness that getting it wrong may have serious consequences!

The one serious consequence from this decision situation is an intermediate condition, prior to the ultimate consequence; it's what happens as soon as the Cooper's "commitment to action" occurs, it's the decision-maker reaches "the point of no return" (Dearlove, 1998). It's like stepping on a banana skin, or a patch of oil, on the pavement, once the step has been taken opportunities for recovery are severely limited and most often do not exist.

And *that* is why a judgement-call-decision needs to be a correctly-made decision. There is, almost always, no second chance, to allow going over and correcting what was done. It's rare that we can hit the rewind button of life and rerun the event to correct whatever went wrong. The rare occasions when that's possible are the result of sheer luck, another mystery-management quality.

Nevertheless, many managers make judgement-call-decisions successfully. How? By practice? By "cheating", falling back on perhaps-ill-remembered incidents from the individual's past? Or, like judges in court, by precedent, from previous cases provided by others? Or is it by some negative-selection reason, because those who are successful are the managers who do not avoid making such decisions?

There's scope for research in this, particularly in the hazardous industries.

AN EXAMPLE OF APPLICATION OF JUDGEMENT

This example is not from industry but from ordinary everyday life, which in this author's opinion makes it more telling - - - certainly for this author - - - and it shows what can happen even with the best of intentions and appropriate knowledge of the situation.

In January this year a group of people had been attending a seminar on one of the Hawaiian islands. After it concluded they set out to celebrate, going to a property owned by one of those attending, and came to a ford across a stream. The property was on the other side, and the owner of the property, who was driving one vehicle, had crossed this ford many times, so he took one group across in his four-wheel drive, then went back for a second group, two in the front seat with him and two in the back seat.

Halfway across the ford the flow across the ford increased very suddenly to a wall of water. The vehicle rolled over and was swept downstream. The two in the rear seat escaped with injuries, the three in the front seat were killed by impact against rocks and by drowning.

We can review how the driver decided to make the second crossing. It was based on experience of having driven across the ford many times to his property, plus looking at the water flow, the general weather conditions, and of course on having made one successful crossing immediately before. His judgement was that another crossing would be safe.

That incident has, through this year, impressed this author with the fallibility of judgement. It's often all we have when making decisions, but it can lead to tragic consequences. In this case one consequence was the loss of my forty-year-old daughter, who was one of the three in the front seat.

CONCLUSION

The conclusion which comes out of the above musing on the hazard every manager faces when making a decision is that there's no way out of the bind managers face.

If the decision's programmed it's a low hazard with low risk, with high probability of being "right", with small consequences if "wrong" for some unexpected reason. But managers are paid to meet unprogrammed decisions, sometimes highly hazardous, with high risk, with potentially serious consequences, and usually with an unknown probability of result.

When the decision is sufficiently highly unprogrammed there are no rational ways to find the right answer. So judgement is needed. Only time can tell whether that produces a satisfactory result. After all, what is the test to be applied to a decision, so we may say whether it was successful? The only test is to look at the result. A good result *suggests* the decision was correct.

Unfortunately, although there are ways of teaching objective, rational, decision-making, there seems to be no way to teach managers how to use judgement. And, equally unfortunately, judgement is often needed in making decisions.

The result which inevitably stems from that analysis (as well as from some historic examples) is that management *per se* is an inherent hazard within a business enterprise. For as long as we have enterprises we need managers. If we could eliminate managers we would remove a major hazard in the system. But having "a manager" has become an essential part of the system, and we cannot imagine an enterprise functioning without management.

The ultimate irony of this circular problem is as well as dealing with all the other hazards in a business (financial, physical, etc) management should recognise and deal with the hazard itself presents.

REFERENCES

Ansoff, H. I. 1968. Corporate Strategy. Penguin Books Ltd. Harmondsworth, England.

Cooper, J.D. 1961. The Art of Decision-Making. Tadworth, Surrey (UK): The World's Work (1913) Ltd.

Dearlove, D. 1998. Key Management Decisions. Pitman Publishing, a Division of Pearson Professional Limited. London.

Drucker, P. 1963. The Practice of Management. W. Heinemann Ltd. London.

Harrison, E. F. 1987. The Managerial Decision-Making Process. Houghton Mifflin Company, Boston.

Massie, J.L. 1971. Essentials of Management. Second edition. Prentice-Hall Inc. Englewood Cliffs, New Jersey.

Mowen, J. C. Judgement Calls. Simon and Schuster Inc. New York. 1993.

Radford, K. J. Managerial Decision Making. Reston Publishing Company Inc. (A Prentice-Hall Company) Reston, Virginia.

Stoner, J.A.F., Collins, R.R. and Yetton, P.W. 1985. Management in Australia. Prentice Hall of Australia Pty. Ltd., Sydney.

Ward, R. B. 1993. *Human Error Among Managers: Observations of Students in a Management Subject.* Second Australian Conference for Engineering Management Educators. Melbourne.

SAFETY IMPROVEMENT THROUGH LEARNING FROM INCIDENTS

Nicholas JL Gardener
Health, Safety, Environment and Risk Manager, Elementis plc

Elementis, a specialty chemicals company, uses global incident reporting and investigation as a key tool for improvements in health, safety and environmental (HS&E) performance. Success has come through attention to the way in which the system was implemented as well as the incident investigation process itself. The system is based around a corporate electronic database deployed at all levels throughout the Company worldwide, There has been a marked and continuing improvement in HS&E performance.

Incident, investigation, root cause, improvement, database

PREAMBLE

Increasingly, as process safety improves, the opportunities for further improvements in occupational health, safety and the environment (HS&E) must come from elsewhere.

Priority must go to preventing the incidents that cause the most suffering, damage or pollution. It is not sufficient however only to work on serious incidents. All incidents matter to a greater or lesser extent. Also, the relatively infrequent occurrence of serious incidents means that if you only work on them there are insufficient data to allow comprehensive action to be taken to prevent other possible serious incidents. The strategy must be to remove the mass of underlying factors, present in more minor incidents that, under other circumstances, could become major.

It is a never-ending journey, which can be likened to breaking down a pyramid shaped iceberg that has the most serious incidents visible at its apex down to unseen or ignored minor near misses towards the base. A model for safety is shown in figure 1.

But to break it down you must first study why the incidents are there. There is nothing fundamentally new about incident investigation. But not all systems are successful. This paper describes how Elementis has implemented a comprehensive incident investigation process, based on a corporate electronic database deployed worldwide, to achieve significant improvements in HS&E performance.

BACKGROUND

Elementis is a specialty chemicals company with operations in the UK, USA, Europe, Asia, Africa and Australia. In the last century managers and employees were continually making efforts to improve, sometimes with great success. However, the process depended mostly on local initiatives. Improvement tended to be restricted either in the scope of problems solved or to the location where the work was done (and often both).

An exception was the increasing adoption of formal management systems for safe working. However, differences in perspectives between the UK and US (in particular) meant that their priorities were different. In the UK greater reliance has been placed on management systems to assure safe working that complies with all legal requirements. The focus in the US is to have compliance systems based more directly on the detail contained in the comprehensive OSHA[*] and EPA[φ] standards.

Clearly the Company must comply with the legal requirements of each country in which it operates. But beyond that the Company recognised that the only acceptable position to take is to conduct its business worldwide with the highest concern for the health and safety of its employees, contractors, customers, neighbours and the general public, and for the environment in which it operates.

Several steps were taken. To raise awareness and ensure that consistent high standards are achieved worldwide, the HS&E team instituted a series of corporate policies for life critical activities such as "Line Breaking", "Working at Heights" and "Vessel Entry". A corporate team checks that these are applied by conducting comprehensive compliance audits.

Another step was to introduce an employee led behavioural safety programme to address the large number of small incidents or hazards directly attributable to the way people work and behave.

The Company also recognised the importance of Product Stewardship. Responsibility towards others, such as those who use our products and the local community, is both ethically correct and one of the building blocks for sustainable development.

Nevertheless improvement in performance was only happening at an evolutionary rate. It was clear that these steps, while necessary, were not sufficient to take the Company to the level of HS&E performance to which it aspires. A step change to a new level of performance was required. An aim was set: reduce the lost time accident frequency by 50% within a year, in preparation for further reductions in all recordable incidents.

The target was set arbitrarily on the basis of being a level that would reflect a systematic change. But, however desirable or essential, an arbitrary target is meaningless unless there is a means to achieve it. The means to achieve the performance improvement was by rigorously correcting the root causes of incidents through greatly improved incident investigation. For further discussion on arbitrary targets see Appendix 1.

In 2001 Elementis plc reduced its lost time accident rate by 53%. figure 2 shows that the improvement is continuing.

IMPROVED INCIDENT INVESTIGATION

REQUIREMENTS OF AN INCIDENT INVESTIGATION SYSTEM
To be effective the root cause corrective action process requires:

*   full reporting of incidents
*   in-depth investigation of each incident to establish the root cause(s) – recognising that there may be more than one root cause per incident
*   development of effective corrective actions (at least one for every root cause)
*   making sure that all corrective actions are carried out

---

[*]Occupational Safety & Health Administration
[φ]Environmental Protection Agency

- checking to ensure that as far as can be ascertained the actions are effective
- learning from incidents including near misses to prevent similar incidents occurring elsewhere.

CONCEPTS FOR SUCCESS

One of the core concepts was that a large number of people should be involved with incident investigations. This is fundamental to demonstrating commitment to improvement and to tap the knowledge and ability that exists throughout the organisation. In any case safety and pollution prevention are everyone's responsibility (though the opportunities and levels vary). So anyone should be able to report an incident. A wide range of people with relevant knowledge, expertise or interest should then be involved with investigation. Plant and other managers must be part of the corrective action process. HS&E personnel have their professional role to play overseeing and participating in the process, and for ensuring that lessons are learnt and applied to advantage elsewhere.

Tangible benefits are achieved from correcting the causes of actual incidents. Intangible benefits accrue from a large number of people at all levels in the organisation developing and using the system collaboratively towards a common goal.

A key feature for success was the way in which the new system has been introduced. This can be summarised as a team approach between corporate HS&E and the businesses, allowing corporate requirements to be met in a way that was compatible with business needs. Central to this was development and implementation of a comprehensive incident investigation database.

DATABASE FUNDAMENTALS

GENERAL

Having a single corporate database in electronic format facilitates the process in several ways:

- it provides the framework in which to report all essential information and guides users in the necessary steps to follow. This is vital because many users are only involved infrequently
- data are consistent, for prioritising actions and for making trend analysis meaningful
- increasing the data capture to include all sites world-wide provides more experience from which others can learn
- using the database collectively acts as a focal point and catalyst, bringing the different businesses worldwide closer together on health, safety and environmental matters.

CHOICE OF DATABASE

The option of taking a proprietary database was discounted.

The company already had a number of Lotus NOTES databases. From this experience it was felt that a Lotus NOTES database for incident investigation could be created relatively rapidly at low cost. Further enhancements would not be constrained by software limitations and licences. Widespread user familiarity with how to use the existing databases would make adoption much easier. The chosen method also provided the opportunity to link easily to the Company's email system and for future integration with other Lotus NOTES databases.

As it turned out there were other advantages. The process of developing the incident investigation system helped create understanding and shared ownership. Audit reports and risk management topics have been added to take advantage of the corrective action routines.

OPERATIONAL DEFINITION
Elementis is a global company. An important recognition was the need for clear understanding between sites in different countries.

The Company adopted US OSHA definitions for "Recordable Injuries and Illnesses". Lost time is recorded both as US definition (all lost time after the day of incident) and UK RIDDOR* (greater than 3-days lost time) An internal Company definition was used for different categories of environmental incident.

DEVELOPING THE DATABASE

PREPARATORY WORK
Sites already had incident reporting systems in place. Although these included sections on investigation and actions, no consistent format was used and the level of detail varied. The process followed was to explain the need for corporate involvement and the benefits a carefully designed common system could bring.

Benefits were broadly that the system would be:

- comprehensive - to guide investigations in an efficient and effective manner
- consistent - for corporate assurance that incidents were investigated properly to a certain standard
  Making the system electronic would add further potential advantages:
- automated notifications and reminders, with remote access possible
- rapid visibility to authorised readers (an issue discussed below)
- search capability and opportunities for trend analysis
- learning through sharing

THE DEVELOPMENT PROCESS
All sites provided copies of their existing incident report forms. Comparisons established the scope of perceived requirements. A draft form was then created to take the best features for a comprehensive, logical workflow. During the process some existing requirements were questioned, other new requirements were added (in particular for analytical purposes).

Senior representatives of each Business reviewed the drafts so that when the process was completed they were satisfied that the system met their needs. Review was in two

---

* Reporting of Injuries, Diseases and Dangerous Occurrences Regulations 1995

forms. By email to give time for reflection and one-to-one comment, and by telephone conference calls to discuss ideas. At the end of this process senior managers in all Business were thus happy to implement the incident investigation system at their sites.

As the benefits of the database for improving health, safety and environmental performance become more and more apparent further enhancements and changes have been proposed. Being in Lotus NOTES means that the system has been able to evolve fairly readily.

Once the template was agreed it was converted into a Lotus NOTES database. A key objective was to get a database up and running quickly to maintain momentum. Enhancements were added later, and as the structure became more complex some parts were re-programmed. The nature of Lotus NOTES made all of this possible. As a result there was minimal lag between agreement to proceed and an operational system.

SOME ISSUES ARISING DURING DEVELOPMENT
There were some issues about the database, which had to be addressed. first, was security and confidentiality. As mentioned above any employee with a Lotus NOTES account can access the database to enter an incident report. This is a fundamental requirement. Other roles require appropriate authorisation, in some cases set by the manager responsible for the incident investigation, in others from predefined lists. Open access to the database generally means that all reports can be read. However, there is the option to make a report private. As people become familiar with the system and confident in the way in which it is used it has been found that there are very few incidents that fall into this category. It is however still available if there are particularly sensitive issues.

Secondly, all incidents are treated equally by the database, regardless of severity. There was some feeling that there ought to be a short version for minor incidents to encourage reporting. This has been resisted on the following grounds. Minor incidents including many near misses are a rich source for learning. The aim is to capture as much near miss and serious condition data as possible. If it is truly of minor importance it should be possible for a small team to deal with it swiftly. But we still want the record and corrective or preventive action.

IMPLEMENTING THE DATABASE
It was recognised at the outset that opening the database to all employees worldwide presented a huge education and training issue. Not just in the logistics but also in the way people respond. Previous work by the author suggested that the approach had to cater for two types of person. Practical people, who would initially take some persuading but would then be model users, and more intuitive types who might adopt more readily but would later need more anchoring to maintain the desired usage. Further description is given in Appendix 2.

Another consideration was that site personnel, who were not involved in the development of the new system, would still need persuading that the proposal was a good idea. The aim was to create understanding and acceptance that it was worth giving up their existing systems, even if their systems still appeared good. The same process had to be gone though with the Businesses managers in the first instance but by implementation they had gone beyond that stage.

Three key steps were therefore taken.

First, a training package was produced. This covered the importance of thorough incident investigation, techniques for investigating incidents and how to use the new database. The training package was supported by comprehensive guidelines on how to enter data into the database. Details of how the database is used are shown in Appendix 3.

Secondly, large numbers of users were trained. Plant managers and HS&E staff, who all have specific roles in the process, were trained. So were the supervisors who need to raise reports. Additionally, and especially for near misses where the intention is that trades union/safety representatives and others can raise a report, these employees were also trained.

Thirdly, continual active support has been provided via telephone, email and in some cases face to face. Queries are encouraged and followed by a timely response. The database is checked routinely to ensure that it is being used fully and correctly. Concerns are followed up where necessary. It was important to recognise that the familiarity gained by the developer and administrator functions led that of the users. Also some users only need to access the database infrequently and may forget what they have been taught. A side effect has been that contact with users has been a fruitful source of ideas for database improvement.

CONCLUSIONS FROM INCIDENT INVESTIGATION

Benefits are being achieved in many ways.

Reports are comprehensive with many root causes found. A key observation, which should be of no surprise to systems thinkers, is the number of incidents where the root cause is due to some management issue, in contrast to an immediate cause seen as being due to "people". Ishikawa cause and effect "fishbone" diagrams have been highly advantageous for investigating root causes.

Analysing the types of incident occurring over the whole company has led to insights and suggested further actions, which would not necessarily be obvious from a single incident. Examples have been learning about slips and trips when using stairs, forklift truck safety, injuries to hands, and office safety.

The incident system has been fully adopted worldwide throughout all Elementis businesses from the large chemical sites to small depots serving mines in the Australian outback. This means that Business managers can keep in touch with the detail of what is happening as and when they wish, wherever they are in the world.

Reliable data are readily available for monthly and annual reports from a single database, with data input at source. Data are summarised in Pareto and control charts to show trends. Information presented covers lost time and other recordable incidents. This helps management to focus on the bigger picture as well as be concerned about single incidents. Also included are charts that indicate the health of the incident investigation process, such as number of reports generated (Figure 3) and times to investigate. This helps indicate the level of preventive actions being taken. Taken together these charts help predict, with risk of error, future short – medium term performance.

As confidence has grown in the database and investigation system, more near misses are being reported. This is encouraged. An initially increasing number of reported near misses is seen as the sign of a mature organisation recognising reality and working to improve, not the sign of deterioration in performance. One ratio used to measure near miss reporting is the number of near misses reported to recordable incidents. Another is the average number of near misses reported per employee. Creating visibility offers the opportunity to work on root causes, as shown in Figure 4. Ultimately, as near misses are tackled, the number reported should reduce, but that may be a long way off. Meanwhile HS&E performance should continue to improve.

APPENDIX 1 NOTES ON CONTROL CHARTS AND ARBITRARY TARGETS

The author uses control charts extensively to track HS&E performance as a means to understand and predict. Charts showed that there was a stable system for lost time accidents and other recordable incidents. As it stood any improvement target would be just a wish.

Those who are familiar with Shewhart's work on control charts[1] and the teachings of Deming[2], may therefore question the logic of setting an arbitrary target for improvement.

The first point to make is that setting a target was recognition that the status quo was unacceptable – something had to be done. The true goal must be zero incidents, but a sign of improvement would be a halving within a year.

A second point is that action would be taken to change the system to a new lower level. Conventionally one would only investigate when a point is outside the 3-sigma control limits (or possibly use a run rule). Action at other times on a stable system is uneconomic. However, the intention was not to try to explain a bad month that was just part of the variability of the system. If you (rightly) want to reduce the number of incidents you need to work on the mean, which is a change to the system. The only way to achieve that is to study the system that gives rise to the incidents (Wheeler[3]). All incidents every month need to be studied to provide the necessary insight.

APPENDIX 2 PERSONALITY TYPE THEORY

Why do some people take readily to incident reporting and investigation while others do not? Previous work by the author[4] investigated the reasons why some people adopt Shewhart control charts while others do not. The results of that study on competent people with proper training and facilities suggested that people then used either their experience or their intuition in deciding how to respond.

A way of showing how the effect of management encouragement on different thinking types is to use catastrophe theory. It should be noted that the dramatically named theory is a revolutionary way of understanding how things change. It does not necessarily mean a disaster!

Catastrophe theory was developed by René Thom and is described in a very simple way by Woodcock & Davis[5]. It describes change in systems having more than one stable state, or following more than one pathway of change. Change may be smooth. Where it is discontinuous a so-called "catastrophe" occurs. The theory provides a way of describing discontinuous transitions.

The mathematics is complex and need not be attempted to gain a qualitative understanding of a process. Elementary catastrophes exist as one of seven possible behaviour surfaces. These can be used as templates. A common one is the cusp catastrophe, which occurs in systems whose behaviour depends on two control factors. Its graph is a 3-dimensional curved surface with a pleat. The catastrophe is the jump from the edge of the pleat another part of the surface.

The particular catastrophe is chosen so that the process behaviour corresponds to some features of the model. One can then study the model to see what other, less obvious types of behaviour it suggests for the process under different conditions.

Figure 5 shows a cusp surface. Every point on the surface represents an equilibrium state.

Changing one or both control factors (Thinking and/or Encouragement) may result in one of two things: a smooth change or a jump to a very different state. The jump may not be instantaneous but, because it goes through an unstable state, it will be brief relative to the time spent in stable states. Note there is no scale. The diagram indicates paths, not measurements.

The following are the main possibilities:

a-b   Intuitive types progressively report and investigate incidents in a tentative manner, which will be affected by the prevailing management attitude.

b-c   With success the intuition is replaced by experience and practical thinking predominates.

d-e-f   Practical types will need a certain degree of encouragement to gain benefit. Once convinced they go straight to full use. Point "e" is the singularity or point of "catastrophe" leading to a new stable level of operation (f).

Once at "c" or "f" stable operation will remain and encouragement can be reduced - up to a point (g). Representing perhaps that active encouragement is no longer necessary. However, if there is no management interest in incident investigation, encouragement is reduced further. The result for decisive types is a jump back in a discontinuous manner via "g" to "h" (non-use). Intuitive types may take a different path and meander more progressively, via, say, f-x.

Note that other paths are theoretically possible. Looking at alternatives may suggest other possibilities for consideration. All paths except those going over the cusp (e-f) or through it (g-h) are theoretically reversible. The path "e-f-g-h" shows that there is a lag (or hystersis) representing inertia to change for those who are guided by experience.

An insight is that "x" represents a critical point. Suppose encouragement from a starting point "a" reaches this point, and no further. As an intuitive type gains experience of incident investigation thinking moves towards a more fixed behaviour. The cusp catastrophe would indicate that diverging paths are possible. The direction will determine whether incidents are reported and investigated (f) or not (h).

APPENDIX 3 USING THE DATABASE

A basic requirement is that incidents should be entered and progressed in a timely manner. All incidents must be entered before the supervisor goes off shift.

The nominated manager for the area where the incident occurred is notified automatically by email. Having checked the content and satisfied themselves on the validity of the report, they must then set up an investigation team (who are notified automatically) and then accept the incident report. If they do not want to proceed e.g. if it is a duplicate of one already reported they "reject" the report. Note they do not delete it. Authority to delete is reserved for database managers. This maintains transparency of what is reported and helps prevent accidental deletion.

Due to the speed of implementation there were some early teething troubles. These were partly due to the database and partly due to the network server infrastructure. Database problems typically arose when the strict discipline for data entry was not followed. For example times had to be entered in 24-hour format, with hours and minutes separated by a colon, with no "return" after the entry. Making this more forgiving by adding selection boxes was a relatively easy cure. Network problems arose because, for speed of response, the database replicates onto the local server for each site. We learnt a lot about replication.

Despite these setbacks users were very supportive, possibly because of the time invested in showing the benefits that would accrue from getting it right.

The investigation team is charged with establishing root causes and developing effective corrective actions (at least one for each root cause). The report is then submitted to an HSE manager and the Plant Manager. This part of the process is strictly controlled. Both names can only be selected from a pre-entered list held securely in the database. The HSE Manager's responsibility is to check that in their professional judgement the actions are appropriate and that from their knowledge there are no other actions that might be included. The Plant Manager is asked to approve the corrective actions based on knowledge of the plant and crucially by approving to indicate a commitment to ensure that the actions are carried out.

In some cases either through the seriousness of the incident or the extent of the corrective action more senior management approval is required. A routine is provided for such eventualities.

A comprehensive corrective action system is included. Those with actions (who have Lotus NOTES email accounts) are notified of their actions. A tracking system shows open actions, separating out any that are overdue, and completed actions. As well as seeing actions against an incident it is easy to see all the actions against an individual person. This visibility helps the individual concerned and their Manager.

The report is not signed off until the Plant Manager (or Senior Manager) is satisfied that the corrective actions have been completed and are satisfactory.

A customised "HELP" section is included with the database A useful feature is a "Message for Today" page that pops up as the first screen whenever a new message is added. This provides a reliable way of e.g. notifying users of changes, alerting them to specific safety issues, or providing feedback on how the database is used.

REFERENCES

1.   Shewhart, W.A., 1931, Economic Control of Quality of Manufactured Product, New York: Van Nostrand Co Inc

2. Deming, W.E., 1986, Out of the Crisis, Cambridge MA:MIT
3. Wheeler, D.J., 1993, Understanding Variation: The Key to Managing Chaos, Knoxville,TN: SPC Press Inc
4. Gardener, N.J.L., 1999, Overcoming Behavioural Barriers to Adoption of Shewhart Control Charts as a Means to Business Development, MA Dissertation, University of Teesside
5. Woodcock, A. and Davis, M., 1978, Catastrophe Theory, London: Penguin Books

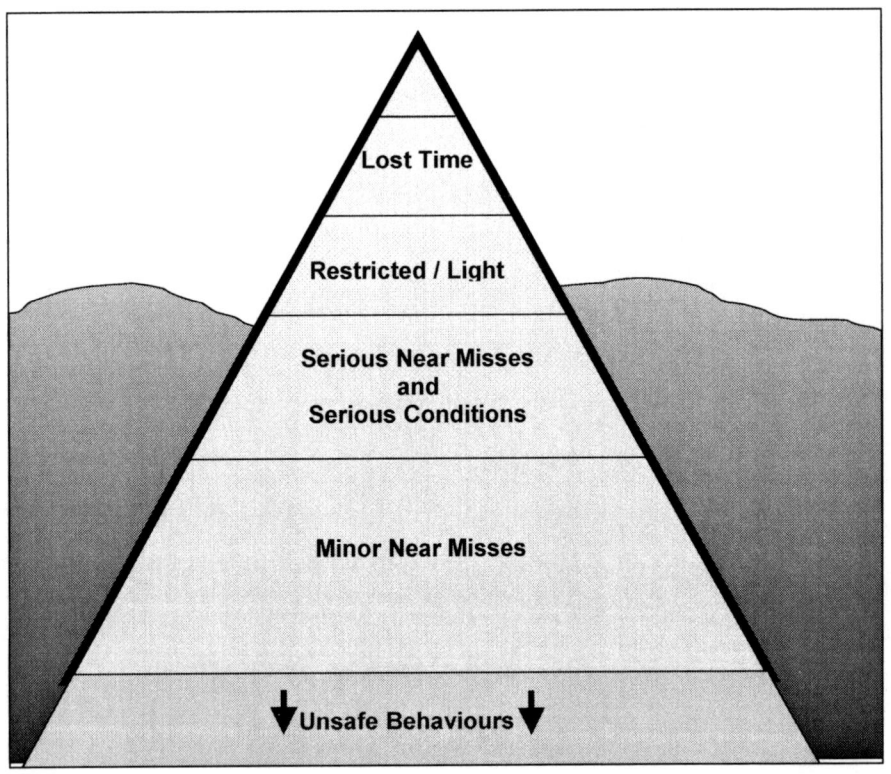

Figure 1. Iceberg pyramid for safety incidents

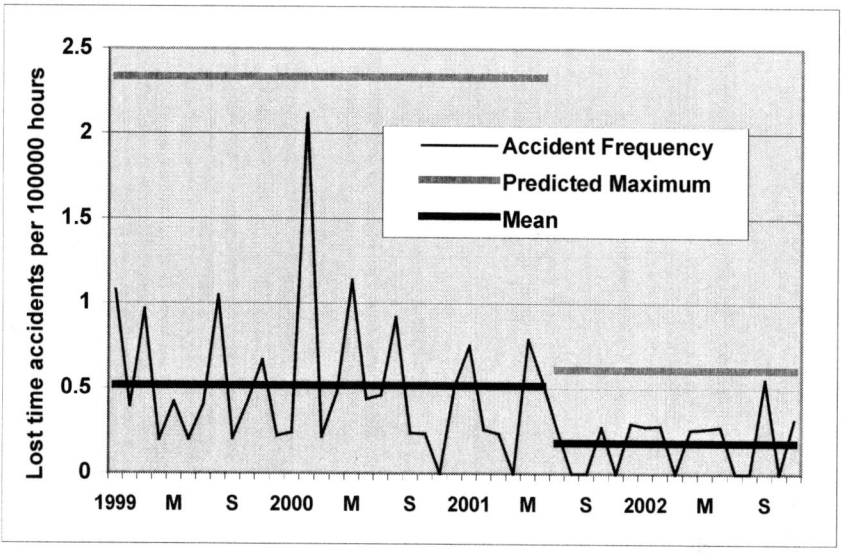

Figure 2.   Greater than 3-day lost time accident (LTA) rate

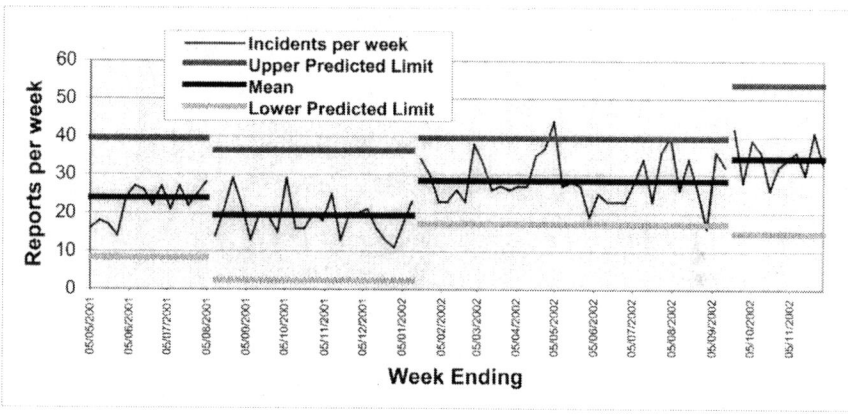

Figure 3.   Incident activity reporting

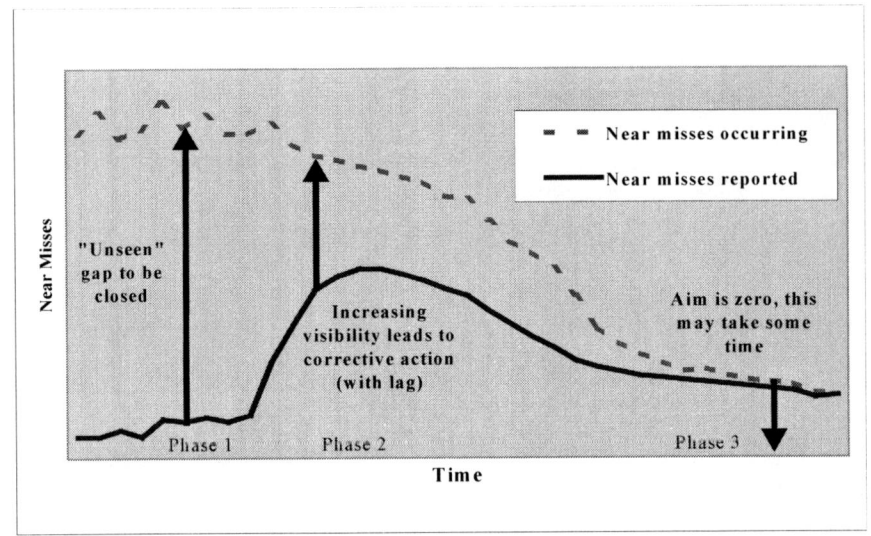

Figure 4. Near miss reporting – evolution

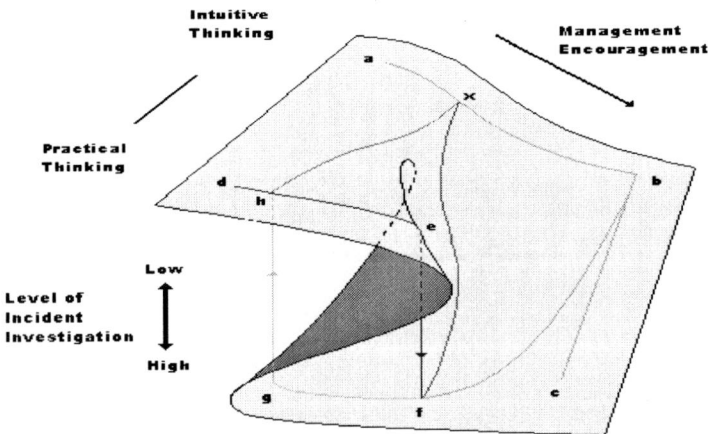

Figure 5. Catastrophe theory cusp surface for incident investigation

NOVEL PROCESS DESIGN METHODS TO ACCESS SAFER PROCESSING OPTIONS

Paul Sharratt[†], Kevin Wall[†] and John Borland[#]
[†]Department of Chemical Engineering, UMIST, [#]BRITEST Limited

The low tonnage organic chemicals sectors are distinct in having particular hazards that derive from the nature of the chemistries and the plant they use. The widespread use of inefficient batch technologies and relatively large inventories of materials are contributors to risk. While in principle, intensive processing techniques can be used to address these problems, adoption is slow. There are many reasons for this, but a central one is the failure of current design methodologies to identify early enough opportunities for more appropriate (and inherently safe) process technologies. The BRITEST™ project has developed a suite of tools that can be used in the early stages of process design to identify opportunities for more intensive and efficient processing. These involve the use of simple qualitative models to deduce favourable operating regimes and conditions. The cost and time requirements are consistent with the short duration and low cost development projects typical of the low tonnage sectors. Further, the qualitative models provide a useful framework to exploit the experimental data frequently collected for hazard assessment (typically calorimetric data). The methods are outlined and illustrated using a case study from a real industrial process.

KEYWORDS: Intensification; Process design; Safety

INTRODUCTION

It has long been recognised that inherent plant safety is improved by reducing the inventory of hazardous materials. This is particularly important in the low tonnage chemicals sectors (intermediates, agrochemicals, pharmaceuticals etc.), where batch processing is widely used. It is typical for batch process designers to slow down chemistry to match the poor processing capabilities of the stirred tank. The result is large volumes of process material that spend extended times (often days) in process, when an equivalent continuous process might need only seconds or minutes for the same transformation.

As well as the inefficiency of batch processes, the complex chemistry they typically involve tends to give rise to a greater scope for unintended reactions. Runaway reaction hazards are a major safety issue[1].

All other things being equal, lower inventories reduce the size of the worst-case incident. Inventory reduction can be achieved through logistical improvements and reductions in the amount of material in storage. It can also be reduced through the "intensification" of the processing operations. This may involve using higher reaction rates (by higher temperature, catalysis, increased concentration) accompanied by higher heat and mass transfer rates. By doing this, Process Intensification (PI) may bring additional benefits beyond inventory reduction:

- Reduced possibility of accumulation of materials in process;
- More rapid response to control actions;
- Better process performance;
- Cheaper plant.

PI is attracting increasing attention worldwide from industry, regulators and academe. However, the uptake of intensive technologies has generally been slow. There are several reasons for this. Firstly, much of the interest has been driven by "technology push" from equipment suppliers and designers. Intensive equipment has been a solution looking for a problem, often applied on a trial basis to look for improvements (rather than as a result of rational argument). Secondly, there is some perception of risk among potential users, who prefer to stick with familiar, tested technologies. Thirdly, intensification has focussed on the more glamorous areas, particularly reaction, rather than the feed preparation and product separation areas. These unglamorous areas often constitute the bulk of capital expenditure and materials inventory in a chemical process, making the potential benefits from intensification of the reaction section alone look rather small.

There have been some attempts at methodological approaches to PI. However, these have in general failed to attract widespread use. Problems with such methods may be:

- That they are not realistic within the resource limitations and business models of the target industry sectors;
- That they do not recognise the important of aligning methods with existing practices in industrial process design;
- That they assume that the solution will be intensive and don't consider the possibility that the existing approaches may be better against several key criteria;
- That they are not credible to industry.

In the low tonnage chemicals sectors, time to market is a key driver, and all process development activities are organised to reduce it. By assuming that a batch stirred tank is to be used, companies can work with standard process development methods and safety screening tools. There is little opportunity to evaluate other processing technologies, and in general laboratories are not set up to do this. Process development is primarily empirical, thus avoiding the time-consuming and expensive activities needed to obtain data for detailed physico-chemical modelling[1].

Under the current low tonnage design paradigm, the assumption that a batch stirred tank will be used immediately focuses attention on the (limited) range of process features that are of importance in the implementation of a batch process – mixing and agitation, heat transfer requirements, phase dispersion etc. Laboratory protocols are designed to deliver representations of batch tank performance in those duties. Many degrees of freedom that in principle exist (contacting pattern, residence times, phases present among others) are not accessible in traditional batch processes. For example, in a two liquid phase system, batch processes cannot deliver counter-current contacting of the liquids. However, it is often those degrees of freedom (rapid mixing, intense heat transfer and so on) that are the key elements of intensive options.

A new set of tools has been developed as part of the BRITEST™ project[2,3,4,5] to support the identification and adoption of innovative (including intensive) processes. The tools attempt to address the problems identified above by adopting a different philosophy - the methods start at the beginning of the problem rather than the end. The methods set out to identify the most appropriate process technology on the basis of the business need and the features of the chemical process. This is self-evidently sensible, and is broadly the practice in large tonnage chemical process design. It is, however, the antithesis of current practice in the batch sectors. The tools have been designed to work with minimal data, yet can identify potentially valuable intensive options very early in design. This maximises the chance that these options can be adopted.

The framework of tools addresses four issues, as illustrated in Figure 1. The order of activities is important. Under the BRITEST™ method, the process concept is developed independently of and before the plant concept. The process concept is developed first to identify the best outcomes that can be delivered, independent of plant constraints. In doing this, the essential processing capabilities (eg the ability to deliver intensive heat transfer or counter-current phase contacting) are identified. This in turn allows the selection of equipment based on its processing capability.

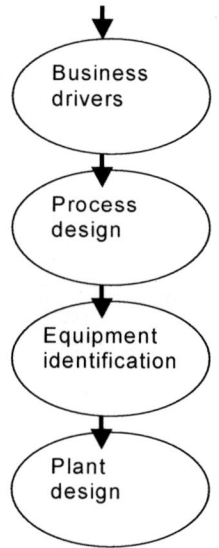

Figure 1.   Overall framework for process and plant design

Of course, this is exactly the type of approach that would be taken in the development of a new continuous, large-scale process. However, the development of such a process would require substantially more time and effort than is available for a low tonnage process. The novel feature of the new methods is the capability to develop process and plant concepts without the need for the detailed quantitative information typically used in continuous process design. This distinguishes the method from other published methods. The latter generally amount to little more than exhortations to think about intensification opportunities early in process development. In this sense, the methods go beyond the rather restrictive philosophy of intensification.

METHODOLOGY

This paper concentrates on the core aspects of process design and plant concept development. The collection and presentation of business drivers will not be discussed here. All that needs to be noted is that the key desired business outcomes of a project are collected and presented in a form that makes them useful to the design technologist. This is important because the method generates a range of options (rather than the very restricted options that arise from the batch paradigm). The process developers need to be able to screen options during design without reference back to the business manager. In what follows, the design of the reaction system will be emphasised in order to illustrate the safety benefits in what is essentially a reaction hazard case study.

The following underlying axioms are used in deriving the approach to reaction conceptual design.

- To obtain the best outcome from a process we need to deliver conditions that in broad terms maximise the rates of the desired processes and minimise the rates of undesired processes.
- Lack of detailed knowledge of the rate behaviour requires the law of mass action to be applied (unless other prior knowledge is available).
- The phases (gas, liquid solid) present in a process represent central degrees of freedom in setting out distinct process options. For a given chemistry (i.e. intended stoichiometric reaction set) there may be several possible phase combinations.

We consider a process to be the provision of an ordered and structured set of conditions applied to the processed materials. This sidesteps the "Unit Operation" concept, and thus the need to assume the type of processing device. We initially seek conditions that should deliver the best process performance, rather than accepting the limitations that are inherent in selection from existing process equipment.

The method involves the collection of two sets of data:

- reaction stoichiometry and associated information about the reactions and other rate processes; and
- information relating to phase transitions.

The first set allows the identification of strategies to improve the outcome of the process in terms of yield and volume efficiency. The second set allows the identification of a set of "phase strategies", or the phases present during the reaction.

The reaction information is presented in the form of a "Driving Force Table", as illustrated in Table 1. Here, each column represents a rate process. Rows represent either factors that influence the rate or key reaction parameters. The symbols "+" and "–" represent a positive or negative influence on rate. Thus, a "+" opposite a species indicates that a higher concentration of the species would be expected to increase the rate of that rate process. It is rare (contrary to popular academic opinion) that designers have detailed and accurate models of reaction kinetics, and frequently the set of reaction pathways will be incompletely understood. Further, the time and cost of detailed kinetic analysis is seldom justified under the pressure of rapid time to market. In contrast, the level of information required to complete the driving force table is realistic in the early stages of process development, and is typically assembled from prior knowledge, the literature and (minimal) experimentation.

The data required to suggest the range of possible phase behaviours are also quite limited – melting and boiling points (estimated if experimental data are unavailable) as well as solubility behaviour (based on molecular polarity for example). Typical data for a set of relevant materials are illustrated in Table 2. Of course, the detailed behaviour would need to be confirmed experimentally; but what is important at the early stages is the identification of possible, distinct strategies that may be broadly favourable to the process outcome.

CASE STUDY

The illustrative case study presented centres on a runaway reaction incident[6]. The process involved the diazotisation of an aromatic amine, followed by decomposition of the diazonium compound in the presence of water to produce a phenol. The case attracted attention from the HSE. Ultimately, the company was allowed to restart the process in a significantly modified batch facility at a cost of £1 million, with further costs associated with the incident reaching £1.7 million.

A set of reactions which represent the chemistry (relying on information from Sykes)[7] is given in Equations 1 to 6 below. Reaction 4 represents the reaction of ion R^+ with a range of other organic species, resulting in higher molecular weight species lumped under the name "tars". Note that this is not a chemically complete or proven representation of the reaction set, but is typical of the state of knowledge that technologists could bring to bear at the early stages of development. The validity of the set of equations would be tested through the experiments during the process development activity. While the BRITEST™ methods assist in the definition of appropriate experiments, this aspect is outside the scope of this paper. It is also important to note that the published information is restricted to avoid revealing confidential details of the process – compositions, detailed conditions etc.

The data used to represent reaction rates was based on some of the calorimetric data presented in the original paper[6]. Such data would be readily accessible at reasonable cost to a competent development team, and the data needs are much less than would be required to elucidate the detailed kinetic behaviour of each reaction.

The driving force diagram that corresponds to the reactions is in Table1.

$$H_2SO_4 + RNH_2 \leftrightarrow RNH_3^+HSO_4^- \tag{1}$$
$$RNH_3^+HSO_4^- + NSA \rightarrow RN_2^+ HSO_4^- + H_2O + H_2SO_4 \tag{2}$$
$$RN_2^+ HSO_4^- \rightarrow R^+ + N_2 + HSO_4^- \tag{3}$$
$$R^+ + \text{organic species} \rightarrow \text{tars} \tag{4}$$
$$R^+ + H_2O \rightarrow ROH + H^+ \tag{5}$$
$$RN_2^+ HSO_4^- + RNH_2 \rightarrow Azo \tag{6}$$

Data concerning the phase behaviour of the system are given in Table 2. The identity of the aromatic amine is not given in the paper, so for the purposes of this study it will be assumed to be solid at room temperature, with a melting point of 120°C. On the basis of these data, a range of "phase strategies" can be developed for the key synthetic reactions. This is done by simple enumeration of the possible combinations of phases for the components present as a function of the relevant variables (such as temperature). Some credible options for the case study are shown in Table 3.

While the paper refers to a process that had been carried out in batch plant, the information in the paper will be used as if it was part of a development process aimed to design a new plant, making no assumptions about the technology to be used.

Given the information in the tables, it is possible to develop processing options and plant concepts for the system. Option selection would rely on the fit with the previously identified business drivers. In the case study, the drivers would include cost issues, but also safety considerations for a reaction type that is known to pose significant hazard.

At least three distinct modes of operation for reaction 1 can be suggested. Any of them could be operated either batchwise or in continuous mode. The identification of "phase options" related directly to some requirements on the processing capabilities of the equipment. For example, the presence of a dissolving solid implies the needs to suspend and disperse the solid. For reaction 1, the main options that arise are as follows.

- Direct mixing of acid and solid amine. This would involve mixing, dispersion of the solid and the removal of heat (the temperature needing to be controlled as the next reaction cannot be allowed to run at too high a temperature). While continuous operation would be possible, the addition of solids into a liquid continuously is a difficult processing step.
- Melting the amine and dispersing it as a liquid in the acid. The amine and acid are mixed and reacted with simultaneous heat removal. Batch operation of this option would have no obvious advantage over the solid amine option, and would raise additional difficulties in heat transfer, but continuous operation in a heat-exchanger reactor is attractive.
- Dissolving the amine in an inert carrier solvent, toluene for example, and the two-liquid phase contacting of the acid and amine solution. The reaction product would dissolve in the acid, leaving the inert solvent to be recycled.

For reaction 1, there are no known side reactions, no and so long as the reaction is completed before the NSA is added, loss of yield issues are not relevant. It would be possible to produce the amine salt solution for storage before use, or equally continuously as needed.

The initiation of reaction 2 by mixing NSA with the amine salt solution has the potential to generate all of the subsequent reactions 3–5. It would be possible to consider suppressing those reactions by running the reaction at low temperature. The practicability of this would depend on the balance of reaction rates and the temperature dependency, but could readily be tested experimentally. Reaction 2 could be carried out batchwise or continuously. Its moderate rate would suggest that continuous operation would be suitable – for example the manufacture of 100 tonnes of diazonium salt per year can crudely be estimated using the published calorimetry data6 to require a reactor volume of a few tens of litres. Given the need to remove heat, the ideal equipment would seem to be another heat-exchanger reactor.

Table 1. Driving force table for the diazotisation synthetic reaction set. Question marks indicate uncertain or missing information

Driving force	Reaction 1	Reaction 2	Reaction 3	Reaction 4	Reaction 5	Reaction 6
H_2SO_4	+	·	·	·	·	·
RNH_2	+	·	·	·	·	+
$RNH_3^+HSO_4^-$	−	+	·	·	·	·
NSA	·	+	·	·	·	·
$RN_2^+HSO_4^-$	·	·	+	·	·	·
H_2O	·	·	·	·	+	·
R^+	·	·	·	+	+	+
N_2	·	·	·	·	·	·
$ROH^\cdot$	·	·	·	·	·	·
organic pecies	·	·	·	+	·	·
Azo	·	·	·	·	·	·
Temperature	+	+	++	+	+	+
PH	·	?	?	?	?	++
Heat of reaction	Moderate exotherm	Moderate exotherm		Strong overall exotherm		Exotherm ?
				For these reactions taken together		
Representative reaction time	Very fast (order of ms?)	Few Minutes	Minutes (?)	Fast – Less than seconds	Fast – Less than seconds	Probably fast(?)

Table 2. Typical initial information on phase behaviour

Species	Melting point/°C	Boiling point/°C	Polarity
H_2SO_4	Low	>>100	High
RNH_2	120	High	Low-moderate
$RNH_3^+HSO_4^-$	High	Decomposes	High
NSA	Low	?	High
$RN_2^+ HSO_4^-$	Decomposes	-	High
H_2O	0	100	Moderate-high
R^+	n/a	n/a	High
N_2	n/a	permanent gas	Low
ROH	c120	high	Low-moderate
organic species	>100	high	Low-moderate
Azo	high	Decomposes?	Moderate

Table 3. Selection of suggested phase strategies for the reactions

Reaction	Phase options
1	Solid amine, liquid acid **or** Molten amine (>120°C), liquid acid **or** Solution of amine, liquid acid **or** Others…
2	Homogeneous liquid phase **or** Gas/liquid (water as vapour >c100°C)
3–6*	Homogeneous liquid phase + gas N_2 **or** Gas water as vapour + gas N_2)/liquid **or** Gas water as vapour +gas N_2)/liquid/liquid (second immiscible phase with ROH)

*Given the reactivity of R^+ it is not likely that its generation and reaction can be separated.

It is likely that the conditions required to drive high selectivity for reaction 5 would be a high concentration of water. This would suggest quenching of the diazonium compound from reaction 2 into water, ideally with a low concentration of the diazonium to minimise the concentration of organic species available for coupling with R+. This would generate aqueous waste and the cost of waste disposal would be balanced against any yield benefits. It is also obvious from Table 1 that slow mixing may increase the yield of crud.

Experimental work would be needed to identify the best conditions (water concentration) and to define the intensity of mixing required.

The equipment for reactions 3–6 needs to mix intensely, remove heat and to decant the phenol phase. Heat removal and mixing may need to be carried out simultaneously, but there appears to be no reason to carry out phase separation in the same equipment. As reaction 6 is suppressed under acid conditions it would not be significant in the conditions envisaged.

PLANT OPTIONS

The process could clearly be carried out in a batch system, as was done by Hollidays both before and after the incident. However, it is evident that a continuous plant using intensive mixers, heat exchanger reactors and continuous phase separation could deliver all of the primary functions identified. The largest volume may be associated with the diazotisation reaction, and even with a residence time of, say, 10 minutes, this would only have a volume of tens to hundreds of litres to give a production rate of hundreds of tonnes per year. One potentially viable continuous configuration is given in Figure 2.

The existing process prior to the incident was in standard batch equipment. As a result of investigations after the incident, a new configuration was developed, the changes centred on the instrumentation to avoid the accumulation or unreacted materials. It is interesting to note that even in the face of serious safety issues, and the possibility of operating the process continuously at low inventory, that the traditional style of process analysis did not result in a continuous process being developed.

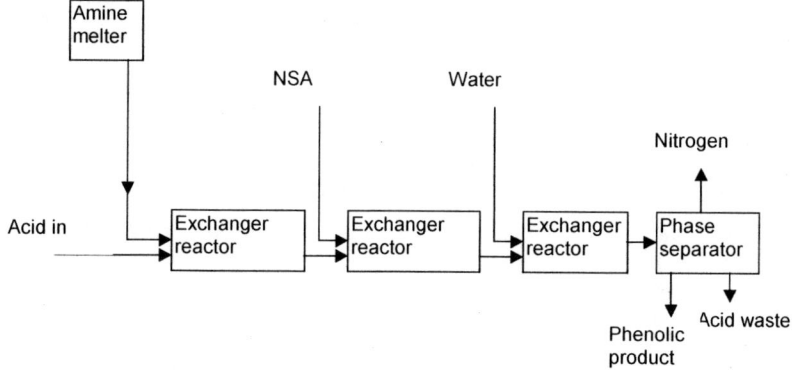

Figure 2. Block diagram representing a continuous process option

DISCUSSION

The analysis above demonstrates that relatively little information, and information of only moderate cost can be used to deduce innovative plant options. The option presented above is one of many that might be suggested, and is chosen to illustrate how easily a radically different approach can be identified. Of course, the identification of a potential processing option would require experimental work to support its development and assessment. However, the early identification of possible options allows timely and appropriate experimentation to be defined. This is a shift away from the prevalent approach in low tonnage process design – that a stirred tank will be used and that experimentation is all carried out to support that decision.

The operation of an approach to design like that in Figure 1 would quickly identify the safety benefits associated with an intensive continuous design. The potential reaction hazards of the process are evident, and would easily have been picked out as important selection criteria. The clear and timely identification of process options that meet such safety criteria inevitably increases the chances of their adoption, as well as potentially reducing the cost and time requirements for implementation of a safe process.

It is important to recognise that while methodologies such as this can improve the outcomes of process and plant design, they do present companies with significant problems in their implementation. Difficulties are not only technology related but also include culture change, risk perception and the need to have a well designed implementation strategy. Nevertheless, the high costs that arise when processes go wrong, as well as the potential benefits in process efficiency, reduced effluent and reduced capital cost provide a strong argument to change.

ACKNOWLEDGEMENTS
The financial support of EPSRC (Grant GR/L65956/01), the UK Department of Trade and Industry and the collaboration of Britest Limited and its sponsor companies is gratefully acknowledged.

REFERENCES
1. Sharratt PN (Ed), 1997, "Handbook of batch process design", Blackie A&P, London, ISBN 0-7514-0369-5
2. BRITEST Limited, 2002, http://www.britest.co.uk/
3. Wall K, Sharratt PN, Sadr-Kazemi N and Borland JN, 2000, "Plant-independent Process Representation", Computer Aided Chemical Engineering 8: European Symposium on Computer Aided Process Engineering-10, Pierucci S., (Ed.), Elsevier Science BV., ISBN 0-444-50520-2
4. Wall K, Sharratt PN, Sadr-Kazemi N and Borland JN, 2001, "Plant-independent Process Representation", Org Proc Res & Dev, 5 (4): 43–437 Jul–Aug 2001
5. Sharratt PN, Wall K and Borland JN, 2001, "Crossing the border - chemistry and chemical engineering in process design", Proc 6th World Congress of Chemical Engineering, Sept 22nd–27th, Melbourne, Australia
6. Partington S and Waldram S, 2001, Runaway reaction during production of an azo dye intermediate, IChemE Symposium Series 148:81–93
7. Sykes P, 1970, A guidebook to mechanism in organic chemistry, 3rd Edn. Longman

A MODIFICATION TO THE K_G METHOD FOR ESTIMATING GAS AND VAPOUR EXPLOSION VENTING REQUIREMENTS

G A Lunn and D K Pritchard
Health and Safety Laboratory, Harpur Hill, Buxton, Derbyshire SK17 9JN.

Measurements of the reduced explosion pressures in vented gas explosions in compact and elongated vessels are reported. Measurements in enclosed vessels of the K_G factor are also reported. The measured explosion pressures are compared to values predicted by the K_G method equation given in NFPA 68. The results indicate that modifications can be made to the K_G method to produce more realistic pressure predictions. The same equation is used in vent area calculations but variants of the K_G factor are substituted in the equation. These variants characterise vented explosions more successfully than the usual values of K_G. This is because the usual measurement of the K_G factor takes place at a late stage in an enclosed explosion and in a relatively small volume and these are inappropriate conditions for characterising vented explosions in which pressures do not reach high values and when vessels have a larger volume. The modified K_G method uses K_G values derived from the present tests and satisfactorily predicts measured reduced explosion pressures for vessels with L/D ratios up to 5.

KEYWORDS: venting, gas explosions, vapour explosions, K_G factor

INTRODUCTION

Explosion relief venting is a technique for mitigating the effects of a gas or vapour explosion in items of industrial plant. It involves fitting into the walls of a vessel weak panels that burst early in the explosion and allow gas and combustion products to vent into the open air. As a result the internal pressure rise is considerably reduced, as is the potential for damage to the vessel and injury to people.

To apply the technique the size of the vent and its failure pressure have to be chosen so that the maximum internal pressure, the reduced explosion pressure (P_{red}), is below that which would cause failure of the enclosure. It may be acceptable to allow some damage to the enclosure, for example bowing of metal panels, provided this does not result in catastrophic failure of the vessel. Incorrectly sized relief vents will result in either the plant not being properly protected or impracticable or unnecessarily expensive vents being fitted.

There are various methods available for sizing explosion relief, ranging from empirical formulae and nomograph type methods to complex mathematical models of the venting process[1,2].

Recently, the Health and Safety Executive has funded a programme of work to generate data on vented explosions that can be used to assess the accuracy and range of application of vent sizing methods and to develop improved guidance where appropriate. This paper reports the analysis of the experimental data in a series of comparisons with vent sizing methods and the development of a modification of the K_G method for vent sizing.

EXPERIMENTAL

The data generated consists of reduced explosion pressures measured during gas explosions in vented compact, elongated and connected cylindrical vessels.

Three compact vessels were used, 2, 6.3 and 20 m^3 in volume. The effects of gas, gas concentration, vent area, bursting pressure of the vent, ignition strength, ignition position and the presence of internal obstacles on the reduced explosion pressure were studied[3].

The elongated vessels were cylindrical, with a diameter of 1.1 m and Length to Diameter (L/D) ratios in the range 5.5 to 11.8, and rectangular with a cross-section of 2.5 m ×2.5 m and Length to Width (L/W) ratios of 3 to 12[4].

The linked vessels were the ones used in the compact vessel tests. The effects of volume, volume ratio, diameter and length of connecting pipe, ignition position and distribution of vent area on the reduced explosion pressures were studied[5].

In addition, the effect of the vessel volume on the K_G factor has been studied. The K_G factor is a parameter, calculated from the equation

$$K_G = V^{1/3} (dP/dt)_{max}, \tag{1}$$

where $(dP/dt)_{max}$ is the maximum rate of pressure rise measured in a contained explosion at the optimum gas or vapour concentration in air (bar sec.$^{-1}$), and V is the vessel volume (m^3).

The K_G factor is essentially equal to the maximum rate of pressure rise at the optimum gas concentration measured under standard test conditions in a 1 m^3 vessel and can be used to estimate vent areas by using either Nomographs or equations. Pressure-time histories from enclosed gas explosions in vessels of 20 litres, 2, 4 and 20 m^3 in volume were analysed[6,7].

The outcome of these studies was a mass of data that quantified the dependency of the reduced explosion pressure on the different variables that can affect it. The analysis in this paper is limited to the compact and elongated vessel results.

DISCUSSION

COMPACT ENCLOSURES

Comparison of Predicted and Measured Reduced Explosion Pressures

The pressure-time histories of explosions in compact vessels exhibited a multi-peak structure, with usually three peaks. The first peak (P_1) is a result of the vent starting to open, and has a value approximately equal to the vent burst pressure. A second peak occurs if and when the rate of volume production due to combustion exceeds the volume rate of outflow through the vent. The second peak pressure, P_2, then occurs when the rate of combustion declines after reaching its maximum value and a situation is again reached where the rate of outflow exceeds the rate of volume production. The third peak (P_3) is acoustically induced and was, in most cases, the largest peak of the three. This peak is capable of causing damage to enclosures if the venting arrangement does not take it into account. In nearly all the HSL tests this large final peak was observed.

The measurements of maximum reduced explosion pressures obtained by HSL for near cubical vessels have been compared to predictions from a number of published vent sizing techniques[3]. In general it was found that the empirical methods, such as those of Cubbage and Simmonds[8] and Rasbash[9], underpredicted the peak reduced explosion pressures. The predictions of the K_G method published in NFPA 68[10], Bradley and Mitcheson's method[11] and Molkov's method[12], on the other hand, were in reasonable agreement with some of the experimental values. These comparisons generally show, however, that no calculation method gives accurate estimates of the reduced explosion pressures under all circumstances.

Modified K_G Factor Approach to Vent Sizing

Of the simpler vent sizing methods, the K_G factor approach from NFPA 68[10] appears to be the most flexible, but it requires modification if it is to be successful. It is clear from the qualifications written down in NFPA 68 that K_G is far from being a universal constant and that it can be affected strongly by alterations to important variables.

In this section a proposal for using a modified K_G method for estimating vent areas is described, based on the measurements of reduced explosion pressures in vented vessels and the K_G factor measurements derived from the enclosed explosion tests performed by HSL.

The original K_G method, based upon the cube root law, was in the form of nomographs for a number of gases (methane, propane, coke gas and hydrogen), that allowed the vent area to be calculated from a knowledge of the vessel volume, the maximum reduced explosion pressure allowed and the opening pressure of the vent. For other gases the vent area could be found by interpolation of the results from the four reference nomographs, provided the K_G value for the gas was known. These nomographs were developed for initial conditions of:

- No initial turbulence in the enclosure at the time of ignition
- No turbulence-producing internal obstacles
- An ignition energy of 10 J or less
- Atmospheric pressure in the vessel prior to ignition.

As an alternative to using the nomographs equation 92) can be used for L/D values of two or less:

$$A_v = [\{(0.1265 \log_{10} K_G - 0.0567) P_{red}^{-0.5817}\} + \{0.1754 P_{red}^{-0.5772} (P_{stat} - 0.1)\}] V^{2/3} \quad (2)$$

The equation is valid for the following conditions:

$K_G \leq 550$ bar m s^{-1}
$P_{stat} \leq 0.5$ bar
$P_{red} \leq 2$ bar and $\geq P_{stat} + 0.05$ bar
$V \leq 1000$ m^3
Initial pressure before ignition of <0.2 bar

When the L/D ratio extends from 2 to 5, the vent area calculated from equation (2) has to be increased. The additional vent area ΔA is calculated from equation (3):

$$\Delta A = [A_v K_G (L/D - 2)^2]/750 \quad (3)$$

and applies when P_{red} is no higher than 2 bar.

K_G values are measured during a standard test but it cannot be considered as a constant that is unchanging with vessel volume. Data in NFPA68 show that the value of K_G for a given gas increases if it is derived from pressure-time traces measured in vessels of increasing volume[10]. The graph in NFPA68 is incorrectly labeled, however; the volume scale should read litres, not m^3.

Nevertheless, the K_G factor method is a relatively simple method to use, even if its straightforward application is not always successful if standard values of K_G are applied under all circumstances. This is because the K_G factor is measured at a point on the pressure-time curve where the explosion overpressure is several bar, well above the pressures likely to be reached in a vented explosion. If the reduced explosion pressure in a vented explosion does not exceed 0.5 bar, say, then it is inappropriate to apply the standard K_G factor, because it is a measure of the rate of pressure rise in circumstances that never occur in the vented explosion. A variation of the K_G factor based on the pressure-time history at a much earlier stage in the closed vessel explosion is likely to be more appropriate. Similarly, as the K_G factor is known to vary with vessel volume, a variation of the K_G factor relevant to larger volumes is more appropriate to vented explosions in larger volumes.

Thus, if a range of K_G factors can be specified, derived from pressure-time histories measured in enclosed vessel explosions, it may be possible to characterise a range of vented explosions with more flexibility than by simply using the standard K_G value.

Based on this idea three variations of the K_G factor have been used in an analysis of HSL's experimental results:

$K_{G(20 \text{ litre})}$ is the K_G factor measured in the 20 litre sphere in a standard test and is shown to characterise vented explosions in small volumes in circumstances where the reduced explosion pressure is relatively high.

K_{GV} is the K_G factor measured in an enclosed vessel with the same volume and L/D ratio as the vented vessel, and is shown to characterise vented explosions in larger vessels in circumstances where reduced explosion pressures are relatively high eg ignition remote from the vent.

$K_{GV(0.5)}$ is the K_G measured at 0.5 bar g in an enclosed vessel of the same volume and L/D ratio as the vented vessel, $K_{GV(0.5)} = V^{1/3} \times (dP/dt)_{(0.5 \text{ bar overpressure})}$, and is shown to characterise explosions in larger vessels in circumstances where the reduced explosion pressures are relatively low eg central ignition.

The values of the K_G variants measured in HSL's enclosed explosion tests are given in Tables 1 and 2.

Table 1. The $K_{G(20 \text{ litre})}$ and K_{GV} values

Gas	20 l sphere $K_{G(20 \text{ litre})}$[1]		2 m^3 vessel K_{GV}	4 m^3 vessel K_{GV}	20 m^3 vessel K_{GV}
	Kuhner*	HSL**	HSL	HSL	HSL
Ethylene	206	158	219	117	132
Propane	104	79	318	52	60
Methane	65	61	61	30	33

* Measurements reported by Kuhner.
** Measurements by HSL.

Table 2.   The $K_{GV(0.5)}$ values for each vessel at a pressure of 0.5 bar gauge

Gas	20 l sphere	2 m³ vessel	4 m³ vessel	20 m³ vessel
	HSL	HSL	HSL	HSL
Ethylene	24	30	32	40
Propane	12	15	16	29
Methane	10	11	12	15

The predicted pressures in the following analysis are obtained by applying equation (2) for the K_G vent sizing method substituted by either K_{GV}, $K_{G(20\ litre)}$ or $K_{GV(0.5)}$, as most appropriate to the experimental conditions and the behaviour of the vented explosion. The effect of vessel length to diameter ratio has been ignored in these calculations.

The reduced explosion pressures are the maximum values measured in each test, regardless of which pressure peak this is. The second peak is combustion controlled, whereas the third peak is influenced by the acoustic resonance of the vessel. Equation (2) is an empirical expression, however, and makes no distinction between these different mechanisms. The test here is whether the worst case pressures can be successfully predicted by Equation (2).

Comparisons between predictions and measurements are shown in figures 1 to 5, in which reduced explosion pressures are plotted against the vent coefficient, $K = (V^{2/3}/A_V)$,

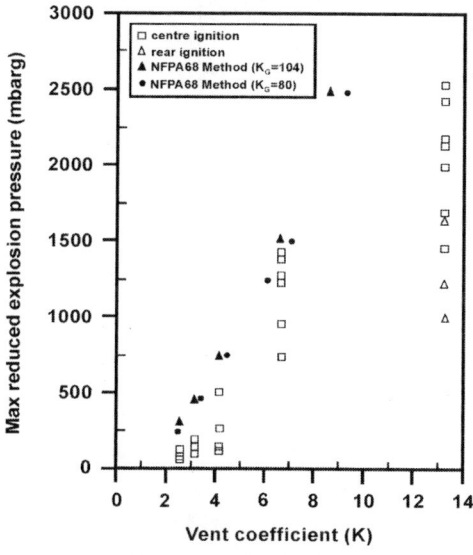

Figure 1.   Comparison of predicted and measured maximum reduced explosion pressures for propane/air mixtures in a 2 m³ vessel

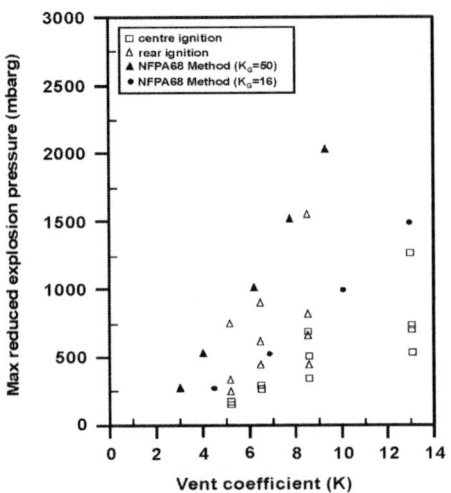

Figure 2. Comparison of predicted and measured maximum reduced explosion pressures for propane/air mixtures in a 6.3 m³ vessel

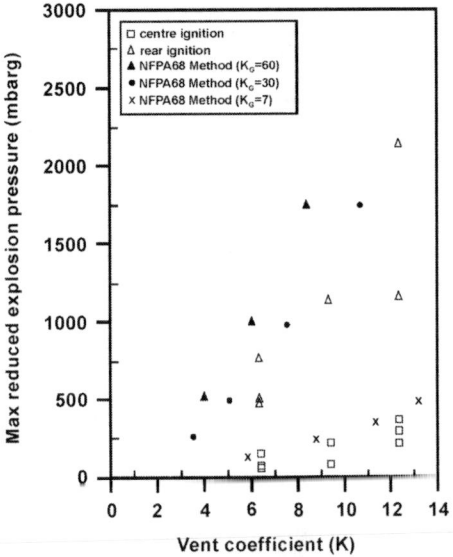

Figure 3. Comparison of predicted and measured maximum reduced explosion pressures for propane/air mixtures in a 20m3 vessel

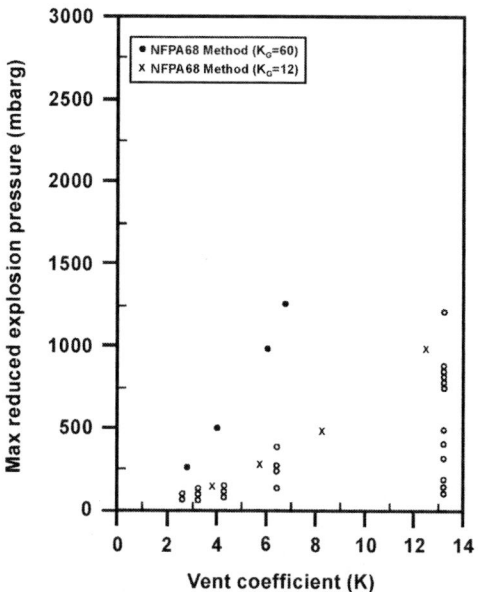

Figure 4.   Comparison of predicted and measured maximum reduced explosion pressures for methane/air mixtures in a 2 m³ vessel

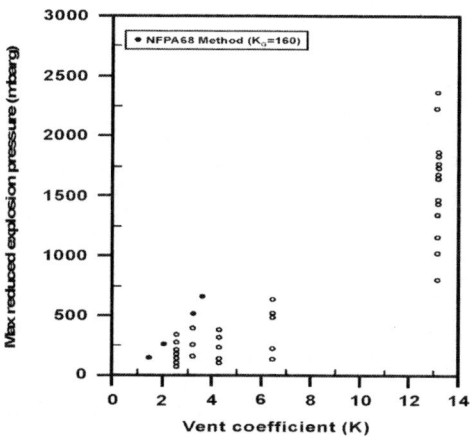

Figure 5.   Comparison of predicted and measured maximum reduced explosion pressures for ethylene/air mixtures in a 2 m³ vessel

where V is the vessel volume and A_V is the vent area. The K_G values used in the predictions are taken from Tables 1 and 2, using the value of K_{GV} or $K_{GV(0.5)}$ measured in circumstances that are closest to the conditions of the vented test. For example, K_{GV} and $K_{GV(0.5)}$ values from 4 m^3 enclosed explosions are used when predicting the vented explosions in the 6.3 m^3 vented tests.

These comparisons show that as the characteristics of a vented explosion are changed by such parameters as vessel volume and the K factor, then the value of K_G that characterises them also alters. If reduced explosion pressures are low then K_G values relevant to the early stages of an explosion are more in tune with the explosion behaviour in the vented explosion. If reduced explosion pressures are relatively high, K_G values derived from the later stages of an explosion are more appropriate.

Figure 1 shows the measured reduced explosion pressures for propane explosions in a vented 2 m^3 vessel and the calculated reduced explosion pressures using a $K_{G(20\ litre)}$ factor of 104 bar m s^{-1} and 80 bar m s^{-1}. The former value is the value recommended by Kuhner, the manufacturers of the 20 litre test apparatus; the latter the value measured by HSL in the 20 litre sphere (see Table 1). Both sets of predictions satisfactorily envelop the measured pressures.

Figure 2 shows measured reduced explosion pressures in propane explosions in the 6.3 m^3 vessel. Differences in these pressures arise depending on the position of the ignition source. For rear ignition, higher pressures are produced compared to when ignition is at the vessel centre. Comparisons are made using K_G values of 50 bar m s^{-1} and 16 bar m s^{-1}. The former is K_{GV} as measured in the enclosed 4 m^3 vessel and the latter is $K_{GV(0.5)}$ as measured at a 0.5 bar g explosion overpressure. Predictions using 50 bar m s^{-1} are a satisfactory envelope for the pressures generated with rear ignition and those using 16 bar m s^{-1} are a satisfactory envelope for the pressures following central ignition.

Figure 3 shows measured reduced explosion pressures from propane explosions in a 20 m^3 vessel. Predictions using K_G values of 60 bar m s^{-1}, 30 bar m s^{-1} and 7 bar m s^{-1} are also shown. The first two of these values are K_{GV} and $K_{GV(0.5)}$ derived from measurements in a 20 m^3 vessel (see Tables 1 and 2). The predictions using 60 bar m s^{-1} are conservative compared to measurement whereas those using 30 bar m s^{-1} are a satisfactory envelope for pressures generated following rear ignition. When the ignition source is at the centre of the vessel, a K_G value of 7 bar m s^{-1} satisfactorily envelops the experimental results.

Figure 4 shows measured reduced explosion pressures for methane explosions in a 2 m^3 vessel compared to predictions using K_G - values of 60 bar m s^{-1} and 12 bar m s^{-1}. These K_G values are the K_{GV} and $K_{GV(0.5)}$ values measured for methane in the 2 m^3 explosion vessels. A K_G value of 12 bar m s^{-1} gives close predictions of the higher reduced explosion pressures in the experimental results and clearly characterises these vented methane explosions.

Figure 5 shows the measured reduced explosion pressures for ethylene explosions in the 2 m^3 vessel. A comparison is made with predictions using a K_G value of 160 bar m s^{-1}, the K_{GV} value measured in the 2 litre vessel. The predictions of reduced explosion pressure satisfactorily envelop the measured values up to 0.5 bar overpressure; at higher reduced explosion pressures, the predictions are much higher than the measurements, although the experimental points are few in this region.

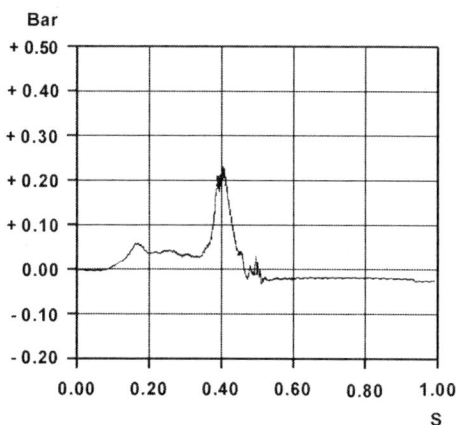

Figure 6.   Pressure-time history: methane/air, 2m3 vessel, K factor = 6.5

The K_G approach to vent sizing implies that for a given K factor and a given K_G, vented explosions should produce the same reduced explosion pressure regardless of the vessel volume. In the present analysis, a K_G of 12 bar m s^{-1} successfully predicts the pressures in methane explosions in a 2 m^3 vessel while 16 bar m s^{-1} does the same for propane explosions centrally ignited in a 6.3 m^3 vessel. The pressure-time trace in Figure 6, for a methane explosion in the 2 m^3 vessel, for a K value of approximately 6.5 and central ignition, can be compared to the pressure-time trace shown in Figure 7 for a propane explosion in the 6.3 m^3 vessel with central ignition. The traces have similar shapes and values of the pressure peak, although the trace from the 2 m^3 vessel is of shorter duration, as would be expected. This similarity indicates that similar types of vented explosion in different vessels and with different gases are characterised by a given value of an appropriate value of K_G.

When the volume of the vented vessel increases, explosions tend to be slower and are characterised by decreasing values of K_G. An illustration of this is shown in Figures 7 and 8.

Figure 7 shows pressure traces for propane explosions in the three vented vessels at a K factor of approximately 6.5 with conditions for the worst case reduced explosion pressures, i.e central ignition in the 2 m^3 vessel, rear ignition in the other volumes. The K_G factors that predict these reduced explosion pressures are 80 bar m s^{-1}, 50 bar m s^{-1} and 30 bar m s^{-1} for, in order of vessel size, 2 m^3, 6.3 m^3 and 20 m^3.

Figure 8 is similar to Figure 7 except that the ignition position is central in all cases. The K_G factors that predict these reduced explosion pressure are 80 bar m s^{-1}, 16 bar m s^{-1} and 7 bar m s^{-1}.

Figure 9 shows a pressure-time trace from an ethylene explosion in the 2 m^3 vessel with a K value of approximately 6.5. Although the reduced explosion pressure is relatively low, the explosion is much faster than with propane and methane under similar conditions.

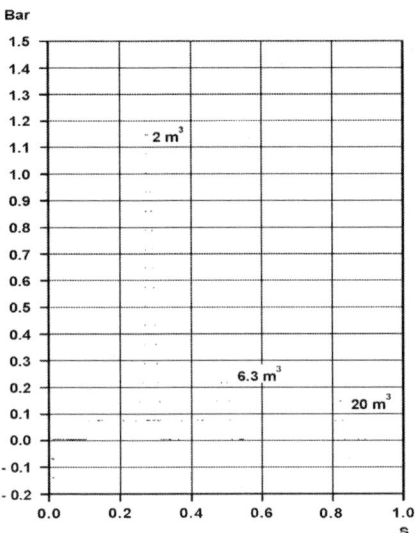

Figure 7. Pressure-time histories: 5% propane/air, central ignition, K factor = 6.5, various volumes

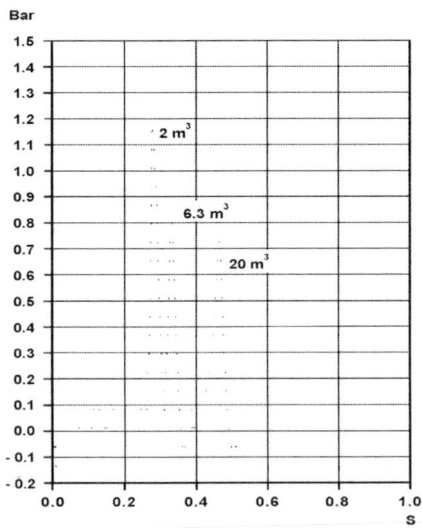

Figure 8. Pressure-time histories: 5% propane/air, rear ignition, K factor = 6.5, various volumes

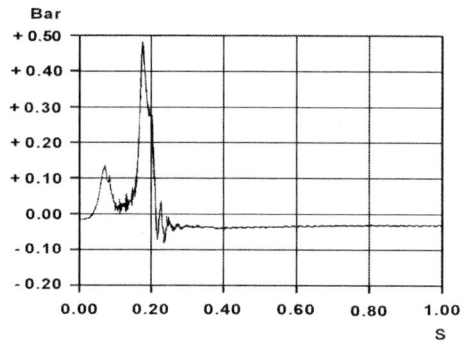

Figure 9.   Pressure-time history: ethylene/air, 2 m^3 vessel, K factor = 6.5

This effect is the opposite of that described in NFPA68, where the K_G factor is shown to increase as vessel volume increases. This is because the K_G factor in NFPA68's data characterises enclosed explosions; but, as the data from vented explosions show, these values cannot be transferred across to vented explosions without modification.

In order to apply the modified K_G approach to other data the analysis of HSL's explosion tests has been summarised as shown in Table 3. The results suggest the trends shown in Figures 10 and 11 can be used to estimate values of $K_{GV(0.5)}$ at intermediate volumes.

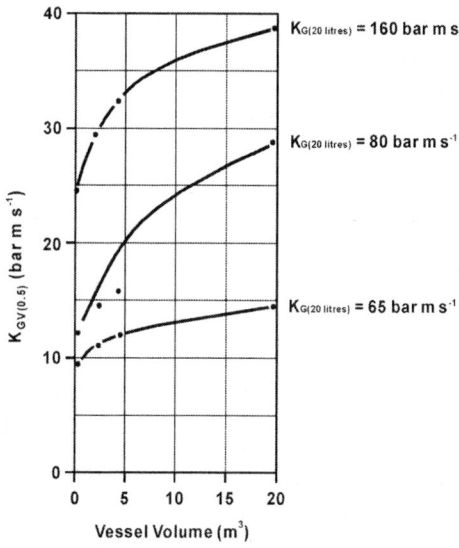

Figure 10. $K_{GV}(0.5)$ values as a function of vessel volume

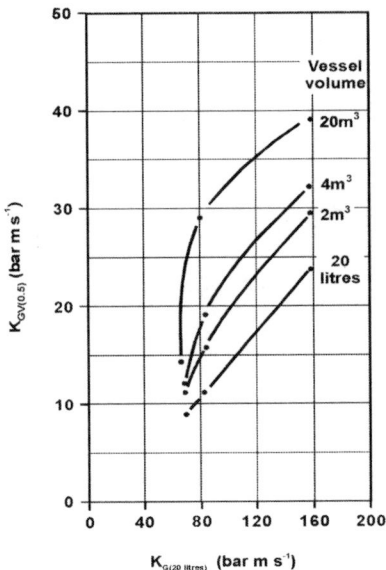

Figure 11. $K_{GV}(0.5)$ values as a function of $K_{G(20\ litres)}$ and vessel volume

Figure 10 shows the effect of vessel volume on $K_{GV(0.5)}$, at three values of $K_{G(20\ litre)}$ based on the HSL determinations from Table 1. For a known volume, three values of $K_{GV(0.5)}$ can be read from Figure 10 corresponding to the three values of $K_{G(20\ litre)}$. This data can then be plotted on Figure 11, which shows how $K_{GV(0.5)}$ changes with $K_{G(20\ litre)}$ and vessel volume, using the lines already there as a guide to connect the three points. For a vessel of known volume and a gas of known $K_{G(20\ litre)}$ value, the $K_{GV(0.5)}$ value can then be read off from the vertical axis.

Because the analysis shows that for the same degree of venting the appropriate value of K_G decreases as the vessel volume increases, data for K_{GV} and $K_{GV(0.5)}$ for vessels of a lower volume can be used if data is unavailable for a specific vessel size. The suggested modifications in Table 3 have been used to calculate reduced explosion pressures for comparisons with published data.

When the scheme given in Table 3 is followed using the K_G values from Tables 1 and 2 that most closely apply to the test conditions, the comparisons with published data listed in Table 4 are obtained. The predicted reduced explosion pressures are either close to or in excess of the measured values. Generally, the predictions for propane are satisfactory, and for methane the predictions mostly exceed the measured values and in those examples where there is an under-prediction the difference is small. Calculations using a K_G factor typical of hydrogen from a 20 litre sphere measurement produce reasonable predictions of P_{red} in these examples.

Table 3. Modifications of the K_G method

Gas reactivity (as measured by the K_G value in a 20 litre test, $K_{G(20\ litre)}$)	Vessel volume, V (m^3)	K_G value to be used for P_{red} prediction
$K_{G(20\ litre)} \leq 65$ bar m s^{-1}	$V < 2$	$K_{G(20\ litre)}$ as measured in the 20 litre sphere
	$2 \leq V$	$K_{GV(0.5)}$ as measured in a vessel with the same volume
$65 < K_{G(20\ litre)} \leq 80$	$V < 6$	$K_{G(20\ litre)}$ as measured in the 20 litre sphere
	$6 \leq V < 20$	Rear Ignition : K_{GV} for a vessel of the same volume Central Ignition: $K_{GV(0.5)}$ in a vessel of the same volume
	$20 \leq V$	Rear Ignition : $K_{GV(0.5)}$ in a vessel of the same volume Central Ignition: $K_G = 7$ bar m s^{-1}
$80 < K_{G(20\ litre)} \leq 160$	$V < 6$	$K_{G(20\ litre)}$ as measured in the 20 litre sphere
	$6 \leq V < 20$	Rear Ignition : K_{GV} for a vessel of the same volume Central Ignition : $K_{GV(0.5)}$ in a vessel of the same volume
	$20 \leq V$	$K_{GV(0.5)}$ in a vessel of the same volume
$160 < K_{G(20\ litre)}$	All V	$K_{G(20\ litre)}$ as measured in the 20 litre sphere

ELONGATED VESSELS

The reduced explosion pressures measured in elongated vessel tests are shown in Figures 12–15, where a normalised distance between the ignition point and vent (d), multiplied by the vent coefficient (K), is plotted against the maximum reduced pressure. The normalised distance d is obtained by dividing the horizontal distance between the ignition point and the centre line of the vent opening by the vessel diameter or width[19].

Figure 12 shows such a plot for the cylindrical vessel results, for propane with a single vent. Examination of the pressure-time profiles and test details for all the tests plotted in Figure 12 found that those tests in which high pressures were generated exhibited strong pressure oscillations, or there was an end vent rather than a side vent. An empirical equation that gives an upper bound that encompasses all the results, apart from the few tests that resulted in exceptionally high maximum pressures, is

Table 4. Measured reduced explosion pressures and predictions

Reference for experimental data	Fuel	Vessel volume V, m^3	Ignition position	Shape	Vent opening pressure P_V, atm	Vent area m^2	P_{red} measured mbar	P_{red} calculated mbar	K_G bar m s^{-1} from Table 3
13	Propane	0.76	C	Cube	0	0.29	48	102	15
13	Propane	0.76	C	Cube	0.095	0.29	178	102	15
13	Propane	0.76	C	Cube	0.095	0.29	178	335	80
14	Propane	35	C	Rect	0	1	1,370	1,700	30
15	Propane	11	R	Cyl	0.05	1.36	90	163	16
16	Propane	4.000	R	Segment	0	563	10	74	30
16	Propane	4.000	R	Segment	0	563	15	74	30
16	Propane	30.4	C	Rect	0.04	0.58	700	745	7
17	Propane	2	C	Cyl	0.008	0.3	130	290	15
17	Propane	2	C	Cyl	0.068	0.3	145	290	15
17	Propane	2	C	Cyl	0.079	0.3	183	290	15
18	Propane	2	C	Cyl	0.03	0.3	170	290	15
19	Methane	0.95	C	Cyl	0.32	0.05	2,000	4,170	12
19	Methane	0.95	C	Cyl	0.16	0.1	1,000	790	12
20	Methane	33.5	C	Rect	0	2.57	150	183	15
21	Methane	49.1	C	Cyl	0	3.46	120	170	15
16	Methane	4.000	R	Segment	0	563	5	4	15
16	Methane	30.4	R	Rect	0.023	2.74	215	146	15
16	Methane	30.4	C	Rect	0.012	1.33	205	508	15

16	Methane	30.4		Rect	0.04	1.33	542	508	15
17	Methane	2	C	Cyl	0.22	0.3	33	227	12
17	Methane	2	C	Cyl	0.65	0.3	104	227	12
17	Methane	2	C	Cyl	0.83	0.3	112	227	12
17	Methane	2	C	Cyl	0.009	0.3	83	227	12
17	Methane	2	C	Cyl	0.002	2.16	30	41	12
17	Methane	2	C	Cyl	0.003	2.16	46	41	12
17	Methane	2	C	Cyl	0.005	2.16	47	41	12
17	Methane	2	C	Cyl	0.016	0.3	42	227	12
18	Hydrogen	0.95	C	Cyl	0.075	0.2	1.250	1.870	637
18	Hydrogen	0.95	C	Cyl	0.135	0.3	490	950	637

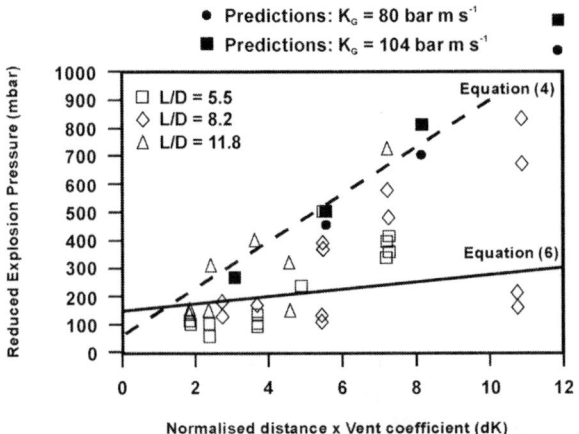

Figure 12. Cylindrical vessel with single vent: propane air

$$P_{max} = P_v + 85 \, d \, K \qquad (4)$$

A similar pattern emerged for the results for methane with a single vent as shown in Figure 13. For methane equation (5)

$$P_{max} = P_v + 70 \, d \, K \qquad (5)$$

provides an upper bound for all but one of the results.

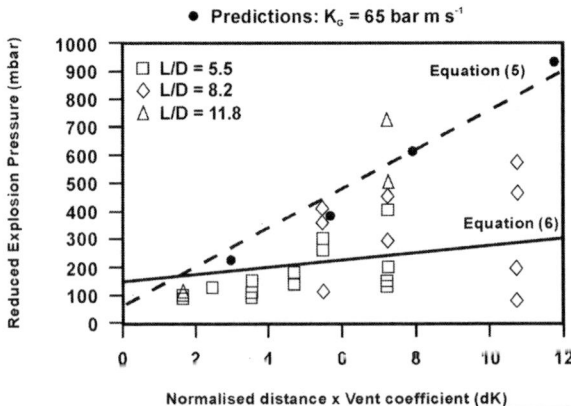

Figure 13. Cylindrical vessel with single vent: methane/air

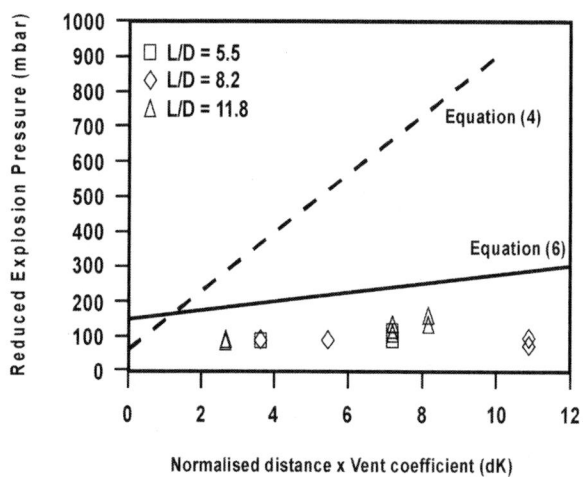

Figure 14.   Cylindrical vessel with multiple vents: propane/air

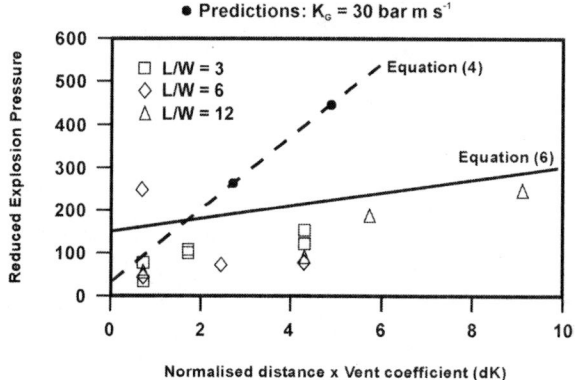

Figure 15.   Rectangular vessel with multiple vents: propane/air

Results obtained for propane with the cylindrical vessel with multiple vents are plotted in Figure 14. For multiple vents the dK value used is that for the vent nearest the point of ignition. Hence the reduced explosion pressure for a given dK should be lower than that obtained for a single vent. In none of the tests with multiple vents were strong low frequency oscillations observed, or a strong third peak. The result is that all the measured maximum explosion pressures lie below the values predicted by an equation published by Tite et al.[19]:

$$P_{max} = 15\ d\ K + 150 \tag{6}$$

The results obtained with the rectangular vessel for propane, with both single and multiple vents, are plotted in Figure 15. With one exception, the measurements lie below the predictions of equation (6). In this test the ignition point was in the middle of the vessel, with a vent on either side. Strong pressure oscillations were generated which resulted in a high maximum pressure.

Figures 12, 13 and 15 also contain predicted reduced explosion pressures from the K_G method derived for compact vessels in section 3.1. These calculations assume rear ignition and full end venting, i.e $K = 1$. No account of L/D ratio is taken in these calculations; the vessel volume is the only parameter involved. The calculations satisfactorily envelop almost all the experimental results, the only exceptions being tests in a vessel with an L/D of 11.8 and the vent positioned close to the ignition source. However, because no end venting experiments were performed above an L/D of 5.5, application of the modified K_G method was not tested beyond this value.

CONCLUSIONS

The K_G factor method for explosion vent sizing published in NFPA68 is a relatively simple method to use, but its straightforward application is not always successful if the standard value of K_G is applied under all circumstances.

The standard value of K_G may characterise enclosed gas explosions adequately but it does not characterise vented explosions well.

A variation of the K_G factor method based on deriving variants of the K_G factor from the earlier stages of a pressure-time history and from pressure time traces in larger volumes is more appropriate to vented explosions.

Three variants of the K_G factor have been used in an extensive analysis of experimental measurements of reduced explosion pressures. When the K_G factor most appropriate to the conditions of the vented explosion is substituted into the vent sizing calculation good agreement can be obtained between experiment and measurement.

For elongated vessels with length to diameter (L/D) or length to width (L/W) ratios in the range 3 to 20 and containing quiescent gas mixtures that are essentially free of turbulence inducing elements, vents should be spaced at regular intervals along the length of the enclosure and ideally there should be a vent at or very close to each end of the enclosure. The modifications to the K_G method shown in Table 3 worked well with elongated vessels with L/D ratios not exceeding 5 with the vent positioned at the end remote from ignition.

REFERENCES

1. Lunn G A. Venting of gas and dust explosions: A review. Institution of Chemical Engineers, Rugby, UK. (1984)
2. Lunn G A and Pritchard D K. Guidance on explosion relief for gases and vapours: a literature review and a proposal for a modified K_G vent sizing method. Health and Safety Laboratory Report. To be issued, 2003.

3. Pritchard D K, Allsopp J A and Eaton G T. Venting requirements for near cubical vessels: Analysis of results. Health and Safety Laboratory Report GE/97/03, 5 June 1998.

4. Pritchard D K, Allsopp J A, Eaton G T and Hedley D. Venting requirements for elongated vessels: Analysis of results. Health and Safety Laboratory Report EC/00/29, 23 June 2000.

5. Pritchard D K, Allsopp J A and Eaton G T. Venting requirements for linked vessels: Analysis of results. Health and Safety Laboratory Report EC/98/48, 3 December 1998.

6. Pritchard D K, Hedley, D and Webber N K. The Nomograph method for the sizing of explosion relief vents for gases and vapours. Health and Safety Laboratory Report EC/98/10, 19 November 1998.

7. Pritchard D K and Hedley D. Venting nomographs based on burning velocities. Health and Safety Laboratory Report EC/99/82, 11 January 2000.

8. Cubbage P A and Simmonds W A. An investigation of explosion reliefs for industrial drying ovens - I. Top reliefs in box ovens. Trans. Inst. Gas Eng. **105**, 470. (1955)

9. Rasbash D J. The relief of gas and vapour explosions in domestic properties. Fire Research Note No 759. (1969)

10. NFPA. NFPA 68 - Guide for Venting Deflagrations. National Fire Protection Association, Quincy, MA, USA. (1998)

11. Bradley D and Mitcheson A. The venting of gaseous explosions in spherical vessels II - Theory and experiment. Combustion and Flame **32**, 237. (1978)

12. Molkov V V (2000) Explosion Safety Engineering: Design of venting areas for enclosures at atmospheric and elevated initial pressures. FABIG Newsletter, Issue No. 27, Article R383, September 2000.

13. Yao C. Explosion venting of low-strength equipment. Loss Prevention **8**, 1–9 (1974).

14. Solberg D M, Pappas J A and Skramstad E. Experimental investigations on flame acceleration and pressure rise phenomena in large scale vented gas explosions. Proceedings of 3rd International Symposium on Loss Prevention and Safety Promotion in Process Industries, Basle, vol 3, pp 16/125–16/1303, 1980.

15. Molkov V V, Aganoff V V and Aleksandrov S V. Deflagration in vented vessel with internal obstacles. Combustion, Explosion and Shock Waves **33**(4), 418–424 (1991).

16. Harrison A J and Eyre J A. The effect of obstacle arrays on the combustion of large premixed gas/air clouds. Combustion Science and Technology **52**, 121–137 (1987).

17. IBExu. Private communication from IBExu to V V Molkov

18. Pasman H J, Groothuisen Th M and Goojier P H. Design of pressure relief vents. Combustion Science and Technology **52**, 91–106, (1987).

19. Tite J P, Binding T M and Marshall M R. Explosion relief for long vessels. Paper presented at Conference on Fire and Explosion Hazards, April 1991, Moreton-in-Marsh.

THE REGULATION OF INHERENT SAFETY

David A. Moore, PE
President and CEO, AcuTech Consulting Group, Chemetica, Inc., 1948 Sutter Street, San
Francisco, CA 94115, dmoore@acutech-consulting.com

INTRODUCTION

Inherent Safety is a worthwhile process risk management strategy to employ, and efforts to implement inherently safer strategies should be given first priority, as is feasible. But regulating the use of inherent safety is proving to be challenging as evidenced by the experiences to date of industrial companies who are under inherently safer requirements as part of the Contra Costa County, California, Industrial Safety Ordinance (ISO)[1,2]. This paper will outline those requirements and explain how the inherently safer requirements were adopted as part of an overall strategy of ISO programs that act in concert to address risk to the community and workers and the issues in achieving compliance.

Inherent safety is gaining momentum in the United States in the minds of the regulators at the Federal, State, and Local Government levels. Inherently safer systems (ISS) requirements of the Contra Costa County, California, Industrial Safety Ordinance (ISO) are a significant new regulation that should be closely monitored by industry[1,2]. While the concept of Inherent Safety has been in the vocabulary of process safety engineers for twenty years, it is only a newly appreciated issue in process safety management. Inherent safety is not widely practiced in an explicit manner in the United States at this time. This is the first process safety regulation in the United States that we are aware of that requires facilities to justify their consideration of inherently safer systems.

Companies in any industry or location should be familiar with the concepts, and concerned with the progress being made in addressing the underlying concerns of the County in enacting the ordinance, the intent of the inherently safer systems requirements, and the merits of any additional requirements. Given that inherent safety is a rather subjective concept, it makes the matter a difficult one to understand.

The ordinance references the definition of inherent safety published by the Center for Chemical Process Safety (CCPS) in its book "*Inherently Safer Chemical Processes: A Life Cycle Approach*[3]". This is significant in that industry guidance is being held as a standard by which companies need to operate.

The intent of this paper is to provide further explanation of the concept of Inherently Safer Systems, to explain the regulatory requirements, to assess their effectiveness, and to make suggestions on how to improve the regulation of Inherently Safer Systems (ISS).

BACKGROUND ON THE ISO

According to the County[4], the Contra Costa County Board of Supervisors passed the Industrial Safety Ordinance because of concerns of the accidents that have occurred at the oil refineries and chemical plants in the County. The effective date of the Industrial Safety Ordinance was January 15, 1999. The ordinance applies to oil refineries or chemical plants that were required to submit a Risk Management Plan to the U.S. EPA and are a program level 3 regulated stationary source as defined by the California Accidental Release Prevention (CalARP) Program. The goals of the ordinance include:

1. To reduce the number of incidents being experienced in the County at the covered stationary sources;
2. To reduce the overall catastrophic risk from the facilities, especially to the public and workers.

It is no doubt believed that by encouraging inherently safer systems be considered, that substantial improvements could be made in realizing the goals mentioned above. Part of the ISO requirements is the need for the regulated stationary sources to consider inherently safer systems when evaluating the recommendations from process hazard analyses for existing processes and to consider inherently safer systems in the development and analysis of mitigation items resulting from a review of new processes and facilities. Contra Costa Health Services completed and issued a Contra Costa County Safety Program Guidance Document on January 15, 2000[4]. This document included a definition of inherent safety and some rules for implementation of the ordinance.

Facilities that are subject to this ordinance are, in total, subject to four safety regulatory programs designed to reduce the potential of an accidental release from a regulated stationary source that could impact the surrounding community[4]. The four programs are the Process Safety Management (PSM) Program administered by Cal/OSHA, the federal Accidental Release Prevention Program administered by the EPA (RMP), the California Accidental Release Prevention Program administered by Health Services (CalARP), and the Industrial Safety Ordinance (ISO) administered by Health Services. All of the above regulations require the identification and analysis of hazards, and meeting these regulations requires a significant effort.

A key difference between the PSM, RMP, and CalARP programs and the ISO program, however, is that the regulated stationary sources are required to consider inherently safer practices for all PHA action items under the current Industrial Safety Ordinance. This requirement is unique and is the topic of this paper. Note that the other three regulations imply that inherently safer systems should be employed, but are non-specific in this requirement.

CURRENT REQUIREMENT OF THE ISO FOR INHERENT SAFETY
The current Industrial Safety Ordinance requires facilities to consider inherently safer systems in the development and analysis of mitigation items resulting from a process hazard analysis and in the design and review of new processes and facilities. The Industrial Safety Ordinance defines inherently safer systems as follows:

"Inherently Safer Systems" means Inherently Safer Design Strategies as discussed in the 1996 Center for Chemical Process Safety Publication (CCPS) "Inherently Safer Chemical Processes"[5] and means Feasible alternative equipment, processes, materials, lay-outs, and procedures meant to eliminate, minimize, or reduce the risk of a Major Chemical Accident or Release by modifying a process rather than adding external layers of protection. Examples include, but are not limited to, substitution of materials with lower vapor pressure, lower flammability, or lower toxicity; isolation of hazardous processes; and use of processes which operate at lower temperatures and/or pressures." County Ordinance Chapter 450-8, §45—8014(g).

SUMMARY OF PROPOSED AMENDMENTS

The Board of Supervisors and the CCHS are analyzing the merits of changing the ordinance or at least updating the Program Guidance Document. A newly proposed guidance document for the ordinance may allow a broader application of inherent safety, and may require that ISS be considered for all new, modified, or existing facilities in a more comprehensive way than just to consider the outcomes of PHA action items. That guidance was in draft form at the time of the development of this paper (September, 2002).

This guidance was felt to be necessary since the first few years of the application of inherently safer systems proved to be a frustration for regulators and industry alike. Industry felt that meeting the intent of the regulation was unrealistic for existing facilities, particularly from an economic standpoint. They felt risks were adequately addressed by strategies other than inherent safety. Regulators seemed to feel that industry didn't take the requirements of the regulation seriously and hadn't made substantial progress in the area of inherent safety. They felt that the continued incidents at the facilities were evidence that inherently safer strategy they had expected was necessary.

This dilemma resulted in a several proposals to stiffen the regulations. There are three amendments being considered by the Board of Supervisors for the Industrial Safety Ordinance. One proposal, the Gerber amendment, which was proposed by one of the County Board of Supervisors, would effectively give the authority for the County Health Services Department to mandate the exact technology and ISS design employed when a company rejects a recommendation from a PHA to implement an ISS proposed change.

The crux of it is that while it is already required to implement the most inherently safer strategy, it appears that CCHS and possibly the Supervisors have doubts that industry is taking the principle to heart. Actually much has been done to implement ISS where possible. Of course not every PHA action item is an opportunity for Inherently Safer Systems (many times various layers of protection are required and are reasonable solutions). This has the potential to force very expensive and significant changes in the name of inherent safety.

Another proposal from the CCC Health Services Department suggests that the Industrial Safety Ordinance be amended in the way inherently safer systems are considered. Based on the way the existing ordinance is written, the reduction of existing hazards at the stationary sources is claimed to be 'minor if any at all' and only a portion of each of the processes is considered using the existing ordinance. Health Services is proposing two changes to the ordinance:

1. That a separate study of existing processes to consider inherently safer systems be performed.
2. That the definition of inherently safer systems as defined by the ordinance be amended to emphasize the inherent and passive layer of protection. The inherent layer considers ways of reducing hazards, while the passive, active, and procedural layers look at ways

of reducing the risks. The amendment proposed by Health Services will look at reducing the hazards at existing process, not just reducing the risk, and will look at the complete processes, instead of only portions of the processes.

BACKGROUND ON INHERENT SAFETY

The history of inherent safety as a documented strategy for loss prevention is rather recent, but the concept is very old. "On December 14, 1977, Trevor Kletz presented the annual Jubilee Lecture to the Society of Chemical Industry in Widnes, England. His topic was "What you don't have, can't leak," and this lecture was the first clear and concise discussion of the concept of inherently safer chemical processes and plants.

Safety professionals agree that this is a good 'way of thinking', and is a best practice in process safety management. The Center for Chemical Process Safety of the American Institute of Chemical Engineers (CCPS) published the book referenced by ISO in 1996 to promote the concept[3]. Inherently Safer Systems is a philosophy that is encouraged in the industry to focus engineers and managers on reducing or eliminating a hazard within a chemical process as a goal as it is feasible. No absolute ground rules exist for accomplishing this – it is a philosophy.

At this time in industry, while it is appreciated, it is not prescribed in any other U.S. process safety regulation we are aware of nor is it widely practiced in a formal, documented way by industry throughout the world. Most companies would recognize it as a philosophy, and would encourage its use where possible.

The ISS method employs four key inherently safer strategies:

1. Minimizing,
2. Substituting,
3. Moderating, and
4. Simplifying

Each of these strategies can accomplish hazard reduction or elimination.[4]

"Hazard is defined as a physical or chemical characteristic that has the potential for causing harm to people, the environment, or property."[3] The concept is based on the belief that if one can eliminate or moderate the hazard, not only is the risk reduced, it may be possible to remove the risk altogether from consideration. Alternatively, an inherently safer system would make the hazard less likely to be realized and less intense if there is an accident.

LAYERS OF PROTECTION AND OVERALL PROCESS RISK MANAGEMENT STRATEGY

Similarly, though, CCPS has published several other publications that sanction the use of other process risk management strategies of Layers of Protection and the use of all four process risk management strategies. In fact, these two concepts are described in the same textbook on inherent safety[6]. The layers of protection concept is that every hazard needs to have a series of layers of prevention, detection, and mitigation systems to either assist in prevention of the incident or in reducing the potential impacts of any event. Key to the concept is the idea of reducing the risk of the chain of events from initiation of an incident to prevention of or reduction of the hazardous outcome.

CCPS also describes four categories of risk reduction strategies:

1. Inherent
2. Passive
3. Active
4. Procedural

These strategies are illustrated in Figure 1.

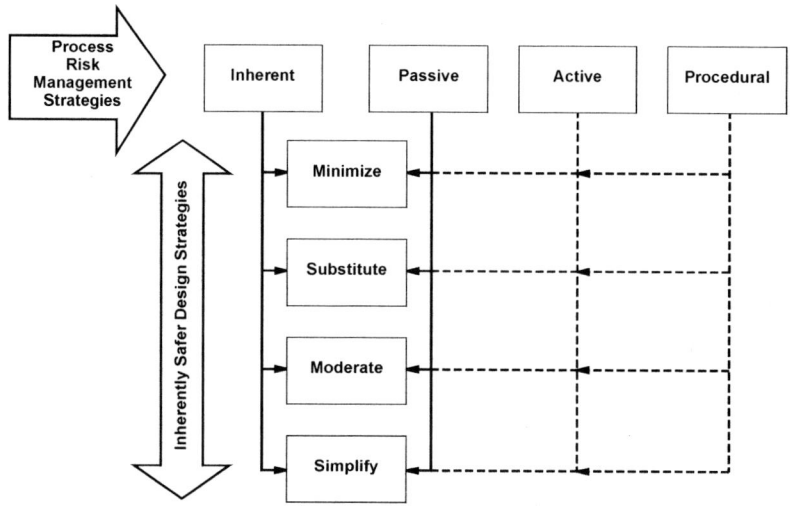

Source: Fig 2.2, CCPS Concept Book *"Inherently Safer Chemical Processes"*[5]

Figure 1.   Process risk management strategies

Figure 2 illustrates the matrix AcuTech developed with abbreviations to illustrate the broad application of inherent safety across all four inherent safety strategies and across all four process risk management strategies. The abbreviations are documented in the worksheet for each inherently safer systems study that is done as a means of ensuring the broad application of strategies is implemented. First preference is given to those in the Inherent column, followed by Passive and then Active and Procedural, due to the degree of reliability of each strategy.

All safety professionals agree that layers of protection (Figure 3) and the four process risk management strategies are necessary and advisable strategies, too, and represent another best practice for process safety management. Inherent safety and layers of protection/process risk management strategies are not opposing strategies – they are complementary strategies in an overall scheme to reduce risk. In fact, the strategies cannot be taken out of context and be effective.

Inherently Safer Systems Matrix

Inherent Safer Design Strategies	Process Risk Management Strategies			
	Inherent (1)	Passive (2)	Active (3)	Procedural (4)
Minimize	MIN1	MIN2	MIN3	MIN4
Substitute	SUB1	SUB2	SUB3	SUB4
Moderate	MOD1	MOD2	MOD3	MOD4
Simplify	SIM1	SIM2	SIM3	SIM4

Figure 2.　Process risk management strategies vs. inherently safer strategies

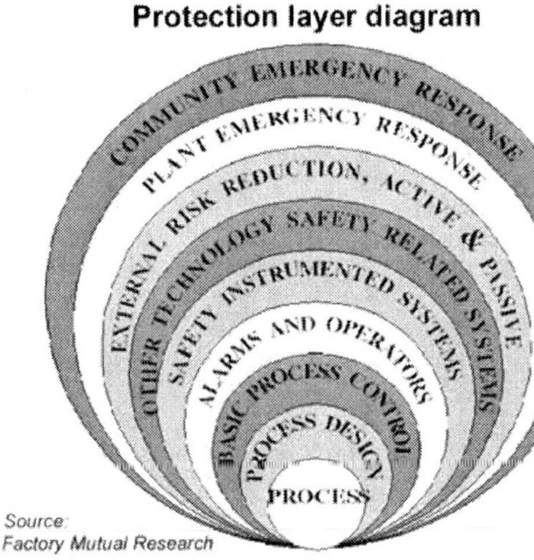

Figure 3.　Layers of protection concept

252

CCHS also recognizes these four strategies in their broad definition of inherently safer systems in their newly drafted ISO guidance, and has encouraged stationary sources to employ all four strategies in an overall risk reduction framework.

OPPORTUNITIES FOR ISS AND LAYERS OF PROTECTION

CCPS states that there are three stages of an accident sequence, the initiation, the propagation, and the termination stages. "Inherently Safer Strategies (and Layers of Protection) can impact the accident process at any of the three stages. The most effective strategies will prevent initiation of the accident. Inherently safer design can also reduce the potential for propagating an accident, or provide an early termination of the accident sequence before there are major impacts on people, property, or the environment"[3]. This description by CCPS says that safety professionals believe that there are many opportunities to employ inherent safety – not only to eliminate a hazard but also to moderate an accident sequence in different ways that, in effect, lowers the risk.

The CCPS book Inherently Safer Chemical Process6 considers the implementation of inherently safer systems over the lifetime of the chemical process. To accomplish this the authors of the book took a broader meaning where the inherent level (the most strict definition of inherent safety) is a subset of their definition, which also include passive, active, and procedural levels. The passive, active, and procedural levels do not reduce or eliminate the hazard, but does reduce the likelihood of a release by reducing the risks.

In other words, inherent safety can be interpreted either narrowly or broadly. "The narrowest interpretation that could be argued from the wording of the ISO definition would be that an ISS reduces the underlying hazard that must be contained and controlled for a Major Chemical accident or Release to be avoided. The broadest interpretation that could be argued from the wording of the ISO definition would be that an ISS reduces the risk of a Major Chemical Accident or Release by reducing the underlying hazard that must be contained and controlled or by improving layer(s) of protection in a way that is permanent and inseparable and not easily weakened or removed from the system. It should be noted that systems that do not address Major Chemical Accidents or Releases, as defined in the ISO, would not qualify for Inherently Safer Systems per the ISO definition."[5]

In either the broad or narrow interpretation case, if the company takes action to employ an inherently safer system, the overall risk is likely to be reduced and the workers, the company, the public, and environment all benefit. The real issue in the end is whether the risk has been reduced sufficiently to a level that is as low as is reasonably practicable. This issue is ultimately more important than whether a company happened to employ an inherently safer system in every PHA action item case or not.

The book explains the dilemma faced by the County ISO regulation on pages 10–11[6]– "There is much discussion about whether or not a particular safety feature in a chemical process is "inherent". Such discussions may arise in part because different people are viewing the process at different levels of resolution, ranging from a global view of the entire process to a very detailed view of specific features of the process. The definition of hazards (an inherent physical or chemical characteristic that has the potential for causing harm to people, the environment, or property) can be applied at any level of resolution."

Based on a review of PHAs and the recommendations associated with them from industry over the past two years, it is clear that this dilemma mentioned in the CCPS book is occurring. In other words, the PHA teams made many decisions that any ordinary company would have made when presented with similar hazards. In some cases, particularly for new processes or for the restart of a crude unit, the PHA teams made many inherently safer decisions. For the most part, the majority of the recommendations involved improving or adding onto layers of protection since the team did not see an opportunity to employ inherently safer systems above existing examples of ISS, and so it was prudent to reduce the risk or improve on the integrity of an existing layer of protection.

This should be encouraged, not criticized. In the end, the reduction of risk is the goal, while it is desirable to eliminate or reduce the hazard itself if possible as a first strategy. The CCPS book explains this concept on page 11 - "For purposes of this text, any improvement in a layer of protection which is permanent and inseparable, and not easily weakened or removed from the system, is considered to be a process safety improvement in an inherently safer direction." This broad definition is not only a move in the right direction; it is a commonplace approach in industry throughout the world and a prudent approach to follow.

BENEFITS AND DRAWBACKS OF ISS

The benefits of employing the inherently safer strategy is that, hopefully, every decision that is made to design, build, operate, and maintain a process attempts to employ an inherently safer system than the original concept or existing process operation. This could result in lower overall risk, since the original hazards may be eliminated or greatly reduced. But it could be that the mere application of an inherently safer solution does not guarantee a safer process. In fact, it is recognized that any change to a process may result in other undesirable effects, even if the change was in the best interest of safety.

Also, an inherently safer change alone does not necessarily constitute a safer plant. If by employing layers of protection the risk can be sufficiently reduced or if it can be reduced to a greater degree than if one or more inherently safer strategies are employed, layers of protection may be a preferred strategy. Even if an inherently safer strategy is employed, the hazard still may be present, and so it may be desirable or necessary to employ layers of protection in addition.

It may be possible to equate the risk of an accidental release to one that has been treated by an inherently safer system if various layers of protection are employed. The concern is that the layers may be unreliable and certainly may be expensive to provide. But the risk could, indeed, be lower with a process employing various layers of protection.

Risk is defined as a measure of economic loss, human injury, or environmental damage in terms of both the incident likelihood and the magnitude of the loss, injury, or damage.[3] The real goal of the ISO is to reduce risk, particularly to the community.

This process works well for existing plants, but in many situations the hazard is not eliminated or reduced. It can be more applicable to new processes or conceptual designs.

LESSONS LEARNED OF THE ISO PROGRAM FOR ISS

In conclusion, depending on the definition of ISS, there were many examples of inherently safer solutions being employed. If the definition is limited to the strictest definition of inherent safety, then only a few of the PHA action items had employed the most inherent strategy for risk reduction. However, other strategies employed resulted in more 'inherently safe' processes since passive, active, and procedural recommendations addressed more inherent approaches.

Lessons learned from the past three years have included:

- Companies found ISS to be economically and technically difficult if not infeasible to accomplish, particularly for existing processes;
- There are different perspectives on what is reasonable and what is feasible when it comes to decisions on the need for implementing ISS;
- The guidance provided to ensure that ISS was being considered consistently and fully was not informative enough, so there was some confusion and an education gap;
- An annual report of the application of ISS showed that the majority of PHA action items did not involve application of ISS (at least not first inherent principles of ISS);
- The public and regulators tend to be impressed with the principles of inherent safety and have high expectations of risk reduction by the approaches;
- The public and regulators often mistrust industry if wholesale technology changes and other first principles of inherent safety aren't applied to achieve major risk reduction;
- This is created by a difference in perception of risk by the public and industry and a misunderstanding or lack of appreciation of the costs of such changes;
- Industry often looks at inherent safety as only the first principles rather than a wider use of the concepts, so this exacerbates the problem;
- Application of inherent safety at only the most purely inherent level (first principles) is often at odds with practical and cost effective risk reduction, especially for existing construction;
- A broader application of inherent safety across all four strategies of process risk management is more practical and may result in novel risk reduction ideas;
- Any move in an inherently safer direction is likely to be a good risk reduction move, so this should be encouraged;
- Inherent Safety is best applied by those knowledgeable of the process with proper training;
- Guidance/training is needed for a team to know how to do this effectively;
- We recommend that ISS be considered as an integral part of the PHA process for efficiency and since it is more widely applied.

CONCLUSIONS

Inherent Safety is a newly appreciated issue in process safety management, and this is the first process safety regulation in the United States that we are aware of that requires facilities to justify their consideration of inherently safer systems. As such, it is important to understand the progress being made in addressing the underlying concerns of the County in enacting the ordinance, the intent of the inherently safer systems requirements, and the merits of any additional requirements. Given that inherent safety is a rather subjective concept, it makes the matter a difficult one to understand, implement, and regulate.

AcuTech does not recommend that the narrowest view of inherent safety be the only one encouraged. Also, AcuTech recommends that inherent safety is integrated into process hazard analysis (PHA) studies to allow for a broader perspective and a day-to-day effort, rather than only a special study once in a while.

AcuTech believes the best approach is to allow those most knowledgeable of the process to justify their design and operating decisions, as the current ordinance requires, rather than to give the regulator the authority to mandate a particular technology against the will of the industrial company. Inherent Safety must be exercised to the fullest, but there are practical bounds on its application.

The Contra Costa County ISO is on the cutting edge of process safety with inherently safer system requirements. It is likely that other regulations may model themselves after this or similar ones and require inherent safety. Already the City of Richmond, California, has adopted the CCC ISO regulation.

Since the terrorist attacks on the United States in September, 2001, there has been a flurry of regulations for homeland security. The proposed U.S. Senate Bill S.1602 "Chemical Security Act of 2001" lists inherent safety requirements as a requirement. This could have very significant consequences if this is emphasized over other strategies for risk reduction from intentional releases. Inherent safety is being focused on as the leading solution for limiting the risk of terrorist acts on chemical facilities by some regulators and environmental groups.

Companies should begin to apply this voluntarily to gain the benefits and prepare for the future business environment.

REFERENCES

1. Section 450-8.016(D)(3) of the County Ordinance Codes as amended by Ordinance No. 2000-20, Contra Costa County, California.
2. "Ordinance No. 2000- Amending Ordinance Code Chapter 450-8, on Industrial Safety, Regarding Inherently Safer Systems and Action Items Identified in Process Hazards Analysis" (Gerber Amendment).
3. Center for Chemical Process Safety, Inherently Safer Chemical Processes: A Life Cycle Approach, Center for Chemical Process Safety, AIChE, 1996.
4. Inherently Safer Systems Health Services Proposal Issue Paper, Contra Contra County Health Services Department, 2001
5. Industrial Safety Ordinance Annual Performance Review and Evaluation December 12, 2000
6. Review of Ultramar Golden Eagle Refinery Inherent Safety Report for Donna Gerber, District III Supervisor, Unwin Co., November 30, 2001
7. Center for Chemical Process Safety, Inherently Safer Chemical Processes: A Life Cycle Approach, Center for Chemical Process Safety, AIChE, 1996.

© 2003 IChemE

APPLICATION OF INHERENT SAFETY CHALLENGE TO AN OFFSHORE PLATFORM DESIGN FOR A NEW GAS FIELD DEVELOPMENT – APPROACHES AND EXPERIENCES

Stuart Chia[1], Kieran Walshe[2], Ed Corpuz[2]
[1]Kellogg Brown & Root (KBR) Pty Ltd, Level 9, 21 Kent Street, Sydney NSW 2001, Australia
[2]BP PLC

In the past, the concept of a "safe design" for an offshore installation was one that had been provided with redundant levels of prevention barriers and mitigation systems. In addition, the lack of engagement of operations personnel in the design process often resulted in modifications of the built platform possibly giving rise to hazards unforeseen by the design team. By having process safety specialists working closely with the design team and operations personnel, the incorporation of the "inherent safety challenge" concepts as a design tool during front-end engineering (FEED) design is currently been applied to a major Greenfield offshore gas Project in South East Asia. The natural gas reservoirs are located in a remote pristine area. The Project faces unique challenges as the gas field is classified as high temperature (up to 130°C) and high pressure (up to 320 bar). The concept had evolved from being initially a manned platform to an unmanned facility. The challenge during FEED was to design an "inherently safe platform". This was taken to be a platform that does not leak or break down, required few people and, as such, is inherently safe as it should not put anyone at risk. The purpose of this paper is to broadly describe the proactive approach adopted for this Project and to relate how the HSE team is championing the inherent safety approach. By involving operations personnel and design team up front, the HSE team is assisting in the delivery of a design that will be safer and more reliable. Inherent safety design examples under consideration for the Project and philosophy are provided.

Inherent Safety, NUI, Design Safety Goals, Hazard Management Plan

FOREWORD

Due to the sensitivities surrounding this Project at the time of preparing this paper, the title of this particular development could not be released. The paper has been structured to be general in nature such that readers can consider and apply inherent safety concepts contained herein to their projects.

INTRODUCTION

At BP, the HSE Policy is quite clear, "Our Goals are simply stated - No Accidents, No Harm to People and No Damage to the Environment". This statement is the guiding principle behind this major gas production development with the expectation that the "offshore facilities will demonstrate world class performance so far as HSE is concerned". This paper details how the Project is currently working towards fulfilling the HSE policy goal in terms of design safety and relays the experience gained thus far.

The Development's natural gas reservoirs are located in a very remote area in South East Asia. The offshore component includes those facilities required to produce, collect, and export natural gas and condensate for delivery to an LNG Plant. The complete offshore Project development will include the design, procurement, fabrication, installation, drilling, completion, pre-commissioning and commissioning for a series of platforms located near shore (approximately 30 kilometres).

Each platform has been designed as an unmanned facility (or Normally Unattended Installation - NUI) in a water depth of up to 60m. The wells will be drilled using a heavy-duty jack-up operating in cantilever mode. No topsides processing is proposed with all production fluids from the wells being routed to shore via individual subsea pipelines, one from each platform, for onshore processing at a Receiving Facility. The design also considers the provision of subsea power and communication cables.

The Project faces unique challenges in that the natural gas fields are at high temperature (HT, up to 130°C) and high pressure (HP, up to 320 barg). In addition, the development will be located in a known active seismic area, sand waves, fast currents and be subject to tidal movements with low visibility water column.

INHERENT SAFETY CHALLENGE

THE NEED TO "THINK OUTSIDE THE BOX"

Past authors (Dalzell) have observed that the formal risk assessment approach and making a case for safety has resulted in an overcomplex and retrospective approach to risk management in design. Hence opportunities for optimising inherent safety may be missed causing risk management to be "reactive" rather than been "pro-active" in developing a safer design.

It is also the past experience of the authors that the identification and active management of technical safety issues for a new facility whether offshore or onshore, is conducted when the final design has been approved for construction, or worse, when the facility is being built. When new and complex technology is involved, this late assessment may find that risk from the operation of the facility is unacceptable, when measured against corporate and regulatory requirements. This may necessitate a re-design, major modification and/ or implementation of expensive hardware changes which ultimately leads to higher project costs, delays in commissioning, and problems in achieving product specifications in operation. There may also be additional operating costs and production interruption costs if the potential problems are not identified *a priori*, especially at the early stage of design.

A pro-active approach avoids or minimises the need to incorporate extensive safety systems to control and mitigate the hazards. Safety systems may themselves introduce additional hazards as they can increase exposure of personnel (e.g. maintenance) to the hazards. The key to providing an inherently safer design is by placing more emphasis on hazard elimination and incident frequency reduction measures (i.e. causation analyses). Dalzell argues that simplicity and inherent safety are complementary.

The approach adopted by the Project Team was to engender a culture whereby the entire team asked, "what's the hazard and what can we do to minimise the risks". This type of questioning leads the team to "not assume that the best design is one that has been achieved previously", that is to "Think Outside the Box".

By encouraging the participation of the designers and operations personnel, the Project aimed to create more opportunities to reduce the hazard likelihood, severity and consequence during the early stages of the Project.

WHAT IS INHERENT SAFETY ?

There are several definitions of inherent safety from eminent leaders in this field. McQuaid defines "an inherently safer approach to hazard management is one that tries to avoid or eliminate hazards or reduce their magnitude, severity, or likelihood of occurrence by careful attention to the fundamental design or layout.

Using Dalzell's definition as the basis, the Project collectively developed a mission statement that defined Inherent Safety as "delivering to Operations, a NUI and pipeline that doesn't leak, collapse or sink and has no one on it to be killed" over the life of the facility. By this definition alone, Operations and Designers were engaged as a single team with the former group assisting in the design process.

It also meant that HSE design issues were an integral part of the discipline engineer's decision-making process rather than been considered separately by safety engineers as has happened in past Projects.

DESIGN SAFETY GOALS AND HAZARD MANAGEMENT

A critical step in the Inherent Safety Challenge was the development and agreement of design safety goals by the Project Team. It was necessary to accomplish this early in the Project so that all team members were aligned in understanding their basis of development and the role of each goal during the FEED phase. HSE roll out workshops were used to assist in this delivery.

The inherent safety goals were expressed as zero base or minimisation base allowing the team to focus on attempting to achieve the Inherent Safety mission. These goals were based upon the generic hazards identified during pre-FEED associated with the gas development Project and unmanned facilities in general such as seismic, vessel collision, weather, hydrocarbon leak and transportation.

With reference to the Inherent Safety mission statement, the overall design safety objective was defined to be "Zero Visits and Zero Manning for the NUI". To achieve this objective, a total of 19 Design Safety Goals applicable to the Project were identified and some of these goals are reproduced below:

- minimising inventory;
- minimising potential leak paths;
- maximising natural ventilation;
- keeping the design and intended operating activities simple;
- minimising helicopter flights;
- minimising processing;
- maximising reliability;
- minimise inspection, maintenance and intervention;
- eliminate specification breaks and relief systems;
- minimising manning requirements.

The goals are used to help develop the design options and assist in identifying and refining the most inherently safer design. The desired outcome is that instead of taking a default set of prevention detection control and mitigation measures, the Project team should actively manage the balance between prevention and cure.

In order to confirm that this approach has been adopted during FEED, the goals will be used as guidewords to review and challenge each of the design options. In this way, transparency is established that describes the progress made in eliminating hazards and, where this is not completely practical, what reductions (in order of likelihood, severity and consequence) are achievable, and what arrangements are to be put in place to control and manage residual risks.

By adopting this hierarchal approach to hazard management, the reader will observe that quantitative risk analysis has not been used to "drive the design". Through the application of inherent safety goals and effective residual hazard management, the risk assessment instead of taking precedence now becomes a tool to verify that we have "got the design right". This is illustrated in Figure 1 whereby the design team will:

- avoid active systems wherever possible;
- passive systems should be used wherever possible; and
- avoid dependence on personnel and third parties to identify, control and/or mitigate a major accident.

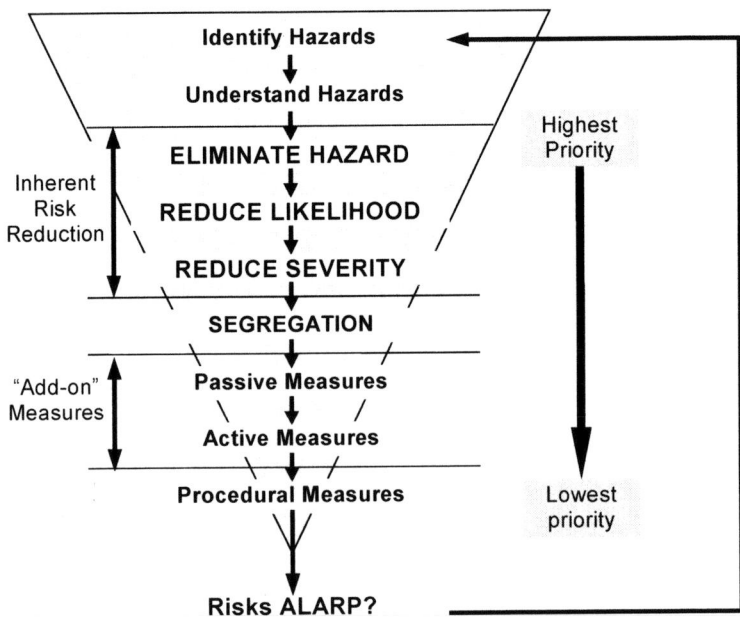

Figure 1.   Risk management approach

MAINTAINING THE INHERENT SAFETY CHALLENGE
Dalzell notes that inherent safety is not a discrete activity carried out by specialists or a series of stand-alone safety studies. It is a living process where both the major decisions and the more modest contributions to inherent safety are documented and communicated to other designers and operators. This is essential so that the reasons for each inherent safe decision that establishes the design limits and operating parameters are well understood and not subject to possible revision or impairment in the future.

During this phase, the "inherent safety challenge" will be maintained through:

- Design Safety Moment at all Project meetings;
- Continuing team presentations which focuses on for example, Lessons Learnt, Achievement against Design Safety Goals;
- Safety Challenge workshops;
- Live Hazard Management Plan database.

With the latter, the Hazard Management Plans (HMP) will be designed to capture all critical inherent safe decisions and proposed measures for any residual hazards. The knowledge and understanding of the design and its hazards gained by the designers will be passed onto the operators via the HMP. The HMP will then be revised in the next phase to ensure that safety and operational management systems have been appropriately cross referenced to the design measures.

THE PROJECT EXPERIENCE

A review of the design safety goals reveals that much emphasis has been placed on "designing out" the threats or causes of a major hazard. At the time of preparing this paper, the Project had commenced FEED and the following are examples of how the Inherent Safety Challenge had been applied with preliminary results.

In developing an inherent safe design, the Project Team is always reminded of the fundamental question, "for any proposed addition or removal of an equipment or process system, does this lead us to achieving our objective of zero manning and/ or zero visits ?".

MINIMISE MANNING REQUIREMENTS – EXAMPLE 1
Issue
During the concept selection phase, it was proposed to develop the gas field through offshore dehydration with process platforms and carbon steel export pipelines. Known as the "Dehydration Case", the process would consist of separation, gas dehydration, and export, together with regeneration of the glycol. In addition, vapour recovery (compressor, vessel and heat exchanger) as well as both condensate, water treatment facilities and gas metering was considered. Accommodation quarters would contain up to 30 Person on Board (POB).

Challenge
In design reviews, the Project Team identified the following:

➤ Could the number of personnel going offshore be minimised?
➤ Could the duration of maintenance to be performed on the platform be optimised?

Achievement

The Project developed the "Minimum Facilities Case" whereby all processing equipment (and test separator) was removed and transferred onshore. The pipeline material was changed to multiphase corrosion resistant alloy (CRA) pipelines.

By designing for the facility lifetime, the original concept of a manned platform became one of an unmanned facility with minimal processing. This eliminated the need to have personnel stationed permanently offshore and minimised operator intervention. This was the design taken forward to FEED for optimisation.

Currently the design has eliminated the Temporary Refuge (TR) that removes the complexity and fabric maintenance requirements. In addition, the removal of the TR would discourage personnel to "rest" on the platform and conducting any "overnight" work.

MINIMISE LEAK PATHS – EXAMPLE 2

Issue

Standard flange joints as a means of mechanical connection for piping systems has been the automatic choice within the industry. However as a result of the high pressure and high temperature that the NUI will operate within, some of the flange sizes will be outside the ANSI range.

Challenge

Flange joints were identified as a likely leak source for hydrocarbons. The project team identified the following:

➤ Could the number of flanges be minimised and a better technology employed?
➤ Could the reliability of equipment be maximised and the maintenance requirements minimised given the environmental conditions it would be subjected to over the design life of the facility?

Achievement

The Project has proposed the use of hub connections for mechanical connections. These connections have had a good history in the North Sea for high pressure and temperature service. The team found that whilst a typical flange design relies on bolt tension for joint integrity, a clamp connector does not directly. The bolts provide the necessary tension for the clamp to act on the hubs and create a seal. Sealing integrity is unaffected by over tensioning the bolts and there are only 4 bolts (which includes 100% redundancy). As a result, the operational maintenance and bolt replacement is significantly quicker compared to a traditional flange.

The smaller overall size of the hub connection (compared to a flange) will assist operations and maintenance from an access viewpoint.

The simple design of the clamp assembly, inherent strength and integrity when installed, coupled with weight and space savings, make it the preferred selection for the NUI pipework. The Project believes that this system offers significant advantages over conventional flanges with less risk of failure requiring less maintenance. This results in less time for personnel to be on the platform.

262

KEEP THE DESIGN SIMPLE – EXAMPLE 3

Issue

A deluge system had been considered to provide protection to personnel evacuating from the NUI in the event of an ignited release and provide asset protection.

Challenge

Any active fire fighting system required equipment (and fuel storage) that would require maintenance and regular testing. The project team identified the following:

➤ Could the system reliability be maximised and the maintenance requirements be minimised?
➤ Is an active system necessary?

Achievement

The design team decided that an active fire fighting system was not required. The team identified that it was more prudent to place more emphasis on technical integrity (i.e. minimise leaks, optimise layout) of the NUI than to develop active systems. In addition, performance standards were to be established such that personnel could safely evacuate from the NUI before escalation could occur. In turn, this ensured the design allowed such egress and evacuation to be safely achieved.

Elimination of this system provided many benefits including

- removal of maintenance requirements (i.e. blocked nozzles) and minimises the need for personnel to visit the NUI.
- removal of congestion on the NUI arising from such equipment.
- elimination of corrosion potential of equipment as the firewater would be saltwater.

MINIMISE RELIEF SYSTEMS – EXAMPLE 4

Issue

Under abnormal condition, the potential existed for the NUI to be exposed to pressure and temperature fluctuations that may require some form of pressure relief.

Challenge

The design team identified the following:

➤ Was there an alternative system to control a process upset other than reliance upon a depressurisation system ?

Achievement

The team decided that topsides piping would be rated to full wellhead shut-in pressure. By setting the design limit to the maximum expected pressure, the NUI has been provided with a robustness and integrity that could withstand all expected pressures occurring during operation, shutdown and maintenance of wells.

This decision will result in eliminating the need to have an emergency depressurising system that includes associated vessels and piping and a potential leak source. This action will also minimise the congestion on board the topsides.

MAXIMISE RELIABILITY – EXAMPLE 5

Issue

Whilst equipment had been minimised, it was still necessary for the NUI to have power for monitoring systems and importantly, the safety systems, active during and throughout production. Equipment reliability, operating conditions (HP/HT) and third party intervention (TPI) were drivers for selection of the power supply for the NUI, not necessarily in that order of priority. The throughput from the producing wells each day is massive such that the cost of deferred production is a risk in itself. This meant that the electrical service will have to be highly reliable and always available.

Challenge

Electrical power supplies have been traditionally associated with rotating machineries. However these engines are maintenance-intensive pieces of equipment as they have moving or rotating components which are subject to wear and tear. The project team identified the following challenges:

➤ Could the need for traditional engines such as diesel for power generation be eliminated?
➤ Could the system be relatively impervious to unwanted TPI which has been an issue in South East Asia where solar panels have often been vandalized or stolen?
➤ Was the alternative system more reliable and effective than traditional power generation systems?

Achievement

The team identified that other alternative power generation techniques such as gas turbines, micro turbines and thermo-electric generators rely on fuel gas to generate the power. Since no processing was envisioned based on the "minimum facilities, normally unattended installation" concept, these technologies do not align with the project goals.

It was also identified that diesel engines introduced emission issues and from a safety perspective required bulk storage on the platform and regular tank filling. This was in conflict with the goal of zero visits by personnel.

In order to fulfil the goals as closely as possible, the Project decided to investigate the use of power transmission from the onshore facility via submarine cables to the NUI. This had the advantages of:

-   removed the need for fuel storage on the NUI;
-   minimised the leak paths associated with traditional systems from fuel piping connections; and
-   maximised the reliability and minimised inspections of the submarine cables as these were static equipment and the cables themselves were of the "install and forget" type of infrastructure.

CONCLUSIONS

By defining a project specific Inherent Safety Mission and accompanying design safety gaols, the Project has been able to focus collectively on developing a design that will be safer and more reliable. The Project has sought to foster a culture by which all on the design team challenges each other in applying inherent safety principles to their everyday engineering activities.

The Project has recognised that to achieve the goal of "zero visits and zero manning" the design must be simple, reliable and robust enough to minimise the maintenance activities and the number of personnel required.

The role of the Project HSE team has been to assist in the championing of the Inherent Safety Challenge by which:

- All hazards will be identified and fully understood.
- Every opportunity to minimise hazards at source will have been identified in time to be implemented where practicable.
- An effective strategy will be implemented to manage each major accident (i.e. design for the incident or ensure that it doesn't happen).

The outcomes of this approach will allow the:

- Project personnel (operations and design engineers) to fully understand the hazards and risk so they may take ownership and manage them effectively;
- Designers and operators to document the hazard management process so that the latter may understand the operating limits, the hazards, the systems to manage them and their responsibilities for safe operation and maintenance
- Provision of knowledge to allow fully effective systems to be provided to prevent and mitigate the hazards.

ACKNOWLEDGEMENTS

The authors would like to acknowledge all personnel in the management, engineering and operations team of the Project for without their efforts and enthusiasm in "taking on the Inherent Safety Challenge", we would not have been able to prepare this paper.

REFERENCES

Dalzell, G., Chesterman, A., 1997, Nothing is Safety Critical, Institute of Chemical Engineers (IChemE) Hazards XIII Conference, United Kingdom.

BP Getting HSE Right, Corporate Documentation.

Chia, S., Raman, J., 2001, Process Safety Approach to Plant Design and Operation, American Institute of Chemical Engineers (AIChE) Spring National Meeting 2001, United States.

A CASE HISTORY - WHOSE RESPONSIBILITY?

G.R. Astbury

Avecia Limited, P.O. Box 42, Blackley Manchester M9 8ZS

A case history of an incident is presented which occurred over 15 years ago in a solvent extraction oil refining Plant, whilst the author was working for a contractor. This occurred after the Plant had been handed over, and was in full production, but the hot oil heating system was not performing correctly. While sampling the hot oil, the oil sprayed out suddenly, burned the operator, and then ignited. Whilst it would be simple to blame the operator, there were consequences which had not been foreseen, and which resulted in the entire Plant being shut down for 12 weeks.

KEYWORDS: Hazard, Fire, Incident, Oil refining, Learning, Safety

INTRODUCTION

This incident occurred after the Plant was handed over to the Operating Company, following commissioning. This was about 16 years ago, prior to the Author joining ICI and the subsequent demerger to Zeneca and the sale as Avecia. The incident resulted in the entire Plant being shut down for a period of 12 weeks after a serious fire. This major shutdown was as a result of a series of small shortcomings in the design, construction, and operation of the Plant, all triggered by a single event of sampling the heat transfer fluid because of degradation of the performance of the heat exchangers in the Plant. Although an investigation was carried out, it concentrated on the immediate cause, but did not review the lessons that could be learned, nor the wider implications of the incident.

THE INCIDENT

The heat transfer hot oil system was not performing at its design rate, so it was decided that it was necessary to determine whether the oil had degraded. The operator had been asked to take a small 500ml sample so that an ASTM distillation[1] could be carried out to determine the fraction of light ends that had formed. There was no dedicated sample valve, so the Operator was instructed to use a 1/2" gate valve which was not intended for taking samples. This had been fitted to the 12" diameter overhead heat transfer hot oil mains to allow for draining the horizontal section after the isolation valves. This 1/2" gate valve was some 4 metres from the ground and was not intended to be used unless the system was cold and the section had to be drained. It was capped to prevent drips of oil which might leak past the gate valve falling onto the work area below. The operator had to stand on a ladder, as it was considered unnecessary to erect scaffolding.

When the cap was unscrewed, the oil sprayed out, and the operator received serious burns to the chest, but was able to climb down the ladder and report to the control room. The operator had checked that the valve was shut prior to unscrewing the cap, and thought that the oil leaking out as he unscrewed it was simply a small quantity of oil which had leaked past the gate. However, as the cap was unscrewed to the end of the thread, it flew off, and the oil sprayed out. It sprayed down onto the mineral wool acoustic insulation forming the stair-well wall, and covered the area beneath the valve, forming a large pool of hot oil, well above its flash point. The acoustic wall had been required as a condition of planning permission to screen neighbouring houses from the noise of the main pumps of the Plant. Shortly after the release the oil ignited as a classic "lagging fire", being hot and dispersed on the mineral wool of the acoustic insulation. The ensuing fire engulfed the north end of the Plant. The main power cables serving the entire Plant entered at the north-west corner and passed along the stair-well wall and then down the centre of the Plant. These cables were PVC covered, single wire armoured PVC insulated cables, so the fire destroyed the cables and shut the whole Plant down. It took 12 weeks of working 24 hours a day to replace the cables and re-commission the Plant. Fortunately, the fire did not spread to any of the major equipment and damage was limited to the stairs and the power and instrument cables.

The initial investigation showed that the 1/2" gate valve, which was on a short set-on branch, had a short stub of welding rod in it. It was not possible to determine when the stub of welding rod had lodged in the valve, but as the valve had been capped on installation, it would not have leaked when the system was commissioned. Therefore, even though the Operator had checked that the valve was closed, there was no way that it was known that there was an obstruction in the valve seat, and only when the screwed cap was removed would the oil come out. It was believed that it was safe to remove the cap even with the main under pressure, as the valve had been confirmed to be closed. In fact the Operator had only checked that he could not close it any more than it was already closed. He did not fully open the valve and then re-close it, counting the turns, so that he knew it was not obstructed. With hindsight, this should have been done, so that it would be known to be closed fully, and only small leakage past the gate would be likely to come out.

PIPE DEBRIS

The stub of welding rod that had been left inside the pipe was there simply because the procedure for cleaning the pipe had not been followed. The 12" pipes had been cut, set up, and welded in a Fabrication Shop on the site, and the 1/2" valve had been screwed onto the set-in branch and seal welded. To prevent potential damage of the valve seat during welding, the piping specification stipulated that the valve should not be tightly closed during seal welding, but left partially open. Thus the procedure had been followed at this point in that the valve had been opened prior to the welding. Once welded, the valve should have then been closed and capped with a screw cap. The entire pipe section was then subject to the regular inspection laid down in the piping specification. The inspection consisted of a visual inspection for cleanliness and freedom from debris, radiography of 10% of the butt-welds, and finally a hydraulic test. As the pipework was to be used on hot oil service, it was also to be drained thoroughly and dried out before installation.

As the weld debris had clearly been left inside the pipe, there was either a failure to apply the appropriate inspection of ensuring that the pipe was free of foreign matter, or contamination after inspection had occurred. In this case, it was not possible to determine whether the welding stub had been in at the start and had not been seen when the pipe was inspected, or whether it had entered during the erection phase whilst the pipe section was being moved from the on-site Fabrication Shop and lifted into the Plant structure.

Whilst such a breach of inspection is rarely so catastrophic, it does underline the need for procedures to be implemented correctly. As the entire construction phase of the project was behind time, it is easy to see that apparently trivial details can be ignored to save time. Hence if it is apparent that procedures are not being followed, then it is necessary to determine whether the reason is that the procedure is unworkable or simply the reason is unknown to those who should follow it. Appropriate corrective action is then required. In this case it is not possible to determine where the fault lay, but the inspection requirement of confirming that pipework is free from foreign matter is a generally accepted good practice, but is often ignored or simply undertaken in a cursory fashion.

THE HOT OIL SYSTEM

After a short period of normal production, it was apparent that the hot oil system was not performing correctly, since the main circulation pump was cavitating, and the heat transfer was not being achieved at the correct operating bulk temperature. The oil temperature had been increased marginally, but this had not improved the Plant performance. It was also apparent that there was some cavitation at the control valve on the largest heat exchanger. This was thought to be due to degradation of the hot oil by producing light ends which were flashing off in the pump and after the control valve. This was the reason for the requirement to sample the oil. The risk of poor performance was not entirely unexpected as the design of the heater had been significantly altered to a non-standard design with an unusually low pressure drop to minimise pumping power. The design for the hot oil system had been altered during the design phase because of the layout of the plant in the existing buildings, in order to save capital costs. The main plant location was constrained by a number of reasons, but the position of the Services Building, electricity Substation and the Control Room were fixed as these were existing buildings, served by existing cables. The Site plan is shown in Figure 1 below.

The existing cable supplying the Services Building from the substation was of a small capacity, and was buried along its entire length and passed under the main access road for the Plant. The installation of a new steam boiler, the hot-oil heater and circulating pumps, and a new air-compressor gave an electrical loading which exceeded the capacity of the existing cable. As all the loads were continuous, there was no possibility of applying diversity, so either the cable had to be replaced or the loads reduced.

The cost of replacing the cable was high, as the existing cable was buried rather than being in a service trench. Hence there was no simple way of running another cable either as a total replacement or as a separate cable running in parallel with the existing cable. Since the total load was only marginally above the continuous capacity of the cable, all the loads were critically examined to see if a reduction was possible. The largest single load was the hot oil main circulating pump rated at 45kW, and was the only load where some reduction might be possible.

The original design flow for the thermal load for the Plant fell between two sizes of pump frame, so the larger pump frame was selected. The standard motor for this pump frame fitted with its largest impeller was 45kW which not only exceeded both the flow and head requirements for the hot oil system, but also overloaded the electric cable supplying the building. Therefore the pump characteristic curves were examined, and it was found that by using a turned down impeller and accepting a lower head, the flow could be maintained and the required motor power could be reduced to only 30kW as the pump would be operating much closer to the best efficiency point. The 30kW motor current was within the capacity of the existing cable, but the available head on the pump was now below that required by the manufacturer of the hot oil heater. Thus there was a possible way of avoiding replacing the cable if the heater could be designed for a lower pressure drop. This was thought to be a distinct possibility, as one of the Contractor's senior staff had previously worked for the heater manufacturer, and was confident that such a design was possible.

The heater manufacturer was consulted about the possibility of designing a special hot-oil heater with a lower pressure drop. The standard heater at that thermal rating and oil flow rate used a three-start coil operating in parallel for the heater elements. This gave too high a pressure drop for the system, so the possibility of using more coils in parallel was investigated. This would result in a slightly larger heat transfer area, but as the velocity would be lower through each coil, the overall heat transfer coefficient would be reduced. Eventually, the heater manufacturer agreed to design a five-start coil heater, which would have a guaranteed low pressure drop at the required flow, thus allowing the smaller pump motor to be fitted, along with a turned-down impeller.

However, as the velocity would be lower, the heater manufacturer would not guarantee the performance of the heater other than the low pressure drop. The manufacturer also warned that as the velocity was much lower in the proposed five-start coil, there was a distinct possibility that the high film temperatures could occur, leading to thermal cracking of the oil. Hence the manufacturer suggested reducing the maximum film temperature from 315°C to 305°C, and a bulk temperature from 285°C to 280°C. The cracking of the oil could lead to deposits of carbon in the inside of the tubes, which would foul the heat transfer surface, leading to higher wall temperatures. As the pressure drop was so low, the risk of flow imbalance was much higher than would be the case for the normal three start coil, so the manufacturer suggested the incorporation of individual thermocouples at the outlet of each coil, so that any imbalance could be detected. In view of the cost of this, it was decided that the option of monitoring would not be purchased. However, it was decided that the oil would be changed on a regular basis to minimise the risk of degradation.

RISK MANAGEMENT

The decision to go ahead and purchase an unknown and unproven design of heater rather than accept the known cost of replacing the cable was taken on the basis that whilst there was a risk of the heater not performing, it was assumed to be manageable. To provide some finance for solving any technical problems which may arise during the commissioning, a sum of money was allocated as an ill-defined contingency, but there was no plan as to exactly what it would cover, or whether it would be adequate. It was decided that the heater manufacturers were being overly pessimistic in their warnings, so the unknown and un-quantified risk was accepted. No proper risk analysis was undertaken to determine the financial, technical and safety implications of the failure of the heater to perform at its required design temperature, despite the reservations of the manufacturer. Methods of risk management in capital projects is discussed elsewhere[2].

However, as there was an known potential for degradation of the hot oil, there was a recognition that complete replacement of the oil would be required, and a suitable 2" connection was provided to allow for emptying and refilling the system. No consideration was given to the provision of a sample point. Clearly the management of the technical risk was ignored. Once it became apparent that the decision had been taken to save capital by not replacing the cable, no follow up was undertaken for contingency plans. From the known potential for degradation of the oil into light ends and carbon deposits in the heater, it was apparent that a sample point should have been provided and a sampling schedule drawn up to monitor the quality of the oil. This omission was one of the root causes of the incident.

The choice of PVC cable for wiring the power into the Plant was made on the grounds of economy, as the risk of loss of the cables due to fire was considered remote. As there were existing pipe-bridges, it was decided to route all the cables along the existing pipe bridges, and provide short new spans to connect to the tank farm and the main Plant. This was seen as being economical as it used already available structures for support. As the Plant was a continuous plant, every pump on the Plant had an installed spare, so that in the event of loss of a pump for any reason, the installed spare would be available to avoid interruption of production. However, the supplies to the duty pumps and the supplies to the standby pumps were not segregated and brought in at opposite ends of the Plant, but were on the same cable run. The use of the single cable run introduced a potential common-mode failure that severance of the cable run for any reason would shut down both duty and standby pumps. Had the standby pump cables been brought in at the opposite end to the duty pumps, a failure at one end of the Plant may have stopped all the duty pumps, but would have allowed continued operation of the standby pumps. The potential for a failure of all the cables entering the Plant at a single point was considered too unlikely to warrant the use of mineral insulated copper clad cable, which is fire resistant. Consequently, the use of the lower cost cable, but using a common cable run, did not achieve the standby availability that had been intended.

LESSONS TO BE LEARNED

- Where procedures are available and required, then it is a management responsibility to ensure that they are followed. Where procedures are not followed, then the reason behind this should be determined, and appropriate corrective action should be taken.
- Project costs should be realistic, and pressure to reduce costs by cutting the scope should be either resisted, or the consequences of the reduced scope should be understood and accepted.
- The use of an ill-defined "contingency" for capital projects should be avoided, and the financial risks should be defined properly.
- Where information or warnings are given, then suitable action should be taken to ensure that the implications are fully considered, and are not simply ignored.
- The need to recognise the implications of operating outside the design parameters of equipment is essential.

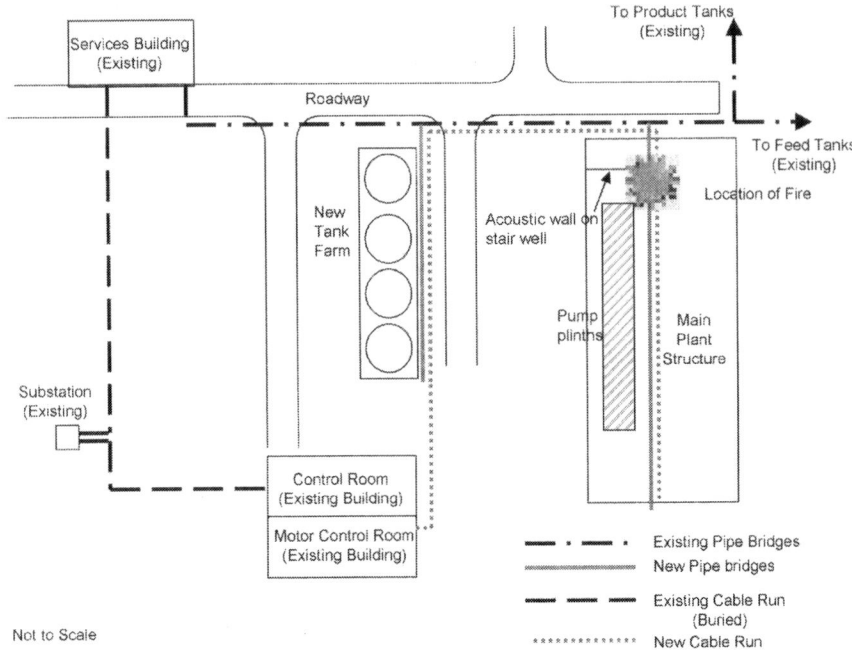

Figure 1.

REFERENCES

1. ASTM Test Method D86-01e1 "Standard Test Method for Distillation of Petroleum Products at Atmospheric Pressure".
2. Youds, P., "Financial risk management - Pharmaceutical capital projects", I.Chem.E. Project Management and Pharma Subject Groups joint meeting, Birchwood Park, Warrington, 13th February 2002.

DESIGN AND PROTECTION OF PRESSURE SYSTEMS TO WITHSTAND SEVERE FIRES

T A Roberts*, I Buckland**, L C Shirvill***, B J Lowesmith**** and P Salater*****
*Health and Safety Laboratory, Buxton, Derbyshire SK17 9JN
**Technology Division, Health and Safety Executive, Merseyside L20 3QZ
***Shell Global Solutions, P.O. Box 1, Chester CH1 3SH
****Advantica Technologies, Loughborough, Leicestershire LE11 3GR
*****Norsk Hydro ASA, Stabekk, 0246 Oslo, Norway

Mitigation of the impact of severe fires on hydrocarbon processing plant is critical to minimising the risk to personnel, reducing damage and limiting capital loss. It is recognised that current experimental data and the associated model validation are mainly confined to the response of vessels containing Liquefied Petroleum Gas. Models for predicting the behaviour of vessels containing multi-component fluids (with or without emergency depressurisation) under severe fire loads exist. However, relatively little validation has been performed.

Currently, industry tends to use the American Petroleum Institute's Recommended Practices 520 and 521 for the design of pressure relieving systems to withstand fire conditions but these are only applicable to some of the less severe, hydrocarbon pool-fire scenarios. Under the auspices of the Institute of Petroleum and with the support of the Health and Safety Executive, interim guidelines have recently been published. They are intended to assist design and process engineers concerned with large, essentially fully enveloping pool fires and jet-fire impingement on pressure vessels and their associated pipework. The guidelines are intended for use primarily for designing new facilities and specifically deal with fires that are more severe than the open pool fires currently covered by API guidance.

This paper considers the guidance in the API recommended practice and the new Institute of Petroleum guidelines and compares the approaches with data from hydrocarbon pool and jet-fire trials on filled propane vessels.

KEYWORDS: Pool fires, jet fires, heat transfer, fire protection, pressure relief, emergency depressurisation, pressure vessels.

INTRODUCTION

Pressure systems are either designed to withstand the highest expected pressure or fitted with means of preventing over-pressurisation. During normal operations, protection of pressurised systems is provided by appropriately sized pressure relieving devices, typically pressure relief valves (PRV) and/or bursting discs, which are designed to automatically limit the maximum pressure within the systems. Process plant, especially that offshore, is usually fitted with an emergency depressurisation (EDP) system, in addition to a PRV, in order to reduce the risk and consequences of failure. In an emergency, the EDP system allows intentional and controlled discharge of the contents of the process plant into the flare or vent facilities.

Whilst PRVs are designed to prevent failure of pressurised systems due to over-pressurisation, it is recognised that the load-carrying capacity of pressure systems will be significantly compromised by exposure to fire. This occurs as a consequence of reduction of the strength of the vessels and pipe-work with increasing temperature. In general, there will be critical temperatures where, for example, vessels will fail, valves will cease to function adequately and/or flange connections will loosen. Pressure systems can also fail due to the increasing pressure exerted by their contents as these to rise in temperature and if the relief system is inadequate. Hence, the ability of pressure relieving and depressurisation systems to safeguard pressurised systems in a fire is critically dependent upon the assumptions made about the type and size of the threatening fire.

Currently, industry tends to use American Petroleum Institute's Recommended Practices 520[1] and 521[2] for the specification of pressure relieving systems to enable pressurised plant to withstand fire conditions. However, these are restricted to some of the less severe hydrocarbon pool-fire scenarios. Under the auspices of the Institute of Petroleum, and with the support of the Health and Safety Executive, interim guidelines[3] have recently been published. They are intended to assist design and process engineers concerned with large, essentially fully enveloping pool fires and jet-fire impingement on pressure vessels and their associated pipework. The guidelines are intended for use primarily for designing new facilities and specifically deal with fires that are more severe than the limited sizes of open pool fires currently covered by API guidance.

This paper considers the approaches in the API Recommended Practice 521[2] and the new Institute of Petroleum guidelines[3] and compares the approaches using data from kerosene pool-fire trials and flashing-liquid propane jet-fire trials on filled, horizontal, cylindrical propane vessels. A brief consideration of EDP systems is also given.

MANAGEMENT OF FIRE HAZARDS

Safe operation does not rely solely on the protection afforded by relief and depressurising systems. Indeed, pressure relief and depressurising systems are safeguard measures that only come into play when other measures aimed at preventing fires have failed. Good design (for example, careful use of flanged connections), elimination of potential ignition sources (e.g. hazardous area classification and permit-to-work procedures; such as for hot work), and early fire detection and response (in the unlikely event that ignition does occur) all play a role.

As shown later, pressure relief and depressurisation do not guarantee that vessels will not fail in a fire, particularly a severe fire. In principle, depressurisation and fire protection can prevent vessels from failing, but in practice the adequacy of these measures must be judged in the context of the likelihood and consequences of vessel failure. For example, a higher level of safeguards will be required on a manned offshore facility or onshore installation with offsite populations in close proximity, than on a small vessel in a remote location.

It is now common practice to carry out a fire risk analysis where measures taken to reduce the risk of a fire occurring, and the effectiveness of mitigation measures should a fire occur, are all included. Depending on the nature and size of the facility, the fire risk analysis may range from a simple assessment through to a fully quantified risk assessment (QRA). For offshore facilities, this would be part of the Fire and Explosion Strategy (FES) outlined in ISO 13702[4]. For onshore installations, this would be incorporated in the process safety management system for the plant and, where applicable, form part of the safety report or demonstration of safe operation to the competent or regulatory authority.

CURRENT INDUSTRY PRACTICE

A survey (under the auspices of the Institute of Petroleum (IP)), of a world-wide range (circa 160 organisations) of operators/owners, consultants and design contractors, indicated that there was no consistent approach or consensus on what criteria to use to design against the risk of severe fires. The main conclusions from the responses and information received were:

(a) There is little consistency in the design methodology used, even within a single company;
(b) Some said they had limited in-house expertise and engaged a specialist design contractor; and
(c) Some applied API RP 521 and assumed that by designing to that code, there was no additional risk.

There appeared to be no industry-preferred methodology and this has been taken as being the justification for preparing interim IP guidelines.

Whilst the recommendations and equations in API RP 521 are generally applicable to a refinery or chemical plant, they were never intended to cover all fire scenarios, especially those that may foreseeably occur on offshore installations. The scope and application does not apply to very large enveloping pool fires or impinging jet fires. Hence, should process plant fitted with protective systems designed to API RP 521 or a similar standard be exposed to severe fires, such systems may be insufficient to prevent failure of the vessel before the inventory has been safely removed. Should the vessel contain a hazardous inventory (usually flammable hydrocarbons), the consequences of failure will be compounded with potentially catastrophic results and escalation of the event.

AMERICAN PETROLEUM INSTITUTE RECOMMENDED PRACTICE 521

As indicated above, industry has traditionally used the American Petroleum Institute's Recommended Practice 521[2] when designing pressure vessels to withstand the effects of fire. When a pressure vessel containing liquid and vapour hydrocarbons is heated by fire, there are two main considerations:

(1) Heat will be transferred to the contents thereby increasing the pressure.
(2) The vessel wall not in contact with liquid may be heated to a point where the vessel fails.

HEAT TRANSFER TO THE CONTENTS

The basic formulae given in API 521 for the heat absorbed by the contents of a vessel engulfed in fire is:

$$Q = 43.2 \ F \ A^{0.82} \text{(with adequate drainage)} \tag{1}$$

$$Q = 71.0 \ F \ A^{0.82} \text{(without adequate drainage)} \tag{2}$$

where Q is the heat absorbed (kW), A is the effective wetted surface area of vessel (m^2) and F is an environment factor.

The effective wetted area of the vessel is defined as the surface area of the vessel in contact with liquid up to a height of 7.62 m (25 ft) above ground level or other surface that could sustain a fire. Effective elevation is based on observations that wind and shape effects limit the contact of the fire with the vessel as the elevation increases. Some companies use larger values for the effective elevation. The philosophy of wetted area is that heat transferred to the liquid will eventually cause it to boil and produce more vapour, whereas heat transferred to the vapour phase will just cause vapour expansion. The environmental factor (F) is an attempt to correct the heat flow for the effect of insulation, water drenching and earth covering. The values used for F and limits of application give rise to most conflicts between codes (see Parry[5]).

HEAT TRANSFER TO THE UNWETTED SURFACE
Heat input from an open fire to the bare outside surface of an unwetted wall may, in time, be sufficient to heat the vessel wall to a temperature high enough to rupture the vessel. API RP 521 gives two illustrative figures. The first figure gives the average rate of heating of steel plates exposed to an open gasoline fire on one side. Observed data (mean temperature versus time) for plate 3.2 mm (1/8 inch) thick are given together with computations for plates 3.2 mm, 12.7 mm and 25.4 mm thick. The second figure illustrates the effect (rupture stress versus time at indicated temperature) of overheating steel (AST A 515, Grade 79).

INSTITUTE OF PETROLEUM INTERIM GUIDANCE
Scandpower Risk Management AS[6] have prepared a guideline for protection of pressurised systems exposed to fire for Statoil and Norsk Hydro. It is based on Norsk Hydro's internal guidance and is primarily concerned with the design of systems fitted with emergency depressuring systems (EDP), although much of the information given is also relevant to other systems. The guidance produced by the Institute of Petroleum (IP) draws heavily on this guideline in relation to EDP. However, the IP guidance[3] has a wider scope in that it is intended to also cover vessels, e.g. storage vessels, which are not normally fitted with EDP systems. The key components of the IP guidance are:

- An outline of the design process;
- fire types and thermal loading;
- Equipment response and failure prediction;
- Protective measures; and
- Areas of uncertainty, which might warrant further study.

The IP guidelines are intended to supplement the existing codes and should be used in conjunction with them. The main differences between the IP and API guidance are in relation to severe fires. These are considered as follows.

MEASURED HEAT FLUXES
Over the past 15 years, there has been a number of joint industry projects designed to generate data for the validation of models for predicting the response to open and confined severe hydrocarbon pool and jet fires. It is not the intention to review this data here but some of the key references are given as background information to the proposed values given later:

- Most of the experimental data prior to 1991 are summarised in reports produced as part of Phase 1 of the Blast and Fire Engineering Project for Topside Structures (Cowley and Johnson[7] and Cowley[8]).

- Data on free and impinging horizontal jet fires were obtained as part of two European Community (EC) projects viz. AA and JIVE. As part of project AA *(Two-phase releases for toxic and flammable substances: Thermal initiation, source terms and fire effects)*, a programme of large-scale steady state ignited jet releases of natural gas and liquefied petroleum gas (LPG) was performed during the period 1988 – 1989. A total of 125 individual experimental data sets were prepared, including 68 experiments in which the fires impacted onto a cylindrical vessel or a pipe. Cowley and Pritchard[9] published some of the work in regard to the thermal impact on structures.

- The JIVE project[10] was concerned with the *Hazard consequences of jet fire interactions with vessels containing pressurised liquids*. Part of the project was concerned with taking 2 tonne propane tanks to failure[11] via a nominal 2 kg s^{-1} flashing liquid propane jet fire. Another part of the JIVE programme of work, performed by Davenport[12], involved measurement of the impact of the flames from the natural gas/butane mixtures on targets.

- In 1995, Gosse and Pritchard[13] studied the heat transfer from vertical natural gas jet fires impacting onto the underside of a flat 20 m by 20 m deck at flow rates up to 3 kg s^{-1}. The effects of exit pressure and orifice diameter, stand-off distance, partial confinement and wind speed were studied.

- Unconfined crude oil jet fires were studied in Phase 2 of the JIP on 'Blast and Fire Engineering of Topside Structures' (Selby and Burgan[14]) and in a subsequent JIP on releases of 'live' crude oil containing dissolved gas and water (Evans[15] et al.).

- A major experimental programme of large-scale compartment fires (Chamberlain[16]) was carried out by Shell research at SINTEF. 0.35 kg s^{-1} propane jet fires were burnt inside a 135 m^3 insulated steel compartment with reduced ventilation to simulate accidental fires in offshore modules. Further work on confined fires was performed in Phase 2 of the JIP on 'Blast and Fire Engineering of Topside Structures' (Selby and Burgan[14]), where a 415 m^3 insulated compartment was also used.

The data from these trials have been used to derive the values proposed later.

HEAT LOADS FROM JET AND POOL FIRES

In order to calculate the heat up of vessels or pipe-work subjected to fire, it is necessary to quantify and understand the thermal load imposed by fires. This thermal load is a combination of convection from the hot combustion products passing over the object surface and radiation emitted by the flame to the object surface. In reality, this is a very complex event and the following issues are relevant:

- The relative proportions of radiative and convective load from a flame will vary depending on the fuel type and location of the object within the flame;

- The total heat loads will vary depending on the fuel type, the size and shape of the object and the location of the object within the fire;

- The heat loads will vary over the surface of the object; and
- The heat absorbed by the object will vary with time.

In most cases, it can be assumed that the flame and the object surface are diffuse grey bodies and that the ambient temperature of the surroundings is low compared to the flame temperature. Using these assumptions, in simple terms for a fully engulfing fire, the heat flux absorbed by the object can be expressed as:

$$Q_{ABS} (kWm^{-2}) = Q_{RAD} + Q_{CONV} = \varepsilon_s \, \sigma \, (\varepsilon_f \, T_f^{\,4} - T_s^{\,4}) + h \, (T_f - T_s) \qquad (3)$$

where ε_f, ε_s are the flame and surface emissivities respectively, σ is the Stefan Boltzmann constant (5.6697×10^{-11} kWm^{-2}K^{-4}), T_f, T_s are the flame and object surface temperatures (K) respectively, and h is the convective heat transfer coefficient (kWm^{-2}K^{-1}).

It is important to note that the thermal loading absorbed by an object in a fire as described by equation (3) will reduce as the object heats up. As can be seen, if the object heats up, T_s will increase whilst T_f is likely to remain much the same, resulting in reduced Q_{ABS}. The important parameters (likely to be unchanging in a steady state fire) are thus T_f and ε_f for radiative loading, and T_f and h for convective loading. Generally speaking, it is these parameters that need to be specified together with ε_s (which may change with temperature) for calculation of the response of an actual section of pipe-work or vessel subjected to fire impact. Therefore, these are the parameters specified in Table 1 for different fire scenarios.

However, researchers most often quote heat flux loadings from fires to objects in kilowatts per square metre, for example, as in the results[14] of the Blast and Fire Engineering Project, Phase 2 and the interim guidance notes issued[17] after Phase I of that project. Without other information such as T_f and ε_f this may be of little value. Furthermore, in experiments designed to assess thermal loading from flames to objects, the "load" actually measured is that absorbed by instruments situated on the object surface, not the surface itself. Typically, calorimeters are used to measure the total heat load and sometimes radiometers are deployed to measure the radiative component. During the experiment, the instrument is deliberately maintained at a constant known temperature. In the Institute of Petroleum Interim Guidance[3], advice is given on how some of the above factors can be simplified for calculation purposes and on the interpretation of heat flux measurements.

PROPOSED VALUES

As mentioned above, different fire types will result in different heat fluxes, for example:

- High pressure releases of fuel, with a significant gas content, will tend to produce a high convective flux.
- Pool fires and jet fires of liquid fuels tend to have a low convective flux.
- Higher hydrocarbons tend to produce more radiative flames.
- Fires in enclosed spaces can result in higher flame temperatures and hence fluxes through restricted heat losses (they can also result in lower values if the air supply rate is too low).
- Very large fires can produce higher flame temperatures and hence fluxes.

Table 1. Typical parameters for pool and jet fires

Fire type	Open pool fires	Severe or confined pool fires	Open jet fires[c]	Confined jet fires
Total incident flux[a] (kWm^{-2})	50–150	100–250	100–400	150–400
Radiative flux[a] (kWm^{-2})	50–150	100–230	50–250	100–300
Convective flux (kWm^{-2})	0	0–20	50–150	50–100
Emissivity[b] of flame ε_f	0.7–0.9	0.8–0.9	0.5–0.9	0.8–0.9
Temperature of flame, $T_f(K)$	1000–1400	1200–1450	1200–1500	1200–1600
Heat transfer coefficient, h $(kWm^{-2}K^{-1})$	0	0–0.02	0.04–0.17	0.04–0.11

Notes: a) Radiative (and total) flux does not take account of emissivity of surface of
   object (that is, quoted values are equivalent to $\varepsilon_s = 1$).
   b) Emissivity is influenced primarily by fuel type and size of the fire. Higher
   hydrocarbons characteristically have higher emissivities. Large fires will
   also tend to produce more luminous flames due to soot production and
   again this will tend to lead to higher values of emissivity. The values
   presented relate primarily to hydrocarbons and values outside this range
   may apply for other fuels, especially if cleaner, essentially soot-free flames
   are observed.
   c) Mixed fuel jet fires with both high velocity gas and a higher hydrocarbon liquid
   fuel tend to produce the highest heat fluxes, producing both high radiative and
   convective components. The lowest overall fluxes are expected from
   pressurised liquid releases of higher hydrocarbons ($\geq C_5$).

Hence, a wide spectrum of fires can be produced with differing heat fluxes depending
on various parameters. Typical heat fluxes produced by hydrocarbon pool and jet fires may
vary as shown in Figure 1, depending on the confinement and severity of the fire. For
simplicity, the following four categories of fire are proposed in the IP Interim Guidelines
for consideration in the design process:

- Open pool fires
- Large or confined pool fires
- Open jet fires
- Confined jet fires

The heat fluxes measured in experiments (see MEASURED HEAT FLUXES
above) are closely related to the initial incident heat flux experienced by an object.
Hence, Table 1 presents values of heat flux (total, radiative and convective) typical of
those initially expected from a fire to an object. The radiative flux given represents
$\sigma\,\varepsilon_f\,T_f^4$ and the convective flux given represents h $(T_f - T_s)$ when T_s is low (nominally

ambient). The table presents ranges of values - emphasising the point illustrated in Figure 1 that fire types are widely varying.

In experiments where direct measurement of heat flux is made, the reported heat flux is the incident heat flux. However, in experiments where the heat flux is calculated from the rise in temperature of the contents the reported heat flux is the absorbed heat flux.

COMPARISON WITH DATA FROM FIRE TRIALS ON PROPANE VESSELS

A comparison, of the heat transfer predictions from the IP guidance and the pressure relief requirements of API, is made with experimental observations from hydrocarbon pool fire and flashing liquid jet-fire trials on filled propane vessels.

EXPERIMENTAL TRIALS

The Health and Safety Laboratory has performed two sets of fire trials on filled propane tanks. Moodie[18] et al. conducted trials to determine the behaviour of a 5 tonne horizontal cylindrical LPG tank engulfed in kerosene pool fires (Figure 2). Five tests were carried out with commercial propane fill levels from 22% to 72%. The kerosene and vessel were contained in a 3.8 m by 6.8 m bund. In fire durations from 11.6 minutes (22% fill) to 31.0 minutes (72%) fill, the peak heat fluxes, corrected for the absorptivity (assumed to be 0.8) of the pipe calorimeters used, was 105 kWm^{-2} (incident heat flux). It was stated that the engulfing fires were fully established within 3 minutes of ignition with heat fluxes of 100 kWm^{-2}.

Roberts and Beckett[19] performed four trials on 2 tonne horizontal, cylindrical LPG tanks engulfed in nominal 2 kgs^{-1} flashing-liquid propane jet fires (Figure 3). Tanks with fill levels from 20% to 85% were heated in a jet fire until they failed (within 5 minutes). A target of similar shape and size to the LPG tanks was fitted with pipe calorimeters at 90° intervals and the mean incident heat fluxes found to be in the range 180 to 200 kWm^{-2}.

COMPARISON WITH PROPOSED VALUES

The Shell HEATUP model[20] was used to derive the parameters, given as ranges in Table 1, for the:

- kerosene pool fire trial on the 72% full tank (typical of an open pool fire of the type referred to in API RP 521); and
- flashing-liquid propane jet fire on the 85% full tank (typical of an open jet fire with a higher-hydrocarbon, more-radiative, fuel).

The values are compared with the ranges of values proposed in the previous section in Table 2. The values used in the HEATUP code gave a good representation of the wall temperatures, pressures and time to failure for the jet-fire impinged vessels. All values used are within the ranges specified in Table 1.

Table 2. Comparison of data derived from lpg tank trials

Fire type	Kerosene pool fire		Open jet fire	
	HEATUP*	Table 1	HEATUP*	Table 1
Total incident flux (kWm^{-2})	75	50–150	170	100–400
Radiative flux (kWm^{-2})	75	50–150	120	50–250
Convective flux (kWm^{-2})	0	0	50	50–150
Emissivity of flame ε_f	0.9	0.7–0.9	0.6	0.5–0.9
Temperature of flame, T_f(K)	1070	1000–1400	1370	1200–1500
Heat transfer coefficient, h (kWm^{-2}K^{-1})	0	0	0.05	0.04–0.17

*ε_s assumed to be 0.65 for the modeling

RELIEF PERFORMANCE

API RP 521 defines a pressure-relieving system as an arrangement of a pressure-relieving device, piping and a means of disposal intended for the safe relief, conveyance and disposal of fluids in a vapour, liquid or gaseous state. A relieving system may consist of only one pressure relief valve or rupture disk, either with or without discharge pipe, on a single vessel or line. A more complex system may involve many pressure-relieving devices manifolded into common headers to terminal disposal equipment. In this section, the API PRV requirements are compared with the PRVs actually used and the level of protection provided in the trials.

The pressure relief valve requirements were sized in accordance with the requirements of API RP 520 and API RP521, which covers calculation of the required pressure relief area for the rate of vapour generation determined. The results are compared with the characteristics of the pressure relief valves actually used in Table 3.

Clearly, the trial conditions are such that drainage for any spilt fuel is inadequate. However, this was taken into account in sizing the installed safety valves, which were adequate in accordance with API RP 520/521 recommendations for such a scenario in both cases. In the 72% fill pool-fire trial, the PRVs opened (14.3 barg) at about the set pressure and the pressure in the vessel did not exceed this throughout the trial confirming the adequacy of the API recommendations. In the case of the 85% fill jet-fire trial, the PRV opened at 18.3 barg (set pressure 17.2 barg), cycled open and shut twice and then remained open until catastrophic tank failure at 24.4 barg. Hence, a PRV sized to API 520/521 cannot be assumed to be capable of keeping the pressure to the set pressure when subjected to a jet fire of the nature used in the trials. This suggests that the combination of heat flux and wetted surface area recommended in API 521 is inadequate for sizing PRVs in a severe fire situation.

Table 3. Comparison of api prv requirements with those used

	5 tonne tank in pool fire		2 tonne tank in jet fire	
Parameter	Adequate drainage	Inadequate drainage	Adequate drainage	Inadequate drainage
Maximum fill level	0.72	0.72	0.85	0.85
Wetted surface area (m^2)	25.7	25.7	15.3	15.3
Heat transfer rate (kW)	577	948	404	664
Vaporisation rate (kgs^{-1})	2.00	3.29	1.51	2.48
Safety valve set pressure (barg)	14.3	14.3	17.2	17.2
API required effective flow area (mm^2)	463	761	296	486
Installed effective flow area (mm^2)	887	887	619	619

Note: Effective flow area is the product of the actual safety valve flow area and its coefficient of discharge.

The pool fire trials were not designed to take the vessel to failure. All the vapour space wall temperatures behaved similarly, increasing rapidly once the fire had become established, but rising less rapidly once venting commenced. In individual tests, large temperature differences (440°C to 610°C for 58% fill) existed at any one time, both across the tank and from end to end. The peak rates of temperature rise (from ambient to 400°C) was roughly 1.25 Ks^{-1} for a vessel with a wall thickness of 12 mm. Figure 1 in API RP 521 suggests a heating rate of 1.75 Ks^{-1} i.e. it is conservative compared to this test. However, for the jet-fire trials, the heating rate with a wall thickness of 7 mm was 7 Ks^{-1}. Interpolation of the plots in API RP 521 suggests a heating rate of around 4 Ks^{-1} i.e. a gasoline fire is much less severe than a jet fire. All the LPG vessels failed catastrophically in the jet fires within 5 minutes, at pressures from 16.5 barg to 24.4 barg and at maximum dry wall temperatures from 704°C to 870°C.

EMERGENCY DEPRESSURISATION
As indicated in the Introduction, API RP 521 is also used for the design of emergency depressuring (EDP) systems. API RP 521 defines a vapour depressurising system as a protective arrangement of valves and piping intended to provide for rapid reduction of pressure in equipment by releasing vapours. The actuation of the system may be automatic or manual. In general, emergency depressuring systems are usually fitted to all offshore, and most onshore, process vessels. However, they are not usually fitted to storage vessels as it is not practicable to remove the large inventories involved. All types of vessel are normally fitted with pressure relieving systems.

For vapour depressurising, API 521 recommends *"reducing the equipment pressure from initial conditions to a level equivalent to 50 per cent of the vessel's design pressure within approximately 15 minutes. This criterion is based on the vessel wall temperature versus stress to rupture and applies generally to vessels with wall thicknesses of approximately 25 mm or more. Vessels with thinner walls require a somewhat greater depressuring rate. The required depressuring rate will depend on the metallurgy of the vessel, the thickness and initial temperature of the vessel wall, and the rate of heat input from the fire."* *"Where fire is controlling, it may be appropriate to limit the application of vapour depressuring to facilities that operate at 17.24 barg and above, where the size of the equipment and volume of the contents are significant. An alternative is to provide depressuring on all equipment that processes light hydrocarbons, and set the depressured rate to achieve 6.9 barg or 50 per cent of the vessel design pressure, whichever is lower, in 15 minutes. The reduced operating pressure is intended to permit somewhat more rapid control in situations in which the source of a fire is the leakage of flammable materials from the equipment to be depressured."*

In the severe fires identified, EDP may not guarantee vessel protection if designed to API 521 because the heat transfer to the contents and dry wall will be higher than assumed. In a review by Roberts[21] et al., an analysis of an example pressure vessel was performed in order to demonstrate the likely thermal and mechanical response of pressurised equipment to a severe fire. It indicated that guidance was required on the behaviour of vessels and their contents in severe fires.

Scandpower Risk Management AS[6] have prepared a guideline for protection of pressurised systems exposed to fire for Statoil and Norsk Hydro. It is based on Norsk Hydro's internal guidance and is primarily concerned with the design of systems fitted with emergency depressuring systems although much of the information given is also relevant to other systems. The guideline reflects the Statoil and Norsk Hydro design philosophy which focuses on fast depressurisation with maximum use of the flare capacity, rather than use of passive fire protection to mitigate the consequences of the fire. It suggests, as a starting point, that for severe fires there should be a pressure reduction to 7 bar within 8 minutes i.e. nearly twice as fast as recommended by API. The guidance produced by the Institute of Petroleum (IP) draws heavily on the Scandpower guideline in relation to EDP.

DISCUSSION

The hydrocarbon pool-fire results suggest that guidance in API 521 works well for the design of systems to resist open hydrocarbon pool fires with incident heat fluxes up to about 100 kWm^{-2}. However, for the flashing liquid propane, jet fires with incident heat fluxes of the order of 180 kWm^{-2}, it appears inadequate. API RP 521 does not offer guidance on jet fires or confined fires but these are not excluded from the scope. As a consequence of this, there is a tendency for the user to follow the code's emergency depressurisation rates and related relief valve sizing recommendations without understanding the limitations in terms of the severity of the fire. This has been raised with API for consideration in their next revision of API RP 521.

The IP guidance allows the higher heat fluxes that occur in severe fires to be taken into account. It categorises the fires into usable data sets that provide the necessary heat transfer properties in order to evaluate the equipment response. It is recognised that in many situations a detailed structural response calculation is not required. However, when it is required, there are no fully validated models available to assist in these calculations as no fire trials have been performed on pressure vessels fitted with EDP systems.

CONCLUSIONS
The main conclusions are that:

- Application of API guidance can lead to a significant under estimation of heat load and hence under-size relief systems in severe fires, which can occur offshore and in some onshore situations.
- The IP guidance provides a more realistic assessment of potential heat loads.
- Validation of the new IP guidance against tests involving pressure vessels incorporating EDP systems is recommended.

ACKNOWLEDGEMENTS
The authors gratefully acknowledge the contribution from other members of the Institute of Petroleum Jet Fire Pilot Group under the chairmanship of S Schuyleman (IP). The PRV sizing calculations were performed by Ms A J Wilday (HSL/HSE).

DISCLAIMER
The opinions and conclusions expressed in this paper are those of the authors and do not necessarily reflect the policy or views of the organisations involved.

REFERENCES
1. American Petroleum Institute Recommended Practice 520, 2000, Sizing, Selection, and Installation of Pressure-relieving Devices in Refineries: Part I – Sizing and Selection, API 2000.
2. American Petroleum Institute Recommended Practice 521, 1997, Guide for pressure-relieving and depressuring systems, 4th Edition, 1997.
3. The Institute of Petroleum, Guidelines for the design and protection of pressure systems to withstand severe fires, reference to be provided.
4. International Standards Organisation 13702:1999, Petroleum and natural gas industries – Control and mitigation of fires and explosions on offshore production installations – Requirements and Guidelines.
5   Parry C F, 1992, Relief systems handbook, IChemE, Chapter 7, 1992, ISBN 0 85295 267 8.
6. Scandpower Risk Management AS, 2002, Guideline for protection of pressurised systems exposed to fire, Scandpower (downloadable from the Scandpower website at http://www.scandpower.com/?CatID=1071).
7. Cowley L T and Johnson A D, 1992, Oil and gas fires – characteristics and impact, OTI 92 596, HSE, 1992.

8. Cowley L T, 1992, Behaviour of oil and gas fires in the presence of confinement and obstacles, OTI 92 597, HSE, 1992.
9. Cowley L T and Pritchard M J, 1990, Large-scale natural gas and LPG jet fires and thermal impact on structures, Gastech 90 Conference and Exhibition, Amsterdam 4–7 December 1990.
10. Duijm N J, 1995, Hazard consequence of jet fire interactions with vessels containing pressurized liquids – JIVE final report, TNO report R95–002, 1995.
11. Roberts T, Gosse A and Hawksworth, 2000, Thermal radiation from fireballs on failure of liquefied petroleum gas storage vessels, Trans IChemE, Vol. 78, Part B, May 2000, pp. 184–192.
12. Davenport J N, 1994, Large Scale Natural Gas/Butane Mixed Fuel Jet Fires, Final Report to the EC. EC Contract STEP-CT90–0098 (DTEE). December 1994.
13. Gosse, A J and Pritchard, M J, 1995, Large Scale Jet Fire Impaction onto a Flat Surface, International Gas Research Conference, Vol. II Exploration and Production, pp. 493 – 504.
14. Selby C A and Burgan B A, 1998, Blast and Fire Engineering for Topsides Structures – Phase 2, Final Summary Report, SCI-P-253, The Steel Construction Institute, ISBN 1 85942 078 8.
15. Evans J A, Exon R and Swaffield F, 2000, Large Scale Experiments to Study Jet Fires of Crude Oil/Gas/Water Mixtures, Report R2961 BG Technology (published internally by HSE as OTN 2000 042).
16. Chamberlain, G A, An Experimental Study of Large-scale Compartment Fires, Trans.I.Chem.E., 72, Part B, pp. 211–219, 1994.
17. The Steel Construction Institute, 1992, Interim Guidance Notes for the Design and Protection of Topside Structures against Explosion and Fire, SCI-P-112/509.
18. Moodie K, Cowley L T, Denny R B, Small L M and Williams I, 1988, Fire engulfment tests on a 5 tonne LPG tank, J. Haz. Mats, 20 (1988) pp. 55–71.
19. Roberts T and Beckett H, 1996, Hazard consequences of jet-fire interactions with vessels containing pressurised liquids: Project final report, HSL Internal Report PS/96/03 (to be published on the HSE/HSL website).
20. Persaud M A, Butler C J, Roberts T A, Shirvill L C and Wright S, 2001, Heat-up and failure of Liquefied Petroleum Gas storage vessels exposed to a jet fire, 10th International Symposium on Loss Prevention and Safety Promotion in the Process Industries, 19–21 June 2001, Stockholm, ISBN 0 444 50699 3.
21. Roberts T A, Medonos S and Shirvill L C, 2000, Review of the response of pressurised process vessels and equipment to fire attack, Offshore Technology Report OTO 2000 051, HSE (downloadable from the HSE website).

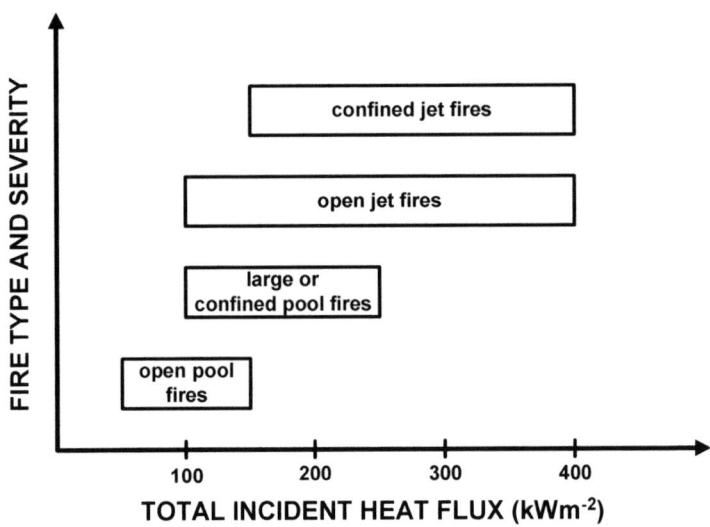

Figure 1. Illustrative heat fluxes for pool and jet fires

(Courtesy of HSL -see reference 18)

Figure 2. Kerosene pool fire trial (with flare from PRV)

(Courtesy of HSL- see reference 11)

Figure 3.   Flashing liquid propane jet fire trial

FIRE AND EXPLOSION HAZARDS OF MEAT & BONE MEAL: STORAGE, TRANSPORT AND PROCESSING

Steven J Manchester

Fire and Risk Sciences (FRS) Division, BRE Ltd, Garston, Watford, WD25 9XX

In order to restore consumer confidence in eating beef following the BSE outbreak, the UK Government instigated the Over Thirty Month Scheme (OTMS). This is a voluntary scheme whereby cattle over thirty months, showing no clinical signs of BSE are slaughtered and their carcasses incinerated or rendered into Meat & Bone Meal (MBM). As a result of this scheme producing huge quantities of MBM, extensive storage areas were required. A number of these stores have experienced fires. BRE was asked to investigate these fire incidents to determine the cause, and to undertake a research programme to assess the fire and explosion hazards in storing, processing and transport. This paper describes the research, testing and consultancy work undertaken by BRE on MBM, which has produced specific guidance to mitigate the hazards.

KEYWORDS: Meat & Bone Meal, spontaneous combustion, dust explosion, calorimetry, fire.

BACKGROUND

As a result of the BSE crisis that afflicted the UK in 1996, the Government instigated the Over Thirty Month Scheme (OTMS) to restore consumer confidence in eating beef. This involved the culling of cattle showing no clinical signs of BSE, once they had reached an age of thirty months. The carcasses were either incinerated or rendered into Meat & Bone Meal (MBM) and tallow. Huge quantities of MBM were produced, which had reached over 470,000 tonnes as of March 2000, prior to the commencement of an incineration programme to destroy the MBM. However, at this time only one incinerator was capable of incinerating the MBM being produced. Large in-door storage areas, mainly old factory and farm buildings, were used to hold the MBM while awaiting incineration, with some stores holding material for over four years before being emptied.

INTRODUCTION

Following a number of fires involving MBM in storage, the Intervention Board (now the Rural Payments Agency) commissioned BRE to undertake research with the following objectives:

- Investigate the potential causes of fire in MBM during bulk transport and storage;
- Assess the hazards and risks associated with fires and dust explosions in the material; and
- Provide guidance on prevention and mitigation of fires and dust explosions with MBM during storage, transportation and processing.

A research programme was devised which focussed on the following areas:

- Site visits to six MBM stores, two of which had experienced fires in the MBM material, and an incinerator process plant handling MBM.

- Investigation into the cause of the fires in the two stores.
- Self-heating investigation on a number of MBM samples from stores around the UK and Northern Ireland.
- Dust explosbility assessments and guidance on preventative and protective measures.
- Toxic analysis of the combustion gases produced from smouldering and flaming combustion of MBM.
- Bomb and cone calorimetry studies on the fire behaviour of MBM.
- CFD modelling studies on the fire plume to predict the deposition of combustion products from a fire in a store and on a vehicle.
- Guidance on fire prevention, fire-fighting and the environmental effects of a MBM fire in a store or on a vehicle.
- Clean-up methodology to be followed after a fire.

In addition a site visit to an incinerator was also undertaken to assess the dust fire and explosion hazards from handling and processing MBM.

MBM IN STORAGE

SITE VISITS AND FIRE INVESTIGATION

The first part of the project was to visit a selection of the MBM stores, including the stores which had experienced fires, to get a first hand knowledge of the conditions within the stores and the potential fire/explosion hazards. This exercise was an essential part of the fire investigation into the two incidents. Discussions with the site operators also provided an invaluable source of information on the fire incidents themselves and the general loading/unloading and storing procedures that were followed at each of the stores.

From the site visits the general conditions of storage were as follows:

- MBM was stored in stockpiles of typically greater than 10,000 tonnes inside old/disused factory or farm buildings.
- The stores contained no temperature controls, although temperature monitoring of the pile was carried out on a regular basis.
- The buildings were generally in good enough condition to keep the MBM dry, some had electricity supplies still functioning.
- Some stores contained piles of MBM which were not compacted, but were piled loosely into the building.

It was also apparent from discussions with the Intervention Board and the store operators that the nature of the MBM varied from store to store. The main reason for this seemed to be due to the different rendering companies employed, which meant that the MBM supplied for storage was not identical in all cases. Variations included the MBM being ground into a fine dust rather than being a mixture of fines and large lumps of bone, larger proportions of blood meal, and different fat contents. All of these variations in the content of the MBM, in conjunction with the different storage volumes and conditions, meant that visiting the stores was essential to ensure a representative experimental programme and hence be able to offer guidance on fire/explosion safety.

From the visits to the two stores which had reported fires it was concluded that both fires almost certainly started from self-heating of the MBM leading to spontaneous combustion. The evidence for this was:

- Ideal conditions for self-heating, i.e. large volumes of MBM stored without compartmentation for very long (3–4 years) periods of time.
- No compaction of the MBM pile in these particular stores, allowing air into the pile.
- The MBM was piled around the building roof supports, allowing air ingress.
- No identifiable external sources of ignition.
- Good security of the sites and the buildings reducing the probability of arson.
- Temperature records from monitoring at these sites and also other stores, showing self-heating of the MBM.
- Experimental evidence of the propensity of MBM to undergo self-heating leading to ignition.

EXPERIMENTAL PROGRAMME

Following the site visits to the stores a detailed experimental programme was discussed and agreed with the Intervention Board in order to fully characterise the fire and explosion properties of MBM and use this data to perform a risk assessment on the storage, transport and processing conditions. Further studies were also undertaken using Computational Fluid Dynamics (CFD) modelling to estimate the environmental consequences of a MBM fire in a store and while being transported.

Isothermal Self-Heating Tests

Self-heating[1,2] is the occurrence of a rise in temperature within a body of material in which heat is being generated by some process taking place within the material at a greater rate than heat is lost to the surroundings. In certain circumstances the temperature rise may increase both in magnitude and rate sufficiently to culminate in combustion; that is, there may be a 'self-ignition' or 'spontaneous ignition'. External heat sources may sometimes be necessary for the occurrence of such ignition but, by definition, these will be less than sufficient in the absence of heat generation within the material itself.

MBM from three stores were chosen for an in-depth investigation into the self-heating properties. The purpose of this part of the project was to provide advice on safe storage temperatures and volumes, and the times required for self-heating to occur. The test method is described in detail by Beever[3].

The materials under test were placed in cubical wire mesh baskets of different side length: 75 mm, 100 mm, 150 mm, and 200 mm. The baskets were filled with the sample and levelled with a straight edge. The material was not compacted into the cube. The filled cube was then suspended in a pre-heated oven and thereafter maintained at a known temperature to within ± 1°C. The centre and surface temperatures of the sample were monitored using 1.0 mm, stainless steel sheathed, chromel/alumel thermocouples. These were connected to a chart recorder and a personal computer, so that the self-heating process could be observed and recorded.

The results from the ignition tests are summarised in Tables 1–3.

Table 1. Summary of test results for Preston store

Test sample	Cube size (mm)	Critical temperature (°C)	Time to ignition (hr)
Preston	75	162	4.5
	100	154	8.1
	150	142	16.1
	200	134	30.3

Table 2. Summary of test results for Alleena store

Test sample	Cube size (mm)	Critical temperature (°C)	Time to ignition (hr)
Alleena	75	171	3.8
	100	162	7.0
	150	150	15.1
	200	144	26.0

Table 3. Summary of test results for Newtownstewart store 2

Test sample	Cube size (mm)	Critical temperature (°C)	Time to ignition (hr)
Newtown-stewart	75	176	3.8
	100	167	6.8
	150	154	15.0
	200	143	28.1

Two types of behaviour were observed;

1. the central sample temperature rising by a relatively small amount ($< 20°C$) above the pre-set oven temperature and then gradually falling back to the oven temperature (sub-critical behaviour, Figure 1), or
2. the central sample temperature rising to a high value of about 300°C, and then gradually falling back to the oven temperature (super-critical behaviour, Figure 2).

The results, when applied to the full-scale MBM storage conditions, showed that the ambient temperature at which self-heating can occur in the samples tested was 46–49°C, with times to ignition varying from 4.9 years to 21.5 years. At the time of the investigation, some of the stores containing the MBM had been storing the material for over four years. It was found that MBM at a temperature of between 67–73°C could lead to an ignition within as short a time-scale as 157 days.

From the experimental results, and the subsequent calculations using Thermal Ignition Theory for the ignition timescale of warm MBM, it is most probable that the fires were due to the MBM being warm when it arrived at the stores from the rendering plants. This, coupled with the non-compaction of the MBM as it was loaded into the stores and the huge volumes resulted in an eventual self-heating leading to a flaming ignition.

Dust Explosion Tests
During the unloading and loading operations at the stores dust clouds were formed by the fine material in the MBM. If ignition sources were present, this could lead to an ignition of the dust cloud creating and unconfined explosion. A series of tests were undertaken on the fine dust component of MBM samples from eleven different stores to ascertain its susceptibility to ignite and the severity of the explosion once ignited. The tests undertaken were:

- Classification
- 5mm layer
- Minimum ignition temperature
- Minimum ignition energy
- Explosion indices (Maximum pressure and Kst)

Details on these tests can be found in reference 4.

The results of the testing showed that MBM can be ignited from hot surfaces at temperature above 280°C when present as a dust layer; and temperatures above 450°C when present as a dust cloud. MBM was not sensitive to ignition from low energy (<500 mJ) electrostatic sparks, but could be easily ignited with higher energy sparks (8–10J). The explosion indices tests showed that pressures as high as 7.6 bar g could be reached in a confined explosion, with Kst values as high as 62 bar ms^{-1}.

These results were then used to determine the risks from identifiable ignition hazards present in a typical store, particularly during the dust creation stages of loading and unloading from transport. Once these had been determined, suitable explosion prevention and protection methods were highlighted.

Toxicity Studies
One of the main concerns of the study was the potential life threat from the combustion products from a fire involving MBM in a store. The aims of this part of the study were to characterise the combustion atmospheres with a view to assessing the toxic hazard. Three MBM samples from three different stores were decomposed in the BRE tube furnace. Two decomposition conditions were studied, a non-flaming thermal oxidative condition at

375°C, representing a slow smouldering combustion (such as may be found during self-heating), and a slightly ventilated flaming condition at 650°C, which may result from an external ignition source or from the development of a self-heating reaction.

The major toxic fire gases were monitored: carbon dioxide, carbon monoxide, oxides of nitrogen, oxygen and smoke optical density. Samples were also taken for the analysis of cyanide, chloride, phosphate/phosphite, sulphide and sulphate anions and for significant volatile organic compounds.

It was found that the yields of major toxic gases were greater under flaming than non-flaming conditions. Under both flaming and non-flaming conditions, the greatest contribution to the lethal toxic potency of the fire effluents was attributable to hydrogen cyanide and organic nitriles (58–78%), with less being attributable to carbon monoxide (4–11%). See Table 4.

Table 4. Contributions to the toxic potency

Material	Test run	LC_{50} Concentration (g/m^3)	Relative contribution to lethal toxic potency (%)			
			from CO	from HCN + nitriles	from organic irritants	from inorganic irritants
Non-flaming @ 375°C						
Preston	T143	7.4	8.5	58.1	23.9	9.7
Alleena	T144	8.9	4.2	55.9	30.2	8.8
Chorley No 2	T146	8.3	3.9	61.2	26.2	7.7
Flaming @ 650°C						
Preston	T142	8.1	10.7	64.9	6.3	11.7
Alleena	T145	9.5	9.2	69.5	7.2	6.9
Chorley No 2	T147	10.0	9.0	77.6	3.0	8.4

The results of the study showed that fires involving MBM are capable of causing dangerously toxic atmospheres in enclosed spaces, such as stores which are filled to capacity, with relatively low masses decomposed.

Cone Calorimetry

One of the identified potential fire hazards of the storage of MBM was the threat of an external ignition source causing a fire. To ascertain the likelihood and consequences of MBM being ignited by an external fire source, a study of the combustion of three MBM samples were undertaken using the Cone Calorimeter. The cone calorimeter is an internationally recognised bench-scale method for assessing the reaction to fire performance of materials[5]. A heat flux of 50kW m^{-2} was used to simulate a large external fire. The test provided data on a wide range of parameters including, ignitability, rate of heat release, smoke production, carbon dioxide and carbon monoxide production.

It was found that MBM is readily ignitable from a large fire source. It ignites within a few seconds exhibiting a rapid burning combustion on ignition, which soon subsides into a slow but steady burning rate. Peak rate of heat release rates were within the range 87–136 kW/m^2 (see Figure 3). The data obtained from the cone calorimetry was also used to assist in the computer modelling work of a fire plume.

Bomb Calorimetry

In order to gain an understanding of the fire hazard posed by charred and burnt MBM material, resulting from a self-heating ignition, its calorific value was determined. The bomb calorimeter apparatus was used for this test and the sample was taken from a charred sample found at one of the stores, after heat measuring probes had identified a hot-spot in the stockpile.

It was found that the material had a gross calorific value of 18.4 MJ/kg, indicating that even though it had already been subjected to self-heating, it still contained a significant quantity of combustible material. As such, burnt MBM should still be regarded as a combustible material and not as "inert waste".

Computational Fluid Dynamic Modelling of the Fire Plume

Studies were undertaken on the behaviour of a fire plume and the potential for the plume to disperse products away from a fire occurring in a store.

Two different scenarios, chosen in discussion with the client, were considered for this study. The first scenario simulated the movement of combustion products of a potential fire inside a typical warehouse building containing MBM. The warehouse chosen for the study was on an urban site, with dimensions of 70m by 80m in plan, 6m up to the eaves and 10m up to the ridge of the building (Figure 4). The second scenario simulated the dispersion of a fire plume, and dry and wet deposition of its combustion products in the surrounding environment from a vent opening of 1.5m by 3.0m in area. The vent opening represents a burned through roof-light as a consequence of a possible localised breach of the roof.

For the first scenario, simulating the movement of combustion products inside the warehouse, a transient simulation with a growing fire was carried out. Assuming that the fire loses approximately 35% of the total heat release rate by radiation, the fire source was modelled as a convective source of heat with peak convective heat release rate of approximately 1.6MW. Initially, apart from a leak around the access point to the building (i.e. the door), the building envelope was considered completely closed. As the fire grew, the combustion products started to accumulate under the roof. When the gas temperatures reached approximately 100°C (which is pessimistic estimate of the softening temperature for plastic roof-light panel) at the roof level, it was assumed that a building breach occurred of 1.5m by 3m in area, simulating a burned through roof light, on the upwind side of the building. At that time, the access door (1.5m wide by 4m high) was opened at the ground level. The model provided predictions of the transient development of the hot layer of combustion products inside the warehouse and characteristics of the fire plume emerging from the breach of the roof.

For the second scenario, the source discharge conditions of the fire effluents at the point of breach from the warehouse fire (convective heat release rate of 1.6 MW) were then used for simulating the dispersion of the fire plume and particle deposition into the surroundings. The fire plume emerging from a single opening resulting from the breach produced an average discharge velocity of 2.5 m/s and gas temperature of 200°C at the vent opening in the roof, thus giving the total discharge mass flow rate of 8.4 kg/s.

In general, a wide variety of local meteorological wind conditions will prevail, which depend upon the local topology and weather conditions. The weather conditions could involve high and low wind speeds, neutral, stable and unstable atmospheric conditions. This study was limited to one meteorological condition corresponding to high wind speed of 9.8m/s under neutral stable condition, which was considered to represent a typical weather condition in the UK.

The predicted results suggest the highest concentrations of the fire effluents to be close to the building, then falling steadily with increasing distance. For an assumed rate of dry deposition of 0.1m/s, the wet deposition generally represents only a relatively small component of the total, around 10%. For a lower assumed rate of dry deposition of 0.01m/s, the wet deposition is comparable to the dry deposition.

GUIDANCE ON FIRE PREVENTION
To reduce the probability of self-heating:

- Keep the storage volumes as small as possible by using a number of small store rather than one large store.
- If a large storage building is used consider splitting the area into smaller compartments.
- Material being placed into the store should be at ambient temperature.
- Material being stored should be kept cool and dry.
- Compact the MBM as it is placed into the store to remove air voids and reduce the quantity of oxygen available for self-heating.
- Monitor the temperature of the MBM pile, ideally at the centre, to enable early detection of the onset of any self-heating.

Buildings and any services such as electricity points must be maintained in good order to reduce the risks of an ignition from an external source. Ideally, buildings should be constructed of non-combustible materials. A high level of security is desirable to minimise the risk of arson. If mechanical equipment is used within the building it must be maintained in good working order. Hot-working procedures, such as welding and cutting should be fully and carefully monitored.

An automatic fire detection system should be installed in each store building to complement the existing practice of manual temperature monitoring of the MBM pile.

FIRE-FIGHTING STRATEGY
Sprinkler systems may be installed if desired, but it is expected, in the absence of any experimental data suggesting otherwise, that they may be of little benefit. They may

extinguish surface fires, but will not be able to penetrate into the MBM pile to extinguish the deeply seated fires created by self-heating.

Due to the risk of production of hazardous gases, smoke and particulates during MBM fires within an enclosed space, and the further risk of subsidence of the MBM pile, it is not advised that store personnel tackle any fires that occur. Personnel should immediately retire from the building and the fire service be called when a fire is detected or suspected. Breathing apparatus must be worn by fire-fighters inside the building.

If the fire is at the surface of the pile water spray may be used to douse the flames at the surface. The hot and burnt material should then be removed and allowed to cool before disposal. If the fire is deep seated within the pile then pumping large quantities of water onto the MBM will not be effective, as penetration will be minimal. Alternative options that may be used include: smothering the fire by limiting the oxygen ingress into the building and using compaction to limit the oxygen ingress within the MBM, directly injecting extinguishant into the pile via probes, digging out and removing all of the hot and burning material, and leaving the smouldering fire to burn.

CLEAN-UP METHODOLOGY
Water used to extinguish the fire will be contaminated and should not be allowed to enter the water table if at all possible. All run-off water should be contained and collected for safe disposal. Removal of burnt MBM within the store should be undertaken, but with care, as digging into the pile may cause re-ignition. Burnt MBM should also be disposed of safely as it will almost certainly contain combustible material. Decontamination using appropriate disinfectants should be undertaken on equipment and apparatus coming into contact with the MBM material, burnt or non-burnt. Some clean up of the surrounding area may also be required if the fire plume resulted in deposits of soot and ash particles on the ground and nearby buildings.

MBM DURING TRANSPORT
MBM is transported at two different stages of the MBM storage/incineration process. It is transported from the rendering plant to the store and then, possibly four years or more later, it is taken from the store to the incineration plant. The results of the experimental tests on MBM, described in the storage section, were also used to assist the fire/explosion hazard identification and risk assessment on transporting MBM.

FIRE HAZARDS
It has been shown that MBM is a combustible material that has the potential to self-heat and can easily be ignited from external fire sources. From the self-heating experimental studies it was found that the risk from self-heating during transport is extremely low. This is due to the short length of time the MBM would be in transit (typically 2–3 hours) and the relatively small volume being transported. This is however for MBM material at ambient temperature. If material were to be transported that was already warm or hot, possibly from the rendering process or from self-heating during storage, then the risk increases significantly. External

fires occurring on the vehicle, possibly as a result of a road traffic accident or vehicle ignition sources, could also ignite the MBM.

Fire prevention methods should include checking the MBM before loading to ensure it does not contain warm or hot material, and ensuring good maintenance of the vehicle to reduce the risk of a fire starting on the vehicle igniting the MBM. All vehicles transporting MBM should carry portable fire extinguishers and have the drivers trained in their use. As a minimum extinguishers should be of the type to fight Class A (solid combustibles) and Class C (electrical). However, if a fire is suspected inside the truck body, i.e. within the MBM load, then the doors should be kept shut and the fire service called. Opening the doors may cause the fire to flare-up due to the ingress of air.

The hazards from burning MBM and the clean-up methodology are generally as described earlier.

DUST EXPLOSION HAZARDS
During transportation of MBM there will be little possibility of a dust cloud being present and hence there is minimal risk from an explosion. However, during loading and unloading of the MBM a dust cloud will be created and care should be taken to minimise the risk of an ignition. Potential ignition sources will need to be identified and measures taken to eliminate or reduce the likelihood of their presence. Methods that reduce or limit the formation of dust clouds during loading and unloading should be explored.

MBM INCINERATION PROCESS

PROCESS OVERVIEW
MBM held in storage is eventually moved to an incineration plant for destruction. On arrival at the plant it is unloaded into a receiver area building of the plant where the MBM is tipped through a metal grid, to break-up any large lumps, and onto a conveyor. The receiver area building above the chutes leading to the conveyor is fitted with local air extraction and dust filter units, as during unloading dust clouds area created within this building.

The MBM is taken by conveyor to a silo before being taken away by screw conveyor feeding another conveyor belt. This conveys the material to the incinerator, which is fed by a screw conveyor. The incinerator operates at temperatures of 1100–1200°C where the MBM is destroyed.

POTENTIAL IGNITION SOURCES
A risk assessment of the process was undertaken and identified a number of potential ignition sources:

- Electrical equipment, particularly within the MBM receiver building, which will need to be rated for use in an explosive atmosphere.
- Burning material. The operator of the site mentioned that on occasions the MBM arrives in a heated condition showing evidence of charring. If glowing embers are

discharged into the process, these may act as a source of ignition, particularly if sucked into the dust filter unit.

• Hot surfaces. All equipment, such as motor surfaces, light fittings, that have hot surfaces which may come into contact with dust clouds or have dust layers formed on them, should be appropriately rated. This rating is based on the results of the 5mm layer ignition and the minimum ignition temperature tests as described briefly earlier.

• Friction heating and impact sparks. Moving parts such as conveyor bearings should be inspected regularly to ensure they do not overheat and any debris particularly metal objects should be removed from the MBM prior to being processed.

• Electrostatic discharge. Although the experimental studies showed that the MBM dust is not sensitive to ignition from electrostatic discharges (minimum ignition energy test), as a minimum precaution the plant should be well earthed.

• Welding or cutting operations. Maintenance or repair work may produce localised heat and sparks that could ignite dust deposits. Work should only be permitted in areas cleared of dust.

EXPLOSION PREVENTION METHODS

To prevent the possibility of an explosion a number of measures can be taken:

• Eliminate as far as possible all potential ignition sources. This should entail ensuring all electrical and mechanical equipment operated in explosive atmospheres is corrected rated, monitoring the condition of the MBM on arrival (i.e. there is no hot or burning material present) and good maintenance of equipment involving moving parts.

• Avoid the formation of dust clouds by local exhaust ventilation and by good housekeeping.

• Correct temperature rating of equipment with hot surfaces.

• Permits to work in areas where flammable materials are present.

• Good quality control of the MBM to eliminate debris, particularly metal objects.

• Keep the storage time of the MBM material in the silos to as short a time as possible to reduce the risk from self-heating, particularly if the temperature inside the silos is above ambient.

EXPLOSION PROTECTION METHODS

A number of techniques may be used to protect against the consequences of an explosion or fire involving MBM.

• Installation of explosion protection systems in plant items where dust clouds are present, for example the dust filter units and silos. The particular type of system chosen will mainly depend on its location and volume. Common methods used are, relief venting, suppression and containment.

• Isolation systems to prevent an explosion from an ignition in one plant item propagating through the process and affecting other equipment.

• Continuous temperature or carbon monoxide monitoring within the silos to detect signs of self-heating and linked to an alarm/fire extinguishing system.

CONCLUSIONS

1. Long-term storage of MBM in large volumes can lead to self-heating and eventual ignition. Care should be taken within the stores as fine dust that forms part of the MBM material is flammable and will ignite if it comes into contact with a source of ignition. Burning MBM produces toxic combustion products that can quickly reach dangerous concentrations, particularly within the store buildings filled close to capacity. Calorimeter studies have shown that MBM is readily ignitable form external heat sources and still contains a high level of combustible material in the charred residues from self-heating. Hence, measures need to be in place to reduce the risk of external fires or other heat sources igniting the MBM.

2. Transportation of MBM also presents a potential fire hazard, though as the quantities being transported as considerably less than in storage, the risk from self-heating is much lower provided the material is not already burning when loaded. There is also a risk of a fire starting elsewhere on the vehicle igniting the MBM, and all practical measures need to be employed to reduce this risk. Loading and unloading vehicles will create dust clouds and care should be taken to ensure no ignition sources are present during these operations.

3. Processing MBM can lead to a number of ignition hazards from the dust clouds generated and from layers of dust on hot surfaces. A formal risk assessment and hazardous area classification will be required in accordance with European Union ATEX Directives[6,7] and UK DSEAR legislation[8].

REFERENCES

1. Bowes, P. Self-heating: Evaluating and Controlling the Hazards. HMSO. 1984.
2. Beever, P. F., SFPE Handbook of Fire Engineering, Society of Fire Prevention Engineers/National Fire Prevention Association, Quincy MA (1988).
3. Beever, P. F., Spontaneous Combustion - Isothermal Test Methods, Building Research Establishment Information Paper IP23/82 (1982).
4. Dust Explosion Prevention and Protection - A Practical Guide. Edited by John Barton. IChemE 2002.
5. BS476-15:1993. Fire tests on building materials and structures - Part 15: Method for measuring the rate of heat release of products. (ISO 5660-1:1993).
6. European Union Directive 94/9/EC Equipment and Protective Systems for use in Potentially Explosive Atmospheres.
7. European Union Directive 99/92/EC Protection of Workers from Explosive Atmospheres.
8. Dangerous Substances and Explosive Atmospheres Regulations. 2002.

ACKNOWLEDGEMENTS

Thanks are due to my colleagues Dr Jenny Purser, Dr Suresh Kumar and Mr Richard Chitty for their contributions to the project and to the Rural Payments Agency (formerly Intervention Board) for funding the work.

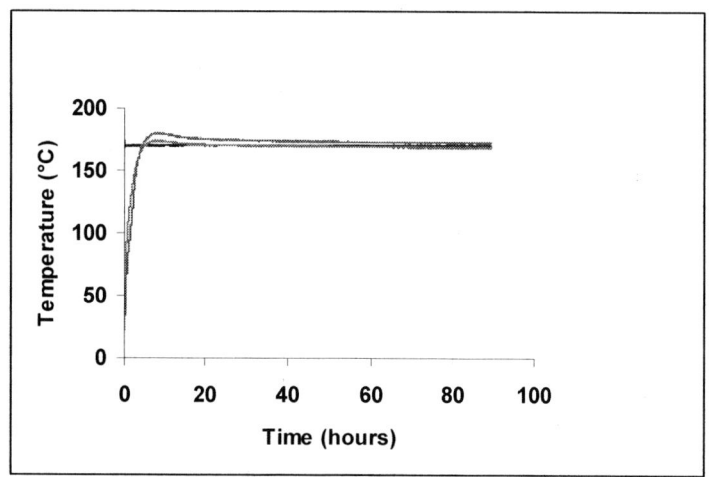

Figure 1. Temperature - time profile at sub-critical temperature of 170°C

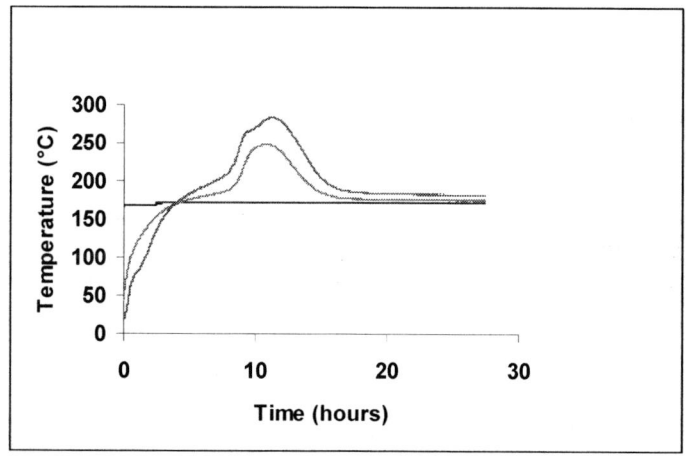

Figure 2. Temperature - time profile at super-critical temperature of 172°C

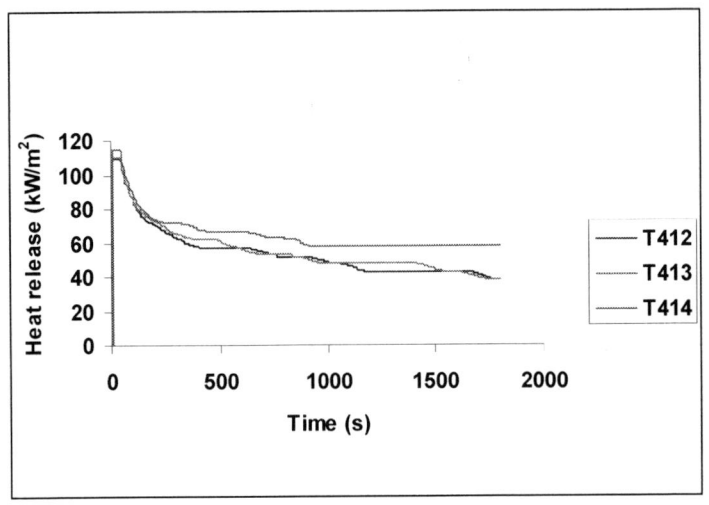

Figure 3.   Cone calorimeter test: heat release vs time

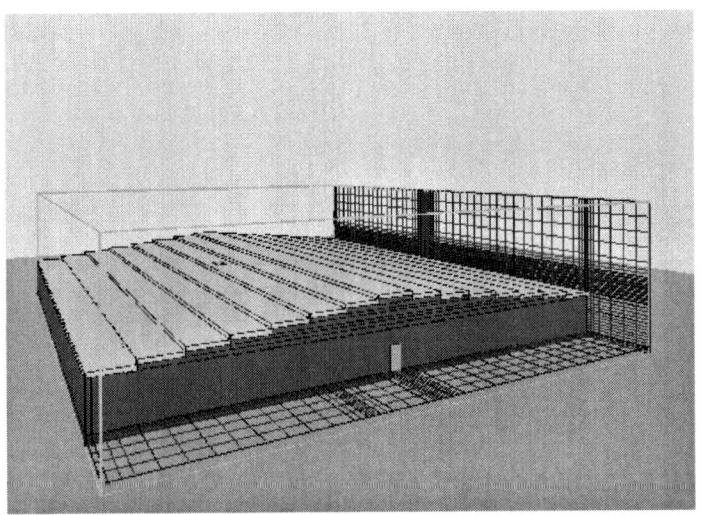

Figure 4.   Typical warehouse used for CFD studies

LESSONS LEARNT FROM FITTING AN INERT GAS BLANKETING FACILITY TO AN EXISTING STORAGE SILO

A Woowat, P Atherton[1], I Kempsell, S Windebank[2], British Nuclear Fuels plc
[1]Sellafield, Cumbria (+44-19467-28333), [2]Risley, Cheshire (+44-1925-832000)

SYNOPSIS

The decommissioning of redundant plant at BNFL Sellafield is proceeding as a priority.

One of the major activities is to remove metallic and organic Intermediate Level Wastes (ILW) from a large reinforced concrete storage silo. The waste has been stored in an air-based atmosphere.

It was decided to fill the silo with inert gas, prior to beginning the retrieval process. The silo had not been designed to retain inert gas. The inerting was completed successfully in early 2002.

This paper outlines how the silo was prepared for inerting and compares the predictions of performance, including anticipated problems, with the actual outcome.

INTRODUCTION

The decommissioning of redundant plant at BNFL Sellafield is proceeding as a priority. One of the major current activities is to improve the waste storage conditions of a large reinforced concrete storage silo located in a congested area of the site. This is being undertaken in advance of waste retrieval operations that will require the construction of new waste retrieval and processing capabilities at a cost of >£100 M.

The silo was built just after World War II at the start of the Cold War to provide waste storage facilities for the cladding from spent nuclear fuel initially arising from the military programme and later from the early civil nuclear programme. Radioactive wastes, generated from 1952 to 1968, were loaded and stored inside the silo within an air atmosphere. More recently, a programme has been initiated to systematically reduce the potential fire risk presented by the waste before waste retrieval can commence.

This paper primarily concentrates on presenting the technical challenges faced in developing a pragmatic solution to reducing the fire hazard by fully inerting the silo, as it was not designed to retain inert gas. Details are provided of the various options assessed, the development work conducted in support of the chosen option, site preparatory works and the plant commissioning carried out to prove the design. Safety and the reduction of risk to a level that is as low as is reasonably practicable (ALARP) has been paramount throughout the works programme and has had a major influence on the design and modification programme.

Substantial improvements of the silo structure and containment have now been achieved and the waste contents were successfully inerted with argon in early 2002. Further work in preparation for retrieval is now underway.

STORAGE PLANT DESCRIPTION

The silo (Figure 1) is a 300 mm thick reinforced concrete structure having external dimensions of approximately 29.5 m x 10 m overall and rising to a height of 18 m above local ground level. It consists of six equal compartments that are full or near to capacity.

Each compartment is divided into two equal sections by a 13 m high longitudinal division wall rising from a height of 1.5 m above the bottom of the silo. The silo is supported on cellular foundations (voids), to elevate the storage compartments above local ground level, with the compartment floors being covered by a layer of gravel and sloping towards a drain located in the corner. A charge hole is positioned at the top of each compartment above an inverted 'V' shaped divider plate, used to divert the tipped waste to the compartment sides. The charge holes connect the silo to a common overhead transfer tunnel running centrally along the silo and through which waste was transported and tipped. Some waste rests on the divider plate and protrudes up into the lower section of the transfer tunnel. Tunnel clearance operations are currently underway to dislodge this waste back into the silo, before sealing the chargeholes to isolate the tunnel from the silo compartments.

REDUCTION OF PLANT RISK (1 & 2)[1]

The key hazard was the potential for a silo fire to be initiated within the bulk waste material. Significant quantities of combustible waste material were sentenced to the silo; principally aluminium, magnox (an alloy of magnesium), uranium, graphite, and organic materials. Due to the reactive nature and degradation products of some of the constituents, there was a small possibility of spontaneous ignition occurring and a fire developing in an air atmosphere if the waste was disturbed. Temperatures and pressures generated during such a fire might, in extreme circumstances, have compromised the silo containment resulting in unacceptable consequences.

The approach taken to reduce the risk posed by the plant is summarised in a FAST diagram (Functional Analysis Systems Technique) (Figure 2). The challenge was to provide a robust means of managing any potential silo fire hazard. Any selected scheme had to be rigorous, satisfy BNFL Company and Regulator safety requirements and represent a solution that ensures the overall risk posed by the plant is ALARP.

Option studies were initiated, but it soon became apparent that only a limited number of options were available to manage the fire risk. It was rapidly concluded that there was a clear requirement to inert the silo contents with an appropriate medium to drive out the air and maintain oxygen levels at a safe limit. Further studies concluded that gas inerting was the best option for providing an inert atmosphere that will not support ignition of the waste and fire propagation.

SELECTION OF INERT GAS (3)

A comparison of the properties for various gases (Table 1) concluded that argon represented the best option for meeting the process, engineering and safety requirements, recognising that the asphyxiation hazard to plant operatives would need to be controlled by engineered and managerial means. Argon was selected over two other cheaper alternatives, carbon dioxide and nitrogen, because it is the only gas available on an industrial scale that could extinguish a burning metal fire. It also has the inert properties to enable permanent blanketing without chemical reaction with the waste mass that might later prove disadvantageous for waste treatment. Nitrogen is less dense than argon making it unsuitable

---

[1]The number(s) in brackets refer to the risk reduction step on the FAST diagram presented in Figure 2.

for dispersing air from within the silo and furthermore, reacts with burning uranium and magnesium. Carbon dioxide on the other hand is heavier than both argon and nitrogen and represents a suitable gas for displacing air. However, it both permits ignition and sustains metal combustion once ignition has occurred preventing it from being selected as a suitable inerting and extinguishing agent.

Other liquid and solid extinguishants were considered, but were rejected for several reasons including inability to ensure rapid delivery to the seat of the fire, and disposal problems once the extinguishant became radioactive.

ASSESSMENT OF ARGON ASPHYXIATION RISK (4)

Argon is a colourless, tasteless, odourless gas that becomes a hazard when present in sufficient concentrations to act as an asphyxiant at oxygen concentrations <18%. The silo compartments would be filled with large volumes of argon gas. There would then be the potential for fugitive argon to seep from the silo into other working areas. Furthermore, the advanced works to prepare for waste retrieval and subsequent processing of waste would require breaking through the silo containment, whilst retaining a fully inert atmosphere within the silo, thereby also presenting a potential asphyxiation hazard. BNFL have therefore developed a robust argon hazard management strategy for the silo that endeavoured to engineer out the hazard wherever possible as a first principle.

Immediately under the structural concrete of the cellular base of the silo is a 75 mm layer of blinding concrete. The ground directly beneath has been well compacted with very few air voids present. The site surrounding the silo has a covering of man made ground between 0.5 m and 8.0 m thickness and comprising of a 0.5 m layer of concrete and hardcore below which is a layer of sand and gravel containing bricks, clay etc. It was expected that this adjacent ground was also well compacted, but might be disturbed in the future during construction work to prepare for waste retrieval operations.

Computational Fluid Dynamic (CFD) modelling of the silo structure was carried out to simulate the leakage of argon from the silo into the atmosphere and surrounding area. The objective was to determine the magnitude of the asphyxiation hazard and to identify the optimum siting for oxygen monitors. A gas dispersion code was used to model the silo and its surrounds to address the dispersion of argon below ground, through the silo walls and from breached argon pipework.

The CFD model indicated that argon concentrations around the silo at ground level and below could present a potential hazard to personnel working in the local area. A second study modelled argon diffusion rates through the concrete silo containment walls and the combined diffusion/convection flows through small cracks in the silo walls. Once again the study revealed that there was a potential argon asphyxiation hazard to personnel that may be located below ground close by the silo. These two studies also demonstrated that concentrations rapidly fell a very short distance from the release site and would therefore, pose no hazard to personnel located outside the immediate vicinity of the silo building. Similar results were obtained from a third study that demonstrated severing the argon distribution pipework running along the north side of the silo could also result in significant argon concentrations. However, this was restricted to a small region local to the breach and the gas was rapidly dispersed as distance from the release site increases.

The conclusion from the CFD modelling work clearly indicated that there might be an argon asphyxiation hazard present within a few metres of the silo once it was inerted. It was clear that the optimum position for locating the oxygen monitors was external to the silo within a service trench that runs alongside the silo. Therefore, the most significant hazard would be from slow seepage into confined spaces, rather than gross leaks from high pressure pipework, the latter release being dispersed quickly and capable of being rapidly detected.

MANAGEMENT OF ARGON ASPHYXIATION RISK (5)

To complement the modelling work, a world-wide review of asphyxiation incidents was attempted. Unfortunately, very few countries made their data available in the public domain. Information was available mostly from the UK and USA process and nuclear industries. It was found that there were not many argon fatalities per se, but in the period 1995–2000, one person died in the UK (Ref. 1) and 23 persons died in the USA (Ref. 2) from Oxygen deficiency due to the presence of argon in work areas. However, there were many asphyxiation fatalities from other causes, some 98 in the UK (Ref. 3) and 195 in the USA (Ref. 2). The vast majority resulted from access to confined spaces such as vessels. The root causes of the argon incidents usually involved inadequate precautions, or safety management arrangements (risk assessment, safe systems of work, training etc.) or emergency procedures. Lessons learnt from these previous incidents and investigations are incorporated into the strategy developed for managing the silo argon hazard. The multi-legged approach to controlling safety in different areas of the plant is summarised in Table 2. It was recognised that these precautions were extensive and exceeded those usually found in most process industry plants. However, this was considered prudent at the time given that the silo represented the largest inerting application to an existing concrete structure on the Sellafield site and possibly the UK.

Management of the argon asphyxiation risk over a period of more than 6 months of active commissioning has resulted in no significant argon leakage being detected outside the operating envelope of the silo. This has been substantiated by an extensive external monitoring regime.

PROVISION OF ARGON SUPPLY SYSTEM (6)

A new modern argon inerting and fire fighting system was designed to provide the silo inerting and fire fighting requirements. Two liquid-argon storage and vaporisation plants are located remotely from each other on the site and are connected to the silo by diverse pipe-routes. The plants comprise of two double skinned liquid argon tanks that provide sufficient quantities of argon to ensure that all compartments and the transfer tunnel remain fully inerted, with sufficient capacity always being available for fire fighting duties. Gas is injected into the compartments at the bottom of the silo and is drawn through each compartment by the silo ventilation system maintaining the oxygen composition of the atmosphere at less than 2% (v/v) oxygen. In the unlikely event of the argon supply failing or during deliberate de-inerting, which may be required to support the pre-retrievals works programme, a segregated argon fire fighting supply is also provided by the argon plants. This injects large quantities of argon into the top of the compartments at such a high rate (capable of delivering 800 Sm^3/hr of gaseous argon for 30 hours) that it would rapidly extinguish any fire detected by the silo sensing instruments.

A supply line from each vaporisation plant feeds argon at its storage pressure of approximately 10 Bar to a pressure-reducing station located on the west wall of the silo. Four sets of regulators facilitate the passage of low-pressure argon to three headers. Each header is capable of serving all six silo compartments, one carrying inerting argon (controlled flow) to low-level connections, one carrying fire-fighting argon to the same low-level connections and one carrying fire-fighting argon to connections on the roof of the silo.

The vaporisation plants ensure an uninterrupted availability of argon for all duties, thereby maximising the reliability of the system. Both plants are on-line continuously supplying the inerting header whilst, normally, each plant serves one fire-fighting header. Thus, there will be two independent fire-fighting supplies available in normal operating circumstances.

The argon plants were installed and commissioned in late 2001, providing all of the silo inerting and fire fighting requirements. Argon is injected into the silo at a rate of 3 to 12 m^3/h per compartment through the single sump located at the base of each compartment.

There were concerns that the argon gas might not rapidly fill and disperse throughout each compartment if the injection points were partially blocked by the gravel/screed covering the silo floor or if the gap between the bottom of the central wall and silo floor was blocked by waste. This concern was investigated in development trials on an inactive model of the silo. It was shown that the forces of gaseous diffusion were dominant within the waste mass, for example, argon rapidly dispersed the trapped air inside an upturned container. Also, it was predicted that the argon would plug flow. The plant results are discussed later in this paper.

The silo ventilation system draws argon from each compartment via the charge hole and transfer tunnel and is set up to maintain the tunnel at a slight negative pressure. This maintains a very small air in-bleed in the upper sections of the silo rather than an egress of contaminated argon out to the environment.

It is worth noting that the existing silo structure and existing argon delivery system (pipework etc.) are capable of meeting the 0.125 g seismic design standard required for existing plants with substantial nuclear inventories. At completion of a current upgrade, the new argon storage plants will be capable of meeting the higher 0.25 g (10^{-4}/y return frequency) standard to cater for potentially better performance by the silo, and to be consistent with internal hazards standards for new plants.

ENSURING DELIVERIES AND QUALITY OF ARGON (7)

The volume of liquid argon required to ensure the full availability of the fire fighting system is approximately 70% of the maximum working volume for a single tank. This volume is maintained in both tanks at all times. Tanker deliveries are required every 6 to 19 days, based on the maximum and minimum injection rates respectively.

The liquid argon purity is <2 ppm (v/v) oxygen. This standard industry grade argon is delivered 'on demand' to the facilities by road tanker from a major UK supplier. Argon is routinely sampled and analysed for oxygen content at the production facility and certified in accordance with the bulk liquid supplier's QA systems, with the road tanker being sampled before and after filling.

The design of couplings on the two argon storage tanks is unique to this particular plant on the site preventing an incorrect tanker load being delivered to the argon tanks. Furthermore, the filling process is performed by the road tanker driver under the supervision of a qualified BNFL operator.

ENSURING SILO CONTAINMENT (8)

A set of R&D trials and supporting test work was performed in support of the silo inerting programme and prior to commissioning the argon inerting system on site. The behaviour of argon gas inside a silo compartment was assessed using a scale model filled with pieces of polystyrene to simulate the waste. Oxygen monitoring equipment distributed throughout the waste measured the oxygen depletion as argon gas was fed into the bottom of the rig and drawn out at the top simulating conditions within the silo. The overall aim was to determine whether the inerting option was viable by:

- Establishing the optimum argon flow rate for inerting the silo and minimising the potential for air retention 'pockets' between the waste.
- Predicting the total volume of argon required to fully inert the silo.
- Determining the optimum location for oxygen monitors in the silo.

A series of tests were conducted in which the argon flow rate was varied and the oxygen concentrations recorded by monitors distributed at various levels throughout the test rig. It was concluded that:

- The minimum argon flow rate required to inert the entire silo to a level <2% oxygen would be 23 m^3/hr, or approximately 4 m^3/hr per compartment.
- Based on this flow rate, the time to inert the silo would be approximately 15 days.
- The argon would behave similar to a liquid when filling the silo from bottom to top similar to a bath filling with water and thus, is better represented by a plug flow rather than a mixed flow model. This important result was the first indication that argon injection at the bottom of the silo was an efficient means of displacing air from the silo and that no pockets of oxygen rich gas would remain. The assertion that argon filling was by plug flow was later verified by two-dimensional CFD modelling of the silo and confirmed during the silo commissioning trials.
- The top section of the silo would be the last to reach the target oxygen concentration and so the top of the silo would be the optimum location for any oxygen monitors.

Concrete porosity tests were performed on actual core samples taken from the silo to investigate the permeation of argon through the concrete silo walls. The results from these tests measured diffusion coefficient in the range 1 x 10^{-4} to 5 x10^{-4} m^2/hr for a non-coated concrete surface. This inferred that argon leakage rates through the 300 mm walls of the silo would be in the range 3.5 x 10^{-4} to 1.25 x 10^{-3} m^3/hr per square metre of concrete. The diffusion coefficient and leak rate was reduced to 3 x 10^{-8} m^2/hr and 2.5 x 10^{-5} m^3/hr per square metre of concrete respectively when a nominal 1 mm layer of sealant was applied to the concrete surface. It was thus concluded that diffusion through the concrete would be very small and any cracks or through wall penetrations would dominate. These results initiated a programme of wall surveys and repairs, culminating in the coating of all external surfaces to enhance argon retention and minimise losses to levels that would support a safety case.

Because of surface cracking of the concrete, other tests were conducted to identify and assess the suitability of surface repair, preparation and sealant materials that could be applied to the external surface of the silo to reduce argon losses through the containment walls. They identified the best primer as a moisture curing urethane, because of its ability to significantly improve the cohesive strength of the concrete substrate. A urethane mastic was the better of two fillers tested, because it was easier to spread over a concrete surface, and a MTM Acothane proved to be the best top coat for filling in minor concrete defects.

The results from these R&D trials increased general confidence in the overall inerting proposal demonstrating that the proposal was practicable and further development of the scheme was worthwhile. They indicated for the first time that full inerting of the silo could be successfully achieved, providing key input information and data utilised during the subsequent design, installation and commissioning of the argon delivery and silo ventilation system.

STRUCTURAL IMPROVEMENTS (9)

A survey and inspection of the external surface condition of the silo walls was carried out in 1994/1995. The walls were washed down and areas of defective concrete and cracks repaired using a high strength polymer modified repair mortar. Following surface preparation, horizontal carbon fibre strips (nominally 100 mm wide by 1 mm thick) were then bonded to the external faces of the silo (1000 mm centre to centre) to strengthen the walls. Once the structural repairs, strengthening and surface repairs were completed, a 2 mm thick coating of mortar was applied before finally sealing the walls with a 1 mm coat of hot applied polyurethane sealant to minimise argon permeability. These measures, augmented by other civil and structural improvements to reduce loads on the silo and strengthen the roof, significantly reduced one of the key project risks.

OXYGEN MONITORING AND CONTROL (10)

Independent oxygen monitoring and argon flow control systems have been installed. There are 20 oxygen analysers. Sample points are distributed throughout the silo and at three levels along the length of the transfer tunnel. These monitors continuously record oxygen levels by extracting gas samples, with the sampled gas being returned back to the silo. Each monitor is set to alarm when oxygen concentrations in a compartment or in the tunnel reach an upper limit of 1.8% by volume.

COMMISSIONING (11)

There were a series of commissioning trials on the silo prior to full and continuous argon inerting. The principal aims were to:

1. Confirm that an inert atmosphere could be established within the silo and tunnel i.e. oxygen levels maintained at <2%. This target was pessimistically based on the minimum oxygen content required to sustain a metal fines fire.
2. Measure and record information and data that would allow the silo to be characterised in terms of argon retention, leakage and oxygen gain.

To achieve this purpose, initially a single compartment (Compartment 2) was inerted. This minimised the potential hazards, while providing sufficient data to determine how best to proceed with inerting the remaining silo compartments. Subsequently, the other five compartments were inerted and the overall objective of fully inerting the silo was achieved.

COMPARTMENT 2 INERTING TRIALS

The expectation, based on the previous R&D Trials and CFD modelling work, was that filling of the silo compartment would be by plug flow. It was predicted that there would be a rapid drop off in oxygen concentration, observed as a sudden fall on the oxygen analysers at the top of the compartment, once the compartment was full. This would imply that there was a narrow interface between the air and argon within the compartment. It was also anticipated that leakage rates through the compartment walls would be low, but there were concerns that the base of the compartment and construction joints might not be well sealed allowing significant argon leakage into the voids.

Inerting of compartment 2 commenced at the maximum design flow rate of 12 m³/hr of argon. Subsequent analyses of the plant recordings showed that oxygen readings initially remained constant at the instrument full scale deflection (fsd) of 5%, but fell rapidly to 2% approximately 40 hours after inerting commenced. This rapid fall corroborated predictions that the filling regime was more akin to a plug flow rather than a mixed flow model, and that the argon was penetrating throughout the waste. The oxygen concentration continued to fall, reaching 1% after approximately 110 hours.

During the inerting of Compartment 2, the oxygen content of the voids was regularly monitored and no appreciable depletion occurred implying that no argon leaked out. Furthermore, monitoring around the silo has confirmed that no detectable leakage of argon occurred from Compartment 2.

COMPARTMENT 2 DE-INERTING AND RE-INERTING TRIALS

A series of further trials were conducted following the successful completion of the inerting trials to provide further important data in support of the safety case. These involved firstly isolating the argon supply and observing the instruments to determine how quickly oxygen levels in the compartment would recover. The oxygen concentration steadily rose from the base line of 1% in a near linear trend over the next 24 hours to approximately 2%, reaching 3% after 40 hours and 5% after 100 hours before exceeding the fsd of the instruments.

A further de-inerting trial assessed what impact the ventilation system had on inerting. It was established that it would take several hours before oxygen concentrations exceeded 2% with the ventilation system operational and more than 24 hours with the ventilation fan stopped.

Finally, the compartment was re-inerted by injecting the argon again at the full flow rate of 12 m³/hr. There was a rapid decrease in the oxygen concentration to <2%, with levels continuing to fall and reaching 1% after approximately 30 hours.

FULL SILO INERTING TRIALS

Full commissioning trials were performed once the Compartment 2 trials had been successfully completed. These effectively repeated the trials performed for Compartment 2.

The results and outcome of the initial inerting trial generally reflected those recorded and observed for Compartment 2. All compartments attained the 2% oxygen concentration level approximately 60 hours after inerting commenced and then continued to fall to 0.5% after 70 hours. Recorded oxygen concentrations from the oxygen analysers located within the silo and tunnel are summarised in a diagram presented as Figure 3. Pressures in the tunnel extract were maintained between –10 and –50 Pa relative to atmosphere. Strong winds did affect maintenance of the pressures and caused the oxygen concentration at the top of the tunnel near to the extract to fluctuate by less than 1%.

CONCLUSIONS

1. Commissioning of the argon inerting system was successfully completed with the structural integrity of the silo and its argon retention capabilities proving to be better than was envisaged, prior to structural improvements. This means that in the event of an argon inerting system failure, the silo would remain in an inerted state (oxygen concentration less than 2%) for a several hours in the vicinity of the waste, even with the ventilation system operative. Argon can be retained for even longer periods of time (days) by switching off the ventilation system. This has enabled simplification of future inerting upgrade requirements, because a less rapid response is required following inert gas failure than was originally envisaged.

2. Commissioning established that atmospheric conditions (pressure fluctuations and wind speed) make fine-tuning of the silo depression to minimise the air in-leakage difficult to achieve. A compromise was implemented that balanced silo inflow, ventilation extract and atmospheric variations by maintaining the silo depression in the range –10 to –50 Pa.

3. Plug flow inerting of the silo compartments occurs during filling.

4. The maximum argon injection rate can be achieved with no sign of any restrictions to flow through the sumps and filter bed through which the argon is injected. It also suggests that there is no bulk free liquid at the base of the compartment as this would act as a resistance to argon flow.

5. No significant leakage of argon via diffusion or through construction joints has been detected into the voids under the silo. Nevertheless, area monitoring will continue to be provided.

6. The low-lying areas outside the silo, where argon could gather, were carefully monitored and no oxygen depletion was detected at any point. This is a function of the good argon retention of the silo and sealing. Whilst vigilance must continue in hazardous areas, it has been demonstrated that the hazard is less onerous than originally envisaged.

REFERENCES

1. Health and Safety Executive (Liverpool) January 2001
2. Occupational Health and Safety Administration. Department of Labor February 2001
3. Annual Report of the Health and Safety Executive 1998

Table 1. Comparison of argon, nitrogen and carbon dioxide properties

Property	Argon	Nitrogen	Carbon dioxide
Density (kg/m^3)	1.78	1.25 Diffusion rate is comparable with air. Would therefore need to change many compartment volumes to displace air and require large purge volumes.	1.98
Reactivity	Very non-reactive	Reacts with Mg, U & Zr at elevated temperatures.	Decomposed by Na, Mg, Li, K, Zr & metal hydrides. Forms weak acid in presence of water vapour that could attack the silo walls and structures.
Prevention of metal fires	Yes - due to oxygen depletion	No	No
Capability for extinguishing metal fires	Yes – any metal	No – once ignited Mg will continue to burn forming nitrides	No – Mg will burn forming oxides, CO & C.
Use as fire fighting system	Yes	No	No
Fulfils compartment inerting requirements	Yes	Yes	Yes
Cost (£/100 m^3)	39	7	17

Table 2. Safety legs providing the hazard management strategy for the silo argon asphyxiation hazard

	Safety legs							
	Engineered			Engineered and ops. management & procedural			Ops. management & procedural	
Activities/Areas	1	2	3	4	5	6	7	8
	Minimisation of argon flowrates	Containment	Access control	Safe working distance	Ventilation (forced)	Detection	Operational procedures	Respiratory protection
Argon delivery/ offloading to tanker	Y	y	Y	y			y	
Argon storage & let down plant		y	Y	y		y	y	
Argon supply to silo		y	Y	y		y	y	
Adjacent to silo	Y	y	Y	y		y	y	
Silo/tunnel roof	Y	y	Y	y		y	y	
Instrument cubicles		y	Y	y		y	y	
Control room		y	Y	y		y	y	
Vent & stacks	Y		Y	y		y	y	
Confined spaces (including tunnel and trenches)	Y	y	Y	y	y	y	y	*

* = As required, but only as a last resort.

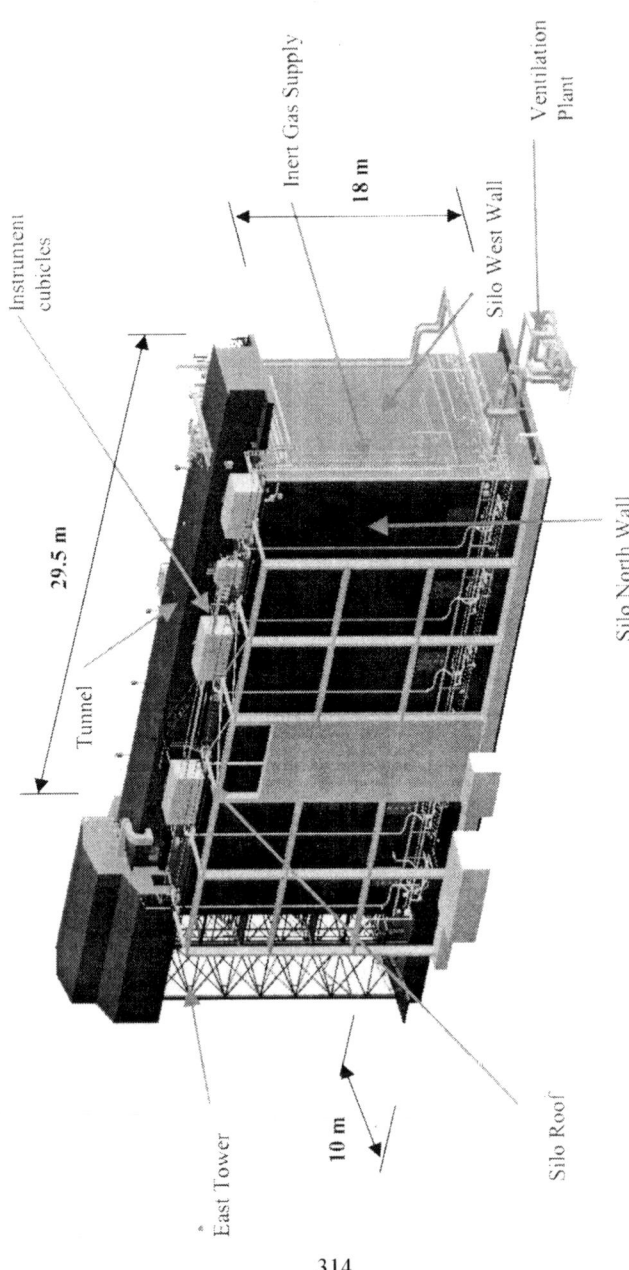

Figure 1. Silo isometric

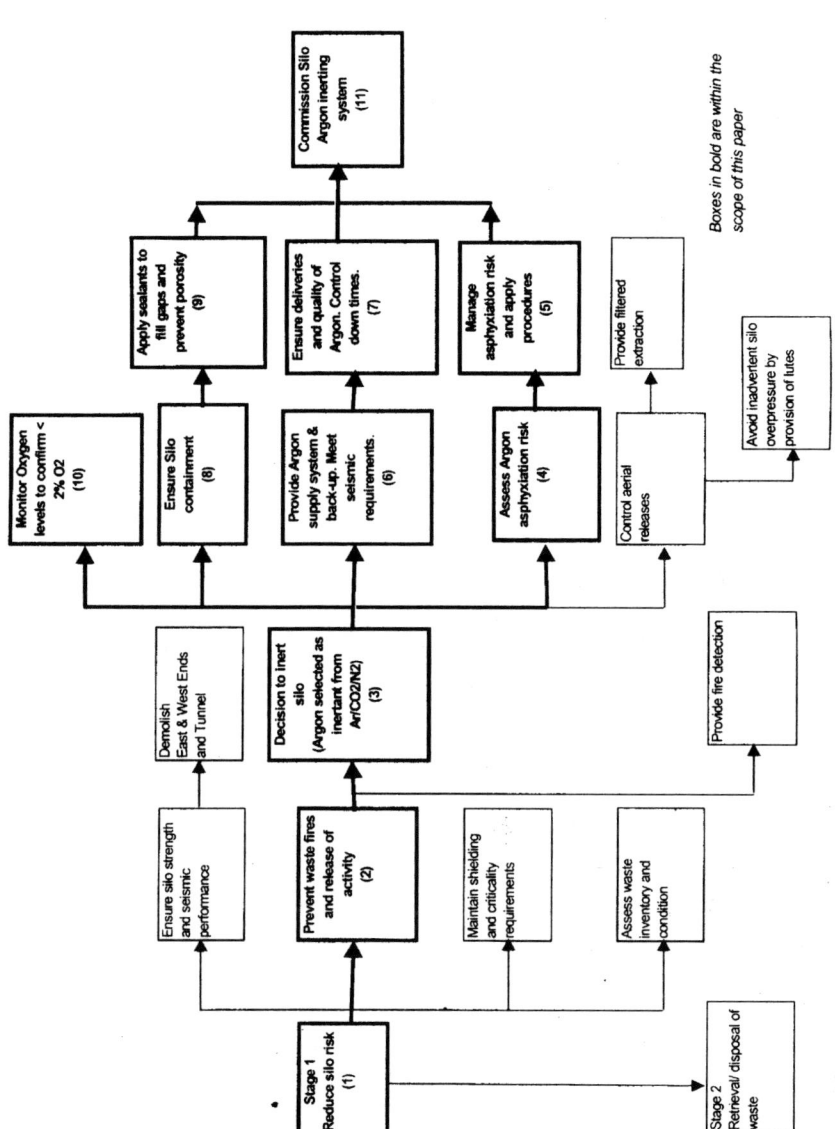

Figure 2. Fast diagram for silo risk reduction philosophy

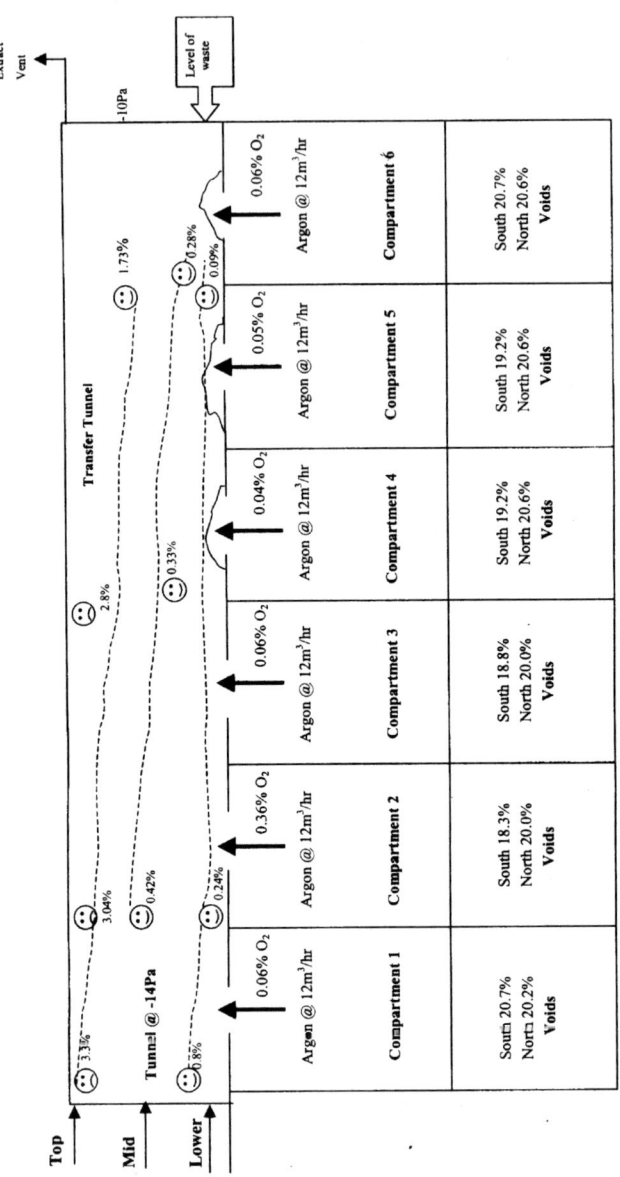

Figure 3. Oxygen concentrations measured during commissioning trials

FIRE RISKS FROM PACKAGED FLAMMABLE DUSTS – HSE FIRE INVESTIGATION AND TEST WORK AT HSL

M Iqbal Essa – MRSC; C.Chem; MICheme; C.Eng; PGDipOH; MIOSH.
HM Principal Specialist Inspector of Health and Safety
Health & Safety Executive, Grove House, Skerton Road, Manchester M16 0RB
Dr. Graham Atkinson – BSc; PhD.
Principal Scientist - Health & Safety Laboratory, Harpur Hill, Buxton, Derbyshire.

In the UK the Health and Safety Executive (HSE) has responsibility for enforcing standards of process fire safety, including the storage of highly flammable solids. Many industrial premises hold solid materials that are flammable. These materials could include natural and synthetic substances such as plastics, rubber, paper and textiles in the form of blocks, sheets, fibres, granules and powders. If these materials are not suitably stored or used they could lead to serious fire hazards. The risk is of heavy losses to the company involved and a threat to the lives of people both on and off site.
This paper discusses an HSE investigation involving a major fire at a rubber crumb processing plant that resulted in a fatality of an employee. The serious nature of the rubber crumb fire and its rapid escalation (which destroyed the company's approximately 70 tonnes of packaged product within minutes) prompted an in depth study at the Health and Safety Laboratory in Buxton. This paper also discusses some intermediary observations made.

INTRODUCTION

In many cases people within industry are not fully aware of the fire characteristics of the materials they use, handle or store within their premises. This is probably largely due to them not fully understanding its properties and the requirements necessary for adequate storage arrangements and additional safeguards that might be needed against a fire.

It is not always easy to predict how a solid flammable material may burn when heated and the resulting fire develops. This is because the behaviour of a burning material largely depends on several factors. These include, the quantity of flammable material involved, how it is packaged and stored in the workplace.

The fire that HSE investigated at a rubber crumb processing factory in Stockport in 1999 established that the packaged rubber crumb fire did indeed occur because of poor understanding about fire hazards associated with the material and because of inappropriate packaging practices.

THE INCIDENT

The fire occurred during the night shift at 00.31 on 25th Nov 1999. There were only 3 people were working the night shift on the site at the time of the incident. Sadly one of them died in the fire.

The fire started in an area where freshly processed rubber crumb was being loaded into 25kg three-ply paper sacks. These sacks were stacked on a pallet accommodating 46 sacks in total to make up an 1150 kg load. The sacks were shrink-wrapped to improve load stability in storage and transportation. A hand held propane gun was used for the shrink-wrapping operation. It was reported that immediately after completion of shrink-wrapping the employee decided to move the pallet into the storage area located at the front of the factory. It was whilst he was moving the load using a fork lift truck that he noticed smoke coming out of the freshly shrink-wrapped load of rubber crumb. He called out to alert his colleagues to the developing fire and raced up to the office, situated on the mezzanine floor, to ring the fire brigade. The fire spread quickly to the neighbouring stock of packaged rubber crumb product. The rate of fire spread and high smoke production hampered attempts to fight the fire with extinguishers and the other two employees were forced to make their escape. Unfortunately, the employee who went to raise the alarm became trapped by smoke and flames.

It was reported that the local fire brigade responded immediately and were at the company's site within 4 minutes of the alarm being raised but by the time they arrived the fire was already raging fiercely.

THE PREMISES
The factory was a single floor corrugated cement sheet building with a pitched roof. The walls were in filled with breeze blocks to a height of around 2.7m and from above about 3.4m from the ground level there were glazed panels. These were protected on the outside by steel mesh. The roof was supported by steel trusses and clad with profiled cement panels.

Small offices were constructed in one front corner of the building with access to a changing room and an office located on the mezzanine floor. The roof of the middle section of the building was at a higher level to accommodate the bucket elevators required for moving the products from the milling and sieving processes. With the exception of the offices, the whole structure was completely open. This openness offered the possibility of free flow of smoke under the roof from one part to another.

THE PROCESS
The main process undertaken at the premises is the grinding of tyre re-tread waste to make different grades of rubber crumb. A range of product grade is produced that vary in the size distribution of the crumb. The mean particle size varies from the finest grades at around 150 micron diameter to over 1mm diameter for the coarsest grades. The processed rubber crumb is used in the manufacture of such products as children's play surfaces.

Raw material is processed via a storage hopper, a loading conveyor and bucket elevator, a box magnet on the conveyor and a rotary drum magnet at the inlet of the bucket elevator remove fragments of wire.

From the storage hopper the material is introduced into one of two independent grinding loops via screw feeders. A second bucket elevator lifts material in the grinding loop on to a sieve. The oversize fraction re-enters the mill whilst product is moved into a conveying duct where a large volume of cooling air is introduced. The product is separated from the cooling air in a cyclone separator and accumulates in a finished product hopper. finer grades are then bagged directly into paper sacks. Coarser grades are sieved again to remove finer fractions before bagging. There are additional magnetic separators to remove any remaining metal fragments in the grinding loop and the product line.

Temperature sensors are fitted under the grinding mill and in the associated elevator feed screws. These sensors are set to trigger an alarm and shut down the plant if the temperature of material in the grinding loop rises above a set temperature. The fire point of vulcanised rubber is around 200°C, so the sensors should prevent overheating leading to a fire.

Tests carried out at HSL following previous HSE visits in 1992 established that the material was explosible. HSE advised the company to afford adequate explosion relief to bucket elevators, hoppers and cyclones.

BAGGING PROCESS

The most significant secondary process undertaken in terms of the fire investigation were bagging of product, stacking of bags on pallets, shrink wrapping of complete pallets load of material and moving of the complete wrapped pallets to a storage area.

Both polythene and 2 and 3-ply paper sacks were used to hold the finished rubber crumb in 25 kg lots. One of the employee stated that 3-ply paper sacks were used in the pallet in which the fire started. The empty sacks used are reinforced with paper tape pre-stitched with natural fibre thread at one end. After filling the sacks are closed with the aid of a hand held stitcher using synthetic fibre thread.

After filling the bags of crumb are flattened by being gentle trampling to assist stacking and stability. The bags are then stacked on to a pallet. Two dabs of a water-based paste are applied between layers of bags to improve the stability of the pallet load.

Some pallets, including the one in which the fire apparently started, were shrink wrapped immediately after being stacked. A 450 gauge polythene cover was put over the load. If the bags had been properly flattened, this cover would reach down to the bottom of the load. A propane fuelled heat gun was used to shrink and tighten the cover around the stack of bags. The shrink-wrapping process depended on the gun being held the correct distance from the polythene cover. If the gun was brought too close to the cover or was left in the same place for too long the cover would melt, creating holes in the cover. The Fire Service inspected several pallets that had escaped fire damage after the fire; holing of the plastic cover caused by close contact with the heat gun was fairly common. The total height of a pallet load containing 46 bags is something over 2 metres in height. This means that it is difficult to monitor and control the heat gun as it is applied to the top surface of the pallet. Finally the completed pallets were moved by a fork lift truck to a storage area to await despatch.

OTHER ISSUES

This investigation also revealed that the standard of housekeeping within the factory was poor. This is in spite of the company's claim that they vacuum cleaned the process area regularly. Accumulations of dust to a depth of 150mm were observed on support beams and electrical apparatus, such as strip lighting. This gave cause for serious concern. The foreseeable secondary explosion hazards of this dust were explained to the company. Appropriate advice to keep such areas free from dust was given.

The level of dust accumulation in and around the electrical motors and their housings was considered a potential fire hazard. Overheating of the motors as a consequence of dust accumulation was explained to the company. Appropriate advice to keep motors and their housing free of dust was thus afforded.

In spite of the earlier advice in 1992 given by HSE on the provision of explosion relief, the dust collection unit and its explosion relief was very poorly positioned within 1m of a desk used by operators. The potential dangerous consequences were stressed to the company. An alterative much safer location for resiting the dust collection unit and the explosion relief was advised. The company agreed to move the dust collection unit well away from the process area into the yard outside.

FIRE DAMAGE

The fire damage was most severe in the front part of the building. Approximately 70 tonnes of fine rubber crumb was destroyed by the fire in this area, together with some other plastic and packaging materials. Above the palletised storage area the cement panels covering the roof in the front part of the building were also completely destroyed. Structural steel in roof trusses at the front part of the building was also badly distorted by the heat from the fire. A wooden staircase leading to the Quality Manager's office and an adjacent toilet on the first floor were also completely burned away.

Heat and fire damage was observed at roof level above the machinery in the middle section of the building. This was consistent with strong flames extending upwards from under the roof on the front section.

IGNITION MECHANISMS

A number of possible causes for the initial ignition were considered but discounted. These included electrical faults, self-heating, malicious ignition, FLT malfunction and process overheating are all considered to be unlikely sources of ignition in this instance.

The propane powered heat gun is however considered an obvious possibility, since the pallet that caught fire had been shrink-wrapped a minute or two before the fire was discovered.

Some basic tests carried out by the Fire Service on empty sacks showed that to ignite the paper sacks required a relatively long exposure to the heat gun. When the heat source was removed the paper sacks generally self-extinguished. Only if the frayed paper end of the sack was ignited did the fire persist when the heat source was removed.

The Fire Service tests also showed that the paper reinforced stitching at the pre-stitched end could also be ignited by a short application of the heat gun. This continued to burn, igniting the end of the sack and the hessian stitching. Eventually the sack was breached and the flammable contents would have started to leak out.

It was concluded that the propane gun could have ignited the edges of the sacks at the top of the pallet during the shrink-wrapping operation. This ignition would have led to a very small but growing fire that might well have been overlooked for a minute or two during the transportation of the rubber crumb loaded pallet.

UNUSUAL FEATURES OF THE FIRE

There were two features of this fire that might have important implications for the safe storage of rubber crumb and possibly other finely divided combustible products.

i) *The fire grew extremely rapidly.* Generally very rapid fire growth is associated with high stacks of expanded plastic foam or baled textiles. In this case the storage was only one pallet high and the product had a high density and was packaged.

ii) *The premises became smoke logged very rapidly.* This is clearly closely related to the rapid spread of flame. Nevertheless, it is significant that smoke began to seriously impede escape within a couple of minutes despite the open character of the building and the fact that the roof extended to a height of over 13 metres in places.

Following the fire the Health and Safety Laboratory carried out a number of tests to investigate the rate and mechanism of fire growth in palletised rubber crumb. The results of these tests have also been used as inputs in an analysis of smoke logging of the factory. It is hoped that the results of these investigation will improve understanding of the fire hazard associated with this kind of product and provide data suitable to undertake realistic risk assessments in other premises.

FULL SCALE TESTS – FIRE GROWTH IN PALLETISED RUBBER CRUMB

Two full scale tests of fire growth in a pallet load of 40 mesh rubber crumb were carried out in the high rack test rig at HSL in Buxton. This rig is essentially a cubical enclosure of height 7.5m that can be conveniently clad with polythene on two sides so that the early stages of fire growth take place in relatively sheltered conditions. If a fire grows to a size greater than around 5 MW the polythene softens and fails, preventing the fire from becoming ventilation controlled.

TEST 1

An array of thermocouples was used to monitor the temperature under the roof deck. In this first test a pallet load of 38 x 25kgs bags was burned on an open steel tray 1.8m x 2.3m x 0.3m (high). This tray was supported on 4 load cells. This arrangement allowed the measurement of the rate of mass loss in the early stages of the fire. After around 5 minutes some material fell from the pallet over the edge of the tray and thereafter effective monitoring of the rate of heat release through measurement of the rate of mass loss was not possible.

The fire was started with a British Standard Number 7 pine wood crib. This has a mass of 125g and produces a peak heat output equivalent to four crumpled sheets of broadsheet newsprint.

When the paper sacks were breached the rate of fire development was extremely rapid. The fine rubber powder ran down from the edges of the sacks igniting and burning immediately. As the rate of heat release increased the updraft of air around the pallet became more vigorous. This flow increased the rate of ablation of the surfaces of the pallet and drove a proportion of the material released upwards, so that it burned as a dust cloud rather than a solid pile. The observed rate of fire development was consistent with the eyewitness accounts of a major conflagration occurring within a couple of minutes of the first signs of fire.

Results of some of the measurements of temperature at ceiling level (7.5 m elevation) in the area just above the pallet are shown in the graph below.

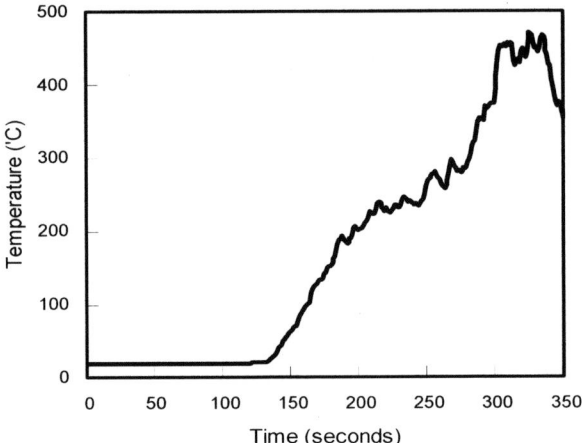

Figure 1.   Ceiling temperatures 7.5m above base of pallet 38 x 25 kgs rubber crumb (40#)

The first signs of raised temperatures at roof level occurred after around 140 seconds. The temperature rises steadily to around 500°C in the next 160-170 seconds. Thereafter there is a fall in the recorded temperature in this area of the ceiling. This is mainly caused by progressive failure of the plastic shielding around the rig and consequently increased plume deflection.

Results of measurements of mass and mass loss rate together with the average temperature are shown in figure 2.

The shifting of load as material pours down from burning bags produces some variations in apparent mass loss but it is clear that the rate of mass loss becomes significant at around 140 seconds after ignition. It rises to a level of around 250g/s after about 250 seconds. The rate of total heat production (both radiant and convective) from styrene-butadiene rubber is around 27 kJ/g so a mass loss rate of 250 g/s corresponds to a rate of heat release of 6.7MW .The proportion of heat production radiated by the highly emissive flames depends on the size of the fire but for a fully burning pallet would probably be slightly over 50%.

In order to compare the rate of fire growth with other types of commodity the rate of mass loss and heat release have been fitted to quadratic "t-squared" curves. If this fit is carried out over the first 210 seconds of the fire a value of 0.28 is obtained for the "a" parameter in the t-squared fire growth expression i.e.

Heat release (kW) = a.t^2 = 0.28. t^2 (t in seconds)
If the fit is carried out over the first 300 seconds a value of 0.18 is obtained for "a".

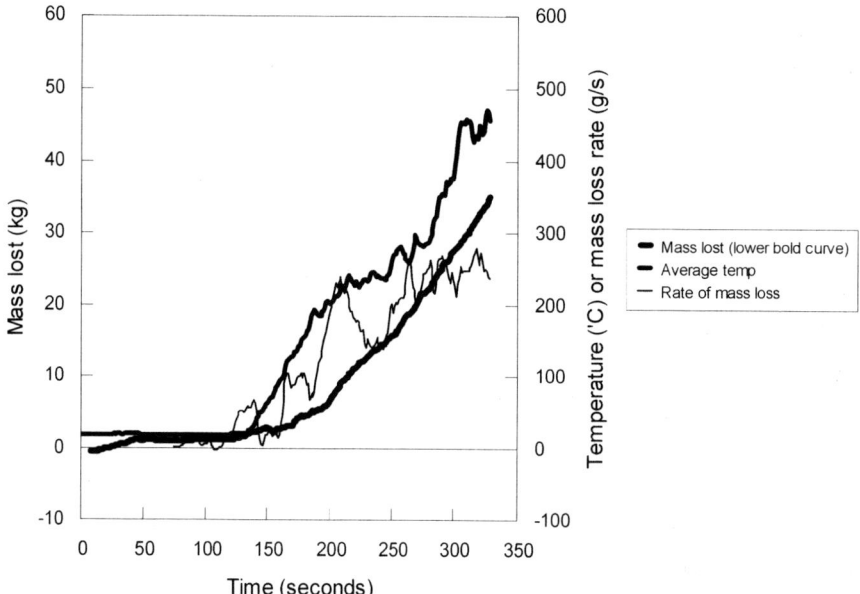

Figure 2. Mass loss, mass loss rate and temperature for 38 x 25 kgs sacks rubber crumb (40 #)

Heat release (kW) = a.t² = 0.18. t² (t in seconds)

The most hazardous category in the American National Fire Prevention Association (NFPA) system of t-squared fires for use in risk assessment is ***ultra-fast***, which is taken to correspond to a value of (a) of 0.18. This means that the rubber crumb fire grew as quickly or more quickly than the most severe case considered in this system of fire risk assessment. Other materials that may fall into this category include, for example, high stacks of polyurethane or polystyrene foam, flammable liquids in plastic bottle and long rolls of paper stored on their ends.

At the site where this fire investigation was carried out, the ignited pallet load of rubber crumb was in close proximity to a number of other such loads. After the initial growth of fire to a few megawatts there would have been very rapid fire growth over exposed surfaces of nearby pallets, driven by the high radiative heat transfer.

TEST 2:

The second test carried out HSL was less successful than the first. In order to improve visibility the plastic shielding (destroyed in the first test) was not replaced. A breeze on the day of the test led to flames being blown away from the ignited face and under the pallet. The fire developed on the sheltered side of the pallet opposite to the point of ignition. The flames were blown away from the top of the pallet reducing the rate of fire spread and thus the rate of burning. However, after a period of 7-8 minutes a large fire developed.

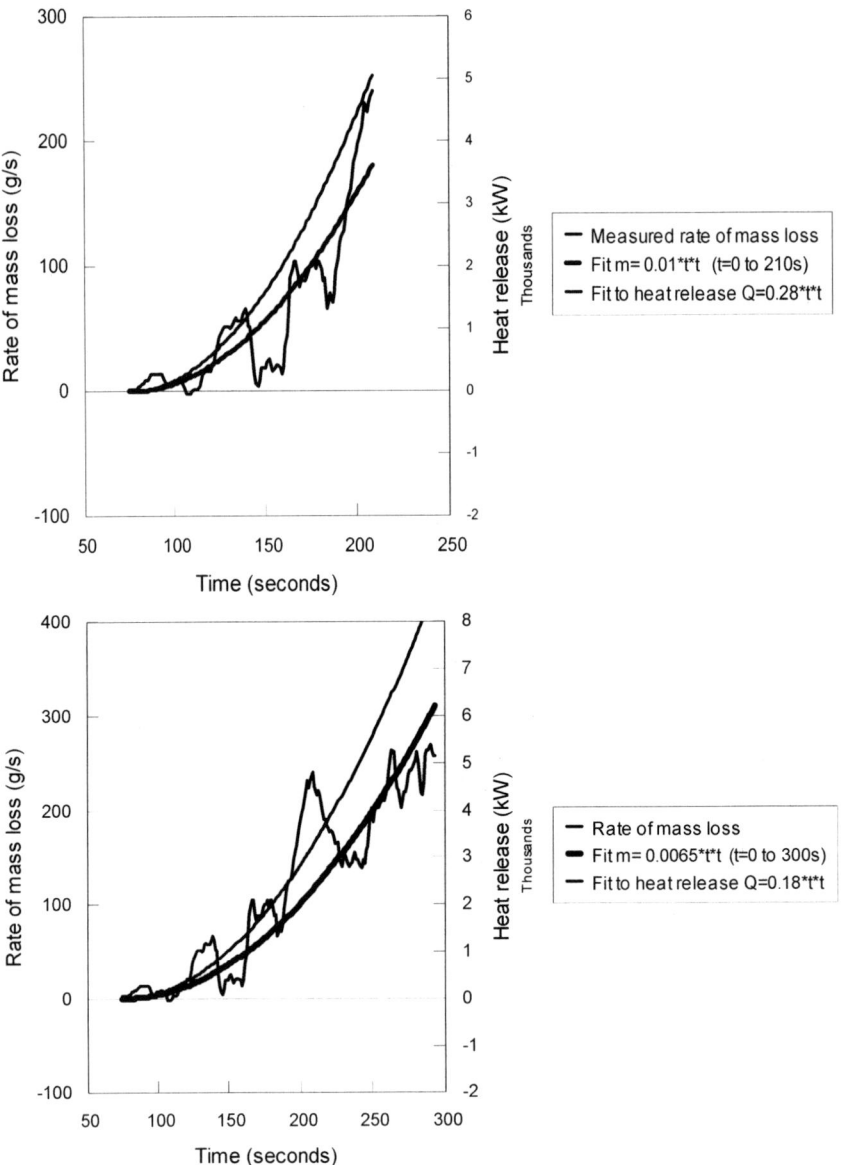

Figure 3. Measured mass loss rate and fits to mass loss and heat release

MEDIUM SCALE TESTS OF FIRE HAZARDS

HSE use a medium scale test to assess the fire hazard of solids stored in bulk. This test classifies materials as HIGH or NORMAL hazard according to two aspects of their fire performance: the rate of development of fire and the potential to generate smoke. The test is general applied to cellular plastics and lightweight textiles in block, roll or baled form.

The medium scale test is essentially a compromise between the full scale testing, which produces immediately applicable results, but is expensive; and bench scale testing that is cheap but produces results that have to be extrapolated to be of use. The hazards of rubber crumb were effectively demonstrated by the large scale tests described above, but it is of interest to see how the materials classified by the medium scale screening test. The test also allows the measurement of smoke yield which is significant in the context of understanding the circumstances that led to the fatality at the incident site in Stockport.

The test is not regularly applied to powdered, high density materials. There is therefore no established method for choosing the sample geometry. The general principle followed for other types of material is to try to reproduce in the test, as precisely as possible, a small section of a storage being assessed. In the case of blocks of foam or rolls of textiles this is generally possible to do satisfactorily. The case of a powder in 25 kg sacks is more problematic. It was observed that the mechanism of burning involved material being dislodged from the surface of breached bags then dispersing and burning as it fell. These processes are difficult to reproduce realistically in a small scale test.

Two sample geometries were chosen in which 5 kgs of rubber crumb was packed in 10 x A4 manila envelopes. These were stacked in different ways in the two tests.

Results from the two tests were broadly similar. The potential of rubber crumb to fuel a very rapidly growing fire did *not* show up in either test. The smoke yield was also low over the period of the test, because only a small proportion of the sample burned in this period. The measurements of smoke and heat production suggest that, if all of the 5 kg sample had in fact burned, the smoke production would have been 900 m^3 ODm1. This is in excess of the limit defining a high hazard material of 400 m^3 ODm1.

The conclusion is that the causes of high rates of fire growth in palletised rubber crumb are probably very difficult to assess effectively in a test of the sort currently used by HSE. A better understanding of the fire hazard of palletised fine combustible powders is required. It may be possible to devise a special smaller scale test to determine whether another material of this sort represents as high a level of hazard as fine rubber crumb.

ANALYSIS OF SMOKE LOGGING

The rapid filling of the building with dense black smoke was probably a key factor in the tragic consequences of the fire at Stockport. The details of the process of smoke logging are quite complex because of the variable roof height, growing fire, uncertain heat transfer, venting of roof panels etc. Nevertheless, it is possible to calculate approximately how long it took for the smoke layer to descend to head height level and what the visibility was at this stage.

BASIC ANALYSIS

The floor to ceiling height in the section of the building where the fire started was 8.64 m. The most commonly used expression for the mass flow of air into a hot layer above a typical fire is based on Thomas, P. et al, Fire Research Technical Paper No.7 -1963)

$$\text{Mass flow} = 0.096 P_{fire} \; \rho_o \; y^{3/2} \; . \; (gT_o/T_f)^{1/2} \qquad \text{(Equation 1)}$$

Where:
P_{fire} is the perimeter of fire plume near the source (m)
ρ_o is the density of air at ambient temperature (1.2 kg/m^3)
y is the height of the base of the hot layer above the fire base (m)
g is gravitational acceleration i.e. 9.81 m/s^2
T_o is the ambient temperature (K)
T_f is the temperature of the gases entering the hot layer.

The perimeter of the rubber crumb pallet fire increases somewhat as the fire progresses and material spills from the pallet; a figure of $P_{fire} \sim 8m$ is appropriate. The initial height of the base of the hot layer is approximately 8.6m. Results from the large scale test undertaken suggest that after the first 60 seconds the parameter $(gT_o/T_f)^{1/2}$ is in the range 1.9 to 2.2, so a value of 2 will be assumed.

With these assumptions Equation 1 gives a mass flow of air into the hot layer of 46 kg/s. Before the raised section of roof is filled there will be further rapid entrainment as hot smoke flows towards and then up into the middle roof space. Such extra entrainment is ignored in this basic calculation and it is assumed that the fire plume is the only source of entrainment into the hot layer. There will also be cooling of the smoke through both radiant and convective contact with the walls and roof of the building. We assume that rapid high temperature radiant cooling lowers the average temperature in the hot layer to around 200°C. In this case the initial rate of increase in volume in the hot layer is approximately 66 m³/s.

A first estimate of the time taken to smoke log the building to a height of 2.7m (floor level in the upper storey) can be obtained by dividing the volume of the building above this level by the initial volume flow rate.

$$T = 6882m^3 / 66 \text{ m}^3/s = 104 \text{ seconds}$$

If it is assumed:

a. The smoke is well mixed.
b. The yield of smoke is 900 m³ ODml / 5kg as in the medium scale tests.
c. The average rate of mass loss during smoke logging is 250 g/s.

The optical density of the smoke (at 200°C) is 0.64. In these conditions a non-luminous object would be completely invisible if it were further than 2 metres distant.

In fact the time to smoke logging would be greater than the result above because as the hot layer deepens the fire plume increasingly re-entrains smoke, so the effective value of y in Equation 1 falls. For example, if the average effective height were to be midway between the roof height and final layer height, the time for smoke logging would be increased to 190 seconds. Where smoke re-entrainment was significant the smoke density would be increased. If the smoke logging time increased to around 200 seconds the visibility would reduce to around 1 metre.

Apart from the neglect of smoke re-entrainment many of the assumptions on which this analysis are based are very crude. The lower surface of the hot layer would not necessarily be horizontal. There might be a number of significant effects that would hasten the contamination of air near the ground; for example, air cooled by contact with the walls flows downward through the hot layer, developing a momentum that takes it below the lower surface of the hot layer. It is also possible that large scale vertical structures driven by the strong flow into the middle part of the ceiling could disturb the lower surface of the hot layer. It would be possible to investigate the smoke logging process in detail but this would require Computational Fluid Dynamics.

CONCLUSIONS FROM ANALYSIS OF SMOKE LOGGING

The conclusion from the basic analysis of smoke logging of the building is that the smoke layer will deepen rapidly, reaching the level of the floor of the first storey after around 200 seconds. This roughly corresponds with the time taken for the burning of a single pallet to reach its maximum severity. This means that anyone attempting to fight the growing fire would begin to perceive smoke close to head height at about the same time as the fire reached a stage when it was clearly uncontrollable. This appears to broadly correspond to the account given by the company.

GENERAL CONCLUSIONS & RECOMMENDATIONS FOR ACTION

1. Storage of palletised rubber crumb represents a high hazard in terms of the potential to fuel a very rapidly developing fire that produces a large amount of dense smoke. It is considered that a lack of awareness of the fire risks associated with this product at the factory may well have been a contributory factor in the death of its employee at Stockport.

2. Steps are necessary to improve awareness of the risks of palletised rubber crumb through advisory and regulatory work by HSE and the Fire Service, and by providing clear warnings when the material is supplied e.g. hazard warning sheets and markings on packaging.

3. The use of a hand-held LPG fired naked flame gun is considered to be totally inappropriate for shrink-wrapping rubber crumb. This is because rubber crumb is a high fire hazard material and as such would fuel a fire producing large quantities of dense smoke very rapidly. To reduce risks alternative cold methods should be used.

4. There is a strong possibility that other finely divided combustible materials may behave in a similar manner to the rubber crumb tested at HSL. This is currently under investigation and the findings will be published at the conclusion of the study being carried out at HSL.

5. If a fire grows rapidly, the time for smoke logging of even large open structures may be very short. Emergency procedures that involve employees contacting the Fire Service before exiting are potentially dangerous. This is especially true when making an emergency call causes someone to lose track of the progress of the fire.

6. Necessary precautions against fire and explosion as set out in HSE's Guidance Note HS(G) 103, entitled "Safe handling of Combustible Dusts", ISBN 0-7176-0725-9. This guidance note can be obtained from good booksellers.

REFERENCES

1. ASTM D93-80 (1980) "Standard test method for flash point by the Pensky Martens Tester"
2. ISO 5660 (1993) "Fire Tests-Reaction to Fire-Rate of Heat Release for Building Products", International Organisation for Standards.
3. Babrauskas, V. (1995) "The Cone Calorimeter", SFPE Handbook –Ed DiNenno, P., Pub. National Fire Protection Association, Quincy MA.
4. BS 2782: Part I Method 141 "Critical Oxygen Index"
5. BS 5306: Part 2 (1990) "Fire Extinguishing installations and Equipment Premises", British Standards Institution.
6. Wharton, R.K. (1981) Fire and Materials, Vol. 5, 2, pp73–76.
7. Foley, M. (1995) Doctoral Thesis, Edinburgh University
8. ASTME 1354 "Standard Test Method for Heat and Visible Smoke Release Rates for Materials and Products Using an Oxygen Consumption Calorimeter."
9. HSE Guidance Note 64 (1991) "Assessment of Fire HazardS form Solid Materials and Precautions Required for their Safe Storage and Use", HMSO.
10. BS 5852 (1982) "Fire Tests for Furniture", British Standards Institution.

BUND DESIGN TO PREVENT OVERTOPPING

Glenn Pettitt,
ERM Risk, 8 Cavendish Sq, London, W1G 0ER.
Peter Waite,
Entec, 17 Angel Gate, City Road, London EC1V 2SH.

Most bunds are designed to contain 110% of the contents of the largest tank in the bund. However, it is now well established through a number of incidents and experimental work that stored materials may overtop the bund wall due to the momentum of the release following catastrophic tank failure. This paper is concerned with experimental work conducted to investigate whether bund walls could be retrofitted to prevent overtopping, avoiding the necessity to extend the bund height by several metres. Such a measure could be used where a risk assessment has shown that the residual risk to various receptors, both people and environmental, is not tolerable, after the effect of preventative measures has been analysed, i.e. bund redesign would only be used in an extreme case. The paper shows the results of the experimental work and demonstrates that the new design successfully prevented overtopping at a 1:30 scale.

KEYWORDS: Atmospheric storage tanks, catastrophic tank failure, bund overtopping, risk assessment, prevention, control.

INTRODUCTION

Modern design standards for bunds surrounding atmospheric storage tanks should ensure that the bunds are able to contain at least 110% of the maximum volume of the largest tank in a bund. This capacity is intended to contain the hazardous liquid, e.g. crude oil, kerosene, should there be a failure of the tank wall or the adjoining pipework. However, in the event of a catastrophic tank failure, or even a large connection failure there is the potential for the released liquid to surge over a bund wall due to the momentum of release. Several incidents have occurred in the past where liquid has been released over the secondary containment[1,2] and theoretical models have been developed to characterise such a release, e.g[3,4,5].

The experiments of Greenspan and Young[3] show that conventional bund walls would need to be almost as high as the initial liquid level to eliminate overtopping due the projection of a 'plume' of liquid with enough kinetic energy (derived from the initial potential energy of the static liquid) to rise over conventional walls.

Bund overtopping is a particular problem when there is a sloping bund wall or dike of low height. Some dikes have a shallow slope of only about 30° from the horizontal and total height of 1.5 m above grade. Following failure from the primary containment overtopping may result in about 50% or more of the contained material being released outside the secondary containment (Greenspan and Young[3]). This may be catastrophic if it affects an environmentally-sensitive area.

A series of experiments has been carried out to investigate the effects of placing a vertical section of wall on top of an existing sloping bund wall to mitigate the effects of overtopping. It was recognised that a vertical wall alone would be unlikely to achieve the desired effect of eliminating overtopping. The second author had observed sea wall tests

329

where overtopping from waves was reduced by the shape of the sea wall, designed to deflect the high velocity wave run-up and any 'splash' back into the sea. Therefore, the experiments included experiments with a horizontal 'lip' at the top of the vertical section on the inside of the wall. The experiments showed that while there was significant overtopping when using a typical bund wall design, overtopping was virtually eliminated when using the horizontal 'lip'.

This paper describes the experimental work carried out, the results and the significance in terms of inherently safer design of bund walls, particularly where the environmental risk may be high.

OBJECTIVE OF EXPERIMENTAL WORK

Bund overtopping has been shown to be a problem in a number of historical incidents. On several occasions, the design of the bund wall or dike was not sufficient to retain the spilled liquid following catastrophic failure of the primary containment; generally, the bund wall or dike was insufficient in height and was sloped so that the escaped liquid could run-up and easily flow over the top. This was particularly illustrated in the Floreffe incident of 1988 when it was estimated that between 40 and 71% of diesel oil from the primary containment overtopped the dike and 750,000 gallons flowed into the adjacent river causing serious disruption to water supplies and the environment[2].

Experimental work has been carried out on a number of occasions[3,4,6–9] to examine the flow of liquids overtopping bund walls following failure of the primary containment. Often the focus has been placed upon how much overtopping may be expected.

The objective of this series of experiments was to investigate if there was the potential to retrofit typical existing bund designs (with sloping walls) with a mechanism to alleviate the overtopping potential. It was not the intention to simply investigate how the height of the bund wall would need to be increased to retain all the liquid, rather if a specific design could be used, i.e. by use of an internal 'lip' on top of the bund wall, similar to a sea wall.

APPARATUS

A model of the bund was constructed in polypropylene to 1:30 scale for typical storage tanks and bunding arrangements. As with other experiments (e.g. Greenspan and Young[3] provide the justification) to study these effects a linear scale was chosen. The "base model" had sloping walls with the top of the wall 57 mm (to model 1.7 m) above the base of the bund. The angle from the horizontal plane was about 35°. The capacity of the bund was about 170% of the maximum scale tank capacity of a prototype tank (27.5 or 35 metres diameter, maximum fill height 11 metres). Alternative bund walls could be fitted to the model along one side to test the effects of different wall profiles:

1. An increased height of 77 mm (to model 2.3 m) above the base of the bund,
2. 77 mm (2.3 m) high sloping wall with an additional vertical section at the top of height 47 mm (1.41 m) with a 'lip' similar to a sea wall of width 19 mm (0.57 m).

A storage tank was represented by a moveable vessel approximately 700 mm (to model 21 m) in diameter, which although not typical of a crude oil or petroleum product storage tank (which have a lower height to diameter ratio) was more convenient to move and was large enough to give more than the maximum head for a liquid release from a typical tank (>10 m).

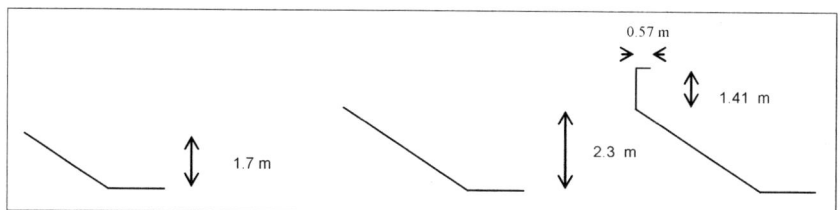

Figure 1.   Bund wall arrangements

The release mechanism consisted of a sliding plate behind a polypropylene block with the required hole cut into it. On the inside of the sliding plate was another polypropylene block with a larger hole than all the test holes. This is shown in Figure 2.

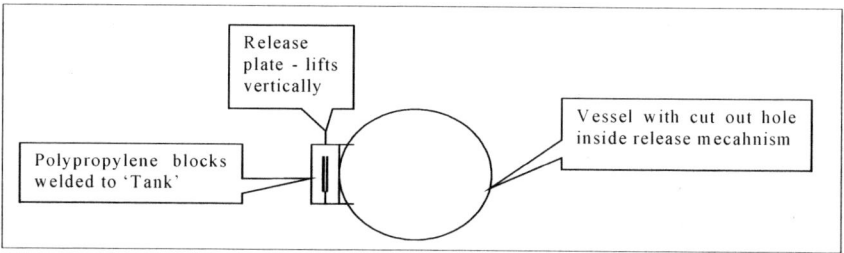

Figure 2.   Test vessel

The design of the release mechanism allowed different holes to be used, in order to model different failure types in the tank wall. Five hole sizes and shapes were used in the tests, as shown in Figure 3.

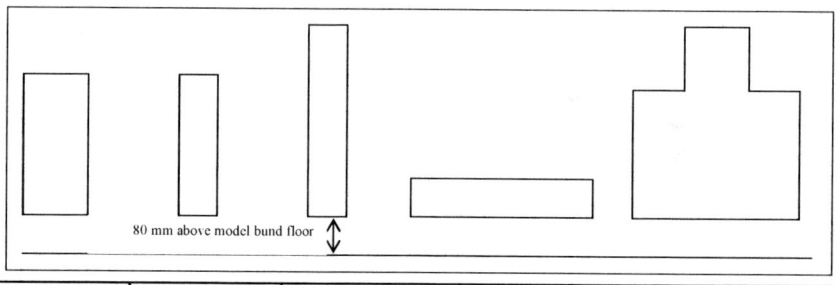

Hole 1	Hole 2	Hole 2 (modified)	Hole 3	Unrestricted
100 × 50 mm	100 × 40 mm	130 × 40 mm	40 × 130 mm	130 × 50 + 2 × 100 × 40 mm

Figure 3.   Holes sizes and shapes

After each test there was some residual fluid remaining in the vessel, below the lower edge of the hole (80 mm above the model bund floor).

The test fluid was water (SG 1 rather than the SG of 0.83 for a typical hydrocarbon fuel) and the flow over the bund wall was collected in rectangular vessels and weighed.

EXPERIMENTAL METHOD

The test runs were arranged to investigate the effects of the different bund walls, different distances between the release point and the bund wall, different heads of liquid and different orientation/shape/size of failure. The combinations are listed in the results.

The tests were performed in order to model a release of kerosene from tanks with various liquid heights. As indicated above in Figure 3, it was not possible to construct the apparatus to demonstrate a failure at the base of the tank.

The tests were filmed on a video recorder and example tests photographed.

The amount of water released was calculated from the geometry of the test vessel. The amount spilling over the bund wall was found by weighing the catch vessels and subtracting the empty weights. Most of the water was released within about 10 to 20 seconds.

Observations on the tests were recorded using the following nomenclature:

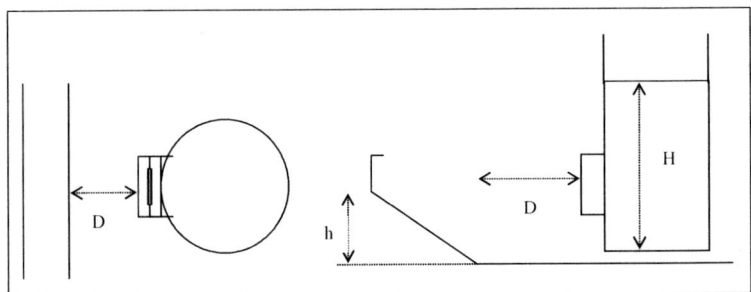

Figure 4. Nomenclature

RESULTS

Records are presented in Table 1 at prototype scale, derived from the model results.

Only two pairs of tests are directly comparable but serve to demonstrate that the two runs which gave the highest proportion of overtopping (0.342 and 0247) with the sloping bund of height 2.3 metres, Tests 2 and 3A showed virtual elimination of overtopping when repeated with the sea wall section in place (Tests 9 and 7 respectively resulted in 0.001 and 0.002 of the release overtopping the bund. This corresponds to less than 10 m³ at full scale, compared with up to around 1000 m³ without the sea wall. The tests were designed to demonstrate the effectiveness of the arrangement rather than to compare the results with and without the additional structure.

Table 1.   Summary of results

Test no.	Liquid head, H, above bottom of hole (m of water)	Distance from bund, D (m)	Hole	Bund wall height, h (m)	Proportion of release overtopping bund
1	6.6	9	1	1.7	0.62
2	8.58	3	1	2.3	0.342
3A	8.58	13.5	1	2.3	0.247
4	6.6	20	1	2.3	0.019
4A	11.4	20	1	2.3	0.103
6	6.6	20	1	2.3 + sea wall	0.001
6A	9	20	1	2.3 + sea wall	0.000
7	8.58	13.5	1	2.3 + sea wall	0.002
7A	9.9	13.5	1	2.3 + sea wall	0.002
8	8.58	9	1	2.3 + sea wall	0.000
8A	9.9	9	1	2.3 + sea wall	0.000
9	8.58	3	1	2.3 + sea wall	0.001
9A	9.9	13.5	1	2.3 + sea wall	0.002
10	6.6	20	2	2.3 + sea wall	0.001
10A	7.5	20	2	2.3 + sea wall	0.002
11	8.58	13.5	2	2.3 + sea wall	0.009
11 Repeat	8.58	13.5	2	2.3 + sea wall	0.000
11A	9.9	13.5	2	2.3 + sea wall	0.000
12	8.58	9	2	2.3 + sea wall	0.006
12 Repeat	8.58	9	2	2.3 + sea wall	0.000
12A	9.9	9	2	2.3 + sea wall	0.000
13	8.58	3	2	2.3 + sea wall	0.018
13A	9.9	3	2	2.3 + sea wall	0.043
14	8.58	3	2 modified	2.3 + sea wall	0.014
15	6.6	20	3	2.3 + sea wall	0.000
16	8.58	23.5	3	2.3 + sea wall	0.000
17	8.58	9	3	2.3 + sea wall	0.000
17A	9.9	9	3	2.3 + sea wall	0.000
18	8.58	3	3	2.3 + sea wall	0.000
18A	9.9	3	3	2.3 + sea wall	0.000
19	8.58	3	Unrestricted	2.3 + sea wall	0.000

Only two pairs of tests are directly comparable but serve to demonstrate that the two runs which gave the highest proportion of overtopping (0.342 and 0247) with the sloping bund of height 2.3 metres. Tests 2 and 3A showed virtual elimination of overtopping when repeated with the sea wall section in place (Tests 9 and 7 respectively resulted in 0.001 and 0.002 of the release overtopping the bund. This corresponds to less than 10 m^3 at full scale, compared with up to around 1000 m^3 without the sea wall. The tests were designed to demonstrate the effectiveness of the arrangement rather than to compare the results with and without the additional structure.

OBSERVATIONS

1. Although the volume of liquid released does not scale to the full contents of a typical tank, the liquid head used is similar to the maximum liquid head in a storage tank when scaled up. The effect of a greater volume (larger tank diameter) would be to reduce the rate at which the head decreased following the start of the release and so increase the duration of the release, the proportion of liquid overtopping the bund would be similar.
2. In Test 6A, only 1 or 2 small drops of liquid splashed over the bund wall.
3. In run 7A, there was initially no overtopping of the bund but after about 10 seconds one large 'splash-over' occurred.
4. For Tests 8 and 8A there was no overflow.
5. For Tests 9 and 9A there was very small overflow, the smallest amount detectable.
6. For Tests 11 and 12, there was a long delay to overflow.
7. It was observed that the silicone sealant (between the floor and the bottom of the bund wall) was being moved by the liquid flow during the test and that this was a possible cause of disruption to the flow causing the delayed splash, i.e. if there is an obstruction on the sloping face of the bund, this may cause some splash over the raised bund wall. The sealant was removed and replaced with a polypropylene weld that solved this problem.
8. The repeat of Tests 11 and 12 then showed no overflow and in Test 12A a small drop splashed over but was not detectable by weighing.
9. Tests 13 and 13A resulted in part of the release jetting directly over the top of the bund and sea wall extension, demonstrating that spigot flow could overtop the bund for a tank close to the bund wall.
10. Tests 11 and 11A were repeated and Test 11 resulted in no overspill but in 11A a small amount splashed over and was just detectable by weighing.
11. Tests 15, 16, 17A and 18A resulted in no overflow. For Tests 17 and 18, there were small drops of overflow but this was not detectable by weighing.
12. The hole restriction was removed for Test 19 to give the maximum possible hole size, but no overflow was observed or measured.

    Still photographs from the tests are shown in Figures 5 and 6.

Figure 5. Bund wall (2.3 m) without 'Sea Wall'

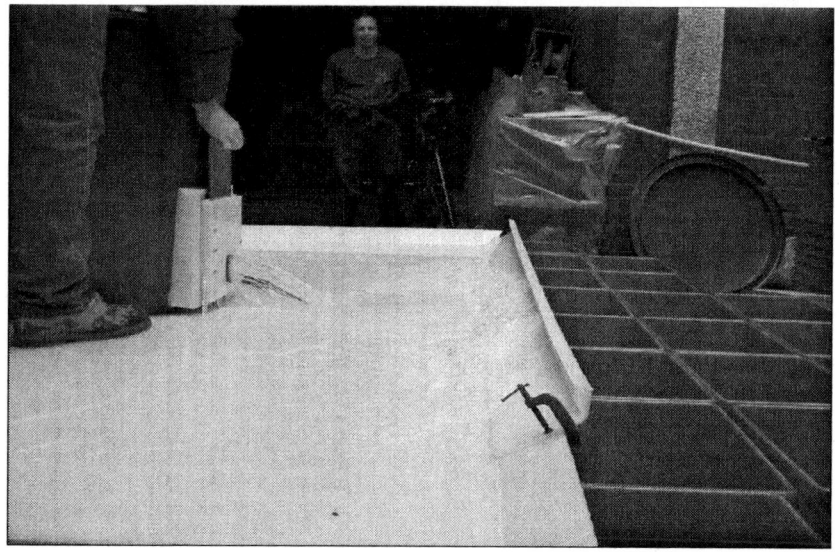

Figure 6. Bund wall with 'Sea Wall'

COMPARISON WITH OTHER MODELS

The experiments showed that when using a typical bund wall design, the results were compared with the model of Michels et al.[4] and shown in Table 2.

Table 2. Comparison of tests 1 to 4A with the Michels model

	Overtopping proportion	Michels et al. prediction
Test 1	0.62	0.25
Test 2	0.34	0.25
Test 3A	0.25	0.12
Test 4	0.02	0.04
Test 4A	0.10	0.04

Thus, it can be seen that the actual amount of overtopping was generally greater than that predicted by the model of Michels et al. This could be due to the fact that the slope in the test runs was shallow and thus it provided a smooth trajectory for the liquid to run up. (Other causes could be the experimental arrangement not allowing complete release to the tank bottom, and the smaller than normal diameter to height ratio). Such a result was expected and, in fact, important, as such an arrangement was seen as a control where it was felt necessary to be in agreement with previous research. However, overtopping was virtually eliminated when using the horizontal 'lip', and this is evident from the videos of the experiments. The work of Greenspan et al.[3] used similar set-ups to investigate the effects of overtopping. Greenspan, however, found that overtopping still occurred. The likely reason for the difference in results here is that Greenspan's experiments were one-dimensional, i.e. the liquid was released down a channel to a wall and spreading across the bund was not modeled. The experimental apparatus used in this work is more realistic as it allows for spreading across the bund two-dimensionally.

The fluid mechanics of the tests have not been investigated theoretically, but it was found experimentally that smooth flow across the bund is important. When there was a disturbance created by the silicone rubber seal between the bund floor and the wall, this caused turbulence in the flow, resulting in water being projected over the sea wall. When the seal was replaced by a polypropylene weld, the flow was more laminar in nature and overtopping did not occur for the majority of the tests. Thus it is likely to be importance not to have such obstacles in actual bunds. The fluid mechanics would need to be investigated further both experimentally and theoretically.

COMPARISON WITH HISTORICAL DATA

Several incidents have occurred in the past where there has been a catastrophic failure of an atmosphere storage tank containing crude oil or petroleum products. Such incidents are well documented by Wilkinson[1]. Following such a failure the tank contents have been released and in some instances the material has been lost outside of the secondary containment due to the momentum from the initial surge. The best example of this was the Floreffe incident[2] of 1988, which is mentioned above. The experimental work conducted here has supported the evidence from past incidents and other experimental work[3,4,6-9] that the effects of overtopping can be catastrophic for typical bund wall arrangements, particularly if the tankage is situated adjacent to vulnerable receptors.

The effects of retrofitting the bund wall with the design discussed in this paper have not been tested in reality. However, it is expected that the effects would be significant and possibly with sufficient design that overtopping would be alleviated altogether. For specific designs, further designs would be desirable, possibly with a larger scale to test the effects up scale-up.

CIVIL ENGINEERING ISSUES

As discussed in a paper by Davies et al.[10] there may be structural failure of a bund wall by the surging of the released liquid. For many current bund walls, where the wall is simply of a vertical construction, the wall would likely collapse, due to the dynamic forces as the released liquid impacted the wall. Hence, even if the wall could prevent overtopping, it may well collapse completely, resulting in a catastrophic release from the secondary containment.

Hence, the bund wall must be designed to withstand the dynamic forces of the surging liquid over a period of many seconds. A retrofit of the design discussed in the experimental point would be pointless if this were to collapse.

DISCUSSION - PREVENTION OR CONTROL?

It should be pointed out that the bund wall design investigated in this experimental work is only to control the effects of catastrophic failures. Its use is only mooted in extreme circumstances where it may be necessary to retrofit existing bunding arrangements for the protection of vulnerable receptors, e.g. Sites of Special Scientific Interest (SSSIs) or populated areas. A risk assessment should be conducted to determine whether it may be necessary to include such a design. In any case, mechanisms of prevention should first be explored to ensure that the risk is as low as reasonably practicable. Such mechanisms may include corrosion prevention, Non Destructive Testing (NDT), hydrostatic testing, etc., where these are set out under robust safety and environmental management systems. Only if the residual risk still deemed as not tolerable and further cost-effective mechanisms for prevention have been exhausted, then mechanisms for control such as retrofitting the bund walls should be considered.

The lessons learnt from previous accidents should be considered in preventing catastrophic tank failures. For example, the Floreffe tank underwent brittle fracture causing it to fail catastrophically. The tank was 40 years old and had been dismantled, transported and then reassembled. It failed on its first refill. The lessons from this and other such failures cited by Wilkinson are important to prevent a reoccurrence.

Davies et al.[10] cite the following reasons for catastrophic releases from storage vessels after inspection of incidents recorded on the MHIDAS database (Major Hazardous Incidents Database):

- brittle failure of primary containment, sometimes caused by rapid changes in ambient temperature,
- failure of tank seams due to fire impingement,

- failure of the tank during the initial filling process,
- boilover of tank contents,
- acts of vandalism or sabotage.

For the construction of new tanks, effective use of land-use planning, together with modern design standards and state-of-the-art methods for accident prevention, should ensure that standard bunds walls that retain 110% of the tank contents should be sufficient.

CONCLUSIONS

1. The experimental work described in this paper has shown that it is possible to design bund walls to prevent overtopping following catastrophic failure, without having to build the walls to extreme heights. By incorporating a design similar to that used for sea walls, i.e. the use of a horizontal 'lip', the surging liquid can be directed back into the bund even for 'unzipping' type releases.

2. Such a bund wall design would need to be of sufficient strength to withstand the dynamic forces of a surging liquid following a catastrophic release. Such forces would likely cause many simple vertical walls to fail and thus the design would require significant reinforcement. The fluid mechanics of the catastrophic release would need to be investigated in detail as it was observed that obstacles in the bund may cause turbulence that may, in turn, cause overtopping to occur.

3. The bund wall design should only be used for retrofitting bund walls after a risk assessment has been carried out. The risk assessment should first explore all mechanisms of prevention and only if the residual risk is still not tolerable should retrofitting then be considered. Such a bund design should not be need for new tanks, where effective use of land-use planning, together with modern design standards and state-of-the-art methods for accident prevention should ensure that the risk of catastrophic failure and overtopping is broadly acceptable or as low as reasonably practicable.

REFERENCES

1. Wilkinson, A, Bund Overtopping - the consequences following catastrophic failure of large volume liquid storage vessels, AEA Technology, SRD/HSE R530, 1991.
2. Prokop, J., The Ashland Tank Collapse, *Hydrocarbon Processing*, 67(5), 105, May, 1988.
3. Greenspan, H.P., Johansson, A.V., An Experimental Study of Flow Over an Impounding Dyke, *Studies in Applied Mathematics*, Vol.64, p.211, 1981.
4. Michels, H.J., Richardson, S.M., Sharifi, T., Catastrophic Failure of Large Storage Tanks, IChemE Symposium Series No.110, 1989.
5. Trbojevic, V.M., Slater, D.H., "Tank Failure Modes and Their Consequences", *Plant/Operations Progress*, 8, 84, 1989.
6. Clark, N. and Savery, J., The Catastrophic Failure of Containment Vessels, 3rd Year Link Project, Dept. of Chem. Eng. and Chem. Technology, Imperial College, London, and the Health and Safety Executive, Buxton, December, 1984.

7. Law, G.D., and Johnskareng, G.R., Containment Provisions and Overflow of 2-Dimensional Catastrophic Tank Failure, 3rd Year Link Project, Dept. of Chem. Eng. and Chem. Technology, Imperial College, London, and the Health and Safety Executive, Buxton, December, 1984.

8. Cleaver, R.P., Cronin, P.S., Evans, J.A., Hirst, I.L., An experimental Study of Spreading Liquid Pools, IChemE, Hazards XVI, Manchester, November 7th–9th, 2001.

9. Thyer, A.M., Hirst, I.L., Jagger, S.F., Bund Overtopping – The Consequences of Catastrophic Tank Failure, J. Loss Prev. Process Ind., 2002.

10. Davies, T., Harding, A.B., McKay, I.P., Robinson, R.G.J., Wilkinson, A., "Bund Effectiveness in Preventing Escalation of Tank Farm Fires", Trans IChemE, Vol. 74, Part B, May 1996.

THE REALITIES OF IMPLEMENTING AN HSE MANAGEMENT SYSTEM IN A JOINT ORGANISATION OF MIXED CULTURES AND LANGUAGES

By Ayoub Hadj-Kouider & Paul Barrett (HSE-Managers)
Sonatrach/Anadarko Algeria Corporation - Groupement Berkine

Groupement Berkine (GB) is the joint operating organisation formed in 1998 between Sonatrach, and Anadarko Algeria Company, a wholly owned subsidiary of Anadarko Petroleum Corporation. GB manages and operates the development assets of the parent companies in blocks 404 and 208 in the Berkine Basin.

Initial contacts between the companies started in 1986 with a Production Sharing Agreement being signed in 1989. Following extensive successful exploration, first oil was produced in 1998. The current production capacity is 285,000 BPD, which is expected to double by 2007.

GB is staffed from a wide range of different cultural backgrounds comprising Algerian nationals together with North Americans and Western Europeans. There are 3 main spoken languages (French/Arabic/English); French and English languages are the working languages.

The cultural mix is compounded further by a range of HSE system implementation challenges. In addition to the language and cultural differences, the rotational nature of the workforce adds to the communication difficulties of the organisation, as well as the normal transitional nature of oil & gas prospecting from seismic exploration, through project development and into operations of more than 10 fields.

KEYWORDS: Safety Management; Algeria; Different Values; Different Cultures; Different Languages

OBJECTIVE OF THIS PAPER

This paper describes some of the ongoing challenges that we face in implementing HSE Management Systems in Groupement Berkine, and some of the activities we have undertaken to implement these changes. The paper is presented in three parts;

1. History of Groupement Berkine
2. Groupement Workforce Relationships
3. The Challenges which face Groupement Berkine
   a. The background from where Algerian Safety Culture has developed
   b. The challenges within Groupement Berkine today
   c. The direction to attain international best HSE practices

The paper is the result of over four years experience gained by the authors, and gives an insight into some of the real issues that are faced on a daily basis.

HISTORY OF GROUPEMENT BERKINE

Groupement Berkine is a non-asset holding, joint operating venture (51%: 49%) between Sonatrach and the Anadarko Corporation. Sonatrach invited Anadarko into Algeria in the late '80s, to assist in boosting exploration effort in the country.

Anadarko was awarded the licences to explore in four blocks in the east of the country, blocks 404, 208, 211 and 245, and established a base camp in Hassi Messaoud, to provide logistic and material support to its seismic and exploration drilling activities. After extensive seismic work and the drilling of a number of exploration wells, 12 fields of significant production potential were discovered from 1993 to 1998.

The first phase of development required a total investment of approximately $200 million, which included the development of the HBNS field to produce 60,000 barrels per day from a single oil & gas separation "train". An EPC Contract was established with Brown & Root/Condor and first oil was produced in May 1998. All produced gas is re-injected back into the reservoir.

In 1999, the agreement for the second phase of development was signed, requiring a further $700 million to design and construct three additional oil & gas processing trains, plus the necessary additional gas compression and water injection facilities to extend production to 285,000 barrels per day. Construction, commissioning and start-up were completed in April 2002. The four trains now produce from two main fields HBNS and HBN, and five satellite fields, in all totalling nearly one quarter of Algeria's oil production. In order to keep pace with developments, five rigs are drilling up the reservoirs, and wells are currently being completed and commissioned at an average of one per week.

The formation of the joint Sonatrach/Anadarko "Groupement Berkine" was initiated in 1998. The Groupement is staffed by employees of both parent companies, together with a number of national and expatriate sub-contractors. The Groupement is also about to embark on a new development in Block 208, which is located some 60 miles to the south of HBNS.

With all the oil & gas developments that are rapidly taking place in the country, the challenges placed upon the workforce and the environment is significant.

WHERE WE ARE LOCATED

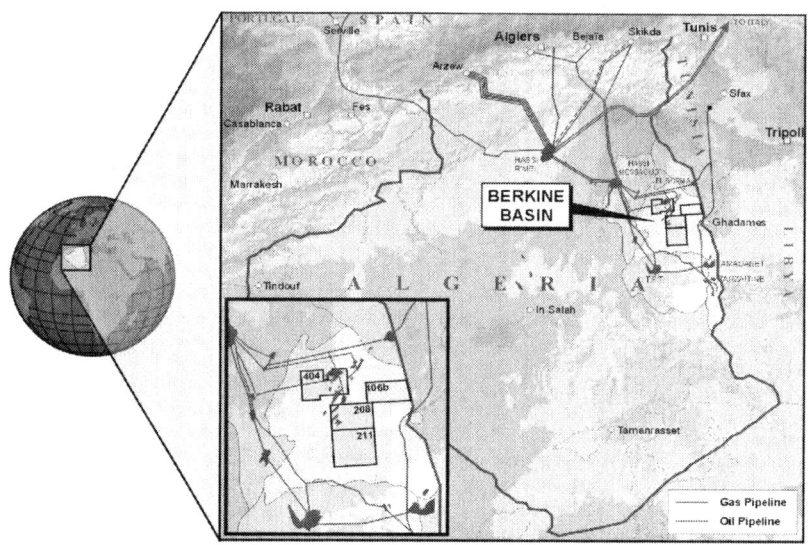

GROUPEMENT WORKFORCE RELATIONSHIPS

This section sets out the underlying complexities associated with the day-to-day working relationships of the Groupement personnel, and describes some of the activities that the Groupement have taken on board to overcome any identified discrepancies. There are many advantages of bringing together different cultures in the workforce; these predominantly result in everyone learning from each other's perspectives.

MIX OF NATIONALITIES

There are approximately 400 Algerian nationals and 160 expatriates employed within the organisation, with around 16 nationalities from as far afield as Canada, United States of America, Belgium, Holland, France, Italy, Trinidad, New Zealand, Australia, Peru, Hungary, South Africa, Germany and the UK.

STATUS

Sonatrach being a State owned company ensures for its staff a fairly stable employment - practically guaranteed - and provides for them and their families an extensive social protection programme. Sonatrach is becoming now more conscious about cost reduction and is undertaking programmes which remind us of what most of the international oil and gas operators have had to undertake through the 1990's of "contracting out" its workforce and services, as part of the cost reduction exercises and downsizing.

As with most joint organisations, there are real and perceived differences in positional status amongst the staff and contractors, and the different levels of management within the organisation. This criterion is held at a higher level of importance amongst some of our colleagues with more traditional values, than others within the Groupement. However, status does make a lot of difference to some people who have only worked in highly denominated and multi-layered organisations, in comparison to those that have experienced the past decade of working elsewhere within "flat management structures".

VALUES

The Groupement Berkine HSE values are set out in the HSE Policy. The policy and the HSE Management System now has the joint agreement of both sets of parent companies' senior management groups, to enable the HSE function to set its course in the same direction as other like minded Oil and Gas Producers' members in the international oil and gas community.

Environmental values are presently being tested in Algeria, and particularly within the Groupement. Historically, the country has yet to establish an infrastructure that is able to manage its domestic and industrial waste effectively. Until fairly recently, the country has had very little motivation or pressure do so. However, with the advent of international loans requiring formal Environmental Impact Assessments (EIA) to be undertaken as part of the commercial agreements, and many senior government officials recognising the commercial and social benefits of good environmental standards, there is a need to establish recycle and regeneration infrastructures across the nation. Groupement has already taken the initiative in conducting EIAs at the start of every field development.

One of the action points raised in our EIA has resulted in the establishment of environmental steering and technical committee groups within the Hassi Messaoud region, to review the methods used, and the criteria set for the treatment of cuttings associated with drilling waste. Groupement Berkine and the other operators are presently working towards better environmental standards in this way.

The differences in the "perceived values" that each nationality brings to the Groupement are sometimes evident. For example, the Algerian culture is very strong on inter-personal relationships. This is less apparent with the Western culture, which is usually focussed singularly on the business in hand.

PROCEDURAL BUSINESS PROCESS VS. PRAGMATISM

Another noticeable difference in working practices that is evident within the Groupement is the impact of some of the more bureaucratic procedural business processes. As we try to modernise and streamline our practices and procedures, we do our best to jointly consider those that are effective and efficient for Groupement. Procedures that have been designed for use within a large single organisation need to be adapted for combined organisations such as GB, particularly when the business performance improves with a lesser administrative workload. Challenging the procedures when they do not work is itself a cultural challenge for many people.

One good example of this process is the development of the Groupement Berkine Permit-to-Work (PTW) Procedures. In 1997, the joint organisation recognised the need to establish safe systems of work ahead of introducing hydrocarbons into the new plant that was under construction. In recognition that it was impracticable to expect to implement comprehensive offshore standards in the Sahara Desert, it was decided to take the framework of a PTW Structure familiar to Sonatrach staff since the 1960's, and develop the procedure to ensure all the key elements of a safe system were included. Experience from the subsequent years of operation has now shown we are in a better position to make the changes to elevate the safety standards to the next higher plane, and as such, the PTW System is presently being augmented.

We need to regularly take stock of all these processes, and work to the solution that takes us forward to the benefit of both parent companies and partners.

OUR "CULTURAL MIX"
As has already been discussed above, we have a wide range of different nationalities, with the rich mix of different cultures that the Groupement has fostered. Locally, we can identify Algerians with more traditional values. Many people who have worked for the national operator for many years will be well versed in the company's methods and procedures of doing business, and may often need convincing of different methods needed to meet Groupement objectives. Many of the younger staff have been raised and educated in a society that is now exposed to the same day-to-day exposure to Satellite TV and the Internet, that is available anywhere else in the world. Contemporary values are much more readily adaptable to working within GB methods.

From an expatriate perspective, there are differences between the "West Texan" and "North Sea Tiger" cultures. When one adds the International "Mercenary" culture - in the financial sense - of the long-term expatriates to the equation, there is certainly an interesting mix of people from totally different backgrounds, adding to the challenge of implementing consistent sets of standards and business processes.

OUR DIFFERENT LANGUAGES

Communication is one of the most difficult aspects of Groupement's business. The HSE function by definition requires good communication, to enable it be an effective factor in improving business performance. Groupement has therefore instigated extensive language training to address this issue.

THE WORK PATTERNS

To amplify the difficulties in communication, this is exacerbated by the complexity in work patterns. The vast majority of people at Groupement Berkine work a phased 28 day on/28 day off rotational cycle. Job handover is also difficult when the back-to-back usually travels to a different national or international airport destination, some people never see their counterpart for many months. This is further complicated by some that work 12 on/11 off, some Algerians work a normal Saturday to Wednesday week. Anadarko works to the international Monday to Friday cycle.

The transient nature of oil & gas work also complicates matters for many people in the management team, particularly in the HSE function. The HSE function covers all aspects of the business, and all departments interface with the HSE managers. The exploration of oil starts with seismic data acquisition at very remote desert locations, which brings its own logistic problems, particularly in emergency response. Exploration and development drilling also takes place remotely, as does the well completion and servicing activities. When we add plant design, fabrication, construction and production activities across the field - any combination of these will be taking place at any given time - across more than 10 oil fields, we start to develop some indication of the scale of task in trying to develop and implement HSE Management Systems at Groupement Berkine.

All of the above factors bring their own challenges to the organisation's success, particularly in communication with one another!

AN INSIGHT INTO THE CHALLENGES WHICH FACE GROUPEMENT BERKINE

On the basis of the foregoing information relating to the multi-cultural mix in the workforce, we can see one dimension of the challenges that Groupement Berkine is faced with. When we add the **background** from where the Algerian Safety Culture has developed from, we can start to recognise the full picture of the **challenges** that face Groupement Berkine today. To conclude our paper, we will present our opinions on the **direction** in which Groupement Berkine will need to go to attain international best HSE practices.

BACKGROUND FROM WHERE ALGERIAN SAFETY CULTURE HAS DEVELOPED.

The background to Algeria's Safety Culture was originally derived from the French Regulatory Model, i.e. based upon the institutions of occupational medicine and the labour inspectorate.

In 1967, the first formal Algerian Safety Legislation was established with a National Charter and an Ordinance upholding French health and safety law, whenever it was not covered by local legislation and rules.

ESTABLISHMENT OF THE ALGERIAN LEGISLATION
In 1970, the National Institute of Hygiene and Safety (NIHS) and the Body for the Prevention of Accidents in Construction (OPREBAT) were established, in a similar context to the French set up. Furthermore, in 1988, a law was introduced with the goal of aligning Algerian Standards to those recognised internationally (Loi No 88-07 du 26 Janvier 1988 relative à l'hygiène a la sécurité et le médecine du travail). This legislation is the most significant, as the responsibility for HSE in the workplace now rests with the Employer – in our case, Sonatrach.

PARALLELS WITH THE EU LEGISLATIVE MODEL (EXAMPLE UK)
Loi No 88-07 in Algerian Law is similar in principle to UK's Health and Safety at Work etc Act 1974. It places the onus upon senior management for the implementation of health and safety policy, and the requirement to provide the necessary resources to prevent accidents in the workplace. By definition, it has similar requirements to the French Health & Safety Law of 1976.

An Algerian Environmental Agency was created in 1993, followed by the Algerian National Safety, Hygiene & Occupational Medicine Council – 1996. These bodies are the Algerian equivalents to the Environmental Agency (and SEPA), and Employee Medical Advisory Service. The following paragraph lists the Regulatory Bodies in place in Algeria.

REGULATORY BODIES
The following organisations are identified as Algerian Government sponsored organisations that have an interest in the health, safety and environmental aspects of work activities throughout the country:

1) Direction de la Protection du Patrimoine (DPP) – an independent organisation whose resources are mainly targeted towards the Oil & Gas Industry. The DPP conducts independent checks and inspections on all new facilities, from design through to commissioning, and grants the necessary approvals to produce oil.
2) Algerian Labour Inspectorate (part of the Public Health Ministry) – there is a legal obligation to report all accidents in the workplace that involve employees. Significantly, there is no obligation to report accidents that involve employees of sub-contractors, as that is deemed the responsibility of the sub-contractor. The "compte rendu" accident reports are submitted to the authority principally as a means of social recompense from the accidents/illness, and not as a means to learn lessons from the event.
3) Le Haut Conseil de l'Environnemment et du Développement Durable (HCEDD) – this is the new Ministry of the Environment that has been set up in Algeria, with the role of setting Environmental Targets for industry. It has not yet been established long enough to give any measurable indication of improvement, however, Groupement Berkine along with several other operators in the region, are now facing more pressure to clean-up the solid, liquid and airborne emissions from our businesses. *(See note above on EIAs)*

The intent therefore, to attain improved HSE Standards in the workplace is reasonably clear within Algeria, however in practice, it is not always effective, as there are insufficient resources allocated to meet the scale of the problem in the country.

MAIN HEALTH, SAFETY AND ENVIRONMENTAL DIFFERENCES TO THE EU (UK EXAMPLE)

At present, there is no independent national enforcement body in existence in Algeria for neither Health & Safety, nor for the protection of the Environment. The Safety Culture is still developing, and apart from locally introduced initiatives in the industry sector (e.g. by the Hassi Safety Council), there is little regulatory pressure to share lessons learned nationally amongst the industry sector, by systems such as Safety Alerts.

The "Regulator" has not been set up as a learned body and does not provide technical advice or guidance; it is purely an administration function which deals in claims for accident compensation only. Under-reporting of accidents is also a major problem nationally, and as with most societies, we have much work to gain everyone's understanding on the need to report accidents and incidents in the workplace.

THE CHALLENGES WITHIN GROUPEMENT BERKINE TODAY

ROTATIONAL NATURE OF THE WORK

Essentially, everyone who works a rotational format lives two separate lives! After 28 days in the desert, Groupement personnel take off for 28 days relaxation and regeneration. This involves periods of adjustment for every individual returning to the field, and returning home.

INTERNAL COMMUNICATIONS

As highlighted earlier in the paper, we have numerous languages and cultures to deal with, as well as the logistical difficulties of installing suitable communications hardware in a desert environment. To make progress in communications in general, Groupement needs to adopt a communication strategy that encompasses the "people" communication issues, as well as the communication "hardware" needs of the business.

AIMS AND OBJECTIVES FROM THE PARENT COMPANIES

We have pointed out some of the differences in values that each parent company embraces. There are different perspectives on one organisation that has its sights set firmly in Algeria, to another that operates internationally, but predominantly in a regulatory environment. Groupement needs to adopt the best of both sets of corporate HSE requirements. Key Performance Indicators (KPIs) are recently evolved management jargon that mean different things to different people. Groupement Berkine has a need to set KPIs that are specific, measurable objectives with the agreement of both parent organisations.

ASSET PROTECTION VERSUS PEOPLE PROTECTION

With effective HSE management systems in place, Groupement can move towards an integrated approach which encompasses asset protection as well as safety of individuals, and the damage to the environment. By adopting hierarchical risk reduction principles – remove the source hazard where practicable; reduce the realised consequences; engineering controls; procedures; personal protective equipment (PPE) etc – we will achieve the win-win situation for the safety, environmental and production KPIs.

ENCOURAGEMENT IN DECISION TAKING, AND DELEGATION

As with all other successful businesses, the management level at which critical decisions are made with regard to the HSE impacts upon the business, needs to take cognisance of the wider impacts these decisions affect. We cannot expect success with a culture of moving issues up the chain of command, which is inherent in much of Groupement's business. Confidence must therefore be developed in placing appropriate delegation levels, to execute these decisions.

ACCEPTANCE OF RESPONSIBILITIES FOR OUR ACTS AND OMISSIONS

Many would argue that human nature itself is the main source of a "blame culture". In our everyday lives, the media rarely assist us in trying to look more positively on the outcome of accidents, hence it becomes exacerbated and even more difficult to change.

One of the most important aspects of making the effort to move away from apportioning blame is gaining a clearer understanding of the responsibility and accountability issues. Groupement itself needs to align itself to a position where we are able to look forward to changes in the way we conduct business.

FOSTERING A POSITIVE REPORTING CULTURE

At present, the legislative driver in Algeria is financial redress for the individual, and not "lessons to learn" for the organisation. There is presently reluctance within Groupement to report accidents, incidents and near-misses. By fostering a positive reporting culture, we can start to gain lessons from human errors and omissions.

MANAGEMENT, EDUCATION AND COMMITMENT

No-one becomes a competent manager overnight. Groupement therefore needs to train and educate all levels of its staff in understanding the key principles and benefits of effective HSE management systems to realise the benefits. As with all successful businesses, Groupement requires the leadership to make the necessary changes to bring about the improvements in HSE performance.

HSE "TIME-LINE"

The following HSE "Time-line" describes a typical development of HSE progression against time. The two graphs plot the progress of HSE within the UK, and the second plot shows the relative Algerian model. The plots are not precise, but indicative of the HSE cultures in existence at any given time for the oil and gas industries. Three distinct phases are identified:

1. "Intervention Safety" – our heritage.
2. Implementation of the HSE Management approach - this is where Groupement Berkine working is today. This is discussed in three contexts; People, Plant and Systems.
3. Behavioural Issues - currently in vogue in EU, for future Groupement Berkine consideration, and not explored further at this point.

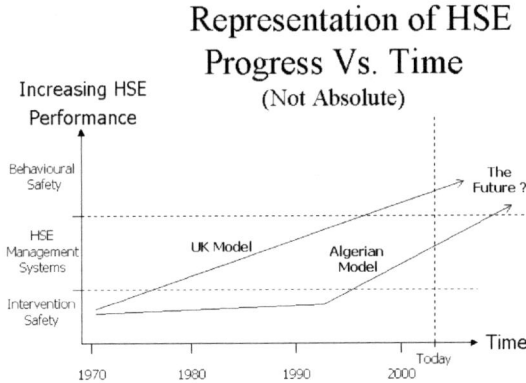

Representation of HSE
Progress Vs. Time
(Not Absolute)

"INTERVENTION SAFETY"
The following activities are examples of our "inheritance":

• We have a relatively large number of fire fighters employed by Groupement Berkine; this is a tradition Groupement has inherited from Sonatrach. The approach for the intervention team is literally to fight the fires when they happen, and hence the balance in resources is traditionally directed towards mitigation of consequences, and less to the prevention of fires. Prevention duties are assigned to the prevention team only. This approach is similar to that adopted in the UK in 1970's.

• Our plants are engineered to international codes and offshore design standards. There is some degree of over-capacity in the engineering integrity of the plant, though there are benefits in constructing robust facilities at remote locations.

• Measurement of "Safety" Performance is still reactive (frequency rates only), Groupement have not developed active indicators as yet. Groupement's own lost-time accident frequency rate (for those accidents that have been reported), is no worse than the national average. However, as stated earlier, we do need to create a "positive reporting culture", to obtain more accurate reactive indicators.

350

IMPLEMENTATION OF THE HSE MANAGEMENT APPROACH

The following HSE management activities are currently being implemented within Groupement Berkine:

PEOPLE

- Training in HSE Management techniques is being undertaken across all disciplines and departments, led by an Algerian Professor in HSE Management. The training is conducted in French and English, and the cross-cultural issues, which relate to all parties are being addressed.
- French and English Language Training is a priority for everyone. As stated earlier in this text, effective communication will be improved by addressing the language barriers.
- HSE information is now integrated into the training provided for Operations and Maintenance staff. O & M staff are now being taught the basic safety concepts required for work on the plant, in addition to Prevention and Intervention staff. The aim is now to integrate safety into the role of line management.
- HSE Inductions are now being undertaken for all new starts and visitors. All the basic requirements for general safety and well-being at the field locations are being addressed.

PLANT

Following on from our stated objective of improving our risk management strategy, the following list of activities gives examples of the improvements we have made:

- Hydrocarbon inventory minimisation – we have installed and commissioned Custody Transfer Meters at our Central Processing Facility, replacing the need for large Crude Oil Storage Tanks. This has required a change in the Algerian Legislation relating to oil movements metering, and has taken over 4 years to implement.
- All new plant has been fully HAZOPed to international standards, and all the actions generated by the study have been closed out.
- In our efforts to change our focus from "putting the fire out, to preventing the fire from occurring in the first instance" – we have made extensive use of the latest Gas Detection Technology throughout our new plant.
- A Computerised Maintenance Management System is in use at the plant, linked to safety criticality of equipment parts. This determines the spares stock as well as prioritising critical maintenance activities.
- Operationally, the performance of the plant and the drilling is good, and losses are maintained at a minimum. Training simulators have been developed and used extensively for training of the operations staff, particularly in simulating plant upsets and emergency conditions.

SYSTEMS

The following activities are a result of the efforts we have made to implement HSE management systems at Groupement Berkine:

- The Groupement Berkine HSE Policy and Management System has now been approved by senior management.

- HSE Dept has direct influence on Groupement Berkine senior management. The HSE function has a direct reporting link to the General Managers.
- The Groupement HSE Plan is continuously being monitored for performance, and developed with experience gained.
- Better communications (both methods and hardware facilities), to enable the correct HSE messages to be relayed to all. We have internet access in the desert, and the same opportunities to relay messages rapidly, as anywhere else in the world.
- Emergency Response capabilities are being improved continuously – ER is not easy with the logistical difficulties posed by a desert environment, and the diversity of different companies within Groupement Berkine. Our biggest challenge is restricted by the lack of an effective medical infrastructure in the Hassi Messaoud/Hassi Berkine region.
- Accident and Incident Reporting is actively being encouraged.

THE DIRECTION TO ATTAIN INTERNATIONAL BEST HSE PRACTICES

The third phase following "Intervention Safety", and "HSE Management Systems" is perhaps, "Behavioural Safety". Groupement Berkine is not ready for this step in direction, though the following considerations may well influence our direction in the future:

- Staff Cost reduction - for example, by the reduction in the number of expatriates in the Groupement workforce. Like all businesses, there will be progressive pressure to reduce the operational expenditure within Groupement Berkine. As our workforce gains in experience and expertise, then the emphasis to reduce numbers will become more apparent.
- Full compliance with regulations - establishment of an HSE Regulatory Body? There are early indications of environmental pressures being applied on the industry, there is no reason to suggest that we may also be more forcefully required to raise safety standards across the industry.
- Outside advantage of looking at the success (or failure) of HSE initiatives used elsewhere in the world, before taking action internally. Groupement Berkine is in a strong position to adopt good practices tried and tested elsewhere.
- Algeria is committed politically to aligning itself with International Best Practices – this is quoted at the highest government levels.

Like everywhere else in the world, we will only succeed with our own senior management's commitment and effective leadership!

CORPORATE MATURITY

In summary, Groupement Berkine is a very new organisation and in terms of maturity in HSE performance - we still have some way to catch up with the major international oil and gas operators - but we are making visible progress!

REFERENCES

1. Professor Sebti Chaabane and Leyla Hayet Mouss, 1998, Occupational health and safety in Algeria, *The Safety and Health Practitioner (IOSH), P23, Dec 98.*

HOW DISTRIBUTION IN HUMAN PROBLEM SOLVING IMPERILS SYSTEMS

J S Busby (University of Bath), E Hughes (Marine and Coastguard Agency),
E Terry (Sauf Ltd), J V Sharp (Cranfield University), J E Strutt (Cranfield University),
M Lemon (Cranfield University)

The distribution of people's problem solving is an important source of failure in complex, hazardous production systems. An analysis was made of about 60 offshore accidents and incidents in an attempt to work out how people distribute their problem solving and how this fails. Inferences were made about the assumptions that were effectively being made in this distribution, and how the system could have been made less vulnerable to such assumptions in the design process. A prompting tool has been developed to help introduce this analysis into risk identification exercises - in both design and operations.

Human error, distributed problem solving, assumptions

INTRODUCTION

When accidents are categorised simply as being 'human error' the natural inference is that something is wrong with the person making the error. They had the wrong intention, a wrong belief, poor memory, a lapse of attention and so on. The environment they were working in plays an important role in inducing, suppressing or helping detect such errors, but error is still located in the mind of the individual. Some of the most influential models of error reflect this position. They associate different types of error with different levels of human performance, for example[1,2].

But this kind of model has been criticised on the basis that it does not help distinguish manifestation and cause: erroneous actions by people are not necessarily caused by erroneous effects within people[3]. Moreover, fairly recent work has emphasised the extent to which people's action and problem solving is determined by the situations they are in[4,5], and the way in which this problem solving is distributed[6]. People get parts of the solutions they need from watching other people's behaviour, custom and practice, organisational procedures and so on. The tools they use often embody other people's knowledge, the routines they follow are often the result of other people's trial and error, and the codes they observe typically have arisen from someone else's activity. This means that when people's problem solving fails (and when, for example, they have an accident) it is important not simply to look at what has gone on in someone's mind. It is important to look at how the distribution of their problem solving has failed. The distribution does not have to be planned, deliberately: it can be emergent[7]. But, whether it is planned or emergent, it can be characteristics of this distribution to which failure is most obviously traceable.

A study has been funded by the HSE to determine how the distribution of problem solving contributes to accident causation. This has involved analysing a set of about 60 accidents and incidents offshore, a substantial proportion of which were hydrocarbons leaks. In almost all cases, people's distribution of problem solving had been entrained in their activity. That is, no-one had explicitly planned this distribution: the way people habitually conducted their tasks meant that distribution was a natural part of what they did.

For this distribution to work out various kinds of implied assumption had to be satisfied - and it was failures in these assumptions that led to failure.

The purpose of the study described in this paper was to determine what these kinds of assumption were, and to find ways of making this knowledge useful to people - the people making the assumptions (operators and maintenance staff largely), people wanting to know how vulnerable a system could be to such assumptions (such as designers), and people needing to make an external assessment or audit of an installation (such as inspectors).

THE UNDERLYING STUDY

METHOD

The study used a secondary data source – a set of 59 offshore incident reports obtained from a database held by the Health and Safety Executive and a small set of 4 reports on major offshore accidents. There are obvious limitations in using such reports as data, but it is important to stress that we were not attempting to diagnose specific failures definitively. The aim of the project was to help people reason about failure in the future, and if our inferences *could* be true they ought to be useful. The reports were analysed in several stages:

- Constructing a simple causal network to represent each narrative report.
- Identifying wherever possible how problem solving had been distributed in the task implicated in the incident and how this distribution had failed.
- Identifying in each case the assumption about the world that was implied in the distribution; (for example, reusing existing solutions typically implies an assumption about the solutions' applicability in different conditions).
- Developing a taxonomy of the implied assumptions.
- Deriving a set of guidelines to express how the design of systems could reduce the potential harm arising from these assumptions.

RESULTS

The taxonomy of implied assumptions had three very general categories at the top level:

- The reasonable system assumption (the assumption that the design and disposition of the technical system was reasonable, in the operator's terms).
- The appropriate organisation assumption (the assumption that organisational systems were complete and fitting for the task in hand).
- The knowledgeable person assumption (the assumption that people involved in the task had proper knowledge of whatever was needed).

Each of these categories then consisted of a set of lower level categories. As our purpose was ultimately to influence system designers, it is sub-categories to the first category that are presented here, in Table 1.

Table 1. Sub-categories of the reasonable system assumption

Assumption	Case example
What is available is what is appropriate	Connection for low pressure line instead made to high pressure line which was the only one available. Implied assumption was that whatever was available at the time was appropriate to the task.
No signals means all is well	Radio used for communication during crane operations. Channel had unknowingly failed and three 'stop' messages not heard by the operator. Implied assumption was that absence of messages meant sender had nothing to communicate - not that channel had failed.
Lapses will not imperil the design	Safety hatches incorporated in design for intermittent tasks. Hatches left open which contributed to capsize. Implied assumption was that the design would not be vulnerable to simple lapses and violations.
Trial and error is not hazardous	Wrong pump in a pair dismantled. Noise masked sound of running pump and poor lighting impeded visual identification. Implied assumption was that you could identify a device by trial and error and would know if you got the wrong one.
Consequences are obvious	Damage caused to sacrificial anodes by pile driving in construction. Damage only obvious during operation. Implied assumption was that if an operation was harmful then the damage would be obvious at the time.
Function follows appearance	System started up with only a blanking plate preventing escape of gas. Possibly fitter thought it would prevent egress of vapour, not just ingress of dirt, because it was solid. Implied assumption was that the solid appearance of the plate meant gas would not escape from aperture.
Things happen in a logical order	During a crane lift slings failed when load was snagged. Operator probably not attentive, expecting that passive slings would not fail before active motors reached limit. Implied assumption was that the properties of the system would follow a natural order.
Ambiguous things do not matter	fitter replaced part of a blowout preventor wrong way round. This then failed when there was a blowout. Implied assumption was that if something could be fitted in different ways then it didn't matter how it was fitted.
Boundaries are obvious	Operators had adopted a ballasting practice which allowed rapid listing. This ultimately contributed to capsize. Implied assumption was that boundaries to safe operation would be obvious.
Redundancy protects systems	Area had to be cleared for radiography. Done both by detection (sending someone to look) and self-detection (making a tannoy announcement). Both failed probably because the other was assumed to be more effective.
Identification cannot go wrong	Drain cut to install break couplings. Second pipe with similar shape also thought to be drain so also cut, but in fact had different function. Implied assumption was that can identify the right objects to work on based on a similar appearance to other objects.
Sequences of actions are not interrupted	Instrument line disconnected during planned maintenance but not reconnected before startup. Implied assumption was that sequences of activity cannot be interrupted or forgotten.

Design guidelines were derived from these assumptions simply by asking how, in general terms, the design *might* help avoid either the assumption arising in the first place or any harm that could arise from it. There turned out to be four main categories of guideline:

- Information - guidelines that help the designer tell the operator or maintainer how or what to do.
- Salience - guidelines that help the designer show the operator what is important at a particular time.
- Restriction - guidelines that help the designer constrain what the operator does.
- Presumption - guidelines that help the designer know what to presume or predict about the operator.

Table 2 shows the guidelines within the last category of 'Presumption'. Most are obvious in the sense that they do not reveal new principles - but the evident failure to follow such guidelines in some cases suggests they are not systematically used.

Table 2.   Sub-categories of the presumption guideline

Presumption guidelines
Determine whether use of equipment requires observers who are then vulnerable to hazards
Expect alarms to be inadvertently left inhibited
Anticipate side-effects of misdirected or excessive force in construction or use
Do not assume that the fitting of foolproofing devices is free of error
Predict the practices that operators will learn to minimise effort and maximise production
Predict that people will omit tasks when they are many and uniform
Expect operators to expect the design to be intuitive and easy to orient
Do not assume users' roles let them perform the functions you delegate to them
Expect the operator to be unpracticed in using emergency controls
Expect operators to expect designs that are reasonable in their eyes and do not expect them to test this
Determine how tasks broken down and allocated to different people could leave the system in a hazardous state
Determine how redundancy in protective actions or devices could be undermined
Anticipate that operators will believe precautionary tests to be comprehensive
Predict how an object would provide hand-holds, steps, and wedges and thereby create a hazard
Predict how the current availability of components or services or could influence operators to use the wrong ones
Do not expect people to notice objects or connections they do not expect to be there
Anticipate that a user might not test a configuration before using it in an environment that will punish incorrect configuration
Determine how interruption of dismantling sequences could be harmful

Assume that precautionary sub-tasks will be forgotten if they are not physically necessary to proceed

Anticipate that people will search for appropriate actions by trial and error

THE PROMPTING TOOL

PRINCIPLES

A tool has been developed in an attempt to make the results of the study accessible and useful to people. The basic principles are these:

- People need help testing their assumptions. The study pointed to a wide variety of assumptions that can imperil hazardous installations, and it was suggested that such assumptions tend to be implied rather than explicit. Both characteristics suggest that without a structure of some kind people will naturally find it hard to test these assumptions.

- The categories of flawed assumption that came out of the analysis should be provided as prompts to people in order to help them examine assumptions. This may be a question of helping operating or maintenance staff test their own assumptions before they are about to engage in some risky process. Or it may be a question of helping designers anticipate the assumptions that operators or maintenance staff could make, in an effort to make the design resilient to such assumptions.

- Links between the categories and accounts of the underlying accidents should be retained in the tool so that users can easily consult examples - examples both of the kind of assumption in question and of how the assumption causes the system to fail. The categories help distil the essential elements from the accident and incident reports, and provide the general type of issue that has to be examined. But they are likely to be too abstract in some cases to be applicable on their own. One of the arguments for providing this kind of tool is that it helps people think about kinds of failure of which they have not had direct experience: it helps people expand their knowledge base, as it were, many times over. But for the kinds of failure of which someone has not had direct experience it is probably necessary to demonstrate how it can happen at the level of specific, concrete events.

STRUCTURE

The structure of the tool is shown in Figure 1. There is an underlying database of accident cases, consisting of reports, causal analyses and very brief, thumbnail sketches of these reports. Overlaid on this database is the category structure of flawed assumptions. The detailed kinds of flawed assumption are then mapped to design guidelines, and there is a category structure above these.

There are two main ways of getting access to this structure. The first, which is the most obvious way for operating staff wanting to test their assumptions, is to work through the categories of flawed assumption. The screenshot in Figure 2 illustrates this. Thus the user selects one of the three coarse categories (for example the 'reasonable system assumption') and then selects, in turn, the lower level categories below this (for example the assumption that 'ambiguous things do not matter'). As each category is selected, the tool displays the thumbnail sketches of the accidents in which this assumption type is implicated.

To give the user a little more structure, a report form can then be brought up which prompts the user to write down, for the assumption type currently selected, any problems they can envisage, any actions required, the responsibility for these actions, their criticality and a deadline. This form, in common with most of the package, can be changed by the user to suit particular needs. From this point (where a particular type of flawed assumption has been selected) the user can also look in more detail at one of the underlying accident reports, or look at the design guidelines that were derived to address the assumption type in question.

An alternative way of organising these assumption types has been provided, based on their frequency. The basic rationale for this kind of organisation is that when people have insufficient time to examine their assumptions against all the 30 or so categories they need to have some rational way of rationing their attention. The tool therefore orders the assumption types according to how many cases implicate each of them. Plainly, the problem with doing this is that frequency is only a partial indication of importance. The fact that the assumption that 'systems are left complete' is the most common does not mean that it led to the greatest losses. And even if the tool had ordered the assumptions according to greatest loss in the past there is no guarantee that this would be reflected in the losses that could occur in the future.

The second main way of using the tool is to consult the design guidelines. Figure 3 shows a screenshot of the window for doing this. The user selects one of the four coarse kinds of design guideline, and can then work through the more detailed guidelines grouped under these. For any one of the guidelines, the user can bring up a list of implementation suggestions, and a form that helps record any thoughts the user has about changes that need to be made to a design. It is important to emphasise that at this level the user is the expert. The tool provides general kinds of failure and general, desirable constraints on a design to reflect these kinds of failure. Converting these into particular features of a particular design is something the tool cannot do. The implementation suggestions are therefore simply possibilities, derived from the particular accidents that the study happened to tackle. They do not provide a checklist for a designer. The user can also, from the design guidelines window, bring up another window explaining the flawed assumption that is linked to the selected guideline.

USES
A distinction was made in the study between uses that were essentially on-line and those that were off-line. On-line uses were to do with direct support of normal activity, while on-line uses were essentially for training - helping people with potential rather than actual problems.

The on-line possibilities are as follows.

- Hazard identification. Hazard identification exercises, however structured, rely on people's knowledge of what can go wrong. There is plainly no guarantee that any particular group of people's experience is broad enough to know of all relevant kinds of failure, so supporting the process is important. A tool like the one developed here synthesises the knowledge available in quite a large number of events. The process is essentially one of inspecting all the categories of assumption and asking 'are we vulnerable to making this assumption?' (in the case of operating staff), or asking 'in what way is the design vulnerable to people making this assumption?' (in the case of design staff).

- Operator participation. Most design organisations involve operators in the design process. But operators sometimes have their own hobby-horses, and they do not necessarily know whether their particular experience and opinion is typical or unusual. A package like this could help operators participate in the design process - giving them an aide memoire and a knowledge of problems that they might personally not have encountered, but that they want to bring to the attention of designers. It seems reasonable to say that involving operators in design does not avoid the need to use a tool of the kind we have developed, and equally that using a tool of this kind does not avoid the need to involve operators in design.

- Aide memoire. Designers often need aide memoires to help them think about all possible problems - especially if these problems are associated with people misusing or misinterpreting the designed system. Each of the assumption types presented by the tool is fairly obvious, once one is told about it. The difficulty is in remembering all the items without any support or structure.

The off-line uses of the tool that we could envisage are as follows:

- Induction training. Empirical knowledge of failure is likely to be most lacking in the least experienced people, so the body of knowledge contained in the tool is likely to be most useful for an organisation's new starters. The essential structure of the tool is a set of cases and a set of generalised concepts (the categories) and one can use the links in both directions. Training could consist of looking at the cases and finding applicable concepts, or working with the concepts and finding applicable cases.

- Toolbox talks. Providing new material for short, general background briefings can sometimes be difficult. The tool could be used to support toolbox talks - for example providing one case and one kind of assumption for each session. People are thereby seeing many cases, over a period, and the message about being attentive to hazards is continually being reinforced.

- Observing others. The tool also provides checklists for observing others. The core of some safety schemes is observing others, and spotting hazardous situations and acts. Sometimes it helps to have some structure to do this - to mark someone's behaviour down as being of a particular kind.

FEEDBACK

The tool has not yet been used to any great extent so it is unclear how well it would support the kind of use that has been suggested. Favourable feedback, so far, has included the following:

- The general principle of linking risk identification strongly to accidents or incidents that have actually occurred is an important one in helping people see that implausible failures do occur. It is very easy for people to be dismissive of the possibility that failures occurring to other people could occur to themselves, and easy to dismiss misled operators as being foolish.
- The tool does seem to address a genuine gap in most risk identification processes, which sometimes lack systematic ways of dealing with the human element. Even highly structured human reliability analysis tools typically say little about the causal mechanisms by which the distribution of problem solving fails.
- The idea of helping people examine assumptions, especially implied assumptions, seems to be reasonable one. Inspecting such assumptions helps people address what have been called the 'pathogens' that contribute to failure[8]. Particular trigger events of an accident - such as an action slip - are unpredictable and hard to forestall, but the 'pathogens' that reside in the system are potentially more open to correction or containment.

Unfavourable feedback has included the following:

- The broad approach of influencing people to test their assumptions suits only cultures in which there is an assumption of empowerment. Fatalistic cultures (whether corporate or national) would probably not be influenced by this kind of work. Fatalistic operators would see the system, rather than their own thinking, as the primary determinant of hazardous outcomes, and fatalistic designers would see operators and operating organisations as the prime determinant.
- The terminology and general tenor of the language used in the package would suit only organisations in which there is a receptiveness to new concepts, particularly new concepts concerning failure and hazard.
- The package does not tell designers what to do at a sufficiently concrete level. Although design guidelines have been derived from the assumptions, and although these have been translated into specific, example implementations, the tool still cannot be applied without thinking hard about how operators or maintenance people will use the design.

ADDENDUM

There is a final part to the tool that is still being developed. A problem for managers at a certain level is less with the question as to whether a particular system design takes account of flawed assumptions, and more with the question of whether a design organisation, as a generality, tends to be good at designing systems that are robust to such assumptions. This problem has been tackled by outlining a capability maturity model. The model suggests that an organisation can sit at one of five broad levels of capability maturity, where the levels are defined qualitatively in terms of well the organisation takes account of the assumptions made by people affected by the designs. It is also possible to define transition rules that show what has to be done to move from one level to the next.

The principle is not in fact to ask which level the organisation in question sits at, since parts of the organisation are likely to be better developed than others. Instead, the user is asked, for each of the five levels, which parts of the organisation have the associated degree of capability, and provide an argument or evidence for it. As with the other parts of the tool the user still has to do thinking: the tool simply provides a structure within which to do it.

CONCLUSIONS

WHY BOTHER?

The difficulty in advocating the use of this kind of tool is that it adds to, rather than replaces, existing processes. The tool's structure does not map directly, for example, to HAZOPs although it could support the conduct of HAZOPs by helping test whether it is likely a human action could lead to some condition and deviation. In fact one could argue that the basic starting point of one's reasoning differs between HAZOPs and the assumptions tool. So, whereas in HAZOP one is starting with deviations and working back to the possibility they could arise in human error, with the assumptions tool one is starting with possible error and working forward to consequences for the system. There is an argument that following both processes would give more assurance than either one alone, but there is obviously a cost in terms of the effort needed.

Nonetheless we felt that there were a number of reasons at least to consider using the assumptions tool. They overlap somewhat, and will not apply in all cases. But they may be compelling on occasion. And even if they are not especially compelling in isolation, it may be that the mass of these reasons together makes it worthwhile using the tool:

- Protecting oneself. The people most imperilled by the assumptions covered in this package were the people making the assumptions themselves. One of the most important reasons for using the tool is to save oneself unnecessary harm.
- Protecting others. Many of the assumptions made by one person imperilled another. So if someone wants to protect colleagues they need to make some effort to inspect their own assumptions.
- Accepting accountability. People are accountable morally and legally for hazards they contribute to, and using the tool should help people meet their responsibilities and demonstrate to others that they are meeting these responsibilities.
- The knowledge arises from harm, or the potential for harm. Some of the cases the tool draws on involved deaths and serious injuries. To ignore the knowledge, unless it is based on flawed inferences, would be to ignore the losses that generated it.
- Observing others. A critical element in the protection of hazardous installations is people's capacity to observe each other. The tool gives people a structure for observing others and testing what they are assuming.
- Common sense limits. We learn how to behave as part of our upbringing and general experience of life. Part of this involves learning what assumptions we can reasonably make about the world. But the world of a hazardous installation is different from the world of home, school and street where we learn our basic behaviour. We cannot therefore depend on our instinctive behaviour in such an installation. We have to know what it is about our assumptions about the world, learned in everyday life, that is misleading or plain wrong.

LIMITATIONS

There are limitations in this work at several levels. First, the basic principle that historical analysis helps one reason about future failures is not watertight. There seem to be some constants in human and organisational behaviour, but new technologies and new laws at the very least change the relative importance of different behaviours and beliefs. Second, the principle of sampling the past in the expectation that the sample will be representative has some obvious limitations. There is no guarantee that our list of assumption types is exhaustive, and in a qualitative study it is very hard to gauge whether one has a good enough sample. Third, the principle of using accident and incident reports as a source of data has some basic problems. The recall of the people involved in accidents can be partial, the process of investigation can be constrained, and the freedom to publish conclusions that are controversial or commercial damaging can be very limited. On the other hand, there is a strong expectation that we learn from accidents and that we can demonstrate this learning. And the complexity of failures is such that simulations and experiments have their own limitations as sources of data.

ACKNOWLEDGEMENTS

Many thanks are due to Bob Miles of the Health and Safety Executive for support and encouragement during this work. The study was funded by the HSE under contract D3916.

REFERENCES

1. Reason, J., 1990, *Human Error*, Cambridge UK: Cambridge University Press.
2. Rasmussen, J., 1983, Skills, rules, and knowledge: signals, signs, and symbols, and other distinctions in human performance models, *IEEE Transactions on Systems, Man, and Cybernetics*, 13: 257–266.
3. Hollnagel E., 1993, *Human Reliability Analysis: Context and Control*, London: Academic Press.
4. Lave, J., 1988, *Cognition in Practice; Mind, Mathematics and Culture in Everyday Life*, Cambridge, UK: Cambridge University Press.
5. Suchman, L.A., 1987, *Plans and Situated Actions*, Cambridge UK: Cambridge University Press.
6. Hutchins, E., 1995, *Cognition in the Wild*, Cambridge MA: The MIT Press.
7. Clegg, C., 1994, Psychology and information technology: the study of cognition in organizations, *British Journal of Psychology*, 85: 449–477.
8. Reason, *op. cit.*

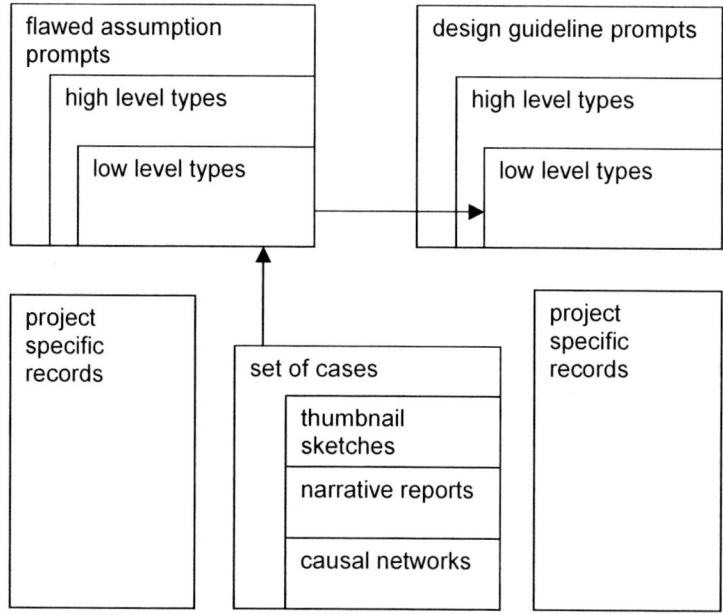

Figure 1. Basic structure of the prompting tool

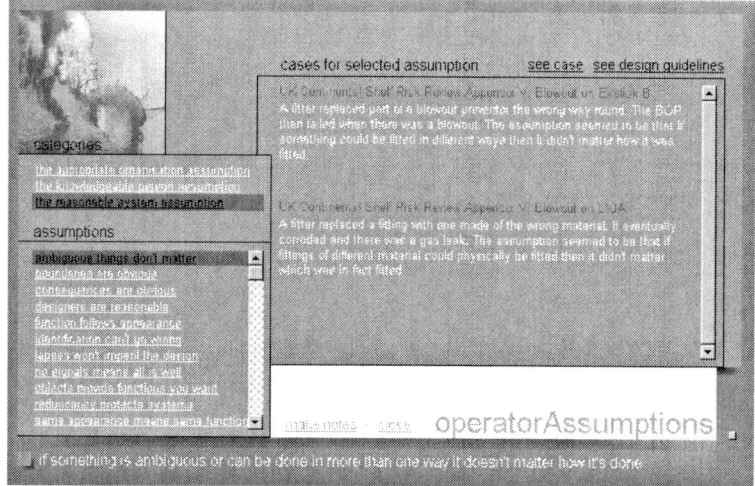

Figure 2. Screenshot on categories of flawed assumption

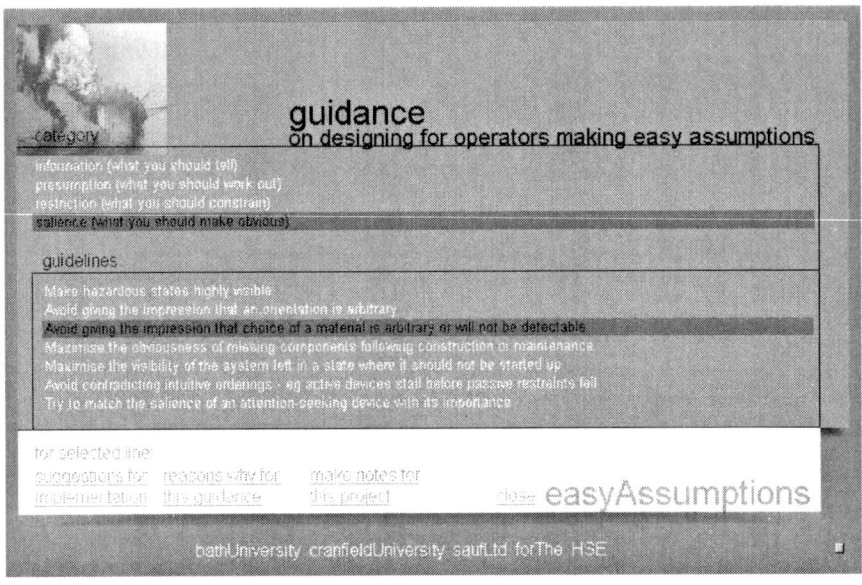

Figure 3. Screenshot on categories of design guideline

ASSESSING HUMAN INVOLVEMENT IN CHEMICAL MANUFACTURING: A HUMAN FACTORS TOOLKIT

Helen Jones*, Steve Shorrock*, Peter Bull[†] and Debby Hallett[†]
*DNV Consulting, Highbank House, Exchange Street, Stockport, Cheshire, SK3 0ET, United Kingdom; [†]Ciba Specialty Chemicals, PO Box 38, Bradford, West Yorkshire, BD12 0JZ, United Kingdom.

The paper outlines a methodology that has been developed for assessing human factors within the process domain. The approach systematically examines how human factors can impact on safety critical tasks. The method comprises six steps that lead the assessment team though task analysis and human error assessment, and then on to evaluate the safeguards in place to prevent a major accident hazard (MAH) and the likelihood of human recovery. The effectiveness of the approach is discussed along with potential improvements that could be made to enable wider application of the approach in the future.

Human factors; process; safety critical; task analysis; manning; safeguards.

INTRODUCTION

Addressing human factors issues has always presented a challenge to the chemical industry, and until recent years, the approach to human factors has often been fragmented or solely reactive. Increasingly, it has been recognised that a systematic, proactive process is more in line with the recent transition from a reactive "fire-fighting" approach to proactive safety management, and that such approaches can also add value by helping to optimise efficiency and productivity. Furthermore, the COMAH regulations now make it clear that human factors must be taken into account in risk management. Although the need for systematic, proactive human factors assessments has been recognised, the approaches available can often appear to be too industry specific, too broad or beyond the reach of non-human factors professionals.

DNV Consulting and Ciba Specialty Chemicals have jointly developed an approach to provide a standard methodology for human factors assessments for Ciba's UK sites. Although the methodology focuses on safety critical tasks, it also provides a framework for a more holistic assessment of how human factors impact on the safety management system. The approach aims to be practical and participative, involving key site personnel with a responsibility for safety.

METHODOLOGY

The methodology integrates and adapts existing tried-and-tested approaches, whilst ensuring that the whole process can be applied with limited training in human factors. The six key steps are shown in Figure 1. These are described in further detail below.

STEP 1: IDENTIFY MAJOR ACCIDENT HAZARD SCENARIOS FROM COMAH SAFETY CASE

The MAH scenarios identified within the existing site safety case should be extracted and a MAH description noted. This description provides the focus for the subsequent stages of the assessment and establishes the activities and site areas to be assessed. The assessment is most effective when applied to MAH scenarios that have been classified as ALARP. Any MAH scenario where the risk has been found to be intolerable are likely to require engineering solutions to reduce the risk before human factors interventions can be applied effectively. The MAH scenarios assessed to be ALARP can be subjectively prioritised according to the level of human action and human recovery involved. These can then be carried forward to Step 2.

STEP 2: IDENTIFY SAFETY CRITICAL TASKS

Once the MAH scenarios have been identified, an inventory of all the related tasks is compiled based on the site procedures. Each procedure identified is then rated on the basis of how frequently the task is performed, its duration, how much operator action/intervention is required, and the likelihood of the worst credible consequence. This rating allows the procedures to be ranked according to their criticality and hence allow the assessors to prioritise their efforts on the most safety critical first. N.B. all the critical procedures should be investigated before repeating the process on the next MAH scenario.

STEP 3: QUALITATIVELY ASSESS CRITICAL TASKS AND POTENTIAL ERRORS

This step consists of five stages: task observation; task analysis; human error assessment; concurrent task analysis and gathering information about the additional roles of shift personnel. This step is performed by a team of assessors including health and safety personnel, site managers and a human factors professional. Most importantly, site personnel with experience of the tasks under investigation must be involved throughout this step.

Observation

The tasks are observed during routine operations to gather information that could not be inferred from the procedures. This includes details of the duration of task steps, the environment in which the task is performed and the roles and responsibilities of the operator that performs the task. Any roles or concurrent tasks that have the potential to impact on task performance should be recorded for use in the concurrent task analysis. Control system prompts (if applicable) and safeguards relating to the task should also be noted.

Hierarchical Task Analysis

A hierarchical task analysis (HTA) is conducted on the task. This allows the task to be broken down into functional steps which can then be analysed for potential human failures. This analysis is vital, as the steps of the procedure may not necessarily correspond to the human actions that comprise the task.

Human Error Assessment

Each task step from the HTA is analysed to determine the potential for human error. Factors that could impact on human performance such as workplace design, tools and equipment, time pressure, workload and training are considered. These factors should correspond to those recorded during the observation, although additional factors (e.g. that could occur during non-routine operations) can be included where appropriate. Once the potential errors have been identified, they are classified as either slips (of action), lapses (of memory), mistakes or violations (i.e. not following the procedure). The consequences of each error are determined, and the safeguards in place and human recovery requirements are recorded.

Concurrent Task Analysis

This analysis has two parts: within task and between tasks. The within-task analysis examines the critical task steps that may be performed whilst simultaneously conducting the primary critical task step under investigation. This demonstrates how operator attention may be split within a critical task.

The between-task analysis examines other tasks that may be performed whilst simultaneously conducting the primary critical task step under investigation. These concurrent tasks may have been recorded during the task observation. However, the operators and site personnel should be consulted to identify all the possible concurrent tasks. This demonstrates which secondary tasks may be attempted during a critical task step, thus requiring the operator to leave the critical task unattended or be distracted from the primary critical task.

If there are a large number of potential concurrent critical tasks, these can be classified into task groups. The concurrent critical tasks or task groups are recorded in a matrix to show the extent to which the secondary task can be conducted according to three categories –

1)   the task can be completed
2)   part of the task can be conducted or
3)   the secondary task cannot be attempted while conducting the primary critical task step under investigation.

An example of the recording sheet is shown in Figure 2. The matrix highlights concurrent tasks where the operator's ability to recover from a problem with the primary critical task may be impaired. If any of the critical task steps highlighted rely on human recovery, then more in-depth investigation may be required to determine whether the operator would be absent long enough to allow the situation to escalate and whether there would be sufficient time remaining for the situation to be recovered.

Shift Information

A list of shift personnel is compiled outlining their job roles and responsibilities. Any additional duties are noted, e.g. first aider, fire crew or emergency warden. This information outlines the number of people available per shift, any dual roles and whether the team members are multi-skilled. This indicates whether individuals must leave tasks unattended when performing other roles. The information also indicates whether sufficient staff are available to cover breaks, etc. The shift information is simply intended to provide additional information to consider when evaluating the safeguards.

STEP 4: EVALUATE THE SAFEGUARDS

Alarm Checklist

The existing techniques for alarm evaluation were found to be relatively complex and suited for separate alarm studies rather than the initial high-level assessment required within this approach. This checklist (Figure 3) was developed to capture the same key issues by asking questions in "everyday" language that allows operators to undertake the assessment with assistance from the human factors professional if necessary. This checklist was developed from industry best practice[1,2,3] and covers issues such as alarm design, alarm inputs and processing, operator interface, operator response required, alarm type, environment, communications, alarm logs, and change management. The checklist is designed to highlight potential weaknesses within the alarm system that might degrade the effectiveness of the alarms as a safeguard. The checklist outlines the standards that must be attained if the alarms are safety critical. Although the checklist is applied to the alarm system as a whole the issues should also be considered in relation to each safety critical task step where alarms have been listed as a safeguard.

Procedure Checklist

The procedure checklist was again developed from industry best practice[4,5,6]. The existing guidance was collated into the same format as the alarm checklist (Figure 4). Once again, the language was screened to avoid obscure terms to ensure that it could be easily applied by operators. This checklist assesses procedure design, language, layout, etc and also the management of procedures within the organisation. Therefore issues such as accessibility, adherence, responsibility for co-ordinating and updating procedures and training can also be taken into consideration. The checklist also explores the extent to which the workforce is involved in the writing and development of procedures on the site. This checklist is designed to determine whether the procedures in place are of a sufficiently high standard to be considered as safeguards and to highlight areas for improvement where necessary. The checklist is applied to the procedures that have been recorded as safeguards in Step 3.

Staffing Assessment Flowchart

The flowchart was based on a HSE staffing assessment methodology[7]. The flowchart examines whether the task is constantly attended by a member of staff. If the task is left unattended at any point, the flowchart leads the assessor to investigate the other safeguards in place to either attract the operator's attention (e.g. alarms) or to control any process upset (e.g. automatic shutdown systems). The flowchart steers the assessor to one of three levels of staffing adequacy:

- Level 0 – staffing or control measures in place are likely to be inadequate. Immediate measures should be taken to improve staffing or control measures
- Level 1 – staffing or control measures may be insufficient. If there is a reliance on trips, slam-shuts or other fail-safe mechanisms, reliability must be justified. Staffing levels should be considered to ensure that essential monitoring, control and incident response activities can be conducted.
- Level 2 – staffing or control measures are likely to be adequate. Monitoring systems should be established to ensure that staffing remains adequate.

The staffing flowchart (Figure 5) is applied to the task steps where training or operator intervention etc have been recorded as a safeguard.

These tools are used to evaluate the effectiveness of the safeguards identified in the error analysis and to provide a rapid "healthcheck" of the alarm system, procedures and staffing levels on site. The tools are not designed to provide a comprehensive assessment of any of the systems in place although they could highlight areas where such assessments would be beneficial. The outcome of each healthcheck assessment gives an indication of whether the reliance on the respective safeguards is justified.

STEP 5: IDENTIFY RISK CONTROL MEASURES

The next step is to improve the safeguards that have been identified as being less than adequate. This may involve, for example, improving the procedures and training, prohibiting certain concurrent tasks, improving alarms, increasing staffing levels or introducing engineering controls to avoid human failures resulting in a MAH. The safeguard evaluation tools should highlight areas for improvement. It is also important to consult with the operators to explore other potential improvements. An action list is compiled to ensure that all the proposed improvements are allocated to the relevant parties, considered and then either followed through to completion or justification provided as to why the improvement should not be adopted. A cost benefit analysis may be required to provide such a justification.

STEP 6: INCORPORATE RELEVANT RESULTS INTO COMAH SAFETY CASE

The results of this team-based process and the improvements adopted are then incorporated into the site's COMAH Safety Case report. The records taken throughout the assessment can be used to demonstrate that human factors issues have been systematically considered.

The six steps must be repeated for each of the MAH scenarios identified in Step 1.

APPLICATION

One of the aims of this joint partnership approach to human factors assessment was to transfer knowledge of human factors to the people who actually deal with these issues on a day-to-day basis on each site. By raising awareness of the issues that can affect human performance, the site personnel can identify issues beyond the scope of this assessment and proactively target new areas for improvement that would be missed by external contractors with a limited knowledge of the site and its operations.

Initially, the assessment team was formed consisting of three health and safety personnel, the site manager, one chemist, three operators and a human factors professional. This resulted in a multidisciplinary team who could provide different perspectives on the issues under consideration. The Ciba team were all experienced in the operations to be assessed. However, in order to enable them to complete the assessment, two days of training were provided in human factors and techniques involved in the assessment.

The methodology was translated into a spreadsheet format to allow the data to be captured effectively and to maintain an overview of the entire approach throughout. This format also provided easy access to all the data captured.

The assessment was conducted on the site over a period of one week. Although this restricted the amount of time available for each step, it ensured that the members of the team were able to participate throughout the assessment. If the assessment had been conducted over a longer period, team members would have been forced to miss sessions due to work commitments and shift patterns. The Ciba team took ownership of the assessment throughout, with guidance and support provided by the human factors professional where necessary. This encouraged the team to use the training that had been provided and to explore the human factors issues on the site.

EVALUATION

The study illustrated how simple, practical and effective methods can be used to assess how human failures contribute to risk. This approach is flexible enough to incorporate existing hazard identification and risk assessment information. For companies where previous evaluations of alarm systems, control systems, staffing levels and procedures have been conducted, these could be included in place of the checklists and flowcharts used in this study. Therefore the flexibility within this approach can avoid the need to revisit issues and perform repeat assessments. However, it is important to ensure that any previous assessments used within this approach are still relevant to the current tasks and current task environment. If minor changes to the site or the tasks have occurred since the previous assessments were conducted, then a partial update may be sufficient. However, if the assessments are outdated then a full re-assessment will be required.

The staffing flowcharts proved to be less effective than the other tools at providing direction for improving low scoring tasks. It was decided that although this flowchart is useful for diagnosing staffing deficiencies, it is not applicable within this methodology as the same information is collected elsewhere within the assessment. This flowchart is to be replaced by a diagnostic flowchart that will allow the assessment team to consider possible improvement measures and select the most appropriate option. Ordinarily, an organisation's first choice when searching for risk-reduction measures will not be to increase staffing levels. Other measures e.g. restructuring job roles may prove to be more effective in reducing the risks. The new flowchart aims to determine whether any benefit would be gained by having the task constantly attended. After all, if human monitoring is not the most reliable safeguard and the consequences are potentially severe, then an alternative solution may be more appropriate.

The concurrent task analysis proved to be one of the most useful exercises within the whole assessment process as this highlighted the situations where a simple human failure could escalate into a major accident hazard due to the human defences being defeated. However, it became clear following the initial assessment that the recording matrix did not include sufficient detail to allow others to interpret how the analysis had been conducted. For example, the coding system states that "P" denotes that part of the secondary task can be attempted whilst conducting the primary task. However, there is no way to distinguish which part of the secondary task can be attempted and whether there is a conflict that prohibits the "other part" (and hence completion of the secondary task). Therefore, the matrix has subsequently been updated to allow a greater level of detail to be recorded.

The method has proven to be ideal for the application it was designed for. However, it is important to recognise that other methodologies may need to be employed subsequently to investigate further the issued identified; e.g. a full alarm handling assessment may be required if the alarm safeguards are found to be inadequate. Therefore it is important to be aware that the checklists used do not necessarily replace existing assessments. Instead they provide an indication of where efforts should be focused for the greatest benefit in terms of risk reduction.

Although this method was originally designed for use within the chemical process industry, it could easily be applied to other industries to facilitate the inclusion of human factors into their safety cases. Although this approach has task analysis and human error assessment as its foundation, it has been specifically designed to be sufficiently flexible to incorporate other assessments and issues where necessary. If the safeguards in place do not take the form of alarms, procedures or manning, but instead are e.g. software-based safeguards, then an appropriate assessment tool could be included to evaluate these.

DISCUSSION

The feedback from the assessment team was positive and the potential improvements uncovered by the process proved the benefits of the systematic approach. However, the approach does require commitment from the organisation in order to succeed. It proved difficult to arrange for all the participants to be available at the same time for the sessions and therefore in some cases two people were trained to fill the same role within the team.

When this methodology is applied to other sites, different safeguards may be noted e.g. ones relating to the DCS control system. Therefore, an interface evaluation would be required rather than simply applying the checklists and flowchart outlined in this paper. Training and competence assurance may also be noted as safeguards. If so, the organisation must be able to demonstrate the training provided meets the specific need identified and that appropriate measures are in place to ensure that the training is effective and competence is maintained. The methodology is still evolving to fulfil such requirements and Ciba are continuing to develop the assessment process.

As with all assessments, it is important that the information gathered throughout the study is utilised effectively. The recommendations improving the safeguards should be fed forward into the design of new process systems and work environments as well as providing solutions to immediate issues.

REFERENCES

1. Engineering Equipment and Materials Users Association (EEMUA), 1999, *Alarm Systems: A Guide to Design, Management and Procurement*. EEMUA Publication No. 191. The Engineering Equipment and Materials Users Association: London.
2. Health and Safety Executive, 2000, *Better Alarm Handling*. HSE Information Sheet - Chemicals Sheet No. 6. March, 2000.
3. Bransby, M.L. and Jenkinson, J., 1998, *The Management of Alarm Systems*. HSE Contract Research Report CRR 166/1998. HSE Books: Norwich.
4. Health & Safety Executive, 1991, *Developing Best Operating Procedures: A guide to designing good manuals and job-aids*. **SRDA-R1**, By The SRD Association, AEA Technology, Cheshire.
5. Human Factors & Reliability Group (HFRG), 1995, *Improving Compliance with Safety Procedures: Reducing industrial violations*. HSE Books.
6. Embrey, D., 2001, Why Don't Real Men Use Procedures? A systematic approach to enhancing compliance to procedures in safety critical systems. *Industrial Safety Management*, 3(3), September 2001.

7.   Health & Safety Executive, 2001, Assessing the safety of staffing arrangements for process
     operators in the chemical and allied industries. Contract Research Report (CRR) 348/2001.
     HSE Books. See http://www.hse.gov.uk/research/frameset/crr/index.htm.

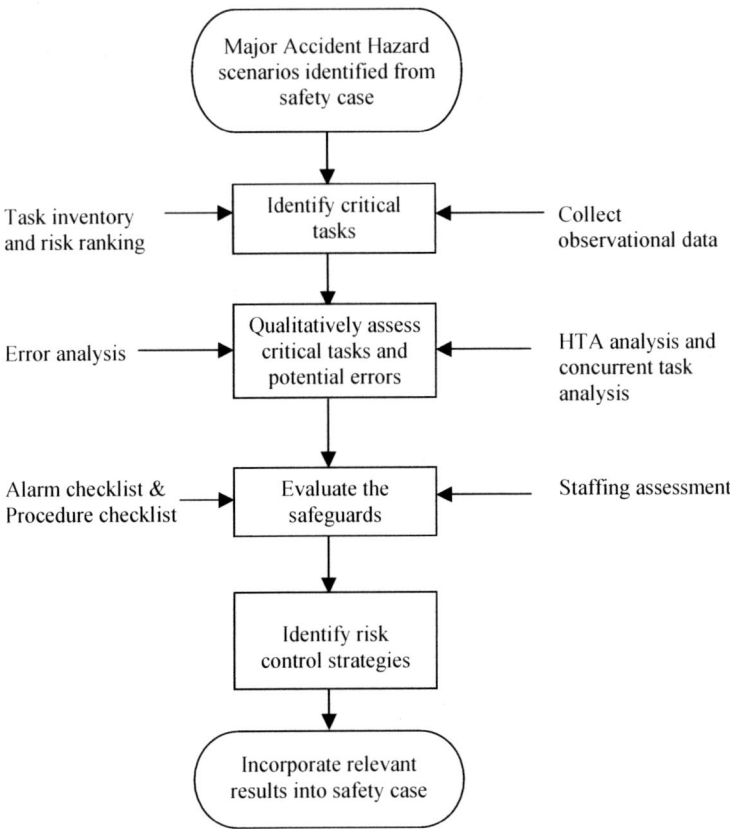

Figure 1.   Six stages of the assessment process

			...which of these tasks can be performed concurrently?											
			Critical (HTA) Tasks											
			Make Monomer						Charge Monomer					
While performing these tasks...	Task No	Task Step	1	2	3	4	5	6	7	8	9	10	11	12
	1	Charge to Vessel	■	A	P	X	X	A	A	A	A	P	A	A
	2	Transfer to Vessel	A	■	P	X	X	A	A	A	A	P	A	A
	3	Add	A	A	■	P	P	A	A	A	A	P	A	A
	4	Sample to lab approva	X	X	S	■	P	X	X	X	X	X	X	X
	5	*Set up suplhuric sotz*	X	X	P	S	■	X	X	X	X	X	X	X
	6	Transfer to Vessel	A	A	P	X	X	■	A	A	A	P	A	A

Key:　　X = None of the task can be simultaneously performed

P = Part of the task can be simultaneously performed

A = All of the task can be simultaneously performed.

Figure 2.　Concurrent task analysis matrix

Auditory alarms		
Can all auditory alarms be heard from all parts of the plant that the operator may be, even when wearing ear protection?	BP	An assessment has been performed to ensure that all alarms can be heard from all parts of the plant. When wearing ear protection, another operator is available to deal with alarms.
	S	No problems have been reported with alarm audibility throughout the plant. When wearing ear protection, another operator is available to deal with alarms. No assessment has been performed.
	P	Certain alarms cannot be heard from certain parts of the plant, or when wearing ear protection.
Are all sound signals easily distinguishable?	BP	Different sounds are used for different alarms (e.g. safety-critical vs. operational) or priorities, and designed according to ergonomic guidance.
	S	A small number of variations are used or a single tone is used where the use of different tones is not essential.
	P	A single tone or no tone used for a large number of different signals. Operator has to search to identify the reason for the tone sounding. OR there are too many different auditory alarms, which cannot be distinguished reliably in operating conditions.

Figure 3.　Alarm checklist

A. Procedure Design		
Is the procedure of an appropriate length?	BP	Procedures are kept as concise as possible whilst still conveying all the necessary information. The length of the procedure has been designed with the context of use in mind.
	S	Procedures are generally of a usable length, but may not be sensitive to the context of use.
	P	Length of procedures makes them very difficult to use. No account taken of context of use.

Figure 4. Procedure checklist

Figure 5: Staffing Level Decision Flow
Diagram 1

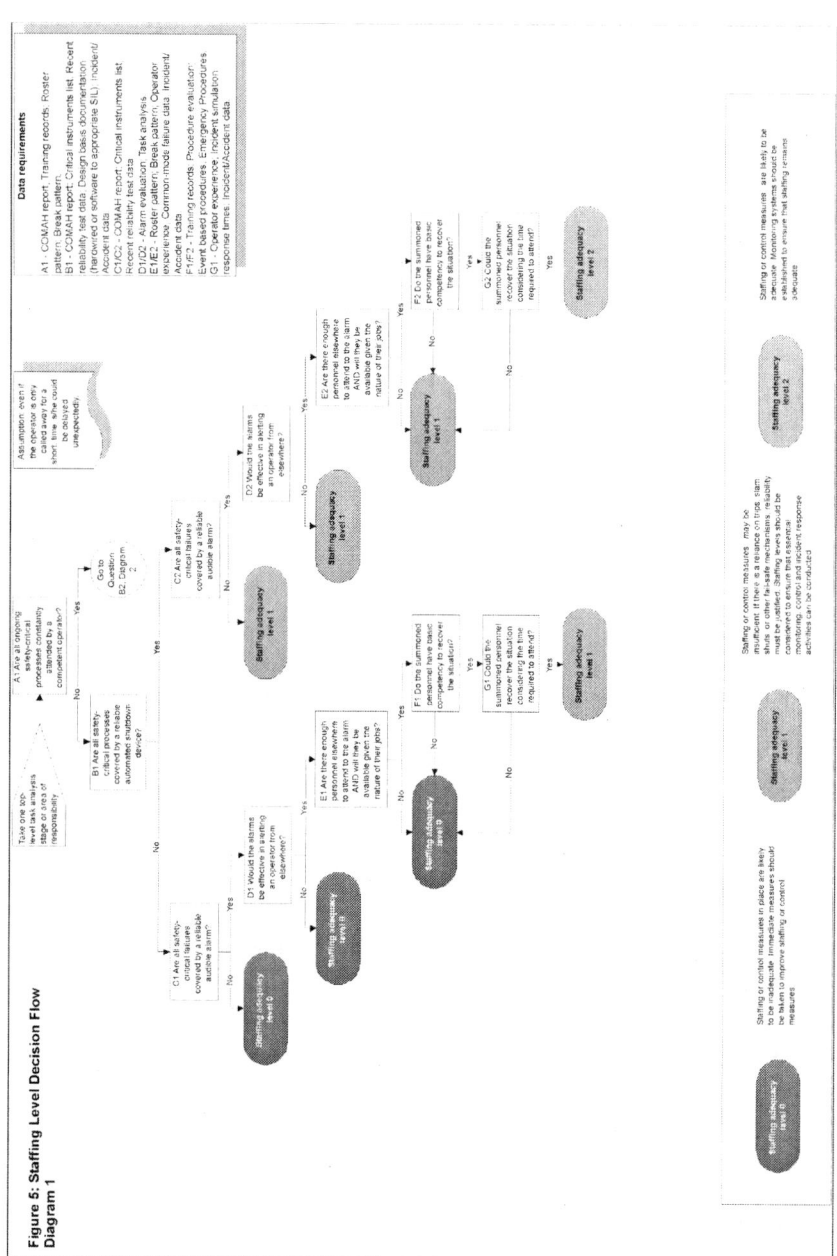

HUMAN FACTORS AND COMAH – THE CHALLENGE OF EXISTING PLANT

Sara Marsden, Mark Bendig, Jonathan Berman
Greenstreet Berman Ltd
Ken Patterson, Martin Colley
Hickson & Welch

The Health and Safety Executive is seeking a higher level of attention to human factors (HF) in the hazardous industries sector. They have issued guidance on the inclusion of HF in COMAH safety reports, focusing on the need to evaluate the potential impact of human error on process safety, and to ensure that appropriate measures are in place to minimise the contribution of human error. One challenge for industry is to identify and implement practical and integrated approaches to the evaluation of existing plant and systems, and plant modifications, and the identification of cost-effective responses to identified HF shortcomings.

This paper describes a systematic HF assessment completed at Hickson & Welch (H&W). The assessment was commissioned by H&W as part of their overall programme for improving process safety management, and was designed to satisfy HSE expectations both in terms of the assessment, and also in terms of enhancement of H&W's existing HF knowledge and competence. The work comprised training sessions, and an assessment involving task analysis, error identification, consequence assessment and recommendations for HF improvements. A designated process (Stage F) was analysed in detail, to provide generic recommendations and a framework for a more simplified future analysis process that could become integral to H&W's approach. The work was accepted by HSE.

The paper highlights practical lessons for undertaking HF analyses of existing plant and for making best use of HF consultants.

Human Factors, COMAH, Assessments, Safety Case

INTRODUCTION

The Health and Safety Executive (HSE) is seeking a higher level of attention to human factors (HF) in the hazardous industries sector. They have issued guidance on the inclusion of HF in COMAH safety reports, focusing on the need to evaluate the potential impact of human error on process safety, and to ensure that appropriate measures are in place to minimise the contribution of human error. One challenge for industry is to identify and implement practical approaches to the evaluation of existing plant and systems, and planned plant modifications, and to identify cost-effective and appropriate responses to identified HF shortcomings.

Following an incident on one of their batch processes, which led to a release, Hickson & Welch (H&W) received an Improvement Notice from HSE that required, *inter alia*, that they take full account of HF. The objective for H&W was both to incorporate HF into their process safety management (PSM) system in an appropriate manner, and to demonstrate that they had done so to HSE's satisfaction.

This paper provides an overview of the issues facing H&W, and how they were addressed. It describes how H&W used HF consultants to enhance their own HF capability, and ultimately to satisfy HSE that they had sufficient in-house HF capability to assure future assessments. Additionally, the paper presents some lessons learned and guidance, both for HF assessments and for the use of HF consultants.

THE BACKGROUND

H&W is a specialty chemicals company operating semi-batch processes. By its nature, their plant tends to operate for comparatively short periods before being reconfigured for the production of a new product. This continuous process of operation, plant redesign, plant reconfiguration means that there is a constant need to assess and assure the safety and operability of the new configuration, and to be able to assure HSE that the assessment undertaken has been done rigorously and effectively and that the error opportunities have been properly identified and controlled. Equally importantly, it is necessary constantly to manage the change process, and to assess error opportunities in the context of previous plant configurations, as well as the current or proposed one.

H&W suffered a runaway exothermic reaction. The immediate cause of the event was difficult to determine, but one root cause was inadequate attention to the types of human error that could arise, and hence to appropriate defences against them that should be put in place.

This incident occurred at a time when HSE was showing increasing interest in HF – a new specific HF unit had been created within the Hazardous Industries Directorate.

HSE served an Improvement Notice that comprised a number of elements. One of these was to carry out an HF assessment of the batch process, to identify clearly the HF shortcomings. A second was to improve H&W's understanding of the HF issues that arose from their batch process operations. A third was to develop an appropriate method for incorporating HF into their PSM system that addresses COMAH safety case requirements, such that H&W could be confident in future assessments. HSE strongly recommended in the Improvement Notice that H&W make use of HF consultants to support the work required to discharge the Improvement Notice.

The Improvement Notice provided a focus for H&W to consider plant, and plant modifications. It also provided a focus on demonstrating proactively that all practical error reduction measures have been implemented.

HSE EXPECTATIONS

The Control of Major Accident Hazard Regulations 1999 (COMAH) came into force on 1 April 1999. HSE COMAH guidance requires that "For hazardous events that could lead to a major accident, the safety report should show that risk-reduction measures have been put in place to reduce the risks to as low a level as is reasonably practicable."

The HSE has set out expectations and principles regarding the HF elements of major accident prevention. These expectations are contained within both the HSE's general Safety Report Assessment Manual, available at www.hse.hid./comah2/index.htm, and in their specific "Human Factors for COMAH safety report assessors" internal guide. In short, the HSE expect that the same systematic, demonstrable and "empirically" based approach that is applied to technical safety be applied to HF. Figure 1 illustrates this approach.

This means that Duty Holders need systematically to identify potential errors, choose and implement safety measures, and manage such measures within their management system. The following quotes from HSE guidance illustrate their expectations:

"The report needs to show that the measures taken and SMS are built upon a real understanding of the potential part that human reliability, or human failure, can play in ... major accidents"

"It should be clear how human factors have been taken into account in the risk assessment."

"The safety report should show what measures are in place to ensure adequate performance by human operators, ..."

"The safety report should also show how human factors have been taken into account in the design of equipment and systems (e.g. usability, tolerance of errors, detectability and recovery)."

"The safety report should show how systems which require **human interaction** have been designed to take into account the needs of the user and be reliable."

All of this emphasises the importance that HSE places on effective management of HF, and how the safety report must provide adequate assurance that HF issues have been properly considered.

When HSE served the Improvement Notice, they emphasised to H&W that they wished to see clearly how H&W was both addressing HF issues to prevent recurrence of the specific incident, and also how their PSM system took proper account of HF.

Generally, HSE has identified a set of 'priority' HF issues of concern to them:

- Organisational change
- Demanning and staffing levels
- Training and competence
- Fatigue from shiftwork and overtime
- Alarm Handling
- Compliance with safety critical procedures
- Safety culture/blame culture
- Communications e.g. shift hand-over
- Ergonomic design of interfaces
- Maintenance error

These span a range of issues far broader than interface design and the working environment – the issues that frequently are considered to be the full range of HF. Instead, this list demonstrates that HSE view HF as being far broader and all-embracing, and that HF issues cover all aspects of human performance at work. The challenge for a COMAH site in preparing its safety case is to ensure that it provides sufficient information to demonstrate that all of these areas have been considered, and appropriate measures implemented.

THE COMMISSION

GSB was approached by H&W to help discharge the Improvement Notice. The assessment was to be carried out on a batch process the company intended to re-introduce to plant – to enable the company to explore how the HF assessment process could be fitted into H&W's existing Process Safety Management (PSM) system. The commission emphasised that the key objective for H&W - although it was not required by the notice - was to ensure that they learned from the process of discharging the Notice. The company was clear that it was not sufficient for them simply to discharge the Improvement Notice, important though that was. Their primary goal was to ensure that they acquired the competence and processes to incorporate HF into their PSM system for use on both future and existing products. This in turn meant ensuring that there was proper 'buy-in' to HF across the company. The HSE had highlighted the need for H&W to understand where there were gaps in their existing HF knowledge – to "know what they don't know" – and then to fill those gaps.

INITIAL APPROACH

In responding to the commission, GSB identified a number of stages to the planned work. Whereas it would be possible for GSB to go to site, carry out an assessment of the relevant batch process, and report on human error opportunities, human engineering discrepancies, and to make recommendations for improving the design of the batch process – this would fail completely to meet the needs of H&W (though it would have discharged the notice).

Instead, it was apparent that perhaps the most important aspect of the support that GSB could offer would be to enhance H&W's understanding of HF, human error, and methods for controlling human error through design and operations. Consequently, the approach adopted was partly to provide formal training at the outset, and partly to use the assessment of the batch process as a form of further on-the-job coaching for selected site staff. The detailed assessment would be used to provide a set of specific recommendations relating to that batch process, but also to draw out generic issues that would be common to the majority of batch processes operated at the H&W site. From this, a more streamlined HF assessment process would be derived, taking account of the detailed recommendations already made. This streamlined process would be planned to be implemented by H&W, and would fit with their existing PSM system.

H&W accepted the proposed approach, and assigned a project manager to act as liaison throughout the work.

TRAINING

The first activity was to provide formal HF training for H&W. Two complementary training activities were planned. The first was a half-day senior management awareness session. The purpose of this session was to enable senior management to understand both the importance of HF, and the manner in which it could be addressed. It was essential that senior management was properly engaged in the subsequent HF assessment processes, and would properly support them. To achieve this, the session was planned to highlight the importance of HF, to highlight the tractability of the issues – human error can be managed, and to highlight the methods and approaches that would be adopted, together with their objectives.

Successful delivery of this session, although only a small part of the commission, was considered extremely important by H&W if they were to be able properly to take ownership of future assessments, and to take proper account of recommendations derived from them.

The second training activity, delivered the following day, comprised a full-day training and awareness course for a cross-section of site staff, including union representatives, to equip them with an understanding of HF issues and solutions, and an understanding of the methods that could be applied. Delegates at this second session included those staff who would receive the more intensive on-the-job training by being involved in the detailed assessment itself – the training session included elements of the processes and tools that would be applied during the detailed assessment. An important element of this training was the focus on practicality of recommendations, and that any HF assessment would be carried out on existing plant, which therefore imposed significant constraints in terms of what could be changed or adapted, and how human error could be controlled.

These two training sessions were delivered back-to-back, ahead of the start of the detailed assessments. Both sessions were well-received, and participants were very conscious of how the value of the session content went beyond the immediate concerns surrounding the Improvement Notice, and extended to all aspects of their normal work.

DETAILED ASSESSMENT OF BATCH PROCESS

Once the training was completed, and the member of H&W staff who would participate in the detailed assessment was comfortable with the planned process, a GSB consultant went to site to work with him.

The batch process selected for analysis was scheduled for implementation shortly after completion of the assessment, and hence it was anticipated that there would be both generic recommendations concerning the overall H&W approach, and batch-specific recommendations that could be implemented immediately to improve the planned process.

The analysis would focus not only on the operation of the planned batch process, but also on the differences between the preceding and following processes, to understand the implications of the change process. Additionally, there was a need to consider maintenance activities, both planned and unplanned, and how they could influence process safety. Whereas H&W considered carefully the opportunities for incorrect assembly of plant when configuring a new batch process, they were less careful about considering the similar opportunities for incorrect assembly following maintenance during a batch campaign. (However, although the opportunities for error might be similar, maintenance during campaigns is either on a "like for like" basis or is controlled by a separate control of change process. The consequences of error here should always be considerably less.)

Because the batch being assessed was still at the planning stage, much of the assessment was based, on the 'batch sheets' – the detailed step-by-step operational instructions.

This assessment required active participation by H&W. GSB worked alongside H&W to complete the task analysis and initial error and consequence analysis. H&W then took this and completed the detailed error and consequence analysis.

The assessment process can be summarised as follows (see also Figure 2):

Task Analysis
Developing and verifying with operational staff, a task analysis of operator actions of the operation, maintenance and emergency/abnormal states of the batch reaction operations. This entailed:

- Identifying key documentation and personnel;
- Reviewing procedures and drawings to develop an initial view of tasks;
- Completing "walkthroughs" of each part of the operation, and interviews with relevant staff;
- Producing a draft Task Analysis – detailed for production, high level for configuration, maintenance and emergency procedures;
- Talking through draft with key staff for confirmation and clarifications.

The purpose of the Task Analysis is to provide the equivalent of a verified "P&ID" for the operator tasks – clarifying the precise tasks, how they are undertaken, and their inputs and outputs. This provides the basis for the subsequent error analyses.

Identifying Safety Related and Safety Critical Tasks
This was a screening of tasks identifying those critical to preventing major accidents. Consequences were allocated to 4 categories according to severity, the top two categories were taken as being Major Accidents as defined in COMAH. Those categorised C & D were not taken forward, but H&W could subsequently decide whether to examine these further for other risks to health or safety (or quality).

The purpose of the screening process was to enable H&W to focus effort only on the critical tasks, and hence to ensure that the HF assessment remained practical in the time available.

Error Analysis
Examining in detail the tasks identified as having potential consequences in the top two categories entailed identifying the sub-tasks (from the task analysis) where human errors have the potential to cause major accidents, reviewing the opportunities for detecting significant errors and the consequences of not doing so.

H&W held several **"what-if"** sessions guided by an error mode checklist. The aim was to identify which errors could lead to a major accident regardless of the probability of error.

Production was analysed to a detailed level of tasks. Plant configuration, maintenance and abnormal states were analysed at a high level due to the very high range of circumstances and potential interactions with processes.

This process provided the basis on which to determine whether error reduction measures were required, and what form they should take.

Identifying Opportunities to Reduce Reliance on Human Intervention
Where significant error modes were identified, engineering and management control measures were reviewed both for prevention and mitigation, as well as for opportunities for detection of error. A view was then taken as to whether these ensure that the human error could not reasonably lead to a major accident hazard.

For those errors for which engineering control measures for prevention or mitigation were not adequate, a view could then be taken as to whether current engineering (and management) control measures could reasonably practicably be improved to prevent (and/or mitigate) errors and their consequences.

HF best practice judgement was applied at this stage in the process.

Identify Opportunities for Human Factors Improvements

Where reliance on human intervention was unavoidable, reviews were made of relevant controls (interface design, procedures, competencies, etc) to identify options for improvement and to show clearly the links between the likely consequences of error and the control measures in place.

Prioritisation and Justification

Review of current risk assessments in light of the above to assist in identifying priorities for action and ensure reliance is justified and can be demonstrated.

IDENTIFICATION OF IMPROVEMENT OPTIONS

Following completion of the detailed error and consequence analyses, GSB worked with H&W to develop recommendations for error reduction and mitigation.

These recommendations fell into two categories. One set was batch-specific. They would allow H&W to improve the design and operability of the planned batch process. The second set of recommendations was generic, and was the reason for undertaking the very detailed assessment. The intention was that subsequent assessments would build upon the detailed generic understanding gained during this work, and hence could be more streamlined.

The level of detail of the assessment did lead to a significant number of issues being raised, and it quickly emphasised the importance of having a process for prioritising recommendations. A further issue that arose during the assessment was the distinction between safety and quality issues. The focus of the study was safety. However, whereas errors that would lead to quality problems without impacting on safety could in theory be discounted, in practice H&W understandably were very keen also to address those. This led to a potentially huge burden of plant modifications, and this burden needed to be managed. GSB's approach was to prioritise with respect to risk, but the site needed also to consider quality and performance issues.

INTEGRATION INTO THE H&W PSM SYSTEM

GSB provided a report that summarised the process undertaken, and the conclusions and recommendations, both batch-specific and generic. GSB also facilitated a meeting with H&W managers to review both the assessment process and the assessment outcomes, in order to help identify how to ensure the integration into the H&W PSM system.

The approach taken, including both the training and the assessment process, supported this integration process. The elements of the training process that were oriented towards achieving management buy-in were considered essential, as was the in-depth coaching of key staff to create an internal 'champion' for the process.

As anticipated, this stage of the overall process required some modification at the time. The complexity of the selected batch process was greater than expected, which increased the time spent on the analysis. H&W needed to make rapid progress on the integration process, and therefore chose to carry out some of the integration work in parallel with the preparation of the GSB report. However, this had the effect of increasing their ownership of the process. Another issue that arose concerned the accuracy of the batch sheets. This was less than originally expected – which both increased the time spent in the analysis, and also influenced the manner in which H&W could expect to carry out subsequent assessments. The batch sheets are aimed at supporting the operators, and do not provide all of the information necessary for assessing potential maintenance errors. However, identifying the inaccuracies was itself beneficial.

SUCCESSES AND ISSUES

A number of particular successes were achieved during the course of the work – some planned, and some that were more unexpected. Similarly, a number of issues arose, either that needed to be resolved as the work progressed, or that adversely affected the progress of the work or its outcome. This section describes some of those successes and issues.

TRAINING AND ASSESSMENT PROCESS

- The training provided at the start of the project was an essential precursor – it enabled H&W to understand the elements of the HF assessment process and their purpose, and also supported buy-in across the company. Furthermore, it provided reassurance that the process that was about to be applied would provide benefit, and hence warranted the H&W effort and resources that would be required.
- The training involved site staff at all levels, from senior managers through to plant operators, and included union representatives. This ensured that the process was seen to be transparent, and was not intended to apportion blame, or to bias the conclusions in favour of 'preferred' solutions.
- The GSB consultant who undertook the detailed task analyses and 'what-if' assessments was not familiar with batch processes. This was both a strength and a limitation. The strength was that it allowed the consultant to ask fundamental questions, and thereby highlighted a number of issues that might otherwise have remained undetected. It also demonstrated that the strength of the analytical process lay, in part, in the manner in which it was applied – that it should address all aspects of the batch process that had the potential for safety-related consequences if there were a failure. The limitation was that, initially, it slowed the process, until an effective relationship was established between the consultant and the H&W assessor. A conclusion is that the 'application' of a naïve view was beneficial, but probably not required throughout the assessment process.
- The assessment was based on the batch sheets for the planned batch process. The development of the detailed task analysis was time-consuming, in part because of batch sheet errors. The process may have been more efficient if the assessment had been

integrated into the PSM earlier, such that a high-level assessment of the proposed batch process could have been undertaken as a screening process.

- The proposed assessment process involved a detailed 'what-if' analysis following on from the task analysis. The task analysis process revealed many issues, and hence was valuable but lengthy. The intention was to use a high-level screening process to determine the safety significance of 'top-level' failures and only to analyse further those that appeared to have such safety significance. In the event, the initial analysis proved extremely time-consuming. It also transpired that the potential blurring by H&W of the distinction between quality and safety issues – because of the wish to address both – made it difficult for the H&W analysts to avoid continually being sucked into a very detailed analysis. Once they had gained familiarity with the screening process, H&W were then able to complete the analysis much more quickly, relying also on their process knowledge.

- The balance of GSB and H&W staff appeared to be essential to the successful assessment, but there is a need to give careful thought to how best to achieve that balance.

- The selected batch process was extremely complex. It was selected for analysis because it was the next batch process planned for a campaign. However, the complexity of the process made it more difficult to stand back and determine how effective the proposed assessment process had been, and how easily it could be incorporated into H&W's PSM system in due course.

- The complexity of the process highlighted the value of having at least two people involved in the assessment – one to carry out the detailed assessment, and one to 'stand-back' and take a view about the implications of the assessment findings.

- Involvement of operators (and the union) was essential to ensure "buy-in" by everyone on site. This also ensured that H&W undertook the bulk of the assessment – both for speed and efficiency of the assessment, and in order that they could determine how best to implement a system that would be effective in the long-term.

- The prioritisation process needed to be pragmatic. When considering existing plant it was apparent that many potential errors were being controlled by administrative defences – and were being controlled effectively. Consequently it would not be appropriate always to recommend plant modifications and hardware defences. However, there was a need carefully to consider whether the administrative defences were robust – this was where the HF consultant added value.

- By providing the one-to-one coaching of H&W staff, and following on from the HF training, it was possible to equip H&W with the skills to develop further their ability easily to assess the adequacy of their administrative controls.

OUTCOMES FROM ASSESSMENT

- The number of quality issues that were raised during the assessment was both a strength and also a challenge. On the one hand it confirmed the value of the detailed analysis process – it highlighted opportunities for enhancing the batch process. On the other hand it created significant concern in respect of the amount of work that would be needed to implement remedial measures.

- The Improvement Notice was successfully discharged, and hence the assessment process could be considered effective. The level of detail in the assessment was greater than was necessary merely to discharge the notice, but it enabled H&W to carry forward some generic messages that could be incorporated into future batch processes. Future assessments would be more streamlined.
- The shortcomings in the batch sheets was a significant finding. In practice, the skill and experience of site staff allowed them to overcome these errors, but it made the process potentially vulnerable. The assessment has allowed H&W to refocus on the validity of the batch sheets.
- The assessment highlighted the importance of the change process – moving from one material to be produced to the next. A number of issues were identified where the error potential was aggravated because of the manner in which the previous batch process was operated.
- The assessment highlighted the importance of providing effective controls over breakdown maintenance activities.
- Integration into PSM.
- The entire assessment was planned as a linear process – the task analysis and error assessments, followed by recommendations, and the development of a process that could be incorporated into the PSM system. In practice it became apparent that, due to time pressures imposed by the Improvement Notice deadline, it was necessary to carry out some steps in parallel – particularly when considering the manner in which PSM was enhanced. It is unclear whether this is a presentation issue, or whether it is important to ensure that the various activities happen in parallel. GSB considered that a strength of their approach was the allocation of an experienced safety management consultant to stand back from the detailed analysis, and work in parallel with H&W to identify how to incorporate the assessment activities into the PSM system.
- H&W need to be able to 'challenge' themselves quickly when developing a new batch process. The assessment undertaken for Stage F was extremely detailed and H&W faced the difficulty of developing a simplified process that would support such a rapid challenge. The generic issues raised in the detailed assessment provide a surrogate checklist to support a more rapid and hence acceptable process.
- H&W had not planned to include HF in their HAZOP process during batch process development. The outcome of the assessment and subsequent discussions led to an acknowledgement that it could beneficially be incorporated at the design stage – to improve their ability to engineer out the opportunities for human error.

LESSONS LEARNED

H&W and GSB have both learned a number of lessons from this work. These lessons cover three areas:

- the assessment process, and the batch design and operating issues that it highlighted;
- how such assessments can and should be incorporated into the company PSM system;
- how the value of the HF consultant can be maximised.

THE ASSESSMENT PROCESS

- The earlier that HF assessments can be incorporated into the design of plant modifications and new batch processes, the more effective will be the error controls that can be implemented. However, there is a need to ensure that they are applied at a point in the design process when the potential human errors can be identified – too early on in the process these may not be readily apparent. In the end a two pass approach may be required.

- A balance must be struck between familiarity with the processes, and hence the efficiency with which the assessment can be undertaken, and 'naivety', and hence the ability to 'challenge' the system. In practice, a significant element of the requirement for challenge was satisfied through the initial assessment undertaken for this project. Those challenges provide a generic framework with which to question future modifications. H&W already had a good pragmatic understanding of potential errors, and the formal assessment provided a more robust framework to support understanding of the underlying error causes.

- An extensive set of recommendations arose from the assessment. The precise recommendations were based, in part, on the application of HF judgement. They also took account of the need to implement pragmatic and practical changes. Consequently, many recommendations focused on enhancing existing checking processes, rather than proposing impractical or costly plant modifications or procedural changes. HF expertise was applied to assure that the proposed changes were likely to be effective in controlling human error, and failures.

- A number of recommendations arose concerning the physical environment. None of them alone was considered fundamental, but together they provide the opportunity to enhance the working environment for operators, and thereby reduce further the likelihood of error.

- Recommendations arose concerning 'hidden knowledge', typically acquired through informal on-the-job training. It was suggested that such knowledge could usefully be formalised – examples included the rules of thumb used by operators for controlling batch temperatures, which if inappropriate could give rise to significant problems.

- Some recommendations concerned plant labelling – the importance for safe operation of the plant was recognised by H&W, and suggestions were made as the where further improvement could be made.

- Recommendations concerned the conceptual design process that preceded modifications to the batch process in order to manufacture a new product. The opportunities for human error in the design process were noted, and H&W has examined the level of independent checking that it applies, in order to be confident that it remains appropriate.

- Correct plant and software configuration was recognised as being critical. Recommendations concerned a review of the strength and independence of the checking processes associated with re-configuring. In particular, the danger of inadvertent failure to change an aspect of the configuration from the previous process was noted. This recommendation also applied to the testing activities that were intended to detect errors introduced in the design and configuration processes.

- Some recommendations concerned batch sheets. These included measures intended to ensure the accuracy of batch sheets, encouragement to operators to question aspects of the process that caused them concern, and also opportunities for making more use of the batch sheets for other purposes such as training, highlighting safety issues, etc.

- A number of recommendations concerned specific ergonomic improvements that could enhance batch processing at H&W (e.g. improve the accuracy with which defined quantities of water or other products were introduced; disable/remove duplicate or redundant interfaces to reduce ambiguity; improve monitoring accuracy; ensuring that ergonomic conventions were followed and that historical maintenance had not inadvertently contravened such conventions – e.g. incorrect alarm colour coding). The generic elements of these recommendations were highlighted.

- Some recommendations concerned opportunities to reduce error likelihood through improved procedural control of product availability. However, the importance of keeping such controls simple, and minimising unnecessary bureaucracy was emphasised in order to have confidence that the procedural controls would be complied with, and hence remain robust.

- A set of recommendations specifically addressed maintenance issues. One recommendation concerned the need to consider how to engineer out the opportunity for maintenance error (e.g. making it not possible incorrectly to assemble a plant item). Another concerned the need clearly to indicate plant item safety significance for that particular batch process. This is important where the safety significance of a particular plant item may change from one batch to the next, and hence maintenance staff cannot know, from the plant item alone, what is its safety significance. Recommendations also concerned maintenance management and planning.

- The assessment also enabled H&W to review its arrangements for responding to abnormal events. It clarified the extent to which operators were sometimes the only line of defence against interlock failure. Furthermore, routine reliance on interlocks during normal operations may reduce the operator's awareness of the criticality of certain operations, and hence of their role in assuring safety.

INCORPORATION INTO PSM

Four complimentary activities exist within PSM where HF has a particular bearing. Each of these was addressed when considering how best to incorporate HF into PSM:

- Process hazard assessment (e.g. HAZOP)
- Engineering (standards) management
- Safety and reliability demonstration
- Other SHE management activities

In the light of this work the following specific recommendations were proposed for H&W to consider.

Process Hazard Assessment
HF should be incorporated at a number of stages, including:

- Stage 2 Activity Risk Assessment (HF HAZOP)
- Stage 3 Detailed Event Assessment (human error analysis, task analysis, human error guidewords/checklists, appropriate human reliability assessment)
- Process hazard analysis to include explicit HF consideration

Engineering Standards
Revisions to current site standards to incorporate HF guidance. Extending the current standards to make them bespoke to H&W processes.

Safety and Reliability Demonstration
Giving greater weight to HF issues in the company's COMAH report, e.g.:

- HF reliability principles (e.g. no single human error shall have a major incident potential)
- Expand principles covering interface design, supervision, competence assurance
- Develop links between human error analysis and HF safety and reliability decisions
  Hickson & Welch hope to be able to describe how they have modified their Process Safety Management System to incorporate the lessons learned in this exercise at a future "Hazards" conference.

General SHE Management Systems
Many aspects of management influence the likelihood of human error, e.g. hiring, placement, training, permitry, change management, organisational structure, supervision. It is important to be able to compare company current practice with best practice/HSE guidance.

USE OF HF CONSULTANTS
H&W commissioned HF consultants for two reasons. One was to support their assessment and to help them understand better the HF issues that affect their operations. The other was because the HSE had clearly indicated that they expected to see external HF consultants involved in the assessment. A number of learning points arose from this:

- "Technology transfer" was an essential element of the work – the ability of the consultants to help H&W acquire new skills, and robust HF assessment processes. This was also important if effective prioritisation of potential error reduction recommendations was to take place.
- H&W and GSB had to work closely together – each brought knowledge and experience that together would enable the development of an effective process. During the work it became apparent that communication between H&W and GSB needed to be strengthened and made more explicit, to allay H&W concerns about the perceived complexity of the process. For example, as planned, H&W continued the analysis using their own resources. It became apparent that they were going further with the consequence analyses than originally intended, and were becoming understandably

concerned about the time and resources required. It was important that lines of communication remained open between H&W and GSB to highlight this concern and correct the process accordingly.

- GSB underestimated the complexity of the Stage F process, because it was not possible to have access to the batch sheets in advance of tendering. This caused GSB initially to propose and commence a process that proved to be too detailed. The more information that can be provided to the consultant at the outset, the more appropriate will be the proposed approach.

- It is essential that both the client and the consultant have a common understanding of the intended deliverables, and their proposed use. In this instance, there would have been benefit from a more detailed discussion of how GSB's report could be used to satisfy HSE. In the event the report structure was more suited to H&W's needs than to HSE's. For future assessments, a report section suited to HSE's concerns (with a focus on the Headings in the Improvement Notice) would be helpful.

- The value of external consultants appears, in part, to lie in the early stage of the assessment process, where they can question and challenge existing arrangements, and propose and co-develop an assessment process. Building on the knowledge transferred during the assessment, the client can carry out subsequent assessments without making such extensive use of consultants.

- There remains a role for the consultants in providing ad-hoc HF advice and guidance in respect of specific issues – assisting prioritisation, assessment of remedial measures, and recommending particular solutions.

- Even with external consultants, H&W needed to put in significant resources. Whilst this was expected – the discharge of the improvement notice was the responsibility of H&W, and could not be transferred to an external consultant – nevertheless, the level of effort to gain sufficiently detailed understanding of the basic processes and HF should not be underestimated. Subsequent assessments should be more manageable.

- There may be merit in setting up an industry peer group to support future assessments and to exchange information on best practice and common issues.

- The involvement of the operators and the union was considered essential. It provided an avenue to ensure that best use was made of operator experience.

- The 'naïve' perspective that can be offered by a consultant is important for ensuring systematic 'challenge' to the 'normal' way of doing things. Additionally, a consultant can stand-back from the process and look for gaps, barriers and other shortcomings.

- The development of an in-house 'champion' for HF also was considered essential. That person could provide robust upward and downward links. It was important that they were a reasonably senior process manager.

- An identified potential vulnerability concerned the expertise that lay within the design team. There was reliance on two people to undertake HAZOPs. Succession management may need to be considered.

GUIDANCE

The following 10 points are recommended as guidance when considering how to incorporate HF assessment into the PSM system for existing plant:

1. HSE has made clear that they consider HF to be of prime importance when considering the adequacy of a COMAH safety case – the need to provide assurance that HF is being adequately addressed is ignored at your peril.
2. Adequate incorporation of HF must start as early as possible in the design process – at the conceptual stage.
3. The Duty Holder requires a good grounding in HF to be able properly to identify HF issues, to prioritise areas for attention, and to identify effective solutions.
4. Independent advice, both to challenge existing practices and to provide guidance on the effectiveness of potential solutions, should be available.
5. The first assessment will be very resource intensive – it is important to ensure that generic messages are taken from this assessment and incorporated.
6. An in-house 'champion' for HF is invaluable.
7. The assessment focus must remain broad, taking account not only of operations, but also of design, commissioning, maintenance, abnormal operations, etc.
8. HF expertise can help to discriminate between alternative solutions, and to advise on the strengths and weaknesses of different error defences, thereby helping to make the chosen defences robust and cost-effective.
9. Existing plant will present a range of HF discrepancies. It is important to be able to distinguish between those that demand attention for safety reasons, and those that may have only economic disadvantages. A screening process will help this.
10. HF consultants are a valuable source of advice, guidance and support, which HSE expects Duty Holders to make use of. However, HF consultants are an adjunct to, and not a substitute for, the application of in-house expertise.

CONCLUSIONS

H&W wished to enhance the manner in which they demonstrated adequate control of HF within their PSM system. They commissioned GSB to assist in analysing a selected batch process, both to reduce the likelihood of human error and hence poor safety on the next planned batch process, and to help them determine how best to improve their treatment of HF.

The assessment and support, carried out over a relatively short period of time, proved invaluable for H&W in helping them better understand HF and how it affects their operations. The process was detailed, and raised many issues. A key requirement was to derive a more streamlined assessment process that could be incorporated reliably into their PSM system. The information derived from the detailed assessment allowed H&W to do this.

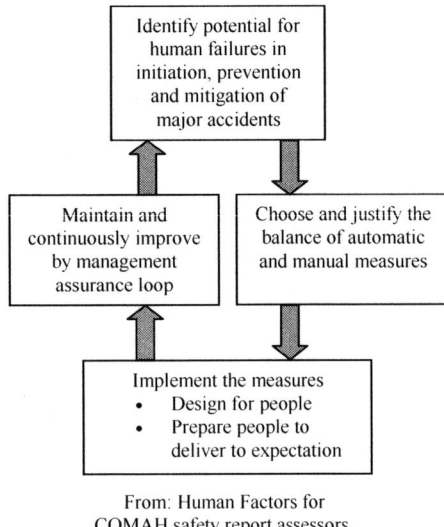

From: Human Factors for
COMAH safety report assessors,
HSE 2001

Figure 1. Human factors and COMAH

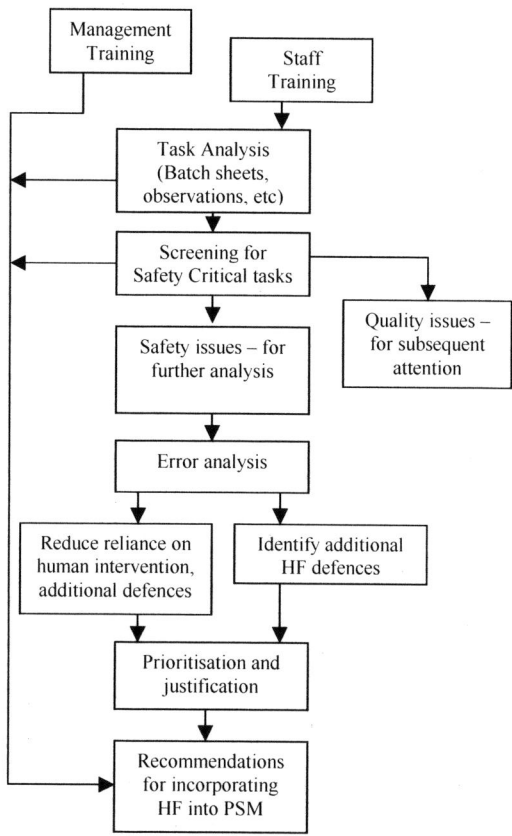

Figure 2. Assessment process

IMPROVING HUMAN FACTORS AND SAFETY IN THE PROCESS INDUSTRIES: 'THE PRISM PROJECT'

Robin Turney,
PRISM Project

SYNOPSIS

To assist the process industries in improving both its understanding of and application of human factors the European Process Safety Centre has taken the initiative in creating PRISM. This is a 'Thematic Network' which aims to create an extensive European forum within which industry, universities, research centres and practitioners can collaborate to improve the flow of practical experience and fundamental knowledge in human factors. The network has been established with financial support from the European Union Department for Research and Development under its Programme for Competitive and Sustainable Growth.

The network, which started in 2001, will last for 3 years and already has the support of organisations from 14 countries in Europe. These include many major chemical producers as well as universities and research organisations. It also has support from the European Union Joint Research Centre, from CEFIC as part of its 'Sustech' programme and from the Health and Safety Executive (UK). It is being co-ordinated by the European Process Safety Centre with support from The Institution of Chemical Engineers.

Since Human Factors is a very broad field four separate 'Focus Groups' have been established within the network covering

- Cultural and organisational factors
- Optimising human performance
- Human factors in high demand situations
- Human factors as part of the engineering design process

Each of the focus groups has already held a seminar and more are planned. In addition they will produce guidance aimed at meeting industrial needs, all of which will be made available on the Internet.

This paper provides a mid-term review of PRISM, describing what is available and how organisations can gain from and participate in its future activities.

Safety, Human Factors, Emergencies, Safety Culture, Alarms.

INTRODUCTION

It is well known that human factors play an important part in most if not all accidents. This can be seen in both the simplest accidents and in those that involve more complex technical interactions. It is also clear that the adoption of the attitude of 'blame the operator' will fail to lead to improvements in safety and will result in the true causes of accidents, for example the pressures which lead operators and others to make mistakes, never coming to light.

An example is the explosion at the Texaco Refinery in Millford Haven in July 1994. The series of events that lead to this explosion can be traced to a severe electrical storm that affected a number of production units. During the following 4 to 5 hours a fire on one of the units was dealt with, parts of the refinery were shut down and attempts were made to bring all units back on-line.

Eventually a combination of failures resulted in a knock-out drum on a flare line being overfilled followed by the failure of the flare line and the release of 20 tonnes of flammable hydrocarbons. This material ignited, causing a major explosion. In its report into the incident (ref. 23 & 24) the HSE identified important human factors issues as well as failures in safety management systems, plant design and construction. The human factors issues included limitations in the display systems of the distributed control system, which made it difficult for the operators to form a clear overview of the state of the whole unit. The operators were further hindered by the alarm systems where there were a total of 2040 alarms, 87% of which were rated as high priority. In the 10 minutes prior to the explosion a key alarm indicating that the knock-out drum was overfilled was submerged amongst a total of 275 alarms which had to be dealt with by the operators (an average of one alarm every 2 seconds). In fact the operating team had been dealing with alarms at a very high rates since the initial lightening strike, 4 hours earlier.

The above incident demonstrates the very important part which human factors play in both the cause and prevention of accidents alongside the other key elements essential for safe operation.

- Plant and equipment which is safe and suitable for its function.
- Effective systems for the management of safety.
- Properly trained and well motivated staff.

High standards of safety can only be obtained if all the above elements are in place.

Looking back over the last 20 to 30 years it is probably in the first two areas that the most important safety improvements have been made in the process industries. Whilst the drive for further improvements in both hardware and management systems will continue the returns are likely to be lower than in the past. It is for this reason that many companies, including those in the PRISM network, are looking at the scope to obtain further improvements in safety through the greater understanding of 'Human Factors'. This is represented in figure 1.

THE PRISM PROJECT

To assist the process industries in improving both its understanding of and application of human factors the European Process Safety Centre has taken the initiative in creating PRISM. This is a 'Thematic Network' which aims to create an extensive forum across Europe within which industry, universities, research centres and practitioners can collaborate to improve the flow of practical experience and fundamental knowledge in human factors. It has been established with financial support from the European Union Department for Research and Development under its Programme for Competitive and Sustainable Growth.

The objective of PRISM is.

"The improvement of safety in the European process industries through raising awareness of, and sharing experience in, the application of human factors approaches.
In addition the network aims to stimulate the development and improvement of human factor approaches in order to address industry-relevant problems in batch and continuous process industries."

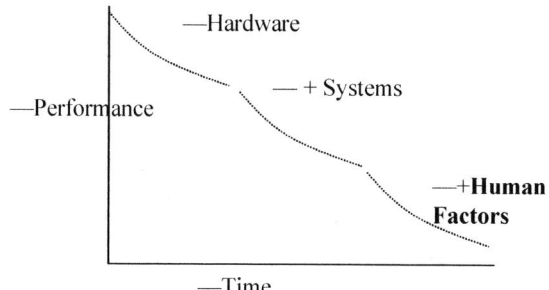

Figure 1. Improvement in safety performance

The network, which will last for 3 years, already has the support of over 100 organisations and individuals from 14 countries in Europe. These include many major chemical producers as well as universities and research centres. It also has support from the European Union Joint Research Centre, from CEFIC as part of its 'Sustech' programme and from the Health and Safety Executive (UK). It is being co-ordinated by the European Process Safety Centre with support from The Institution of Chemical Engineers.

This paper provides an overview of progress and achievements of the project at its half way stage.

So how does PRISM operate? It is recognised that the field of Human Factors is a very broad one and for this reason four separate 'Focus Groups' have been established within the network.

These cover
- *Cultural and organisational factors*
- *Optimising human performance*
- *Human factors in high demand situations*
- *Human factors as part of the engineering design process*

CULTURAL AND ORGANISATIONAL FACTORS (FOCUS GROUP 1)
This Focus Group is examining topics such as:

- the influence of cultural factors (e.g. national, organisational, and site culture)
- effective behaviour modification programmes
- safety implications of team working (benefits and pitfalls)

This Focus Group is being lead by The Keil Centre (UK) and John Ormond Management Consultants (UK) as Principal Contractors, in conjunction with Solvay (Belgium) and Lyondell (Netherlands) as End-User Advisors. The role of the End-User Advisers is the important one of ensuring that the work of the focus group will be of practical relevance to industry. The topics and deliverables will include:

- A general guide on the application and value of behavioural modification programmes.
- A report on cultural factors in safety management and a guide on identifying and managing key factors
- Information and guidance on the safety implications of team working and their management.

The first Focus Group 1 seminar was held in Edinburgh in January 2002 with the theme 'Improving Safety: Cultural & Organisational Factors'. Approximately 70 delegates from 11 European countries attended to take part in a programme which included both presentations and group work sessions. The presentations included inputs from industry, academics and consultants.

In a paper, 'Benchmarking offshore safety culture' Mearns (1) outlined the results of a 2 year study sponsored by the HSE and involving 14 off-shore operating and support companies. During the study a number of indicators or measures of safety culture were used. In the second year of the study only 1 company showed a reduction in the measure of its safety culture, 3 were about the same and 5 showed an improvement. More importantly the study demonstrated a positive correlation between the measures of safety culture and the lost time injury rate, with low injury rates being associated with adoption of 'Best management practice' More detailed analysis also showed significant differences between year 1 and year 2 in those practices rated as most important by the operators.

In a case study, Joyner (2) explained why bp's Dalmeny and Hound Point tanker loading facility had embarked upon a project to assess and improve the existing strong safety culture. The site had in recent years undergone significant changes with the introduction of new technology, organisational changes including 'enhanced teamwork'. The results had been deemed successful but had been disappointing in safety, with 2 serious incidents occurring early in the year 2000. The study identified a number of areas for improvement, including the need for more face to face communication (not just e-mails), a simpler way of reporting minor incidents and near misses and a need to involve all staff, including management, in the observation and correction of unsafe acts. Of particular interest was the way in which different safety subcultures were found within different occupational groups on the site, which suggests safety culture improvement activities need to be tailored to local circumstances. Overall it was considered that the exercise had increased two-way communication as well as understanding and had provided a sound basis for safety improvement plans.

Having established that there is a link between Safety Culture and safety performance ways of measuring safety culture are necessary. Two techniques have been shared within PRISM. Lardner et al (3) described the 'Safety Culture Maturity' model. The essence of the model is the definition of 5 stages of Safety Culture maturity. These range from Emerging, through Managing, Involving, Cooperating, to Continually Improving. An essential element of the approach is that, for sustained improvement an organisation needs to assess the level which it is at and then make improvements step-wise through the various levels. Trying to jump steps may lead to failure because the organisation is not mature enough to take the new ideas on board. The model is based on 10 elements of safety culture which are assessed in workshops involving front-line personnel and management from across the whole organisation. Byron (4) separates Safety Culture from Safety Climate which he defines as *Tangible outputs of the safety culture as viewed by an individual or group at a particular point in time.*. The HSE has developed a tool designed to provide an assessment of the safety climate through a 71 statement questionnaire, which is distributed throughout the organisation (The questionnaire is available from the HSE website). Byron has emphasised the way in which a survey can reveal important differences in opinion between different groups within an organisation. An example was provided where the workforce strongly agreed with a statement that 'the permit to work system introduced unnecessary delays' whereas the management & supervision disagreed. Further investigation showed the root cause to be a change in the organisation which had reduced supervision resources and introduced the delays in the issue of permits which were causing the concern.

Many organizations have adopted new working methods over recent years with many moving towards self managed work teams. Lardner (5) called on the results of four recent case studies from oil exploration, chemical and offshore gas maintenance industries to illustrate the gains and "tripwires" of teamworking. These showed that the Self Managed Teams had more involvement in risk assessment, safety auditing, monitoring safety indicators, plant design and other key safety issues. Communication, and knowledge of plant and processes, improved and there was greater involvement in planning and problem solving. On the other hand, major changes to roles and responsibilities required careful planning and retraining. The jobs can become more demanding, and the assumption that safety is "always someone else's responsibility" is a potential problem.

Corpe (6) led an interactive sessions which introduced delegates to Smarteams, an internet-based team development resource which was developed specifically for the upstream oil and gas industry. Workshop sessions demonstrated the relevance of this resource to teamworking in all industries, and explained how health and safety benefits had been achieved through participation in teamworking initiatives. Whiting (7) outlined the work which had won a team from Bradwell Power Station the NUMEX 2000 Award for excellence in maintenance in the nuclear industry. They described how a project to refurbish maintenance workshops was used as a vehicle to enhance teamworking and ownership of safety. This project achieved impressive improvements in commercial and health and safety performance within existing budgets

Wright (9) presented some results from a study which identified best practice in involving employees in health and safety. He distinguished between different degrees of employee involvement, and highlighted the health and safety benefits obtained by a number of organisations through effective employee involvement.

In addition fishwick (8) described the approaches being made to small to medium sized organisations in order to prepare the guide on 'Application of Behaviour Improvement Programmes'. At the time of writing this guide is being reviewed by PRISM members.

OPTIMISING HUMAN PERFORMANCE (FOCUS GROUP 2)
This Focus Group is examining topics such as:

- task design
- procedures
- ergonomics
- training
- staff selection
- shift work

This Focus Group is lead by DNV (UK & Norway) as Principal Contractors, in conjunction with Chinoin (Hungary) and Huntsman Polyurethanes (Belgium) as End-User Advisors. The topics and deliverables will focus on:

- The production of best practice guidance for the process industries on task design, including consideration of both cognitive and physical ergonomics issues.
- Best practice guidance on man-machine interface and human computer interaction.
- Best practice guidance for the process industries on the production of effective procedures.
- Guidance on human factors considerations in the training of staff.

Both focus group 2 and focus group 3 will consider many of the issues highlighted by the Texaco incident.

Focus Group 2 held a seminar in Budapest in March 2002.

Although a great deal of time and resources are devoted to the development of procedures in the process industries, less attention is paid to how to ensure compliance once they have been developed. In his paper, Embrey (10) quotes data from the power generation industry which indicates that 46% of all incidents are caused by failure of personnel to follow instructions. When such incidents arise it is important to look for the underlying system failures which may have generated the situation in which the violations are encouraged or condoned. CARMAN (Consensus Approach to Risk MANagement) actively involves the operators in the development of procedures, considers the level of support required for different levels of competence and the interface between training and procedures. The approach has been applied to a major oil & chemical site following a series of dangerous near misses. Following the use of the new approach for over 3 years, surveys have shown significant improvements in a number of areas including a 52% reduction of those who saw the procedures as unworkable

In a verbal presentation, Labudde, shared approaches used by DuPont. Compliance with standard procedures is seen as critical to safe operation and significant time and effort is devoted to training. When a facility is acquired from another operator detailed retraining of staff is undertaken, an exercise which can, in some cases, involve a stoppage of production to ensure that the new approach to safety is adopted. The emphasis does not stop at this point and on-going training is provided to all staff, figures from one site showing 7% of the total manpower hours being devoted to training. DuPont do not see this as a cost but as an investment in safety and product quality.

For plants covered by COMAH (Seveso II), the demonstration that employee training has been carried out effectively is extremely important. Bull, (11) reported on the way in which

Ciba Speciality Chemicals have used a Systematic Approach to Training (SAT) developed by the US Department of Energy. A comprehensive computer system had been implemented to manage the large amount of information involved, and to maintain up-to-date records.

PRISM presentations have also addressed the effectiveness of training. In a paper presented to Focus Group 3, Schaafstal, (12) reported on studies into the effectiveness of the training of Weapons Technicians in the Royal Dutch Navy. Results showed that newly qualified technicians were able to solve only 40% of the problems presented. This lead to the redesign of the training course to incorporate a structured approach to trouble shooting, the total length of the course being increased from 6 to 7 weeks. Results were encouraging with the proportion of problems being solved rising to 86%. Following this there was a complete redesign of the course with a greater concentration on trouble shooting, the total duration being reduced to 4 weeks. With this course the proportion of problems solved increased again to 95% indicating an approach which obviously warrants wider consider.

At the time of writing best practice guidance for small to medium sized operations on 'Training ' and 'Preparation of Procedures' are available to PRISM members for comment.

HUMAN FACTORS IN HIGH DEMAND SITUATIONS (FOCUS GROUP 3)
This Focus Group is examining topics such as:

- diagnosis of process upsets
- cognitive (alarm) overload
- emergency response
- control room layout
- abnormal situation management

This Focus Group is lead by TNO (Netherlands) as Principal Contractor, in conjunction with ATOFINA (Belgium) and BASF (Germany) as End User Advisors. The topics and deliverables will focus on:

- workload, stress and work environment
- information (alarm) overload
- emergency procedures
- application of virtual reality to human factors
- resource deployment
- the human in the emergency loop

A seminar was held in Soesterberg in May 2002.

The UK Heath and Safety Executive have recognised the importance of ensuring that the manning levels in critical situations are sufficient to ensure the safe management of upset and emergency conditions. Contract research carried-out for the Hazardous Installations Directorate of the HSE has lead to the development of an socio-technical assessment methods to determine areas where the level of manning may be insufficient. This method has been described in a number of places and application in a number of different situations has confirmed its value. This approach has been shared within PRISM by Conlin. (15)

Neerincx & Passenier (16) have developed a model for mental load under high demand situations. This takes into account the level of information processing, the time occupied by the tasks and the number of time set switches (changes to tasks being undertaken). In addition to other applications the model has been found to be useful when applied to high demand situations in the control of ships of the Royal Dutch Navy. (17)

Within the PRISM group there has been a high level of interest in this model together with a recognition that for wide application, more guidance will be needed on the high and low levels of the parameters used in the model.

One of the problems common to all control room situations is 'Alarm Overload'. Herbaux (18) described a practical approach to 'Alarm Management' being applied by Atofina. The approach covered, the establishment of an alarm philosophy, the identification & treatment of 'bad actors', application of advanced techniques such as alarm masking, alarm grouping or the replacement of individual alarms by overviews. The application of this approach to an ethylene plant led to a reduction in the monthly number of alarms from 19000 to 14000. Whilst this level is still considered high the proportion of alarms caused by instrument problems was reduced from 40% to 25% and the number of disabled alarms from 50 to 15. This was seen to represent major progress involving a significant change in mentality whilst it was still recognised that there is scope for further improvement.

High demand situations are not restricted to process plants or naval applications but cover a wide range of industries. Northern Ireland Electricity has recently reorganised its 24 hour emergency processes. This had involved the introduction of new IT support systems, the creation of new job roles, and the development of a revamped incident management centre and emergency response plan. A paper by Hamilton (19) described how human factors integration techniques were applied in the accomplishment of these changes and the lessons learnt from this work.

Other papers presented at this seminar covered, Display Development for Manned Space Missions (20) and Identifying Critical Situations (21). In addition workshop sessions considered future developments in high demand situations and a number of demonstrations were arranged showing the ways in which virtual reality techniques are able to provide assistance.

Within Focus Group 3 a separate sub group lead by Polytechnica Milano are studying the use of Virtual Reality techniques to improve the understanding of human factors. A number of meetings have been held and a survey produced on 'Human Reliability Methods for Safety Assessment & Risk Management' (22) as well as a state of the art report on the use of Virtual Reality. Currently proposals for further work on this topic are being developed.

HUMAN FACTORS AS PART OF THE ENGINEERING DESIGN PROCESS (FOCUS GROUP 4)

The concept of this Focus Group is to take direction/outputs from research and other Focus Groups and consider them in relation to typical engineering design processes in place at both major process industry companies and at SME's.

Current practice is being reviewed with the objective of producing guidelines on how to take human factors into account as part of the engineering design process. In doing this account will be taken of know-how and experience from nuclear, oil and gas and other process operations.

This Focus Group is being led by the Technical University of Berlin (TUB), in conjunction with ExxonMobil (Germany) and Snamprogetti (Italy) as End User Advisers.

SMALLER ORGANISATIONS

Whilst the human factors are clearly seen to be of value in large organisations an important objective of the project is to find the way in which human factors issues are addressed in smaller companies. From this it is hoped to establish what, if any, barriers prevent the application of good practice and to find ways to increase application. To do this contact has been made with a number of smaller companies through national associations of chemical manufactures, including the UK Chemical Industries Association.

Preliminary results indicate a much lower level of interest in the topic and less appreciation of the potential value of human factors. In addition many small organisations appear to be averse to obtaining help from consultants in this field.

CONCLUSIONS

At the time of writing the PRISM network is a little over half-way through its 3 year life. Although much work has still to be accomplished the following conclusions can be drawn.

- Human factors continue to contribute to accidents across all industries,
- Within the process industries improvements in hardware and management systems have lead to significant improvements in safety over recent years. Further improvements in these areas are likely to be much more limited.
- An improved understanding of Human Factors offers the opportunity for further significant benefits to safety.
- Leading companies in the process industries already show a high degree of interest in Human Factors and recognise the value it can provide in improving safety and business performance.
- This interest is shared by a number of smaller companies although the majority have still to be convinced of the value.
- New approaches to Human Factors, such as the assessment of safety culture and techniques to assist in the assessment of manning levels are proving to be of value to industry.
- The PRISM network is meeting its objective in providing an opportunity to share information and experience on Human Factors across Europe.

REFERENCES

1) Merans K, Fin R, Whitaker S, Benchmarking Offshore Safety Culture, Focus Group 1 Seminar, 2002, PRISM Website.

2) Joyner P, Towards a More Mature Safety Culture, Focus Group 1 Seminar, 2002, PRISM Website.

3) Lardner R, Amati C, Lee S, Safety Culture Marurity, Focus Group 1 Seminar, 2002, PRISM Website.

4) Byrom N, A Tool to Assess Aspects of an Organisation's Health & Safety Climate, Focus Group 1 Seminar, 2002, PRISM Website.

5) Lardner R, Teamworking & Safety, Focus Group 1 Seminar, 2002, PRISM Website.

6) Corpe J, Grange D, Effective Teambased Working with 'smateams',. Focus Group 1 Seminar, 2002, PRISM Website.

7) Whiting D, Minter N, Ruffle B, A Case Study in Team Working & Safety, Focus Group 1 Seminar, 2002, PRISM Website.

8) Fishwick, T, Culture & Organisatioal Factors, Focus Group 1 Seminar, 2002, PRISM Website.

9) Wright M, Involving Employees in Health & Safety: Practical Guidance, Focus Group 1 Seminar, 2002, PRISM Website.

10) Embrey D, A Consensus based Approach to Risk Management, Focus Group 2 Seminar, 2002, PRISM Website.

11) Bull P, Establishing Effective Training Arrangements, Focus Group 2 Seminar, 2002, PRISM Website.

12) Schaafstal A, Structured Troubleshooting: A New Perspective on Training & Maintenance, Focus Group 3 Seminar, 2002, PRISM Website.

13) Kassa O, Safety Critical Operations & Accident Barriers, Focus Group 2 Seminar, 2002, PRISM Website.

14) Hughes G, Issues in Task Design, Focus Group 2 Seminar, 2002, PRISM Website.

15) Conlin H, Assessing the Safety of Process Operation Staffing Arrangements, Focus Group 3 Seminar, 2002, PRISM Website.

16) Neerincx M, Passenier P, Envisioning High Demand Situations, Focus Group 3 Seminar, 2002, PRISM Website.

17) Grootjen M, Neerincx M, Passenier P, Cognitive Support for Ship Damage Control, Focus Group 3 Seminar, 2002, PRISM Website

18) Herbaux J, Alarm Management, Focus Group 3 Seminar, PRISM Website.

19) Hamilton I, Skelton M, Human Factors Integration for 24 Hour Response Business Process, Focus Group 3 Seminar, 2002, PRISM Website.

20) Wolff M, The International Space Station, Focus Group 3 Seminar, 2002, PRISM Website.

21) Logtenberg T, Identifying Critical Situations, Focus Group 3 Seminar, 2002, PRISM Website.

22) Cacciabue P, Survey of Human Relaibility Methods for Safety Assessment and Risk Management, 2002, PRISM Website.

23) The explosion and fire at the Texaco Refinery, Milford Haven, 24 July 1994, Health and Safety Executive, HSE Books1997.

24) The Explosion and Fire at the Texaco Refinery, Loss prevention Bulletin, Issue 138, December 1997, IChemE.

APPENDIX I

EXCHANGE OF INFORMATION AND EXPERIENCE

The value of the network lies the opportunity it provides to share experience in the field of human factors as well as the guides and reports which will be produced. This sharing of experience is being accomplished in a number of ways.

Plenary Meetings

There will be two plenary meetings of the entire Network membership during the project. The first plenary meeting was a two-day event was held in June 2001 at the CEFIC offices in Brussels. At this meeting, focus groups outlined their plans of work and organised breakout sessions in order to gain contributions and feedback from all network members. The conclusions from these sessions are being incorporated into the workings of the network.

The final plenary meeting will be held in Prague in June, 2004 as part of the International Symposium on Safety & Loss Prevention in the Process Industries.

Focus Group Seminars

Each of the Focus Groups plans to hold four meetings in total during the project. For each focus group two will be traditional "physical" meetings, eight in total for the whole project, and two will be on-line meetings, hosted at a PRISM web page.

The meetings are open to all who register with the network and the on-line meetings are open to anyone with access to the Internet.

Virtual Conference

A key planned dissemination activity will be a major international Internet-based conference on Human Factors. This event will be used to communicate good practice and roll-out the deliverables at the end of the project. Papers and documents will be down loadable from the conference web site and presenters will be available for on-line discussion. This event will be open to the worldwide online public.

EXPANDING THE MEMBERSHIP OF PRISM

It is the belief of all members of PRISM that work on human factors offers the opportunity for further significant improvements in safety in the process industries. To achieve this there is a need to discuss and share experiences, both good and bad, and find ways to overcome any problems. The network already contains some of the leading consultancies and research organisations in the field of human factors together with operating companies with safety records which are amongst the best in the world.

More members from the process industries, universities and consultancies will be welcome and it is hoped that the combination of meetings in different European countries, together with the use of the Internet will enable everyone with an interest in human factors to participate in one way or another.

More information on how to become a member of PRISM and be kept up-to-date on its future activities can be obtained from the PRISM web-site www.prism-network.org

LESS STRESS = MORE PERFORMANCE

Lardner, R CPsychol; & Wilson, D. MRSC CChem
The Keil Centre Ltd, Edinburgh & BP Grangemouth

Work-related stress is a significant occupational health issue. The UK Health and Safety Executive's guidance on work-related stress requires organisations to adopt a risk assessment / management approach to tackling stress at source. This paper describes an innovative employee involvement project, which used a simple stress risk assessment method to prevent or mitigate key work-related stressors associated with process plant commissioning. The project's straightforward approach to addressing work-related stress won a 2002 European Safety and Health Agency Good Practice Award for preventing work-related stress.

Work-related stress; risk assessment; risk management; good practice

THE SCALE OF WORK-RELATED STRESS

In 2000 the Health and Safety Executive announced its 'Securing Health Together' occupational health strategy[1]. 'Securing Health Together' aims to achieve by 2010:

- a 20% reduction in work-related ill health to both workers and the public
- a 30% reduction work days lost to work-related ill health
- an opportunity for everyone not working due to ill health to rehabilitate back to work or gain access to work as appropriate

An important aspect of achieving these targets is addressing the topic of work-related stress, as HSE estimates stress-related illness to be responsible for the loss of 6.5 million working days each year[2], with one in five workers reporting their job is "very" or "extremely" stressful[3].

RISK ASSESSMENT AND WORK-RELATED STRESS

Recent HSE guidance[4] recommends that the 5 risk assessment steps[5] be followed when tackling work-related stress. However, there are some difficulties in applying the risk assessment methodology, originally developed for physical hazards, to the examination of psychosocial hazards. A risk assessment for physical hazards is based on quantifying the hazard-harm relationship in order to gauge associated risk. Attempts to quantify this relationship for psychosocial hazards are problematic, as this relationship is influenced by individual differences. These difficulties can be illustrated through an example. Workload (potential hazard) may be perceived as positive and stimulating when at a certain optimum level, but can become a source of harm if too high, leading to feelings of stress and tiredness, or if too low, leading boredom and frustration. The main problem is that what is considered to be an optimum level will depend on the individual. Moreover, the harmful consequences of excessively high or low work demands may only be revealed after long periods of time, when the person's physical and mental health deteriorates, and may not be readily linked to the original hazard.

These difficulties have led some to argue that a risk assessment methodology is not a 'fruitful' method for the assessment of psychosocial risk. However, the opposing argument is that, compared to other stress management techniques, the risk assessment approach to stress is likely to be more effective, as the source is being addressed rather than the symptoms. The argument is that the influence of individual differences is not sufficient reason reject a risk assessment approach, but simply a significant factor to be taken into account when designing and implementing a stress risk assessment method. For a comprehensive review of the stress/stress risk assessment literature, an excellent free downloadable report is available[6].

WORK-RELATED STRESS AND OTHER RISKS TO BUSINESS PERFORMANCE
A stress risk assessment involves identifying the main work-related stressors (sources of stress) affecting workers, and taking steps to prevent or mitigate their effects. However, the potential benefits of tackling work-related stressors extend beyond benefits for individual psychological health. Work-related stressors are typically conceptualised as causing harm to psychological health. The degree of overlap between common work-related stressors and recognised root causes of accidents was therefore examined. This was achieved by comparing two root cause analysis models used in the UK offshore oil and gas industry[7,8] and HSE's human factors guidance[9] with the common work-related stressors. This established that approx. 70% of common work-related stressors are also potential root causes of accidents. For example, "lack of training to do my job" is a recognised work-related stressor. Not being properly trained is also a recognised potential root cause of accidents. Moreover, lack of adequate training could cause other types of harm to a business, for example leading to customer dissatisfaction, poor quality work etc.

Establishing a link between promoting psychological health, preventing accidents and managing other business risks is likely to increase the relevance and uptake of a stress risk assessment.

Figure 1 below indicates the degree of overlap between 40 common work-related stressors, root causes of accidents and other types of business risks.

CONTEXT OF THIS PROJECT
BP's Grangemouth petrochemicals-manufacturing complex employs around 2000 people. The complex stabilises crude oil and gas piped from offshore platforms, refines and exports crude oil, and manufactures petrochemical products. It is a major contributor to the UK economy, and is a COMAH site. The Technology Scale-Up Group is responsible for demonstrating the new ATC chemical process technology.

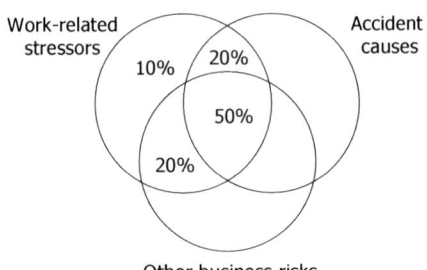

Figure 1. Different types of hazards

PROBLEM

During 2001, the site's Technology Scale-Up Group was about to commission an extension to the ATC demonstration plant. This involved expanding the operator team from five to fifteen. A new plant with a new control system meant that all the team were facing fresh challenges, especially those coming from outside pilot plant operations.

The commissioning was being conducted against a backdrop of site staff reductions, leading to considerable job insecurity. From previous experience, the ATC project team recognised that plant commissioning can be a very demanding time for all concerned, as the commissioning team strive to overcome the inevitable technical challenges. Moreover, the ATC team wanted to do their utmost to avoid the new technology they were developing being associated with anything less than an excellent health and safety record. The ATC team had already taken steps to eliminate physical hazards and improve safety culture. They now wished to take action to prevent avoidable stress by identifying and mitigating any work-related sources of stress arising from the commissioning project.

SOLUTION

The stress prevention project began by forming a team of 6 people, which included the project manager and representatives of the two main groups working on the project: the day support team of engineers, project leaders and chemists and the shift operations team who operate and maintain the plant. The aim was to involve a cross-section of the workforce in cooperating to prevent work-related stress.

This project team met for a briefing on the nature, signs and symptoms and causes of stress, provided by a local Chartered Psychologist. The team then brainstormed all the likely sources of stress that might arise from the commissioning project. These were compared to a set of 40 common work-related stressors, and this generic set was expanded to include the stressors specific to the project. The day and operations team members then worked separately to prioritise their stressors, as these may differ due to their distinct role and tasks.

The stressors were sorted according to whether they were (a) relevant to the project (b) currently well controlled (c) likely to cause stress. For each of the two groups, this yielded a set of "top five" stressors on which there was a consensus about their potential to cause harm. Examples of the "top-five" stressors from each group included high workload, job insecurity, and demands from others for unnecessary detail, and pressure from senior managers.

The team then worked together to complete the risk assessment process. They identified how or why each stressor caused harm, and shared ideas about what organisational and individual actions would mitigate the effects of the top-five stressors, and identified relevant, practical control measures. This involved discussing some very personal issues, such as the effects of job insecurity on other family members.

"It was a really open debate between all involved: day and shift teams" – Shift Technician

The proposed control measures were recorded and later shared and endorsed at a second workshop attended by the whole ATC team of 25 staff.

The most striking example of a stressor which was effectively identified and controlled, was "unnecessary detail". This stressor, specific to the demands of plant commissioning, referred to the effects of other people not specifying the amount of technical detail they required, and the timescales involved. As a result, staff worked long and hard to promptly produce detailed technical information, which was often not required.

At the second workshop, where the stress prevention project team's "top-five" stressors were shared and endorsed by the whole ATC team, a phrase was coined which became the watchword for dealing with unnecessary detail. This phrase – "the minimum requirements" – is now used by all team members to challenge others on the level of detail and deadlines attached to work they require. Adoption of the "minimum requirements" concept has had lasting benefits for managing workload amongst the ATC team. It has also changed how some team members deal with work-life balance issues. For example, Kenny Fraser is a shift technician with a young family, who at the time of the stress prevention project was also studying part-time for a chemical engineering degree. Kenny's degree design project was at a crucial stage, leading to considerable personal stress as he tried to reconcile the demands of work, family and study. For Kenny, the "minimum requirements" concept was a breakthrough:

"The minimum requirement did not just impact on one specific thing … it changed my thinking in respect of my university design project. I was getting into too much detail, and I realised I could take some pressure off myself by making assumptions and estimates, which then allowed me to make good progress, get a bit more sleep and feel better at work. I found it was a win-win…" – Kenny Fraser, Shift Technician

EFFECTIVENESS OF THE RESULTS

The stress prevention project was evaluated to judge its effectiveness by interviewing a sample of people from the ATC team, including the project manager, day and shift team members. These interviews focused on how the project had impacted upon their perceptions and personal experience of stress, and whether & how their behaviour or the behaviour of others had changed. Their quotations have been used to illustrate the effectiveness of the project.

In summary, this relatively simple, low-cost stress prevention project was conducted by as cross-section of employees, with minimal external input. The project's design and execution exceeds the requirements of UK legislation and regulatory guidance on preventing risks to health and safety arising from psychosocial hazards at work.

Framing stressors as a hazard to be controlled, just like the more familiar process and chemical hazards, was a logical extension to existing risk assessment processes, and opened a mature debate about otherwise delicate topics such as the effect of management style on others.

The project normalised discussion of stress and stressors amongst the team, and facilitated team spirit and open communication.

" It was enormously beneficial for team bonding, and discussing human issues, which we don't do often" – Ruth Robinson, Technologist

In the opinion of the external Chartered Psychologist who facilitated part of the project, this project is unique in (a) demonstrating considerable management foresight & leadership by taking action to prevent work-related stress, before any stress problem arose, (b) its use of a very simple yet robust employee involvement process added to existing risk assessment processes and (c) how lasting behavioural changes have been achieved in the ATC team, thus enhancing their existing health and safety culture.

COSTS / BENEFITS

The project costs were low. Apart from staff time, the main cost was external input from a Chartered Psychologist to explain the nature and symptoms of stress.

The project manager firmly believes that the project led to a deeper level of communication, sharing of feelings about work, and enhanced trust. This belief is also held by team members, for example:

"the whole team is now more open about confronting the issue of stress, and more likely to support and challenge colleagues under stress" - Technologist

"... my opinion is that it has helped to head off undue stress arising from the project – I'd firmly advocate that others do something like this – being proactive rather than reactive" – Stress Prevention Project Team Member

Others have commented on the benefits beyond the workplace:

"... one thing I found useful was highlighting the need to get the home / work balance right" – Shift Technician

Since the stress prevention project, the ATC team have had a perfect health and safety record, with no stress-related absence.

CONCLUSION

This project demonstrated that a proactive approach to anticipating potential stressors associated with a future events is possible, and how opening up a mature debate on such topics can bring benefits which extend beyond individual psychological health, as many work-related stressors are also potential accident causes or otherwise harmful to a business. The project's simplicity and low cost means this approach is particularly suited to SMEs, who may be constrained by shortage of time, money and internal expertise.

The project won a European Safety and Health at Work Agency 2002 Good Practice Award, which were based on the theme of "Preventing psychosocial hazards at work, especially stress". The awarding agency commended how successful the intervention had been in taking a preventative and holistic approach to stress at work at the beginning of the project. They said "... it shows how at the design stage, future hazards can be identified and removed or reduced. The effects should be sustainable. It also illustrates how employees can be involved in the risk assessment and management process. The cost was low and the method straightforward so this initiative would be appropriate for others including SMEs. The initiative also illustrates appropriate use of an external expert – the chartered psychologist – to assist the team in the risk assessment process"[10].

PRACTICAL STRESS RISK ASSESSMENT TOOLS

In conjunction with industry, the stress risk assessment method described, and two other stress risk assessment methods, have been developed into StressTools, a simple software package designed for use by non-specialists to identify and manage the risks associated with work-related stressors.

FURTHER INFORMATION

European Safety and Health Agency's Good Practice Awards http://osha.eu.int/ew2002
HSE's stress web-site – www.hse.gov.uk/stress
StressTools – www.keilcentre.co.uk

REFERENCE

1. HSE (2000) Securing Health Together : A long-term occupational health strategy for England, Scotland and Wales. HSE Books; Sudbury MISC 225
2. Health and Safety Commission (2001), *Strategic Plan 2001-2004* www.hse.gov.uk/ action/content
3. Smith, A; Johal, S, Wadsworth, E; Davey Smith, G & Peters, T (2000) *The Scale of Occupational Stress: the Bristol Stress and Health at Work Study* HSE Contract Research Project 265/ 2000
4. HSE (2001) Tackling work-related Stress: A manager's guide to improving and maintaining employee health and well-being HSG 218 HSE Books; Sudbury
5. HSE (1999) *5 steps to risk assessment* HSE Books, Sudbury INDG 163
6. European Agency for Safety and Health at Work (2000) *Research on Work-related Stress* ISBN 92-828-9255-7 downloadable free at http://agency.osha.eu.int/publications/ reports/203/en/index.htm
7. HSE (2000) *Factoring the Human into Safety* Offshore Technology Report OTO 2000/036
8. BP Amoco (undated) *Incident Investigation: root cause analysis training manual*
9. HSE (2000) *Reducing Error and Influencing Behaviour* HS(G)48
10. European Agency for Safety and Health at Work (2002) *Prevention of Psychosocial Risks and Stress at Work in Practice* ISBN 92-919-012-0

THE ATEX DIRECTIVES – A ROUTE MAP FOR COMPLIANCE WITH THE UK REGULATIONS

John Walkington – Senior Lead Consultant ABB Process Industries
Eric Gilchrist – Principal Consultant ABB Process Industries

WHO & WHAT IS 'ATEX'?

There are two ATEX ('Atmospheriques Explosives') Directives that are applicable to equipment in potentially explosive atmospheres and these will be implemented throughout the European Union after 30 June 2003.

The Explosive Atmospheres Directives (ATEX 95 & 137) are new legislation covering the requirements of employers to protect both staff and local communities from the risk of an explosive atmosphere. An explosive atmosphere can be one in the form of gases, vapours, mists or dusts, which can ignited under certain operating conditions by a source of ignition being, electrical, mechanical, static, hot surfaces, etc.

Directive 94/9/EC, known as ATEX 95, allows the free movement of goods throughout the EU by harmonising the technical and legal requirements for products that will be used in potentially explosive atmospheres. In the UK, its requirements are implemented by 'The Equipment and Protective Systems Intended for Use in Potentially Explosive Atmospheres Regulations 1996' (EPS). The Directive is supported by a set of guidelines published by the European Commission in 2001 to help with the interpretation of the Directive and there is a set of DTI guidelines on the implementation of the UK Regulations.

After 30 June 2003, which follows a six and a half year transition period, manufacturers and users of equipment, protective systems and safety devices will have to comply with this Directive. The directive applies to both electrical and mechanical equipment and protective systems for use in potentially explosive atmospheres. Also covered are components and devices for use outside potentially explosive atmospheres but which are necessary for, or contribute to, the safe functioning of equipment and protective systems in such atmospheres. Equipment already in use at 30 June 2003 may continue to be used providing the risk assessment required by Dangerous Substances and Explosive Atmospheres Regulations (DSEAR) indicates that it is safe to do so. However a piece of equipment certified before 30 June 2003 under superseded provisions, can be repaired using an identical part or with a new ATEX compliant component, which may not be identical, be still used without the need to bring the equipment into conformity again after this date, provided that the equipment is not changed substantially. What this means is that equipment placed on the market before the Directive came fully into force is not caught by the Directive unless it is changed into a "new" product by substantial modification.

This Directive means that manufactures in the European Union (EU) member states are, for the first time, set to work to common agreed standards. For products, which are in the higher risk categories, for example where an explosive atmosphere is likely to be present for a considerable periods of time, or where the consequence of an explosion would be particularly severe, the manufacture will ask a Notified Body to conduct the EC type Examination procedure to ensure that the design of the product conforms to essential health and safety requirements. These Notified Bodies (NB) are not peculiar to ATEX. EU

Member States are responsible for notification of bodies to carry out tasks relating to the conformity assessment procedures for all "new approach" Directives, which includes ATEX. Therefore from 30 June 2003 products could have been certified by any of the Notified Bodies appointed by the EU Member States and therefore not only the three UK appointed Notified bodies we have been used to. So it will be important that any certificates and any associated documents are available in the language of the country in which it is to be used. This will be especially relevant for products designed for lower risk situations principally, category 3, where manufactures are entitled to self declare conformity to the Regulations.

EN 13463-1 "Non – electrical equipment for potentially explosive atmospheres, Part 1: Basic method and requirements" was published in January 2002. The standard requires the manufacture to carry out a formal documented hazard analysis that identifies and lists all of the potential sources of ignition by the equipment and the measures to be applied to prevent them becoming effective. If equipment is designed and constructed according to good engineering practice and the ignition hazard assessment ensures that the equipment does not contain any effective ignition sources in normal operation, the equipment can be classified as Category 3 equipment solely on the basis of this standard. Therefore as this standard requires the manufacture to provide a copy of this ignition hazard assessment they have carried out on their equipment to the user this will help the user in the preparation of their 'risk assessment document' called for by DSEAR. For older equipment or equipment imported directly by the end user from outside the EU member states this ignition hazard assessment may need to be carried out by the end user.

Directive 99/92/EC, known as ATEX 137, ensures that workers enjoy a minimum level of protection from potentially explosive atmospheres. In covering safety and health protection of employees, it places duties on the employer. The employer should demonstrate that explosion risks have been determined and assessed with places classified into zones. Appropriate management systems should be in place including training of workers and control of work. The workplace and work equipment, including warning devices, should be designed, operated and maintained with due regard for safety. In the UK the requirements under this directive are implemented in the 'Dangerous Substances and Explosive Atmospheres Regulations' (DSEAR). These Regulations came into force on the 9 December 2002 with the ATEX part coming into force on the 1 July 2003. An Approved Code of Practice will support these Regulations, which is due to be published early in 2003.

Workplaces used for the first time after 30 June 2003 and also all modifications to existing plant after this date should comply. Existing workplaces already in use at this date will also need to comply no later than 30 June 2006.

Essentially the Directive calls for the following activities to be undertaken and documented accordingly:

♦ Risk assessments for operating plant are to be made up to date covering hazard identification and likelihood of risk potential.

♦ Zoning of plant to the new categories has been duly undertaken.

♦ Identification of zoned areas and contents of pipes and containers are adequately labeled.

416

♦ Emergency and incident handling mechanisms are in place and understood.
♦ Use of appropriately ATEX 'CE' marked equipment following the end of June 2003 is installed maintained and repaired in accordance with the Product Directive 94/9/EC requirements.
♦ Operations staff are fully informed and trained in all of the above!

DANGEROUS SUBSTANCES AND EXPLOSIVE ATMOSPHERES REGULATIONS (DSEAR)

UK regulations implementing both the flammable/explosive elements of both the Chemical Agents and ATEX 137 Directive will come into force during 2002 and will underpin the June 2003 ATEX Directive enforcement date.

The essence of DSEAR is to protect the safety of employees and others from those dangerous substances that can cause explosions or fires during the working day. The new DSEAR regulations will deal with explosions and fires that are caused by dangerous substances that generate explosive atmospheres whether they are gases, vapours or dusts. The new 17 part regulations will replace certain older UK safety legislation in the process e.g. The Highly Flammable Liquids & Liquefied Petroleum Gases Regulations 1972.

The new Regulations will augment the existing Health And Safety at Work Act requirements for employers to provide competent management and personnel as well as policies and procedures for correct maintenance activities.

A major consequence is the requirement for all existing hazardous areas to be re-assessed for safety risk and for mechanical sources of ignition and heat energy to be identified and added to the existing instrument / electrical sources.

With the advent of DSEAR there are new additional constraints for the User on how he identifies process hazards and operational risks, designs (with particular inclusion of formal risk assessments), maintains and repairs the hazardous area equipment in service. Many operators will have already covered hazardous areas for instrument / electrical equipment, but there is now a need to risk assess the mechanical equipment as well.

Similar in nature to the key attributes of ATEX, DSEAR covers the following main topic areas:

♦ The identification of dangerous substances used within the operation and the inherent safety risks associated with the operation of the plant processes.
♦ The requirement of a formal risk assessment to be carried out prior to the introduction of any equipment, protective systems or components into potentially flammable atmospheres.
♦ Safety measures are taken to eliminate, control and mitigate the identified risks.
♦ Area classification and zoning of the operating plant is undertaken.
♦ New requirements for the design and project process of hazardous area installations.
♦ Marking and Signage appropriate to the zoning category is put in place.
♦ Consideration to the issue of non-static clothing to operations staff working within these zoned areas.

♦ A person competent in explosion protection undertakes verification of new installations before use.
♦ Information is to be generated and made available to deal with incidents, accidents and emergencies.
♦ Full instructions and training is given to staff including actions to be taken in the event of an emergency.

SO HOW DO I COMPLY?

ABB address this question by offering to outline one approach to compliance covering the requirements of the DSEAR regulations. In most cases the implications for industry are not too onerous. Responsible operators with mature safety management policies and systems in place to comply with current legislation will be well placed to offer evidence and justification in support of DSEAR requirements.

KEY AREAS OF CONSIDERATION WILL BE:

♦ The majority of the existing guidance to hazardous area management will still be valid from the safety viewpoint and hence there should be no need to start from scratch with regard to DSEAR compliance.
♦ The view from industry that complying with ATEX in a logical structured way is not a huge burden in terms of cost.
♦ Current zoning calculations may need to be reviewed because of the statutory link to EPS 1996 and the selection of equipment (Particularly Mechanical items).
♦ The current documentation under the Management of Health and Safety at Works Regulations 1999 (MHSW) may need revising in line with DSEAR wording. This should be seen as not too much of a change to the existing documents, however the requirement for formal area classification (risk assessments) and zoning diagrams will now need to be a formal part of the existing safety management procedure set.
♦ It is expected that the required 'Risk Assessment document' will not require a re-write of the safety documents, which already exist for the plants. It should mostly contain the references to the whereabouts and mapping interpretation of the DSEAR requirements to the existing current safety management system documentation.
♦ Control of flammable atmospheres will need to be very evident within the supporting safety management and risk assessment documentation.

THE ABB STEP APPROACH

DSEAR will apply to a wide range of businesses. Business premises will normally include all industrial and commercial premises, (Note - offshore facilities and domestic dwellings are excluded in the Regulations) where a dangerous substance is present or is liable to be present during the working day. Here ABB offer as part of their project management, consultancy and equipment methodologies a step approach for compliance.

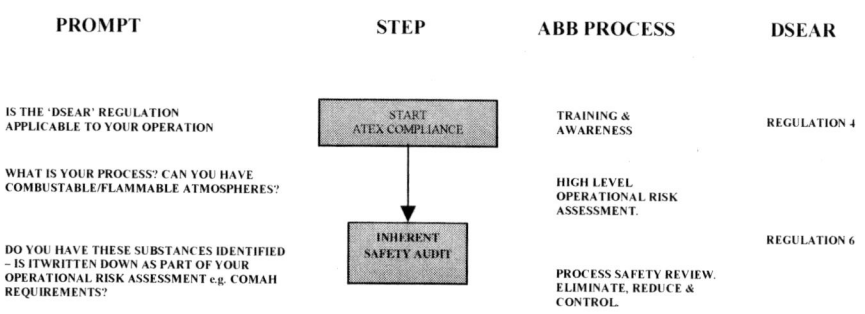

PROMPT	STEP	ABB PROCESS	DSEAR
IS THE 'DSEAR' REGULATION APPLICABLE TO YOUR OPERATION	START ATEX COMPLIANCE	TRAINING & AWARENESS	REGULATION 4
WHAT IS YOUR PROCESS? CAN YOU HAVE COMBUSTABLE/FLAMMABLE ATMOSPHERES?		HIGH LEVEL OPERATIONAL RISK ASSESSMENT.	
DO YOU HAVE THESE SUBSTANCES IDENTIFIED – IS IT WRITTEN DOWN AS PART OF YOUR OPERATIONAL RISK ASSESSMENT e.g. COMAH REQUIREMENTS?	INHERENT SAFETY AUDIT	PROCESS SAFETY REVIEW. ELIMINATE, REDUCE & CONTROL.	REGULATION 6

STEP 1

♦   Identify and assess the fire and explosion risks of dangerous substances used within the operating plant. Note the Chemicals Hazard Information and Packaging for Supply Regulations 2002 (CHIPS) will help in this process as this will automatically indicate the substance as assessed is a dangerous substance under DSEAR. Properties of materials are to be agreed and documented.

♦   Apply safety measures to eliminate or reduce the risks from the use of these substances to be as low as reasonably practicable. Dusts will need to be part of this process as they have previously not been considered in as much detail as gases and liquids until now. Note that the elimination process has not been as emphasised in existing legislation and forms a crucial part of this risk assessment process. Reduce risks further by control and mitigation measures (Often involving substitute substances, with say a higher flashpoint, where possible).

♦   (Overall aim) To ensure that employees and public are protected from fires and explosion. Remember this is the real detail for the high level steps.

Once the safety measures are understood and no further improvements can be made to the operation the plant can be put forward on this basis of safety for an area classification risk assessment meeting. However bearing in mind that any changes made to the operation as part of the inherent safety review in themselves do not introduce other or increase the potential safety risks to the plant.

STEP 2

♦   Undertake an area classification risk assessment meeting of the plant operation to prevent and provide protection against explosions covering control / electrical and mechanical potential sources – including friction or heating, presence of foreign bodies and static discharge.

| PROMPT | STEP | ABB PROCESS | DSEAR |

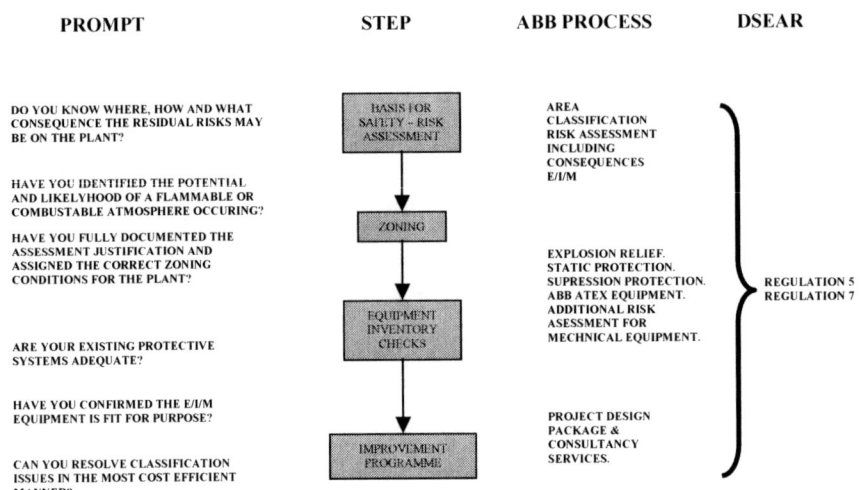

- Prevent formation of explosive atmospheres, avoid ignition of explosive atmospheres, and mitigate effects of explosion so as to ensure health & safety of staff and others.
- Take certain steps in addition to the above requirements so that overall precautions for one explosion in one factory cannot be seen to affect another business nearby.
- Classify the operating plant into zones with corresponding equipment categories.
- Generate design change or improvement programmes for the plant to decrease the zoning category so as to make installation, inspection and repair costs as effective as possible.
- Select equipment to EPS 1996 Regulation requirements under ATEX 95.
- Before first use of workplace – verify the installation safety case and design process with a competent person – normally as part of the commissioning checklists and seen as a trade-off from excessive detail and what is required for safety.
- Co-ordinate safety controls and measures in shared workplaces.
- Provide operating and safety information, instructions and training to employees including information to ensure safe working in operations and maintenance.
- Provide systems and procedures for emergencies - Explosion relief, static controls, fire detection, procedures, for emergency services and response processes in place). Note this should be in conjunction with the relevant fire authorities that should contribute and have ownership of the requirements.

Once agreed and the risk assessment and design package phase is completed with the plant put into service, the ongoing maintenance, inspection and repair attributes of the regulations are required.

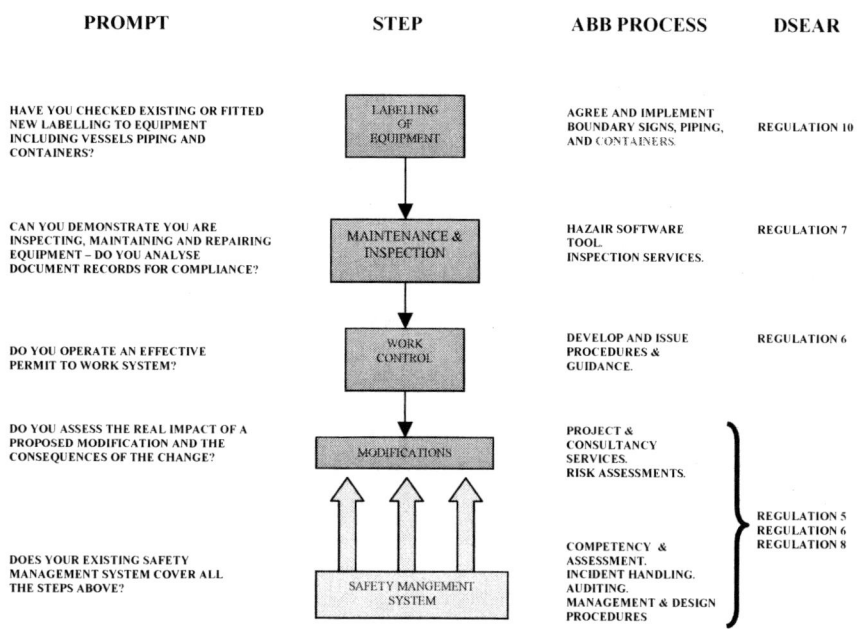

PROMPT	STEP	ABB PROCESS	DSEAR
HAVE YOU CHECKED EXISTING OR FITTED NEW LABELLING TO EQUIPMENT INCLUDING VESSELS PIPING AND CONTAINERS?	LABELLING OF EQUIPMENT	AGREE AND IMPLEMENT BOUNDARY SIGNS, PIPING, AND CONTAINERS.	REGULATION 10
CAN YOU DEMONSTRATE YOU ARE INSPECTING, MAINTAINING AND REPAIRING EQUIPMENT – DO YOU ANALYSE DOCUMENT RECORDS FOR COMPLIANCE?	MAINTENANCE & INSPECTION	HAZAIR SOFTWARE TOOL. INSPECTION SERVICES.	REGULATION 7
DO YOU OPERATE AN EFFECTIVE PERMIT TO WORK SYSTEM?	WORK CONTROL	DEVELOP AND ISSUE PROCEDURES & GUIDANCE.	REGULATION 6
DO YOU ASSESS THE REAL IMPACT OF A PROPOSED MODIFICATION AND THE CONSEQUENCES OF THE CHANGE?	MODIFICATIONS	PROJECT & CONSULTANCY SERVICES. RISK ASSESSMENTS.	
DOES YOUR EXISTING SAFETY MANAGEMENT SYSTEM COVER ALL THE STEPS ABOVE?	SAFETY MANAGEMENT SYSTEM	COMPETENCY & ASSESSMENT. INCIDENT HANDLING. AUDITING. MANAGEMENT & DESIGN PROCEDURES	REGULATION 5 REGULATION 6 REGULATION 8

STEP 3

♦ Provide policies procedures and guidance information to ensure safe working in operations and maintenance.

♦ Ensure adequate training is given to all relevant staff including a programme of future refresher training. Consideration should be given to recognised competency schemes such as COMPEX training for installation and inspection tradesmen.

♦ Mark zoned areas with Ex sign where possible – Only where necessary – Provide suitable signage on the main access thoroughfares to a zoned area and identify and label piping and containers that contain recognised dangerous substances. Particularly review the permit to work details to ensure hot work such as welding is reviewed in light of any new changes.

♦ Update the general zoning site map from the specific project documentation map – reclassify the area for specific zones.

♦ Implement a programme of equipment inspection and repair. This should be documented to provide reporting analysis on equipment performance.

♦ The management of change procedures will need to be changed to augment the need for compliance with DSEAR.

♦ Ensure there is a clear policy communicated to staff for the reporting and remediation of accidents, incidents and emergencies including a programme of regular practice.

CONCLUSION

THE PRACTICAL REALITY:-

♦ It's going to be Law.

♦ For organisations who already have mature systems conforming to the requirements of MHSW, two thirds of the existing plant operational documentation should be available to demonstrate compliance to DSEAR.

♦ It could cost end users a lot of money if not properly implemented. Especially on the amount of new labelling required when considering piping and vessels.

♦ Mechanical risk assessments are now required – Implementation of forthcoming standards and learning will take a little time to bed in.

♦ Equipment for mounting in Zone 1 or 0 could become a lot more expensive – (Particularly for Mechanical Items!).

♦ An explanatory mapping document (Risk Assessment Document) describing how the ATEX Directive has been complied with is required.

REFERENCES

The Dangerous Substances and Explosive Atmospheres Regulations 2002.

The Chemicals (Hazard Information and Packaging for Supply) Regulations 2002.

The Equipment and Protective Systems Intended for Use in Potentially Explosive Atmospheres Regulations 1996.

The Management of Health and Safety at Works Regulations 1999.

Hazardous Area Inspection and Repair Software (HAZAIR).

DSEAR; EARLY EXPERIENCE OF IMPLEMENTING THE NEW REGULATIONS, CONTROLLING THE STORAGE AND USE OF DANGEROUS SUBSTANCES

Alan Tyldesley
Principal Specialist Inspector, Health and Safety Executive, Bootle, Merseyside

The Dangerous Substances and Explosive Atmospheres Regulations were made in October 2002, to implement the fire and explosion aspects of the Chemical Agents Directive, and the Explosive Atmospheres (ATEX) user directive. They have allowed the most extensive changes to the health and safety law relating to flammable substances in 30 years. Extensive supporting guidance has been provided by HSE, but most of the significant new requirements are not yet fully in force, and their impact is not yet clear. This paper discusses issues that havearisen during the development of the new regulations and supporting guidance.

Hazardous area classification, dangerous substances, ATEX, notified bodies

INTRODUCTION

Ever since the Health and Safety at Work Act was passed, and HSE was established, in 1974/5, there has been a planned programme of replacement of health and safety law. Two principal driving forces were set out by the Robens report that preceded the HSW Act; first that health and safety law should apply to all places of work, and not just factories, and secondly that modern legislation should be of a goal setting type, not providing prescriptive solutions that could be rapidly overtaken by technological progress. Earlier attempts to update the law relating to storage and processing of flammable materials have foundered, for reasons that need not detain us now, but new regulations, titled Dangerous Substances and Explosive Atmospheres Regulations (DSEAR) 2002 represent the last big package of revision of pre 1974 legislation, made under the Factories Act 1961.

ANOMALIES OF OLD LEGISLATION

Taking a broad view we would have to say that there certainly are some curiosities about the law to be repealed. Why should you need a licence from HSE to store Calcium Carbide, but not other products with similar hazards? The minimum flash point of solvents used for dry cleaning has been specified precisely since 1949, but the regulations take no account of the manner in which the solvent will be used. Given the range of standards for ignition protected electrical equipment, is it really necessary to keep an absolute ban on installing a motor within a ventilation duct drawing flammable vapours from a process?

External events of course have shaped our scope for action, and almost all new H & S legislation has its roots in European law, and DSEAR is no exception. Not all our European colleagues are quite as steeped in the risk assessment approach as HSE, and so the directives that come from DG Employment, the EU commission department that deals with H & S law, sometimes have more specific requirements in than we might wish, and the EU parliament also has a significant voice in drafting legislation.

So this paper will look at some of the issues that have caused most discussion during the drafting of the regulations, the supporting Approved Codes of Practice, and as a consequence of the consultation phase. In particular, these have been the formal requirement for hazardous area classification to be applied in areas where it has never been done before; the interaction of the ATEX product directive and DSEAR; the hierarchy of control principles and the changing responsibilities of the local authority Petroleum Licensing Officers.

HAZARDOUS AREA CLASSIFICATION

Hazardous area classification (HAC), as a technique has been round for a very long time, with a gradual development of the principles, both for zoning, and for the construction of ignition protected electrical equipment over the last 50 years. So in a real sense, the law this time is catching up with good practice from the chemical and oil industries, and laying down that the same framework should be used more widely. In doing so, it has brought to light assorted grey areas, where the logic for our approach is not as rigorous as we might like, and has exposed some inconsistencies in earlier advice.

Hazardous area classification began as a system for deciding where fixed electrical equipment needed special designs to control the ignition risk, and we are all familiar with the terms flameproof, and intrinsically safe, even if they are sometimes used incorrectly. It is obvious that handheld, or mobile equipment can also create a risk, and many sites place controls over the use of vehicles, cameras, grinding equipment and power saws around their sites, because they need to control ignition risks. Overcautious zones and a strict application of the rules about what can be used where may seem to lead to anomalies. An everyday example is the petrol filling station, where people drive into one side of a pump, while another vehicle is filling up on the other side. We build the pumps to a high standard, they will be ATEX category 2 in future , but cannot be too surprised that some customers fail to understand why should not use their mobile phones, leave the car radio on, or fill up a unsuitable containers! In the same sort of way, we have in the past specified LPG cylinder stores as zone 2, but have been willing to allow in normal delivery vehicles, full of ignition sources.

MINIMUM QUANTITIES

Considerable unease has been expressed by those new to HAC, who have read the regulations, and wondered if they will have to fit explosion protected lighting in their shops or warehouses. It has not helped that almost all technical committees who have looked at the problem, have shied away from discussing the issue of package sizes or minimum quantities; consequently we have been asked 'do I still need to assign zones where quantities of material liable to be spilt will be very small?'

Any attempt to pin this down has to be surrounded by caveats, as we know that even 100g of petrol, evaporated and dispersed inside a fuel tank can kill someone foolish enough to apply a welding torch to the outside. Similarly, there is a world of difference between storing a few small tins of acetone, and using an acetone rag to clean up sticky polymer on an item of machinery. You might not formally draw up a zoning diagram for places where small quantities are stored but the need to handle it safely and avoid igniting the vapours is the same.

AREA CLASSIFICATION FOR THE DUST HANDLING INDUSTRIES

The extension of the concepts of hazardous area classification to dust handling plant creates some tricky problems, and sometimes it seems like a complication that will do little to improve plant safety. If we look at the issue of dust clouds, and explosions, we find from any examination of the incident history, that almost all explosion incidents start within the handling system. There are some exceptions, for example, the General Foods incident at Banbury in 1981 started with bursting of a conveying system caused by pneumatic pressure, and a recent fatality in the chemical industry started with a fire under a Flexible Intermediate Bulk Container containing a plastic additive. This caused the bag to melt and fail so that a large dust cloud escaped, which ignited and engulfed an operator nearby.

Faulty or unsuitable equipment does ignite dust clouds causing explosions, and there are many examples from grinding, blending and conveying processes. We cannot however, make high speed powerful equipment as safe as 'ia' intrinsically safe electrical equipment, and so regardless of whether the inside of your plant is described as zone 20 or 21, additional protective features are likely to be needed.

Fires are of course much more common than explosions, and can be just as devastating in the damage they cause, but the risks to people are quite different. Fires involving dusts in layers commonly grow quite slowly at first, and are likely to give those in the premises time to tackle the incident, if only someone is alerted in time. The consequence of this, is that we need to ask ourselves, how we should zone places where dust layers may collect over an extended period, yet no dense cloud ever exists, unless perhaps a primary explosion rocks the building? By this time, worrying about the control of ignition sources is a bit late.

These sorts of arguments have been rehearsed in various committees, and the footnote to the zone definitions found in DSEAR was the result of a last minute compromise by those who were writing EN 1127 part 1, 'Explosion Prevention and Protection, basic concepts and methodology'. The same issue came back during the writing of the European dust area classification standard, EN 50281 part 3, and has been settled by the text in annex B.

My own view is that you should decide if there is a need to prevent fires as well as explosions. If so, then no matter whether you choose to describe a place where dust layers form, but dust clouds are unlikely as zone 22 or unclassified, the equipment selection criteria are essentially the same. With the long delay in availability of anything published elsewhere in English, HSE is intending to plug the gap, with some industry specific leaflets, but it is, for instance, difficult to pitch something for the woodworking industries correctly for both a large chipboard factory, and a small jobbing joinery shop.

SOPHISTICATED OR SIMPLE APPROACHES TO AREA CLASSIFICATION?

A recurring comment from the consultation exercise on DSEAR, was that existing guidance on hazardous area classification was written from the standpoint of high hazard industry, and was difficult to apply to lower hazard industries. There is some validity in this, and back in the 1980s, the complexity of the topic was recognised, and a more rigorous analysis made clear a need for work that cut across the boundaries of individual engineering disciplines. The consequence was a unique cooperative exercise between the I Chem E, the I Mech E, the I Gas E and the I.E.E which resulted in the book by Cox, Lees and Ang. No one would claim that this was easy reading for a small company, worried whether to make their process area zone 1 or zone 2. For them a much cruder system of classification may be appropriate, while recognising that this might in turn throw up difficulties.

Much more recently in Aug 2002 the Institute of Petroleum launched an update of their code on the subject, which provides sophisticated risk assessment arguments to produce hazard radii from release sources, which may be used to set zone boundaries. A particular advance is the methodology provided to address releases from high pressure sources. The driving force for a new approach to the topic was a wish to move away from broad brush zoning, to something more closely linked to the reality of the situation. Potentially there are significant capital equipment cost savings, at the expense of more detailed analysis at the outset. However, some of the low hazard industries want to specify no zoned areas, but also avoid the costs of paying anyone consultancy fees to justify their position. Perhaps sector specific advice could help here, but the range of special groups looking for help is rather long.

CHEMICAL LABORATORIES

Area classification for chemical laboratories was a particular special case raised during consultation. Here we are faced with some uneasy realities. A huge range of electrical laboratory equipment is supplied in a form that is not ignition protected, and some of it is of considerable size. Heating mantles for glass reactor vessels are available up to 50 l in size, ovens for drying off samples can easily be up to 200 l in size. Smaller scale glass equipment can be very fragile, which laboratory worker has never broken a glass tap, or pipette?

No one wants to zone university laboratories where the cost of chemicals looms large, and all work is done at less than 10 g scale. Where however, do we draw the line as the scale goes up? Should I be allowed to distil 5 or 10 l of ethyl acetate on the open bench, from an isomantle? How do I make a system handling unstenched hydrocarbon gases, supplied from a high pressure cylinder, via plastic or rubber hose, sufficiently certain not to leak to justify no zones? If I carry out my larger experiment within a fume cupboard, will the forced draft encourage fire growth if a liquid spill is ignited? Would a large spill flow out of the front in any case? Current guidance from the Royal Society of Chemistry is of limited help in coming to a sensible basis for justifying either a policy of no hazardous areas, or defining where these should be specified.

INTERACTION BETWEEN THE TWO ATEX DIRECTIVES

The interaction between the ATEX equipment directive (EC/94/9) and DSEAR exposes the difference in origins and purpose between the two pieces of European legislation. The product directive concerns the single market, and the need to remove safety issues as an excuse for barriers to trade. Consequently it gives countries little flexibility in implementation. If a product meets the defined safety criteria, it must be acceptable for sale anywhere in Europe, and conversely, if the product does not meet the criteria, it is not acceptable.

To complicate matters, the product directive applies not only to the act of placing something on the market, but also to putting it into service. This has some sense, as it makes clear that as an end user , you cannot sidestep the safety standards, either by direct import of substandard equipment from the outside the European Economic Area, or by building it for your own use. There is no element of risk assessment here, nor an option of preventing the explosion risk by other means.

In contrast, European health and safety legislation specifies minimum standards for worker protection; firstly to give fair weight to the wishes of working people to have a safe workplace, and secondly to prevent industry in one country undercutting another, by permitting low safety standards. Individual countries have the right to enact or retain stricter law if they wish. DSEAR does have built in the concepts of risk assessment, and in particular the wording of schedule 3 allows some flexibility in the otherwise mechanistic link between ATEX equipment category, and the zone in which it will be used. We have debated whether this should only allow ATEX equipment of a lower category to be used (e.g. category 3 in a zone 1), whether it extends also to using safe but not ATEX certified equipment in a zone 2. In fact there are so many and varied possibilities , it seems unwise to be too dogmatic about this.

NOTIFIED BODIES

The European Commission recognises that the whole single market rests to some degree on trust. Can we trust self certified products to meet all the relevant safety criteria, or when third party testing is specified, will all the test houses adopt similar standards? Without this trust, even if legislative hurdles preventing free trade are removed, commercial considerations and bias will remain to distort the market. This is of course not simply an ATEX question, and similar legislation in other fields has spawned a huge growth industry of testing.

There are currently about 1000 notified bodies, test houses with the right to issue certificates of conformity, and this number is set to rise to 1250 when the construction products directive is fully in place. The consequence of this is pressure from Brussels, first to look at the competence of notified bodies some of whom actually have very little work, and secondly to press countries to report back about the nature and amount of enforcement activity they undertake for this single market legislation.

No one wants to see a league table of competence among such notified bodies, but this is a touchy subject for many countries. Equally touchy for many are the systems they have for enforcing this product legislation. HSE has the responsibility for enforcing the ATEX equipment rules, but there are many other competing pressures for inspectors' time. Experience shows that taking effective action against substandard equipment made outside the UK takes considerable time and a strong case. We need to be clear that the resources involved bring safety benefits.

A particular concern is the difficulty of expecting notified bodies to test equipment where there is no previous tradition of independent testing in most countries, so available test equipment and expertise is very limited. This is the case for explosion suppression systems, dust explosion barrier devices, explosion vent panels and doors, and probably other items classed as autonomous protective systems.

HIERARCHY OF CONTROL MEASURES

Regulation 6 sets out a hierarchy of control measures, none of which, of themselves are contentious. So employers must look at reducing quantities of dangerous substances; minimising releases; controlling releases at source; preventing formation of explosive atmospheres; avoiding ignition sources; avoiding adverse conditions that create danger; and segregating incompatible substances.

Many people can see situations in which some of these steps are not possible, or when taken to a logical conclusion would increase the risk. For example, if your business is selling fuel, you cannot sensibly avoid storing some, and if you store less, that may mean more deliveries and more transfer operations. Similarly, you cannot avoid releasing material if your business is painting by spray gun or brush. You can control the release at source by local exhaust ventilation, but new hazards are created if the exhaust stream is fed to a thermal oxidiser instead of dispersing through a high vent stack.

Preventing the formation of explosive atmospheres might seem a good idea, but if this is done inside process plant by displacement of air with inert gas, new hazards arise if workers ever need to enter the enclosed space.

Controlling ignition sources is all very well, but clearly does not apply to plant for controlled combustion, and in this case if too much fuel gas is released before the ignition source is applied an explosion may follow.

So we can find exceptions and arguments against all these types of control measures. That does not make them invalid, and all we are really providing is a framework for the risk assessment. Can you replace your cleaning solvent with something safer? Can you make the LEV on your paint booth more effective, reducing not only the safety risks but the health hazards? Can you prevent the formation of an explosive dust atmosphere, not by inert gas, but by using a paste or pelletised product?

So, whatever the wording of the actual regulation, it is often going to be the case that all the options need to be properly considered, and that within a single plant different options may be appropriate at different locations. At least we do now have a legal framework that is comprehensive.

PETROLEUM LICENSED STORES

The old Petroleum Consolidation Act of 1928 used a system of licensing to ensure safe keeping of petrol and other products deemed as similar. This did not extend to use of the product, and this was highlighted by Flixborough as that site needed a license to store cyclohexane, but the large amounts in the process were not subject to the licensing.

Licensing officers employed in the main by the local authorities look after conditions at more than 20,000 bulk or container stores for petroleum spirit, and a further 12,000 licenses are issued for dispensing petrol at non retail sites.

It is clearly possible to argue that it is more efficient to have a single body enforce all aspects of health and safety at a single site, and HSE doubted that a licencing regime was appropriate for simple storage sites. As with the ATEX product legislation, finding the correct enforcement regime is an issue of resources: how much safety benefit is there from a given level of activity?

There are no absolute answers to this question, it is really a matter of political judgement. In a slight change from the proposals HSE presented in the consultation document, the Health and Safety Commission decided that for the time being licensing would remain for workplace dispensing of petrol. This reflected concerns about the safety of the public at retail sites, and evidence of poor standards at some non retail sites with very low throughput. The non retail group includes some rather diverse premises, from taxi firms and farmers, through to vehicle builders who put a gallon of petrol in cars coming off the production line. Storage of petroleum spirit in containers will however no longer be subject to a licensing regime, but covered by DSEAR in the same way as other substances.

CONCLUSIONS

As the requirements that come directly from the ATEX user directive, such as area classification of new plant do not come into force until June 2003 for new plant, and transition arrangements extend this for existing plants, it really is rather early to assess the impact of the regulations, except where diligent employers wishing to be ahead of the game have already raised questions with HSE. The targets set by the Government on HSE, in the Revitalising Health and Safety programme have caused HSE to concentrate on issues that create the most accidents and ill health, and fire and explosion hazards are not one of the priority programmes. Consequently these new regulations will have to take their place alongside other priorities when your inspector next calls.

It would be good to think that this was the end of the change process, but a major review of all the fire legislation is under way, and a consultation exercise from the Office of the Deputy Prime Minister was undertaken during 2002. A prime objective is to make the legislation simpler to understand, so that all involved may concentrate their efforts to best effect. HSE is actively involved in the review.

EXPLOSION SAFETY DOCUMENT FOR THE ATEX 137 DIRECTIVE – NEW NAME FOR A FIRE AND EXPLOSION HAZARD ASSESSMENT?

Dr. Richard L. Rogers, Dr. Bernd Broeckmann, Nigel Maddison*
INBUREX Consulting GmbH, Hamm, Germany,
* INBUREX (UK) Ltd, Glossop, Derbyshire

This paper describes the requirements contained in the ATEX 137 Directive which includes an obligation of an employer to prepare an Explosion Safety Document. A risk assessment based approach is described whereby the potential hazards are identified on the basis of an area classification exercise to determine the extent and occurrence of potentially explosive atmospheres. This is followed by the selection of appropriate equipment and the identification of possible ignition sources arising from the plant operations. The requirements of ATEX 137 will be discussed and compared to the traditional fire and explosion hazard assessment currently carried out by many firms where potentially explosive atmospheres are likely to be present.

Explosion protection document; ATEX 137, DSEAR, Explosive atmospheres, Area classification; Ignition sources

INTRODUCTION

In December 1999 the European Parliament and Council agreed and passed the so called ATEX 137 Directive or Directive 99/92/EC which sets out minimum requirements for the safety and health protection of workers potentially at risk from explosive atmospheres. It is the 15th individual Directive within the meaning of Article 16(1) of Directive 89/391/EEC, the base Directive setting out measures to improve the health and safety of Workers, and must be enacted in the member states by the 30th June 2003.

For the purposes of this Directive, 'explosive atmosphere' means a mixture with air, under atmospheric conditions, of flammable substances in the form of gases, vapours, mists or dusts in which, after ignition has occurred, combustion spreads to the entire unburned mixture. The Directive does not apply to:

- areas used directly for and during the medical treatment of patients;
- the use of domestic gas burning appliances;
- manufacture, handling, use, storage of explosives or chemically unstable substances;
- mines or mineral-extracting industries;
- means of transport by land, water and air however, means of transport intended for use in a potentially explosive atmosphere are included.

The Directive sets out a set of obligations of the employer and includes two Annexes, the first provides the definitions for the classification of places where explosive atmospheres may occur – the so called 'Zone Definitions'. The second Annex specifies the minimum requirements for improving the safety and health protection of worker potentially at risk from explosive atmospheres and includes the criteria to be used for the selection of equipment and protective systems in the different Zones.

The ATEX 137 Directive is the 'User' Directive which corresponds to the ATEX 95 Directive (Directive 94/9/EC) for manufacturers which sets out essential health and safety requirements for equipment for use in potentially explosive atmospheres.

EMPLOYER OBLIGATIONS OF DIRECTIVE 99/92/EC

The Directive sets out specific obligations that the employer has to fulfil as follows

- Prevent and protect against explosions.
- Carry out an assessment of the explosion risks.
- Ensure safe working conditions including the provision of instructions, training supervision and technical measures.
- Duty of Coordination subcontractors/visitors.
- Classify the areas where an explosive atmosphere may occur in to Zones including where appropriate the marking of entry points into such areas.
- Select appropriate equipment.
- Prepare an explosion protection document.

The requirements set out in the Directive apply to new work places used for the first time after 30.06.03. Existing workplaces must comply with the requirements set out in the Directive by 30.06.06.

The detailed employer obligations set out in the Directive are summarised in Appendix I.

MINIMUM REQUIREMENTS SPECIFIED IN DIRECTIVE 99/92/EC

Annex II of the Directive sets out the minimum requirements for improving the safety and health protection of workers potentially at risk from explosive atmospheres. The requirements are divided into organisational and explosion protection measures and the degree to which they have to be implemented depends on the risk assessment:

ORGANISATIONAL MEASURES

The employer must provide those working in places where explosive atmospheres may occur with sufficient and appropriate training with regard to explosion protection. In addition work in hazardous places must be carried out in accordance with written instructions issued by the employer and a system of permits to work must be applied for carrying out both hazardous activities and activities which may interact with other work to cause hazards.

EXPLOSION PROTECTION MEASURES

Where appropriate i.e. following the risk assessment which has been carried out and documented in the explosion protection document, the following explosion protection measures have to be implemented·

- Any hazardous releases of explosive atmospheres (intentional or non-intentional) must be diverted to a safe place or safely contained or rendered safe.
- Prevention of ignition hazards must take account of electrostatic discharges and workers must be provided with suitable clothing.

- Only equipment and connecting devices which are safe should be used.
- Measures must be taken to minimise the risks of an explosion and, if an explosion occurs to minimise the explosion effects.
- Optical/acoustic warning and evacuation before explosion conditions reached.
- Provision and maintenance of emergency escape facilities.
- Emergency power supply/manual override of automatic processes.
- Overall explosion safety must be verified by a competent person before use.

ADDITIONAL REQUIREMENTS OF THE ATEX DIRECTIVE 99/92/EC COMPARED TO THE CHEMICAL AGENTS AT WORK DIRECTIVE 98/24/EC

Both the ATEX Directive and the Chemical Agents at Work (CAD) Directive come under the base European Directive for improving the safety and health of workers at work. As with all the subordinate Directives covering individual hazards, both these Directives use a risk based approach to evaluate the actual hazard and to specify the preventative or protective measures required. Thus many of the requirements in the main provisions of the ATEX 99/92/EC Directive are similar to the safety requirements contained in the in Chemical Agents at Work Directive 98/24/EC. Such as the requirements to prevent and protect against explosions and to assess explosion risks. Additionally, many of the minimum requirements in Annex II of ATEX are either implicit in CAD or represent good safety practices, some of which are already widely taken, and are equally applicable to other risks from dangerous substances.

However, some requirements in ATEX require substantially more than CAD or are not appropriate to risks other than those arising from explosive atmospheres. These requirements concern:

- the classification and zoning of places where explosive atmospheres may occur;
- the selection of equipment for use in those places;
- the marking of places where explosive atmospheres may occur;
- the verification of overall explosion safety before new workplaces are used for the first time;
- the provision of appropriate work clothing in explosive atmospheres, and
- coordination where employers share a workplace.

CORRELATION BETWEEN ATEX 137 DIRECTIVE FOR 'USERS' AND ATEX 95 FOR MANUFACTURERS

The Directive 94/9/EC or so called ATEX 95 Directive sets out specifications for both electrical and non-electrical equipment to be used in potentially explosive atmospheres. The aim of the Directive is to facilitate free trade within the member states and it provides a classification scheme and essential safety and health requirements for equipment of different categories which must be followed by manufacturers. This Directive also comes into force on 30.06.2003.

A comparison of the requirements of the different Directives for manufacturers and for employers is given in Table 1. It can be seen that there is a direct link between the two

Directives in that the 3 Categories of equipment specified in the ATEX 95 Directive correspond to the 3 Zones used in the ATEX 137 Directive for the classification of areas where explosive atmospheres are likely to occur.

Thus in Zone 2 or 22 an explosive atmosphere is unlikely to occur in normal operation and equipment of Category 3 may be used i.e. equipment that does not pose an ignition source in normal operation, whereas in Zone 0 or 20 where an explosive atmosphere can be present continuously equipment of Category 1 must be used i.e. equipment that does not present an ignition source even under conditions of rare malfunctions.

Table 1.   Requirements of manufacturers and employers ATEX directives

Manufacturer requirements 'ATEX 95' - 94/9/EC	Employer requirements 'ATEX 137' - 99/92/EC
Definition of area of use of equipment, specification of equipment group/category	Determination of Zones in plant Select appropriate equipment
Category 1:	Zone 0/20
Category 2:	Zone 1/21
Category 3:	Zone 2/22
Comply with essential safety and health requirements or relevant standard	Comply with installation and maintenance requirements
Carry out a risk/ignition hazard assessment of equipment	Carry out a risk assessment of the work place, duty of Coordination
Prepare Conformity documentation	Prepare an Explosion protection document
Appropriate quality control	Regular updates

EXPLOSION PROTECTION DOCUMENT

The Explosion Protection Document required by Directive 99/92/EC is intended to demonstrate that the employer has complied with the requirements of the Directive. Thus it should demonstrate that explosion risks have been determined and assessed and show that adequate prevention and/or protection measures have been taken.

Where explosive atmospheres may occur in such quantities as to requires special precautions the Explosion Protection Document must show how the areas have been classified in zones and that the minimum requirements of the Directive have been applied, in particular that the safe operation and maintenance of work place and equipment is ensured and that there are appropriate arrangements for the safe use of work equipment. In addition the Directive requires that the Explosion Protection Document is produced before work starts and is revised following changes.

It is clear that the Directive only requires the production of an Explosion Protection Document when the quantity of explosive atmosphere is hazardous. Thus the use of a small quantity of flammable solvent in an office does not require the production of such a document.

The Directive allows the Explosion Protection Document to be either a separate document or part of a combined safety report and as a whole must cover both technical and organisational or procedural aspects of explosion prevention and protection. Thus for the majority of process plant, it is recommended that these two aspects be separated and dealt with in different documents.

TECHNICAL CONTENT OF THE EXPLOSION PROTECTION DOCUMENT

- Description of process & plant
- Fire and explosion characteristics of materials
- Occurrence of flammable atmospheres - Zoning
- Identification of possible ignition sources - selection of equipment
- Risk assessment – i.e. discussion and justification of the measures taken
- Preventative and protective measures specific to this process/plant
- Technical/organisational measures

It can be seen that the technical aspects of the explosion protection document mirror the contents of a traditional fire and explosion hazard assessment which the majority of manufacturers using flammable/explosible materials already carry out.

PROCESS/PLANT DESCRIPTION
Depending on the nature of the operation i.e. batch/semi-batch or continuous and the consequential frequency of change in materials and or operations the technical aspects of the Explosion Protection Document will need to be written either specific to a plant or to a process. This section should therefore include a short description of the essential aspects of the process or plant sufficient to identify the objectives, key operations and major plant items.

FIRE AND EXPLOSION CHARACTERISTICS OF MATERIALS
A list of the substances used in the process and their fire and explosion properties should be given in the Explosion Protection Document. In order to carry out the assessment of the possible risk present in the process and in particular the likelihood of ignition, the required characteristics go beyond just whether a material is capable of forming an explosive atmosphere. Typical characteristic that will be required are shown in Table 2. It should however be recognised that this does not imply that the properties listed need to be measured for each substance.

In the many cases the data is either readily available in the literature or typical limit values may be used e.g. the Minimum Oxygen Concentration of typical hydrocarbons in Nitrogen lies between 9-10 vol.%.

OCCURRENCE OF FLAMMABLE ATMOSPHERES – ZONING
Directive 99/92/EC includes definitions for the classification of hazardous places in terms of zones on the basis of the frequency and duration of the occurrence of an explosive atmosphere. The 3 Zone concept used for many years for the specification and selection of electrical equipment for explosive gas and vapour atmospheres and more recently specified for explosive dust atmospheres is now embodied in European legislation.

Table 2. Flammability and explosion characteristics

Gases/liquids/vapours	Dusts
Flash point	Burning behaviour
Auto ignition temperature	Explosibility
Explosibility limits	Minimum ignition temperature (Cloud)
Minimum oxygen concentration	Layer ignition temperature
Density (relative to air)	Minimum ignition energy
Explosion characteristics	Explosion characteristics
- Flame speed	- Maximum pressure
- Maximum pressure	- Rate of pressure rise
- Rate of pressure rise	Thermal stability of bulk powder
Maximum experimental safe gap	

The Directive includes a definition of a hazardous place as a place in which an explosive atmosphere may occur in such quantities as to require special precautions to protect the health and safety of the workers. For the Zone definitions Zone 0, Zone 1, and Zone 2 are used to denote explosive atmospheres containing gases, vapours or mists while Zone 20, Zone 21 and Zone 22 are used to denote explosive atmospheres containing dusts. The same definitions are used for the different Zones of all explosive atmospheres as shown in Table 3. In addition the Directive includes a definition of normal operation and requires that hazards arising from deposits or layers of dusts must also be considered.

A methodology for area classification of places containing explosive atmospheres is already available in European standards, EN 60079-10 for both gases and vapours [3] and EN 50281-3 for dusts [4] though it should be recognised that the Directive does not specify that the methodology contained in these standards has to be used.

Table 3. Zone definitions of places containing explosive atmospheres

Zone	Duration and frequency of explosive atmosphere
0/20	Present continuously, or for long periods or frequently
1/21	Likely to occur in normal operation occasionally
2/22	Not likely in normal operation, but if so, only for a short period

Layers, deposits and heaps of dust must be considered as any other source which can form an explosive atmosphere.

"Normal operation" - situation when installations are used within their design parameters (i.e. including start up and shut down).

As mentioned the concept of area classification has been used for many years for the selection of electrical equipment. However the requirement of the Directive that areas or places containing explosive dust atmospheres together with the need to document the results the areas classification exercise in the Explosion Protection Document will often lead to a reappraisal of the Zone classification. This will be driven by the need specified in the Directive to only use equipment, both electrical and non-electrical of a specific category in a particular Zone.

The area classification methodology used in the European standards for both gases and vapours and for dusts is based on the concept of 3 'sources of release' namely continuous, primary and secondary which depending on the presence of openings and more importantly on the degree and efficiency of ventilation are used to determine the nature and extent of the Zone. An 'exact' evaluation is theoretically possible using numerical simulation but in practice the assumptions involved do not justify this approach and usually estimations of the extent of the Zones are made based on experience. Thus:

Zone 0/20 – continuous source of release occurs almost always only inside equipment or dust containment or transport systems or near the surface of flammable liquids.

Zone 1/21 – primary source of release usually occurs outside equipment for example around sampling points and filling/emptying points without extract, in the immediate vicinity of access doors frequently used when an explosive atmosphere is inside. It may also occur inside equipment for example inside extraction systems or in dust containment equipment such as some silos and filters where a dust cloud is only occasionally formed.

Zone 2/22 – secondary source of release for example around seals, fittings, valves, relief valves, outlets from vents, in the vicinity of access doors/openings infrequently used, storage areas (due to possible breakage) around filling/emptying points with extract/ventilation and areas with dust layers that can form explosive dust clouds.

The extent of the Zones is dependent on the degree of ventilation of extract but in many cases the extent of a zone around the source of release may be taken as 1 m for dusts and 1 m to 3 m for gases and vapours though in the case of gases and vapours heavier than air these may extend much further. It is clear that in the case of dusts good housekeeping is essential to remove deposits of dusts as soon as they are formed. Where this is effectively carried out the area may be classified as non-hazardous provided any layers are removed by cleaning before an explosive dust/air mixture is formed.

IDENTIFICATION OF POSSIBLE IGNITION SOURCES
A complete list of possible ignition sources is given in Table 4 [5].

Ignition sources can arise not only from the equipment used but also from the mode or type of operation carried out, thus a careful assessment of the likelihood of them occurring during the process is needed.

Static electricity, specifically mentioned in the Directive as a possible ignition source that must be considered, is generated whenever movement occurs. However provided all conducting parts of the plant and process materials are earthed and consideration has been given to the use of non-conducting materials, it is seldom an effective ignition source [6].

Table 4. List of possible ignition sources

Common	Special
Hot surfaces	Lightning
lames and hot gases (including hot particles)	Stray electric currents (e.g. from Cathodic protection systems)
Mechanical sparks	Radiation - High frequency, Optical, Ionising
Electrical equipment	Ultrasonics
Static electricity Chemical reaction (thermal instability)	Adiabatic compression and shock waves

SELECTION OF EQUIPMENT

As mentioned above, Directive 99/92/EC specifies criteria for the selection of equipment and protective systems. Thus unless the explosion protection document based on a risk assessment does not state otherwise, equipment and protective systems for all places in which explosive atmospheres may occur must be selected on the basis of the categories set out in Directive 94/9/EC. In particular, the following categories of equipment must be used in the zones indicated, provided they are suitable for gases, vapours or mists and/or dusts as appropriate:

- In zone 0 or zone 20, category 1 equipment.
- In zone 1 or zone 21, category 1 or 2 equipment.
- In zone 2 or zone 22, category 1, 2 or 3 equipment.

In the future the choice of equipment that is safe for use in a particular zone should present few problems as non-electrical as well as electrical equipment corresponding to the different categories becomes available. However for existing plants, it must be demonstrated that equipment currently in use is safe. The basic method and requirements developed and contained in the European Standard on non-electrical equipment for potentially explosive atmospheres EN 13463-1 [7] can be applied by users to decide whether a particular piece of equipment is safe to use. These concepts together with the ignition protection methods are further described in another paper in this symposium [8]. The basis of these concepts is the link between the frequency of occurrence of an explosive atmosphere and the conditions of operation when an ignition source is likely to occur as shown in Table 5. It is clear that this can be generally applied to decide whether any piece of equipment or operation is suitable for a particular zone.

As currently with electrical equipment, the choice of suitable equipment also depends on ensuring that the maximum surface temperature remains below the ignition temperature of the explosive atmosphere. For gases and vapours equipment is classified into classes T1 to T6 depending on the maximum surface temperature and this can be compared to the auto-ignition temperature of the gas or vapour. For explosive dust atmospheres the maximum allowable surface temperature is the lower of either 2/3 rds of the ignition temperature of the dust cloud or the layer ignition temperature – 75 K.

RISK ASSESSMENT
The Directive specifies that the Explosion Protection Document should include an assessment of the risks associated with the work being carried out. In the majority of cases this needs to be no more than a discussion the occurrence and frequency of the presence of an explosive atmosphere and the likelihood of occurrence of an effective ignition source and justification of the measures taken. These will need to take into consideration the ignition energy and or ignition temperature required to ignite the particular atmosphere.

Table 5.   Relationship between Zones and Categories of Equipment

Explosive atmosphere	Zone	Category	Ignition source
Continuous	0/20	1	None with rare malfunction
Occasional	1/21	2	None with malfunction
Not likely	2/22	3	None in normal operation

For more complex plants it is often useful to use a more structured approach to ensure that all the possible sources of hazards have been identified and appropriate measures taken to reduce the risk to an acceptable level. A methodology has been developed for equipment and unit operations for use in potentially explosive atmospheres as part of a European research project, the RASE Project [9]. This uses a risk matrix to evaluate the final risk and the final report of the project includes worked examples of the application of the methodology for spray drying operation, powder pneumatic transfer, paint spray booth, an exhaust system for gas turbine and an oil seed extraction unit.

PREVENTATIVE, PROTECTIVE MEASURES SPECIFIC TO THE PROCESS/PLANT
This section of the Explosion Protection Document should include a summary of the measures which have been applied to prevent the occurrence of an explosion or to protect against the effects of an explosion. A variety of measure are available and include the avoidance of ignition sources, the prevention of the formation of an explosive atmosphere, for example by inert gas blanketing, or where necessary protection measures such as:

- Explosion proof construction
- Explosion venting
- Explosion isolation
- Explosion suppression

The summary should include specific technical and or organisational measures, for example earthing of drums, which are necessary to ensure safe operation.

SAFETY MANAGEMENT ASPECTS OF THE EXPLOSION PROTECTION DOCUMENT

Many of the organisational requirements which need to be demonstrated under the ATEX 137 Directive will or should be already be documented as part of a safety management system. Written procedures should be in place for common activities which include:

- SHE Policy/Responsibilities
- Management of change
- Permit to work
- Procedures for visitors/subcontractors
- Instructions/training
- Frequency of review etc.

Safety management aspects will not be further covered here. Further details of setting up such a system can be found in HSE [1] and EPSC [2] publications.

IMPLEMENTATION IN THE EUROPEAN MEMBER STATES

European Directives have to be implemented in member states regulations within the time period specified in a particular Directive, in the case of the ATEX 137 Directive 99/92/EC this date is the 30th June 2003. According to the information available to the Authors in September 2002, the national regulations or their drafts for the implementation of the ATEX Directive are as yet only available in Holland, Germany and the United Kingdom, though all states contacted indicated that they intended implementing the requirements in the Directive by the Deadline date.

The way the Directives are implemented is left to the individual member state and there are often major differences. Thus, for example in the UK the ATEX 137 Directive is being implemented together with the Chemical Agents at Work Directive in one set of regulations known as the DSEAR Regulation or Dangerous Substances and Explosive Atmospheres Regulation. In contrast, in Germany two different laws are being used to implement the ATEX 137 Directive. Where an employer has a plant which uses or produces an explosive atmosphere then the plant has to comply with the so called "Betriebssicherheitverordnung" (Plant Safety regulations) which includes all the requirements of the Directive including the necessity to produce an Explosion Protection Document. The safety of workers is in part covered by the "Gefahrstoffverordnung" (Dangerous Substances Regulations) and the requirements of the ATEX 137 Directive are in essence repeated in this regulation which also includes the requirements of the Chemical Agents at Work Directive.

As mentioned previously the ATEX 137 Directive specifies minimum requirements and the European Treaty allows member states to set more stringent requirements. From the perspective of international companies it is to be hoped that there will not be major differences though it is rumoured that one member state is considering requiring the installation of Category 2, instead of Category 3, non-electrical equipment in Zones 2 and 22! Small variations are to be expected particularly in view of the different ways that the Directive is implemented. Thus in Germany current plants have to comply with the complete requirements including the production of all the documentation by 2005 and not 2006 as specified in the Directive.

CONCLUSIONS

The new ATEX 137 Directive 99/92/EC places an onus on employers to ensure the provision of safe working conditions and equipment where potentially explosive atmospheres can occur. The minimum requirements specified in the Directive are no more stringent than current good practice already being used in the process industries. The documentary requirement of the Directive i.e. the production of an Explosion Protection Document, can be meet in most cases by a combination of the fire and explosion hazard assessment together with reference to procedures and structures covering organisational aspects which are part of a safety management system.

The Directive provides flexibility for employers to continue using existing equipment and practices provided they are safe and allows the use of equipment not classified under Directive 94/9/EC in the future again provided that its safety has been demonstrated.

APPENDIX I: SUMMARY OF THE EMPLOYER OBLIGATIONS OF DIRECTIVE 99/92/EC [11]

PREVENT AND PROTECT AGAINST EXPLOSIONS

The employer has to take technical and/or organisational measures appropriate to the nature of the operation, in order of priority and in accordance with the following basic principles:

- the prevention of the formation of explosive atmospheres, or where the nature of the activity does not allow that,
- the avoidance of the ignition of explosive atmospheres, and
- the mitigation of the detrimental effects of an explosion so as to ensure the health and safety of workers.

Where necessary, these measures have to be combined and/or supplemented with measures against the propagation of explosions. They have to be reviewed regularly and, in any event, whenever significant changes occur.

ASSESSMENT OF EXPLOSION RISKS

The Directive requires employers to assess the specific risks arising from explosive atmospheres, taking account the following:

- the likelihood that explosive atmospheres will occur and their persistence,
- the likelihood that ignition sources, including electrostatic discharges, will be present and become active and effective,
- the installations, substances used, processes, and their possible interactions,
- the scale of the anticipated effects.

In addition the Directive requires that explosion risks shall be assessed overall and specifically mentions the problem of connections and/or openings to places where explosive atmospheres may occur.

GENERAL OBLIGATIONS
Once the risk assessment has been carried out the Directive requires employers to implement the findings to ensure safe working conditions with appropriate supervision and technical measures.

DUTY OF COORDINATION
The Directive requires that where workers from several undertakings are present at the same workplace, each employer shall be responsible for all matters coming under his control. However in addition it requires that the employer responsible for the work place coordinates the implementation of all the measures concerning workers' health and safety and shall state, in the explosion protection document, the aim of that coordination and the measures and procedures for implementing it.

PLACES WHERE EXPLOSIVE ATMOSPHERES MAY OCCUR
The Directive requires that the employer shall classify places where explosive atmospheres may occur into zones in accordance with the definitions given in Annex I. Following the area classification the employer has to ensure that the minimum requirements set out in Annex II of the Directive are implemented and, where necessary, that the entry points to the areas have to be marked.

EXPLOSION PROTECTION DOCUMENT
The employer has to ensure that a document, called the 'explosion protection document', is drawn up and kept up to date. The document has to demonstrate in particular:

- that the explosion risks have been determined and assessed,
- that adequate measures will be taken to attain the aims of the Directive,
- those places which have been classified into zones in accordance with Annex I,
- those places where the minimum requirements set out in Annex II will apply,
- that the workplace and work equipment, including warning devices, are designed, operated and maintained with due regard for safety,
- that arrangements have been made for the safe use of work equipment.

The explosion protection document shall be drawn up prior to the commencement of work and be revised when the workplace, work equipment or organisation of the work undergoes significant changes, extensions or conversions. The employer may combine existing explosion risk assessments, documents or other equivalent reports.

SPECIAL REQUIREMENTS FOR WORK EQUIPMENT AND WORKPLACES
This section of the Directive sets out the dates by which new and existing work places have comply with the Directive and in particular when it is necessary to use equipment that has been design and manufactured according to the so called 'ATEX 95' Directive. In essence new work places that are used for the first time after 30.06.03 have to comply with the Directive whereas existing work places have to comply by 30.06.06.

It should be recognised however, that the Directive allows the unlimited continued use of existing equipment and also the installation of new equipment that is not manufactured according to the ATEX 95 Directive provided the Explosion Protection Document shows that it is safe.

REFERENCES

1. Health and Safety Executive (1997) *Successful Health and Safety Management*, HS(G)65, (second edition), ISBN 0 7176 12767.
2. European Process Safety Centre (1994) *Safety Management Systems: sharing experiences in process safety*, IChemE, ISBN 0 85295 356 9.
3. EN 60079-10 Electrical Apparatus for explosive gas atmosphere – Part 10: Classification of hazardous areas.
4. EN 50281-3 Electrical apparatus for use in the presence of combustible dust - Part 3: Classification of areas where combustible dusts are or may be present.
5. EN 1127-1 Explosive atmospheres – Explosion prevention and protection Part 1: Basic concepts and methodology.
6. CENELEC report R044-001 Safety of machinery – Guidance and recommendations for the avoidance of hazards due to static electricity.
7. EN 13463-1 "Non-electrical equipment for potentially explosive atmospheres Part 1: Basic method and requirements".
8. R.L. Rogers, 2003, *Development of European standards for non-electrical equipment for use in explosive atmospheres* This conference.
9. R.L. Rogers, B. Broeckmann, S. Radandt, K-H. Grass, J-P. Pineau, C. Loyer, N. Worsell, C. Schwartzbach and K. van Wingerden; *Risk assessment of equipment for use in explosive atmospheres: The RASE Project* 10th Int. Symp. on Loss Prev. and Safety Promotion in the Process Industries 2001 Stockholm, Sweden.
10. The RASE Project – Final report can be downloaded from www.safetynet.de.
11. The full text of European Directives can be downloaded from the European Union web site http://www.europa.eu.int/eur-lex/.

THE EC 'SAFEC' PROJECT: ATEX MEETS IEC 61508

Jill Wilday*, Tony Wray** and Simon Brown***
*Health and Safety Laboratory, Harpur Hill, Buxton, UK
**Health and Safety Laboratory, Broad Lane, Sheffield, UK
***Health and Safety Executive, Stanley Precinct, Bootle, UK

Some types of electrical equipment intended for use in potentially explosive atmospheres rely on so-called 'safety devices' to reduce the likelihood of the equipment presenting a source of ignition which could cause an explosion. Examples of safety devices are motor protection circuits (to limit temperature rise during stall conditions) and pressurisation systems (to prevent ingress of an explosive atmosphere into an electrical equipment enclosure). The EC SAFEC project had the objective of producing a methodology for deciding how to determine the requirements for safety devices to achieve compliance with the ATEX Directive (94/9/EC). Candidate control system standards for categorising the safety devices were EN 954 and IEC 61508 (now EN 61508).

For simple safety devices, EN 954 is sufficient. However, more complex safety devices, particularly if programmable, are better thought of as safety-related systems, and IEC 61508 is appropriate. This requires that the safety device is specified in terms of a safety integrity level (SIL). Three approaches were used to calibrate the SIL required in the different ATEX equipment categories: use of individual risk criteria; use of accident statistics; and estimation of the SIL for a generic design of pressurisation equipment. Further case studies tested the proposed methodology for determining the SIL for a diode safety barrier, a level detection device and both pressure and temperature safety devices. The SAFEC project and its results are described, particularly in terms of how to determine the SIL required in a particular application.

KEYWORDS: IEC 61508; ATEX; Safety integrity; Risk assessment

INTRODUCTION

Some types of electrical equipment intended for use in potentially explosive atmospheres rely on so-called 'safety devices' to reduce the likelihood of the equipment presenting a source of ignition which could cause an explosion. Examples of safety devices are motor protection circuits (to limit temperature rise during stall conditions) and pressurisation systems (to prevent ingress of an explosive atmosphere into an electrical equipment enclosure). The approval and certification of electrical apparatus for potentially explosive atmospheres requires that, where such safety devices are used to reduce the risk of explosion, an assessment be made of their suitability for the intended purpose from a functional safety viewpoint. This needs to be expressed in terms of some measure of confidence that the devices will be able to maintain a required level of safety in accordance with the requirements of the EC ATEX Directive[1], CENELEC standards for electrical apparatus for use in potentially explosive atmospheres[2-9] and relevant standards for safety-related electrical control systems.

CENELEC identified the need for research to determine whether existing and proposed standards in the field of safety-related control systems are suitable for this purpose, and to develop a methodology which will provide the required support for the approval and certification process. Research proposals on this topic were invited under the Standardisation, Measurement and Testing (SMT) Programme and the SAFEC project (contract SMT4-CT98-2255) was selected for funding. It ran from January 1999 to May 2000.

The partners in the SAFEC project were the Health and Safety Laboratory of the Health and Safety Executive (HSL) in the UK (the project coordinator), the Deutsche Montan Technologie (DMT) in Germany, the National Institute for Industrial Environment and Risks (INERIS) in France and the Laboratorio Oficial J.M. Madariaga (LOM) in Spain. The SAFEC partners worked cooperatively with the members of CENELEC Technical Committee 31, Working Group 09 (WG09), which is drafting a standard on "Reliability of safety-related devices" with the intention that the SAFEC results be utilised by WG09 in this standard. Several joint meetings were held.

The SAFEC project comprised six tasks:

1. Derivation of target failure measures (all/HSL).
2. Assessment of current control system standards with reference to the target failure measures from Task 1 (HSL).
3. Identification of safety devices currently used with reference to CENELEC standards (LOM).
4. Study of a selection of safety devices identified in Task 3 (INERIS).
5. Determination of a methodology for testing, validation and certification (DMT).
6. Production of a final report[10] (all/HSL).

This paper concentrates on the choice of control system standards and the calibration of target failure measures for safety devices according to those standards. This work was particularly carried out during tasks 2, 4 and 6.

REQUIREMENTS OF ATEX DIRECTIVE

The ATEX product Directive[1] defines two Groups of application of electrical equipment, each of which has Categories of electrical equipment according to the level of protection required:

*   Group I comprises mining applications where the flammable material is methane (firedamp) or flammable dust:
    *   Category M1 means that the equipment is required to remain functional in an explosive atmosphere.
    *   Category M2 equipment is intended to be de-energised in the event of an explosive atmosphere.

- Group II comprises other applications where equipment is to be used in a potentially explosive atmosphere:
  - Category 1 equipment is intended for use where explosive atmospheres are present continuously, for long periods of time or frequently (referred to elsewhere as Zone 0 and/or 20).
  - Category 2 equipment is intended for use where explosive atmospheres are likely to occur (referred to elsewhere as Zone 1 and/or 21).
  - Category 3 equipment is intended for use where explosive atmospheres are less likely to occur, and if they do occur, do so infrequently and for only a short period of time (referred to elsewhere as Zone 2 and/or 22).

The ATEX product Directive fault tolerance requirements can be summarised as follows:

- A fault tolerance of at least 2 is required for the means of protection of Category 1 equipment.
- A fault tolerance of at least 1 is required for the means of protection of Category 2 equipment.
- No fault tolerance is required for the means of protection of Category 3 equipment.

The SAFEC project regarded a 'safety device' as a part of the equipment, which has an autonomous safety function with respect to the risk of explosion.

CHOICE OF CONTROL SYSTEM STANDARDS

Task 2 of the SAFEC project, carried out by HSL, included a review of existing control system standards, with reference to the requirements of the ATEX product Directive[1]. Since safety devices are defined as having an autonomous safety function (or controlling function), it was expected that control system standards might be useful in defining the requirements for safety devices. There are two standards which provide guidance on the design of control systems for use in safety-related applications: EN 954-1[11]; and IEC 61508[12] (now also published as EN 61508).

A discussion of the relative merits of the two standards for this purpose has been published previously[13]. EN 954 can be used for simple safety devices, e.g. non-programmable electrical interlocks, especially where the appropriate CENELEC standard refers to EN 954. However, it was recognised that some existing CENELEC standards make reference to EN 954 in cases where nowadays it would be more appropriate to refer to IEC 61508, particularly for complex or programmable electronic safety devices (such as a pressurisation control system using a programmable logic controller). Therefore, it was proposed[10] that any industry-specific standard for complex and programmable safety devices should be based on IEC 61508 but have an additional requirement, based on fault tolerance, which will ensure that the fault tolerance requirements of the ATEX Directive are met. However, it was also recognised that some safety devices may already be fully specified within relevant CENELEC standards, e.g. references 2–9. In these cases, it may not be necessary to further specify the safety device in terms of IEC 61508 or EN 954.

In considering the requirements for safety devices according to these two standards, it is useful to define the equipment under control (EUC), according to IEC 61508, as that part of the equipment (as defined by the ATEX Directive) which is not the safety device. See Figure 1.

CALIBRATION OF REQUIRED IEC 61508 SAFETY INTEGRITY LEVELS

INTRODUCTION

IEC 61508 defines safety integrity levels (SIL) for safety functions by taking into account:

- quantified reliability of the safety function (see Table 1). The quantified analysis of a system deals with the random hardware failure rate;
- qualitative reliability. The techniques used to design, maintain, etc. the system throughout its lifecycle must be sufficient to ensure that the rate of systematic failures is less than the random hardware failure rate; and
- architectural constraints, based on fault tolerance and fail-to-safety characteristics. These put a ceiling on the safety integrity level (SIL) that can be claimed for any particular system in order to ensure that uncertain reliability calculations, e.g., where reliability data are sparse, do not lead to an inflated SIL (see Table 2).

Table 1. Quantitative reliability requirements of IEC 61508

SIL	Probability of failure on demand (for low demand rate operation)	Frequency of failure (per hour) for continuous operation
4	10^{-5}–10^{-4}	10^{-9}–10^{-8}
3	10^{-4}–10^{-3}	10^{-8}–10^{-7}
2	10^{-3}–10^{-2}	10^{-7}–10^{-6}
1	10^{-2}–10^{-1}	10^{-6}–10^{-5}

Table 2. Architectural constraints of IEC 61508

Safe failure fraction	Hardware fault tolerance		
	0	1	2
For type A safety-related subsystems			
<60%	SIL1	SIL2	SIL3
60%–<90%	SIL2	SIL3	SIL4
90%–<99%	SIL3	SIL4	SIL4
>99%	SIL3	SIL4	SIL4
For type B safety-related subsystems			
<60%	not allowed	SIL1	SIL2
60%–<90%	SIL1	SIL2	SIL3
90%–<99%	SIL2	SIL3	SIL4
≥99%	SIL3	SIL4	SIL4

The early stages in the IEC 61508 lifecycle involve carrying out hazard and risk assessment and allocating safety requirements to relevant safety functions. It was necessary within SAFEC to define or calibrate the SIL required for each ATEX equipment category. A target SIL requirement applies to a particular safety function, not to a safety device. The safety function may be implemented by a range of technologies and each may achieve a part of the required risk reduction. This is illustrated in Figures A.1 and A.2 of Part 3, Annex A of IEC 61508[12], on which Figure 2 is based.

In calibrating the required SIL, a useful hypothetical concept was a safety function protecting against a case in which there is a source of ignition in normal operation. However, this would not, of course, be a practical design for equipment intended for use in potentially explosive atmospheres.

Three approaches were used to calibrate the SILs required:

- Use of individual risk criteria to determine the necessary risk reduction;
- Use of accident statistics to attempt to determine the SIL for existing equipment;
- Estimation of SILs of safety devices within existing equipment.

These are discussed in more detail in the following sections.

INDIVIDUAL RISK

A convenient quantitative definition of hazardous zones, in terms of the time that flammable gas would be expected to be present, is given by Table 3[14]. In all cases, the probability of occurrence of a flammable atmosphere corresponds to the worst-case probability for the particular zone. It should be noted that these values have not been well accepted in all industrial sectors so, although they have been considered by CENELEC working groups, they have not been incorporated in standards.

Calculations of required risk reduction and hence SIL are shown in Table 4 for a range of risk criteria. The criteria range from intolerable (10^{-3} per year) to broadly acceptable (10^{-6} per year)[15]. A criterion of 10^{-5} per year has been used in previous work by the Institute of Petroleum[16].

Table 3.   Probability of an explosive atmosphere being present

Zone	Quantitative assumption (hrs/yr)	Probability of occurrence (%)
0	> 1000	100
1	< 1000 and > 100	10
2	< 10	0.1

Table 4. Coarse estimate of integrity requirement based on risk tolerability criteria

					Unit
Criterion for probability of death	10^{-3}	10^{-4}	10^{-5}	10^{-6}	per year
Number of workers/members of the public present[a]	0.2	0.2	0.2	0.2	
Required risk reduction:					
Maximum possible failure frequency, assuming a continuous source of ignition, Zone 0	0.57	0.057	0.006	0.0006	per 10^6 hrs
Maximum possible failure frequency, assuming a continuous source of ignition, Zone 1	5.7	0.57	0.06	0.006	per 10^6 hrs
Maximum possible failure frequency, assuming a continuous source of ignition, Zone 2	570	57	5.7	0.57	per 10^6 hrs
Equivalent safety integrity requirement:					
SIL required to achieve target[b], Zone 0	SIL2	SIL3	SIL4	SIL5[c]	
SIL required to achieve target, Zone 1	SIL1	SIL2	SIL3	SIL4	
SIL required to achieve target, Zone 2	SIL1[d]	SIL1[e]	SIL1	SIL2	

Notes to Table 4:

[a]This assumes 20 deaths per 100 explosions involving pressurisation systems[16, 17]

[b]This is the SIL of the overall safety function and includes all protection measures/devices. It is based directly on the maximum allowable failure frequency of the safety function, from the rows above, and assumes continuous operation of the safety function with the SIL taken from Table 1.

[c]SIL5 is outside the range of achievable SILs considered by IEC 61508; however, SIL 5 has been used here in order to make the table more meaningful.

[d] and [e]SIL1 represents the minimum integrity requirement of IEC 61508 for a system defined as being safety-related; therefore, SIL1 must apply to these positions.

ACCIDENT STATISTICS

Discussion with a UK manufacturer of pressurization systems has indicated that about 18,000 such systems have been put into service in the UK over the past 20 years. Assuming a life expectancy in the region of 8 years, this suggests an average of about 6,000 systems have been in use over this time. The partners were not aware of any explosions resulting from the failure of a pressurization system. Therefore, this sets a lower limit on the integrity of pressurization systems over the past 20 years, as shown in Table 5, below. The values in Table 5 were calculated on the assumption that, if no explosions occur over N operating hours, then a reasonable assumption is that the probability of an explosion occurring in the next N operating hours is 0.5.

Table 5 suggests that the integrity of existing pressurization systems is:

Table 5. SIL indications from accident records

	Assumed zone of operation[a]			
	Zone 1H[b]	Zone 1L[b]	Zone 2	Units
Period of study	20	20	20	years
Number of systems in use in the UK over this period	6,000	6,000	6,000	
Total operating period	1,051,920,000	1,051,920,000	1,051,920,000	system-hours
Probability of gas presence[c]	0.032	0.0032	0.00032	
Operating period with gas present	33,661,440	3,366,144	336,614	"gas" hours
Number of known explosions	0	0	0	
Indicated dangerous failure rate for each system	0.015	0.15	1.5	per 10^6 hrs
Indicated SIL for the overall safety system[d]	SIL3	SIL2	SIL1	

Notes to Table 5:

[a]The data in each of the columns have been calculated on the basis that all systems were used in the single specified zone.

[b]For the purpose of these calculations, Zone 1 has been split into two regions.

[c]It would be inappropriate to use the worst-case probabilities for the presence of flammable gas in the calculations in this particular table, as we must use an estimate of the actual probability. Without any prior knowledge of the distribution of this probability, the logarithmic mean of the range of probabilities covered by each (sub) zone has been used. This is: Zone 1H - 3.2%; Zone 1L - 0.32% and Zone 2 - 0.032%.

[d]This is the average SIL of the total configuration of safety-related systems. The pressurization control system (e.g., purge and shutdown systems) will contribute to this SIL together with other systems, e.g., the air supply.

    SIL1, if they have been mainly used in Zone 2;

    SIL2, if they have been mainly used at the lower end of Zone 1, or

SIL3, if they have been mainly used at the upper end of Zone 1.

However, as the probability of gas in the majority of Zone 1 environments will probably lie near the lower end of the zone (i.e., Zone 1L as shown in Table 5) with few at the upper end (shown as Zone 1H), Table 5 should not be considered to indicate that existing pressurization systems are able to achieve SIL3.

ESTIMATION OF SIL FOR SAFETY DEVICES ON EXISTING EQUIPMENT

Again, it can be assumed that existing certified electrical equipment is of adequate integrity, given that there is no history of explosions which have been initiated by certified electrical equipment. Therefore the SILs of existing safety devices can be assumed adequate. SILs for the following safety devices have been estimated (on the basis of random hardware failures only) during the SAFEC project[10] and results are given in Table 6 below.

- Two safety functions within a pressurisation system.
- Diode safety barrier.
- Level detection safety device.
- Pressure and temperature safety devices.

An example is provided by one of the safety functions for the pressurisation system. A generic design of pressurisation equipment was provided by a manufacturer (see Figures 3 and 4). One of the safety functions was to purge the enclosure prior to power being allowed to the equipment within it. The estimation of SIL took account only of the quantitative reliability aspects and the calculation is summarised in Table 6. Reliability data from Smith[18] was used.

Table 6. Determination of failure rate of purging delay function

Component	Failure mode	Failure rate, etc.	Unit	Comments
Contactor K	Energized state. Assumes power circuit correctly fused.	0.400	per 10^6 hrs	Assume 10% failure to open
RY2	Energized state	0.030	per 10^6 hrs	Armature. 10% failure to open.
Discriminator A	Output high	0.120	per 10^6 hrs	Bipolar linear
Capacitor C	Reduced capacitance	0.300	per 10^6 hrs	Assume aluminium electrolytic.
Circuit board	Ignored as de-energized = safe state	0.000	per 10^6 hrs	

452

Diode D	Short circuit	0.006	per 10^6 hrs	Assume 15% to short-circuit
Resistor Rb	Short circuit/ reduced resistance	0.000	per 10^6 hrs	Not credible
Resistor Ra	Open circuit/ increased resistance	0.002	per 10^6 hrs	Assume 50% to drift
Flow sensor AND Pressure sensor	Contacts-closed-β-factor of 0.05 assumed	0.050	per 10^6 hrs	
Overall failure rate: Function 2 (λ)		0.908	per 10^6 hrs	
Proof test interval, T (six months)		4,383	hours	
Probability of failure on demand ($\lambda T/2$)		1.99	$*10^{-3}$	
Safety integrity level of Function 2		SIL2		

Because the frequency of access to the pressurized cabinet is likely to be significantly less than the proof test interval, at first sight it may be assumed that failures of the purging function are unlikely to be revealed by the proof tests. However, this does not take into account that there may be no gas present when the pressurized cabinet is opened, and that the person opening the pressurized cabinet will be able to smell the flammable gas (unless this is, for example, hydrogen) at a level well below the lower explosive limit. If these are taken into account, a demand on the purging function (i.e., when the cabinet has been opened in the presence of flammable gas) occurs less often than the proof tests as is shown in Table 6, which determines the explosion rate from the failure rate of the purging function.

SUMMARY OF ESTIMATIONS OF SIL

A summary of the results of the above calculations for the purpose of calibrating the target risk reduction (SIL) requirement are given in Table 7. It can be that there is a good degree of convergence between the different methods of calibrating the target risk reduction requirements for the different hazardous zones. The approach of the SAFEC project has been to find targets which are in line with published risk tolerability criteria and are also achievable by existing safety devices. The lack of any history of explosions ignited by certified electrical equipment strongly suggests that current designs of safety devices are adequate.

It is proposed that the target risk reduction requirements, for the safety function of protecting against a hypothetical case in which there is a source of ignition in normal operation, be defined according to Table 8.

It is very important to note that these target risk reduction requirements refer to the safety function and not to the safety device. The safety function may be partly achieved by design features of the certified electrical equipment other than the safety device. Indeed, for certified electrical equipment, such design features will usually be present to prevent there being a source of ignition during normal operation. The necessary risk reduction can be allocated between available safety systems, including the safety device (see Figure 2).

Table 7. Summary of calculations for calibrating target risk reduction requirement

Description of method	Target risk reduction requirement		
	Zone 0	Zone 1	Zone 2
Use of individual risk criteria	SIL 3	SIL 2	SIL 1
Use of accident statistics applied to pressurised systems		SIL 2 or SIL 3	SIL 1
Estimated SIL for pressurisation system. Turn off equipment if pressurisation fails.		SIL 2 or SIL 3 (Note a)	
Estimated SIL for pressurisation system. Purge before allowing power onto equipment		SIL 2 (Note b)	
Estimated SIL for diode safety barrier	SIL 4		
Estimated SIL for low level detection system			SIL 1 (Note c)
Estimated SIL for pressure safety device		SIL 2 (note d)	
Estimated SIL for temperature safety device		SIL 2 (note e)	SIL 2 (Note e)

Notes for Table 7
(a) SIL 3 is possible given a suitably reliable air supply.
(b) The overall integrity could be increased by suitable operating procedures, such that SIL 3 may also be possible.
(c) The assumed application was within an LPG tank. This will usually be non-flammable (above UFL) and will therefore correspond to Zone 2.
(d) This could be increased given a suitably reliable air supply (see 5.4.1)
(e) The temperature safety device is assumed to be on a motor intended for use in either Zone 1 or Zone 2.

Table 8. Proposed target risk reduction requirements for the hypothetical case of protecting against an ignition source during normal operation

Hazardous zone	ATEX equipment categories	Target SIL requirement
0 or 20	1	SIL 3
1 or 21	2	SIL 2
2 or 22	3	SIL 1

Table 9 gives the proposed SIL requirements for safety devices as a function of the hazardous zone and the fault tolerance of the equipment under control.

Table 9. Proposed IEC 61508 safety requirements for safety functions

Hazardous area	Zone 0 Zone 20			Zone 1 Zone 21			Zone 2 Zone 22	
Fault tolerance requirement of ATEX Directive	2			1			0	
Equipment (EUC) fault tolerance	2	1	0	1	0	−1	0	−1
SIL of the safety function that the monitoring or control unit is providing	-	SIL 2	SIL 3	-	SIL 1	SIL 2	-	SIL 1
Resulting equipment category (under ATEX) of the combination	category 1			category 2			category 3	

CALIBRATION OF REQUIRED EN 954 CATEGORIES

It was concluded above that simple safety devices should meet the EN 954 category, which achieves the relevant ATEX fault tolerance requirement. A suggested definition of "simple safety device" is one which is simple enough that all the failure modes can be identified.

EN 954 has 5 categories for describing control systems:

- Category B has a fault tolerance of 0;
- Category 1 has a fault tolerance of 0;
- Category 2 has a fault tolerance of 0 but has automatic monitoring;
- Category 3 has a fault tolerance of 1, and
- Category 4 has:
  - a fault tolerance of 1 with automatic monitoring, **or**
  - a fault tolerance of 2 or more.

It therefore follows that the mapping between ATEX equipment categories and EN 954 categories for the safety devices is as given in Table 10. (Note that the addition of a safety device with a fault tolerance of zero to equipment with a fault tolerance of zero gives an overall fault tolerance of one.) In Table 10, the category of the safety device depends on the fault tolerance of the EUC.

Table 10. Proposed EN 954 requirements for simple safety devices

Hazardous area	Zone 0 Zone 20				Zone 1 Zone 21				Zone 2 Zone 22	
Fault tolerance requirement of ATEX Directive	2				1				0	
Equipment (EUC) fault tolerance	2	1	0	1	0	−1	0		−1	
EN 954 category of the safety device	-	B, 1, 2, 3 or 4	3 or 4	-	B, 1, 2, 3 or 4	3 or 4	-		B, 1, 2, 3 or 4	
Resulting equipment category (under ATEX) of the combination	ATEX category 1				ATEX category 2				ATEX category 3	

Note that a fault tolerance of "−1" implies that the equipment would be incendive in normal operation, without the intervention of the safety device

CONCLUSIONS

1. Safety devices, as defined under the ATEX Directive[1] have an autonomous safety function. They include implementation in a number of technologies and can be specified in a number of ways:
   - Devices which are already fully defined in CENELEC standards, e.g. references 2–9.
   - Simple safety devices, which can be defined according to EN 954[11].
   - More complex devices, which are generally electric/electronic/electronic programmable in nature and can be defined according to IEC 61508[12].
2. Proposed requirements for safety devices specified under IEC 61508 or EN 954 have been derived and are given in Tables 9 and 10, respectively.

REFERENCES

1. Directive 94/9/EC of the European Parliament and the Council of 23 March 1994 on the approximation of the laws of the Member States concerning equipment and protective systems intended for use in potentially explosive atmospheres, Official Journal of the European Communities, 19/4/94
2. EN 50014 Electrical apparatus for potentially explosive atmospheres. General requirements
3. EN 50015 Electrical apparatus for potentially explosive atmospheres. Specific requirements for the protective mode "o" oil immersion.
4. EN 50016 Electrical apparatus for potentially explosive atmospheres. Specific requirements for the protective mode : pressurised apparatus "p".

5. EN 50017 Electrical apparatus for potentially explosive atmospheres. Specific requirements for the protective mode : powder filling "q".

6. EN 50018 Electrical apparatus for potentially explosive atmospheres. Specific requirements for the protective mode : flameproof enclosure "d".

7. EN 50019 Electrical apparatus for potentially explosive atmospheres. Specific requirements for the protective mode : increased safety "e".

8. EN 50020 Electrical apparatus for potentially explosive atmospheres. Specific requirements for the protective mode : intrinsic safety "i".

9. EN 50028 Electrical apparatus for potentially explosive atmospheres. Specific requirements for the protective mode : encapsulation "m".

10. Wilday, A J et al., (2000), "Determination of safety categories of electrical devices used in potentially explosive atmospheres (SAFEC) Final Report", www.safetynet.de/EC-Projects/40.html.

11. BS EN 954-1: 1997, Safety of machinery - Safety-related parts of control systems - Part 1. General principles for design., BSI Standards, ISBN 0 580 27466 7.

12. IEC 61508 Functional safety of electrical/electronic/programmable electronic safety-related systems, Parts 1 to 7, 1998 (also published as BS EN 61508).

13. Wilday, A J, and Wray, A M, "Safety categories for safety devices used in electrical equipment for use in potentially explosive atmospheres", Proceedings of International Conference on Explosion Safety in Hazardous Areas, 11–12 November 1999, Institution of Electrical Engineers, UK

14. Area Classification Code for Petroleum Installations (Part 15 of the Institute of Petroleum Model Code of Safe Practice in the Petroleum Industry), Institute of Petroleum/John Wiley, 1990

15. The tolerability of risk from nuclear power stations, HSE/HMSO, 1992.

16. A. W. Cox, F. P. Lees & M. L. Ang, "Classification of hazardous locations", Institution of Chemical Engineers, 1990

17. BIA, "Dokumentation Staubexplosionen, Analyse und Einzelfalldarstellung", Report 11/97, 1997

18. Smith, D J (1993), Reliability, maintainability and risk - Practical methods for engineers, Fourth edition, Butterworth Heinemann, 1993, ISBN 0 7506 0854 4.

ACKNOWLEDGEMENTS

The work described was carried out within the EC SAFEC project (SMT4-CT98-2255) with the support of the European Commission. The input of the other partners and of the members of CENELEC TC 31/WG09 is gratefully acknowledged. Information on pressurised systems was kindly provided by Mr A Owler of "Ex" Certification Outsource Ltd.

DISCLAIMER

The views expressed in this paper are those of the authors alone and are not a statement of HSE or HSL policy.

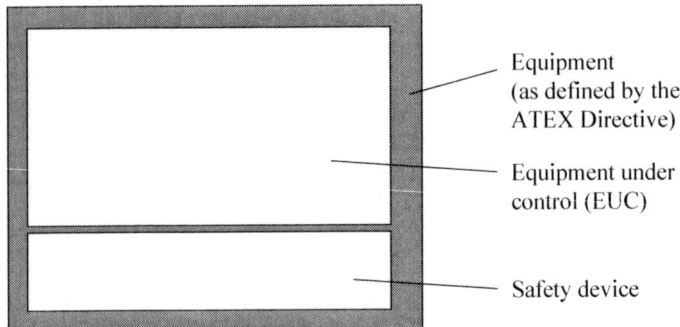

Equipment
(as defined by the
ATEX Directive)

Equipment under
control (EUC)

Safety device

Figure 1. Definition of terms

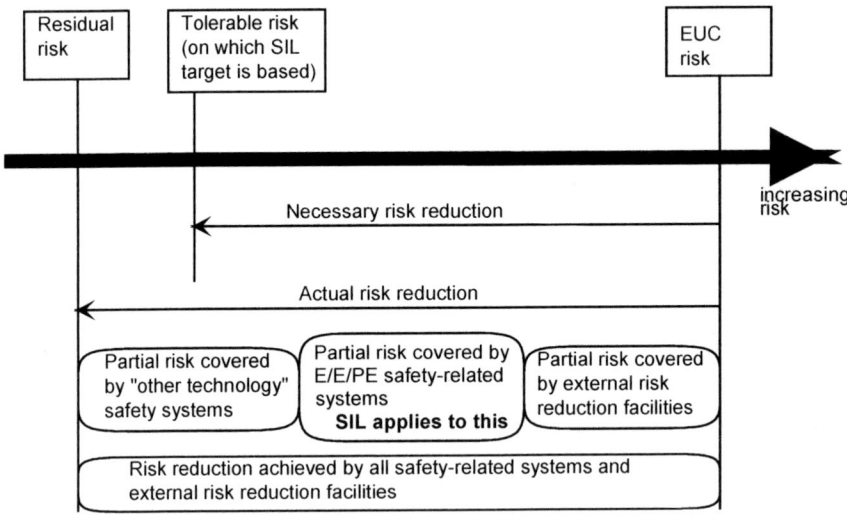

Figure 2. Risk concepts from IEC 61508

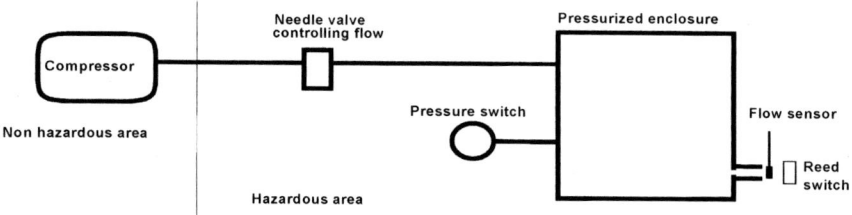

Figure 3. Generic design for a pressurisation system: air-flow diagram

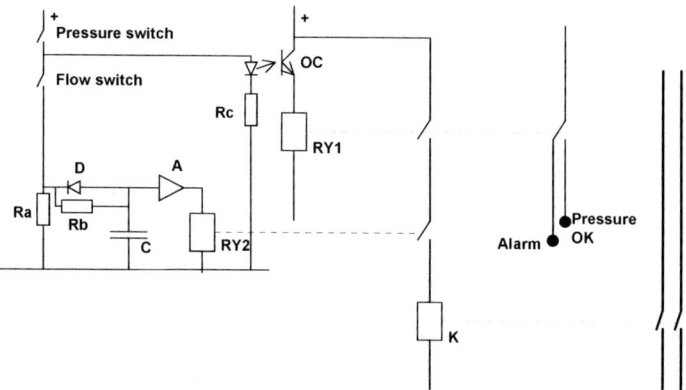

Figure 4. Generic design for a pressurisation system: electrical diagram

DEVELOPMENT OF EUROPEAN STANDARDS: NON-ELECTRICAL EQUIPMENT FOR USE IN EXPLOSIVE ATMOSPHERES

Dr. Richard L. Rogers

INBUREX Consulting GmbH, Hamm, Germany. email: richard.rogers@inburex.com

The European standards body, CEN, has a mandate to produce standards in support of Directive ATEX 100 which specifies requirements for both electrical and no-electrical equipment for use in potentially explosive atmospheres. This work is being carried out in the technical committee CEN/TC 305 with working group WG2 being responsible for standards for non-electrical equipment. The paper describes the development and current status of the standards being produced and the reasons for the differences between the standards for electrical and non-electrical equipment. In particular the requirements contained in the European standard EN 13463 part 1 "Non-electrical equipment for potentially explosive atmospheres -Part 1: Basic method and requirements" with its specification that manufacturers must carry out an ignition hazard assessment for the classification of equipment are elaborated.

Explosive atmospheres, Standards, ATEX, Non-electrical equipment

INTRODUCTION

In order to help manufacturers of equipment and protective systems intended for use in potentially explosive atmospheres meet the Essential Safety Requirements relating to EU Directives 89/392/EC [1] (machinery directive) and 94/9/EC [2] (ATEX 100A) mandated standards for non-electrical equipment are being prepared by CEN/TC305 Working Group 2. *EN 1127 – 1 Explosive atmospheres – Explosion prevention and protection Part 1: Basic concepts and methodology* [3] is a type A standard and sets out the overall philosophy of explosion prevention and protection. The standards currently being prepared by WG2 are type B standards and set out the requirements for specific means of prevention and protection which can be used for different types of equipment.

The first of these standards provides the basic method and requirements and is the core document which sets out the philosophy, concepts and requirements for all the standards in this series while parts 2 to 8 give the requirements for the specific 'types of ignition protection' which may be used.

TYPES OF 'IGNITION PROTECTION MEASURES'
STATUS OF THE STANDARDS

Table 1 lists the current standards in the EN13463 series and their current status. In addition a subgroup of WG2 is preparing a type C product standard on the construction and specifications for fans working in potentially explosive atmospheres.

Although there appears at first sight to be a degree of similarity between the standards being developed for non-electrical equipment and those currently available for electrical equipment, there are in fact major differences. These arise because in normal operation the majority of non-electrical equipment does not constitute an ignition source whereas the converse is often true for electrical equipment.

Table 1.   Status of the EN 13463 Standards as of September 2002

EN 13463 Part No:	Title	Status
1	Basic method and requirements	Published Jan 2002
2	Protection by flow restricting enclosure (fr)	Public Enquiry (ends 22.03.03)
3	Protection by flameproof enclosure (d)	Public Enquiry (ends 22.02.03)
4	Protection by inherent safety (g)	Draft
5	Protection by constructional safety (c)	Vote (ends 08.10.02)
6	Protection by control of ignition sources (b)	Public Enquiry (ends 22.02.03)
7	Protection by pressurisation (p)	Draft
8	Protection by liquid immersion (k)	Vote
	Safety requirements for ignition protected fans	Public Enquiry

EN 13463-1 PART 1 BASIC METHOD AND REQUIREMENTS

EN 13463-1 "Non-electrical equipment for potentially explosive atmospheres Part 1: Basic method and requirements" was finally published in January 2002. The standard provides a methodology for classifying equipment into the different Categories of Group I and Group II. The standard requires the manufacture to carry out a formal documented hazard analysis that identifies and lists all of the potential sources of ignition by the equipment and the measures to be applied to prevent them becoming effective. Examples of such sources include hot surfaces, naked flames, hot gases/liquids, mechanically generated sparks, adiabatic compression, shock waves, exothermic chemical reaction, thermite reactions, self ignition of dust, electrical arcing and static electricity discharges.

If equipment is designed and constructed according to good engineering practice and the ignition hazard assessment ensures that the equipment does not contain any effective ignition sources in normal operation, the equipment can be classified as category 3 equipment. Similarly where the ignition hazard assessment ensures that the equipment does not contain any effective ignition sources during foreseeable malfunctions or rare malfunctions, the equipment can be classified as category 2 or category 1 equipment respectively.

SCOPE OF EN 13463-1

Part 1 of the standards for non-electrical equipment for potentially explosive atmospheres specifies the basic method and requirements for design, construction, testing and marking of non-electrical equipment intended for use in potentially explosive atmospheres of gas, vapour, mist and dusts. It is valid for atmospheres having pressures ranging from 0.8 bar to 1.1 bar and temperatures ranging from −20 °C to + 60 °C. It also includes atmospheres that can exist inside the equipment if, for example, the external atmosphere can be drawn inside the equipment by natural breathing produced as a result of fluctuations in the equipment's internal operating pressure, and/or temperature.

The standard may also be used for the design, construction, testing and marking of equipment intended for use in atmospheres outside the validity range stated above, but in this case, the ignition risk assessment, ignition protection provided, additional testing (if necessary), manufacturer's technical documentation and instructions to the user, shall clearly demonstrate and indicate the equipment's suitability for the conditions it may encounter.

METHODOLOGY FOR CLASSIFYING EQUIPMENT INTO DIFFERENT CATEGORIES

The standard contains a methodology which will enable equipment manufacturers to classify non-electrical equipment into the different categories. This is shown diagrammatically in figure 1.

Equipment Intended for use in a Potentially Explosive Atmosphere

All equipment is intended for use in a potentially explosive atmosphere needs to be assessed to determine if there are effective ignition sources present in normal operation, or if there are sources that might become effective if faults occur. In performing this assessment, account has to be taken, not only of the equipment's moving parts, but also the equipment's enclosure. This is because some enclosure materials pose an effective ignition risk in normal operation irrespective of their contents . For example, enclosures made of light alloys or plastics, which can give rise to either thermite ignition when struck by rusty steel, or by electrostatic discharge when rubbed by other materials or contacted by flowing liquids. It is for this reason that even relatively innocuous mechanical mechanisms have to be assessed. The requirements of this standard do not apply to a piece of equipment with no potential ignition sources.

Ignition Hazard Assessment

The ignition hazard assessment, described in Section 5.2 of the standard, and explained below, is used both to identify potential sources of ignition in a piece of equipment and also to determine whether any applied protective measures render these non-effective.

Eliminate the Ignition Source(s)/prevent them becoming Active/or Apply Protective Measures

These include measures to ensure that the ignition source does not arise, measures to ensure that the ignition source cannot become active, measures to prevent the explosive atmosphere reaching the ignition source or measures to contain the explosion and prevent flame propagation. The various protective measures and the corresponding different types of ignition protection are applied in addition to the measures described in this standard.

The application of a protective measures is designed to make the ignition source non-effective. The ignition source may or may not be eliminated depending on the protective measure or type of ignition protection applied.

Whether a protective measures has to be applied and how many depends on the result of the ignition hazard assessment and the desired final classification of the equipment. For example:

- No protective measures are required and a piece of equipment may be classified as Category 3, 2 or 1 if the ignition hazard assessment shows that it has no effective ignition source during normal operation, expected malfunctions or rare malfunctions respectively, provided that the requirements of this standard are applied.

- A single appropriate protective measures will be required for a piece of equipment which in normal operation has no effective ignition source but has an effective ignition source during expected malfunctions in order for it to be classified as Category 2. The requirements of this standard and of the chosen ignition protection standard have to be applied.
- Two appropriate protective measures will be required for a piece of equipment which has an ignition source both in normal operation and during expected malfunctions in order for it to be classified as Category 2. The requirements of this standard and of both the chosen ignition protection standards have to be applied.

A similar logic applies for the classification of Category 1 equipment. The use and limitations of the types of ignition protection are given in the individual ignition protection standards.

Is the Equipment Capable of Igniting the Explosive Atmosphere during Normal Operation/Expected Malfunctions /Rare Malfunctions?
The classification of a piece of equipment into Category 3, 2 or 1 depends on the result of the ignition hazard assessment of the equipment together with any protective measures that are applied. Potential ignition sources may have been rendered non-effective, by means of one or more appropriate types of ignition protection. This depends on the category of equipment that is required and whether effective ignition sources occur in normal operation or during expected malfunctions or rare malfunctions.

IGNITION HAZARD ASSESSMENT
The standard requires that all equipment and all parts of it shall be subjected to a formal documented hazard analysis that identifies and lists all of the potential sources of ignition by the equipment and the measures to be applied to prevent them becoming effective. Examples of such sources include hot surfaces, naked flames, hot gases/liquids, mechanically generated sparks, adiabatic compression, shock waves, exothermic chemical reaction, thermite reactions, self ignition of dust, electrical arcing and static electricity discharge.
       Consistent with the requirements of the ATEX 100a Directive the standard specifies that protective measures/types of protection shall be considered and/or applied in the following order:

- ensure that ignition sources cannot arise
- ensure that ignition sources cannot become effective
- prevent explosive atmosphere reaching the ignition source
- contain the explosion and prevent flame propagation

The ignition hazard assessment document will differ according to the different equipment groups and categories of equipment in a particular group. The standard requires that the results of the ignition hazard assessment shall include as a minimum, information on all potential ignition sources, the measures which have been applied to prevent the sources becoming effective, and the ignition protection used.
       The manufacturer has to record the results a defined tabular format and the hazard assessment report must be included with the required technical documentation which

demonstrates compliance with the standard. An example of the form for recording the results for equipment group II is shown in table 2.

Two examples of completed ignition hazard assessments are included in the standard to help manufacturers and users of the standard.

Table 2.   Defined format for recording the Ignition Hazard Assessment (given in EN13463-1)

Potential ignition source				
Normal operation (Cat. 3)	Foreseeable malfunction (Cat. 2)	Rare malfunction (Cat. 1)	Measures applied to prevent the source becoming effective	Ignition protection used

MAXIMUM SURFACE TEMPERATURE

To maintain consistency with standards for electrical equipment, the temperature classes T1 to T6 are used to classify the maximum surface temperature for Group IIG equipment. The safety margin to the minimum ignition temperature of the potentially explosive atmosphere as required by EN1127-1 has been included in the defined maximum surface temperature of the equipment so that the temperatures are directly comparable with those for electrical equipment.

ELECTROSTATIC HAZARDS

The standard includes requirements to prevent the occurrence of electrostatic hazards. These apply to any non-conductive parts of the equipment exposed to the explosive atmospheres and susceptible to electrostatic charging. The requirements are based on the recommendations in the CENELEC report [4]

Occurrence of Highly Efficient Charge Generating Mechanisms (Propagating Brush Discharges)

Where propagating brush discharges can arise following highly efficient charging of non-conductive layers and coatings on metal surfaces the standard requires that they shall be prevented in both Group I and Group II equipment from occurring by ensuring that the breakdown voltage across the layers is less than 4 kV.

For Group IID equipment to be used only in the presence of dust atmospheres with a minimum ignition energy of greater than 3 mJ propagating brush discharges can also be prevented by ensuring that the thickness of the non-conducting layer is greater than 10 mm.

Occurrence of Brush Discharges

The occurrence of brush discharges are prevented by the requirement that the projected surface areas conductive materials shall be so designed that under normal conditions of use, maintenance and cleaning, danger of ignition due to electrostatic charges is avoided.

For Group I equipment of both Category M1and M2 this requirement applies when the surface area projected in any direction of more than 100 cm².

For Group II equipment the standard requires that it shall be so designed that under conditions of use, maintenance and cleaning, danger of ignition due to electrostatic charges is avoided. Three means are provided for satisfying this requirement:

- by suitable selection of the material so that the insulation resistance of the enclosure does not exceed 1 GΩ at (23 ± 2) °C and (50 ± 5)% relative humidity,
- or by virtue of the size, shape and lay-out, or other protective methods, such that dangerous electrostatic charges are not likely to occur. For category 2G equipment this requirement can be satisfied by using the test provided in Annex C of the standard.
- or by limitation of the surface area projected in any direction of non-conductive parts of equipment liable to become electrostatically charged as shown in the table 3:

To prevent incendive brush discharges, the thickness of layers or coatings of plastic (non-conductive) solids on earthed metal (conducting) surfaces which can become charged in Group IIG equipment shall not exceed 2 mm in the case of gases and vapours of Group IIA and IIB or 0.2 mm in the case of gases and vapours of Group IIC.

There is no need to prevent brush discharges and hence no restriction on the thickness of layers or coatings of plastic (non-conductive) solids on earthed metal (conducting) surfaces which can become charged in Group II equipment intended for use on potentially explosive dust atmospheres with a minimum ignition energy of greater than 3 mJ.

Table 3. Permitted maximum projected areas for non-conductive parts of equipment liable to become electrostatically charged

Category	Permitted area cm^2			
	Dusts (MIE < 3 mJ)	IIA	IIB	IIC
1	250	50	25	4
2	500	100	100	20
3	No limit *	No limit*	No limit*	No limit*

*unless the intended use of the equipment can result in frequent incendive discharges occurring in normal operation, in which case the criteria for Category 2 equipment shall apply.

These values may be multiplied by 4 if the exposed flat areas of plastics are surrounded by conductive earthed frames.

CONTENTS OF THE 'IGNITION PROTECTION MEASURES STANDARDS

PART 2: PROTECTION BY FLOW RESTRICTING ENCLOSURE (fr)

Protection by flow restricting enclosure is "... a type of ignition protection which, by sealing of the means of an enclosure, reduces the probability of ingress of a surrounding explosive atmosphere into the enclosure to an acceptably low level so that the concentration inside the enclosure is below the lower explosive limit. " (Definition from prEN 13463-2).

Experience has shown that even simple enclosures can prevent a surrounding explosive atmosphere from reaching ignition sources inside them. flow restricting enclosures are such simple enclosures, which will prevent, with adequate probability, the atmosphere inside the enclosures becoming explosive if the atmosphere outside the enclosure becomes explosive rarely and for a short duration only. For this reason their use is restricted to the fulfilment of the requirements of Group II - Category 3 equipment.

An explosive atmosphere surrounding an enclosure can penetrate it mainly due to the influence of three mechanisms, namely: ventilation, equalisation of pressure differences between the inside and outside (breathing), and diffusion.

If such an enclosure is effectively sealed, but not necessarily gas-tight, it can be assumed that ventilation and diffusion will not cause a significant short-time exchange of atmosphere. Under these conditions, an exchange of the external and internal atmospheres through the seals will only take place if there is a pressure difference across them. Such pressure differences may be caused by changes in temperature and will result in the enclosure "breathing" but will not cause a significant flow of atmosphere into or through the enclosure

PART 3: PROTECTION BY FLAMEPROOF ENCLOSURE(d)

Since its conception, protection by flameproof enclosure has been developed to allow many kinds of continuously sparking equipment to be used safely in places where a potentially explosive atmosphere exists. For electrical equipment, this type of protection is well known for protecting power arcing components and is defined and described in EN 50018. As the electrical equipment standard contains the generic testing, verification and marking requirements, unnecessary duplication of the requirements in this non-electrical equipment standard is avoided by cross reference to the electrical standard. In this standard, only those differences necessary for the purpose of providing protection for non-electrical equipment are written in full.

The basic principle of ignition protection by the use of a flameproof enclosure, is that gases, or vapour, may enter the enclosure through the cover joints/flanges and if an explosive atmosphere inside the enclosure ignites, neither the enclosure will be deformed significantly, nor flame transmitted through the joints/flanges to the explosive atmosphere outside. For this reason the enclosure has to be both robust and have dimensionally controlled cover joints/flanges with maximum allowable safe gaps appropriate for the types of explosive gas/vapour likely to occur inside the equipment

EN 50018 does not consider explosive atmospheres formed by dusts, except for Group I, category M2 electrical equipment, where its associated General Requirements document - EN 50014, states that flameproof equipment designed, constructed and tested for use in explosive atmospheres of firedamp (explosive mine gas consisting mainly of methane) needs no alteration, or further testing to allow it to be used where a coal dust cloud is present.

The flame proof enclosures standard for non-electrical equipment has been extended to allow the concept to be used in potentially explosive dust atmospheres. The concept used for protecting equipment against dust cloud ignition in this standard for both Group I, Category M2 mining equipment, and Group II, Category 2G and 2D non-mining equipment is the testing of an enclosure in a gas/air mixture. This is because it introduces an acceptable

467

safety factor against ignition and it allows a much more simple method of testing and verifying its explosion protection properties.

Examples of non-electrical types of equipment that can be protected by flameproof enclosure include equipment with potentially hot rubbing surfaces exceeding the ignition temperature of the atmosphere surrounding them, e.g. friction clutches and brake linings.

PART 4: PROTECTION BY INHERENT SAFETY (g)

The concept of this standard is to define limits for energy and relative speed between movable components of equipment below which it is impossible to form ignition sources. Two methods to realise this are:

- Definition of limits for energy and relative speed preventing the formation of ignitable sparks or hot surfaces in any potentially explosive atmosphere. Mechanical equipment which is small enough or designed to make it impossible to exceed these limits may be used in any potentially explosive atmosphere without additional hazard assessment.

- Definition of acceptable limits for energy and relative speed in dependence on the probability of occurrence and type of explosive atmosphere. The higher the impact energy and the relative speed between movable parts the higher is the probability of generating ignitable sparks or hot surfaces. The idea of defining limits is that the increase of the probability of generating ignitable sparks or hot surfaces has to be related to the decrease in the probability of occurrence of explosive atmospheres and their ignition energy. This then provides a relationship between the category of the equipment and its use. Mechanical equipment which is designed to make it impossible to exceed limits defined for a given zone may be used as equipment of the respective category in this type of explosive atmosphere without additional hazard assessment.

This concept is being developed into a draft standard in the working group.

PART 5: PROTECTION BY CONSTRUCTIONAL SAFETY (c)

The type of protection "constructional safety" is defined as a "type of ignition protection in which constructional measures are applied so as to protect against the possibility of ignition from hot surfaces, sparks and adiabatic compression generated by moving parts.", (Definition from prEN 13463-5)

Mechanical (non-electrical) equipment has been used for decades in potentially explosive atmospheres. Effective ignition sources were most frequently avoided by application of sound engineering principles so that the probability of creating high temperatures or mechanical sparks which could act as an ignition source was reduced to an appropriate level. These measures are part of the safe construction of the equipment based on sound engineering principles and provide ignition protection without additional protection measures thus, this equipment is "constructionally safe".

The standard provides requirements relating to materials for external enclosures and exposed equipment parts, ingress protection, gaskets and sealing arrangements, lubricants/coolants/fluids, moving parts, bearings, power transmission systems, clutches and couplings, brakes and braking systems; springs and absorbing elements and conveyor belts.

PART 6: PROTECTION BY CONTROL OF IGNITION SOURCES (b)
The type of protection "control of ignition sources" is defined as "a type of ignition protection applied to one or more potential ignition sources in non-electrical equipment, whereby integral sensors detect impending operation likely to cause an ignition and initiate control measures before a potential ignition source becomes an effective ignition source. The control measures applied may be either automatic, or manual." (Definition from prEN 13463-5)

Many types of non-electrical equipment intended for use in potentially explosive atmospheres of gas, vapour, mist and/or combustible dust, do not contain an effective ignition source in normal operation. However, there is a risk that an ignition source might arise in such equipment if the moving parts suffer a malfunction or an abnormal operation occurs.

An example of this is turbine, having high speed rotating blades fixed to a shaft, supported on rolling element bearings, inside a stator. In normal operation no ignition capable frictional ignition sources should be present. However, because the clearances between the rotor and stator are very small, malfunctions such as the collapse of a shaft bearing, distortion of a rotating blade, build up of foreign material on a rotating blade, etc. could cause the clearance to be reduced and frictional sparking, or hot surfaces, to occur.

To prevent potential ignition sources from becoming effective during normal operation, malfunction and rare malfunction, it is possible to incorporate sensors into the equipment to detect impending dangerous conditions and initiate control measures at an early stage of deterioration before the potential sources are converted into effective sources. The control measures applied, may be initiated automatically, via direct connections between the sensors and the ignition control actuators, or manually, by providing a warning to the equipment operator (With the intention of the operator applying the ignition control measures e.g. by stopping the equipment).

In this standard, the incorporation of such sensors and their associated automatic/manual ignition control measures, to prevent potential ignition sources becoming effective ignition sources, is known as protection by "Control of ignition source 'b' ".

This type of ignition protection, and the devices used to achieve it, can take many forms. In practice, they may be mechanical, electrical, optical, visual or a combination of all of these. Some examples of simple mechanical sensor/actuator devices are fusible plugs centrifugal speed governors, thermostatic valves, pressure relief valves (using springs or weights), etc.. Although this standard deals with the ignition protection of non-electrical equipment, it nevertheless has to take account of the fact that an increasing amount of non-electrical equipment makes use of electrical sensors to detect and initiate the ignition control measures. It is therefore impossible to produce a non-electrical equipment protection standard without making reference to the use of electrical sensors and their associated ignition control actuator circuits.

Some examples of combined electro-mechanical sensor/actuator devices are temperature, flow and level monitoring/control devices, optical pulse counters, that sense abnormal rotational speeds, vibration sensors, that detect abnormal vibration, from e.g. rolling element bearings, before they fail, conveyor belt alignment devices, power transmission belt tension devices, wear detectors on clutches, that detect unacceptable wear likely to cause frictional heating by incorrect engagement of the clutch.

Such sensor/actuator control devices may be either, continuously active in normal operation of the equipment (e.g. to control the temperature of category 3 equipment), or be arranged so that they only to detect abnormal operation (e.g. to detect impending dangerous over-temperature in category 2 equipment).

Integrity of the Protection System – Functional Failure Rates (FFR)
As malfunction of any of the above sensors/actuator control devices, may result in failure to apply the appropriate ignition control measure, they must be considered to be safety related parts of the equipment. This ignition protection standard therefore calls for them to be assessed and suggests a minimum quality for such devices in the form of a Functional Failure Rate (FFR) that the equipment manufacturer must attempt to achieve.

Thus, to meet the requirements of this standard, the non-electrical equipment manufacturer has to perform both the ignition hazard assessment (Required by EN 13463-1), and additionally, a risk evaluation, to determine the Functional Failure Rate (FFR) necessary to ensure that the sensors/ignition control actuators function when they are called upon to contain the ignition risk within tolerable limits.

The Functional Failure Rate (FFR) is defined as a level of risk reduction to be aimed for by the equipment manufacturer as a result of an evaluation of the ignition risk, caused by the failure of a sensor or ignition control actuator to perform its intended function, at the same time as a potential ignition source in the equipment converts into an effective ignition source in the presence of an explosive atmosphere. Three classes of FFR are defined depending on the probability of occurrence of the above three events occurring simultaneously, i.e. the occurrence of the ignition source, the failure of the control system and the presence of an explosive atmosphere, the latter being defined by the category of the equipment. FFR1 is defined as a low probability of all three events occurring simultaneously; FFR2 has a foreseeable probability and FFR3 has a high probability.

Criteria used in the Different Functional Failure Rate Levels
One of the main difficulties facing a manufacturer in using control of ignition sources as a means of protection is the selection and classification of integrity of the control system to be used as there are currently no defined criteria for non-electrical control systems. The standard therefore specifies criteria which have been based on the concepts used in various European Standard. Thus *EN 954-1 "Safety of Machinery – Safety related parts of control systems : Part1: General principles for design."* written by CEN/TC/114 to assist machinery manufacturers, describes 5 categories (B, 1, 2, 3 and 4) that can be applied to assess the quality of the safety related parts of machinery control systems. Although not specifically written for the purpose of assessing ignition control devices, some of the principles described in that standard have been used in the development of the criteria used in prEN 13463-6.

In the case of electrical control systems, the International Electrotechnical Commission Standard IEC 61508 "Functional safety of electrical/electronic/ programmable electronic safety-related systems", was written by IEC/65A to assist manufacturers of safety related systems. It contains the requirements for four Safety Integrity Levels (SIL 1, 2, 3 and 4) that can be applied to describe the quality of the safety related parts of a control system. Following the recent publication of the seven parts of IEC 61508, some member state test authorities have announced their intention to offer a service for checking such safety related components and protective systems and provide manufacturers with an attestation of its Safety Integrity Level (SIL) rating. In addition IEC/TC/44 Committee has recently started work on a document that is the equivalent of EN 954-1 for electrical/electronic and programmable controlled machine safety. This is based on IEC 61508 and was circulated in September 2000 as IEC draft 44/292/CD. When published this latter document will give more definitive guidance on the SILs of safety related parts of machines.

At the present time however, most sensors and ignition control actuators used for the purpose of this standard will not have been assessed or given a SIL rating, and in addition these are not applicable to non-electrical control systems. Thus in order to provide a common classification of the control systems to be used for ignition protected equipment and in order to easily link these to the 3 categories of equipment the standard specifies three "Functional Failure Rate " levels.

Application of a Functional Failure Rate (FFR) to Different Categories of Equipment
The likelihood that a hazard will occur increases from Function Failure Rate class 1 to 3 and this is reflected in more stringent requirements for the control system for FFR1 to FFR3, i.e. control systems of class FFR3 have therefore a higher reliability. Thus suitable sensors and/or ignition control actuators for use with a FFR of 1 are characterised by well tried components, having a proven history of reliability, assembled and installed in accordance with any relevant standards, adopting well tried safety principles, able to withstand expected influences during operation of the equipment and checked for failure to perform their intended functions at each periodic maintenance check on the equipment, while FFR3 systems have to be so arranged that a single fault on a sensor, or ignition control actuator, does not cause loss of the ignition protection and any such fault is immediately detected at the time it arises. The following table shows the link between the likelihood of occurrence of an ignition source (identified by the manufacturer during the ignition hazard assessment described in EN13463 Part-1), the desired category of the equipment and the resulting Functional Failure Rate class.

For Category 3 Non-electrical Equipment
This equipment, by definition, does not contain sources of ignition in normal operation. To meet this basic requirement, it will not therefore usually be necessary to apply additional Control of Ignition Source 'b' protection to cater for abnormal operation of the equipment. The exception to this, is equipment that has to be controlled by some device as part of its normal operation. For example, a speed control device fitted to ensure that a rotating part of a machine maintains the correct speed in normal operation. In this case, the speed control device can be interpreted as an "Ignition control actuator" as described in this standard.

Table 4. Minimum FFR requirements for a single sensor/ignition control actuator used to protect Group II equipment

Ignition source	Category 3	Category 2	Category 1
Foreseeable in normal operation	FFR 1	FFR 2	FFR 3
Foreseeable during malfunction	Not relevant for category 3	FFR 1	FFR 2
Foreseeable during rare malfunction	Not relevant for category 3	Not relevant for category 2	FFR 1

It is also be possible to fit a Control of Ignition Source 'b' device to normal industrial equipment, thereby convert it from equipment that is not intended for use in a potentially explosive atmosphere to a type that meets the definition of category 3 equipment.

In the above cases, the probability of the control actuator failing at the same time as an explosive atmosphere occurs will be rare and consequently FFR1 has therefore been assigned by this standard to those ignition control sensors/actuators used to protect category 3 equipment.

For Category 2 Non-electrical Equipment
This category of equipment needs to be protected against ignition sources occurring in normal operation and also with foreseeable faults on the equipment. In this case, the probability of an ignition source developing in the equipment at the same time as its associated ignition control actuator is faulty and an explosive atmosphere is present is higher than for category 3 equipment. Nevertheless, the mid-range functional failure rate will suffice. FFR 2 is therefore been assigned by this standard to those ignition control sensors/actuators used to protect category 2 equipment.

For Category M2 Non-electrical Equipment
Category M2 equipment needs to be ignition protected and suitable for the severe operating conditions of use found in gassy mines, but it is intended to be de-energised if an explosive atmosphere occurs. The probability of an ignition source developing in the equipment at the same time as its associated ignition control actuator is faulty and an explosive atmosphere is present is therefore higher than category 3, but not as high as category 2 because of its intended short time exposure to an explosive atmosphere. A mid-range functional failure rate will however suffice. FFR 2 has therefore been assigned by this standard to those ignition control sensors/actuators used to protect category M2 equipment.

For Category 1 and Category M1 Non-electrical Equipment
Category 1 equipment needs to be ignition protected in normal operation, also with foreseeable faults and rare faults applied to the equipment. Category M1 equipment needs to be ignition protected to a very high level, also be suitable for the changing conditions in mines and continued use in an explosive gassy mine atmosphere.

The definitions and requirements for both categories of equipment also include reference to such equipment being either safe with more than one fault applied, or double ignition protected. Thus a commensurate high functional failure rate class is needed. FFR 3 is therefore been assigned by this standard to those ignition control sensors/actuators used to protect category 1 and category M1 equipment.

A flow diagram shown in figure 2 is provided which illustrates this procedure.

PART 7: PROTECTION BY PRESSURISATION (p)

As with the standard on flameproof enclosures, the electrical standard on pressurisation is being extended to cover the situation of equipment for use in potentially explosive dust atmospheres. A draft is in preparation.

PART 8: PROTECTION BY LIQUID IMMERSION (k)

The type of protection "liquid immersion" is defined as is a "type of protection in which potential ignition sources are made ineffective or separated from the flammable atmosphere by either totally immersing them in a protective liquid, or by partially immersing and continuously coating their active surfaces with a protective liquid in such a way that an explosive atmosphere which may be above the liquid, or outside the equipment enclosure cannot be ignited." (Definition prEN13463-8).

Certain types of non-electrical equipment, intended for use in potentially explosive atmospheres of gas, vapour and/or dust, have their potential ignition sources rendered ineffective by either submersing them in a protective liquid, or by continuously coating them with a flowing film of protective liquid. In some equipment, the protective liquid is provided solely for the purpose of preventing the potential ignition sources from becoming effective. In other equipment, the protective liquid serves additional purposes, such as lubricating and/or cooling moving parts, or as in the case of hydraulic systems, for transmitting energy. In some equipment, the protective liquid may be the actual process liquid itself.

Examples of the kinds of equipment utilising this type of ignition protection are oil immersed disc brakes, diaphragm and other submersible pumps used for delivering flammable liquids, oil filled gearboxes, fluid couplings etc. In all of the above, ignition protection is achieved by the fact that protective liquid prevents the surrounding explosive atmosphere from coming into contact with the ignition source(s) by continuously coating, and/or lubricating and cooling the moving parts.

A similar type of ignition protection, known as oil immersion "o", has been used for many years for electrical equipment, where, in addition to the above, the liquid also acts as an electrical insulating medium. It is for this latter reason that this standard cannot be applied to electrical equipment, because it allows the use of liquids that conduct electricity.

SAFETY REQUIREMENTS FOR IGNITION PROTECTED FANS

This draft standard specifies the basic methods and requirements for design, construction, testing and marking of complete fan units intended for use in potentially explosive atmospheres in air containing gas, vapour, mist and/or dusts. Such atmospheres may exist inside, outside or inside and outside of the fan. A draft standard is available.

CONCLUSIONS

The development of standards for non-electrical equipment for use in potentially explosive atmospheres is progressing satisfactorily with the important first standard on Basic method and requirements now published. The standard includes many novel aspects when compared with the equivalent standard for electrical equipment and requires that the manufacturer carries out an ignition hazard assessment. Requirements are given for both equipment groups I and II and for the different categories of equipment. In addition specific requirements are given for non-conductive parts of equipment to protect against the hazards of electrostatic charging.

ACKNOWLEDGEMENTS

The author acknowledges the work by the other members of CEN TC/305/WG2 without whom this standard would not have been possible.

REFERENCES

1. Machinery Directive 98/37/EC
2. ATEX100a (Equipment for use in potentially explosive atmospheres) Directive 94/9/EC
3. EN 1127 – 1 Explosive atmospheres – Explosion prevention and protection Part 1: Basic concepts and methodology
4. CENELEC report R044-001 Safety of machinery – Guidance and recommendations for the avoidance of hazards due to static electricity

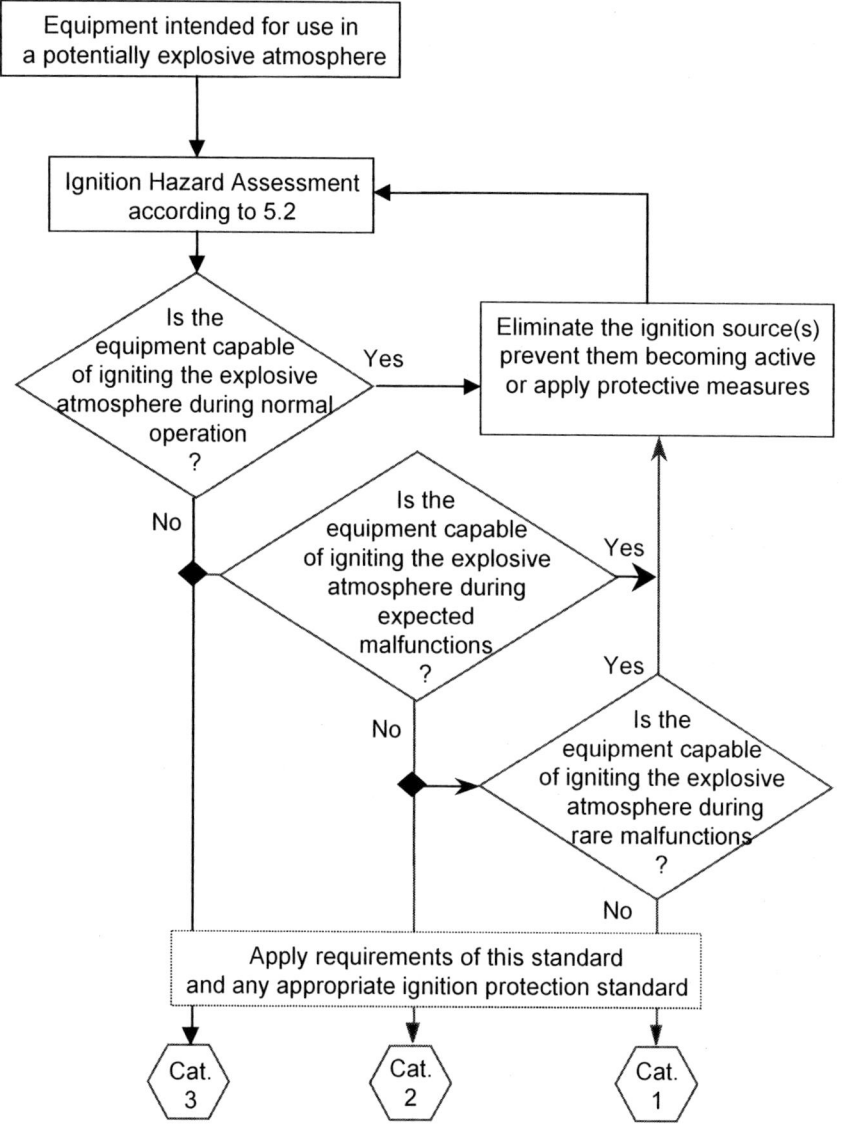

Figure 1. Methodology for classifying equipment into the different categories of group II

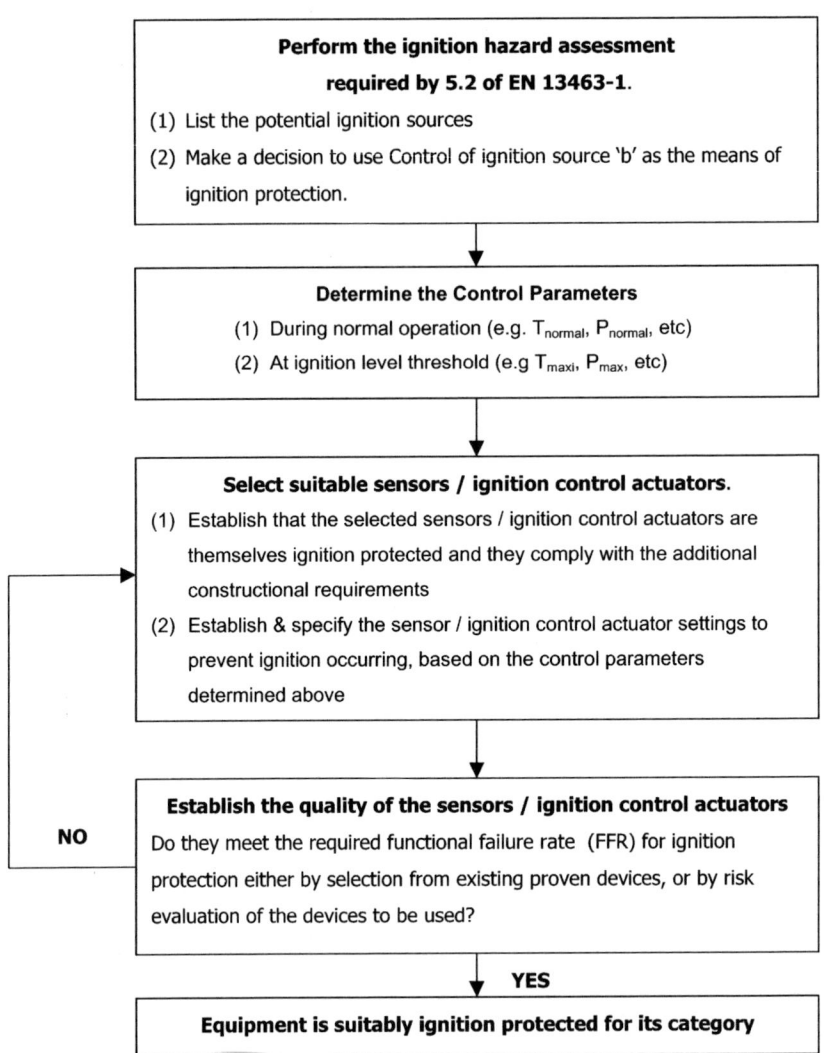

Figure 2. Flow diagram showing steps in the design of control of ignition sources protection

OPTIMISING THE LAND-USE AROUND TRANSMISSION PIPELINES

FK Crawley (Fellow)*, I Lines**, J Mather**,
*WSAtkins Glasgow G3 7LD and University of Strathclyde, G1 1XJ
**WSAtkins Warrington WA3 7WA

SYNOPSIS

This paper shows that the present more simplified approach to risk assessment for high-pressure hydrocarbon pipelines is somewhat conservative. The outflow modelling is more complex that might be predicted by the traditional, simple outflow equations due to two-phase formation in gas pipelines and froth formation in oil pipelines. The consequence models for jet fires, historically based on flare type model, are shown to be conservative. Taking these two points together the magnitude of the hazards can be overstated. The failure rate data for (underground) pipelines show that the greatest causes of massive releases are due to mechanical faults and third party intervention. When these effects are dealt with through design standards and policing of the pipe corridor the full bore failure rate of oil and gas pipelines is very low, 2×10^{-5} per kmyr which indicates that the frequency of the hazard is probably significantly over stated.

Taking all of the results together, both failure data and modelling, and knowing the specifics of each case, the assessed risks for land-use near Transmission Pipelines will be reduced, so releasing land for use in strategic areas.

INTRODUCTION

There are many potentially hazardous materials transported in pipelines at high pressure within the United Kingdom. Failure of the pipeline could affect a large population if it is close to residential houses. As a result there is careful attention given to land use close to such pipelines. However, in the United Kingdom, there is a high population density particularly near industrial areas and there is a conflict between the need for development of land for industry or housing and the need for the maintenance of the safety of the public at large.

Example of the piping in use are 17000 kms carrying high pressure Natural Gas, 200 kms carrying "live crude oil", 200 kms carrying high vapour pressure LPGs and 1000 kms carrying high pressure Ethene. There are, in addition, other pipelines carrying refined hydrocarbon products for both civil and military uses.

Inevitably there is pressure to develop land closer to the pipelines. The requirements for land-use planning are generally given in the Third Report of the Advisory Committee on Major Hazards (ACMH 3) [1] and the Health and Safety Executive's approach to risk criteria are discussed in Risk Criteria for Land-use Planning [2]. In this there are general duties placed on the Planning Authorities and the Health and Safety Executive. The first has to take the decisions taking into account the potential gains and risks to the local community and the second gives the assessment of the risks in order to make a Safety Judgement. ACMH 3 highlights this clearly, "Decisions where safety is involved often present a dilemma for the planning authorities. In many cases the authorities have to weigh up the advantages which a proposed development might bring against the disadvantages that more people might be at risk. The decision is less difficult when the risk is very great or very small, but in many cases fall between these extremes" [paragraph 111, verbatim]. However ACMH 3 then states further on in paragraph 111 "When a planning application is being considered a balanced view should be taken of all aspects including social and economic factors and not just health and safety".

It is at this point that there may be some conflict between the planning requirements and the risks defined by more simple risk models. As a result more sophisticated risk models and assessment are required. However many risk screening models are of necessity relatively simple, quick to use and also introduce some conservatism, which may preclude development in an area that might be assessed as "safe" if more sophisticated models were used. This paper discusses those areas where the conservatism may arise and shows that there has to be a better understanding of the outflow characteristics of the fluids, the consequence modelling and of the appropriate failure rates for piping. It is clear that the outflows for hydrocarbons in particular are complex, they can not be treated as simple fluids and the modelling must take into account the physical properties of the fluids at the transport conditions. The consequence models, particularly for fires, have to be re-examined to ensure that they represent the reality pertaining to the conditions of the assessment. Finally the failure rate data have to be carefully examined on a case-specific-basis to ensure that they are robust and relevant.

A significant percentage of this paper is based on work carried out following the Piper Alpha disaster as it was essential that the oil industry understood the out-flow characterises of both oil and gas pipelines in great detail when determining the need for sub-sea isolation valves (SSIV) and the fluid properties temperature, flow and phase. This work can now be applied to on-shore pipelines.

OUT FLOW CHARACTERISTICS

The out flow characteristics from high pressure pipelines are too complex for them to be reduced to simple single or two phase models. In the case of high pressure hydrocarbons operating in the dense phase above the critical pressure the fluids enter the two-phase regime following pressure loss resulting from loss of containment. In turn the compression wave velocities fall to a fraction of the gaseous values, the fluid is released as an aerosol and in some conditions it may pass through the phase diagram and be released as a cold gas. This in turn reduces the out flow rate but extends the duration. In the case of liquids with elevated vapour pressures transported at a pressure greater than the vapour pressure the fluids "boil" on loss of containment and are released as a froth or aerosol.

In the case of high-pressure hydrocarbons such as high-pressure Natural Gas and Ethene the temperatures of the fluids and the piping may fall to levels which might compromise the integrity of isolation valves. Further the reduction in the compression wave velocity in the two-phase fluids can result in the crack propagation velocity being greater then the compression wave velocity and there is then a "running crack" which will propagate from the source to the end of the line. In the worst-case scenario this will result in a crack starting in a benign zone of the pipeline route propagating into a sensitive area. The effect of this is that the contents of the pipeline will be released as a line source and not a local point source.

The outflow characteristics from severed high vapour pressure oil pipelines, which were modelled by the computer code OLGA [3], are discussed in Oil and Gas Pipeline Failure Modelling [4]. In this typical sub-sea oil pipeline with length 100 km and diameter 0.6 m (24 inches) was modelled with different vapour pressures (compositions are given) and static heads. In the paper it is shown that the pressure, the diameter, the length and the properties of the fluid dictate the outflow characteristics. This indicates that the simple assumption, that there will be no flow from a live crude oil pipeline provided the static head of a column of fluid is greater than the vapour pressure of the fluid, is totally erroneous. The out flows of the oil were first dictated by the stored energy in the fluids and the pipeline (line pack) and then by the physics of the system. Initially, for a severed line (full bore rupture, FBR), the pack was some 25 to 50 tonnes; thereafter the vapour pressure of the fluid and the static head of the fluid dictated the out flow. In the case in question the modelling was carried out as a single fluid in a pipeline with the two ends at different elevations representing different riser heights. (The riser is the vertical section of pipeline which connects the pipeline on the seabed to the platform itself.) The vapour pressure was then varied by adjusting the composition and the temperature. The initial loss of containment resulted in the lower molecular weight components "coming out of solution", this in turn resulted in the mean density of the column of fluid falling from about 825 kg/m^3 to about 275 kg/m^3 and as a result the imposed head at the bottom of the column would be below the vapour pressure of the fluid. The nett result was that the fluids in the 100 km reservoir of liquid in the horizontal section of the pipeline would release the lower molecular weight gases, which would then drive the "bubble pump". In reality the physical properties of oils would result in a foamy system and not discrete bubbles. The final result is that the reservoir of fluids would release about 10 % of the line content at a rate of 500 kg/s falling to 250 kg/s. The outflow with time will be regulated by two–phase pressure drops and not the traditional single phase as well as the residual driving gas reserve in the pipeline. The result is that flows will arrest at some time after the initial rupture. Figures 1 to 4 show the typical outflow characteristics. The outcome of this analysis is that any severed pipeline either under water or under ground carrying high vapour pressure oil has the potential to release a significant percentage of its total content over some minutes and only when the true static head imposed by the line contours exceeds 3 times the vapour pressure of the oil being transported is non-continuous release from a severed riser, or pipeline, realistic.

The small-bore releases were not modelled in detail. However it is clear that the line pack must be released before the pressure in the line reaches atmospheric pressure. Due to the surface chemistry the frothing action will ensure that there is still some of the line content discharged, but that will be dictated by the position of the hole and the line contours. Theoretically, for a long horizontal oil line transporting fluids with an elevated vapour pressure, nearly all of the fluids will be released following rupture but at a rate declining with time. This has some significant implications for a Pipeline Integrity Monitoring System (PLIMS) that uses mass balance and pressure transients to detect loss of containment. In the theoretical example in [3] the line capacity is about 300 kg/s and for a 25 mm hole at an operating pressure of 70 bar the outflow will be about 39 kg/s, if on the seabed, and 54 kg/s, if on land. Of this total outflow about 3 to 5 % will boil off. The impact of small releases with breach sizes of less than 2.5 mm suggests that there will be more environmental risk than human risk and that it may be some time before the initial defect is detected by the PLIMS and the pipeline is shutdown. As first the pressure profile may not change significantly and also the mass balancing may not be sufficiently accurate to detect such small releases.

The outflow from a dense phase gaseous pipeline is equally complex. Once the fluids enter the two-phase regime the standard rules break down. The initial compression wave velocity will be about 380m/s for Natural Gas and about 310 m/s for Ethene but these could fall to well below 100 m/s. The reduced compression wave velocity in the two-phase fluid, about 5 to 10 % wet, forms this choke. See figure 1. The flow characteristics are then determined by the flow through the moving choke inside the pipeline and the frictional pressure drop. Once again using the code OLGA [4], a study [5], based on a typical North Sea gas pipeline operating at a pressure of 170 bar, the pressure at the exit of the pipeline choked at about 20 bar (compared to the hydrostatic pressure of 10 bar). The moving choke within the pipeline resulted in fluid velocities of about 175 m/s, falling with time, and the gas temperatures which fell to about 200K within 60 seconds. (The wall temperatures were about 5K warmer.) The pressure and temperature at the closed end of the pipeline does not change for some time; in the case of a long pipeline it could be many minutes. In effect the gas at the closed end of the pipeline is "dynamically isolated" from the rest of the pipeline and does not start to depressurise until the choke reaches it. This is to be seen in figures 5 and 6. A similar effect is to be found with Ethene [3] where the temperature falls as low as 169K. The nett effect is that once the moving choke is established the frictional effects greatly reduce the out flow within seconds. In this study the out flow fluxes fell to 5750 kg/s/m^2 in 30 seconds. In the Piper Alpha Disaster [3,6] the outflows fell, both by calculation and the fireball modelling, from a peak instantaneous value of 7000 kg/s to about 1000 to 2000 kg/s within 15 seconds depending on the assumed wall roughness. The later flows give fluxes of between 3500 and 7000 kg/s/m^2, which are similar to study [5] using OLGA [4]. Modelling the flames seen in the Piper Alpha Report [6] shows that the outflows some 2 hours into the event were about 50 to 100 kg/s and even after some 24 hours the flows were a few kg/s. This demonstrates the difficulty with depressurising long pipelines.

Clearly the modelling of the outflows for high pressure gases, be they hydrocarbon based or other, is very complex and requires more sophisticated analysis with time than the simple traditional flow models. The implications of this analysis is that the outflows will have a longer duration but at a lower absolute rate. If the initial concern is for radiation levels, as is to be expected the simple modelling will over state the risk. If the initial concern is for the safe dispersion of the gas the high initial out flow will generate a trench and thereafter the jet effects will result in jet dispersion.

The outflow characteristics from other piping systems where there are no complex phase effects can be modelled by the traditional out flow models, but the "pack" created by the strain energy in the fluid and the pipe walls must still be taken into account.

CONSEQUENCE MODELLING
The consequence modelling of hydrocarbon systems in particular requires an understanding of the dispersion and combustion processes. For other fluids there will be particular special

features such as toxicity. More particularly there is the need to understand the failure modes of the piping itself.

First the material properties of the pipeline must be understood. In the case of dense phase fluids the drop in the compression wave velocity is such that it can be below the crack propagation velocity in the pipeline wall, which could result in a "running crack" in high pressure pipelines if crack arresters are not fitted. Also pipelines, which are highly stressed and have a low "toughness", are not tolerant of small defects, which could run into a total rupture. Conversely, lines which are less highly stressed, can tolerate quite large defects without the risk of total rupture. Fearnehough [6] suggests that, provided the ratio of the imposed to minimum yield stress does not exceed 60%, running cracks are unlikely for corrosion, and for "man-made defects" the value falls to 30%. However in the case of dense phase fluids the compression wave velocity influences the situation and the maximum stress will always be at the crack tip resulting in the potential for a running crack. The implications of this are quite significant. Highly stressed pipelines may spontaneously rupture and pipelines transporting dense phase fluids may split and the crack may run from a "notionally safe" zone to a zone where there may be a significant population. In particular it has some significance for the Offshore Gas Industry where tests on scored pipelines on land and under water produce different responses [8].

The effects modelling of a sub-sea release of oil or gas are somewhat different to those on land. The effects will be very much location dependent. In the case of on-shore releases the effects will be one of:

- Dispersion
- Pool fires
- Jet fire
- Fire-ball

Dispersion without ignition is taken as a safe event other than for toxic or aggressive materials. However the fires are different. The geometry of pool fires and fireballs are fairly well understood but the geometry of jet fires, particularly those with high liquid contents are less certain. The traditional approach to a jet fire has been to model it as a flare using the work of Chamberlain [9]. This has been proved to work on flares but the results are less applicable to the releases in question. Examination of video footage of test liquid jets [3] and an analysis of real measurements on blow-outs [10] show that the predictions of the geometry are less certain. In addition the aeration of the fire has an effect on the Surface Emissive Powers (SEP), "optically thin" flames such as Methane have low SEPs with emissivities about 0.1 and flames which have soot as the radiating species have higher SEPs with emissivities approaching unity. The literature gives a whole range of SEPs ranging from an impossible value of 1000 kW/m^2 to less than 50 kW/m^2. The theoretical maximum SEP based on an energy balance round a flame is about 500 kW/m^2 but in reality non-idealities result in a maximum nearer 350 kW/m^2. The traditional way of assessing the amount of heat radiated from a flare has been to assume that a fraction of the total combustion heat, designated as "F", is radiated. This ranges from 0.1 to 0.35. Various references for values of SEPs are quoted in Reference [3] but some rules can be derived from the observation of flare flame colours. For methane the

SEP is about 75 kW/m^2 and for ethene it is 300 kW/m^2. A value of 100 kW/m^2 would be applicable to Natural Gas. Reference [10] gives some very useful real time measurement of SEPs around a blow out. Not surprisingly the values vary round the flame surface and are influenced by the local aeration such that the highest values are to be found on the upwind side. For the release of a well fluid with a gas to oil ratio of 0.10 to 0.15 wt/wt the SEPs were 150 kW/m^2 at the upwind side giving an overall SEP which would not exceed 150 kW/m^2. These figures are quite consistent with the SEPs from very high-pressure jet sources quoted in Reference 3. It is proposed that the SEP for a Natural Gas type torch fire is 75 kW/m^2 and for an oil based torch is 150 kW/m^2. Likewise the accepted SEP for a fireball is 250 kW/m^2.

The flame geometry for the only well (17A) that was well documented in Reference [10] shows that the predictions from Chamberlain are somewhat in error for blowouts. This is quite to be expected, as the mass fluxes at the source will be well outside the experimental boundaries. The results are compared in table 1.

Table 1. Results of analysis of well 17 a based on references [9] and [10]

Observed frustum length (m)	Observed maximum diameter of frustum (m)	Assessed percentage of heat radiated based on flame Shape (%)	Calculated frustum length (m)	Calculated maximum diameter of Frustum (m)	Percentage of heat radiated based on graph in reference 8 (%)
45	17	11.1	65	27	28

It will be noted that the sizes of the flames in Reference [10] are smaller than that predicted in Reference [9]. Not only are the predicted flame dimensions longer and wider but also the shape is fatter, also the amount of heat radiated is less. Taken together, the distance to a tolerable heat flux would be over stated by a factor of 2 by Reference [9]. (Although there is data from other wells in Reference [10], including the release rates, the size of the source is open to interpretation so were not included in this analysis). The value for the percentage of heat radiated in this assessment is very much in line with that experienced in real blowouts.

There is always a difficulty in predicting the SEP for a flame as it is affected by the physics and chemistry of the combustion process. The physics are the nature of the source and the mixing/aeration of the jet and the chemistry is the pyrolysis of the heavier molecules leading to soot or the primary radiating species in the flame, and thereby the reduction of the total heat that can be released. While Chamberlain predicts a radiated heat release for Methane of 15 to 29% the real value for the design of flare stacks is nearer 15 % for low velocity flares and less than 10% for high velocity flares using the Coanda effect [11]. Further observed jet Methane flames are almost translucent giving radiant heat releases of well under 10 %.

The modelling of fireballs is relatively simple. The fireball is taken to be a sphere and then the sphere is assumed to sit on the ground. This has to be reviewed against the likely flame characteristics. The generally accepted parameters of Diameter (D_{Metres}) and Fireball Duration ($t_{seconds}$), for fireballs containing less than 35,000 kgs of fuel, are given [12] as:

$$D_{Metres} = 5.8 \ M^{1/3}$$
$$t_{seconds} = 0.45 \ M^{1/3}$$

These can be solved to produce a burning rate per m^2 of flame surface, this is 0.021 kg/m^2s. In the same manner it is possible to assess the burn rate for pool fires in the range 0.008 $kg/m^2/s$ for small fires rising to 0.014 $kgs/m^2/s$ for larger fires. It is now possible to assess the stand off distance should the release of fuel be ignited as a fireball. The dimensions have been compared to the photographs in the Piper Alpha Report [6] as shown in Table 2.

Table 2. Comparisons of observed and assessed fireball dimensions for Piper Alpha

Observed fireball diameter (m)	Outflow from fireball dimensions (kg/s)	Calculated outflow from line dynamics (kg/s)
340	7600	7000
180	2000	1000 to 2000 *

*Dependent on the roughness factor used

Clearly the model is consistent and can be used to assess any fireball knowing the outflow of fuel. This has been used for past offshore designs to assess the separation distances of "occupied structures" from Riser structures.

However for the full bore leakage from a severed oil pipeline, which has a vapour pressure greater than 101 kPa, there will be a fair measure of rain out and the fuel will burn as a mixture of a fireball and jet fire with the appropriate SEPs. The residue, which may be as high as 75%, will then burn as a smoky pool fire with a SEP of 75 kW/m^2 or lower.

FAILURE RATE DATA

It is evident that failure data for pipelines have to be examined with care to ensure that the data are both robust and relevant to the specific environment and fluid. Process piping may have a benign fluid and a harsh external environment and vice versa. For example the low temperature hydrocarbons such as Propane at a temperature of less than 273 K is a non-corrosive fluid but the external corrosion (corrosion under the insulation) can be rapid if the engineering is poor. Possibly the most secure source of oil pipeline failure data is CONCAWE where there are both annual reports and a 25-year summary [13]. In this report there are 341 recorded incidents of leakage in 520,717 pipeline years. The data have been examined in some detail and are summarised below in Table 3.

Table 3. Break down of main causes of transmission pipeline CONCAWE (Note the values sum to 345)

Cause	Number of incidents
Mechanical (construction)	32
Mechanical (materials)	56
Operational	25
Corrosion (internal and external – cold oil)	51
Corrosion (external – hot oil)	37
Corrosion (internal all lines)	18
Natural Hazards	14
Third Party	112
Third Party (accidental)	86
Third Party (incidental)	18
Pump Stations (included in the above)	43

However one of the major causes of leakage is Third Party (112) and again one of the sources of internal corrosion is to be found in Hot Oils. When the results are examined to determine the main causes of leakage in the pipeline itself the overall, cold oil corrosion rate both internally and externally is 10^{-4} per km per yr. It should be noted that crude oil can and often does contain traces of water so the internal environment of an oil lines may be more hostile than that of, say, gas pipelines. Further, while construction faults can be expected even in the best-regulated country, material defects are amenable to good engineering. More particularly for pipeline alone, not including fittings, the mechanical failures were only 35 of which 6 were "dents", 5 were "weld faults" and 13 were faulty materials (11 more were classified as "other" and "above ground"). This suggests that the Mechanical Failure Rate should be 0.5×10^{-4} per km per yr.

The third party causes are the largest common cause of leakage as shown in Table 4.

It is assumed that the reference to drilling means drilling into the line to tap the fluids, that is with the intent to steal.

Table 4. Causes of third party damage to pipelines

Cause	Number
Other	7
Ploughing	26
Digging	32
Bulldozing	11
Drilling	7
Unallocated	3

More particularly, well over 50% of the third party damage is due to trenching and farming activities. There is also clear evidence of the third party damage being both Company and Country related and the spread is a factor of 5 by Company and 3 to 4 by Country. The likely Third Party Failure Rate for the UK should not exceed 0.5×10^{-4} per km per yr.

The final oil pipeline failure rate for the UK should be as follows:

Table 5. Failure rate for oil pipelines in the UK based on CONCAWE data

Cause	Rate (per km per yr)
Corrosion	1×10^{-4}
Mechanical	0.5×10^{-4}
Third Party	0.5×10^{-4}
Total	2×10^{-4}

The CONCAWE report does show that overall there is a downward trend in failures per kmyr over the last 25 years (60% over 25 years) although the average age of the pipeline is increasing. The report also suggests that larger diameter pipelines, as might be expected, are less vulnerable to damage. A previous analysis of CONCAWE annual reports for the years 1982 to 1991 [3] gave an indicative total failure rate for oil pipelines of 2.62×10^{-4} per km per yr. Bearing in mind the larger data base and the downward general trend in failures these two numbers match well. However it is not only the failure rate but also the failure rate spectrum that counts. Possibly one of the more reasonable conclusions of the CONCAWE report is that the use of On Line Inspection Vehicles (OLIV) can go a long way to preventing leakage. It is possible that the downward trend shown in the 25 year analysis is to some degree explained by their usage.

The CONCAWE reports do not give explicit failure sizes and some deduction is required. If Full Bore Rupture is taken as resulting in greater than or equal to 1000 m³ of material released, only 9 failures of this size or greater were of in pipelines giving a rupture rate of 1.8×10^{-5} per km per yr, averaged over the 25 years, which is similar to Reference 3. Moreover only two of these large events were due to corrosion the rest being due to impact, construction or material deficiencies. The final spectrum of smaller leaks is impossible to assess from the data but a reasonable judgement suggests that the spectrum would be 1/3 of 25 mm and 2/3 of 10 mm. If the improvements in leakage rates are sustained a further factor of 0.7 could be applied for pipelines built in recent years [13].

Table 6. Suggested leak spectrums for pipelines in UK

Leak size	Frequency per km per yr
FBR	1.8×10^{-5}
25 mm	6.1×10^{-5}
10 mm	12.1×10^{-5}

It is worth comparing the failure rate data for Gas Pipelines quoted by Fearnehough [7].

Table 7. Leak spectrums for UK gas pipelines [7]

Leak size	Frequency per km per yr
Over 80 mm	7.5×10^{-6}
20 to 80 mm	2.2×10^{-5}
O to 20 mm	2×10^{-4}

A further set of Oil Pipeline failure data is to be found [13].

Table 8. Leak spectrums for UK sub-sea oil pipelines [14]

Leak Size	Frequency per km per yr
Over 80 mm	4.43×10^{-3}
20 to 80 mm	5.04×10^{-4}
10 mm	1.7×10^{-3}

These data include fittings and such events as trawl board and anchor impacts. Clearly the spectrum is totally different and this data set is neither relevant nor applicable to the on-shore pipelines.

DISCUSSION

The modelling of out-flows from hydrocarbon pipelines is exceedingly complex and requires the use of sophisticated modelling. The flows will be both smaller and longer duration than that predicted from the standard out-flow equations. This results in the instantaneous worst-case hazard being over stated. By the use of more complex calculations it can be shown that out-flows are significantly less and therefore the perceived risk is lower.

The consequence of a leaking pipeline, other than the environmental impact, is likely to be a pool fire, a torch type fire or a fireball. The modelling of pool fires is robust, provided that the SEP can be assessed with accuracy, but the jet type fires assessed from the effects models are also complex and at present the oft used correlations for oil leaks, and also liquid oil product leaks, significantly over state the potential hazard. The resultant fireballs for ruptured oil pipelines will not involve total vaporisation and a significant proportion will burn as a smoky pool fire. In particular the effects of an ignited oil based leak will be over stated if it is not analysed in the correct detail. For jet fires of the gaseous form the present correlations are robust. The jet fire models previously used can be recalibrated for the releases of high momentum releases such as LPG and high vapour pressure oils.

The failure data for on-shore pipelines show that the failure rate is very low and that the most significant causes of leakage are mechanical specification and third party impact. If these are managed properly and there is a good monitoring program for corrosion the full bore rupture leak frequency could fall to less than 1×10^{-5} per km per yr. As the risk to the public is dominated by the full-bore leak frequency there is much to be said for monitoring and policing.

The result of this analysis is that there are good reasons for examining the risks near pipelines in the area of the public sector in more detail particularly if there are both social and financial implications. The analysis given supports this conclusion.

CONCLUSIONS

1. Land use planning near to developed areas creates a potential conflict between economic requirements of the many and the safety of a few.
2. Out-flow modelling from high-pressure hydrocarbon pipelines is complicated by their physical properties and complex models are required.
3. The consequence modelling, for jet fires in particular, still requires more research. Present models could over estimate the effects considerably.
4. Failure rate data are dominated by third party damage, which is both company and country sensitive.
5. More monitoring, both using OLIVs and routine survey or patrol of pipe corridors could reduce the risks in sensitive development areas.
6. Failure rate data for pipelines shows a general improvement with time.
7. Combining the better failure rate data with the outflow and consequence modelling, may free land for development, allowing a small but beneficial encroachment of local development, particularly at the approaches to industrial sites.

REFERENCES

1. Advisory Committee on Major Hazard Third Report. HMSO London 1984 ISBN 0 11 883753 2.
2. Risk Criteria for Land-use Planning in the Vicinity of Major Industrial Hazards. HMSO London 1989 ISBN 0 11 885491 7.
3. Bendiksen, K. H. et al The Dynamic Two-phase Flow Model OLGA: Theory ands Application, SPE Production Engineering. 6: pp171 – 180.1991.
4. Crawley F K, Lines I and Mather J Oil and Gas Pipeline Failure Modelling Trans Part B IChemE Vol81 Part B1 (to be published).
5. Confidential Study.
6. The Piper Alpha Disaster Hon Lord Cullen HMSO 1992.
7. Fearnehough G D, The Control of Risk in Gas Transmission Pipelines. The Assessment and Control of Major Hazards Symposium Series 93, Manchester 1985. IChemE Rugby.
8. Unpublished Study carried out as a Joint Industry Project.
9. Chamberlain G A, Development in Design Methods for Predicting Thermal Radiation from Flares Chem Eng Res Des Vol 65 July 1985. IChemE Rugby.

10. Study Kuwait Scientific Mission, Volume 2 Technical Report, July 1992. OTI 96 641, HSE Books.
11. API RP 521 Guide for Pressure Relief and Depressuring Systems American Petroleum Institute.
12. Major Hazards Assessment Panel (MHAP) Thermal Radiation Monograph. IChemE Loss Prevention Bulletin No 82 1982.
13. Western European Cross-country Oil Pipelines 25-year Performance Statistics Report 2/98 CONCAWE, Brussels 1998.
14. The Update of Loss of Containment Data for Offshore Pipelines. OTH 551 1996 (PARLOC 96) HSE Books.

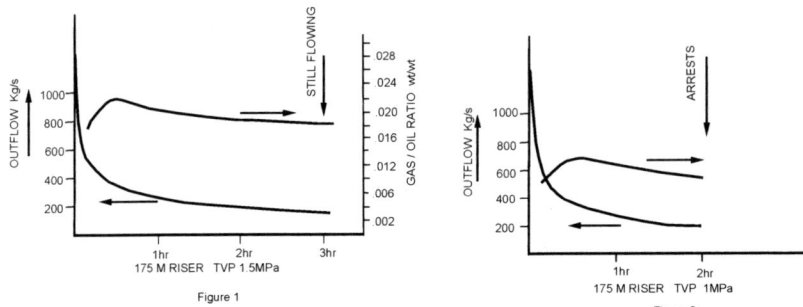

Figure 1

Figure 2

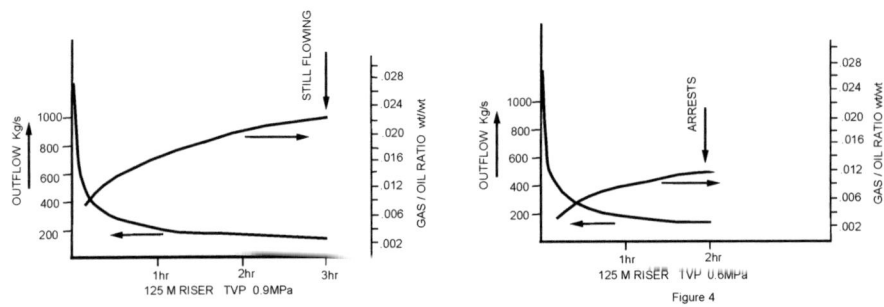

Figure 3

Figure 4

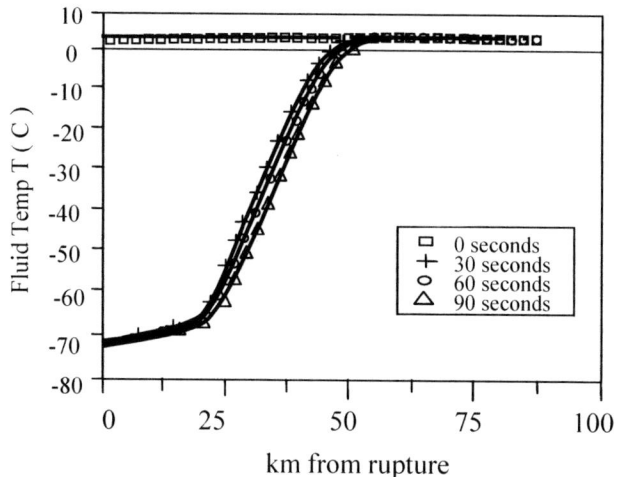

Figure 5. Temperature profile at 30's intervals after rupture

Pressure vs Pipe Section
At various intervals after rupture

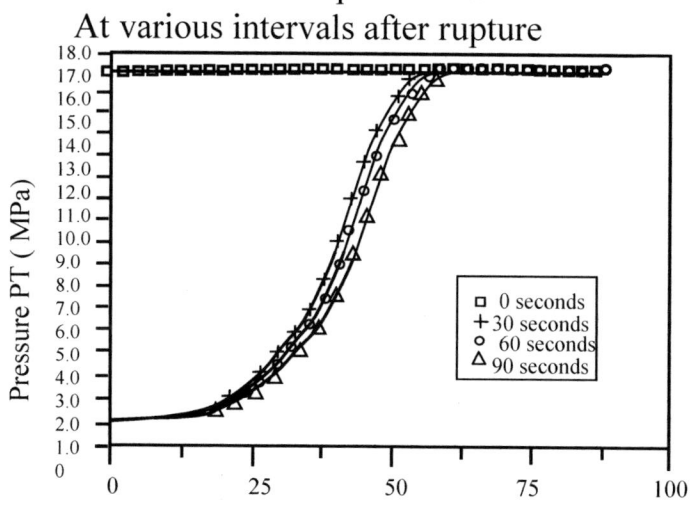

Figure 6. Pressure profile at 30's intervals after rupture

LINKING AN ACCIDENT DATABASE TO DESIGN AND OPERATIONAL SOFTWARE

Dr. John Bond. C-MIST

Abstract

The potential advantages of a linkage between an accident database and the software used in procedures for design, hazard and operability studies and in risk assessment are described.

An established software for the design of equipment is taken and the potential for linking it to an accident database demonstrated by simulating a design of a distillation column. This linkage will allow the design engineer to immediately learn lessons from past accidents and to incorporate them in the design at an early stage. A risk assessment procedure will be carried out using a "Risk Assessment Methodology" developed for the training of personnel. This procedure uses keywords and an accident database to identify hazards and to help in the provision of Control Measures. It will also assist in identifying hazards that may not be in the knowledge of the person carrying out the assessment. A HAZOP procedure software will also be used to show how linking with an accident database will assist in establishing a variety of possible scenarios and assist the team in reviewing the hazards that can occur.

The use of an accident database in combination with other software for design and operational work establishes a more user-friendly system for incorporating lessons learnt from accidents.

KEYWORDS: Accident database, design, risk assessment, HAZOP.

INTRODUCTION

Accidents that have occurred in the past have been investigated, often with associated research work, and have been adopted in engineering standards. Subsequent design work has used these standards to the benefit of the public, the employees and the company. Such a situation occurred after three very large crude oil carriers (VLCC) marine tankers were subject to explosions in December 1969. The *Mactra* was ripped open for 500 feet but reached port, the *King Haakon* was severely damaged but also reached port. The *Marpessa* caught fire and sank. In each case empty oil tanks were being cleaned by high pressure water washing equipment. Extensive research showed that ignition of flammable gases in the tanks was caused by induced electrostatic charges from the water washing equipment (1–5). This research work resulted in the adoption of inert gas systems on VLCCs and adopted in the International Convention for the Safety of Life at Sea (SOLAS). The marine chemical tankers were not required to have the inert gas systems due to research work (6,7) defining the maximum size of equipment to avoid the generation of incendive static charges. This was also incorporated in the SOLAS agreement. The research work of two major oil companies was shared with all companies, incorporated in the SOLAS agreement and became the standard for marine tanker design to the advantage of all companies, the crews of the tankers and the public.

However, due to a variety of reasons, not all lessons learnt are incorporated in standards or shared. Another marine example is in the carrying of 98% formic acid. An accident (8) on a marine chemical tanker occurred after discharging formic acid. The tank was washed with sea water followed by fresh water. A seaman entered the tank to mop out water and was asphyxiated by carbon monoxide. The ship's crew did not know that formic acid would decompose and that even if they had ventilated the tank and tested with an oxygen meter they would still have to test for residual carbon monoxide. The Material Safety Data Sheet at the time did not have this information but the information is shared in the Accident Database (9).

In order to be fully aware of hazards, engineers must consult an accident database to be certain that he is learning lessons from past accidents. This can be done by separately looking up in the database each and every piece of equipment that he is using and each chemical. Even then he may miss some hazards. Just as a person does not look up the spelling of words in a dictionary when he is writing a report, so an engineer may not always consult a database to identify hazards. The linking of an accident database to a software used in the design, risk assessment or review of design would make the process of learning lesson from accidents that more easy and user friendly. It could work in the same way that a dictionary is in the background of a word processor. Indeed, a facility for ensuring that a procedure adhered to the risk assessment could be likened to the grammar checker in a word processor, but that is for another paper.

The use of an accident database such as the IChemE Version 4 in conjunction with software for design, for risk assessment and in the review of a HAZOP flow sheet will be demonstrated.

DESIGN OF A DISTILLATION COLUMN

It is particularly important to identify the hazards at an early stage of the project as the capital expenditure rises as the project proceeds. Any modifications that may subsequently be necessary are more expensive the further you are into the project. With this in mind the design engineer must be aware of all the accidents involving the substances and equipment he is working on. If he is, say, designing the distillation column for ethylene oxide and had the Accident Database in the background of his design software the accidents associated with this equipment and chemical could be triggered. In this case he could find a message appearing on his screen as shown in Figure 1.

If the records for the 5 accidents are viewed, it will be seen that all involved explosions in the distillation column and were caused by:

- Leak from flange or weld
- Reaction in the insulation with water
- Auto-oxidation catalysed by rust with heating from an insulation fire

The lessons learnt are also given and can be incorporated in the design from the start.

- Reduction in the number of flanges or flanges to be left uninsulated. Areas of possible leak should be inspected and tested regularly.
- Upper part of reboiler must be covered. Avoid condensate backup.
- Positive purge of inert gases from shell.
- Ensure minimum heating temperature.
- Insulation non-absorbent and test for glycol formation.

- Avoid piping with no flow of ethylene oxide vapour or stagnant lines.
- Remove any rust from pipework.

Noting these lessons, and others, at this early stage of the design will ensure a saving of time at the HAZOP stage as well as making it unnecessary to make modifications later.

RISK ASSESSMENT
The stages of risk assessment for an individual task are:-

1. Identify the hazards for each task
2. Assess the risk associated with the hazard
3. Devise suitable control measures
4. Reassess the risk to ensure that it is reduced to an acceptable level
5. Produce a method statement

This process requires an experienced person or a team of people with an experienced chairperson to control the operation. Access to an accident database is also vital for the various stages to be fully assessed.

In the analysis that is done it is essential that all the reasonable hazards that are identified as possible are recorded and actions noted, even if the team decide the risk is so low that no action is necessary. Clearly a hazard that is so unlikely need not be recorded, e.g. being hit by a meteorite.

An accident database is an essential tool for identifying hazards and particularly if attached to the Risk Assessment Form software. Consider the risk assessment of a maintenance operation for the removal of a submerged pump from a vessel. Most hazards can be identified by an experienced maintenance person but other hazards can be identified from an accident database. The words maintenance and pump can be highlighted as shown in Figure 2 and the database questioned.

A flag could then come on the screen showing the accident given below.

No. 13152   Date. Unknown
Source: Loss Prevention Bulletin 078
Location: UK
Injured: 1   **Dead:** 0

Abstract:
A submerged pump on a horizontal vessel containing molten phthalic anhydride at 160°C was removed for repair at the workshop. To prevent fumes of phthalic anhydride leaving the 700 mm diameter manhole, a piece of jointing was placed over the hole with a piece of plywood to hold the jointing down. The electrical connections for the motor had been removed before the pump had been taken out. Nevertheless, a repair to the electrical connections was carried out by the electrician. During this work the electrician stepped onto the plywood which broke under his weight. The electrician managed to save himself by spreading out his arms. He managed to pull himself out without assistance, suffering only from shock

Lessons
Temporary covers over holes must be substantial. It is good practice to have a spare blank adjacent to the pumps for use when the submerged pump is removed.

The hazard of an open hole would be recognised and the lesson learnt. The control measure would be put into the Risk Assessment form for a blank to be bolted over the manhole even though it is outside the walkway.

After identifying the hazards, the next operation is assessing the risk. The risk is generally defined as:

RISK = PROBABILITY OF OCCURRENCE x SEVERITY

Table 1 shows the typical probability and severity tables used in a qualitative approach.

Table 1.   Typical table used for establishing probability and occurrence

Rating	Probability of occurrence	Severity of occurrence
1	An unlikely/unknown occurrence. Very unlikely to occur during the operation/facility or process.	Scratches, minor burns, bruises or abrasions. Minor injury 1 person.
2	A remotely possible but known occurrence. Unlikely to occur during the life of operation/facility or process.	Minor injury, laceration requiring stitches, secondary degree burns or severe bruises. Minor injuries 2–10 people.
3	An occasional occurrence. Likely to occur once during the life of operation/facility or process.	Major injury to one person, broken bone, amputation, third degree burns. Major injury 1 person, Minor injuries to >10.
4	A frequent occurrence. Likely to occur from time to time during the life of the operation/facility or process.	Death or permanent severe disablement of one person. Major injury <5 people.
5	A highly likely occurrence Likely to occur repeatedly during the life of the operation/facility or process.	Multiple deaths or multiple severe permanent disablement. Major injuries >5 people or fatality.

These are, however, open to considerable variation of interpretation depending on the person carrying out the risk assessment and even with a group there can be considerable variations. A more definitive interpretation of the rating points is required. Table 2 gives a fuller interpretation of the ratings and should lead to less disagreement amongst the team.

Table 2. A more definitive table for establishing probability and occurrence

Rating	Probability of occurrence	Severity of occurrence
1	An unlikely/unknown occurrence. The team or person carrying out the Risk Assessment has never heard of such a hazard taking place either at his place of work or in another company, nor in any accident database, the media, his training etc.	Scratches, minor burns, bruises or abrasions. Minor injury 1 person.
2	A remote but possible occurrence: The team or person who is carrying out the Risk Assessment has heard about this hazard occurring in another company but not at his place of work, i.e. the hazard has not been experienced by the team or the person carrying out the Risk Assessment but it is reported in an accident database or the media.	Likely to result in minor injury, laceration requiring stitches, second degree burns or severe bruises. Minor injuries 2–10 people.
3	An occasional occurrence: The team or person who is carrying out the Risk Assessment has seldom experienced such a hazard at their place of work but it has occurred at another company and is found in an accident database.	Likely to result in major injury to a few persons, broken bone, amputation or third degree burns. Injuries reported in accident database or media reports. Major injury 1 person. Minor injuries to >10.
4	A frequent occurrence: The team or person who is carrying out the Risk Assessment has frequently experienced such a hazard at their place of work and it is reported frequently in accident databases.	Likely to result in death or permanent severe disablement of one or more persons. Reported in accident databases or media with these conditions. Major injury <5 people.
5	A highly likely occurrence: The likelihood of the hazard taking place is very high and there is a greater chance that the hazard will take place than not.	Likely to result in multiple deaths or multiple severe permanent disablement. Reported in accident databases or media with these severe conditions. Major injuries > 5people or fatality.

Table 3 gives the risk ranking bands

Table 3. Table for establishing risk bands

Severity					
	5	10	15	20	25
	4	8	12	16	20
	3	6	9	12	15
	2	4	6	8	10
	1	2	3	4	5

Probability

Risk bands 12–25 Highly hazardous and highly likely event. In all cases the potential severity is too high to allow the operation to continue. Operation in this risk band be eliminated, avoided or totally replanned.

Risk bands 5–10 Within this band severity and probability are high and the work cannot be carried out until risk is reduced to an acceptable level. Mitigating the hazard can be via the provision of written procedures or work instructions, supervising the work, isolation or limiting exposure.

Risk bands 1–4 Within this band it is acceptable to carry out the work but with Appropriate personal protective equipment, warning signs, barriers, tannoy announcements etc to mitigate the initial risk.

In risk assessments the combination of the accident database with the software for the relevant forms provides a useful and user-friendly system for help in identifying and assessing the risks. Software for writing out Work Permits could be treated in a similar manner.

HAZARD AND OPERABILITY STUDY

The Hazard and Operability Study is clearly another area where an accident database should be linked to the software for carrying out the study. Take the case where a gasoline pipeline is being considered as the Node Point and "More Flow" is being considered. A possibility for more flow could result from rupture of the pipeline from internal or external corrosion. The pipeline metal is considered satisfactory for internal corrosion but the case of underlagging corrosion is considered a possibility due to its temperature of 40°C. The Accident Database is consulted by the keywords 'pipeline' and 'corrosion' and the accident shown in Figure 3 would be brought up onto the screen.

The lessons learnt from this accident could then be incorporated into the main HAZOP document and control measures adopted from the lessons as shown in Figure 4.

CONCLUSIONS

An appropriate accident database with lessons learnt and a keywording system can be combined with modern design, risk assessment and Hazop software to provide a powerful tool for design and operational engineers to improve safety and hence reduce costs. The linking of the software is a logical step.

ACKNOWLEDGEMENTS
The author would like to thank Hyprotech (recently merged with Aspentech) the manufacturer of software for equipment design; C-MIST risk assessment consultants; and Rowan House hazard study consultants; for permission to use copies of their screens for this paper.

REFERENCES
1. Van der Meer, D. 1971. Pr. 3rd Conf. Static Electrification (Inst. Phys., Bristol) pp. 153–157.
2. Van der Weerd, J.M. 1971. Pr. 3rd Conf. Static Electrification (Inst. Phys., Bristol) pp. 158–173.
3. Smit, W., 1971. Pr. 3rd Conf. Static Electrification (Inst. Phys., Bristol) pp. 178–183.
4. Vos, B., 1971. Pr. 3rd Conf. Static Electrification (Inst. Phys., Bristol) pp. 184–192.
5. Van der Meer, D et al. 1973 2nd Int. Conf. Static Electricity Special Colloquim: Problems on Tank Washing in Very Large Crude Tankers (Dechema, Frankfurt).
6. Jones, M.R.O and Bond, J. Chem Eng Res , IChemE. Vol. 62 Sept 1984.
7. Jones, M.R.O and Bond, J. Chem Eng Res , IChemE. Vol. 63 Nov. 1985.
8. Bond, J. Loss Prevention Bulletin 056, pp. 24–25.
9. The Accident Database. Version 4. The Institution of Chemical Engineers.

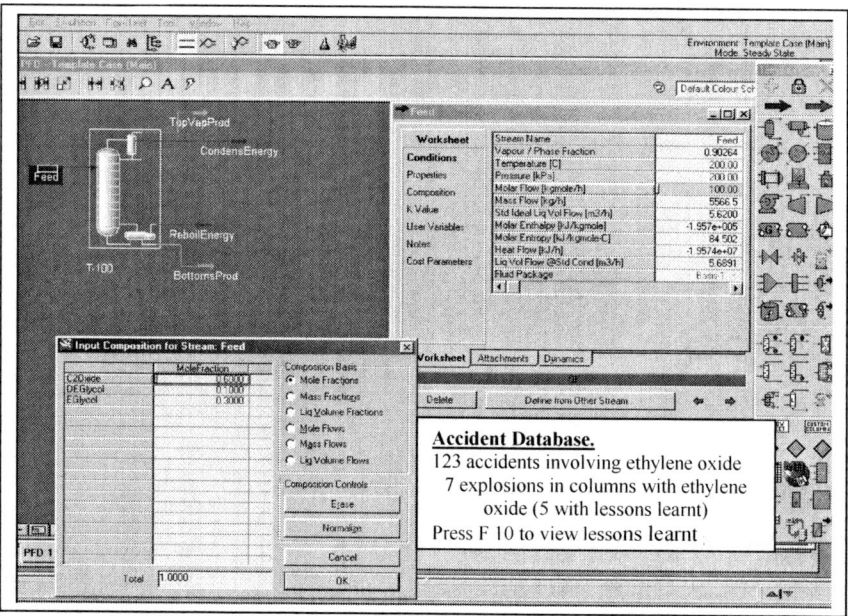

Figure 1. Hyprotech screen for design of distillation column for ethylene oxide and showing accident database warning

Risk Assessment Form

Site Location/Project:		
Work Activity/Task:	REMOVE SUBMERGED PUMP	
Reference No:	MAINTENANCE TASK	
Assessment carried out by:	Assessment checked by:	
Name Date	Name .... Date	

Risk = Probability x Severity of consequence ; R = Risk, S = Severity, P = Probability

	HAZARDS	Risk R = P x S		
		P	S	R
1	Working on vessel top – falling			
2	Scaffolding requirements			
3	Lifting pump			
4	Open manhole			
5				

Figure 2. Part of a risk assessment form

No. 8720 Date. 09 March 1983
Source: Loss Prevention Bulletin136 21–23, Western Mail 1983, 10 March. Loss Prevention Bulletin. 083, 13-14 I.Chem.E.

Abstract: A major fire occurred on a gasoline treater unit in an Olefins complex. Prior to the incident the unit was operating at reduced rates. A fire occurred as a result of the ignition of a mixture of raw gasoline and hydrogen, which was released as a result of a rupture in the horizontal section of the 20 cm diameter, ferretic steel, insulated and clad feed-line between the preheater and the reactor. The impact energy from the violent movement of pipework associated with the failure was sufficient to cause the ignition, which occurred almost immediately after the failure. The fire was isolated and controlled within ten minutes and was eventually extinguished after ninety minutes, by which time some 40 tonnes of hydrocarbons had been consumed. Two operators were injured in the incident one severely and one superficially. Prompt action by the plant and local emergency services quickly contained the main fire and the secondary fires and prevented any further escalation. Fire damage was extensive with preliminary estimates for the rehabilitation costs of £1.25 million (1983). [fire - consequence, gas/vapour release, reactors and reaction equipment, processing, injury]

Lessons
External corrosion of carbon steel and low chrome ferretic steels beneath insulation has been a cause for concern for many years and has resulted in a number of serious incidents from sudden failure of high pressure containment. Although many of the reasons for under-lagging corrosion are well understood it is unfortunate that in many incidents, involving under-lagging corrosion, the known lessons learned are rarely put into practice or maintained. It is useful therefore to use this incident to remind ourselves of the lessons about under-lagging corrosion and the precautionary measures to be taken to prevent such occurrences:

1. The ingress of water through inadequate water proofing, or through damaged cladding, must be minimised or if possible eliminated.
2. The use of absorbent insulating materials must be avoided whenever possible, with the effect of such materials on corrosion rates fully taken into account.
3. Operations above ambient, and particularly in the temperature range 77–115 degrees C are susceptible to under-lagging corrosion.
4. Line preparation prior to insulation - i.e. painting, must be carefully monitored and maintained during construction and any maintenance phases.
5. She presence of chlorides originating from lagging material, process materials, plant, or environment conditions - sea air -- may also accelerate corrosion.

Since it is virtually impossible to guarantee that water will not permeate insulation, added protection can be achieved by metal coating systems.
Further protective measures to prevent under-lagging corrosion include:

1. Careful consideration at the design and construction phases of new plant to the potential problem with adequate consideration given to weather proofing around protrusions in insulation such as pipe supports and nozzles etc.
2. Facilitate ease of removal/replacement of cladding for inspection purposes
3. Actual operating conditions may differ from design specifications thereby increasing the possibility for corrosion
4. The possibility of leaks from small bore piping, steam tracing, with sampling systems another potential source of corrosion

A comprehensive maintenance inspection programme will reduce the likelihood of, and potential consequences from under-lagging corrosion. In the development of such a programme the following points are to be considered:

1. Materials of construction, and insulation material.
2. The age of the plant including a review of previous inspections undertaken and to follow up any actions.
3. The effect of normal operating conditions on the rate of corrosion.
4. Intermittent operations and the time out of service of any plant equipment.
5. The potential hazards from any loss of containment, flammable, toxic, corrosive etc.
6. The integrity of any steam/electric tracing systems.
7. Pipe diameters and pipework configuration - in particular to reduce or prevent areas for water build up.
8. The possibility of internal corrosion to be investigated.
   Plant layout and operating environment (i.e. close proximity to cooling towers, sea, rivers etc).

Figure 3.   Accident database record relevant to the scenario raised in HAZOP case

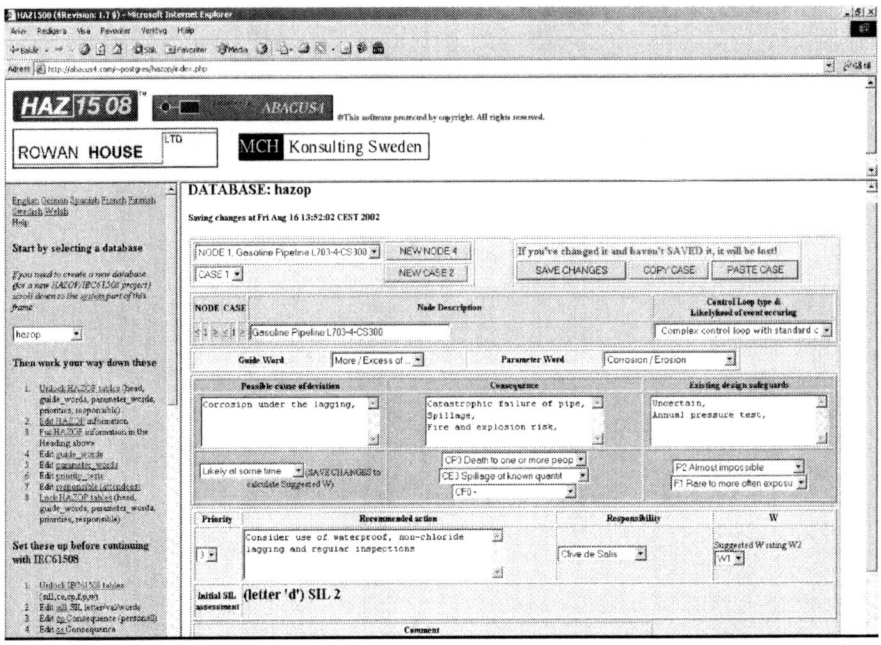

Figure 4. Screen from Rowan house HAZOP software with entry from accident databas

UNDERSTANDING MAJOR ACCIDENT HAZARDS –
THE CUTTING EDGE OF COMMON SENSE

Graham Dalzell, BSc FI Mech E, Safety Consultant, Safety Analysis, BP Exploration & Production Ltd, Dyce and Stuart Ditchburn, CEng BSc M Inst MC, Principal Consultant, Det Norske Veritas Limited, Aberdeen.

INTRODUCTION

During the last decade, hazard identification, risk assessment and management has become a universally used tool in the offshore and petrochemical industries. Huge improvements have been made in the way that these industries now manage their business, leading thankfully to a greatly reduced likelihood of a Major Accident. But the process has now matured to the point where it is doubtful that it is capable of delivering much more in terms of added value.

Whilst it is certainly true that the introduction of a suite of goal setting regulations has help drive this vast improvement in hazard management, it is also the case that it has led to an unwelcome side effect. The detailed hazard evaluation and risk analysis now required by regulation, has necessitated the extensive use of risk analysts, invariably provided by consulting organisations. The level of detail demanded by the regulations has also led to an enormous amount of 'hazard information' being required supporting the Safety Cases and Safety Reports

Thus we have a situation where valuable hazard management information is often 'locked up' in the Safety Case, Safety Report and associated studies. Also, all the 'hazard understanding' is most likely in the hands and minds of the specialist risk analyst, who do not usually play an active part in the day to day operations where risk is most effectively managed

So the search is on to find new ideas or refinements of the process which can be used to 'ratchet up' the value gained in terms of tangible and worthwhile risk reduction. Unsurprisingly, cost is a significant factor in determining which new ideas and alternative approaches are viable; so cost effective methods are liable to be taken up more readily.

This Paper proposes one solution in the form of an integrated Major Accident Hazard Management System (MAHMS), which is currently being developed and implemented on bp Trinidad & Tobago's oil and gas operations in the Caribbean.

CONTENTS

THE PROBLEM
Hazard Understanding – in the hands of the Specialists

- The Safety Case Customer
- Access to Hazard Information
- Relinquishment of Major Accident Hazard Risk Management Responsibility
- Incoherent Risk Management

A SOLUTION
Understanding Major Accident Hazard Management – a Common Sense Approach

- Wholistic Approach to Risk Management
- Managing all Levels of Risk as a Continuum
- Pragmatic and Qualitative Approach
- Critical Measures (People, Processes & Plant)
- Major Accident Hazard Understanding
- Major Accident Hazard Information

THE PROBLEM: HAZARD UNDERSTANDING – IN THE HANDS OF THE SPECIALISTS

Under the UK's present 'Goal Setting Regulations' there is a compelling need for duty holders to demonstrate in their installation Safety Case that they have:-

- Identified all the hazards involved in their undertaking which may pose risks to personnel
- Evaluated the risks arising from those hazards
- Identified and put in place the means to manage the hazards and reduce the risks
- Made a demonstration that those risks are as low as reasonably practicable

These tasks necessitate an analysis in some considerable detail, often of a quantitative nature. Most duty holders engage specialist consultants to perform the analysis, using Quantitative Risk Analysis (QRA) techniques. Often, the same risk analysts conduct the numerous supporting studies required to identify the hazards, determine the characteristics of the risks arising from the hazards and recommend measures required to reduce the risks to as low as reasonably practicable (ALARP). The risk analysis concentrates purely on risks to personnel as required by the regulations. Risk to the environment or production in terms of business interruption, are largely ignored. Some operators use specialist risk consultant to provide a service covering the ongoing management of the Safety Case and the associated risk analysis and risk management activities that are entailed.

Before going further we must emphasise that we are not advocating an end to the above approach. Specialist risk analysts have and continue to deliver exceptional service in this area, where not even the largest operators have much, if any, 'in-house' expertise. Indeed it would be difficult to perceive how duty holders would be successful at achieving the goals of the Safety Case Regulations without such specialist assistance. This is a process that will need to continue for the foreseeable future in the UK Sector at least.

However, the above situation has introduced an unwelcome side effect, which has four symptoms:-

(i)   The Safety Case 'customer' tends to be viewed as the Regulatory Body (HSE), rather than the 'workforce' which we believe was the spirit if not the intent of Lord Cullen's Recommendations;

(ii)   All the 'hazard understanding' is in the hands and minds of the specialist risk analyst, who do not usually play an active part in the day to day operations where risk is most effectively managed;

(iii) There is a major disconnect between the management of occupational health and safety risks and the management of major accident hazard risks, the latter often been relinquished to the specialists; and

(iv)   Environmental risks and risks to business interruption, if considered at all are not being managed wholistically and with the same rigour as safety risks to personnel.

So what are the effects of the above symptoms and how are they detracting from the effectiveness of risk management on the facility?

SAFETY CASE CUSTOMER

A notion that the Safety Case was written for the HSE, is quite widespread throughout the workforce. This means that there is a great deal of apathy towards the Safety Case process, both in terms of its preparation, maintenance and the usefulness of the information it contains. This is most acute in those lower levels of the workforce, who paradoxically are perhaps those personnel at the greatest risk.

Those who do make a point of consulting the facility Safety Case, more often than not find that:-

• the information it contains is difficult to interpret and understand,

• the risk information is couched in terms and values which are almost impossible to appreciate; and

• the information is of little practical use in their daily activity.

If net result is to produce a Safety Case with the primary aim of satisfying the regulatory requirements, it may be of very little use in the day to day management of risk at any level.

ACCESS TO HAZARD INFORMATION

Those personnel in the duty holders organisation who are responsible for managing risk, invariably do not have the information or knowledge that they require to discharge their responsibilities effectively. This information, which we are calling 'hazard understanding', is more than likely available somewhere but is often 'locked-up' in the mass of documents and specialist studies which support the Safety Case.

Even when the relevant information is found, it is not usually presented in a form that is readily appreciated or re-usable directly. To achieve this often requires the safety specialist/risk analyst to interpret the information they have to hand to reproduce it in a useable form. This takes time and because the specialists are not part of the operations team, the information is not readily available when required. There is also a risk that if the risk analyst is not familiar with the day to day operations, that his/her interpretation of the data is inappropriate. Information is therefore often supplied either too late or in an inappropriate form to be of effective use.

RELINQUISHMENT OF MAJOR ACCIDENT HAZARD RISK MANAGEMENT RESPONSIBILITY

The management of occupational health and safety has latterly been much improved and continues to improve throughout the industry. However, there is a serious disconnect between the management of occupational safety, health and indeed environmental risk management and the management of major accident hazard risks. Most line management tend to assume that the responsibility for the latter is that of the risk management specialists and not part of their role.

It is also the case that causes of major accidents could originate as occupational safety related incidents, because the barriers designed to prevent incidents or limit escalation are ineffective. This is often due to them being inhibited as part of the work being carried out, without any consideration of what other temporary measures could be put in place to provide equivalent protection.

This situation is of great concern. Firstly, it raises a doubt that major accident hazard management is being performed with the same rigour as that for occupational safety. Secondly, the obvious benefits of addressing the management of occupational and major accident hazard risk as a continuum are not being realised.

INCOHERENT RISK MANAGEMENT

Major Accident Hazards (MAH), put personnel, production, capital investment and corporate reputations at risk. In most cases, they also pose a threat of environmental damage. Figure 1, below illustrates the way risks are related. It also shows the links that should exist between the management systems although they are often independent.

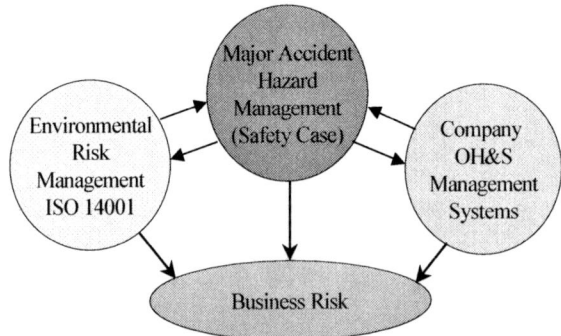

Do these links exist in practice?

Figure 1.   (In)Coherent risk management

It seems sensible and convenient therefore to consider all three types of risk together, yet this rarely happens. This appears to be due to the fact that the legislative drive is purely on safety risks to personnel. Most organisations however regard preventing environmental damage as having equal importance with safety. Also, although not stated, it is implicit that any operator will also rank production as having high importance, often equal with safety and protecting the environment. We should not be reticent to accept this fact as the basic function of an operators' business, nor feel we are in any way reducing the importance of safety by doing so. If the business is successful, then there is more likely to be more resources devoted to safety and environmental risk management.

In summary, whilst a Safety Case represents a sizeable investment, the situation discussed above prevents duty holders from maximising its value.

A SOLUTION: UNDERSTANDING MAJOR ACCIDENT HAZARD MANAGEMENT – A COMMON SENSE APPROACH

The following proposal describes how one Operator intends to overcome the difficulties discussed above by adopting a common sense approach called Major Accident Hazard Management System (MAHMS).

MAHMS is currently being developed and implemented for the Operators offshore and onshore oil and gas operating assets in Trinidad & Tobago. Although this is an operating environment where there is as yet very little regulation, we believe the principles employed are equally viable in a tightly regulated operating area like the North Sea. These principles of MAHMS are that it:-

- Adopts a wholistic approach to risk management, dealing with safety, environment and business interruption risk in one integrated system, considering the complete lifecycle of the operation.

- Manages all levels of risk, from occupational health & safety to major accident hazards, as a continuum, on a day-to-day basis and with the same rigour.

- Adopts a pragmatic and largely qualitative approach to risk assessment and management, only resorting to detailed quantitative analysis where it is not obvious that critical measures have reduced residual risks to ALARP.

- Recognises that measures in place to manage safety, environment or business risks, can be categorised as people, processes or plant; each of which have a performance standard to state functionality, performance and survivability expectations.

- Provides information on major accident hazards, sufficient to impart an understanding about the characteristics of the hazards or Major Accident Hazard Events to all levels of the workforce; thus enabling everyone to understand their role in MAHMS and carry it out effectively.

- Makes MAHM information continuously available in a form that can be readily understood and used in every day management of risks.

The above principles should be embedded into the key business management processes such that they become a fully integrated part of the management system. This is illustrated in the MAHMS Process Map, Figure 2.

What follows is a more detailed description of how each of these elements is intended to work in practice.

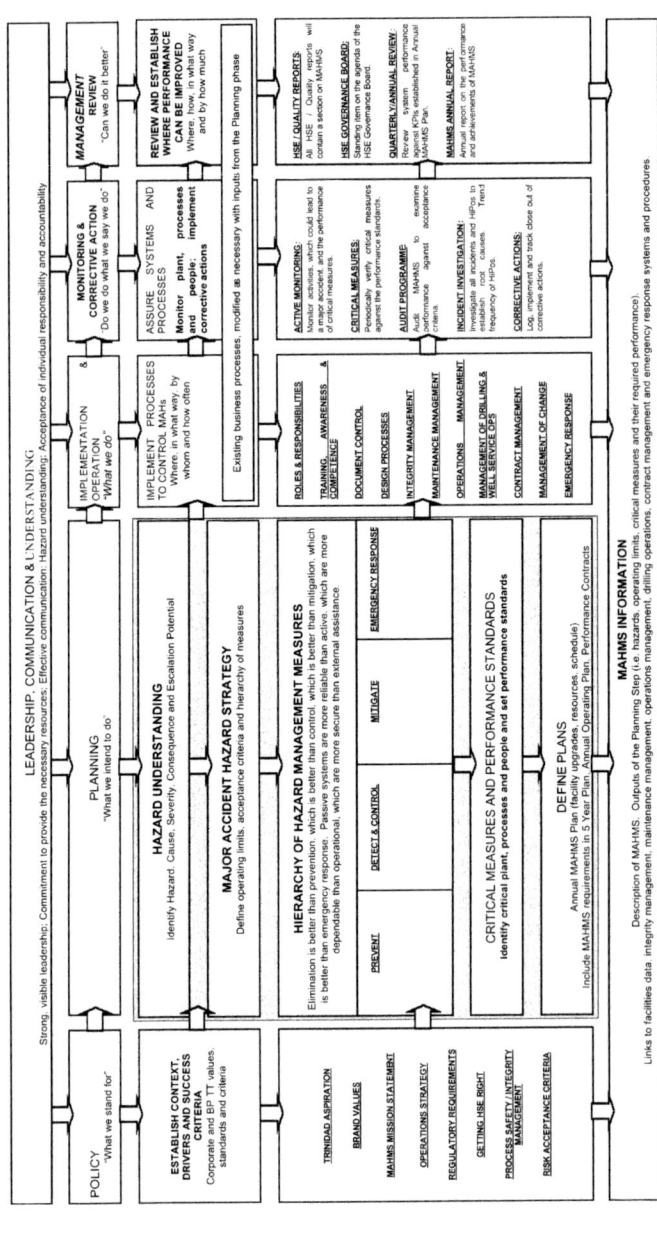

Figure 2. MAHMS process map

Revision 2: Dated 2nd November 2001

THE WHOLISTIC APPROACH TO RISK MANAGEMENT

Major Accident Hazard Management (MAHM) is a wholistic, structured approach to minimising the likelihood and reducing the consequences of a Major Accident Event, throughout the lifecycle of the operation. Here the lifecycle includes concept selection, detail design, construction, installation, hook-up, commissioning, operation, modification, de-commissioning and abandonment. This is shown in Figure 3, below:-

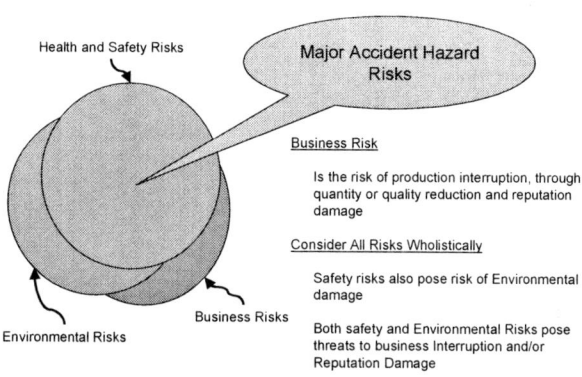

Figure 3.   Wholistic risk management

Increasingly, there is a move towards voluntarily considering and managing risks of environmental damage through emissions or accidental spillage. This is largely as a result of societal pressure to do so, a realisation amongst employees that it is the right thing to do and Operators recognising that they should be adopting an environmentally responsible attitude when conducting their operations.

A business needs to be successful in order to thrive. A successful business generates revenue, some of which can be invested back into operations to support amongst other things, the wholistic management of risk. Where investment monies are limited, the activities that have the best financial justifications are more successful at securing funding. Although accepted as a high priority, spending on safety and environmental risk management alone has very often to vie with many other demands on a limited operating (OpEx) budget. Spending on safety and environmental protection measures does not normally show a tangible financial payback. Combining business risk management with safety and environmental risk makes good business sense and effective use of limited resources, providing a more robust business justification for allocation of monies.

MAHMS is unashamed and blatant at taking this approach to securing its funding.
MAHMS also considers risk management throughout the lifecycle of the operation. For a
new facility this means eliminating hazards or reducing the likelihood of major accident
hazard events by:-

- Ensuring concept selection evaluates options to reduce risks to personnel, the
  environment and business interruption; for example:-
  - Adopting Normally Unattended Installation designs wherever possible.
  - Maximising spatial separation of the process plant, control rooms and
    accommodation by adopting bridge-linked multiple jacket designs
  - Building-in robust integrity management features and arrangements
- Taking full advantage of the opportunity to achieve an inherently safe design by
  incorporating as many measures to avoid or prevent the hazards as possible
- Adopt a layout of the facility, plant and hazardous areas, which minimises the effects
  and escalation potential during a major accident hazard event.

For an existing facility this means reducing the consequences of a major accident hazard
event by:-

- Selecting critical measures which control and mitigate the effects of the event
  effectively whilst minimising the cost and difficulty of installing the measures or
  modifying plant.
- Reduce safety risks by segregating people from the hazardous operations or areas
- Placing a greater emphasis on the effectiveness of emergency response measures to protect
  people and the environment from the worst effects of a major accident hazard event, and
  ensure that personnel can muster, evacuate or escape and be rescued and recovered safely.

*MAHMS adopts a hierarchy of measures, placing emphasis on avoiding and preventing
MAH events.*

MANAGING ALL LEVELS OF RISK AS A CONTINUUM
MAHMS utilises the variety of sound workplace risk assessment tools employed and
extends their use beyond occupational safety and health risk management, into the realm of
major accident hazard prevention. The basic tools and techniques are the same regardless of
the hazards being managed. Figure 4 shows how risks can be regarded as a continuum.
     MAHMS achieves this by providing the information to impart a thorough
understanding about the causes of major accident hazards, and then coaching users of
workplace risk assessment tools to use the information to assess:-

- Whether the work being planned could initiate a major accident hazard event
- In what ways could the work being planned go wrong to cause or threaten to cause a
  major accident hazard event
- Whether the work will disable or interfere with any critical measures designed to
  prevent, control, mitigate or provide emergency response to major accident hazard
  events; and if so what additional temporary measures will be required to provide
  equivalent protection.

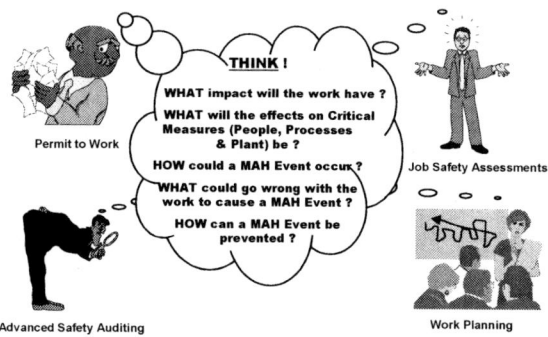

Figure 4. Risk management as a continuum

Some tools in common use where the above process can be applied are:-

Table 1. Common workplace risk analysis tools

Tools	Methodology
Job Safety Analysis	• Provide Hazard information to impart MAH Understanding • Provide guide-words and checklists of Facility MAH to ensure all potential MAH events are covered • Provide checklists to ensure all critical measures for each MAH are considered
Combined Operations or SIMOPS HAZOP	• Ensure a MAHMS specialist attends HAZOP Meeting • Ensure all potential work conflicts are considered • Develop a Combined Ops/SIMOPS Matrix to provide guidance on allowable simultaneous operations
Advanced Safety Auditing	• Provide Hazard information to impart MAH Understanding • Engage in on-the-job conversations about job planning and risk assessment • Extend the conversations to cover MAH avoidance and prevention considerations

PRAGMATIC & QUALITATIVE APPROACH
One thing must be borne in mind in the pursuit of perfection in terms of risk assessment and management, that is the resources (operators' personnel, professional assistance and funding for critical measures) is and always will be finite.

This means we must be *realistic* in what can be achieved, whilst at the same time having a goal in mind which will achieve levels of acceptable risk. The MAHMS approach is to *evaluate risk, primarily in a qualitative* way where it is obvious that risk are being adequately managed and only resort to *detailed quantitative risk analysis* to show how the levels of risk compare with the Operators' Risk Acceptance Criteria.

Risk assessment is the evaluation of the likelihood and consequence and the judgement of its acceptability. It should be applied to evaluate the risks to individuals, the environment and business interruption arising from each hazard, and to the cumulative risk on each asset/facility both individually and its contribution to the business unit as a whole. Work is also necessary to demonstrate compliance with the Operators' Risk Acceptability Criteria.

Where the initial assessment determines that further risk reduction is required, it should follow the standard hierarchy of risk reduction measures shown in Figure 5, below.

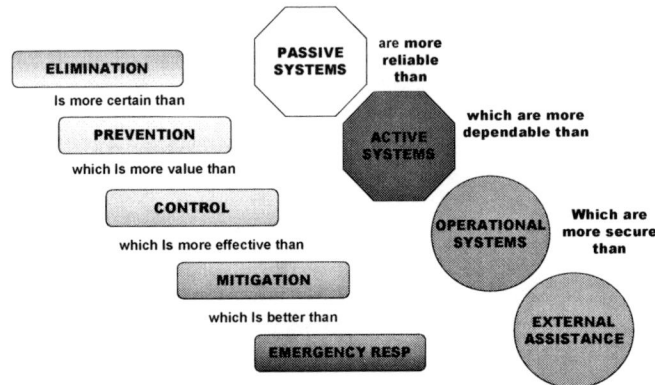

Figure 5.   Hierarchy of risk reduction measures

Qualitative Risk Assessment

This is the use of informed judgement of an experienced group of people to assess the tolerability of risks and the adequacy of measures to prevent or control them. It is essential that this team has an adequate understanding of the cause and consequence, rather than applying guesswork and perception. The group should include people who have operating experience and the eventual responsibility for managing these risks.

Hazard/Risk matrices are the preferred qualitative documentation tool shown in Figure 10. These should be appropriate for major accident hazards and calibrated to align with corporate risk criteria when used for an overall assessment.

Quantitative Risk Assessment

This is the numerical quantification of the totality of risks to life on a facility. It may calculate both individual and societal risk and be used to determine if further investment to reduce risks is warranted. It should integrate all of the information about the hazards and their potential for escalation to a major accident to deliver both the overall risk figures and a picture of the spread of those risks.

The input data to quantitative analysis should be based on realistic likelihood of initiating events taking into account both historical data, actual site conditions, and an assessment of the long term effectiveness of proposed prevention measures. The likelihood and severity of the consequences should take full account of the realistic performance and reliability of the critical measures to eliminate, prevent, control, mitigate and provide emergency response.

Risk Assessment Methodology

For the above approach to work, we need some means of making a qualitative assessment of the risks (safety, environmental and business risk), which can also link qualitative levels of risk to quantitative risk figures if required.

The following methodology describes one possible approach. This methodology is for rapidly ranking the major accident hazard (MAH) on onshore and offshore facilities, and has been adapted for use on the Major Accident Hazard Management System (MAHMS) for Trinidad.

The methodology is used where a multi-discipline team of operations, technical and specialists in hazard assessment carry out Major Accident Hazard Identification (MAHID) for each asset/facility, identifying the principal major accident hazards. In some cases, it may be better if these are grouped so that all hazards in one area could be considered together, e.g. a cellar deck.

Using the experience of the group and specific knowledge of the arrangements, manning and condition of the facilities, the likelihood and safety, environment and business consequence can be assessed and qualitatively ranked according to tables 2 to 5, given below. The risk ranking is the event likelihood and the highest of the safety, environmental or business consequence categories.

These rankings can be assigned a quantitative 'score' according to values given in the risk-ranking matrix. These scores for all the MAHs, can then be added to give an overall risk picture for the facility, denoting the contribution from each MAH. Note that the scoring system is exponential; i.e. the method recognises that one increment of frequency represents a factor of 10 and similarly, one increment of consequence is also a factor of 10. As a result, a risk ranked as "C III" has 100 times the relative risk level of a "B II"

It should be noted that this methodology only provides relative risk-rankings. To achieve numerical risk values, these scores must be 'calibrated' against a detailed quantitative risk analysis. Until this is done, they must not be cross-related to any specific assessment of risk, such as individual risk, probability of any given event, or other established numerical criteria.

Using this approach allows the summation of all of the results from each of the hazards on one facility to give an overall figure that may be compared with the others hazards of facilities.

Example

On a three jacket installation, the following hazards were identified, ranked and scored:

- Gas riser release with escalation to adjacent high-pressure risers which could then affect the accommodation; Ranking B III; Score 10^{-7}.
- Fuel gas leakage at the generators under the accommodation exploding within the engine enclosure but not affecting other areas; Ranking C II; Score 10^{-7}.
- Gas leakage from compressors ingested into and exploding within a local control room with four occupants; Ranking D III; Score 10^{-5}.
- Manifold oil fire in a cellar deck with smoke affecting the top deck and bridge escape route to other jackets; Ranking C II; Score 10^{-7}.
- Separator fire engulfing the accommodation in smoke; Ranking C III; Score 10^{-6}.
- Gas leak and explosion in the cellar deck damaging the riser ESD valves and allowing simultaneous release through the manifolds; Ranking B IV; Score 10^{-6}.

Overall Score; $10^{-7} + 10^{-7} + 10^{-5} + 10^{-7} + 10^{-6} + 10^{-6} = \underline{1.23 \times 10^{-5}}$

Table 2. Likelihood category

Category	Description
E	Incident happens several times per year in BP Indonesia/Likely to continue to occur
D	Incident occurs several times per year in BP/Incident may occur in BP Indonesia at some time
C	Incident has occurred in BP/Incident unlikely to occur in BP Indonesia
B	Incident has occurred in major accident hazard industries (e.g. oil and gas, petrochemical, chemical)/Incident very unlikely to occur in BP Indonesia
A	Never heard of in the world

Table 3.   Consequence category (Safety) | Signifies a Major Accident Hazard |

Category	Description
V	Potential for more than 30 fatalities (e.g. potential loss of entire drilling rig or large installation with immediate fatalities and survivors having to escape to sea in an uncontrolled manner)
IV	Potential for between 10 and 30 fatalities (e.g. an event on a drilling rig or large installation with immediate fatalities and which may require controlled evacuation or a helicopter ditching with loss of all onboard)
III	Potential for between 2 and 10 fatalities (e.g. an event that does not escalate with the potential for immediate fatalities only or loss of an entire satellite crew)
II	Potential for a single fatality or serious injury
I	Potential for first aid or medical treatment only, possible lost time injury

Table 4.   Consequence category (Environment)

Category	Environmental description	Socio-economic description
V	Potential to change ecosystem or activity leading to long term (+10 years) damage and poor potential for recovery to a normal state (e.g. significant damage to a fragile ecosystem)	Potential for long term loss or change to users or public finance (e.g. long term impact on fishing or tourist industry)
IV	Potential to change ecosystem or activity leading to medium term (+2 years) damage but with likelihood of recovery within 10 years	Potential to cause financial loss to other users or the public (e.g. cause a temporary suspension of fishing in the area)
III	Potential to change ecosystem or activity in a localised area for a short time, with good recovery potential. Similar scale of effect to existing variability	Potential to cause a nuisance to other users or the public
II	Change which is within scope of existing variability but can be monitored and/or noticed	Potential to affect behaviour but not a nuisance to other users or the public
I	Negligible effect (unlikely to be noticed or measurable against background activities)	Negligible effect

Table 5.   Consequence category (Business)

Category	Description
V	Potential for major cost or revenue impact across the whole of the Indonesia and/or loss of reputation impacting on future viability of BP's Indonesia operations. (e.g. major contract violation, impact on Indonesia gas supplies for thirty or more days) **Total Company Losses >$US 1billion**
IV	Potential for major cost or revenue impact to BP Indonesia (e.g. contract violation, impact on Indonesia gas supplies for more than 3days or significant loss of oil production for prolonged period) **Total Company Losses $US 100M – $US 1billion**
III	Potential for major cost or revenue impact on Performance Unit (e.g. contract violation, impact on Indonesia gas supplies for less than 3 days or significant loss of oil production for a period of days/weeks) **Total Company Losses $US 10M – $US 100M**
II	Potential for minor cost or revenue impact
I	Negligible cost or revenue impact

Consequence		A	B	C	D	E
	V	10^{-6}	10^{-5}	10^{-4}	10^{-3}	10^{-2}
	IV	10^{-7}	10^{-6}	10^{-5}	10^{-4}	10^{-3}
	III	10^{-8}	10^{-7}	10^{-6}	10^{-5}	10^{-4}
	II	10^{-9}	10^{-8}	10^{-7}	10^{-6}	10^{-5}
	I	10^{-10}	10^{-9}	10^{-8}	10^{-7}	10^{-6}
		A	B	C	D	E
				Likelihood		

Figure 6.   Hazard risk-ranking matrix

Note that when assessing consequences, likelihood and corresponding risk level, credit should be taken for passive safeguards, e.g. the layout of the facility, drains and bunds, passive fire protection, escape routes, etc.

The three colours in the matrices are the classic gradings; red; unacceptable and must be improved; yellow; significant risk and risk reduction measures must be evaluated and implemented if reasonably practicable; and green, risks are low but existing management systems must be maintained. The scoring is simply a means of describing the relative risk according to the position in the matrix. The authors acknowledge the contribution of Chris Rawlings for the concept of an exponential scoring system.

Major Accident Hazard Definition (for bpTT)
An accidental event that has the potential to lead to:

- the death of 2 or more people
- long term or widespread damage to the environment (>2 years)
- major costs or loss of revenue (>$US 10 M)

Having determined the likelihood and safety, environmental and business interruption consequences of each hazard in the Hazard/Risk Matrices, the hazard management strategy can be formulated.

CRITICAL MEASURES (PEOPLE, PROCESSES & PLANT)
The Design and Construction Regulations, DCR, within the UK Goal Setting Regulations require that items of plant that are Safety Critical are identified with a view to their performance being subject to regular maintenance, testing and assurance, which we know as 'verification'.

The above requirement detracts from the possibility that safety critical measures could also be people; in that the roles they play in an emergency for example, are crucial to the effectiveness of muster, evacuation or escape rescue and recovery. It also neglects the possibility that certain processes could be safety critical; for example corrosion monitoring and control is a management process which is often crucial to maintaining integrity of the pressure envelope.

Since MAHMS also takes a wholistic view on risk, the notion that measures are only safety critical no longer applies. MAHMS therefore refers to *critical measures* as being the *people, processes or plant* which eliminate, prevent, control, mitigate or provide the arrangements for muster, evacuation or escape and the facilities for rescue & recovery. These critical measures may also provide a role in either reducing the likelihood and/or the safety, environmental or business consequences of a major accident hazard event.

Within reason the bigger the range and depth of measures the better, since this provides strength in depth in terms of preventative or control barriers. Prevention or control of a major accident hazard event, should not normally rely on only one barrier and should never rely solely on either a people or process critical measure, since both these measures are highly vulnerable to human factors effects.

In order to determine priorities, for repair, maintenance, testing and other assurance processes; it may sometimes be necessary to identify the relative importance of a range of critical measures. The following provides some guidance on how this may be achieved.

Not all critical measures have the same criticality (importance). One common system should be used to determine the relative importance of widely different systems relying on competencies (people), providing hydrocarbon containment assurance (processes) or physical safety systems (plant).

There are numerous factors that dictate the importance of critical measures. These are:-

- **Their relative positions in the hierarchy of measures (eliminate, prevent, control, mitigate or provide emergency response).**
- **Whether the measure is passive or active; automatic or manual**
- **Whether there are human factors dimensions and potential for human error**
- **The required levels of availability and reliability (performance specification)**
- **Whether there is duplication or redundancy**
- **Whether the measure is an emergency response measure crucial to effect successful muster, evacuation, escape, rescue or recovery.**
- **Whether failure of the critical measure would initiate a major accident hazard event**
- **The extent to which the measure is reliant on or interacts with other measures**
- **The critical measures vulnerability to the effects of a MAH event.**
- **Whether the critical measure is providing the only barrier, which prevents a MAH event. (This situation should normally be avoided by design).**

The criticality of any given measure therefore depends upon the above factors. The diagram of Figure 7 below, illustrates this:-

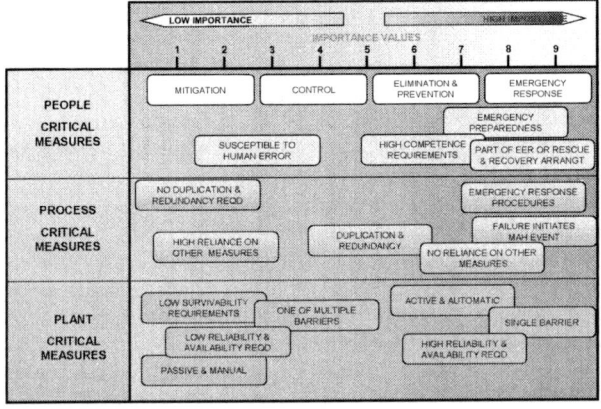

Figure 7.   Importance of critical measures

The importance of any given critical measure can be determined by a summation of points values of all the factors which apply to that measure. Doing this for all critical measures, will provide a relative importance ranking and hence form the basis for prioritising effort and resources to install, maintain, repair, test and examine them.

CRITICAL MEASURE PERFORMANCE STANDARDS

A Performance Standard is a clear unambiguous statement of what a critical measure is required to do, in terms of its performance parameters. The critical measure performance parameters are defined for people, process and plant measures as shown in Figure 8, below:-

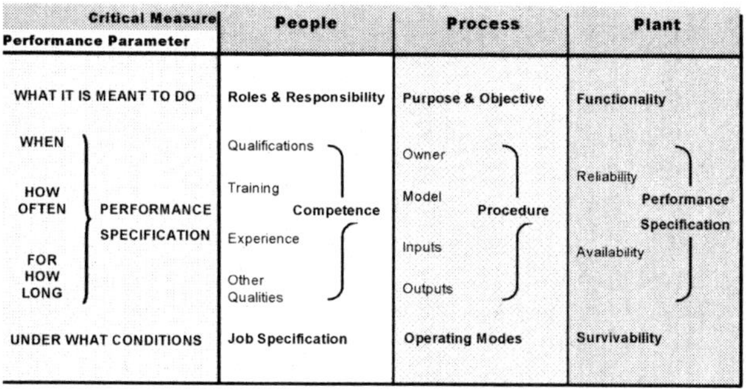

Critical Measure Performance Parameter	People	Process	Plant
WHAT IT IS MEANT TO DO	Roles & Responsibility	Purpose & Objective	Functionality
WHEN, HOW OFTEN, FOR HOW LONG (PERFORMANCE SPECIFICATION)	Qualifications, Training, Experience, Other Qualities (Competence)	Owner, Model, Inputs, Outputs (Procedure)	Reliability, Performance Specification, Availability
UNDER WHAT CONDITIONS	Job Specification	Operating Modes	Survivability

Figure 8. Critical measure performance parameters

What the measure is <u>required</u> to do, and not what it actually does, is defined by the *functionality* of the measure. This should be a very simple statement giving a clear unambiguous definition of what the measure is for. This will be the role in the case of a people critical measure, the purpose in the case of a process or the function in the case of a plant critical measure. The functionality statement should not go into details about how the role, purpose or function of the critical measure is achieved.

How the critical measure is required to perform, how long and how often it will be required to do it, is described in the *performance specification* part of the performance standard. It may be for example that a crucial role in emergency response will be performed in isolation (eg Lifeboat Coxswain) or as part of a team (eg. a member of a fire-fighting team). How long the role is to be performed could be determined in terms of the expected duration of events, or the physical limits of an individual to remain on duty. These latter factors will help determine how many personnel should be trained and in place to take on such crucial roles. Finally, the performance specification will stipulate when (how often) those able to conduct the roles should be available to do it. For most emergency response roles this will be continuous (ie. 24 hours per day, 7 days per week). Where this is the case, then this dictates that a crew manning a facility will need to have as a minimum the crucial roles covered at all times.

Similarly for plant critical measures, where a measure is required to be continuously available (eg a life boat), and this cannot be achieved because of the need to maintain and service that measure, then there is a need for duplication and redundancy. This could also apply to a process critical measure (eg crane driver competence assurance), which is only continuously available so long as only competent crane drivers are allowed to operate the cranes.

Lastly, the performance standard defines under what conditions the critical measure is required to continue performing, and for how long; in other words its **survivability**. This is particularly relevant to plant critical measures. For example, an emergency shut-down valve and actuator will be required to survive the effects of explosions and fires for long enough for the valve to close; and once closed remain able to effect a seal in a fire.

Once a Performance Standard has been written to unambiguously define all the factors discussed above, it provides the unequivocal benchmark against which the performance assurance activities can be judged. If at any time a critical measure is found to be failing to meet its performance standard, then action needs to be taken to rectify the problem and put such temporary arrangements or limitations in place until the measure is satisfactory.

It may be that in the light of more information or operating experience, that the performance standard requirements are found to be either more onerous than required or not good enough. In this case, the performance standard can be revised, approved and re-issued under rigid document control processes. It is not acceptable however, to revise a performance standard just because the critical measure concerned cannot meet it.

MAJOR ACCIDENT HAZARD UNDERSTANDING
One of the primary objectives of MAHMS is to identify the hazards and assess the risks posed by the operation. From this information an "understanding" of the major accident hazards can be imparted to members of the organisation working at each operating level. The level of detail of the understanding and thus hazard information requirements will be different as dictated by their operating level, roles and responsibilities and job function.

Table 6, over the page illustrates this:-

The following describes the hazard information needs and the hazard understanding to be provided at each operational level.

Hazard/Risk Picture – Senior Management
The Senior Management in an organisation need to understand the Major Accident Hazards arising from their operation, in order to make sure that the risks introduced by the hazards are managed adequately. The Senior Manager has ultimate **accountability** for the risk assessment and management activity and provide **leadership** to the business in this task.

Understanding will be provided at this level by means of hazard/risk pictures. These will be presented in the form of simple pie charts showing a qualitative risk analysis of Business Unit risks by "Hazard" and by "type of operation".

Table 6.   MAHMS hazard understanding & information

	Hazard understanding & information	Risk assessment & management
Level 1 **Accountable**	Have the ultimately **Accountability** for assessment and management of risk and	**Evaluating risk**, to show that they are being adequately managed and show how the levels of risk

SENIOR MANAGEMENT	provide the *Leadership* to ensure MAHMS is successful:- • By understanding the hazards of the operation and their contribution to the overall risk picture. • By providing visible and unwavering leadership and commitment to MAHMS • By being responsible for creating the conditions under which the right HSE Culture is in place and engenders the right attitude amongst the workforce for MAHMS to succeed.	compare with the BP Corporate Risk Acceptance Criteria:- • Through risk pictures which show the relative levels of risk across the BU • By the qualitative assessment of risk to show the risk distribution by activity and Performance Unit/Asset and Facility • Through appropriate quantitative risk analysis to show how the significant risks compare with the bp Corporate Risk Acceptance Criteria
Level 2 Responsible OPERATIONAL MANAGEMENT	To be able to decide the *priorities* and supply the *resources* to manage the business as efficiently as possible whilst reducing and maintaining MAH risks to as low a level as possible:- • By having a sufficiently detailed understanding of the principal risk drivers and how they influence the likelihood and consequences of a MAH event in their operation. • Through knowing how changes to the operation, organisation and resources could have a detrimental effect on risk management arrangements.	Determine for each MAH a *risk management strategy* to reduce the safety, environmental and business interruption risk within each Performance Unit:- • Through conducting a qualitative Risk Ranking exercise for all MAHs which shows the event likelihood, and Safety, Environmental and Business Interruption consequences • By identifying the Key Risk Drivers which influence likelihood and/or Safety, Environmental or Business Interruption consequences • By deciding the most appropriate strategy to avoid or reduce the likelihood and reduce the consequences to tolerable levels.
Level 3 Facilitators	Instinctively know what is important about day to day operation, so that they can	Determine a range of *critical measures* to be deployed which enable the Asset/Facility

FACILITY MANAGEMENT (OIMs OSMs & SUPERVISORS)	make ***informed judgements*** about the criticality of work and assign the priorities and resources accordingly:- • By understanding the role of critical measures (people, processes and plant) in MAHMS. • By playing a key role in the day to day management of critical measures to ensure their sufficiency and effectiveness through assurance processes like active monitoring and verification.	Management to effectively manage their MAH risks:- • Through Risk Registers detailing the Anatomy of all Reasonably Foreseeable MAH Events • By putting in place the Critical Measures (People, Processes & Plant) required to manage risk to a Reasonably Practicable level
Level 4 Implementers INDIVIDUALS (The Wider WORKFORCE)	Be fully aware of how his/her ***acts or omissions*** could lead to a MAH event, either by triggering the event directly or by rendering ineffective one or more of the critical measures. • Through workforce involvement in MAHMS Hazard Identification and Risk Analysis to gain the appropriate levels of Hazard Understanding • By having Training in terms of MAHMS awareness integrated into a structured Induction process for new starts and refreshers for all personnel • By having access to MAHMS information through an on-line MAHMS Web site Database, readily accessible for all who need to identify and use information.	Determine the ***Performance Standards*** required for critical measures to facilitate the management of their maintenance, availability and provide Performance Assurance on a day to day basis:- • By preparing a Performance Standard for each Critical Measure (People, Processes & Plant), which describes what it is meant to do; when, how often and for how long it is meant to do it and under what conditions it should remain effective • The Performance Standard will provide the benchmark against which Performance Assurance activities will be carried out.

Figure 9 below shows a typical Hazard/Risk Picture (NB: risk values shown are for demonstration only).

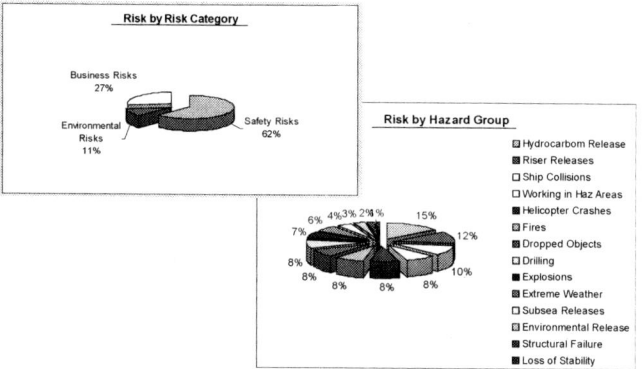

Figure 9.   Hazard/risk ranking by risk category and hazard group

Hazard/Risk Ranking Matrices – Asset Managers
Asset Managers require sufficient MAH understanding to be able to appreciate the priorities and resource requirements are for MAHMS to be effective. Thus their primary role is to determine *priorities* and provide *resources* for MAHMS activities in their part of the organisation.

The relevant level of understanding will be provided to the Performance Units by means of both the risk pictures described above; and Hazard/Risk Ranking Matrices. The Matrices will be the output from qualitative risk analysis and will list the Hazards in descending order of risk contribution. They will also identify the principal drivers influencing likelihood and consequence for each different MAH. This information will be presented in the form of:

- Simple pie charts showing risk profile between different facilities/assets
- Hazard/Risk ranking matrix for each facility/asset showing the principal risk drivers

Figure 10 illustrates a typical Hazard/Risk Ranking Matrix (NB: risk values shown are for demonstration only).

Hazard Registers – Facility Management
The Facility Management Team (Delivery Managers, OIMs, Onshore Shift Managers and Supervisors), need to instinctively know what is important about day to day risk management on the operation, so that they can make *informed judgements* about the criticality of work and assign the priorities and resources accordingly.

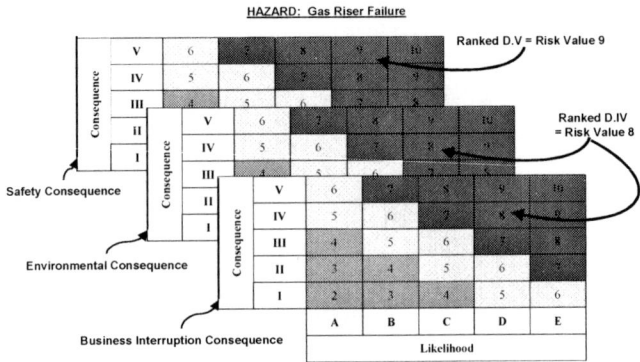

Figure 10. Hazard/risk ranking of a MAH

MAHM understanding will be provided to facility operations and maintenance personnel through the Hazard/Risk Pictures and Rankings described above, and the hazard register. Hazard registers are where the detailed information on each hazard will be documented. The information will be gleaned through Major Hazard Identification (MAHID) exercises to be held for each facility.

The layout and content of a hazard register is illustrated in Figure 11.

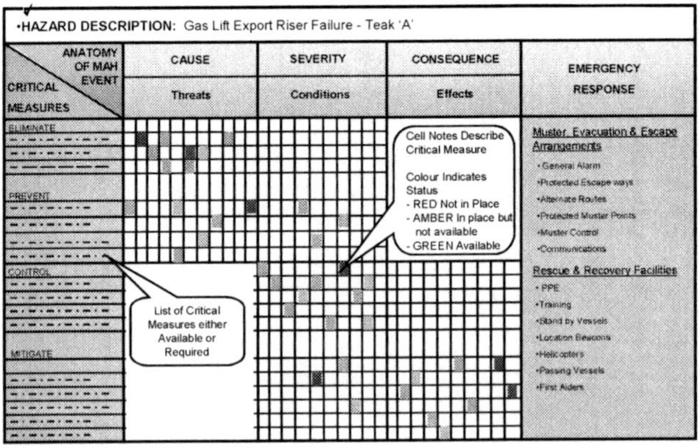

Figure 11. Layout and content of a hazard register

In particular, the information presented in the Hazard Register will provide the means for supervisors to apply "active monitoring" to manage their MAHMS Performance Assurance responsibilities on a day to day basis.

Anatomy of a Major Accident Hazard – Operations Workforce
Apart from their other MAHMS Roles and Responsibilities, individuals have a need to understand Major Accident Hazards sufficiently well to be fully aware of how his/her *acts or omissions* could lead to a MAH event. This could either be by triggering the event directly or by rendering ineffective one or more of the critical measures.

As well as all the sources of information described above, a detailed understanding of the relevant Hazards, will be supplied through various communication tools all describing the Anatomy of each Major Accident Hazard. The Anatomy of a Hazard is illustrated in Figure 12 below:-

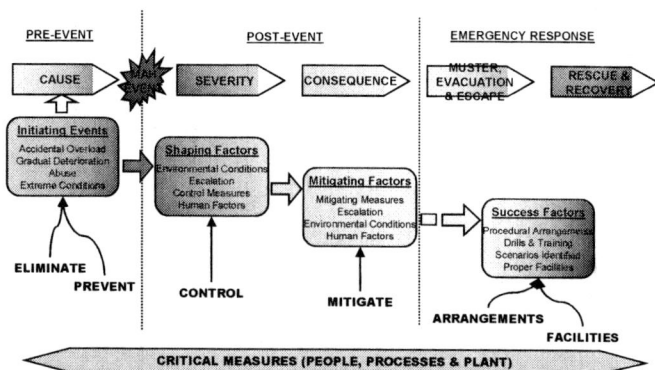

Figure 12.   Anatomy of a major accident hazard event

MAJOR ACCIDENT HAZARD INFORMATION
The Major Accident Hazard Information required for each of the operating levels described above, will be continuously available to all individuals in the organisation through the MAHMS on-line web based database. It will be presented in a form suitable for them to understand MAH sufficient to fulfil their individual requirements.

Figure 13 shows an overview of the information obtained from the above analyses, and illustrates how it will be presented in order to be usable to the various personnel involved in day to day risk management operations.

The type of MAH information available is typically as follows:-

Causes and Likelihood
There should be a detailed understanding of the types of causes, whether due to human factors, procedural malfunction or failure of plant or the structure, and the likelihood for those causes to occur on the facility. For example the possibility of corrosion, the number of corrosion sites and the corrosivity of the process fluids, are all factors which determine the likelihood of corrosion led failures of the pressure envelope.

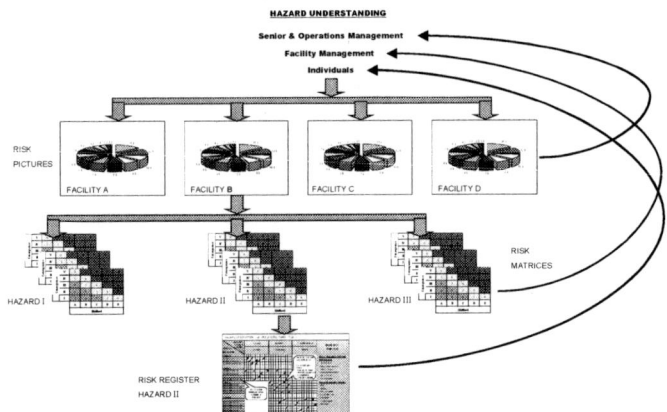

Figure 13.   Overview of MAHMS database

There should be a formal process, such as a MAHID, to confirm that all causes have been identified. It must be applied to all major hazards arising from any aspect of the operation, failure of plant or structures, not just process plant. For example, all of the means whereby a floating marine structure could be overloaded or damaged should be rigorously examined. The level of further analysis of the identified causes should be determined, so that they may be effectively managed.

Severity
Where hazards have the potential to cause a major accident, the severity should be quantified in relevant terms such as the type of hazardous event, the energy released, the size and the area affected, intensity of noise, heat and smoke, location and duration. The quality of analysis needed will depend upon the potential impact of the event, the risks and the type of information needed to make informed decisions.

Immediate Effects and Escalation Potential
Where hazards have the potential to escalate into a major accident, the effects of the initial event on people, processes and the plant, structure or safety systems that may fail leading to further escalation, should be determined. The quality of that analysis required will depend on the type of event, the underlying potential for significant escalation, and the type of information needed to make informed decisions. Tools will be required to map these escalations and methods to assess the effectiveness or vulnerability of safety systems and emergency response facilities.

Consequence
Where hazards have the potential to result in a major accident, all of the routes that lead to that consequence should be identified and mapped. The sequence, timing and characteristics of the event progression should be determined. The quality of the analysis required will

depend on the likelihood of that escalation, the overall potential loss of life and facility damage, and the type of information needed to make informed decisions.

Risk

A picture of the relative risks to the overall facility from individual hazards should progressively develop as the assessment progresses. This is essential information for the decisions in the hazard management process. The quality of risk analysis required for individual hazards will depend upon their contribution to the overall risk. Major accident hazards may each be subjected to a qualitative risk analysis that covers both the initial event and its potential for escalation. Where specific hazards make a dominant contribution to the overall risk, or the overall risks cannot easily be reduced to tolerable levels, a more formal process of Quantitative Risk Assessment (QRA) may be applied.

MODELLING RELEASES OF WATER REACTIVE CHEMICALS

Quinn, D.J. and Davies, P.A.
ERM Risk, Suite 8.01, 8 Exchange Quay, Manchester M5 3EJ, England

Water reactive chemicals are commonly used in many manufacturing processes, particularly in the soap and detergent industry. The hazardous nature of these chemicals is recognised in the Approved Supply List where they are generally attributed risk phrases such as R14, 'Reacts violently with water' and R29, 'Contact with water liberates toxic gas'. This paper is principally related to chemicals which liberate toxic gas on contact with water. Pool models typically deal with boiling and/or evaporating liquids. However, few are available to model the behaviour of water reactive pools. There are a number of properties which influence the behaviour of the pool such as: pool diameter, composition and temperature. The detailed interaction of these and other time dependent pool properties requires a 'complex' computer model which, depending on the spill size, may take some time to generate results. Clearly this may be impractical if many release sizes require modelling. This paper details a simplified method for calculating toxic gas evolution rates from releases of water reactive chemicals. This method has been used as the basis for calculating consultation distances around major hazard sites with the results shown to closely match those calculated by the Health and Safety Executive.

KEYWORDS: Water reactive chemicals, dispersion modelling, oleum, sulphur trioxide, phosphorus oxychloride, thionyl chloride, chlorosulphonic acid

INTRODUCTION

A number of chemicals react violently with water liberating toxic gas and/or flammable gases, that can present a major accident hazard. This paper focuses on those water reactive chemicals (WRCs) that liberate toxic gases on contact with water, though the method can be adapted and applied to those that liberate flammable gases.

Where significant quantities of WRCs are stored and processed on a chemical facility, it may be necessary to model the consequences of a spillage in order to determine harm extents. The time dependent complex nature of the pool and its properties makes detailed calculation time consuming and complicated. These detailed methods may be suited to 'one off' calculations, but are considered less suited to calculating consequences from many releases.

Numerous studies have employed a method developed by ERM Risk, based on that described in a monograph published by the IChemE[1]. The model has been modified, adapted and refined according to the studies it has been used for. Therefore, it is felt that the model is versatile, robust and consistently constructed to account for case specific detail. The model has been used to replicate results calculated by the UK Health and Safety Executive (HSE) and has been adapted to model other WRCs.

OTHER MODELS FOR WATER REACTIVE CHEMICALS

For releases which are entirely indoors in perfectly 'dry' conditions (i.e. no surface water), the method described by Pettit et al[2] can be employed. However, perfectly 'dry' conditions are unlikely unless special measures are in place as described later. In addition, unlike the model proposed here, this method assumes that the release is for a fixed duration of 1 hour (though it can be adapted for 30 minutes duration since typically, HSE assume that the release has been controlled after 30 minutes). Furthermore, the limited egress of toxic gas from the building is not accounted for. Essentially, the model assumes that a column of air, of specified height above the pool contains some moisture according to the humidity. This column is continually moving, according to the building air change rate, providing a continuous supply of moisture for the WRC to react with. The evaporation rate from the pool is not calculated. Instead it is assumed that sufficient WRC is evaporated to react with the moisture in the column of air above the pool.

A more detailed model for the behaviour of water reactive chemicals is described by Kapias[3]. This is an extension to the models describing the behaviour of the water reactive materials, oleum and sulphur trioxide[1, 4].

DESCRIPTION OF THE MODEL

The model developed by ERM can model indoor and outdoor releases of WRC, accounting for toxic gas evolution by wind driven evaporation and reaction with water. Site specific characteristics can be incorporated. For example, the minimum pool depth can be increased according to the ground type, or the forced ventilation in a building can be considered to account for limited egress from the building. A suitable decision process has been derived to determine the minimum pool depth, water thickness, a building's air change rate and wind speed, all of which have an important affect on toxic gas evolution rates which can significantly affect results. The model is versatile and has been used successfully in a number of studies.

EVOLUTION FROM A POOL
Pool Size Calculation
The release is assumed to form a circular pool of uniform depth.

Unbunded Effective Pool Radius (UEPR)
The UEPR is calculated from the volume, using Equation 1.

$$r = 6.85V^{0.44537}$$ Equation 1

where

r pool radius (m)
V volume released (m³)

Equation 1 has been regressed from the formula for a spreading pool[5], based on a 5 mm minimum depth, which is usually related to smooth concrete surfaces. This leads to the 'worst case' (i.e. maximum radius pool), and is understood to be the method typically employed by HSE to calculate pool radius.

If required, the volume released may be calculated from the liquid density and the mass released.

$$V = \frac{M}{\rho}$$

Equation 2

where

M mass released (kg)
ρ liquid density (kg/m³)

Effective Bund Radius (EBR)
Spill size may be limited if contained within a bund. Typically, bunds are a rectangular shape, having a certain floor surface area. The effective bund radius is equal to that of a circle whose area is equal to the bund floor surface area.

Actual Effective Pool Radius (AEPR)
Clearly the AEPR cannot be larger than the effective bund radius (EBR). The following applies:

- if UEPR > EBR, then AEPR = EBR;
- if UEPR < EBR, then AEPR = UEPR.

Wind Driven Evaporation
The wind driven release rate is calculated using the Sutton correlation[6] for evaporation from a free liquid surface in the atmosphere, adapted as described by Grint and Purdy[7].

$$E_{w,WRC} = 1.684 \times 10^{-6} \left(\frac{mP}{T} \right) U^{0.78} r^{1.89} Sc^{-\frac{2}{3}}$$

Equation 3

where

$E_{w,WRC}$ wind driven release rate of WRC (kg/s)
m molecular weight (kg/kmol)
P vapour pressure of the evaporating WRC (N/m²)
T temperature of the evaporating WRC (K)
U wind speed at a height of 10 m (m/s)
r pool radius (m)
Sc Schmidt number of the evaporating WRC (no units)

The Schmidt number for sulphur trioxide is 1.1[1].

Note that Equation 3 calculates the wind driven release rate of the evaporating WRC. It is judged that the WRC will completely react with atmospheric moisture according to the general form of chemical reaction detailed below.

The quantity of toxic gas can be calculated from the mass of water and the stoichiometry of the reaction. In general, the chemical reaction is of the form:

$$WRC + xH_2O \rightarrow ySO_2 + zHCl$$

where

WRC   water reactive chemical
x number of moles of water required per mole of WRC

y number of moles of sulphur dioxide produced per mole of WRC
z number of moles of hydrogen chloride produced per mole of WRC

The stoichiometry is summarised in Table 1.

Note that the reaction of sulphur trioxide with water yields sulphuric acid mist and sulphur trioxide vapour which subsequently reacts with atmospheric moisture. Sulphur trioxide reacts with water according to the equation:

$$SO_3 + H_2O \rightarrow H_2SO_4$$

Table 1. Stoichiometric summary for selected water reactive chemicals

WRC	x	y	z
Chlorosulphonic acid	1	0	1
Phosphorus oxychloride	3	0	3
Thionyl chloride	1	1	2

It is judged that the vast majority of the evaporating WRC will react with water to form toxic gas(es). Therefore, for the purpose of modelling wind driven releases, it is assumed that all the WRC evaporated by wind reacts with water and there is sufficient atmospheric moisture to facilitate this. The evolution rates of toxic gases can be calculated:

$$E_{w.HCl} = \frac{36.5z}{m} E_{w.WRC} \qquad \text{Equation 4}$$

$$E_{w.SO_2} = \frac{64y}{m} E_{w.WRC} \qquad \text{Equation 5}$$

where

$E_{w.HCl}$ evolution rate of $HCl_{(g)}$ from wind driven release of WRC (kg/s)

$E_{w.SO_2}$ evolution rate of SO_2 from wind driven release of WRC (kg/s)

Evaporation Due to Reaction With Water

It is assumed that all the surface water covered by the pool reacts. Therefore, the mass of water available for reaction is calculated as follows:

$$M_{r.H_2O} = \pi r^2 \rho_{H_2O} h_{H_2O} \qquad \text{Equation 6}$$

where

$M_{r.H_2O}$ mass of water available for reaction (kg)
r pool radius (m)
ρ_{H_2O} density of water (kg/m³)
h_{H_2O} depth of water (m)

The quantity of toxic gas can be calculated from the mass of water and the stoichiometry of the reaction. In general, the chemical reaction is of the form described above.

Assuming that there is complete reaction with water, limited by the quantity of available water (i.e. WRC is in excess), the mass of toxic gas produced from the reaction is calculated:

$$M_{r,HCl} = \frac{36.5z}{18x} M_{r,H_2O}$$

Equation 7

where

$M_{r,HCl}$ mass of HCl produced in the reaction (kg)

It is assumed that the mass of toxic gas produced, as a result of reaction with water, is released over a 3 minute period. Therefore, the release rate is calculated as follows:

$$E_{r,HCl} = \frac{36.5z}{18 \times 180x} M_{r,H_2O}$$

Equation 8

where

$E_{r,HCl}$ HCl evolution rate from the reaction with water (kg/s)

Overall Average Evolution Rate

Evolution of toxic gas is assumed to continue for 30 minutes. In most cases, it is understood that HSE assumes that releases from pools last for 30 minutes, by which time it is assumed that onsite emergency action will have controlled the evaporation from the pool. For many releases, it is judged that the pool will reach its maximum radius within 3 minutes after the start of release. Therefore, after 3 minutes the pool will not contact additional surface water. Evaporation due to reaction with surface water is assumed to continue for 3 minutes (180 s), with wind driven evaporation only, continuing for the remaining 27 minutes. Increasing the time for reaction with water tends to decrease the overall average toxic gas release rate. It has been found that the release rates calculated for a 3 minute reaction time with water are broadly in line with those calculated using other models. However, the reaction time with water can be altered to reflect the characteristics of specific releases.

The overall average toxic gas release rate is calculated using the root mean square relationship:

$$E_{av} = \left\{ \frac{E_{r,T}^2 t_r + E_{w,T}^2 (t - t_r)}{t} \right\}^{1/2}$$

Equation 9

where

E_{av} overall average evolution rate (kg/s)
t_r duration of pool reaction with water (s)
t duration of release from pool (s)

CALCULATION OF EGRESS FROM A BUILDING

For releases in buildings, the toxic gas egress will be limited according to the volume of the building and its air change rate. The release rate from the building is calculated using a mass balance, accounting for the gas release rate into the building and its egress rate from the building. During the period where toxic gas is released from the source, assuming perfect mixing, the release rate from the building is:

$$m_{out} = m_{in}\left(1 - e^{-\frac{q_{air}t}{V_B}}\right) = m_{in}\left(1 - e^{-kt}\right)$$ Equation 10

After the release from the source has ceased, assuming perfect mixing, the release rate from the building is:

$$m_{out} = m_{out,max}e^{-\frac{q_{air}(t-t_{max})}{V_B}} = m_{out,max}e^{-k(t-t_{max})}$$ Equation 11

where

m_{out}	release rate of toxic gas from the building (kg/s)
m_{in}	release rate of toxic gas within the building (kg/s)
q_{air}	building ventilation rate (m³/s)
V_B	building volume (m³)
k	building air change rate per second (s^{-1})

The release rate from the building can be calculated at appropriate times and entered into a suitable transient release dispersion model.

GUIDANCE ON RELEASE SPECIFIC PROPERTIES

The consequence analysis is dependent on release and site specific properties. For example, the available water thickness, building air change rate and 'indoor' wind speed. These properties are not amenable to accurate measurement and are likely to be unique to the particular storage arrangement. However, ERM has developed some guidance, based on literary sources and judgement.

WATER THICKNESS

It should be noted that although it is less likely that surface water will be available for reaction with indoor releases, some consideration of its presence is recommended. For example, surface water may accumulate indoors from process water, condensed steam or be transported inside buildings on the surface of vehicles that have been out in the rain. Judgements must be made as to how much water is likely to be present. Furthermore, WRC can react with the water content of substrates such as concrete.

For releases outdoors, it is assumed that water is available to react with any spillage. Although it is recognised that 'special' acid resistant and water repellent tiles may limit the amount of water available for reaction, it is understood that HSE would be unlikely to attribute any credit in their calculations. Therefore, even for indoor releases which are more likely to be in 'dry' conditions, it is understood that HSE would still assume a small amount of water. The scheme summarised in Table 2 may be used as guidance on determining the mean water depth.

Table 2.   Guidance on mean water depth

Description	Recommended mean water depth
Well sealed building or room featuring locking doors, limited openings (if any).	0.5 mm
Sealed road tanker offloading building with roller shutter door.	1 mm
Building or room with frequently opened doors and/or windows.	1.5 mm
Building or room with process water and/or condensed steam frequently present.	1.5 mm
Outside in 'dry' conditions	2 mm
Outside in 'wet' conditions	5 mm

BUILDING AIR CHANGE RATE

When estimating a suitable air change rate for a building, the following characteristics need to be considered.

- Is there forced ventilation?
- If so, what height above the ground are the ventilation intakes?
- Is there a noticeable draft?
- Is the building well sealed (double glazing)?
- Are windows and doors typically closed?
- Is the building occupied much of the day by numerous amounts of people?

For example, 5 and 3 air changes per hour (ACH) have been judged suitable for a sulphur trioxide road tanker offloading building and bulk storage building, respectively. Note that buildings used as laboratories, may have a higher air change rate than for a similarly constructed building used for other purposes. This is largely due to the presence of forced ventilation, in the form of fume cupboards or systems to help prevent flammable and/or toxic atmospheres developing. Laboratories and areas having forced ventilation typically have a minimum air change rate of 12 ACH and could be as high as 40 ACH. For such buildings, it may be that the ventilation intakes are sufficiently elevated that toxic gas is unlikely to enter the building via the forced ventilation system.

WIND SPEED WITHIN BUILDING

The wind driven evaporation from a pool inside a building is determined by the assumed wind speed. Although the inside of a building is somewhat shielded from wind, there will be some air movement which will facilitate evaporation by the 'wind driven' mechanism. In general, an upper bound for operator comfort is considered to be 0.5 m/s. Therefore, in practice lower wind speeds may be experienced. Similarly, wind speeds of 5 m/s or greater, are generally felt as a draft. Therefore, judgement must be used for specific buildings. The following examples are based on judgements made for different types of building:

- Open-sided storage building, 'dutch barn' – 2 m/s.
- Road tanker offloading building with roller shutter door, typically closed – 1.5 m/s.
- Typically unoccupied bulk storage building – 0.25 m/s.

DISPERSION MODELLING

The calculated overall average evolution rate or the time dependent release rates from a building are the principal entries to a suitable gas dispersion model. Gas dispersion models are widely described and are included in consequence modelling software such as BP Cirrus[8]. Therefore, gas dispersion models are not described in detail here.

The toxic gases which may be produced when WRCs contact water are typically denser than air at ambient temperature. However, these toxic gases are typically produced in an exothermic reaction which tends to produce 'warm' gases. Therefore, the gases tend to be more buoyant and hence, a neutral, 'Gaussian' dispersion model is often judged most suitable.

The HSE has published a series of reports on the toxicology of materials in relation to major hazards. These reports are primarily directed at establishing toxicity values for land use planning in the vicinity of industrial facilities. Toxicological data are reviewed to derive a measure referred to as the Dangerous Toxic Load (DTL).

The DTL is associated with a level of impact which is designated the Specified Level Of Toxicity (SLOT) where the SLOT would "...*cause severe distress to almost everyone, many [would] require medical treatment, some [would] be seriously injured and highly vulnerable people might be killed.*" Therefore, to judge the extent of impact to persons from airborne toxic releases, the distance to the SLOT concentration is calculated.

Depending upon the properties of the material, the relationship between concentration and DTL can be linear or non linear:

$$DTL = c^n t \qquad \text{Equation 12}$$

where

c   concentration (ppm)
n   concentration exponent (no units)
t   exposure duration (minutes)

The downwind distance at which a person outdoors would receive a dangerous toxic load is calculated. Selected dangerous toxic loads are detailed in Table 3.

The reaction of sulphur trioxide and oleum with water yields sulphuric acid mist. Similarly chlorosulphonic acid yields hydrogen chloride gas. However, other chemicals, for example phosphorus oxychloride and thionyl chloride, can produce more than one toxic component. These substances require further dispersion modelling techniques as follows.

Phosphorus Oxychloride
There is evidence to suggest that, although phosphorus oxychloride ($POCl_3$) reacts violently with water, a significant proportion may remain unreacted[3]. Phosphorus oxychloride has the

risk phrase R26, 'Very toxic by inhalation'. The reaction of $POCl_3$ with water produces hydrogen chloride gas ($HCl_{(g)}$) (risk phrase R23, 'Toxic by inhalation'). Therefore, it is judged that $POCl_3$ is 'more toxic' than the toxic gas produced when it reacts with water. However, it is judged that the majority of the $POCl_3$ will react with atmospheric moisture, on account of its affinity to reaction with water.

Table 3.   Selected dangerous toxic loads[9]

Substance	Dangerous toxic load (ppm^n minute)	Concentration exponent 'n'
Hydrogen chloride	23,700	1
Sulphur dioxide	4.655×10^6	2
Sulphuric acid mist	13,000	2
Phosphorus oxychloride	2,880	1

Thionyl Chloride

On contact with water, $HCl_{(g)}$ and $SO_{2(g)}$ are produced. The release rates of $HCl_{(g)}$ and $SO_{2(g)}$ can be calculated. The two components have different dangerous doses, therefore, to calculate a cumulative dose, the release rates are 'standardised'.

In general, at a certain downwind distance, the concentration is proportional to the release rate. That is, double the release rate will produce double the concentration at a particular distance downwind. For example, if a concentration of 100 ppm is calculated at a distance 400 m from the source for 1 kg/s release rate, then a 2 kg/s release will result in a 200 ppm concentration, at that same downwind distance.

The dangerous toxic loads (DTL) for $HCl_{(g)}$ and SO_2 are 23,700 ppm.minute and 4.655×10^6 ppm^2.minute, respectively. The corresponding 30 minute SLOT concentrations for $HCl_{(g)}$ and SO_2 are 790 ppm and 394 ppm, respectively. Therefore, the following two cases yield approximately the same downwind dispersion distance:

1. a release rate of 1 kg/s of SO_2, modelled to an end point corresponding to a concentration of 394 ppm; and
2. a release rate of 2 kg/s (790/394 × 1 kg/s) of SO_2, modelled to an end point corresponding to a concentration of 790 ppm.

Therefore, as guidance a release rate of SO_2 can be scaled up (by multiplying by 2) such that its toxic load characteristics can be summed with those of $HCl_{(g)}$.

REFERENCES

1.   IChemE. Richard Griffiths (Editor). (1996). Major Hazards Monograph. Sulphur Trioxide, Oleum and Sulphuric Acid Mist.
2.   Pettitt, G, Bains, G and Dutton, T. (2001). Institution of Chemical Engineers Symposium Series No. 148. Hazards XVI – Analysing the past, planning the future.
3.   T Kapias and R F Griffiths. Environmental Technology Centre, Department of Chemical Engineering, UMIST. (2001). REACTPOOL: A new model for accidental

releases of water reactive chemicals. (2001). Health and Safety Executive Contract Research Report 331/2001.

4. Kapias and R F Griffiths. Environmental Technology Centre, Department of Chemical Engineering, UMIST. (1999). Modelling the behaviour of spillages of sulphur trioxide and oleum. Health and Safety Executive, Contract Research Report 217/1999.

5. Committee for the Prevention of Disasters (TNO). (1997). Methods for the calculation of physical effects - due to releases of hazardous materials (liquids and gases). 'The Yellow Book'. CPR 14E.

6. Sutton, O. G. (1953). Micrometeorology. McGraw-Hill.

7. Grint, G and Purdy, G. (1990). Sulphur trioxide and oleum hazard assessment. Journal of Loss Prevention in the Process Industries, 3: 177-184.

8. BP Research and Engineering Centre. Sunbury-on-Thames, England. BP Cirrus, 6.2.

9. Health and Safety Executive, Hazardous Installations Directorate. Assessment of the Dangerous Toxic Load (DTL) for Specified Level of Toxicity (SLOT) and Significant Likelihood of Death. http://www.hse.gov.uk/hid/haztox.htm.

A SIMPLIFIED RISK-BASED APPROACH FOR ANALYZING HUMAN FACTORS

David A. Moore, PE, President and CEO, AcuTech Consulting Group, Chemetica, Inc.
1948 Sutter Street, San Francisco, CA 94115, dmoore@acutech-consulting.com

INTRODUCTION

While industry agrees that human factors issues are critically important to process safety performance, the lack of a universally adopted, practical approach to address human factors has hampered progress in the practical application of human factors in process safety. Most process safety management programs do not formally address human factors, other than possibly in a superficial way during traditional PHA studies. As such, human factor risk issues are not addressed in a fully comprehensive approach. For the industry to embrace human factors more than they currently are, more practical guidelines are required and additional information is needed for industry to understand how to expend their efforts on this cause.

This paper presents a format for a task-based analysis approach that can be introduced into an existing PSM structure at a process facility.

PRACTICAL METHODS TO ADDRESS HUMAN FACTORS HAZARDS

PSM is a management systems approach, and does not include details on the methods recommended to identify and address risk. This is particularly true for human factors. For the latter issue, it is imperative to develop a methodology for internal teams to routinely and systematically address risks posed by human factors. There are a variety of techniques available, but we believe that many of them are not practical enough to be commonly used or accepted in industry. PHA teams would benefit from a superior approach for addressing human factors during the PHA, and there are other opportunities for special studies to be conducted then as required.

The basis of the approach is to assimilate human factors into the PHA activities of the organization in a more explicit way than is customarily done.

It is only a matter of time until regulators are more interested in human factors than they have shown to be in the past. OSHA and EPA merely gave mention to the topic when promulgating the PSM and RMP regulations. No doubt human factors are an area ready for significant growth and more attention in the very near future.

DEFINITION OF HUMAN FACTORS AND HUMAN ERROR

In order to properly manage human factors, it has to be clear what is involved. With such a fuzzy definition of human factors in industry presently, there could likely be confusion. Accepted definitions of human factors are:

1. A discipline concerned with designing machines, operations, and work environments so that they match human capabilities, limitations, and needs.[i]

2. [E]nvironmental, organizational, and job factors, and human and individual characteristics which influence behavior at work in a way which can affect health and safety.[ii]
3. Departure from acceptable or desirable practice on the part of an individual that can result in unacceptable or undesirable results.[iii]

Regarding process safety, human factors are a collective issue of prevention and mitigation of catastrophic releases of highly hazardous materials through various human factors considerations.

As it relates to process environments, it is recognized that management decisions and programs, and operational procedures, training, and actions can all contribute to human errors. In addition to these parameters, consideration should be given to incorporating human factors into inherently safer design practices and to improvements in the work environment to reduce the number and likelihood of situations to produce error.

OBJECTIVES AND SCOPE OF THE EFFORT
To properly address human factors it is necessary to implement a program that takes a management systems viewpoint to the problem. Included with this management system for human factors is the need for a means to identify and analyze human error likely situations. The corporation should develop a strategy for the implementation of a process safety program to address human errors more carefully. This may include the following objectives:

1. To assist the corporation with the development of a human factors approach as a supplement to and to be integrated with the existing process safety management systems;
2. To assist the corporation to develop a specific approach to conduct a Process Hazard Analysis (PHA) with an emphasis on human factors;
3. To develop all supporting training materials and provide the training for the successful implementation of the program.

The scope of the implementation of such a process is as follows:

TASK 1 – ORIENTATION AND PROJECT SCOPING
The first goal is to meet with the appropriate parties in the corporation to scope out the project and to determine goals, schedule, and other administrative details. Included in Task 1 are the following:

1.1. To become familiar with the corporation's Process Safety Management (PSM) program and PSM metrics by reviewing PSM documents, policies, and procedures and through discussions with process safety personnel;
1.2. To become familiar with the corporation's perception of human factors and human error issues affecting the corporation through a review of incidents and discussions with process safety personnel;

1	Absolute risk comparison approach	Setting safety goals to a defined risk level, such as no more than X events per year caused by human error
2	Benchmarking approach	Comparison of loss statistics and setting goals based on performance of peer industrial companies, i.e., no worse than the average performer
3	Relative risk reduction approach	Setting goals to reduce risk from where it exists at the time specific to the company, i.e., 20% improvement per year
4	Idealistic approach	Zero incidents or no events attributable to human error

Figure 1. Human factors program risk reduction goal strategies

1.3. To determine program goals per the strategies in Figure 1 and in positioning the program for management commitment and approval. Management commitment is essential for incorporating human factors issues, especially because many of the program suggestions are not explicitly required under the law. A full explanation of the benefits is essential, along with the costs involved. An ongoing management system should be put in place for implementing and supervising the program, ensuring its quality, measuring its success, and providing ongoing training so that expectations under the program are understood. The management system should also include written procedures with designated roles and responsibilities, program requirements, implementation schedule, communications procedures, documentation requirements, and technical procedures;

1.4. To develop a draft of the overall project approach based on the findings of Tasks 1.1 and 1.2 and to review it with the corporation and revise it as necessary. The recommended approach is illustrated in Figures 2 and 3.

1.5. To define a project schedule with milestones and responsibilities and resources required;

1.6. To develop a project budget.

TASK 2 – HFPHA APPROACH DEVELOPMENT

Figure 5 illustrates an overall HFPSM (Human Factors for Process Safety Management) approach that includes an Element 6 for human factors consideration in hazards analysis and risk assessment. The team should further refine a generic HFPHA (Human Factors for Process Hazards Analysis) methodology to be specific for the company and discuss how this approach could be implemented within the corporation. The methods developed would be a supplement to and integrated with the existing process safety management systems implemented throughout the corporation.

The philosophy is that risk reduction is justified where the time, expense, and effort required to reduce the risk is commensurate with the level of risk reduction achieved. The underlying basis of the goals of the program to reduce human error risk is a choice of the company management.

Step	Task	Purpose
1	Problem Definition	Define the scope, magnitude, and nature of the human factors problem
2	Scope, Objectives, Goals Definition and Commitment	Define the scope, objectives, and goals of a human factors-oriented risk reduction program and obtain management commitment to the program goals
3	Approach Definition	Define the actual methodology of the approach to be followed
4	Pilot Application and Evaluation/Modification	Test the approach on a limited problem, evaluate the success, and modify as required prior to widespread rollout
5	Policy Implementation	Determine the necessary organizational policies required and obtain employee agreement on the merits of the program
7	Procedures and Specifications Preparation	Develop the necessary procedures and engineering design specifications required to ensure human factors are considered in design and operation
6	Training	Train all individuals who must implement or be subjected to the program. Both general orientation and specific procedural training is required.
8	Rollout of Approach	Implement the approach in a priority order
9	Measurement and Audit of Initial Success and Evaluation/Modification	Obtain feedback on the program's effectiveness and acceptance of the program, and respond to concerns; modify as necessary
10	Oversight and Continuous Improvement	Monitor the program on a periodic basis to ensure that it is functioning per plan, and effective in meeting goals

Figure 2. Human factors implementation approach

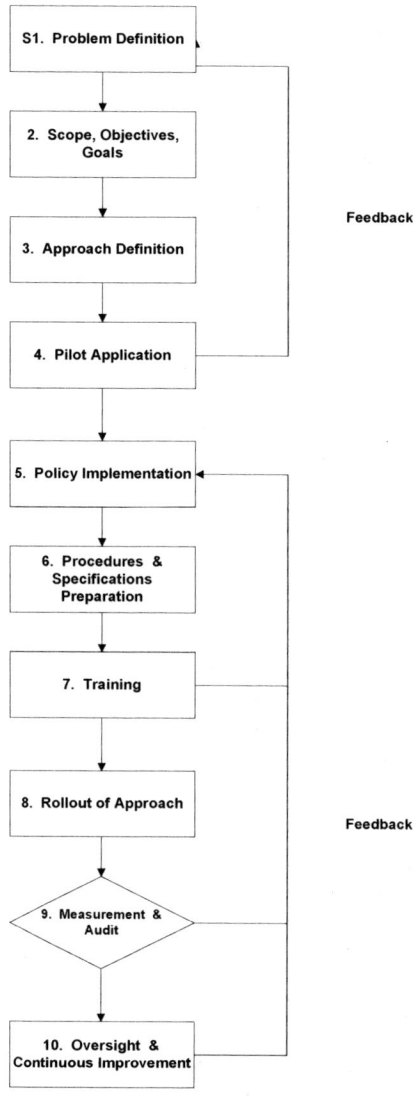

Figure 3. Human factors program implementation approach

Element	Description
1	Management knowledge and commitment
2	Written human factors policy
3	Management system for implementing the human factors program
4	Employee knowledge and involvement on human factors
5	Training on human factors issues
6	Incorporating human factors into hazards analysis and risk assessment
7	Human factors in process design and process change
8	Incident investigation and human factors root cause assessment
9	Consideration of human factors in written work procedures
10	Measurement and auditing of the human factors program performance

Figure 5.   Human factors management system

TASK 3 – HUMAN FACTORS PILOT STUDY

Next, it is recommended that the team field tests the approach as a pilot test and evaluates its effectiveness before widespread implementation. For Task 3, the team will:

3.1 Develop a pilot test protocol and all necessary forms, documents, or reports needed for the evaluation program

3.2 Based on the Human Factors Process Safety Management System, the team will review and evaluate the Process Safety Management programs, work processes, and management systems for the pilot process. This may be accomplished through worker surveys, interviews, onsite inspections, job observations, and audits.

3.3 Develop all supporting training materials and provide the training for the successful implementation of the program at the pilot site at a particular site.

3.4 Conduct certain agreed-upon human error analysis studies to determine the utility of different methods for analyzing risk. Included may be such methods as are listed in Figure 6. To accomplish this, the team will review documents, interview operators, engineering personnel, contract personnel, and management as required, and will conduct a site survey for human factors issues. One of the essential human factors program requirements is that it be incorporated into ongoing hazard analysis and risk assessment efforts. Human factors program development will require that procedures be adopted to conduct the analyses, as well as tools and technical approaches (documentation formats, checklists). In most cases, human factors considerations may be incorporated into existing PHA studies, but for selected studies, specific human factors methods should be adopted. Large facilities or groups of facilities may also consider expediting the typical 5-year PHA cycle in order to review human factors more quickly at areas where human error is likely, or when the consequences of an event are especially high. In developing a workable program, facilities may consider starting with a pilot study before widespread implementation takes place in order to refine procedures and improve long-term implementation efficiency.

3.5   Prepare a written report of findings and recommendations based on the results on the study and recommending forward approaches.

Method	Description
1. HFPHA - HAZOP or What if for Human Factors	Conduct a typical PHA however include specific deviations and checklist questions related to active and passive human error, latent conditions leading to human error, and an analysis of the response of operators to abnormal situations. Conduct the methods as part of the PHA studies typically done on process systems, however in more detail. Use the procedures as the basis of the analysis, rather than primarily the P&ID. In particular, study startup, shutdown, maintenance, and emergency procedures. Examine control schemes, critical human-dependencies, and areas of high consequence with few reliable and effective safeguards. Focus on the highest risks that are mostly influenced by human actions and at least selectively apply the methods where justified.
2. HFTA - Task Analysis (Facility Survey, Job Observation, Interviews, Detailed Procedure Review, Hazard Analysis, Risk Ranking, Integrity Level determination, Human Performance Improvement Measures)	Conduct field observations of the operation of a process and an onsite observation to ensure human factors specifications and design considerations are met and to identify hazards. Involve operations in the analysis, including a discussion of concerns and risks they perceive. Combine this with a task-wise detailed step analysis that identifies the purpose of the step, the criteria for success, the safe operating limits, the indicators for exceeding those limits, and the possible hazards of exceeding those limits. Identify all means to prevent, detect, and mitigate the hazard, including management systems, procedures, training, facility design changes, operational changes, or additional safeguards. Rank each risk based on a scale or likelihood and severity. Determine the Integrity Level of the human element of the system, and the potential means to improve the reliability of that level.
3. HFPRO - Procedures and Training Reviews using Performance Influencing Factor Analysis	Review operating procedures against the hazards analysis to determine if they coincide and if sufficient guidance is provided to prevent or control the identified hazards. Identify the performance influencing factors for every operating step to ensure they are understood, documented, and managed.

Figure 6.   Qualitative methods for human factors hazards analysis

543

STEPS IN THE HFPHA PROCESS

The HFPHA is a task-based analysis of targeted facilities that are more likely to cause significant consequences should error-likely conditions result in human errors. In this way, resources are focused on the areas of greatest significance based on risk.

The steps in the process are described in Figure 7 and the worksheet for conducting the HFPHA method is shown in Figure 8.

The steps in the process are:

DATA GATHERING –
Information/data needed:

1. Process safety information (PSI)
2. Written procedures
3. Field survey results
4. Interviews of the humans (to confirm or not confirm the written procedures)

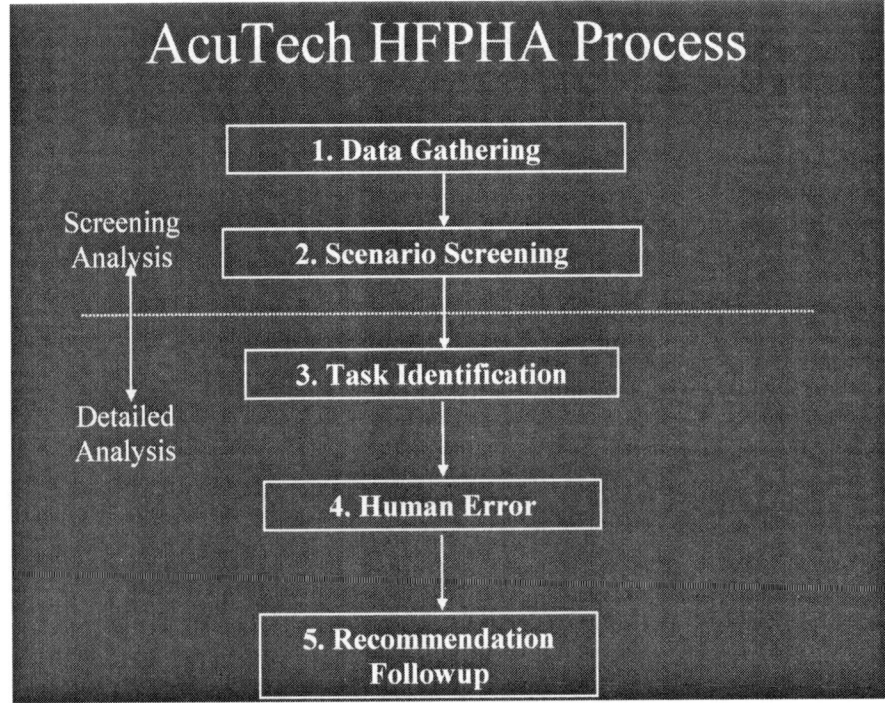

Figure 7.

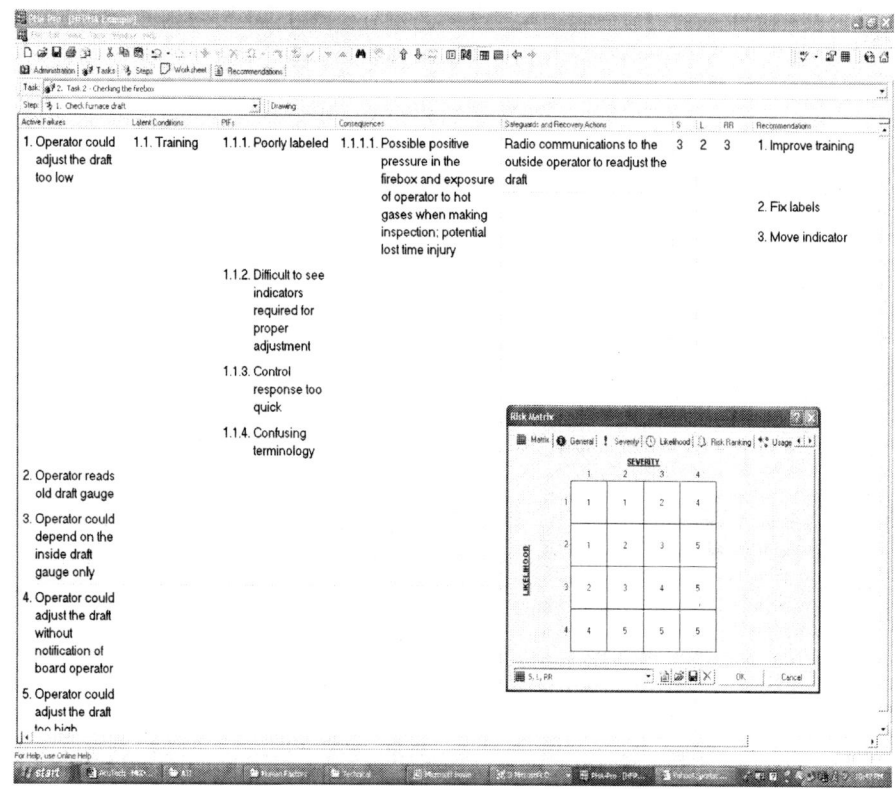

Figure 8. HFPHA worksheet

5. Identify possible Performance Influencing Factors (PIFs) (from checklist)
6. Identify possible latent conditions for the tasks/subtasks to be analyzed (checklist)

Performance Influencing Factors may include:

1. Task, Equipment, and Procedural Characteristics
   a. Feedback mechanisms available?
   b. Hardware interface factors:
      i. Job aids
      ii. Labeling
      iii. Color coding
2. Physiological/Psychological Stressors:
   a. Fatigue
   b. Climate extremes
   c. Movement repetition

      d.   Distractions
      e.   Sleep deprivation
      f.   High task overload
      g.   Threats
      h.   Negative reinforcement
      i.   Lack of rewards, recognition, or benefits

3. Review written documentation and interview someone knowledgeable of the procedures for the following modes of operation:
      a.   Normal startup
      b.   Startup following temporary shutdown
      c.   Startup following emergency shutdown
      d.   Normal operations
      e.   Emergency operations
      f.   Alternative operations
      g.   Temporary operations
      h.   Normal shutdown
      i.   Emergency shutdown

4. Have operations walk through the steps for each process above. Pay particular attention to:
      a.   the relevancy and accuracy of the operating procedures and training;
      b.   the actual practice vs. documented procedures;
      c.   any undocumented steps to avoid hazards;
      d.   communications issues during the above steps between field and control room staff;
      e.   perceptions of authority for shutdown vs. rules;
      f.   depth of rules vs. general training and knowledge required for operations.
      g.   discussion of system human factors issues.

5. Use a checklist to review issues by:
      a.   visual walkaround to observe physical layout and design and to identify any human factors (design-related) issues;
      b.   interview operations to obtain feedback of design and policy/procedure/ management/staffing issues that may affect human error;
      c.   discuss near misses and human errors that have occurred;
      d.   discuss training and supervision received on operations, hazards recognition, near miss or actual events;
      e.   discsuss feedback mechanisms for any staff concerns with human factors issues.

SCENARIO SCREENING -

Review the following sources to determine which processes deserve consideration:

1. Existing process PHAs (assuming high quality)
2. Management of change PHAs or safety reviews or hazard reviews
3. Risk assessments
4. Incident investigations (particularly near misses)
5. PSM audit results
6. Any other safety-related analytical activity

TASK IDENTIFICATION
1. What humans and their activities interact with the process? Review:
    a. SOPs (normal, startup, shutdown, etc.)
    b. Emergency operating procedures
    c. Inspection/Testing/Preventive Maintenance (ITPM) tasks
    d. Sampling/lab activities
    e. Unloading/loading activities
    f. Others
2. Select those tasks for further analysis from above that have contributed to the screened scenarios

HUMAN ERROR ANALYSIS
Analyze the tasks using the worksheet format provided in Figure 8.

RECOMMENDATIONS & FOLLOW-UP
Use the normal method(s) of collecting, managing, and documenting the resolution of PSM-related recommendations

CONCLUSION

The HFPHA method is a practical approach to focus resources on human factors issues and human error-likely situations on a risk-basis. It has the following advantages:

1. Relatively simple
2. Includes a screening step to reduce overall workload and focus resources on a risk-basis
3. Uses an analytical framework and documentation methods (i.e., PHA, recommendations resolution) that are already familiar to plant personnel
4. Uses ranking tool that is already familiar to plant personnel
5. Includes consideration of latent conditions, PIFs, and recovery actions
6. Can be applied to any level of human activity (task or subtask)

REFERENCES

1. Lorenzo, D.K. (1990). *A Manager's Guide to Reducing Human Errors*. Washington, DC: Chemical Manufacturers Association, Inc.
2. Reducing Error and Influencing Behaviour, HSG48, HSE (1999)
3. Bea, Holdsworth, and Smith, "Human and Organization Factors in the Safety of Offshore Platforms", a paper presented at the 1996 International Workshop on Human Factors in Offshore Operations
4. 20 CFR 119.119 Process Safety Management of Highly Hazardous Chemicals, OSHA
5. 40 CFR Part 68 Risk Management Plan, EPA
6. CPL2-2.45a, PSM Compliance Directive, OSHA
7. Section 450-8.016(B) of County Ordinance 98-48, Contra Costa County, California, December 3, 1999.

8.   AcuSafe   News,   February   2000,   http://www.acusafe.com/Newsletter/Stories/
     0200CCHS-Written-Human-Factors.htm
9.   CCPS (1994), *Guidelines for Preventing Human Error in Process Safety.*
10. Meshkati, Najmedin, "Human Factors in Process Plants and Facility Design" Chapter
     6, Cost-Effective Risk Assessment for Process Design (1995)
11. Reason, J., Managing the Risks of Organizational Accidents (1998)

ENDNOTES

[i]Lorenzo, D.K. (1990). *A Manager's Guide to Reducing Human Errors.* Washington, DC:
Chemical Manufacturers Association, Inc.
[ii]Reducing Error and Influencing Behavior, HSG48, HSE (1999).
[iii]Bea, Holdsworth, and Smith, "Human and Organization Factors in the Safety of Offshore
Platforms", a paper presented at the 1996 International Workshop on Human Factors in
Offshore Operations.

UPGRADING AN ALKOXYLATION FACILITY: THE VALUE OF CALORIMETRIC STUDIES

Dr. Richard L. Rogers and Dr. Klaus Hermann
INBUREX Consulting GmbH, Hamm, Germany

This paper describes the approach used during the upgrading of an Alkoxylation facility which resulted in the removal of the reactor bursting discs thus avoiding the need to install dump tanks. Inherently safe operating conditions, now ensured by a high integrity process control system, were determined by extensive calorimetric studies to measure the heats of reaction and degree of accumulation during the process. The experimental methods used to characterise the highly exothermic desired and runaway reactions which show non-ideal temperature-pressure behaviour and, in the case of ethylene oxide, are also highly toxic are described. The evaluation of the results obtained and justification for the decisions taken are elaborated in the paper.

Alkoxylation, ethylene oxide, venting, inherent safety, calorimetry

INTRODUCTION

The Alkoxylation facility before the upgrade consisted of 6 reactors (volumes up to 6 m³)with associated feed vessels for Ethylene Oxide and Propylene Oxide and was operated under complete manual control. Reactions were carried out using a semi-batch mode i.e. pre-charging of the various substrates with the subsequent addition of the Ethylene or Propylene Oxide at a rate commensurate with available cooling. The reactors were fitted with a safety valve set to open at 1 bar below their design pressure (this ranged from 8 to 10 barg) and a bursting disc with a set pressure equal to the design pressure. The relief vents from both the safety valves and bursting discs (diameter up to 150 mm) vented direct to the atmosphere above roof level. The Ethylene Oxide and Propylene Oxide feed system was also fitted with a hard-wired trip system which automatically stopped addition above a set temperature and pressure and also prevented reverse flow back into the feed vessels.

The operating procedure used was to pre-charge the substrate and catalyst at ambient temperature, inert the reactor with Nitrogen gas followed by heating to the desired operating temperature. Ethylene or Propylene Oxide was then charged with cooling to keep the reactor at the process temperature.

An additional safety measure had been implemented following a runaway reaction which occurred in the 1980's due to the rapid reaction of accumulated Ethylene Oxide which had been added before the reaction had initiated. The operating procedure was changed so that an initial aliquot of Ethylene or Propylene Oxide was charged and the reaction temperature and pressure observed to determine whether the reaction had initiated. This was shown by an increase in temperature due to the exothermic reaction and a reduction in pressure due to the reaction of Oxide. Following initiation of the reaction, cooling was applied and the main charge of Ethylene or Propylene Oxide was added at a rate to keep the reactor at the process temperature.

The degree of accumulation was unknown but the resulting runaway reaction was sufficient to cause the bursting disc to open and the reactor contents, including unreacted Ethylene Oxide, were released to the environment. Luckily no injuries were sustained and the released Ethylene Oxide was dispersed to below hazardous levels before the site boundary. A second release of Ethylene Oxide occurred when the bursting disc on one of the 6 m^3 reactors prematurely failed during production at a pressure of 5 barg (set pressure 9 barg). Again the Ethylene Oxide released was rapidly dispersed and there were no injuries and hazardous concentrations were not detected.

Following a review of the Alkoxylation facility and future production requirements in 1999 the company decided that the plant did not meet the current technical safety standards (absence of dump or quench tanks on the relief lines) and, in its then current single shift operating mode, would not meet future production demands. A decision was therefore made to both upgrade the facility and increase production.

OPTIONS FOR SAFE OPERATION

It was decided to increase production by working with multiple shifts and to upgrade the safety of the plant to current standards by fitting a dump tank on the relief lines. However, although the relief system had functioned satisfactorily once in the past, it was uncertain whether the bursting discs currently fitted were large enough to cope with the worse case runaway which could occur. The initial aliquot of Ethylene or Propylene Oxide which was charged was different for each process and the quantity had been empirically set based on plant experience of the amount required to ensure the reaction would start. In addition no information was available on the degree of accumulation during the course of the reaction. An experimental program was therefore started to obtain the thermal and pressure data required to specify the required vent area and to design the complete emergency relief system.

During the course of this initial investigation it became clear that the mainly manual process control system used was insufficient to limit the worse case scenario such that the existing bursting discs would be large enough to cope with a possible runaway. The cost of modifying the reactors and fitting of a suitably large dump tank system was such that an alternative basis for safe operation was needed.

The initial reaction hazard data obtained allowed a HAZOP of a typical process to be carried out. As a result of this it became clear that it would be possible to use an automated process control system to limit the hazard potential of the processes. In addition, provided sufficient thermo-chemical data was available to characterise the reactions, the fitting of appropriate safety trips would allow the degree of accumulation of Ethylene or Propylene Oxide to be controlled such that any runaway reaction would remain within the design parameters of the reactors. This option would remove the need for the emergency vents and the associated relief stream containment system. In addition, the automated control system would allow the plant to be run with fewer operators allowing a multi-shift system to be implemented without additional personnel. An extensive calorimetric study was therefore undertaken to obtain the required reaction data.

HAZARDS OF ALKOXYLATION REACTIONS

Alkoxylation reactions are highly energetic with typical heats of reaction of ca. −100 kJ/mol and, as in this case, are usually carried out industrially in a semi-batch reactor mode with the controlled addition of either Ethylene or Propylene oxide under pressure. A review of the hazards associated with reactions involving Ethylene Oxide has been published by Gustin [1]. The major reaction hazards arise from the runaway reaction of unreacted, accumulated Ethylene or Propylene Oxide which can lead to rapid temperature and pressure increases. A further hazard arises from the possible initiation of a decomposition reaction of Ethylene Oxide in the gas phase. Gas phase decomposition is usually avoided by inert gas blanketing to ensure that the concentration of Ethylene Oxide in the gas phase does not exceed 50 vol.% to 60 vol.% [2].

Although Ethylene Oxide is highly reactive, alkoxylation reactions are prone to delayed initiation, i.e. the reaction does not start immediately on addition of the Ethylene or Propylene Oxide though once the reaction starts it proceeds rapidly. The initiation delay is dependant on the substrate, catalyst and operating conditions used and can be affected by minor impurities in the substrate.

In order to assess the possible hazards of such reactions and to develop safe operating conditions, it is necessary to determine the possible accumulation that could occur which, in the event of a mal-operation, would result in a runway reaction. The liquid concentration of unreacted EO or PO can theoretically be calculated from the partial pressure, however mixtures of substrates containing EO or PO are known to show highly non-ideal behaviour and the calculated concentrations can have errors of over 100%. Figure 1 shows a comparison of the calculated ideal and real vapour pressures of mixtures of Ethylene Oxide and 1-Hexanol. The activity coefficients needed to calculate the real vapour pressures of the mixtures have been determined using the modified UNIFAC group interaction method [3]. It can be seen that at a typical process pressure of 7 atm and temperature of 125 °C, the calculated mass fraction of EO in the mixture assuming ideal behaviour is 0.17. In comparison the mass fraction of EO calculated assuming real behaviour is 0.34. It is clear that a much lower concentration of EO in the mixture at a particular temperature and pressure will be assumed if ideal mixture behaviour is used for the calculation. The higher concentrations of EO which would actually be present will result in a more violent runaway in the event of a mal-operation.

Non-ideality estimation methods such as UNIFAC rely on correlations derived from measured mixture vapour pressures. Unfortunately there is relatively little data available for EO/PO systems and such methods can only be used with confidence for safety assessments if the results can be checked against experimental data. Vapour pressures of mixtures containing EO or PO are however difficult to measure due to the high reactivity of the oxides even in the absence of catalysts which are usually used in industrial processes. This is shown in Figure 2 which compares the vapour pressure measurements obtained from mini-autoclave experiments of Propylene Oxide and a mixture of a long chain Fatty Acid and Propylene Oxide without catalyst. It can be seen that even at low temperatures a reaction takes place resulting in a reduction of the total pressure. It is therefore not possible to obtain reliable vapour pressure values from simple mixing experiments.

CALORIMETRIC MEASUREMENT OF THE DEGREE OF ACCUMULATION

The alkoxylation of various substrates by EO and PO under pressure was therefore studied to determine the degree of accumulation under industrial operating conditions and mal-operations. A Chemisens RM2 isothermal reaction calorimeter was used with piston pumps for the controlled dosing of the EO/PO. The small reactor volume (200 ml) minimised the quantity of highly toxic EO which needed to be used during the experiments.

As is often the case in fine chemical manufacturing a large number of different substrates are used in the alkoxylation facility. In order to minimise the number of experimental tests, 5 representative substrates where chosen on the basis of their chemical structure and reactivity for the detailed study. Thermal stability tests on the substrates and products involved in manufacturing operations had shown that there was no hazard from exothermic decomposition reactions and that they were stable to temperatures well above those used in the processes.

Experiments were carried out with EO and PO at temperatures and pressures used in the process and also at lower temperatures and differing feed rates to simulate the effect of mal-operations. A typical result for the ethoxylation of a short chain alcohol is shown in figure 3. It can be seen that the rate of reaction, shown by the heat flow curve, closely follows the pressure of EO in the reactor which is representative of the concentration of EO in the liquid phase. For the normal process the rate of reaction can be controlled by the rate of addition of Ethylene Oxide and matched to the available cooling. The degree of accumulation of EO can be calculated from the heat released and the amount of EO added at any point in the process and is represented in figure 3 as the percent EO present compared to the total EO charged. The accumulation of EO increases with increasing pressure in the reactor and in this case reaches a steady value of ca. 20% equivalent to ca. 5 wt% EO in the total reaction mixture. Once addition of EO is stopped, the pressure rapidly reduces due to the further reaction of the EO present. The total heat release of the reaction measured was -113 kJ/mol EO which corresponds well to the typical literature values of -100 kJ/mol EO.

Figure 4 shows the isothermal heat flow calorimeter result for the reaction of Propylene Oxide with a long chain Fatty Acid. In this experiment once the reaction had initiated, the rate of addition of PO was adjusted as in the plant process such that the total pressure remained constant. It can be seen that the heat flow trace shows two peaks indicating a multi-step reaction mechanism. Di Serio et al [4] have proposed such a mechanism for the Ethoxylation of Fatty Acids which can be used analogously to explain the behaviour with Propylene Oxide.

The initial reaction of the Fatty Acid with Propylene Oxide can occur by a catalytic or non-catalytic step where M is the catalyst:

$$RCOOH + PO \rightarrow RCOO(PO)H \quad \text{non-catalytic}$$

$$RCOO^-M^+ \rightarrow RCOO(PO)^-M^+ \quad \text{catalytic}$$

The species $RCOO^-M^+$ is poorly active as a catalyst, therefore the non-catalytic reaction mainly occurs. The catalytic reaction which leads to further alkoxylation only occurs after the complete monopropoxylation of the Fatty Acid substrate. This is confirmed by the heat flow trace observed which shows a minimum after the addition of one mole equivalent of Propylene Oxide. The further catalytic polypropoxylation reaction, which is slower, then occurs by the catalytic route:

$$RCOO(PO)^- M^+ + PO \rightarrow RCOO(PO)_2^- M^+ \quad \text{propagation}$$

As with the Ethoxylation reactions, the degree of accumulation of PO correlates with the maximum pressure and in this case reaches a maximum of 27% equivalent to 6 wt% PO in the reaction mixture. The pressure drop and further reaction of PO once the addition is stopped is slower than in the case of the Ethoxylation reactions.

In addition to providing the necessary information about the degree of accumulation of EO or PO during the reaction, the isothermal heat flow calorimeter experiments allowed the course of the alkoxylation reactions to be followed during the addition of EO or PO, providing valuable information about the reactivity and solubility of EO and PO in the substrates under the industrial conditions. This data allowed the process operating conditions to be optimised with a resulting improvement in plant utilisation.

CHARACTERISATION OF THE RUNAWAY REACTION

It is clear from the highly exothermic heat of reaction that an unlimited accumulation of EO or PO would lead to a runaway reaction resulting in temperatures and pressures which could burst the reactor. A safety concept based purely on the provision of emergency relief venting is therefore not feasible for such reactions and a high integrity trip system is needed to limit the amount of EO or PO which can be charged. The isothermal heat flow calorimeter experiments which were carried out allowed the specification of the minimum temperature and maximum pressure limits which are required to prevent an unlimited accumulation of EO or PO but which at the same time allow economic operation of the plant.

The measurement of the degree of accumulation during the process and the overall heat of reaction allowed the 'worse credible case' conditions to be identified i.e. the maximum accumulation which would occur when EO or PO was added within the temperature and pressure limits set by the trip system. These conditions were then used in adiabatic runaway reaction experiments as part of the safety assessment for the plant. Tests were carried out using a Phitec adiabatic reaction calorimeter modified to allow rapid addition of EO or PO during the experiment. Initial tests were carried out with the addition of the maximum accumulated amount of EO or PO to the substrate at the process temperature. These showed that although the temperature of the mixture increased as expected, the pressure in the test can decreased due to the reaction of the EO or PO.

Additional tests therefore were carried out with the addition of greater amounts of EO and PO. A typical result for the runaway reaction of Nonyl Phenol with EO in the presence of the catalyst is shown in Figure 5. During the addition of the Ethylene Oxide (17 wt%) over 5 min the temperature initially decreases due to the cooling effect of the added EO. The amount of unreacted EO remaining at the end of the addition was calculated from the temperature and specific heats of the reactants together with the heat of reaction, in this case the unreacted amount of EO was determined to be 13%. Once addition was complete the temperature increased due to the exothermic reaction however, even with this high concentration of EO, the pressure did not increase, it initially remained constant and then decreased.

Further adiabatic tests were carried out to identify possible conditions which could lead to a runaway reaction with increasing pressure. It was found that this would only occur if much higher concentrations of EO were present which could only be simulated by the addition of the EO at a low temperature such that it did not react. Figure 6 shows the adiabatic calorimeter result of the addition 25 wt% EO to Nonyl Phenol. The EO was added at a temperature of 88 °C and the mixture cooled to a temperature of 72 °C at the end of addition. As the rate of reaction was insignificant at this temperature the test cell was heated using a 'Heat – Wait – Search' mode to 81 °C and then 101 °C at both these temperatures negligible reaction occurred. On increasing to the process temperature of 120 °C the runaway reaction started and reached a maximum of 260 °C. The pressure in the test cell at the start of the runaway was 11 bara and increased to a maximum of 14 bara before rapidly decreasing.

The tests confirmed that it is possible to obtain a runaway reaction which results in increasing pressure and reinforced the importance of ensuring that EO or PO is not added at low temperatures at which high amounts can accumulate.

NEW SAFETY CONCEPT

The calorimetric studies confirmed the high energy release of the alkoxylation reactions and showed that once initiated the reactions proceed rapidly. During the normal process the maximum accumulation of EO or PO which occurred within the temperature and pressure limits specified and controlled by a trip system, was less than 7 wt% based on the total reactor contents. Tests on the runaway reaction of mixtures containing accumulated EO or PO showed that in such cases the maximum pressure of the reactors would not be exceeded.

It is clear that the degree of EO or PO accumulation depends markedly on the temperature and pressure and provided these are controlled within set limits 'unsafe' accumulation can be prevented. In addition the maximum fill level of the reactors must be controlled to prevent over-pressure due to thermal expansion of the reaction mixture following the temperature increase due to the runaway reaction.

In addition to the process control system, the reactors were therefore fitted with safety trips to stop addition of PO or EO in the event of mal-operations. A HAZOP study was carried out to identify possible mal-operations and their consequences. As a result the main trips fitted to prevent accumulation and which stop addition were:

1. Low/High reactor temperature
2. Low/High reactor pressure
3. High fill level
4. Low pressure difference between feed oxide vessel and reactor to prevent back flow into feed vessel.
5. Low feed vessel pressure
6. High feed vessel temperature
7. Low cooling water pressure
8. Agitator failure
9. Lower pressure in oil seal stirrer bearing (must be always greater than reactor pressure)
10. High temperature in oil seal bearing

The trip system removed the need for the bursting discs on the reactors and these were removed and replaced with safety valves design to cope with service overpressure and fire scenarios. The relief lines from the safety valves discharged into a water reservoir already existing as part of the safety concept for the Ethylene Oxide storage facility.

CONCLUSIONS

The calorimetric studies carried out to characterise both the desired reactions and credible runaway reactions allowed an alternative safety strategy for the alkoxylation facility to be developed. This was based on a safety trip system. The safety analysis of the new process control system and safety trips showed that there was no need for emergency relief thus avoiding the necessity to install a new containment system to cope with the consequences of a runaway reaction caused by accumulation of EO or PO. In addition the automatic process control system fitted allowed the plant to be operated with fewer personnel. This enabled a multi-shift system to be implemented with the same number of workers thus meeting the increased production requirement.

The study demonstrates the benefit of re-examining the safety concept when an upgrade is carried out and the value of calorimetric data in developing inherently safer operating conditions.

ACKNOWLEDGEMENT

The authors are grateful to the firm Dr. Th. Böhme KG Chem. Fabrik GmbH & Co. for permission to publish the results presented in this paper.

REFERENCES

1. Gustin, J.L., 2000, *Safety of Ethoxylation Reactions*, Hazards XV – The Process, its Safety and the Environment, Symposium Series No. 147 IChemE, Rugby, UK.
2. Britton L. G.: *Use of Propylene Oxide Versus Nitrogen as an Ethylene Oxide Diluent*, Process Safety Progress Vol. 19, No. 4, pp. 199–209.
3. Gmehling, J., Lohmann, J., Jakob, A., Li, J., Joh, R. 1998, Ind. Eng. Chem. Res 37, 4876–4882.
4. Di Serio, M., Di Martino, S., Santacesario, E., 1994, *Kinetics of Fatty Acids Polyethoxylation*, Ind. Eng. Chem. Res. 33, 509–514.

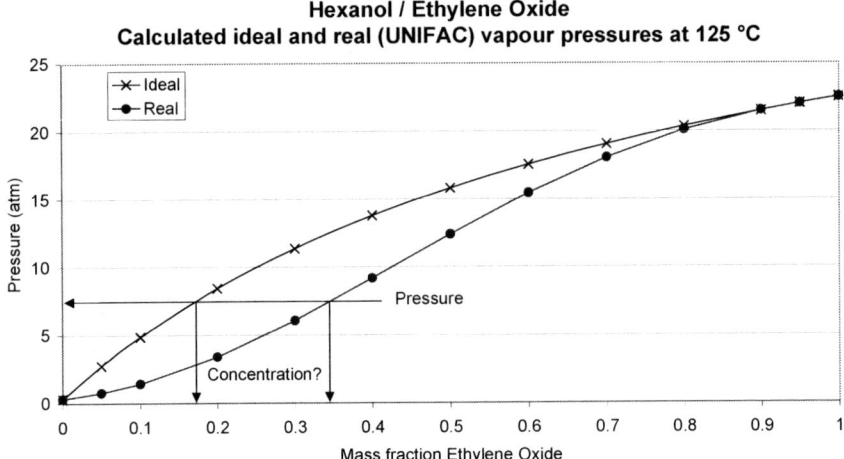

Figure 1. Comparison of calculated ideal and real vapour pressures of mixtures of hexanol and ethylene oxide

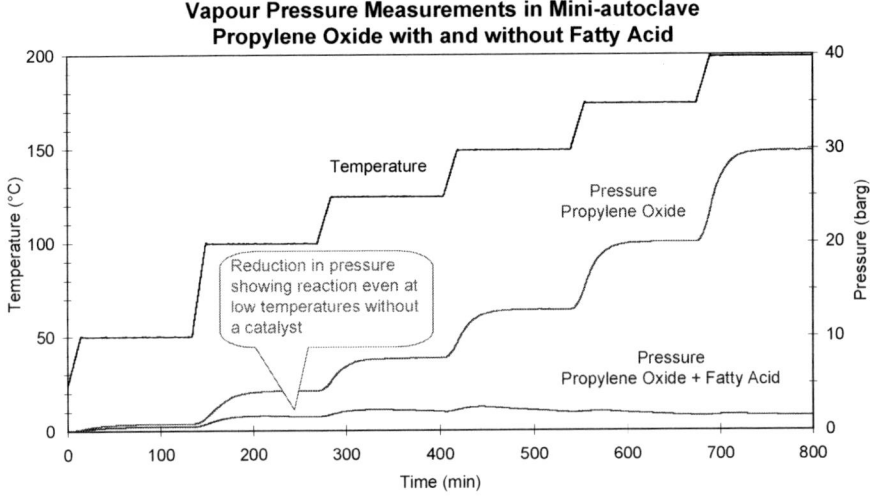

Figure 2. Comparison of vapour pressures of propylene oxide and a mixture of propylene oxide and fatty acid at different temperatures

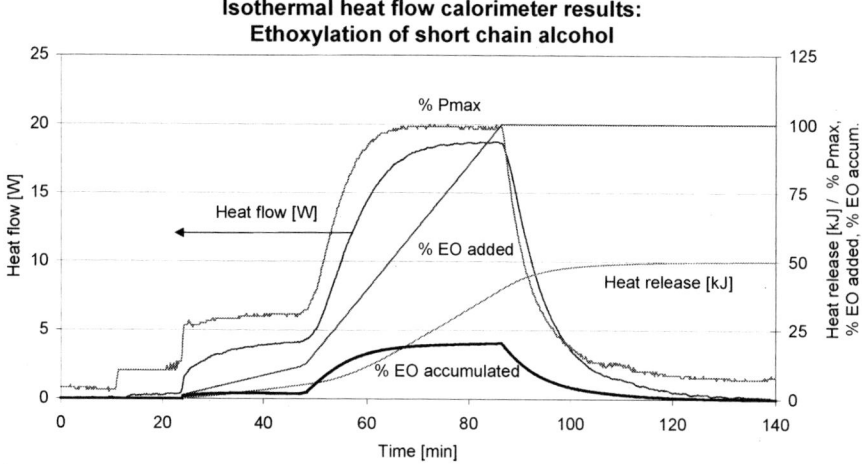

Figure 3. Isothermal heat flow calorimeter results of the ethoxylation of short chain alcohol

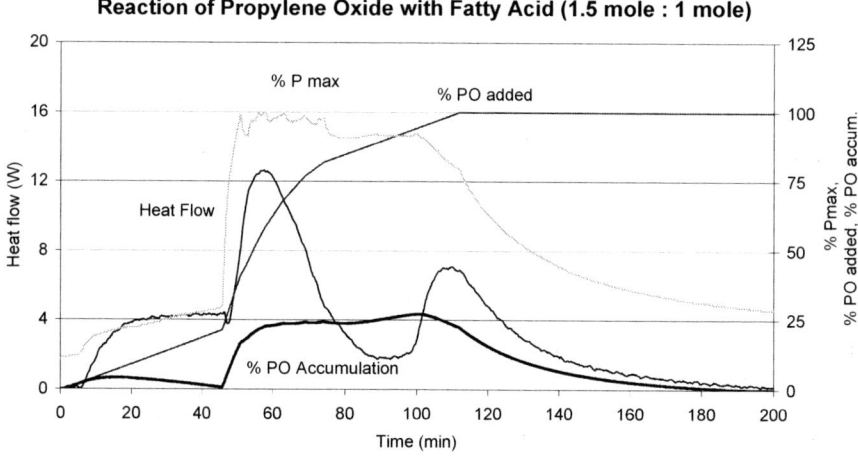

Figure 4. Isothermal heat flow calorimeter results of the reaction of propylene oxide with fatty acid

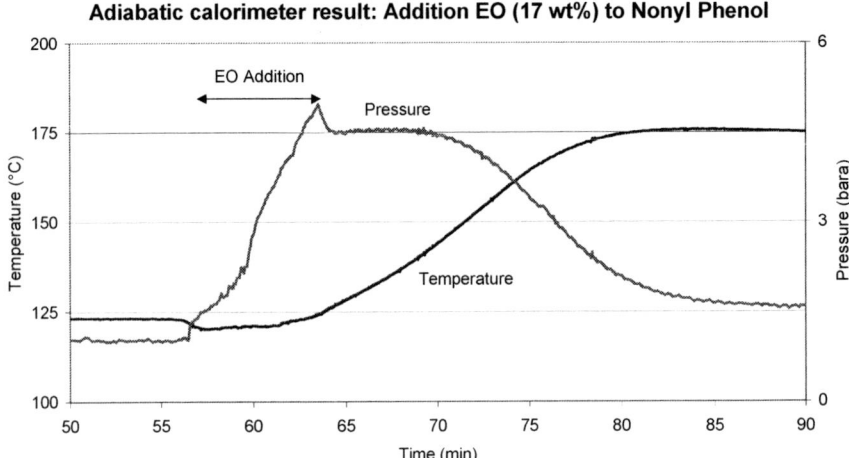

Figure 5. Adiabatic calorimeter result showing runaway reaction of ethylene oxide (17 wt%) with nonyl phenol

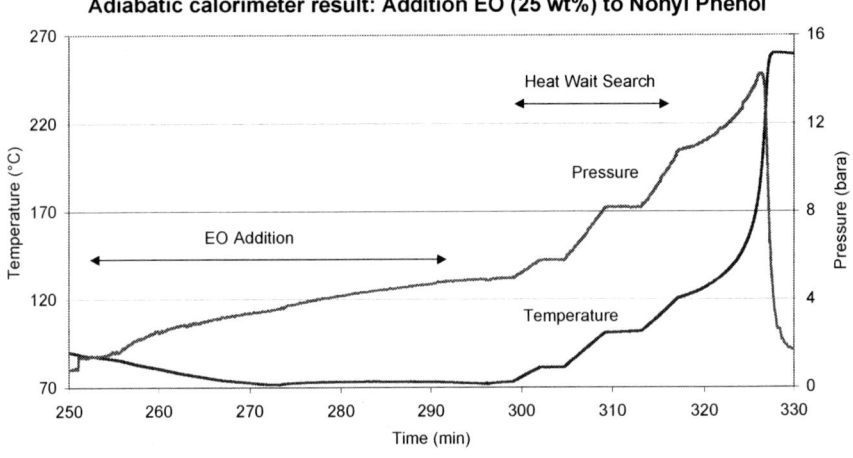

Figure 6. Adiabatic calorimeter result showing runaway reaction of ethylene oxide (25 wt%) with nonyl phenol

PRESSURE RELIEF OF LIQUIDS CONTAINING SUSPENDED SOLIDS

Derick McIntosh, Simon Waldram (Hazard Evaluation Laboratory Ltd, 50 Moxon Street, Barnet, Herts. EN5 5TS) and Janet Etchells (Health and Safety Executive, Magdalen House, Stanley Precinct, Bootle, Merseyside L20 3QZ)

Abstract

The multi-phase venting of vapour, liquid and solids has been studied experimentally on the 1 and 10 litre scales. In non-reacting systems, the depressurisation profiles of superheated water or water-glycerol mixtures were studied on their own and with added glass particles. The particles were both solid and hollow, with specific gravities both greater and less than 1. Similar depressurisation experiments were made during the runaway reaction of acetic anhydride and water, both with and without solids. Relief set pressures were between 3 and 5 bara. Solids concentrations were up to 16% v/v. Nozzle to median solid diameter ratios were between 6 and 500. Experimental design techniques were used to study the effects of many factors efficiently. Depressurisation profiles with and without solids present were compared: in general the solids had little statistically significant effect. There was limited evidence that the less dense solids could increase depressurisation rates slightly. Tests with runaway reactions have highlighted some difficulties in comparing systems with and without solids present. In all cases, the vented fluids were less concentrated in solids than those in the reactor. This may have important implications for design and sizing of pressure relief vents.

INTRODUCTION

The field of two-phase liquid/vapour venting has been the subject of considerable research in recent years[1]. The situation will often arise, however, when there are solids present in the discharge stream. The solid phase might be a heterogeneous catalyst, (e.g. a platinum group metal on a porous carbon support particle, for a hydrogenation reaction,) a partially dissolved reactant or a solid product that is crystallizing as the reaction proceeds. There is little guidance about how to allow for the presence of such solids when sizing a pressure relief system, although some initial work has been carried out by Beyer and Steinbach[2]. Some preliminary information, based on recommendations of The Design Institute for Emergency Relief Systems (DIERS), is given in reference 3. However this has not been validated experimentally and the authors point out that these methods may not apply if the solids are not carried over at the same velocity as the liquid. There may also be problems if the solids affect the flow from the reactor or cause fouling in the relief system. HSE, in collaboration with a consortium of companies has, therefore, sponsored a project to investigate the problems further on the laboratory scale, and to identify the main issues involved. This paper describes the main findings, which will be published in an HSE Research Report[4].

EXPERIMENTAL PROGRAM

An experimental programme was devised to examine the effects of the addition of suspended solids on a two-phase vented system. To examine the effects of the solids, the venting profiles from tests with the addition of solids were compared to those without. Initially non-reacting systems were examined, using superheated water or water/glycerol liquid mixtures, and then reacting systems were examined, using acetic anhydride and water. Experimentation was carried out on both the 1 litre and 10 litre scales.

There are a vast number of potential variables that could be studied in this project, e.g. solid concentration, solid diameter, solid density, nozzle diameter, fill level, stir rate, relief pressure etc. In these circumstances it was important to plan experiments, execute the experimental work and process and analyse the results in a structured and efficient manner. To achieve this, factorial experimental design techniques have been used: see reference 5 for an introduction to this topic. When the effects of many factors are to be examined then very large experimental programmes are required. In these cases fractional factorial designs (e.g. half or quarter) can be used to reduce the required number of experiments (e.g. by factors of 2 or 4). However, a compromise must be made: such designs lose the ability to discriminate clearly between the effects on the response of combinations of factors.

As an example, a full level factorial design involving 6 factors would require 2^6, or 64, experiments. Statistical analysis of the results enables the effect on the response variable to be estimated not only for every factor, but also for all possible combinations of factors. A half factorial design would only involve 32 experiments, and hence half of the experimental effort, without significant loss of response data. Additionally, replicate centrepoint experiments are commonly included to give an indication of the experimental error, and this is used in the statistical analysis.

Planning such experiments, and interpreting the results from them, can be tedious and complex but the use of standard software packages avoids many of these difficulties. In this project the "Design Expert" software was used, see http://www.statease.com.

APPARATUS

A schematic diagram of the apparatus is shown in figure 1. The test rig consisted of a reactor vessel, a vent line and a catch tank. The 10 litre reactor was a purpose built, baffled, stainless steel vessel, rated to 20 bar. Thermocouples, a pressure transducer, a bleed valve, a magnetic drive coupling, an overpressure relief valve and the vent line were incorporated into the reactor top plate. Four electric rod heaters were used and entered through the bottom plate. Two Rushton type impellers and a baffle system were used for the agitation of the mixtures.

The reactor was connected to the catch tank via a 12.7 mm (½") vent line that incorporated a pneumatically operated ball valve. When open this gave full bore unobstructed flow. The vent line was connected to the reactor via a fitting, designed to hold a variety of different sized nozzles. The automatic ball valve was sited just downstream of the nozzle. The catch tank was placed on a balance to allow continuous measurement of the vented mass, as a function of time.

For the tests with reacting systems, it was important to minimise discharge of vapours. The vented catch tank was replaced with a sealed vessel and downstream sparged quench tank. To aid condensation of the vapours, approximately 8 kg of water were placed in the catch tank prior to the test. This worked well and there was very little mass loss from the system overall. A solenoid bleed valve, fitted between the catch tank and the quench tank, was opened after venting to prevent the cooling vapours in the catch tank creating a vacuum and sucking back liquid from the quench tank.

Control and data logging were achieved using HEL software. This allows data acquisition at a maximum logging rate of 10 points per second. Other than during the venting stage of the experiment, the data was only logged every 20 seconds.

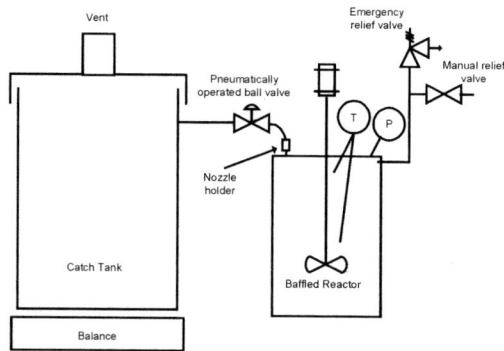

Figure 1. Schematic diagram of the 1 and 10 litre test apparatus

EXPERIMENTAL PROCEDURES
The same general procedure was used for all the tests using non-reacting fluids:

- Charge the materials and seal the reactor.
- Pull a vacuum in the reactor vessel[1]
- Allow the reactor contents to reach the relief pressure/temperature.
- Turn off the heating and wait 6 seconds.
- Open the relief valve and vent the reactor until the temperature dropped to 101°C or the pressure reached 1.05 bara.
- Close the relief valve and allow the reactor to cool.

The 6 second delay was incorporated to allow the data logging rate to be changed to the maximum rate without any delays caused by the software controls. Once initiated, the experimental procedure was fully automated and computer controlled to allow reproducible experimental sequences.

---

[1]Note that in the case of the 1 litre tests, a vacuum was not drawn, rather the vessel was heated with the vent valve open until the liquid reached its boiling point. In this manner the water was degassed and the reactor was purged of air.

Non-reacting test series were carried out by depressurising water and water/glycerol mixtures, both with and without the addition of solid particles. Glycerol was added to vary the liquid density (liquid to solid density ratio) and the liquid viscosity. The solids used were spherical glass particles in size ranges from 4–45 μm to 250–425 μm. Additionally, low density hollow glass particles were used. These have a density of 0.6 kg m^{-3} and are in the size range 0–65 μm. This range is not too dissimilar to the 4–45 μm solid particles, and comparisons have been made between the results from tests using these two sets of particles.

In the case of reacting systems, using acetic anhydride and water, the acetic anhydride was added to the reactor and heated to 50 °C under sealed conditions. The water was then added, the reactor was re-sealed, the mixture heated to 80°C and the heaters were turned off. The exothermic reaction was then allowed to proceed naturally until the venting condition of 3 bara was reached. At this point the reactor vent valve was opened and the reactor vented to the catch tank. The reactor was re-sealed when the pressure dropped to 1.05 bara.

TESTING ON 1 LITRE SCALE

Several series of tests were carried out on the 1 litre scale. Water and water/glass mixtures were charged to the reactor and heated to a temperature of 152°C (5.06 bara) before venting. The data obtained showed that there was very little difference in the depressurisation profiles between 700 ml of water and 700 ml of water plus up to 16% by volume of solids. Indeed, the profiles from tests with the same volumetric fill levels were very similar. There was limited evidence that the presence of the hollow particles resulted in a very slightly quicker depressurisation.

Two full factorial test series, with 4 factors, using a 2 mm and a 5 mm nozzle, were performed. The data analysis on the times taken for the pressure to drop in small increments throughout the depressurisation showed that there was no statistically significant effect due to the addition of solids.

A potential problem that was identified on the 1 litre scale was that the repeatability of the test was poor, particularly when comparing many identical tests carried out over an extended time period. Tests on the 10 litre scale were more reproducible. For this reason, experimental work concentrated on the 10 litre scale.

NON-REACTING EXPERIMENTS ON THE 10 LITRE SCALE

A series of tests were carried out by depressurising water and water/glycerol mixtures, both with and without the addition of solid particles. A 6 factor, half factorial experimental design was used. The following factors were examined: solid concentration, solid diameter, nozzle diameter, fill level, stirring speed and glycerol concentration. The relief pressure was fixed at 5 bara.

In order to examine the effects of the particle density, a further series of tests were planned to complement those already performed. The low density glass has a similar diameter range to the 4–45 μm solid glass. Many of the tests already run in the previous series could be incorporated into a new experimental design, so that only 4 new tests had to be performed, all of these were with the low density glass particles. A 5 factor, half factorial design was analysed, with the factors being: solid concentration, solid density, fill level, stir speed, and nozzle diameter. The effect of glycerol was not studied in this series.

Some factors, e.g. fill level and nozzle diameter, are already known to affect the depressurisation. As this project is aimed at investigating the effects of solids, the prime interest was to identify those experiments in which solid diameter, or solid concentration, combined with the other factors affected any aspect of the depressurisation profile.

The response variables chosen for the analysis of this data are the times taken for incremental depressurisation, in steps of 0.1 bar down to a pressure of 4 bara, then in 0.2 bar steps to 3 bar, and thereafter in 0.5 bar steps. The results showed that the nozzle diameter has an effect on the times throughout the depressurisation, and the fill level has an effect down to 2.5 bara. This would be expected. Additionally, almost throughout there was the effect of an interaction between the fill level and the nozzle diameter. The stirrer speed was identified as having an effect in the early stages, down to 4.7 bara.

The first design showed that the solid concentration had an effect on the initial depressuriation, with the incremental depressurisation times between 4.5 bara and 3.0 bara reducing with increasing solids concentration (ie increasing the solids, reduced the time for depressurisation). This may be simply because at a fixed reactor fill level the liquid contents reduce at increased solid concentration, and the amount of vapour produced as the reactor contents cool to atmospheric conditions will therefore become smaller at high solids concentration. This may affect the transition point between single and multi-phase flow. Otherwise the presence of the solids had no statistically significant effect on the depressurisation.

It is interesting to note that there was no effect on any of the response variables due to the amount of glycerol added. This suggests that the viscosity and density difference between the water and water/glycerol mixtures had no effect on depressurisation for the range of variables studied.

The analysis of the second experimental design showed similar findings to the previous design in that neither the solid concentration nor the solid density had a statistically significant effect on the response variables. Fill level and nozzle diameter had the main effect on the response factors studied.

REACTING SYSTEM

Several properties were desirable when choosing the reacting system. The reaction needed to be proceeding at a relatively high rate during venting so as to promote multiphase flow and to test venting under demanding conditions. Additionally, the 10 litre vessel has a high thermal inertia (ϕ factor) and therefore a high fraction of the heat of reaction would be absorbed by the vessel itself during slow periods of an exotherm.

The reaction of water and acetic anhydride to produce acetic acid was eventually chosen for several reasons – in particular, the reaction mechanism is relatively simple, and the flammability hazards are minimised by using an aqueous reaction. The stoichiometry of this reaction is 1:1, but examination of vapour pressure data indicated that a slight excess of water would generate a higher pressure, and hence the reaction was run with a 3:2 water:acetic anhydride mole ratio. Perhaps more importantly, this vigorous exothermic reaction has been involved in a several publicly reported incidents[6,7,8]. Leigh and

Krzeminski[8] quote an incident where water had entered a storage tank containing acetic anhydride and 15% acetic acid at ambient temperature. The resultant overpressure and vessel rupture killed one person and injured 20 more.

This series of tests was initially planned as a factorial design, with factors of fill level, glass diameter and glass concentration. The high, low and centrepoint values used in the factorial design are given in table 1. Note that reactant volume and glass concentration are independent factors. Combinations of these at high, centrepoint and low values will determine the actual reactor fill level.

A 10.5 mm diameter constriction was installed on the reactor. This was chosen following small scale PHI-TEC tests, and was predicted to give significant overpressure during the venting under the conditions of the reaction. A relief set pressure of 3 bara was used.

In order to give the same thermal inertia, as tests run with glass particles, the tests run at 0% suspended solids concentration were, in fact, run with 15% v/v of 3 mm diameter glass beads present. The diameter of these beads was so large as to ensure that they always settled out in the bottom of the reactor and were not entrained in the vented flow. This was thought to be a good method of keeping the heat capacity of the system constant and at the same time the glass ballast would not affect the two-phase venting rate from the reactor.

Additionally, two tests (one at each fill level) were run using the hollow glass beads (density 0.6 kg m^{-3}) for comparison with the tests with the 4–45 μm solid particles (density 2.5 kg m^{-3}).

RESULTS

Table 2 gives the conditions for each test. The test numbers in table 2 have been sorted in terms of reactant volume and have been given a letter designator for clarity.

The main conditions of the tests are given in Table 2. Figures 2 and 3 give the respective pressure and temperature profiles during venting for the tests using 5000 ml of reactants +15% v/v solids. It can be seen that the tests A and B (both with 3mm glass beads), which had two-phase flow gave very much higher overpressures than the tests with three-phase flow: tests C and D resulted in virtually no overpressure. This was not expected, as in the non-reacting systems, the presence of solids had very little effect on blowdown profiles.

Figure 4 shows the self-heat rate profiles for the tests starting from the temperature at which the electrical heating was turned off (80°C). It is clear that tests A and B (with 3 mm glass beads as thermal ballast) reached much higher heat-rates before venting commenced at 3 bara. As the test recipes are exactly the same, there must be an effect due to the diameter of the solids. Transient heat transfer calculations confirm that the large glass beads absorb less of the heat from the reaction, due to their smaller external surface area and low thermal diffusivity[9]. This will become increasingly apparent at high runaway reaction rates. The fact that glass is a relatively poor conductor of heat will also have an effect, with smaller glass particles heating up more quickly and more uniformly than larger particles.

A similar effect is likely when φ factors are considered: at very high self-heat rates the observed φ factor may be much closer to unity than expected from simple calculation, simply due to the fact that the heat transfer to the reactor vessel cannot occur at a sufficiently high rate for the reactants and vessel to be at an essentially identical temperature at any time. This means that, during fast thermal runaways, the phi factor may approach unity irrespective of the thermal mass of solids that may be present or of the reactor vessel itself. As a consequence, large diameter inert solids result in a much more vigorous reaction compared with the same mass of smaller glass particles.

Table 3 gives the mass balance information and maximum temperature and pressure data for the tests with 5000 ml of liquid. Again the effects of the glass diameter can be seen. The finest particles, which also result in the lowest heat rate due to reaction, result in a greater mass retained in the reactor. The lower heat rate and hence lower rate of vapour generation should give a lower superficial velocity and would be expected to result in less carryover.

Similar observations can be seen in figures 5 to 7 for the tests with the higher fill level. The data is also summarised in table 4. It is interesting to note from figures 3 and 6 that there is slight variation between the temperatures at the start of venting even though the pressure was identical each time. This could be attributable to experimental errors. However, the lower fill level does appear to result in slightly higher relief temperatures. One possible explanation is that the density of the acetic anhydride and acetic acid mixture falls with increasing temperature and the compression of the head space gas may be a significant factor. At higher fill levels the compression effect would be greater and venting would occur slightly earlier in the runaway. The data from a test with 6500 ml of reactants but without glass solids is also shown, and the significant effect of adding glass can be seen on the reaction rate.

The data for the repeat centrepoint tests are shown in figures 8 to 10 and in table 5. These tests show excellent repeatability in terms of blowdown profiles and self heat rates: the lines on the graphs virtually overlap each other.

NOTE ON MIXING

Table 6 shows the mass balance data for the glass particles in each test. In the case of an ideally mixed system, with no slip between the solid and liquid phases, it could be expected that the glass concentration in the reactor and the vented fluid would be the same. Table 6 gives the fraction of the initial liquid and glass masses that were drained from the reactor. In every case, there is an increased concentration of solids in the reactor following venting. It is likely that there was some accumulation of solids towards the bottom of the reactor with the solid glass, particularly with the larger solids. The nature of the reactor vessel makes it impossible to make a visual inspection and qualitative judgement of the stirring efficiency. However, based on visual observation on the 1 litre scale, at similar tip speeds, it was believed to be good at the centrepoint and high stirring rates.

It is very interesting to note the data from the tests using the hollow solids. From previous observations on the one litre scale, these particles are very easily mixed with the liquid, and float when not agitated. Therefore, with poor mixing, any increase in solid concentration would be towards the top of the reactor. It would be expected that as this will vent first, and that a greater fraction of the glass would be vented than that in the reactor as a whole. If the mixing was always perfect, then an equal fraction of solid and liquid would remain in the reactor and the catch tank. The data in table 6 shows that this is not the case, and a greater fraction of the liquid was vented. It is clear therefore that the liquid is vented preferentially even though the solids are lighter than the liquid. If the solids are taking part in the reaction, then this will need to be considered in the design of the relief system, particularly if it means that the reactants could be concentrated, either in the reactor, or in the downstream containment or disposal system.

CONCLUSIONS
From the testing and analysis described above the following can be concluded.

A. Non-reacting tests
- a. Pressure and temperature profiles versus time during venting were in general not influenced to a statistically significant extent by the presence of solids.
- b. There is limited evidence that, under some circumstances, the presence of solids can increase depressurisation rates, particularly at intermediate times. It is likely that the presence of solids may promote even more vapour bubble nucleation and promote bubbly or homogeneous rather than churn turbulent flow.
- c. During venting, liquid is discharged preferentially to the solids. This was observed for both naturally floating and sedimenting particles, i.e. both less and more dense than the fluid in which they are suspended.

B. Reacting systems
- a. Adding inert particles to a reacting system increases the phi factor and hence reduces the reaction runaway rates. This effect is highly non-linear with particle mass.
- b. Depressurisation of reacting systems containing inert solids was highly reproducible on the 10 litre scale.
- c. The temperature of inert particles suspended in a liquid whose temperature is changing rapidly may lag behind the fluid temperature. This means that the effective phi factor can change during the course of a fast runaway. For this reason, large inert particles appeared to accelerate the runaway reaction relative to the same mass of smaller particles and this can lead to larger overpressures during venting.
- d. Heat transfer limitations to the body of a large or massive reactor vessel during a fast runaway may mean that the average temperature in the reactor body is much lower than in the reacting fluid. In an analogous way to item c this can lead to a shifting value of the phi factor as a reaction proceeds. The reaction may become much faster than that anticipated from small scale studies at an analogous phi factor.
- e. Direct comparison of the results from the reacting system tests with, and without, solids is very difficult because of the change in the phi factor (and hence reaction rate) and the ability of inert solids to accelerate a runaway reaction, see items a, and c.
- f. Production of vapour and preferential flow of the liquid (relative to the solid) in the vent discharge will enhance the solid concentration in the reactor. This will then alter the phi factor and hence the runaway rate. If the solids are participating in the reaction this may also affect the reaction rate per unit volume.

As a general conclusion these preliminary studies show that, for the ranges of variables studied, small diameter inert solids have little influence on the rates of depressurisation achieved. Although it was outside the scope of this project, there are several issues that would need to be carefully considered before calculating the required vent areas in the case of three-phase flow. Any effects of concentration of the solids during venting must be considered in selection of the calorimetric test methods used for relief line sizing. Solid deposition and downstream fouling may also be issues.

REFERENCES

1. Fisher, H.G., Forrest, H.S., Grossel, S.S., Huff, J.E., Muller, A.R., Noronha, J.A., Shaw, D.A. and Tilley, B.J: "Emergency relief system design using DIERS technology, The Design Institute for Emergency Relief Systems (DIERS) project manual," 1992, AIChE, NY. ISBN 0-8169-05668-1.
2. Beyer, R. and Steinbach, J: "Source term characterization for three-phase venting scenarios." Paper presented at the 10th International Symposium on Loss Prevention and Safety Promotion in The Process Industries, Stockholm, June 19th –21st, 2000.
3. Etchells, J. and Wilday, J: "Workbook for chemical reactor relief system sizing," 1998, HSE Books. ISBN 0-7176-1389-5
4. McIntosh, R.D. and Waldram, S.P.: "Reactor pressure relief of fluids containing suspended solids" HSE Contract Research Report, project reference 4187/R05.101.
5. Davies, L., Efficiency in research, development and production: the statistical design and analysis of chemical experiments, 1993, Royal Society of Chemistry, ISBN 0 85186 137 7
6. Barton, J. and Rogers, R.L.: "Chemical reaction hazards," second edition, 1997, I.Chem.E. ISBN 0-85295-341-0
7. I.Chem.E, Industrial accidents database, 2002, available from http//www.icheme.org.uk.
8. Leigh, W.R.D., and Krzeminski, Z.S.: "The uncatalysed reaction of acetic anhydride with water," Chemistry and Industry, pp778-779, April 28, 1962
9. Welty, J.R., Wicks, C.E. and Wilson, R.E.: "Fundamentals of momentum, heat, and mass transfer." John Wiley and Sons, New York 1976. ISBN 0-471-93354-6

ACKNOWLEDGEMENT

This work was co-sponsored by the Health and Safety Executive, Great Lakes Fine Chemicals and Syngenta. The inputs from Mr Graham Arthur (Syngenta), Dr Caroline Ladlow (Ciba Speciality Chemicals), Dr Allan Timms (Great Lakes Fine Chemical Company) and Ms Jill Wilday (Health and Safety Laboratory) are gratefully acknowledged.

The views expressed in this paper are those of the authors and should not necessarily be taken as those of the Health and Safety Executive.

Table 1.   Variables and values used in factorial design for the reacting system

Variable	High	Centrepoint	Low
Reactant volume (ml)	6500	5750	5000
Solid concentration (% v/v)	15	7.5	0*
Solid diameter (µm)	150–250[2]	70–110	4–45

*15% of 3 mm glass ballast was added in these cases

Table 2.   Data from 10 litre blowdown tests with the reacting system

Test Number	Reactant volume (ml)	Glass diameter (µm)
A	5000	3000
B	5000	3000
C	5000	4–45
D	5000	150–250
E	5750	70–110
F	5750	70–110
G	6500	3000
H	6500	4–45
I	6500	150–250
J	6500	250–425[3]
K	6500	0–65 (hollow glass)
L	5000	0–65 (hollow glass)

Note: the tests with 3000 µm glass should be treated as 0 % solids (i.e. 2-phase venting with no solid carryover)

---

[2]This value was chosen as the slightly larger particles (250–425 µm diameter) resulted in no solid carry over.

[3] This test gave no solid carryover and is included for comparison only

Table 3. Data obtained from acetic anhydride/water tests with 5000 ml of reactants

Test	Glass diameter (μm)	Glass carryover to catch tank (g)	Glass carryover to catch tank (ml)	Liquid retained in catch tank (g)	Liquid retained in reactor (g)	Maximum pressure (bara)	Maximum temperature (°C)
A	3000	0	0	2425	2880	5.24	169
B	3000	0	0	2523	2045	5.07	168
C	4–45	44.69	17.9	1048	4050	3.05	145
D	150–250	20.36	8.1	2449	2736	3.08	145
L	0–65 (hollow glass)	69.78	116.3	1707	3508	3.09	144

Table 4. Data obtained from acetic anhydride/water tests with 6500 ml of reactants

Test	Glass diameter (μm)	Glass carryover to catch tank (g)	Glass carryover to catch tank (ml)	Liquid retained in catch tank (g)	Liquid retained in reactor (g)	Maximum pressure (bara)	Maximum temperature (°C)
G	3000	0	0	3479	2693	4.39	162
H	4–45	299.0	119.6	2989	3619	3.10	147
I	150–250	24.58	9.8	4190	2443	3.92	157
J	250–425	0	0	3471	2898	4.67	165
K	0–65 (hollow glass)	219.56	365.9	3811	3048	3.47	151
M	No glass	N/A	N/A	3350	3571	7.68	195

Table 5. Data obtained from centrepoint cases

Test	Glass diameter (μm)	Glass carryover to catch tank (g)	Glass carryover to catch tank (ml)	Liquid retained of catch tank (g)	Liquid retained of reactor (g)	Maximum pressure (bara)	Maximum temperature (°C)
E	70–110	19.92	8.0	3116	2767	4.05	159
F	70–110	19.82	7.9	3063	2939	4.05	159

Table 6. Selected mass balance data from acetic anhydride/water tests

Test	Glass diameter (μm)	Initial glass charge (g)	Glass carryover to catch tank (g)	Glass remaining in reactor (g)	Fraction of initial glass remaining in reactor	Fraction of initial liquid charge remaining in reactor
A	3000	1875	0	1868	1.00	0.54
B	3000	1875	0	1870	1.00	0.38
C	4–45	1875	45	1782	0.95	0.76
D	150–250	1875	20	1806	0.96	0.51
E	70–110	1078.1	20	1042	0.97	0.45
F	70–110	1078.1	20	992	0.92	0.48
G	3000	2437.5	0	2421	0.99	0.39
H	4–45	2437.5	299	2078	0.85	0.52
I	150–250	2437.5	25	2354	0.97	0.35
J	250–425	2437.5	0	2362	0.97	0.43
K	0–65 (hollow glass)	585	220	346	0.59	0.44
L	0–65 (hollow glass)	450	70	383	0.85	0.66

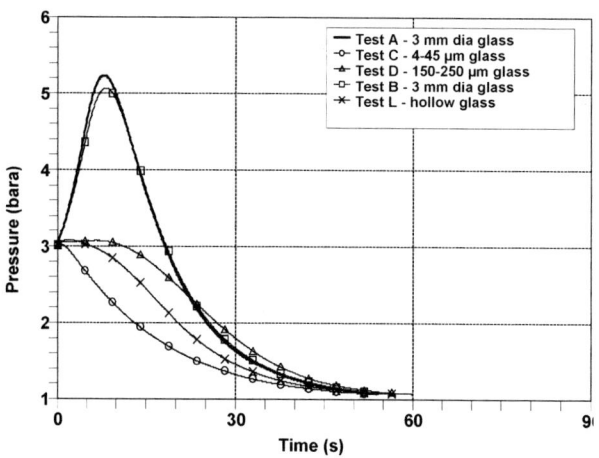

Figure 2. Pressure profile for venting of reacting system tests at low fill level
10 litre scale, 5750 ml charge (5000 ml reactants + 15% by volume glass), water and acetic anhydride (mole ratio 1.5) 9 mm nozzle, 200 rpm stirring

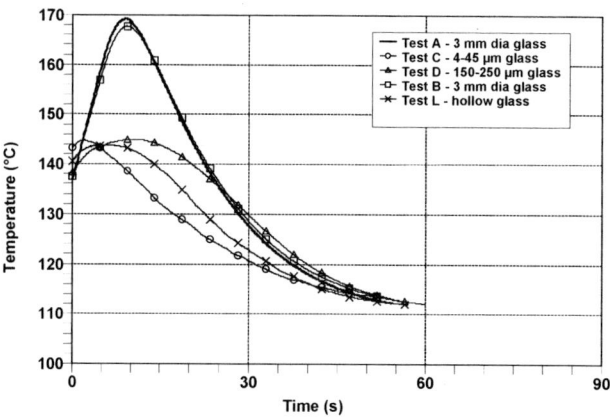

Figure 3. Temperature profile for venting of reacting system tests at low fill level
10 litre scale, 5750 ml charge (5000 ml reactants + 15% by volume glass), water and acetic anhydride (mole ratio 1.5) 9 mm nozzle, 200 rpm stirring

571

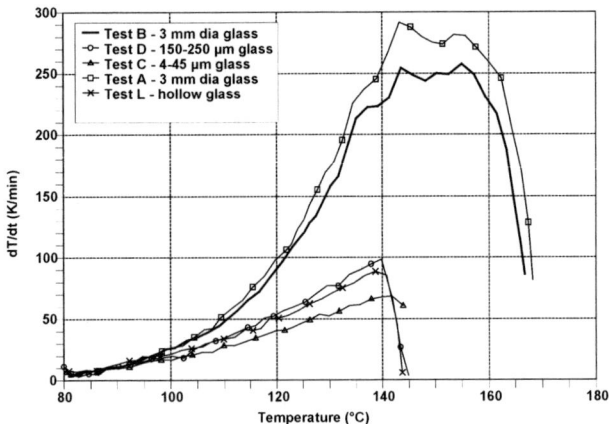

Figure 4.   Heat-rate profiles of reacting system tests at low fill level
10 litre scale, 5750 ml charge (5000 ml reactants + 15% by volume glass), water and acetic anhydride (mole ratio 1.5) 9 mm nozzle, 200 rpm stirring

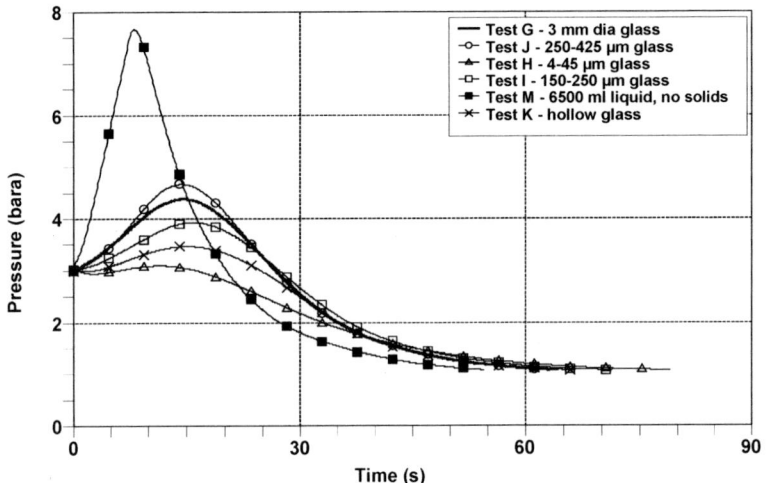

Figure 5.   Pressure profile for venting of reacting system tests at high fill level
10 litre scale, 7475 ml charge (6500 ml reactants + 15% by volume glass), water and acetic anhydride (mole ratio 1.5) 9 mm nozzle, 200 rpm stirring

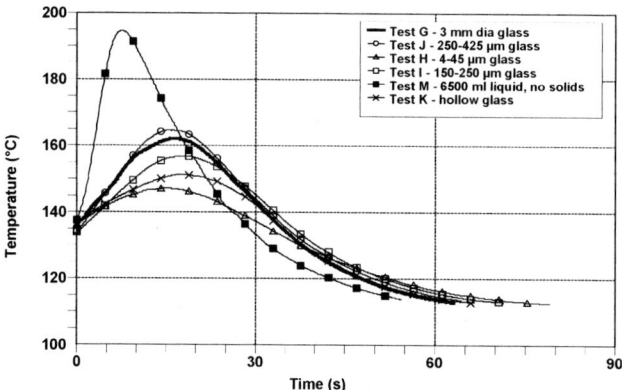

Figure 6. Temperature profile for venting of reacting system tests at high fill level
10 litre scale, 7475 ml charge (6500 ml reactants + 15% by volume glass), water and acetic anhydride (mole ratio 1.5) 9 mm nozzle, 200 rpm stirring

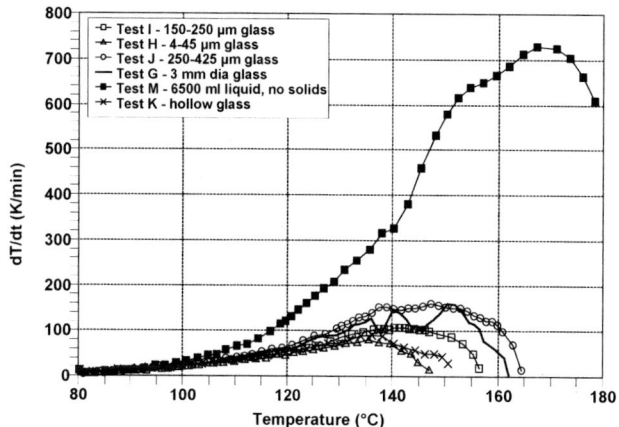

Figure 7. Heat-rate profiles of reacting system tests at high fill level
10 litre scale, 7475 ml charge (6500 ml reactants + 15% by volume glass), water and acetic anhydride (mole ratio 1.5) 9 mm nozzle, 200 rpm stirring

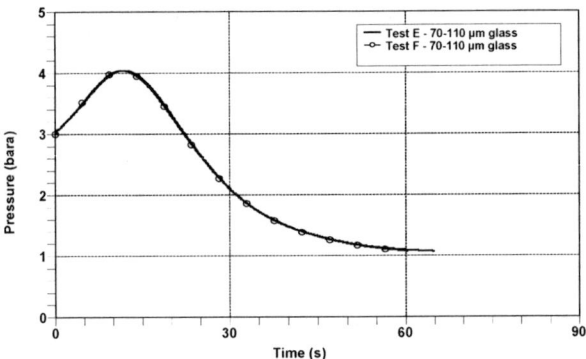

Figure 8. Pressure profile for venting of reacting system tests at centrepoint
*10 litre scale, 6210 ml charge (5750 ml reactants + 8% by volume glass), water and
acetic anhydride (mole ratio 1.5) 9 mm nozzle, 200 rpm stirring*

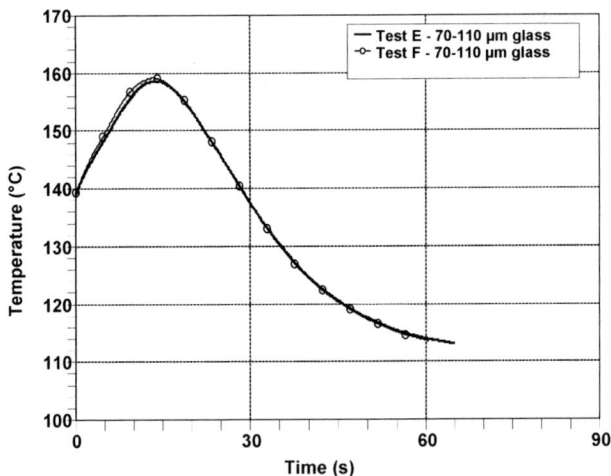

Figure 9. Temperature profile for venting of reacting system tests at centrepoint
*10 litre scale, 6210 ml charge (5750 ml reactants + 8% by volume glass), water and
acetic anhydride (mole ratio 1.5) 9 mm nozzle, 200 rpm stirring*

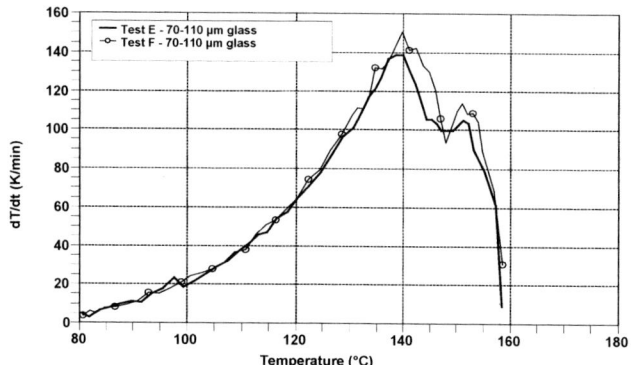

Figure 10. Heat rate profiles for reacting system tests at centrepoint
10 litre scale, 6210 ml charge (5750 ml reactants + 8% by volume glass), water and acetic anhydride (mole ratio 1.5) 9 mm nozzle, 200 rpm stirring

ALKOXYLATION RUNAWAY REACTION INCIDENT AT BAKER PETROLITE, HARTLEPOOL

S. Gakhar*, D. Carr[†]
*ABB Ltd, Belasis Hall Technology Park, Billingham, Cleveland, TS23 4YS
[†]Baker Petrolite, Tekchem Works, Brenda Road, Hartlepool, TS25 2BQ

On 12th October 1999, a runaway reaction occurred in a 22 m³ reactor during an alkoxylation reaction using Propylene Oxide. The subsequent pressure increase caused the bursting disc to rupture and a few tonnes of Propylene Oxide was released to atmosphere. The release occurred despite the presence of a knockout pot designed to capture released material. Although the incident caused no injury and only minimal plant damage, it did result in substantial costs both in terms of investment in corrective measures, and business loss. The incident and its causes are described along with improvement measures implemented to prevent recurrence and provide better protection against overpressure and environmental release. Improvement measures included improvements to process control, a re-design of the pressure relief system of all reactors in the plant, and the installation of a quench tank. The methodologies used to design the improvement measures are described. The paper also describes a later incident which provided a full scale test of the new quench tank.

KEYWORDS: Propylene Oxide, Ethylene Oxide, Runaway Reaction, Quench tank, Pressure Relief, Discharge and Disposal

INTRODUCTION

Baker Petrolite (part of the Baker Hughes group) operates an upper tier COMAH site near Hartlepool, United Kingdom. They employ approximately 85 people on a site covering approximately 11 acres. On the site, they manufacture a range of specialty chemical products. Many of these products are synthesised using semi-batch alkoxylation reactions i.e. involving Ethylene Oxide (EO) and/or Propylene Oxide (PO).

On 12th October 1999, a runaway reaction occurred. Whilst nobody was injured, the incident resulted in a significant release of Propylene Oxide to atmosphere despite the presence of a knockout pot. The incident highlighted a number of deficiencies both in the design and operation of the reactor and the pressure relief system. This paper describes the incident and its causes, the improvements made to prevent further recurrence, and the re-design of the relief and discharge and disposal system.

DESCRIPTION OF THE REACTOR PLANT

Prior to the incident, Baker Petrolite had six stainless steel reactors of varying capacity in their reactor plant building. Most of the reactions carried out in these reactors involved Alkoxylation reactions, although other products were also made. Four reactors have top entry agitators and two reactors have external re-circulation loops. Each reactor was protected from overpressure by a relief system comprising of a graphite bursting disc and a vent line. Each vent line was routed independently to a common knockout pot (V7). A Photograph of V7 is given in figure 1.

V7 was a vertical vessel of approximately 27 m^3 in volume and atmospheric pressure rated. It was located inside the reactor building and contained approximately 2000 kg of water (intended to provided some degree of quench and dissolution of EO or PO) maintained at a constant level by a ball cock level control device. Each of the six reactor vent lines entered the knockout pot separately. The outlet from the knockout pot was 0.6 m diameter and routed vertically to atmosphere via the roof of the building in which the reactors were housed. The discharge height was approximately 10 m above ground.

Reactors were mounted on load cells and as such the relief vents were linked to the reactor via a sliding joint arrangement originally designed by engineering consultants. This allowed sufficient relative movement to ensure effective operation of the load cells. The incident occurred in reactor V6 which is one of the reactors with an external re-circulation loop. V6 has a volume of 22 m^3 and a design pressure rating of 8.6 barg. The graphite bursting disc had a burst pressure of 7.7 barg at 20°C, and a diameter of 0.3 m. The vent line from V6 to V7 had a diameter of 0.4 m.

DESCRIPTION OF THE INCIDENT
The incident occurred at approximately 16:50 hours on 12th October 1999. The product being made was a polypropylene glycol in reactor V6. This involved the reaction between a low molecular weight Polyol and Propylene Oxide. The process steps involved up to and including Propylene Oxide addition are summarised below

1. Add starting Polyol
2. Add 45% Potassium Hydroxide solution (catalyst)
3. Perform pressure test to approximately 60 psi
4. Heat to approximately 110°C to dehydrate
5. Sample to check moisture
6. Apply nitrogen padding to reactor
7. Add Propylene Oxide at controlled rate

The Propylene Oxide was pumped to the reactor from a storage tank and the addition quantity was controlled to maintain the temperature between 110–115°C. If necessary, cooling was applied by use of water at ambient temperature: normally done automatically, it was under manual control due to external circumstances. On the day in question, the steps as far as the pressure test had been completed as normal and PO addition started about 0830 hours. The temperature of the reaction mix just prior to PO addition was between 80 and 90°C. Normally, the PO addition continues uninterrupted until all PO has been added. However, on this day, a number of interruptions to PO addition occurred one due to filling of the PO storage tanks and another due to the installation of new gas detectors in the reactor building.

After all interruptions were over, PO addition was resumed although at this point the temperature of the reaction mixture had dropped to 60°C. Over the next 2 hours, the temperature of the reaction mix increased from 60°C to 100°C and the pressure from 2.8 to 4.1 barg at which time the operator stopped PO addition and applied cooling. The pressure started to fall so the operator isolated the cooling and left the control room to carry out other duties. A few minutes later, a loud bang was heard. Immediately following the bang, a spray of liquid was seen leaving V7 vent on the reactor plant roof directing towards 400 and 700 tank farms. Eyewitness reports indicated release duration of approximately 60 seconds. Subsequent inspection showed that reactor V6 bursting disc had ruptured.

CONSEQUENCES AND INVESTIGATION OF THE INCIDENT

Fortunately, nobody either on or off-site was injured. However some storage tanks were splashed by released material. In addition, some liquid reached a branch railway line serving industrial users and which runs close to the Eastern edge of the site boundary.

A mass balance indicated that approximately 4775 kg of liquid and vapour was released to atmosphere. This included approximately 1300 kg of water from the knockout pot. This indicated that about 3475 kg was lost from reactor V6. The water in V7 was removed probably by a combination of vaporisation and displacement. The amount of pure PO released to atmosphere was difficult to determine. Subsequent calorimetry on a similar process indicated that in order to reach the disc burst pressure, approximately 30–40% of the PO charged would need to accumulate and much of this would have relieved to V7 due to its volatility. This would have amounted to a few tonnes. It was difficult to predict how much pure PO was released to atmosphere as some may have reacted with water in V7 to form Propylene Glycol. However, due to the short contact time and design of V7 (dip tubes into water only) only a fraction is likely to have reacted so much of the PO remaining in the reactor was assumed to have been released to atmosphere.

The amount of material released from V6 was about 30% by weight of the total. This indicated that homogeneous two-phase flow almost certainly didn't occur although some liquid was entrained in the venting stream. There was no visible damage to V6, V7 or the vent line from V6 to V7. In addition, V6 passed a subsequent pressure test indicating no distortion of flanges etc. Although this didn't necessarily indicate that the relief system was adequately sized i.e. design pressure not exceeded, it did demonstrate that the relief event didn't subject V6 to a pressure beyond its yield. Lack of any visible damage to V7 was perhaps more surprising due its atmospheric design pressure rating. This was probably attributable to the fact that the discharge line from V7 was large in diameter i.e. 0.6 m, straight and short in length, thus giving minimal back pressure. Figure 2 shows the internals of V7 after the incident. The original water level can be clearly seen from the watermarks on the dip pipes.

CAUSE OF THE INCIDENT

The cause of the incident was almost certainly due to accumulation of PO at the end of the addition period prior to the incident. The reaction rate is known to be slow at 60°C resulting in accumulation of PO. Nevertheless some reaction occurs at 60°C and this slowly heated the reaction mix to about 100°C over a 2 hour period accelerating all the time but without consuming much PO. In addition, because little reaction was evident, PO continued to be added, further increasing accumulation. At 100°C, the reaction rate is much faster and a large amount of the PO would have suddenly reacted as if the process was batch. Clearly, isolation of the cooling just prior to the incident was also a significant contributory factor although by this time the amount of accumulation would likely have made the event unrecoverable by normal cooling alone.

Although not the direct cause interruptions to PO addition meant that the reactor was allowed to cool to 60°C prior to resuming PO addition. There was a low temperature trip (softwired) set at 60°C preventing addition of PO below this temperature.

IMPROVEMENTS

The incident was reportable under RIDDOR and as a result The Health and Safety Executive were notified. An improvement notice was issued which required consideration of

1.   Prevention of such an incident happening again
2.   Review of adequacy of pressure relief system
3.   Review of discharge and disposal arrangements

This incident clearly showed that the knockout pot (V7) was ineffective and steps were taken to commission the design of a replacement.

PREVENTION

With semi-batch alkoxylation reactions such as these, safety is often based upon prevention of accumulation of unreacted EO or PO[1]. Accumulation can occur in a number of ways

1.   EO or PO added at too low a temperature
2.   EO or PO added too fast
3.   Agitation failure
4.   Catalyst not added
5.   Combinations of the above

Some controls were already in place at the time of the incident to avoid these scenarios. These included

- Low Temperature trip to prevent addition of EO or PO below 60°C
- High pressure trip to indicate accumulation of EO or PO although this was shown to be ineffective for PO due to the lower vapour pressure of PO
- Agitator sensing interlocked to EO/PO feed systems
- Exotherm detection on some reactors

Adiabatic dewar calorimetry was carried out on a similar PO process to determine the effect of accumulation of PO on peak reaction pressure and rate of temperature and pressure rise. The results are presented in figure 3. The minimum reactor design pressure is 8.6 barg and the disc burst pressure is 7.7 barg on all reactors. This indicates that provided the accumulation is not allowed to increase beyond about 30%, the disc burst pressure shouldn't be exceeded. Subsequent isothermal calorimetry indicated that addition of PO at 60°C and at the then maximum allowable feedrate of 3500 kg/hr, resulted in accumulation levels at least as high as 40%. Using DIERS methodology[2,3], the required vent diameter at 40%

accumulation was calculated to be 0.7 m. The vent diameter on V6 and other similar sized reactors was 0.3 m and it was not practical to increase to 0.7 m due to space restriction on the reactor top dish.

A detailed risk assessment was then carried out for the process involved in the incident. One of the outcomes of this assessment was to review the low temperature trip settings and maximum allowable feedrates of EO and PO. Further calorimetry work was undertaken. At 95°C, isothermal calorimetry showed that addition of PO or EO at a maximum rate of 2000 kg/hr would result in a maximum accumulation level of about 15% i.e. 15% of the reaction heat was released at the end of the addition of PO/EO. From figure 3, this indicated a peak pressure of 2.6 barg for the PO product tested, which is below all reactor design pressures. The equivalent maximum pressure in the case of EO (based upon a typical process) was 7.8 barg, again below the design pressure of all the reactors, although not below the disc burst pressures. The adiabatic results for EO at 15% accumulation are shown graphically in figure 4.

As a result, the low temperature trip was increased to 95°C. In addition, restrictor orifices (ROPs) were fitted to the EO and PO feed systems to restrict feedrates to 2000 kg/hr. Whilst it was shown that 2000 kg/hr was acceptable as a maximum feedrate, 1500 kg/hr has been adopted as a maximum operating rate to provide an additional margin of safety particularly for the EO processes and to keep peak pressure below disc burst pressure. A margin of safety was also required to account for the fact that the PO and EO processes tested were being assumed to be representative of a range of PO and EO reactions carried out. However, the two selected were considered to be worst case representations of the PO and EO processes respectively. The ROPs, low temperature trips and agitator sensors and interlocks are clearly safety critical in terms of preventing accumulation to levels whereby a relief system upgrade was not required.

PRESSURE RELIEF SYSTEM

Based upon the improvements made to the process control to prevent accumulation, it was decided that the relief system would not need to be designed to deal with runaway reaction scenarios. Work is still ongoing to study the required safety integrity levels of all safety critical systems according to the requirements of IEC61508. Most of the other non-alkoxylation reactions carried out on the plant were shown not to present a high potential for runaway. This was because many of the reactions were operated semi-batch and it was shown that even at room temperature, accumulation was insignificant. Again, process control is used as the primary basis of safety.

However, the relief systems would need to be able to deal with the most demanding of other non-reactive relief cases. It was shown that the most demanding of the non-reactive relief cases was external fire (as is often the case). Relief calculations showed that the existing vents had sufficient capacity to cope with all vapour flow due to fire boil up. The vents also had sufficient capacity for homogeneous two-phase flow in the event of external fire. However, it was considered that for alkoxylation reactions at least, homogeneous two-phase flow was unlikely as demonstrated by the incident itself. Two-phase flow calculations were performed using DIERS methodology.[2,3]

DISCHARGE AND DISPOSAL

One of the most important improvement measures was how to prevent further releases of PO or EO to the atmosphere. Clearly V7 was ineffective as a knockout pot as it contained very little of the relieving fluid during the incident. Because of the high volatility of EO and PO, it was decided that in the event of pressure relief, EO or PO would flash to vapour and thus a knockout pot or catch tank would be ineffective in preventing atmospheric release. However, EO and PO are highly soluble in water, so it was decided to design and install a quench tank. A quench tank has a number of advantages over a simple knockout pot in that it would

1. Achieve partial if not complete absorption of EO or PO preventing atmospheric release
2. Cool the mixture thereby reducing the likelihood of any on-going reaction

The ideal solution would have been to install a dedicated quench tank for each reactor. However, this would not only have incurred very high cost and leadtime, but more importantly, insufficient space was available for more than one tank. However, use of a single tank meant that the relief systems would need to be combined prior to entering the tank. This presented a few issues which needed consideration and which are discussed later in the paper.

Detailed bench scale quench tests were not performed. Whilst such tests are normally required to prove that a particular reaction will quench[1] (and recommended by the authors), their omission in this case was justified for the following reasons

1. Following involvement of the HSE, time was very tight to complete the design. Although not a key factor where safety is concerned, the company was keen to ensure continuity of supply to key customers.
2. The primary purpose of the quench tank was to absorb EO or PO rather than to achieve full reaction quenching. The ability of a particular design to do this would be less easy to demonstrate by small-scale tests.
3. Due to the process control improvements made, it was unlikely that runaway reactions on PO processes would need quenching as they would be contained.
4. The worst credible EO runaway following improvements also wouldn't exceed the reactor design pressures, although did approach the disc burst pressure. The need to quench EO reactions was therefore considered a possibility. However, because the remaining EO quantity available to react would be small, it was considered that only a small proportion of the reaction mix would need quenching.
5. In the case of external fire, it was also estimated that much of the EO or PO would have reacted prior to the disc bursting, again reducing the emphasis on the need to quench.
6. Many alkoxylation reaction products are highly soluble in water. Thus it was likely that water would be an effective cooling medium.

In addition, to compensate for absence of quench test data and to satisfy the HSE, some pessimistic assumptions were made in the design of the quench tank

1. Heat balance assumed that all the contents of the largest reactor (18000 kg) could be released to the quench tank at maximum EO reaction temperature of 200°C

2.   Heat balance took into account heat of solution of EO in water
3.   Heat balance assumed that unreacted EO would continue to react in the quench tank

The heat balance assumed a final temperature below 80°C where it was known that the normal reaction at least, would slow down appreciably. In addition, the 80°C was below the atmospheric boiling point of the most concentrated EO solution likely to form in the quench tank. The heat balance showed that 35000 kg of quench fluid was required. The quench fluid was a 25% solution of Glycol in water to prevent freezing. This is a very important consideration with quench tanks and provision needs to be made to sample the quench fluid to check Glycol levels. Another important factor was how to obtain a suitable vessel for the quench tank within a very short timeframe. A second hand pressure vessel had become available ex Chlorine duty. This was a horizontal vessel of approximately 74 m³ in volume, with a design pressure of 13 barg. This was sufficiently large to contain all the quench fluid and provide adequate freeboard height to allow for liquid swell and to prevent droplet entrainment.

The possibility of simultaneous relief (i.e. relief from more than one reactor at the same time) was considered. Simultaneous relief would have required a quenchtank of at least twice the volume i.e. at least 150 m³. This would have required construction of a new tank at increased cost and leadtime. Whilst simultaneous relief could not be totally discounted, it was considered that the frequency was sufficiently low such that installation of a larger quenchtank was not justified on reasonably practicable grounds. The next challenge was to design the tank internals i.e. the internal sparger. Factors to consider here were:

- Selection of appropriate hole size e.g. to minimise blockage and water hammer risk
- Determination of number of holes
- How to ensure even distribution of relieving liquid and vapour amongst the holes
- How to maximise contact time and area with the quench fluid

To design the sparger for the tank, guidance given in the CCPS guide on Pressure Relief and Effluent handling[5] was used. This was found to be a very useful and practical source of information. The final design consists of a central manifold made of 0.76 m diameter pipe, with twelve 0.15 m diameter pipe branches. The holes were drilled into the branches uniformly around the circumference to even out reaction forces as much as possible. In total, 1540 holes of 10 mm diameter were used. The final sparger arrangement inside the quench tank is shown diagrammatically in figure 5.

A sparger with side arms was used to maximise contact area with the quench fluid. A single central pipe would probably have been adequate but this wouldn't have as good use of all available water volume and would thus be less efficient. The sparger was positioned as low as possible within the tank to maximise depth of submergence and thus maximise the likely contact time for heat and mass transfer. However, the sparger had to be positioned sufficiently far above the base to permit the side branch length to be long enough to fit on the required number of holes whilst achieving minimum hole spacing in line with recommended guidance. All in all, the design process was iterative.

The relief line from the reactors entered axially primarily to facilitate installation of the 0.76 m central manifold into the tank. The side branches were fitted onto the central manifold within the tank. This required construction of a new manway on one dished end of the vessel. The design pressure rating of the tank needed to be maintained. Special supports were designed to secure the sparger arrangement within the tank to withstand likely reaction forces. The tank itself rested on saddles and it was shown that the friction generated by the weight of the tank would be sufficient to withstand axial reaction forces as the vented liquid first came into contact with the quench fluid. However, additional restraints were fitted to provide greater security. The final installed quench tank is shown in figure 6.

Another factor to consider was the effect of the nitrogen padding in the reactor. The nitrogen wouldn't be dissolved within the quench fluid and would need to disengage from the liquid. Because the reaction products were effectively surfactants, it was considered likely that some foaming would occur during sparging. It was considered possible that the foam would prevent disengagement of the nitrogen gas without carry over of foam out of the new tank. Whilst release of foam to atmosphere did not carry a high risk, it was considered unacceptable to risk any release to atmosphere. For this reason, a second tank was installed to catch any foam or liquid which might get entrained by the nitrogen. Installation of this tank was an accepted option for the Health and Safety Executive. A photo of the second containment tank is given in figure 7.

The next step was to design the final atmospheric stack from the second containment vessel. It was decided to adopt a pessimistic approach and design the stack on the basis that no EO would be absorbed by the quench fluid. Dispersion modelling was performed using PHAST version 6[6]. The modelling considered the stack height and diameter required to avoid a potentially flammable cloud impinging upon a nearby elevated light. It was also necessary to avoid ground level concentrations both on and offsite which could cause acute harm to people. The concentrations of interest were the Lower Flammable Limit (LFL) for flammable effects and the Immediately Dangerous to Life and Health (IDLH) concentration for toxic effects. For EO, these concentrations were 30000 parts per million (ppm) and 800 ppm respectively. To provide an added margin of safety, the flammable cloud at LFL/2 was modelled. Weather conditions 1.5F and 5D were used which are present 70–80% of the time in the area.

After consideration of wake effects of nearby tank farms, it was decided to install a stack of 0.6 m diameter, 15 metres high with a 0.3 m diameter tip to increase vertical velocity and aid dispersion. This gave a comfortable clearance of the LFL/2 plume envelope from the light and other obvious ignition sources. Also, at no point would the ground level concentration have exceeded the IDLH concentration. The final stack design is also shown in figure 7 in relation to the elevated light.

MECHANICAL DESIGN OF PRESSURE RELIEF SYSTEM

The new position of the quench tank required a re-design of the pressure relief pipe work routings. It was also necessary to link all the relief pipework together prior to entering the quench tank. Because a choke would likely occur upon entry to the quench tank, a relief event in one reactor would create backpressure within the vent network and impose backpressure on other reactor discs. Also leakage from one disc could allow product to accumulate on the vent side of another disc. There were therefore several challenges in the re-design of the vent system pipework. Some of these included

1. Routing of pipework to minimise stress e.g. due to thermal expansion
2. Design of pipework supports for reaction forces
3. Design of flexible connections to allow load cells to function
4. Prevention of discs rupturing in reverse direction
5. Provision of a method to inspect discs without removing vent pipework.

The most difficult challenge was in the design of the flexible connections. Bellows joints were not capable of withstanding the reaction forces, particularly in the event of two-phase flow. The final design used a vertical sliding joint arrangement, which provided just enough movement to allow correct function of the load cells at the same time providing sufficient strength along its axis to withstand at least 20 tonnes force. An example of the final design is given in figure 8. Prevention of discs rupturing in the reverse direction required consideration of two factors

1. Static pressure in vent network during venting
2. Reflected pressure when vented fluid first contacts stationary water in spargers.

Due to the complexity of the system and the wide range of likely relief scenarios and products, it was difficult to evaluate such pressures with confidence. However, the existing graphite discs had a reverse pressure capability of only a fraction of the forward burst pressure. Therefore it was decided to replace the discs with stainless steel reverse acting discs of a design to provide a higher reverse pressure strength than forward burst strength. This was certainly adequate to handle the maximum vent static pressure which would not exceed forward burst pressure. It was estimated that the fluid static pressure when it first contacted the quench fluid would be about 4 barg. The reverse pressure capability of the new discs was in excess of 8 barg which was more than twice the static pressure at the point of quench fluid contact. This was considered adequate in the absence of a more complex analysis. In order to inspect the discs in situ, a manhole was installed as shown in figure 8.

SECOND RUNAWAY REACTION INCIDENT

In January 2002, control was lost of another reaction on the reactor plant. This was not an alkoxylation reaction. The bursting disc on the reactor ruptured and several tonnes of hot material was released as a two-phase flow to the new quench tank. Whilst this was not a reaction for which the tank was primarily designed, it did provide a full-scale test of the new system which performed well. In particular

*   No material was carried beyond the quench tank
*   No other discs burst in reverse
*   The flexible joints performed well
*   Pipework and supports withstood reaction forces
*   The sparger holes didn't block despite the very high viscosity of this particular reaction product (very much higher than a typical alkoxylated product)

The good performance of the system was not surprising bearing in mind the safety factors built into the design to account for uncertainties (uncertainties are always present with such designs). However, there is now a higher degree of confidence that the system will perform when required to do so. Also, the system design is being adopted as best practice within the company and designs are underway for a similar system at another UK site.

LESSONS LEARNT

A number of general lessons can be learnt from this incident. Some of these are highlighted below

1. Incidents cost money. Due to the use of a common manifold, all reactors had to be shut down during installation resulting in lost production, in addition to the cost of the plant improvements.
2. Semi-batch reactions can be as inherently unsafe as pure batch reactions, unless the correct operating conditions to prevent accumulation are known. Calorimetry is required to determine such safe operating conditions. The cause of the incident was directly related to inadequate understanding of the reaction thermochemistry.
3. Process control can often be used effectively as a means to reduce relief demands due to runaway reaction and thus reduce required relief size. However, consideration needs to be given to the required safety integrity required.
4. Designing pressure relief systems is not just a case of determining a correct vent diameter. Close attention to the mechanical and discharge and disposal design is also required. This is particularly true when linking several reactors into a common header.
5. Discharge and disposal Systems e.g. Knockout pots, are unlikely to work effectively unless properly designed.

REFERENCES

1. Gustin, J.L., 2000, Safety of Ethoxylation Reactions, *Hazards XV – The Process, its Safety, and the Environment*, Symposium Series No. 147, Institution of Chemical Engineers, Rugby, UK.
2. Fisher, H.G. et al, 1992 Emergency Relief System Sizing using DIERS Technology, The Design Institute for Emergency Relief Systems (DIERS) Project Manual, AIChemE, ISBN 0 8169 0568 1.
3. Etchells, J., Wilday, J., 1998, Workbook for Chemical Reactor Relief System Sizing, CRR 136/1998, HSE Books, ISBN 0 7176 1389 5.
4. Singh, J., 1996, Safe Disposal of Vented Reacting Fluids, CRR 100/1996, HSE Books, ISBN 0 7176 1107 8.
5. CCPS, 1998, Guidelines for Pressure Relief and Effluent handling Systems, AIChemE.
6. Process Hazard and Safety Tools (PHAST) version 6, Det Norsk Veritas.

Figure 1.   V7 knockout pot

Figure 2.   Inside of V7 after incident

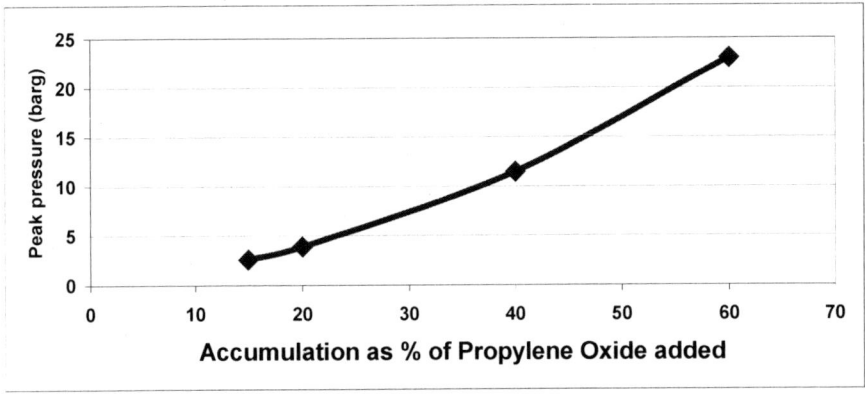

Figure 3. Peak reactor pressure as a function of propylene oxide accumulation

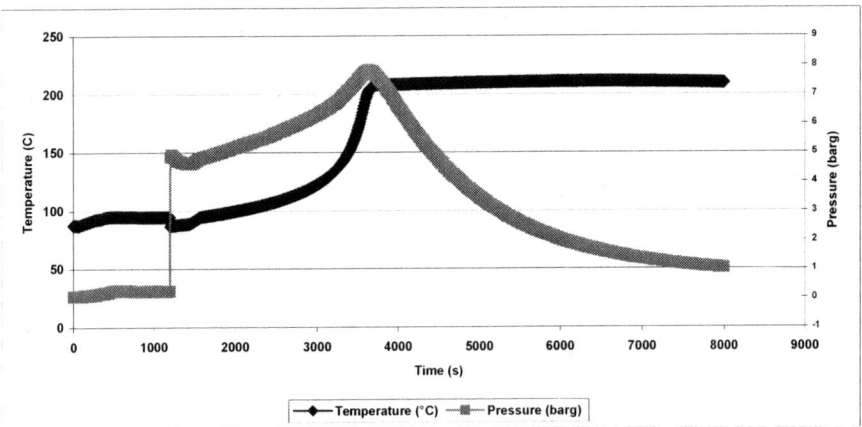

Figure 4. Adiabatic dewar data for a typical EO reaction at 15% accumulation

Figure 5.   Sparger inside new quench tank

Figure 6.   New quench tank

Figure 7.   Secondary containment vessel and vapour discharge stack

Figure 8.   New sliding joint arrangement on pressure relief pipework

AN EXPLOSION ACCIDENT – CAUSES AND SAFETY INFORMATION MANAGEMENT LESSONS TO BE LEARNED

Tzu-Lien Tzou[a], David W. Edwards[a] and Paul W.H. Chung[b]
[a]Department of Chemical Engineering, Loughborough University, Leicestershire, LE11 3TU, UK;
[b]Department of Computer Science, Loughborough University, Leicestershire, LE11 3TU, UK

A terrible accident occurred at a process plant in Taiwan on May 18, 2001. The plant was destroyed by a series of explosions that resulted in the death of one man, 112 injuries and extensive damage. The accident was caused by the ignition of a leak of mixed flammable vapours from an out-of-control exothermic batch reactor, which produced water-born acrylic resin. Most of the victims, who were employees of nearby factories, were cut by glass splinters and other debris that rained down over an area of radius 200 metres around the plant.

This accident reveals that both the process plant and the neighbouring factories did not handle safety information properly. This paper describes the accident, the discussions and the conclusions from the viewpoint of safety information management. The lessons learned from this accident include the importance of information management and the need for using a safety information management system throughout the plant life cycle. A research project at Loughborough University is investigating safety information management and its measurement and is developing a prototype tool for use in this area. This project is also described briefly in the paper.

KEYWORDS: accident, explosion, safety information management

INTRODUCTION

There is no doubt that batch polymerization reactors for the manufacture of polymers and resins should be considered inherently risky systems. Their main purpose is to react acrylic monomers to form high molecular weight acrylic resins via free radical exothermic polymerization. An analysis of industrial incidents in the UK involving thermal chemical reactions in batch or semi-batch reactors has shown that 47.8 percent of these 134 incidents were related to polymerization reactions[1]. Because acrylic monomers are highly reactive and are capable of undergoing fast polymerization that can generate substantial heat and pressure if not controlled suitably and correctly [2,3,4], most companies making these polymers are aware of the importance of safe handling of chemical material to avoid runaway reactions. However, do they have enough information to do so, particularly when the reaction equipment has been successfully operated over a long period of time, say twenty years or more?

During the afternoon of May 18, 2001, a 6-ton reactor at the Fu-Kao Chemical Plant ruptured during a runaway polymerization reaction, leading to a terrible accident. The plant was destroyed and nearby factories were seriously damaged by a series of explosions and fires that resulted in the death of one man, 112 injuries and extensive damage. The accident was caused by the ignition of a leak of mixed flammable vapours from an out-of-control exothermic batch reactor. Most of the victims were cut by glass splinters and other debris that rained down over an area of radius 200 metres around the plant.

This paper presents the initiating and root causes of the accident. In particular, it focuses on those things that were wrong or deficient with respect to managing safety information. This accident demonstrates that both the plant and the neighboring factories did not manage safety information properly. The plant did have some information about the hazard held somewhere within its organization. However, this knowledge was 'inactive' because it was simply filed and no one knew about it. Most of the neighboring factories had no knowledge of the hazards existing in the facilities close to them. This accident provides a typical case history of how safety information management could have prevented the occurrence of such an accident or, at least, have reduced its effect. This case is also interesting because it illustrates the importance of information exchange in an industrial park.

Safety information is the basis of safety management for the entire plant life cycle, from design, through fabrication, construction, operation, and maintenance to decommissioning of the process plant. There should be systems in place to accumulate safety information and communicate it to employees and to the public who need to use this information. Many organizations and software houses have created commercial products to support safety information management for clients. However, there is not enough guidance on how to improve safety information management for the process industry. The industry needs a new paradigm for measuring the performance of safety information management. Therefore, a safety information management audit tool (SIMAT) for the process industry is being developed at Loughborough University.

THE PROCESS INSTALLATION

The accident happened at the Fu-Kao Chemical Plant which is located in an industrial park in northern Taiwan. The plant is a medium-sized manufacturer of polymers and resins used in the composites and coatings industry. The plant was situated in a three-floor building, divided into three areas: raw materials with 7 storage tanks, production with 7 reaction units and product storage area.

A unit, called Reactor A, was in the production area. The reactor produced water-born acrylic resins through the reaction of acrylic monomers in solvent and using organic peroxide initiators. Because of the highly exothermic polymerisation reaction, the solvents, methyl alcohol and isopropyl alcohol (IPA), were used in the reaction to remove heat through a condenser. The initiator, Dibenzoyl peroxideb (BPO), which could then thermally decomposed to form primary free radicals that reacted with monomers grow into long polymer chains.

The Reactor A system was composed of an agitated 6-ton capacity vessel, an overhead condenser, a pure water feed tank and pump and emergency cooling water tank. Heating and cooling of the reactor was achieved using the same external dimpled jacket with manually operated valves. The operators determined the degree to which these valves were opened based on their experience of running the process. The timing of switches from heating to cooling or vice versa was also based on operator experience. The overhead condenser provided additional cooling capacity during reactions operating under reflux. The pressurized (3.5 kg/cm^2) pure water tank which supplied water as a reactant in normal operation as well as acting as a killing system when a runaway reaction happened, was connected to the reactor via a manually operated valve. Further cooling was achieved by spraying water on to the outer wall of the reactor. However, since Reactor A was designed and installed in the plant about twenty years ago, it did not have an emergency relief disposal system attached. Figure 1 shows the Reactor A system.

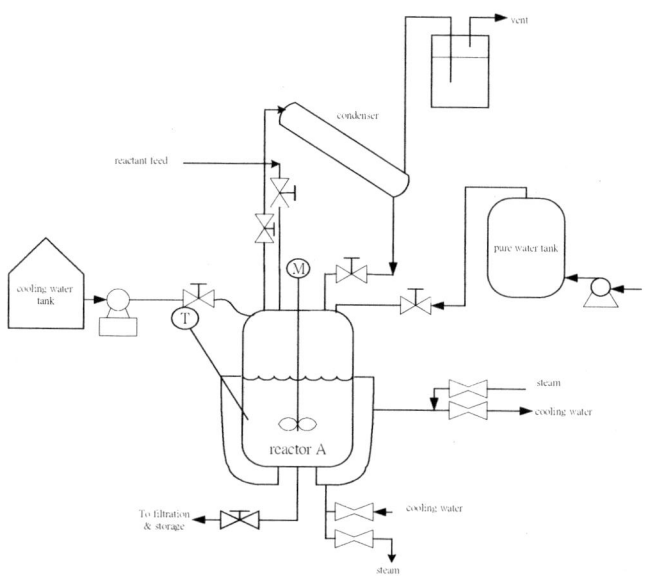

Figure 1. Simplified process flow diagram of Reactor A system

The operating procedure for making a batch in Reactor A was:

1. Add 1293.2 kg of methyl alcohol and 373.3 kg of IPA to the reactor.
2. Start the agitator.
3. The desired amount of several kinds of acrylic monomers, including 172.8 kg of acrylic acid (AA), 1500 kg of methyl acrylate (MA), 32.6 kg of methacrylic acid (MAA) and 20 kg of acrylonitrile (AN), are added to the reactor in sequence.
4. Pump 2130 kg of pure water and add 5.6 kg of BPO into the reactor.
5. Supply steam to heat the reactor between 60–65°C using the reactor jacket, and continue mixing until the exothermic polymerisation is initiated while the reactants in the reactor are boiling.
6. Stop the steam flow, vent the jacket and coils, and circulate cooling water under manual control and allow the temperature to gradually rise to 70°C within 70 minutes.
7. Stop circulating the cooling water, stop the agitator and maintain the polymerisation reaction for about 4 hours.
8. When the batch is completed, the acrylic resin is transferred to a product adjustment tank. If necessary the reactor is cleaned, then the next batch is started.

The Reactor A process had been in use for over twenty years. The Reactor A system was staffed with only one experienced operator. The plant manager did not order anyone to assist the operator in normal circumstances but the operator could ask for assistance when needed.

ACCIDENT DESCRIPTION

On May 18, 2001 at approximately 8:00 A.M., the operator of Reactor A began a batch of acrylic resin just like every other normal day. He stated in interviews that he fed the required amount of solvents, acrylic monomers and initiators in the morning. The steam valve to the reactor jacket and coil was opened to begin heating the reactant at 10:00 A.M. while the agitator continued to operate. The operator left to have lunch at 0:10 P.M. and came back at 0:40 P.M. During lunchtime there was no one monitoring the reactor. Soon after the operator came back from lunch the reaction temperature reached 65°C. So, the operator stopped the steam and started the cooling water at 0:50 P.M. He stated in interviews that he had heard the sound of the cooling water flowing through the piping, an indicator he used to check cooling water flow. Approximately 5 minutes later, the operator reported that the reaction temperature had extraordinarily increased to 80°C and it was out-of-control. The operator tried to add extra cooling water from the valve on the pipe that connects the reactor to the condenser but he failed. At 1:10 P.M. the flashing reactor contents were ejected upward from the reactor. Meanwhile the emergency alarm was started and the plant manager instructed all the plant employees to evacuate immediately. At 1:20 P.M. the first explosion happened and set fire to the plant. Moments later there were several further explosions with accompanying fires. The fire was extinguished by the Fire Bridge in two hours.

During the accident, the ambient temperature was about 23°C. It was a clear day, with light winds mostly from the northwest. These winds blew plumes of reactant, products, and smoke off the plant site. The fallout was mainly to the south of the plant. The vapours spurted out and spread as far as 1 kilometre downwind from the plant.

After the explosions the resulting pressure wave and fire completely destroyed the plant (see Figure 2) and extensively damaged the surrounding structures of neighbouring factories (see Figure 3), resulting in the death of one man and 112 injuries. The man who died was an employee of a factory next to the plant. He was killed by the flame/blast while he was working outside the factory building very close to the Fu-Kao plant. Most of the injuries were to employees of nearby factories or residents around the industrial park. They were cut by glass splinters and other debris that rained down over an area of radius 200 metres around the plant, or/and felt dizziness and nausea after smelling the odours. Whereas, responding to an emergency alarm, most of the plant workers were evacuated about 5 minutes before the first explosion occurred and most escaped serious injury.

Many odour complaints were received from community members, during and following the accident. Air monitoring was performed by the Environmental Protection Agency and Industrial Technology Research Institute. Tests were negative for acrylonitrile, methanol, methyl acrylate and acrylic acid.

Figure 2.   Explosion at Fu-Kao chemical plant

The accident totally destroyed the plant and seriously damaged 46 nearby factories including 16 high-tech companies that produced IT-related products. These factories had to stop their production for at least 5 days and some of them took more than 30 days to recover. The total property loss was estimated to be US $12m and the total loss due to business suspension was estimated to be US $50m.

INITIATING CAUSE OF THE ACCIDENT
An official committee of technical experts was formed whose members came from the Institute of Occupational Safety and Health and the local HSE authority in Taiwan. The committee concluded that the initiating cause of the accident was that the operator failed to recognize and act promptly when the reaction temperature was too high, leading to an uncontrollable runaway reaction. The committee briefly explained the accident thus:

1.   The first explosion was a small vapour cloud explosion (VCE), which was caused by the ignition of a leak of mixed flammable vapours from the out-of-control exothermic 6-ton batch reactor, while the reaction temperature was extraordinarily increased to 80°C and the cooling water system failed.

2.   According to evidence gathered on-site, such as the agitator pump was blown far away and the condenser and decanter had unfolded outwards but the vessel was not blown away (see Figure 4), the explosion location was assessed to be 1–2 metres above the reactor.

3.   The second and main explosion was due to the combustion of 100 kilograms of BPO stored on a shelf 10 meters away from Reactor A. The first explosion heated the stored BPO over its unreturned temperature of 65°C. Then the BPO was decomposed at 104°C and exploded after 0.1–0.3 second. Even with a small amount of BPO the peroxide explosion has a huge effect, because its maximum explosion pressure rise $(dP/dt)_{max}$, 900-1100 bar/sec, is similar to explosive combustion and it is 2 to 3 times greater than either an organic vapour or gas explosion.

4.   Several of the explosions that followed were believed to be BLEVEs, since most storage tanks in the plant had exploded upwards and the contents of these tanks were completely burnt.

(3a)

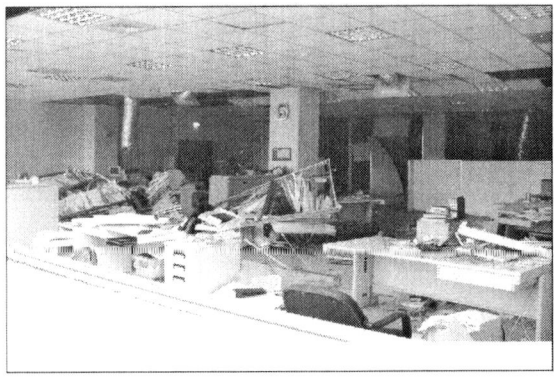

(3b)

(3c)

Figure 3. Explosion effects on neighbouring factories

Figure 4.   Post-accident photo of the 6-ton reactor

The Ministry of Economic Affairs of Taiwan also funded a research institute to analyse the causes of the accident. The research carried out a consequence analysis and calorimetric tests, and concluded the causes of the accident were briefly[5]:

1. The initiating cause was a vapour cloud explosion due to a runaway reaction in the 6-ton reactor. The blast mass was estimated to be equivalent to 1000 kg of TNT.
2. During the runaway the temperature had risen rapidly from 60°C to about 170–210°C and the maximum temperature rise rate might have reached 192 K/min.
3. The three most probable failure scenarios (with cooling failure) were identified as follows:
    *   Normal recipe;
    *   Normal recipe with 50% methyl alcohol added (solvent undercharged);
    *   Normal recipe with double charge of BPO (initiator overcharged).

The two investigations have different views as to the cause of the most significant explosion, there are BPO explosion or VCE. These differences have not yet been reconciled at the time of writing. However, both investigations agreed that the main initiating causes of the accident were:

1. The reactor cooling system was insufficient to safely control the exothermic polymerisation reaction. The operator controlled a process temperature manually only by opening and closing steam and cooling water valves. The runaway happened because the operator delayed operating the valves.
2. Inadequate process design led to operating procedures requiring that the process run in the temperature range (65–70°C), which is too near the temperature at which the reaction could become uncontrollable (80°C). Thus, unusual variations in the operator's responses to the batch temperature or delays in adjusting the valves could result in overheating or undercooling, with the result that the heat generated by the reaction would exceed the cooling capacity of the reactor.
3. The reactor was not equipped with safety equipment, such as a quench system or a reactor dump system, to stop the process and avoid a runaway reaction.

ANALYSIS OF THE CAUSE FROM THE VIEWPOINT OF SAFETY INFORMATION MANAGEMENT

Kletz[6] describes accident investigation as being like peeling an onion. The outer layers deal with ways of avoiding the hazards, while the inner layers are concerned with the underlying causes, such as weakness in the management. Many accident investigation tools and guides also emphasize the need to 'drill down' the causal factors into Process Safety Management (PSM) programme. For instance, The Center for Chemical Process Safety of the AIChE suggests a 'multiple-cause' systems-oriented investigation that focuses on root cause determination, integrated with an overall PSM[7]. Therefore, when we examine the performance of the plant with respect to the twelve elements of the CCPS's PSM model, as reproduced in the Table 1, we find all 12 to be weakly and insufficiently represented. This is a sure sign that integral safety was not in the minds of top management and it explains the root cause of this accident.

Table 1. Assessment of the PSM performance at the Fu-Kao Chemical Plant

PSM element	Performance
1. Accountability: objectives and goals	No overall safety goal and objective for the plant
2. Process knowledge and documentation	Limited in operating procedures and maintenance records
3. Capital project review and design procedures	No critical reviews
4. Process risk assessment	No activity
5. Management of change	No activity
6. Process and equipment integrity	Very limited on equipment fabrication, initial inspection and testing
7. Incident investigation	No near-miss reporting and investigating
8. Training and performance	Insufficient
9. Human factors (error assessment, task design, ergonomics)	None
10. Standards, codes, and laws	Partly covered
11. Audit and corrective actions	No activity
12. Enhancement of process safety knowledge	Poor activity

Examination of the PSM performance of the plant revealed the poor documentation and safety information management (SIM) of the plant, which should be analysed in greater depth. Reviewing data, information and the investigation papers related to the accident, and interviewing with the operator and manager of the plant, led to the following main findings and discussion points with respect to SIM:

1. For the Fu-Kao Chemical Plant

    • The plant maintained information on the manufacturing process, such as process and instrument diagrams, design codes and standards, a simplified process-flow diagram and Material Safety Data Sheets (MSDS) for most of the explosives it used. Workers at the plant did not use and were not aware of most of the written safety programmes and documentation. They were aware of the MSDSs and of the information they contain; however, they were not aware of any specific hazards associated with the explosive materials.

    • In several previous instances, the operator had reported to the manager that the process temperature rose at a faster-than-expected rate and exceeded the upper limit specified in the operating procedures, in spite of the operator's efforts. The temperature of these batches eventually returned to within the operating limits. These operator reports of significant deviations in controlling batch temperature were vital safety information but the management did not investigate the causes of these events. This shows that the plant simply put safety information away and waited for another deviation to happen. Unfortunately, this time there was not only a deviation but also a catastrophe.

    • The Safety Information File and MSDS, which the plant used as the basis for a Process Hazard Analysis (PHA) conducted in 1995, noted the desired exothermic synthesis reaction, but did not include information on the exothermic runaway reaction. The inadequate PHA did not provide sufficient safety information to help revise the weakness of the safety equipment and management of the plant.

    • Similar accidents had happened in 1997 in the USA[8] and in 2000 in Taiwan. The local HSE authority sent a leaflet to inform the plant to learn from these accidents. This kind of information can be an ideal training material and should be used for PHA. However, the manager ignored the information and put it in an archive cabinet only.

    • In this accident although a manually operated emergency shutdown system was available, the operator did not know what was the 'exact time' to activate it, since there was no emergency operating instruction.

    • The plant also did not effectively implement the requirements of its internal PSM programme. The PHA conducted for the process and the operating procedures (batch sheet) did not address the consequences of potential deviations such as excessive heating and the runaway operating range. The batch sheet did not list the actions operators should take to correct or avoid deviations.

    • During the first interview of the investigation, the Fu-Kao manager said that the reactor was equipped with safety equipment in the form of a dump system to shut down the process in case of a runaway reaction emergency. However, according to the accident site and the blueprint of the reactor, there is no dump or quench system in the equipment. This revealed that the manager already had information and knowledge that the plant should add desired safety devices to the reactor but he did not act.

- The last inspection by the local HSE authority was on February 2001. The result of the inspection was sent to inform the plant to carry out a prompt review of its operating procedures and training on hazardous material handling. The authority also suggested the plant reviewed its PHA to assess the safety equipment of the process reactor. The plant manager did not act or respond to this information. After a couple of months the accident happened.

2. For the neighbouring factories
   - The Fu-Kao Chemical plant was built earlier than most of the other factories in the industrial park. Some of the new 'high-tech' facilities were only built a few years ago. Most of these factories did not know what was happening in neighbouring sites and had no information about the hazards existing in the facilities close to them. So most of the buildings in these factories were constructed using a lot of glass, which acted as flying knifes injuring victims in the accident. Furthermore, these factories had no efficient emergency response plans for evacuation, rescuing victims nor did they have personal protection equipment. Fortunately, the accident did not spread toxic materials to hurt people in these factories.
   - If materials that are not there cannot leak, people who are not there cannot be killed[9]. This accident would have had a much smaller impact if the neighboring factories had not been allowed so near the plant. It is, of course, much more difficult to forego high-profit 'high-tech' industries than chemical industries. Nevertheless this must not stop collection of information and risk assessments to prevent accidents or, at least, reduce their effects.
   - In this 300-factory industrial park there is no information exchange system nor emergency plan for the whole park. This accident also showed the need for companies to collaborate with local authorities and emergency services in drawing up plans for handling emergencies to reduce consequences and to prevent a 'domino effect'.

SAFETY INFORMATION MANAGEMENT (SIM) RESEARCH AT LOUGHBOROUGH

According to the findings and discussion above, inadequate SIM was a critical factor contributing to the accident. Tuner[10] writes "Disasters equal energy plus misinformation" and King & Hirst[11] modify the equation to "Disasters equal energy and/or toxic substances plus misinformation or rejection of information" to emphasize the importance of handling information in preventing disaster. Kletz[12] also states that industrial accidents occur because organisations have no memory and do not use the knowledge that is available. There have been too many experiences in which even simple data were not noted because they were relatively hard to find[13]. Furthermore, 'safety information' has been treated as one of major components not only in PPCS's PSM programme but also in other PSM models published by U.S. Chemical Manufactures Association, the American Petroleum Institute, the U.S. Environmental Protection agency and the Occupation Safety and Health Administration. Thus, every company should ask itself not only if it has enough safety devices, alarm systems and so on, but also if its organizations have an adequate SIM system to improve understanding and communication, which will enhance awareness of hazards in the workplace and result in a better, safer working environment.

Safety information flow in a process plant should link all elements together. The SIM system should ensure that safety information is kept current throughout process changes, equipment maintenance, and other normal activities. Maintaining information requires appropriate linkages between the process safety information management system and the four other elements of the PSM system: capital project review, management of change, process equipment integrity, and process risk management[14]. These linkages are depicted in Figure 5. As changes to the facility or process are reviewed and implemented, these changes should be reflected in the process safety information. Specific responsibility should be assigned and resources allocated for ensuring that this occurs.

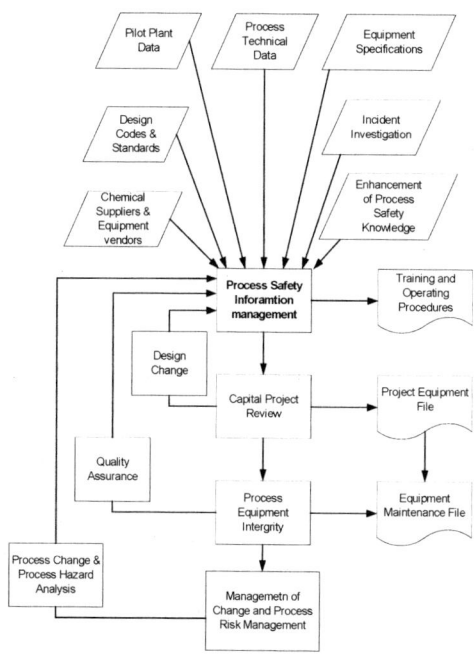

Figure 5.   Process safety information management linkage

Furthermore, during the last 10 years, there has been a revolution in information handing technology. For example, many types of documents are now created, revised and stored using computers and the Internet[15,16]. The opportunity and challenge today is to make the best use of new technology to more effectively manage the increased volume of information. Many process plants might want to developed SIM systems that specifically addressed their requirements; it can still be a frustrating and bewildering task. Learning how to properly acquire, utilize and apply safety information can be considered the very core to successful safety programmes. The main problem is how can we ensure that a SIM system in current use or a newly developed one is suitable for the plant. Therefore, an information audit, which is an established management methodology, must be used to evaluate the current information environment[17].

Nevertheless, evaluating a practical process industry SIM system poses a number of questions, including:

1. Why and how does a SIM system fail?
2. What are the scope, function and model of a good SIM?
3. How to make SIM available proactively during all interactions with the plant?
4. How to audit SIM?

These questions are still not answered. Therefore, Loughborough University is doing a SIM research project to explore these issues. The principal aim of this research is to investigate and analyse major process industry accidents in relation to inappropriate SIM and to provide a tool to audit not only the integrity of a SIM but to evaluate the effectiveness of SIM performance. This will be realised through the following three objectives:

1. identify and classify inadequate safety information by conducting causal analysis of past incident reports;
2. identify the important proactive indicators of SIM performance;
3. develop a SIM audit tool (SIMAT) and implement it on site to demonstrate its ease of use and its accuracy and usefulness.

CONCLUSIONS

The initiating cause of the Fu-Kao accident was that the operator failed to recognize and act promptly when the reaction temperature was too high, leading to an uncontrollable runaway reaction. The reactor was not equipped with adequate safety equipment to stop the process and avoid such a runaway reaction. However, inadequate SIM was one of the root causes of the accident. Learning from accidents dictates that every process plant should establish a programme that ensures that reactive chemical process safety information and operating experience are collected and shared with all relevant internal unit and external organizations. Unfortunately, past research and investigation of accidents did not focus on this issue.

Successful SIM will help reduce operational risk, avert litigation and regulatory fines, cut management costs, and ensure reputation. A computer-aided SIM may reduce accidents by 60% in one year[18]. However, good performance in this field requires a consistent prevention programme built upon shared experience. Lack of a common approach to SIM often makes this difficult. Moreover, industrial facilities today are faced with the challenge of collecting, storing, managing and evaluating an immense amount of safety information. There are potentially many information management problems: not getting the right information; spending more time organizing and finding information than analysing it; the length of time it takes to reinterpret an information set after new information is collected; no 'big picture' of site conditions. It can be simply summarised that one has too much information but not enough useful information.

SIM is a primary element in a Process Safety Management programme. Many organizations and software houses have created commercial products to support safety information management for clients. However, how to measure SIM performance is still unclear, and there is not enough guidance on how to improve SIM for the process industry. Therefore, the ongoing research project at Loughborough University, to develop a SIM audit tool (SIMAT), will focus on examining SIM performance in order to establishing and maintaining good SIM practices.

ACKNOWLEDGEMENTS
The authors acknowledge the assistance of Mr Hong-Chun Wu, a researcher at the Institute of Occupational Safety and Health in Taiwan, in the gathering of information and material for this paper.

REFERENCES
1. Barton, J.A. & Nolan, P.F. (1989). Incidents in the chemical industry due to thermal-runaway chemical reactions, Hazards X: process safety in fine and speciality chemical plants. Symposium series No.115, 3-8. IchemE, Rugby, UK
2. Lees, F. P. (1996). Loss prevention in the process industry (2nd ed). Butterworth Heinemann, Oxford, UK.
3. Klein, J. A. & Balchan, A. S. (1996). Safety formulation and manufacture of acrylic resins, CCPS international conference and workshop on process safety management and inherently safety processes, Orlando, Florida, UAS.
4. CSB (1998). Investigation report: chemical manufacturing incident, Report No. 1998-06-I-NJ, U.S. Chemical Safety and Hazard Investigation Board, Washington DC, USA.
5. Kao, C. S. & Hu, K. H. (2002). Acrylic reactor runaway and explosion accident analysis, *Journal of Loss Prevention in the Process Industries*, 15, 213–222.
6. Kletz, T. (1988). Learning from Accidents. Gulf Professional Publishing, Oxford, UK.
7. CCPS (1992). Guidelines for investigating chemical process incidents. AIChE, New York, UAS.
8. Gromacki, M.(2000). Acrylic polymer reactor accident investigation: Lesson learned and three years later, CCPS International Conference and workshop on process industry incidents, October, Orlando, Florida, USA.
9. Kletz, T. (1998). What went wrong?: case histories of process plant disasters (4th ed). Gulf Professional Publishing, Oxford, UK.
10. Tuner, B.A. (1978). Man-made disasters. Taylor and Francis. London, UK.
11. King, R. & Hirst, R. (1998). King's safety in process industries (2nd ed). Arnold, London, UK.
12. Kletz, T. (1993). Lesson from disaster: how organizations have no memory and accidents recur. IchemE, Rugby, UK
13. CCPS (1995). Guidelines for process safety documentation. AIChE, New York.
14. CCPS (1993). Guidelines for auditing process safety management systems. AIChE, New York, UAS.

15. Megill, K. A. & Schantz, H. (1998). Document management: new technologies for the information services manager. Power-Saur. UK.
16. Laudon, K.C. & Laudon, J. P. (2000). Management information system (6th ed). Prentice-Hall Inc., New Jersey, USA.
17. Henczel, S. (2001). The information audit: a practical guide. Power-Saur. UK.
18. Yoon, H. & Moon, I. (2000). Industrial applications of safety information management systems, *Hydrocarbon Processing*, September 2000, 84-A-D.

SAFETY CASE IMPLEMENTATION – AN AUSTRALIAN REGULATOR'S EXPERIENCE

Geoff Cooke
Principal Safety Analyst, Major Hazards Division, WorkSafe Victoria
Rob Sheers
Director, Major Hazards Division, WorkSafe Victoria

Each state in Australia individually regulates safety with the State of Victoria having a population of 5 million people. In June 2000 Victoria enacted the Occupational Health and Safety (Major Hazard Facilities) Regulations. These Regulations were made following the 1998 incident at the Esso Longford gas plant to further promote the safe operation of major hazard facilities in the state. As a result, approximately fifty Major Hazard Facilities had an obligation to develop and submit a Safety Case by 30 June 2002 to the Regulator, WorkSafe, (a division of the Victorian Workcover Authority) in accordance with the Regulations. The Regulator then has six months to assess the Safety Case, verify findings and make an appropriate licence decision regarding the operation of the Major Hazard Facility.

This paper discusses a Regulator's experience and observations in Safety Case preparation, assessment, and licensing.

KEYWORDS: Safety Case, Implementation of Regulations, Major Hazards, Licensing Major Hazard Facilities, Victoria, WorkSafe, Australia

INTRODUCTION

On 25 September 1998, an explosion and fire at Esso Longford Gas Plant killed two workers and left substantial parts of the State of Victoria without gas for 10 days. A Royal Commission[1] investigated the incident and recommended the Victorian Government establish a Safety Case regime for major hazard facilities (MHFs) in the State. The Government implemented the recommendation through the Occupational Health and Safety (OH&S) Regulator, WorkSafe Victoria. After a period of intense stakeholder consultation, the *Occupational Health and Safety (Major Hazard Facilities) Regulations 2000*[2] (the Regulations) were introduced in June 2000, 19 months after the Longford incident.

The 50 MHFs range from four refineries operated by multi-national companies to privately owned warehouses storing material for the chemical industry as shown in Table 1. Many facilities were constructed over 30 years ago and some have changed ownership several times. The initial approach to the Safety Cases varied widely from a simple warehouse adjusting from prescriptive dangerous goods regulations to large multi-national complex facilities that were unresponsive to local regulation.

Two years after the incident at Longford, the Government had established Major Hazard Regulations with a dedicated WorkSafe Division to regulate these facilities. This paper discusses the subsequent three years when Safety Cases were prepared by Industry and assessed for licensing. The first section deals with major issues in chronological order with the second section identifying good practice from Safety Cases.

Table 1. Classification of MHFs into industry sectors

Industry Sector	Number
Chemical Manufacturer-includes explosives	21
Chemical User- includes water treatment and paper manufacturers	5
Warehouse	7
LPG and Gas Distribution	5
Petroleum Refining	4
Bulk Terminal	8
Total	50

THE REGULATIONS 1998 - 2000

The Regulations were based on the 1996 Australian National Standard for Control of Major Hazard Facilities[3] which in turn was based on a draft of the Seveso II framework. The desire to quickly implement the Regulations required Industry and the Regulator to meet tight timeframes. Industry was required to submit their Safety Case within two years or by 30 June 2002 with WorkSafe allowed a further six months to assess and licence the facilities. The licence was to ensure a rigorous approach to safety and had a maximum term of five years.

Some parts of the National Standard were outside the power of an OH&S Regulator. Land use planning and environment were excluded, but the Regulations did include additional mandatory requirements such as worker involvement in the preparation of the Safety Case and performance monitoring of major hazard controls.

The Regulations are performance-based and mainly draw their power from the *Occupational Health and Safety Act 1985 (OHS Act)*. Parts of the Regulations do prescribe mandatory processes, for example the requirement to prepare an emergency plan in conjunction with the emergency services, or prescribe Safety Case content such as the requirement to include a management of change procedure. The major parts of the Regulations are shown in Table 2.

The major elements of the Regulations in Part 3 *Safety Duties of Operators* and Part 5 *Consulting, Informing, Instructing and Training* are shown in Figure 1. The Regulations outline a process for preparing the Safety Case, the safety role of employees, the requirements to review the Safety Case with an emphasis on the safety management systems and control measures.

PERFORMANCE STANDARDS UNDER THE REGULATIONS

Industry and Unions were consulted extensively during the preparation of the Regulations, but were uncertain of their performance rather than prescriptive nature, particularly the standards required for the licence. After Longford, the community and government had the resolve to refuse to licence a facility that could not systematically demonstrate adequate control of its major hazards and stakeholders wanted predictability of the outcomes.

Table 2.   Major parts of the regulations

Part	Contents
Part 3	*Safety Duties of Operators* describes the requirement for safety management, identification of major incidents, safety assessment, control measures, emergency planning, and the safety role of employees in undertaking the Part 3 requirements.
Part 4	*Safety Case* describes the contents required in the Safety Case document. This includes 'demonstrations' in relation to the adequacy of control measures and the safety management system.
Part 5	*Consulting, Informing, Instructing and Training* describes the Operator requirements for consulting, informing, instructing and training with Health and Safety Representatives (HSRs), employees, non-employees, the local community, and municipal councils.

The licence standard could not be defined before the licensing phase. WorkSafe's view was that the standard was set by all stakeholders and as such could not be determined before the licensing process. This dilemma of a defined standard seems to be a common problem with the first cycle of performance-based regulations. In the case of Victoria, there was a clear expectation that improvements in the control of risk at MHFs would be the greatest possible and any appearance of a minimum standards approach would have drawn criticism.

THE FIRST YEAR 2001

PLANNING AND RESOURCING

To meet the tight timelines, the Regulations required Industry to produce a project plan for the preparation of the Safety Case to be monitored by the Regulator. About 25% of MHFs planned to deliver their Safety Case ahead of the 30 June 2002 deadline with the remaining MHFs planning to submit immediately before the deadline. Emergency services personnel were also seconded to WorkSafe and formal coordination requirements would apply across all government agencies regulating safety at MHFs.

Although requiring considerable administration by Industry and the Regulator, this project plan did prove a useful tool in ensuring the tight timeframes were met. Early Safety Cases were delayed by up to three months but all Operators delivered their Safety Case by the deadline. The timeline for implementation of the Regulations is shown in Figure 2.

As the project plans were being received in late 2000, WorkSafe finished recruiting the Safety Analysts for the Major Hazards Division (MHD) Safety Case team. The Major Hazards Division was well resourced with one Safety Case officer per eight MHFs. The Safety Case Officer's role was to ensure regulatory compliance during the preparation of the Safety Cases and then conduct Safety Case assessments. To ensure independence, the assessor had to be an officer that had not been involved in the preparation of the Safety Case.

It quickly became apparent that to gain sufficient resources to meet the tight timeframes that the Division would compete for resources with Industry and with other state regulators. The shortage in expertise to regulate and consult for Safety Cases seems to be a common problem with the first cycle of this type of regulation. If all states had started simultaneously, expert resources would have been a crippling issue for both Regulator and Industry.

ADVICE AND GUIDANCE

In recognition of the timeframes and shortage of expertise, WorkSafe implemented a range of support programs for the two years which included two briefings for Chief Executives of MHFs to ensure their support for the process, 25 days of seminars for MHF Operators and HSRs, the preparation of 17 detailed guidance notes and three Exemplar Safety Cases to share early lessons across Industry. WorkSafe found the guidance work extremely time consuming. Generally, Industry feedback was that the seminars were the most useful guidance but sought specific examples that applied to their particular industry sector. Industry also wanted written guidance to include examples and for the guidance to be provided earlier.

COMPLIANCE AND INTERVENTION

During the first year of the Regulations, Industry and the Regulator both observed that the level of safety compliance assumed for the introduction of the Regulations was lower than expected. Many companies with good safety records were surprised at their own gaps in compliance. Rectifying these regulatory gaps found in the Safety Case process started to increase the overall cost of implementation of the new Regulations.

At this point a team of dedicated OH&S Inspectors was also provided for the MHFs to primarily ensure compliance with other safety legislation. The allocation of more compliance inspection resources to higher hazard industries was logical but at this point there was a risk of diverting Operators of MHFs from their Safety Case. The issues were resolved by prioritising what should be reviewed so that the two teams supported each other.

At this stage WorkSafe categorised the MHFs into six industry sectors and started comparing their OH&S performance and progress toward achieving the goals of the Regulations. About 25% of the MHFs had difficulty preparing their Safety Case and required a stronger regulatory intervention strategy. Further resources were focussed on these facilities with a system of escalating the Regulator's concerns to more senior management if progress did not improve. 65% of facilities required an advisory strategy with normal monitoring and 10% required little attention or an observation strategy. The number of MHFs in each category at the end of the first year is show in Figure 3.

Most of the seven MHFs in the Warehouse Industry Sector were assessed as progressing slowly. These MHFs had less formal business systems than other MHFs and fierce competition reduced the possibility of working collectively. The Regulator and Industry associations tried a variety of initiatives to educate and assist this sector but all proposals were unsuccessful. With hindsight, more assertive intervention was needed during this period to ensure the warehouses rectified their performance issues.

At the end of the first year, implementation was reviewed by a consultant. The major findings were that the Regulator had an aggressive approach that was disconcerting Industry; the cost was starting to exceed that forecasted and there was heightened concern over the standards that would be set for licensing. The report did indicate that there was no doubt that all stakeholders were fully committed to delivering quality Safety Cases by the deadline.

THE SECOND YEAR 2002

FEEDBACK

WorkSafe ensured that when administering the Regulations it was objective and independent and determined it would not fulfil the role of consultant for industry. WorkSafe was also concerned that it should not partially approve the Safety Case in advance or give specific advice that it would later assess. This approach of monitoring progress rather than providing direct and specific feedback caused Industry some frustration. WorkSafe's view was that the adequacy of the Safety Case was only apparent when most of the document was completed. Feedback at each stage of preparation would have given a false impression that may have dissuaded the Operator from revisiting early stages of the Safety Case as the process developed.

WorkSafe did respond to the need for feedback by introducing the Pilot Safety Case assessment. A Pilot assessment was an onsite review by WorkSafe of a vertical slice through the Safety Case that included sections from hazard identification, safety assessment through to control measures and relevant safety management system elements. Pilot assessments were well received by Operators and led to immediate changes. Some of the common improvements included clearly linking the elements in the process so that omitted items were easily detected, identifying blind spots in methodologies or work that did not challenge fundamental assumptions.

Industry feedback was that Pilot Safety Case assessments should be conducted earlier than the programmed 6–12 months before submission of the Safety Case. WorkSafe thought the Pilot Safety Case assessments were worthwhile both from the resources and the training benefit to WorkSafe as well as in anticipating the types of Safety Cases to be submitted. Other types of feedback, such as commenting on drafts, were not undertaken as much more time would be consumed by iterative exchanges of comments between the Regulator and Operator. The pilot assessment of documentation with the monthly on-site feedback and clarification meetings with the Operator was thought to be a more efficient and effective use of time.

During the second year, the first feedback on the effect the Regulations on the workforce was received. The involvement of HSRs from the beginning of hazard identification and throughout the process, rather than at the end in a training or review role, was found to be particularly effective. Feedback through the Unions indicated that, at some sites, their members were starting to see a fundamental turn around in safety at the site. This is an important lesson for Safety Case development which can often be largely a technical process. Involving experienced workers added to the development of the Safety Case and facilitated implementation. Overall, worker involvement extended outside the Safety Case and led to more active involvement in safety at the MHF.

THE ASSESSMENT FRAMEWORK

During the second year WorkSafe began preparing its Safety Case Assessment methodology. A consultative group was formed with Unions and Industry to review documents being prepared. The first obstacle encountered was application of the legal framework formed by the Regulations to a strongly operational and technical process. The legal framework gave little discretion on the prescriptive contents of the Safety Case document including the process for its development. Such prescriptive requirements were less important than many other performance-based aspects of the Safety Case which had a more direct and specific impact on the safe operation of the facility. The final assessment framework proved to be more extensive than first intended. At this point, Industry was concerned that as assessment and licensing approached, administration of the Regulations appeared more legalistic.

The first draft of the assessment framework was available 12 months before the deadline for submission of Safety Cases. The framework was then trialled on early submissions to give a final form after a further six months. This timing was too late to contribute to the development of Safety Cases which further compounded the uncertainty of the standard that would apply for the licence.

THE EARLY SAFETY CASE SUBMISSIONS

Ten of the simpler facilities, such as water treatment plants using chlorine, submitted Safety Cases which enabled WorkSafe to test and develop its assessment framework. The assessment framework generated information for the licence decision, possible safety improvements and the inspection plan for the post-licence period of up to five years. The legal framework had emphasised the legal licence tests and reduced the emphasis on safety improvements and post-licence inspection plans. A strong vision in WorkSafe that major incidents would not be prevented through a minimum compliance approach had retained these important elements during the assessment.

Soon after commencing assessment of the Safety Cases, WorkSafe realised that the six month maximum assessment period might not allow an Operator the necessary time to rectify simple non-compliances that could lead to a licence failure. After consultation with stakeholders, the Regulations were amended to allow WorkSafe to extend the assessment period to 12 months for specific MHFs with non-compliances.

SUBMISSIONS AT DEADLINE

In June 2002, two years after introduction of the Regulations, all 42 MHFs had submitted their Safety Cases. From the original 50 MHFs, about 10% had decided to reduce their inventory and deregister as a MHF. An example of this category is a chemical salvage company which decided the recovery of arsenic did not warrant the expense of the Safety Case.

Of the remaining submissions, 50% of MHFs had their assessment period extended to 12 months. This group was reasonably well forecast from the progress monitoring during preparation of the Safety Case. These facilities can be divided into five groups in decreasing order of safety significance, as shown in Table 3.

Overall, WorkSafe was satisfied with progress made against the ambitious timings in the first two years. More development work was undertaken during the assessment period than intended, but this work supported substantial improvement at facilities. Feedback from Operators supported the overall view that the ambitious timings were met whilst achieving the objectives.

GOOD PRACTICE AND PITFALLS IN PREPARATION OF SAFETY CASES

Many of the Safety Case project plans showed that operators intended to use generic qualitative, semi-quantitative, and quantitative hazard and safety assessment methodologies. As work progressed the methodology became more specific for the facility. In many cases the benefit of worker involvement led to analytical or quantitative methods being modified to also show results qualitatively to allow employees to remain involved throughout the process. Managers of Safety Cases also reported the need to conduct both 'top-down' and 'bottom-up' approaches to comprehensively deal with their hazards. WorkSafe found that Operators who solely conducted top-down methods had serious omissions whist a solely bottom-up approach led to a poorly structured Safety Case that was difficult to understand.

EMPLOYEE INVOLVEMENT

Employee involvement tended to be focussed around workshops such as hazard identification, safety assessment, and demonstration of adequacy of control measures. Employee involvement was particularly strong during the hazard identification and safety assessment workshops. The attendees at hazard identification workshops included HSRs and shop floor employees with 'hands-on' operational and engineering knowledge of the area of the facility under consideration. The workshops were often facilitated by a safety consultant or company safety professional.

Table 3. Category of facilities extended to 12 months

Group	Number	Comment
Systems below industry standard	2	Chemical manufacturers
Safety performance below norm	1	Complex facility
Industry sector performance issues	11	Largely warehouse and LPG/Gases
Document issues - comprehensiveness	2	Chemical Manufacturers
Minor methodological problems	2	Chemical Manufacturers

The majority of the hazard identification workshops were structured brainstorming sessions that focussed on a specific area, or activity at the MHF. The hazard identification workshop methodology improved as operators gained greater skills in application of the methodology. For example, hazard studies that started out as a check on engineering codes and standards evolved into a systematic assessment of site-specific operational and engineering activities. This led to the identification and adoption of new and additional procedural controls that would not normally arise from a simple compliance check against codes and standards.

CHALLENGING THE NORMS

Hazard identification often showed very divergent views on the importance of the hazards. The workshops were able to combine the experience of the workers with the analytical approach shown by managers and engineers. The workshops also showed how different people operating under the same procedures had quite different understandings and perceptions of the hazards.

Operators have used the safety assessment to critically review their management of safety and the reliability of their assets. Such a review gave a clearer picture of what safety at the facility really depends on. In the wake of these critical reviews, generic systems have seemed inadequate and the need to start tailoring safety management systems has become necessary. Many facilities have now tailored their systems to meet the specific needs of their facility.

International companies face a particular problem with tailoring. They must maintain consistency across their facilities yet allow individual facilities to tailor their systems. WorkSafe's observation is that rigid systems imposed externally were not highly regarded by the workforce. They tended to reduce the possibility of a fundamental review during safety assessment and overall resulted in less site ownership of safety at the facility. There is a clear need for international systems to achieve both global consistency and allow for flexibility at the local level.

Other safety improvements that resulted from MHF operator's challenging their norms include improvements to emergency plans to deal with specific scenarios, the increasing effectiveness of existing controls and the adoption of new controls. Tracking and monitoring tools also evolved from the Safety Case process, such as hazard registers that capture findings and actions arising from workshops and other hazard studies. Most importantly, many operators have shown the most significant findings from their Safety Case in databases to allow easy linking and manipulation to support their safety.

SAFETY IMPROVEMENT PLANS

Many Operators produced action plans that identified planned improvements as they progressed through the process. The plan often identified the priorities for the next two years and the critical actions to be completed before the licence decision was made. Initially these plans caused some consternation to the Regulator because they had to be checked to ensure that necessary controls were not listed for future implementation and that interim controls met the standard of 'so far as is practicable'. But these action plans provided a clear perspective of how the operator is managing their safety and contributed significantly to the assurance that the maximum term of the licence should be granted.

COMMON PITFALLS

A number of pitfalls were observed when preparing Safety Cases. Some Operators under-resourced the Safety Case, or had difficulty in translating performance-based Regulations into practice, or appeared to be focused on the document rather than the practicalities of their facility's safety. Some Operators became over-reliant on consultants and contractor support or held unrealistic expectations on the effectiveness of control measures or excluded potential major incidents too early in the process. Many of these pitfalls were not apparent until the several months of assessment were conducted and often came to notice when the Safety Case documents were being tested for implementation.

COMMON GOOD PRACTICE

From the Regulator's perspective, good Safety Cases were produced when Operators used a systematic approach and clearly identified their basis for safety. These Operators planned and implemented a robust risk assessment process, involved the right people and underpinned the Safety Case process with a sound safety management system. The resulting Safety Case described the current situation to the desired goals rather than simply the desired goals.

SAFETY CASE ASSESSMENT AND LICENSING

After a Safety Case is submitted, WorkSafe has six months to make a licence decision for the majority of cases. The licence decision is made on the four tests of whether the document is complete, whether the Operator meets its Part 3 duties, whether the Operator meets its training, consultation and informing duties under Part 5, and whether the Operator has the 'ability to operate safely'. All tests must be met to be licensed.

TECHNICAL ASSESSMENT

The assessment and licensing involves a number of steps. The 'desk-top' technical assessment of the Safety Case ascertains whether the Operator is complying with the Regulations and is operating safely. Experience has shown that the desk-top study needs to review documentation not included in the Safety Case, particularly the safety management system. For example, assessors sample site-based manuals and safety management information on intranet and computer based systems.

The desk-top review has also involved considerable "clarification" of information in the Safety Case which is not understood by assessors or appears to be incomplete. It is not uncommon for clarification responses to exceed 10 substantial documents. This clarification assists the Operator to identify areas that cannot be understood by a third party. In general, the amount of time spent by the Regulator on assessment is more a function of the quality of the Safety Case than the complexity of the facility.

After the majority of the desk-top study is complete, the Safety Case is verified by the OH&S inspectors using audit techniques. Verification targets implementation, the safety management systems, possible weaknesses and critical control measures. One of the main findings from verification was that engineering control measures were in place but systems which relied on people were not well implemented. Training had clearly lagged the preparation of the Safety Case.

The desk-top assessment, verification and the regulatory history of the site are combined into a report which is reviewed by the Operator before the licence decision. The findings report aims to present a balanced view on compliance of the Safety Case document and safety compliance at the facility. Positive findings are reported as well as negative findings. Where deficiencies are apparent the report discusses options on how the deficiencies can be best addressed to secure a safety outcome. The preference is to extend the assessment period to rectify the non-compliances, but some licences contain conditions or limited terms. At the conclusion of the assessment of a large facility, up to 800 hours of assessment may have been conducted.

The report includes a whole-of-government approach with the emergency services, gas safety regulator, fire services, electrical regulator and environmental regulator being involved in the assessment and verification of the Safety Case. This involvement has not reduced the requirements of the other Regulators. Industry is advocating that the next step should be to produce only one document for all safety regulation.

SAFETY CASE ASSESSMENT OBSERVATIONS
Some common observations have occurred in many of the Safety Cases. The Safety Case document should not simply represent the final or desired situation at the facility. For instance, transition to new systems should be defined and planned and be apparent in the Safety Case. The verification of the Safety Case will show when transition has not been adequately addressed. The demonstrations have also often challenged the Operator. In some cases the Regulator may effectively make the demonstration for the operator when attempting to understand the Safety Case. The use of clarification and the ability to extend the assessment period have allowed the additional time for the Operator to make the demonstration if necessary.

LICENSING AND THE FUTURE BENIFITS
Victoria is rare in that the government has chosen to licence MHF facilities. The dangers of this approach were that assessment would focus solely on legal requirements including minimisation of legal liability. If such a legalistic approach had been adopted then technical and operational requirements at the facility could be overlooked and improvements in safety and the targeted inspection plans of the Regulator would not be identified. These dangers have been overcome with some unexpected benefits from licensing. The time limits on both Industry and the Regulator have forced resources to be committed and decisions to be made to achieve the outcomes, which avoids the risk of an iterative exchange of documents and correspondence without conclusion. Most importantly, it establishes a cycle where after the document is approved, the focus is clearly on the site and not the document.

A benefit expected from the Regulations is that measurement of improvement in control of risk should be possible because of the requirement for performance monitoring. Major Hazard Regulation is costly and often comes under community scrutiny and should show measurable results. The Regulatory requirement for performance monitoring of control measures should allow the development of measures which will monitor the overall performance of MHFs in Victoria.

CONCLUSIONS

After the incident at Longford, Victoria set out to install a Safety Case regime across 50 facilities in minimum time. Although the first licence cycle is incomplete, both the Regulator and Industry believe that substantial change has been accomplished in the two and half years since the Regulations were introduced. Table 4 outlines the effect on the industry sectors.

From the Regulator's perspective, the change has been possible because of the resources committed by stakeholders, the use of tools in the Regulations such as the Safety Case project plan and the successful use of the legal framework for large scale performance regulations. The community resolve to use sanctions contained in the Regulations and the tight timeframes have undoubtedly contributed to achieving the objectives. Missed opportunities included finding effective interventions for industry sectors that were not systems-orientated and bringing forward the publishing of much of the guidance by twelve months.

Table 4. The regulation's safety effect on industry sectors

Industry sector	Effect of major hazard regulation
All Sectors	Engagement of at least the HSR in the fundamental safety issues at site.
	The use of databases to link and implement the complex information from the Safety Case.
	A common understanding of the hazards at the MHF.
	Engagement of other government agencies to contingency plan for specific emergency scenarios rather than generic.
Chemical Manufacturer	About half of this group were at a high standard and used the opportunity to test the latest corporate safety techniques.
	The second part of the group has reviewed the balance between their procedural and engineering controls.
Chemical User	This group have established and tailored systems so that their safety does not rely on a highly experienced workforce.
Warehouse	The warehouses now manage their safety systematically rather than by exception.
	Warehouses whose operator also manufactures chemicals were at a much higher standard.
LPG and Gas Distribution	This sector has overcome a compliance with prescriptive standards approach.
Petroleum Refining	The refineries have assessed 30 years of modifications and reviewed their asset integrity to reach a comprehensive approach.
Bulk Terminal	Greater focus on emergency and incidents.

Industry did recognise the community resolve to regulate major hazards and has made a fundamental review of their safety. Industry found that some of their safety was of a lower standard and not as integrated as expected. Once started, Industry found the process self-sustaining and many are intending to gain corporate benefit from their development work outside Victoria. A number of Operators have commented that they would recommend this form of Safety Case because of its inclusion of workforce experience, the requirement to fundamentally review their safety, and the focus on implementation.

The Regulator, Industry and Unions must maintain and improve the standards over the next five years. The nature of the Regulatory framework and its relatively small size in Victoria does allow opportunities for innovation. These developments will be posted on WorkSafe's website at www.workcover.vic.gov.au.

REFERENCES

1. Longford Royal Commission 1999, The Esso Longford Gas Plant Accident - Report of the Longford Royal Commission, Government Printer for the State of Victoria, Melbourne. The report can be purchased from the Information Victoria website at www.bookshop.vic.gov.au/infovic/anonymous/home.asp for AUS$77.00.
2. Occupational Health and Safety (Major Hazard Facilities) Regulations 2000: Statutory Rule No. 50/2000 can be viewed at www.dms.dpc.vic.gov.au (Victorian Statute Book/2000 Statutory Rules).
3. National Occupational Health and Safety Commission, 1996, National Standard for the Control of Major Hazard Facilities [NOHSC:1014(1996)], Australian Government Publishing Service, Canberra. The National Standard (and the associated National Code of Practice) can be viewed at www.nohsc.gov.au/OHSInformation/NOHSC Publications/#1.

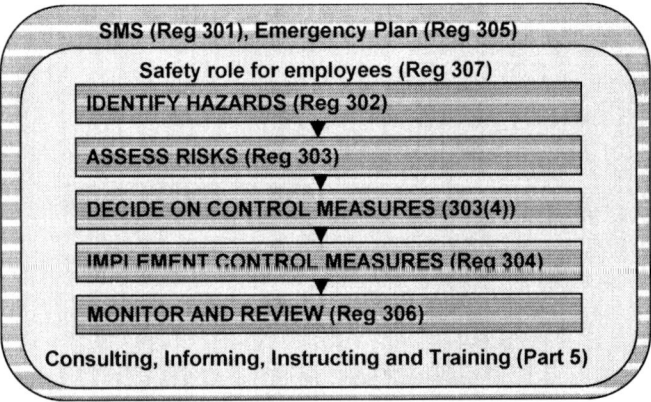

Figure 1.   Regulations in part 3 safety duties of operators and part 5 consulting, informing, instructing and training

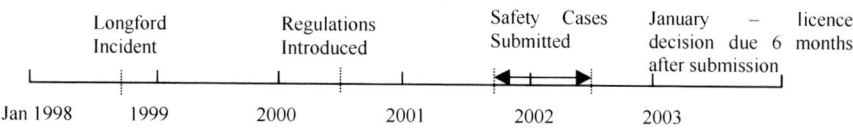

Figure 2. Timeline for implementation of the major hazard facilities regulations

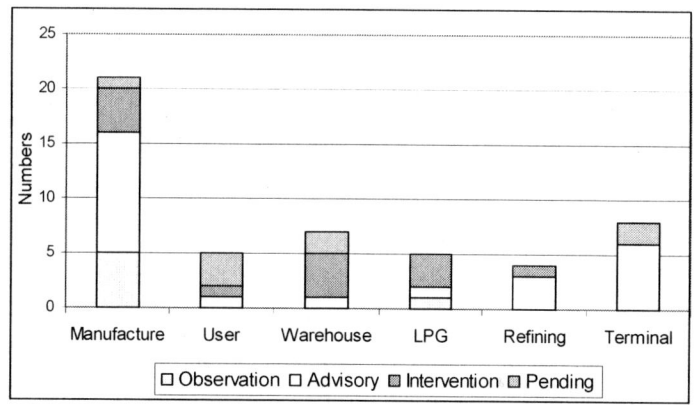

Figure 3. Oversight strategy for MHFs June 2001

MAJOR RISK AVOIDANCE–FULFILLING OUR RESPONSIBILITIES – COSTS AND BENEFITS OF ROMANIA'S SAFETY INTEGRATION INTO THE EUROPEAN UNION

Prof.Dr.ing.Darabont Alexandru, Prof.dr.ing.Diatcu Eugeniu,Ph.D. Kovacs Ştefan, Apostol George Claudiu
INCDPM Bucharest and Hyperion University, Romania

The European Union documents regarding risks (Seveso II directive, other EU safety and health directives etc.) are novel to Romanian specialists. In the view of Romania's integration in the EU, safety issues emerging from the local implementation of these directives raise various important structural and financial issues. This paper focuses on providing a valid methodology that could be engaged for comparative analysis between both different local economic sectors and a broad range of industries within other candidate states for entry into the EU. It does so by presenting the operational analysis of a recent survey on the identification of necessary costs for compliance with the European Safety Directives, that has engaged more than 10000 various economic enterprises. Its findings were examined in relation to collected data on Romanian occupational accidents and diseases. The outcome both advances the development of a cost effective integration model based on the optimisation of expenses and touches upon practical implementation of safety issues encountered locally.

KEYWORDS: Risk, safety, integration into EU, integration study

INTRODUCTION

Romania's integration into the European Union does not simply refer the advantages offered by the European market but also the responsibilities that come with providing a qualitative shift in the living standard. As better life means better work conditions, safety and health assurance in the workplace constitute basic elements in the improvement of working environments. The close observation and implementation of the European Union social directives will be beneficial in both complying with European Union provisions and improving the actual local economy safety level. As such, the evaluation of the necessary costs for implementation is a vital task.

Assessing such costs means:

-To compare the compliance of our safety legislative system to its European counterpart.

-To estimate the real implementation costs of EU directives into Romanian economy, at its basic (enterprise) level;

While the majority of Romania's work safety documents are straightforwardly adapted from or developed in the spirit of EU trends, the cost evaluation of the local implementation of EU directives is more difficult. This paper presents some of the most significant aspects of such an undertaking by devising and testing methodological approaches and fine tuning an implementation cost model.

METHODOLOGICAL ASPECTS OF THE STUDY

The aim was to obtain a realistic mid-term forecast of the EU safety directives implementation action expressed by overall costs at national level. To obtain this forecast we worked with the safety managers of various enterprises who were requested to participate in our survey.

From the beginning we realised the necessity to work with statistically significant samples of Romanian enterprises. Our samples were built around economic activities as described in the NACE code (see Table 1). Their selection was according to:

1. Economic activity domains.
2. Financial strength – they had to have significant resources that allow that the investment they were prepared to make in order to comply with EU provisions would not disturb the financial health of the enterprise.
3. Size of the enterprise.
4. Regional representativeness: we sought to capture a uniform geographical distribution of the sampled enterprises.

The chosen enterprises were diverse in size (ranging from unit samples-industries where one unit covers near one hundred percent of the economic activity and ending with samples having thousand of units, like those in commercial services). Our time frame was fixed between 2002–2006 in order to produce an overview of the implementation process.

A number of extrapolating criteria for the whole Romanian economic activity were also considered:

1. The domain's participation in the national Internal Gross Product (IGP);
2. The domain's participation to yearly production;
3. The contribution made by the sampled enterprise to the domain.

Table 1.   Major sampled activities

Sample activity no.	Main sampled economic activities
1	Agriculture
2	Wood and hunting activities
3	Fishing
4	Extractive industry
5	Processing industry
6	Production, transport and distribution of electric and thermal energy
7	Building and maintenance activities
8	Trade
9	Transport and storage
10	Postal activities
11	Other services
12	Health and social assistance

The intention to produce objective assessments was-somewhat hampered by external factors, such as the managers' belief that the forecasted implementation sum should be provided by the enterprise itself. Other managers speculated that the forecasted sum would be given by the European community in order to do the improvement. These assumptions were accompanied by a more concrete obstacle: an uncertainty at the enterprise level dealing with the difficulty in forecasting the amount of money for safety and health improvement when the work force can not always be paid regular wages.

Following this latter aspect, instead of requesting the approximate costs of implementation we offered the choice of assessment on a 0...10 scale where 0 was the minimum value (no money needed for the implementation of a specific directive) and 10 was the maximal value-1 cost unit = aprox 10 million Euro.

This scale approach has the advantages of:

1. Eliminating psychological barriers for the safety managers.
2. Using standard data collection procedures, processing and interpretation, that take into account the necessity of objective validation;
3. Providing an accurate national overview of the safety levels of various economic activities, and the possibility of comparison between various activities.

Anticipating the gap between the actual safety level of our pilot enterprises and their expressed needs we also used the 0...10 scale to assess the safety level of each enterprise taking into account:

1. The amount of incidents, accidents and professional diseases present;
2. The operating safety policies and
3. More recent safety assessments.

Using this approach we computed a reality coefficient using a formula such as:

$$Rc = \sim Abs\ (Exp\text{-}S) \tag{1}$$

Where:
Rc = reality coefficient;
Exp = forecasted sum for the sampled enterprise on a 0..10 scale;
S = safety state for the sampled enterprise on a 0..10 scale.

FINDINGS

Our findings are significant for the financial safety needs of Romanian economy.

Figure 1 shows the implementation cost repartition on main sampled economic activities.

Samples 4–7 refer to core domains: number 5, for instance represents the processing industry that includes the bulk of industrial activities, starting with machines and equipment and ending with beverages and clothing. Domains 8 (Trade) and 9 (Postal) have had no significant safety investments made so far.

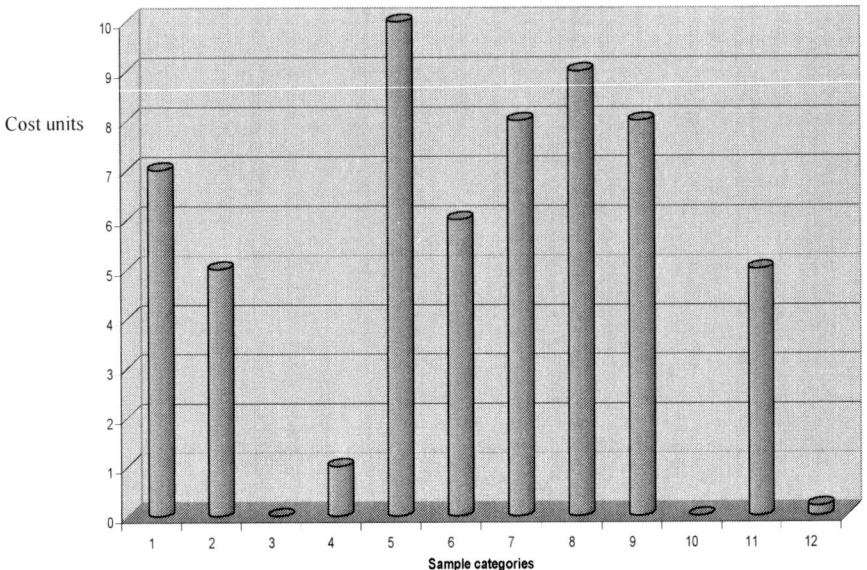

Cost repartition on sampled activities

Figure 1.　Implementation cost repartition on main sampled economic activities

The next graph shows previsioned costs versus safety problems.

Analysing safety problems we took into account the occupational accident statistics between 1996 and 2000, and transposed them on a 0...10 scale in order to make useful comparisons.

A closer look at the comparisons between these categories of analysis reveals that there are:

1. A *"normal"* region where the forecasted costs are comparable to perceived safety problems, showing that the management is aware of existing safety problems and willing to improve the safety. This refers to number 5, 6 and 7 samples.
2. A *singularity* represented by the extractive industry where most occupational accidents occur, explained by:
    -the amount of safety investments made in 1996–2000.
    -the transitional state of our economy and the actual or foreseeable closure of unprofitable mines, pits and other extractive units.
3. *"Abnormal"* regions where safety problems are minor but the forecasted sums are large.

Further analysis distinguishes between:
-economic activities and industries that are able to support the implementation costs by themselves;
-economic activities and industries that are not able to auto finance the implementation costs.

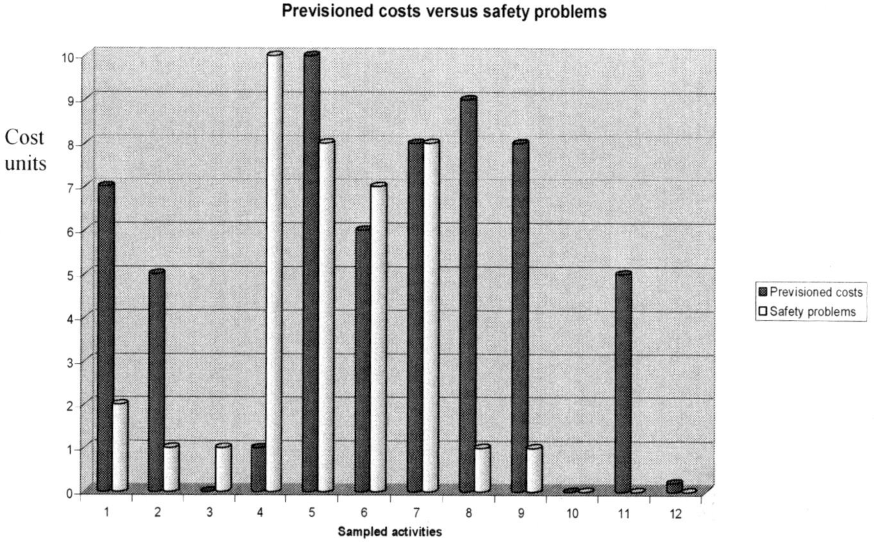

Figure 2. Implementation cost repartition on sampled activities compared with safety problems

Concerning the latter, there are two particular ways to deal with this situation: If the specific activity or industry is significant, it must be assisted in order to implement the EU safety directives by the Romanian government, possibly through EU funding. If it is not, the implementation issue should be postponed.

Figure 3 shows the breakdown of implementation costs on major specific EU safety directives.

It is apparent that the general safety directives (minimal safety conditions at the workplace (89/654), the usage of technical equipment (89/655) and of PPE's (89/656) and the protection against chemical risks at the workplace (98/24/CEE) are the most needed in the Romanian economy. In the above table the need is expressed in cost figures

The breakdown on specific cost categories is presented in Figure 4.

The cost breakdown considering the size of sampled enterprises is presented in Figure 5.

Of interest here is the co-existence of the big enterprises (over 250 employees) with small and medium enterprises (between 50 and 249 employees).

IMPLEMENTATION MODELS

The difficulties in forecasting the resources needed for implementation point to the need to develop viable cost models. Ours has the advantage of dealing with all the above mentioned aspects and with inherent uncertainty[1]. The model is presented in Figure 6.

The model's cost effectiveness relates to progress already done towards the improvement of work conditions and safety in Romania.

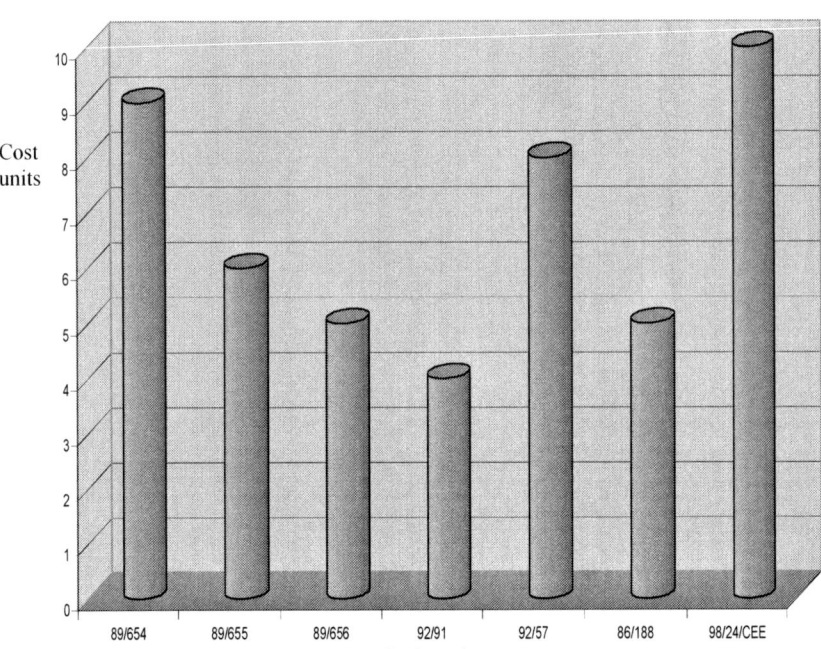

Figure 3.   Implementation cost figures distributed on the significant* EU social directives

In developing this model our focus was on the results of the implementation. We classified the results according to their time span as:

-*immediate:*

    -the compliance with EU documents regarding risks and major risks;

    -a 25% reduction of major hazards;

    -the 75% decrease of the major events covered by Seveso II Directive;

    -a 25...50% improvement of work conditions on workplace;

-*mid-term:*

    -a 50...75% reduction of major hazards for safety and environment;

    -a 75...100% decrease of the major events covered by Seveso II Directive;

    -a 50...75% improvement of work conditions on workplace;

-*long-term:*-could be defined as reaching similar safety levels as other EU countries.

The quantitative and qualitative components of our model are presented below.

---

* From the cost point of view

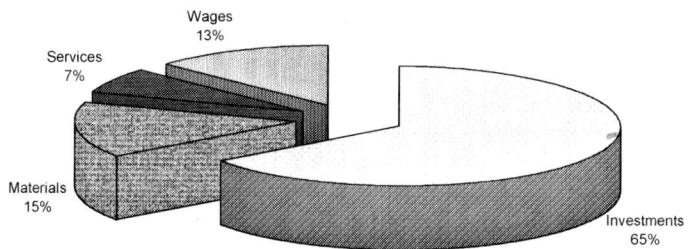

Figure 4. Implementation cost categories

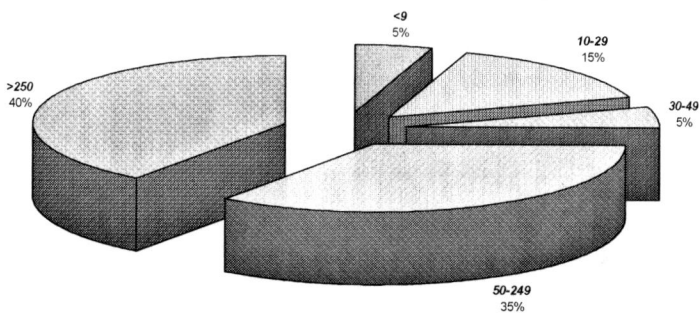

Figure 5. Implementation cost distribution considering the size of sampled enterprises

The quantitative component GEIA/SACOMO was developed based on algorithmic cost modelling, as most recent cost integration models. Here cost is analysed using mathematical formulas linking costs or inputs with metrics to produce an estimated output.

This component is presented in Figure 7.

Algorithmic models generally provide direct estimates on effort or duration.

Our quantitative model is two-levelled. The first level- *General Evaluation of the Implementation Action (GEIA)* is a global evaluation of the implementation effort.

Effort prediction models take the general form:

$$Eff = p*S*e \qquad [2]$$

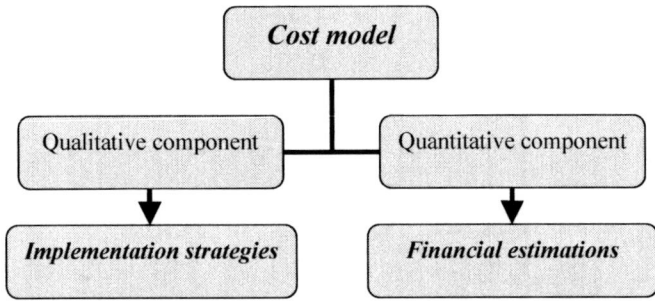

Figure 6.   Components of the cost model

where

-p = productivity constant;
-S = size of implementation;
-e = economy coefficient of scale.

However, if this evaluation is not satisfactory, the second level *Safety Constructive Cost Model (SACOMO)* detailed at the enterprise/activity/industry level, can be observed.

In SACOMO we considered three stages:

-*basic implementation*-at the enterprise level;
-*intermediate implementation*-at the activity level;
-*detailed implementation*-at the industrial domain level, idea taken also as the basis of our qualitative component which is presented next.

The two basic equations of the SACOMO model are presented below:

$$MM = a*KDSI*b \tag{3}$$

$$TDEV = c*MM*d \tag{4}$$

where

MM^2 is the effort (usually expressed in man-months); in this case is possible to express the effort in enterprise-months implying the duration of the full implementation at a specific enterprise level;

$KDSI^3$ is the "engine" of implementation; on our three levelled scale, KDSI is different for small, medium and large enterprises; KDSI also depends significantly on resources available (e.g. internal resources, e.g. personnel and experience and external resources, such as money and knowledge of restrictions imposed).

$KDSI = a_k*f(Ri,Re,Rs)$ [5] where Ri are internal resources; Re are external resources and Rs are restrictions; KDSI could be expressed more exactly[4] by the following formula $KDSI = ak*(qi*Ri+qe*Re-qs*Rs*ets)$ [6]

where ak is the coefficient for the enterprise size,

$q_{i,e,s}$ are the weights for resources and restrictions, t is the implementation time and s is a probabilistic constant;

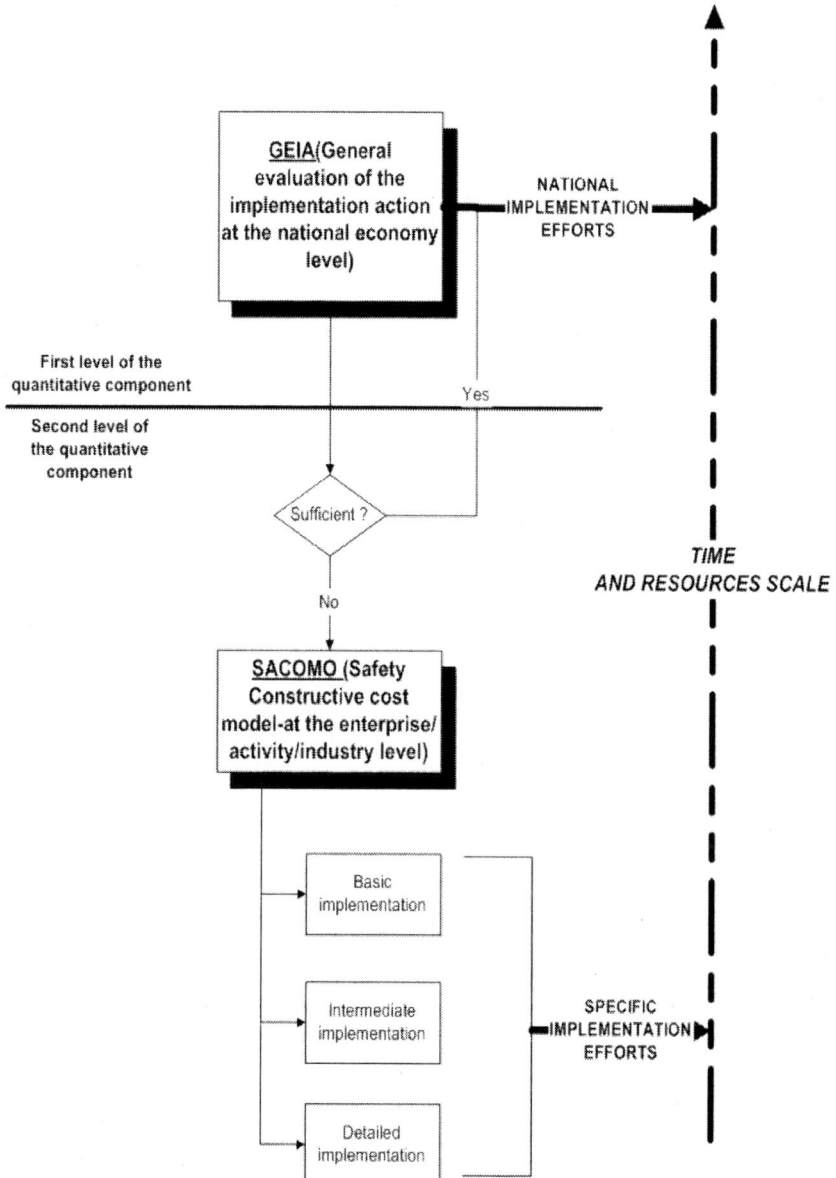

Figure 7. The quantitative component (GEIA/SACOMO)

TDEV is the development time; coefficients a, b, c and d are dependent upon the "mode" of development which could be:

-organic[5], for implementation in small enterprises with no significant safety problems;
-semi-detached, for implementation in medium enterprises, with normal safety problems;
-embedded, for implementation in large enterprises, with significant safety problems.

Table 2. Model coefficients

Development mode	A	B	C	d
Organic	3.2	1.05	2.5	0.38
Semi-detached	3.0	1.12	2.5	0.35
Embedded	2.8	1.2	2.5	0.32

The formulas used in the model arise from the analysis of statistical data[6]-the accuracy of the model could be improved later by its calibration to specific development environments.
The real model has 33 influence factors that are driving its precision towards 85–90%[7].
Some preliminary results of the model are presented in Table 3.

Table 3. Preliminary results of the model application

Result parameter	At the modelled enterprise level	At the modelled activity level	At the modelled economy level
MBIC (Model based implementation costs)	~10.000 CU[†]	~ 2.1 million CU	~ 3 billion CU
MBITF (Model based implementation time frame)	6–12 months	15–24 months	24–48 months
MBB(Model based benefits-% incident reduction)	~ 45–70%	~ 65%	~ 62.3%

It was interesting to see that by increasing the implementation expenses in our model, the time frame needed for the full implementation was reduced to a specific threshold after which the increase expense had no effect on the time frame.

The qualitative component of our model derives implementation strategies from the bi-polar integration[8] process, taking into account a cost-effective approach. HSE-AIM (Health, Safety and Environment Assurance Implementation Model) describes the steps that must be taken in order to maximise the efficiency of implementation process, isolating the non-viable enterprises and promoting the profitable ones. It is a qualitative part of the model because develops strategies and not figures.

[†]The cost unit takes into account the report between the financial strength of the national currency and the economic trends for the forecast period.

The qualitative component[9] considers implementation as a bottom-up process[10]. Therefore, the implementation costs are minimal, maximising the efficiency.

It is also a Darwinist model[11]. The market forces combined with safety barriers take out the enterprises that are not fit for implementation[12].

The idea behind this component is straightforward. On the first step of the model, if an enterprise qualifies for the implementation of EU safety directives (i.e. is a profitable one) it could get, and eventually return resources in the form of funds and specific knowledge.

If the enterprise does not qualify for the implementation, a waiting policy for the enterprise is recommended with two possibilities:

-If the enterprise becomes bankrupt – then no implementation is needed; or
-If the enterprise progress towards a profitable state - then it qualifies for the implementation. The waiting policy must isolate the enterprise through "a sanitary frontier", so that the enterprise should not develop a major occupational or environmental incident/accident in the waiting period.

At the second level of the implementation model[13], industries or specific activities must be analysed as a whole. Thus, if the activity or industry is profitable or has a profitable horizon for further national development it must get resources for implementation, resources that are expressed by funds (for example, for safety compliance assessment) and expertise. At this level, specific knowledge must be replaced by specific expertise from the EU developed countries so that all the regulatory mechanisms for the specific activity or industry at the national level should be fine-tuned to the EU pace.

At the third level, if the economy is healthy (from more than the safety point of view) a reaction occurs towards the basic level, implying a top-down safety determination derived from the bottom-up safety implementation. This component is a reactive one, the enterprise having a direct reactive link with the economy as a whole.

CONCLUSIONS

The safety integration costs of Romania into the European Union could be seen as moderate. In making this claim the economic level, the actual safety state and also the progress being made in the last years regarding safety improvements have been considered. Our past experience and our integration model show that while the profit making enterprises could support the costs, the ones that are not profitable must be closed[14]. Some success stories of foreign owned enterprises on Romanian soil[15] showed that if the investors are interested in obtaining quality products, the human safety and health is implicit. Partly, these costs must be supported by the state, eventually through foreign support. This financial support must be safety and profit oriented.

What are the benefits of the safety integration of Romania into European Union? First, the improvement of work conditions – with important social implications like the rise of productivity and the slow down of immigration. Romania could become a safe and environment friendly country in no more than a decade. More important, perhaps, is that, with appropriate aid, Romania could become a developer and exporter of safety, the experience we provide being representative for many developing countries.

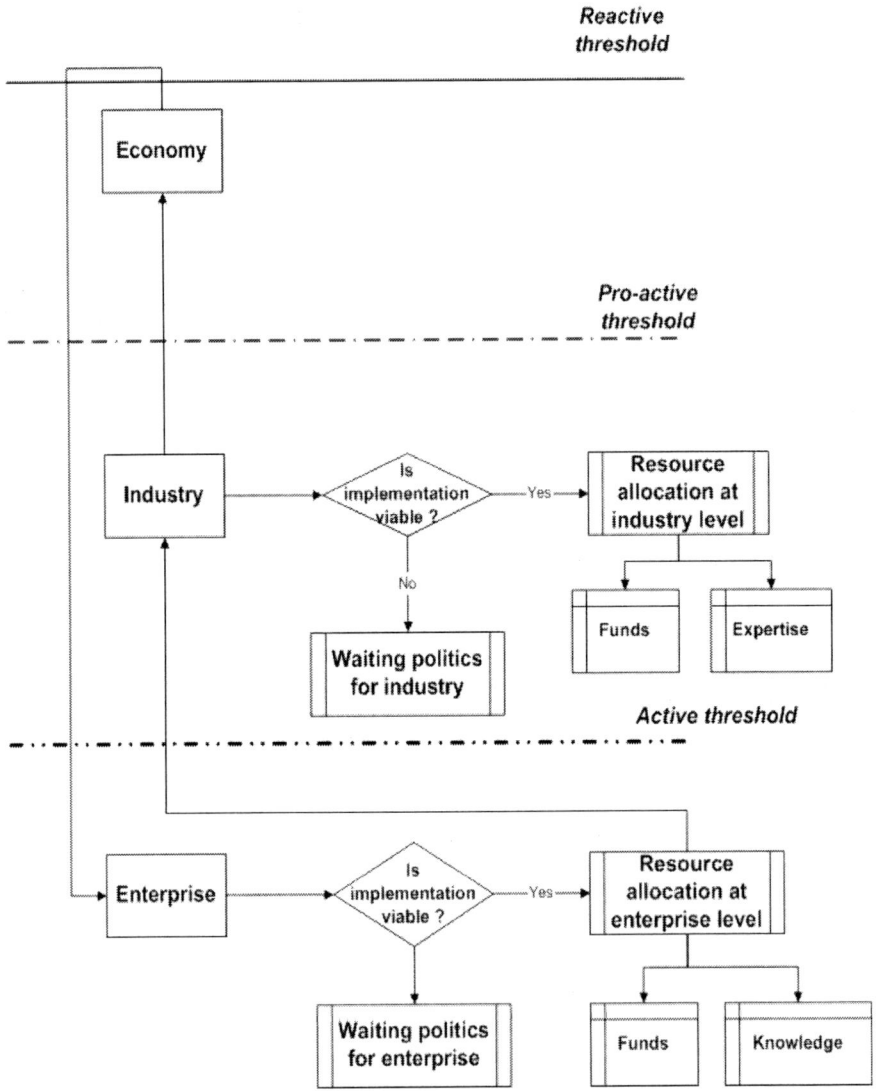

Figure 8. HSE-AIM qualitative component

In our opinion, now, our main safety necessities should be not so much oriented towards money but towards obtaining sound safety expertise. In this respect, we launched the idea of development of a European safety knowledge sphere[16].

This knowledge structure will offer, at reasonable costs, through formal and informal means, the necessary information to safety specialists. This transfer will be beneficial for all the involved parties in assuring:

- -the preservation and further usage of safety individual and corporate memories across Europe;
- -knowledge dissemination to all the interested parties;
- -joint expertise development;
- -an informal immediate warning system regarding incidents that could develop into major occupational accidents or environmental catastrophes;
- -collaboration and team working to solve safety problems across Europe;
- -access to the most modern and efficient assessment and prevention techniques;

The safety knowledge sphere could be developed mainly using Internet facilities. Some of the expertise exists already in the form of various networks like SafetyNet, HarshNet, and Prism.

The next step will be the development of formal and informal local knowledge nodes (big, medium and small enterprises that have safety experience to share, universities, research institutions, safety service firms, non-governmental organisations and trade unions, etc.) that will act as developers and disseminators. This will allow the development of such a network as an informal, non-centralised structure.

REFERENCES

1. Klir G.J, Wierman M.J., 1998, Uncertainity based information, Heidelberg, Physica Verlag
2. Ash P., 1998, The reliability of job evaluation rankings, in Journal of Applied Psychology, nr.32, pg.313–320.
3. Klir G.J., Folger T.A., 1998, Fuzzy sets, uncertainty and information, Engelwood Cliffs, Prentice Hall, N.J.
4. Patnouck R, Sallasar N., Algorithmic integration models, in Algorithmic Models Handbook, Neuwen-Tolland, Liege ed., 2001.
5. Zhang Z. and others, 1999, Perspective based usability inspection-an empirical validation of efficacy, in ESE International Journal, vol.4, no. 1.
6. X.Cunz, A quantitative approach in risk and safety modelling, in University of Maryland Safety Journal, no.12/1, pg.10–14.
7. Barry Boehm, 1981, Software Engineering Economics, Prentice Hall, N.Y.
8. Doolan E.P., 1992, Experiences with Fagan's inspection model, in Practice and Experience, vol.22, no.2, pg. 173–182.
9. Briand C., Basili V., Hetrmanski C., 1993, Developing interpretable models for identifying high risk components, in IEEE Transactions on Software Engineering, vol SE 9, #6, pg.652–663.
10. Kwarecki K, 2002, Schichtarbeit und Gesundheit, in Proceedings of the XVI th World Congress on Safety and Health at Work, Wien.
11. Barry Boehm, 1996, Ada and Beyond-study chair, National Academy Press.
12. Jorgensen K, 2002, Risikoanalyse, Ergebnisse und umsetzung aus der sicht der behorden, in National Working Environment Authority Journal, no.22, pg.24–29.

13. Taghi M., Khosgoftaar M. and others, 1999, A comparative study of ordering and classification of fault-prine modules, in *EMPIRICAL Software Engineering Journal*, vol.4, no.2, pg. 159–186.

14. Darabont Al., Kovacs Şt., 2002, Risk approach-a fuzzy approach, *in Proceedings of the XVI th World Congress on Safety and Health at Work*, Wien.

15. Safety implementation into a car part enterprise, 2000, case study in the Romanian car part enterprise at Satu Mare, BMW Internal Document, pg. 15–45.

16. Proposal no. 112 made by INCDPM to the Expression of Interest inside the FP 6 Programme.

EXPLORING THE ROLE AND CONTENT OF THE SAFETY CASE

Helen Conlin[1], Philip G. Brabazon[2], King Lee[1]
[1]Parsons Brinckerhoff Ltd.
[2]University of Nottingham

It is a requirement to provide a safety case for all of the Major Accident Hazard (MAH) industries. The prime purpose of the safety case is for the Dutyholder to demonstrate to the Regulator there are effective means for ensuring safe operation in accordance with a goal setting safety regulation regime. This paper will take forward ideas presented in Expert Advice to the Ladbroke Grove Rail Inquiry Part 2[6] in regard to the principles of a safety regulation regime and the role and content of the safety case.

Currently, with local variations within different major hazard sectors, it is common for a Safety Case (SC) to describe an organisation's technical systems and processes and its Safety Management System (SMS). These sections are accompanied by a risk assessment which attempts to demonstrate that through these measures and perhaps further identified control measures, risk has been reduced to As Low As Reasonably Practical (ALARP). This paper criticises the current norm, accusing it of producing unbalanced documents that fail to present a complete and strong argument for why the organisation's arrangement lead to continuing safe operation.

The completeness and strength of an argument can be rated by comparison to theories and models of argument construction e.g. Toulmin et al 1979[39]. The strength of an argument is limited by the quality of knowledge and science used as the basis. In the field of safety the quality of science varies depending on the context, for example more confidence can be placed in the understanding of the science of materials and structures than in the understanding of organisational behaviour. Consequently, in general, the conditions under which hardware components and structures fail can be forecast with greater certainty than the performance of a team of people tackling a complex problem. The benefit of using formal argument structures is that constituent parts of an argument are made explicit, increasing the visibility of incomplete and weak arguments. Validation of the Safety Case would entail appraisal of the constituents of the argument.

This paper explores the application of an argument model, commenting on the viability and benefit of its adoption. This paper also considers the degree to which the proposed approach is implemented in other fields such as insurance, aerospace and defence.

Safety Case (SC), argument, Toulmin, Safety Management System (SMS), Regulator

THE BENEFITS OF COMPLETE AND STRONG ARGUMENT IN SAFETY CASES

This paper proposes that the SC should set out an argument for why an activity is as safe as it needs to be. That those involved in the operation of a MAH facility should make a case for how their activities lead to continuing safe operation of that facility and form a plan for strengthening the argument. This is consistent with the requirements of a goal setting safety regime.

The term 'argument' is being applied in place of 'demonstration' which is used in the regulations defining the Safety Case/Safety Report regimes[12,13,16,18,20,19,20,21,22,23,24]. This is due to there being no dictionary definition of demonstration which reflects the meaning implied in such safety regulations. The regulations imply the need for a reasoned explanation rather than a scientific proof of safety. Scientific proof is the available definition of demonstration, however this is not achievable within the current understanding of safety science.

A concern is that the present SC process is drifting from the intention of the SC regime. Benefit has been gained from SC's, but for further benefits the SC process needs a prod. A SC regime puts the onus on the Dutyholder to find ways of operating safely. When brought in to a sector the SC regime stimulates considerable thought. A criticism is that the reflection can stagnate, and a reason for this is that the regulations are not pushing Dutyholders to produce SC's with complete or strong arguments. The approach of demanding Dutyholders to put forward a convincing argument for why they are safe and to seek to continually strengthen the argument is a remedy.

FORMS OF ARGUMENTS

For centuries, philosophers have believed that arguments can either be explained by absolute means or by relative means. These positions briefly are:

*   Positivism and the belief that there is only one absolute logic and one form of approach to rational understanding (i.e. truth) called absolutism.
*   The counter-position known as relativism with many logics.

Toulmin's model for argument[39] sits outside both these positions. Using either of these methods according to Toulmin[38] is irrational to the modern argument. First, Toulmin claims that by using a relative method, no standards for the claims are made because the analysis of the argument is only relative to that particular argument. Additionally, Toulmin believes that absolutism or foundationalism is irrelevant in the modern era. He claims absolutism is irrelevant for several reasons. First is the fact that this absolute logic is based in mathematics and geometry. Therefore the concepts which are contained in them are field dependent. Hence, Toulmin argues that there is no room for these viewpoints in other areas of logic. Secondly, Toulmin believes that due to there being a definite grey area in some arguments, they do not allow for the absolutism position that answers are either correct or incorrect. The overall problem that Toulmin expresses about absolutism is that its rules are so strict that it just does not apply to modern reasoning.

Certainly the field of safety science is unsuited to the use of absolute logic and the analyses of SC arguments must be capable of comparison within a regulatory framework.

THEORIES OF ARGUMENT

Argument theories and models can provide assistance for constructing a SC argument for continued safe operation and for validation of such an argument.

An argument involves putting forward reasons to influence someone's belief that what you are proposing is the case[11]. An argument has at least two components, a point and a reason(9):

*   making a point (or statement) by
*   providing sufficient reason (or evidence) for the point to be accepted by others.

These elements are related and the movement can go either way to form the argument:
- a movement from either a point to a reason

    or
- from evidence to conclusion (the point)

The movement from one to the other can be supported by other components called inferential devices. These are rules or principles which permit the making of a claim on the basis of some evidence (or warrants). These components are explored further below. For the purposes of constructing a SC argument the movement would be from a point (activities lead to continuing safe operation) to the reasons for this with support from evidence.

Some theories/models of argumentation and argument analysis are introduced below.

CORAX

Corax[2] outlined the concept of argument in two areas. First, he believed argumentation is practical. Second, he contended that probability is a factor that affects this concept. Corax detailed what he called 'practical disputation' or argument. Additionally, Corax incorporated probability into his definition of argumentation. Unlike hard sciences that advocate certainty, he asserted that argumentation was a process of debate wherein exploration is encouraged as opposed to certainty. In fact, the term 'reasonable doubt' used in our judicial system stems from this Greek idea. Hikins[10] writes, 'Corax, as far as we know, was the first to notice the importance of probability in public argument and discuss it in a book on rhetoric. From this point on, the concept of probability persists as a central term in rhetorical history to the present day.'

TOULMIN

Various models of argument have been developed, the leading one being the Toulmin et al (1979) model[39]. According to this model, an argument has several constituents - claim, grounds, warrants, backing, qualifier and rebuttals. The model may be used for constructing an argument and for analysing arguments. The argument components are defined as follows:

- claim: 'the assertion put forward publicly for general acceptance' – proposition;
- grounds: 'the specific facts relied on to support a given claim' – data or facts;
- warrants: 'statements indicating how the facts are connected to the claim' – explanation of how the data supports the proposition;
- backing: 'generalisations making explicit the body of experience relied on to establish the trustworthiness of the ways of arguing in any particular case' – credentials or general information in support of the explanation;
- qualifier: 'phrases that show what kind and degree of reliance is to be placed on the conclusions, given the arguments available to support them' – the strength of the claim; and,
- rebuttals: 'the extraordinary or exceptional circumstances that might undermine the force of the supporting arguments' – under what conditions may the claim not be true, counter examples.

Toulmin et al. say the first four elements need to be present for an argument to be *sound*. The last two are required for an argument to be *strong*.

The Toulmin structure of argument is illustrated in Figure 1. A challenge can be made to any or all elements. Is the claim justified? Are the evidence, warrant and backing justified? Additionally, we can ask whether the claim stands up to major challenges? Is it sufficiently robust? The validation of a SC argument would ask such questions, a further method of analysing argumentation is introduced in the next section.

FISHERS METHOD OF CRITICAL READING

Fisher (1993)[7] provides a systematic technique for reading analytically which allows evaluation of any argument to be done by analysis of its formal structure. Words that are used to structure an argument are the focus for the analysis. Words such as *thus* and *therefore* are highlighted because they are used to link evidence with claims and suggest inference, reasons and conclusions. Fisher's approach provides a systematic set of procedures for the analysis and subsequent evaluation of an argument, for example, one procedure seeks to extract the conclusions and reasons of an argument.

Fisher's method is based on what he calls the assertability question, it questions both the premises and the conclusions of an argument. The main assertability question is: what argument (what *you* need to believe) or evidence (what *you* would need to know) would justify the acceptance of the conclusion? Note, this question is not attempting to establish truth, it is about establishing justified reasons for accepting an assertion[9]. This analysis method may lead to the problems associated with a relative analysis of argumentation in that due to the analysis only being relative to a particular argument there is a lack of standards being applied to the claims.

REVIEW OF CURRENT SITUATION

The sequence of implementation of MAH industry regulations is summarised below:

- Nuclear, 1971 [12,13];
- CIMAH, 1984 [14];
- Offshore, 1992 [16,20,22];
- Railway, 1994 [18,24];
- Pipelines, 1996 [21];
- COMAH, 1999 [23].

Throughout this time the Regulator's approach to the structure and content of the SC has evolved. The Nuclear regulatory regime[12,13,15,17] is based upon the Regulator granting a license to operate, part of this exercise requires production of a SC by the Dutyholder. The nuclear site license has conditions attached which define areas of nuclear safety which the licensee should pay attention to ensure safe operation of the site. Some conditions impose specific duties whilst others require the licensee to devise and implement adequate arrangements in particular areas. Schedule 14[17] on Safety Documentation states that: *'the licensee shall make and implement adequate arrangements for the production and assessment of safety cases consisting of documentation to justify safety during the design, construction, manufacture, commissioning, operation and decommissioning phases of the installation'*. Note the use of 'justify' rather than 'demonstration'. Supporting the Nuclear

Site Licence conditions[17] are the Safety Assessment Principles for Nuclear Plants[15]. These principles define how the Dutyholder will be assessed for safety, they do not specify content, rather they provide a benchmark for safety both in general terms through the fundamental principles and in specific areas such as Equipment Qualification and Reliability. The approach used within Nuclear SC's is similar to other MAH SC's, it is the rigour and robustness of the 'arguments' used to justify safety which differs.

The use of the word demonstration is most apparent within the COMAH regulations[23] and guidance[25,26] and is explored further below.

DEMONSTRATION WITHIN COMAH

The COMAH regulations[23] require that a number of demonstrations relating to SMS are provided, major accident scenarios are identified and the necessary measures have been implemented, the measures have adequate safety and reliability and an onsite emergency plan has been developed.

COMAH safety report guidance defines demonstration as 'to show, justify or make the case/argument through the information given'[26]. In particular this requires:

- A sufficiently rigorous and systematic process;
- A link between the measures taken and the major accident scenario;
- Provision of prima-facie evidence that the necessary measures have been taken.

Part of the problem limiting the successful achievement of the level of demonstration required by COMAH may be that the document is described as a 'safety report' and not a 'safety case'.

CURRENT PROBLEMS

Industry has difficulties in understanding and meeting the requirements of demonstration. This was highlighted as a problem in the early days of COMAH[4]. However, it has continued to be a reported problem[5].

The HSE have suggested a certain structure for the safety report and its content[25,26]. This structure may not be most appropriate to present a logical safety argument.

Current COMAH safety report argument is often restricted to one section and can be incoherent and disjointed due to excessive cross referencing to descriptive sections of the document plus there is limited analysis e.g. whether the SMS is a good one and why it was chosen compared to other available models. Does this arise due to the prescriptive nature of the guidance for safety report content? Is it a problem due to inadequacies of the previous CIMAH regime and attempts by Dutyholders to 'bolt-on' sections for a new regime such as COMAH?

The main element of the report which contains safety arguments is the risk assessment section which tries to identify all major accident scenarios and assess them. Most other sections of safety reports are descriptive. However, often the link between risk assessment and control measures is weak. Information on safety control measures tends to be descriptive rather than showing that they are fit for purpose. Additionally the risk assessment generally starts at a low level and does not question why, for example, a particular process route has been selected and discuss alternatives that were considered and rejected e.g. alkylation process within oil refining and alternatives to the use of hydrofluoric acid.

Appreciating that risk assessments take many forms, it is considered that applying the forms of qualitative or quantitative risk assessments typically used for analysing engineering systems for analysing an organisation would be very demanding and prone to error. Simplistically, it is routine in a risk assessment of an engineering system to decompose the system into its constituent components and to determine how the states of the components can combine to cause the system to fail. Applying a similar approach to analysis of an organisation has not proven successful. The use of benchmarking through comparison with one or more 'best practice' models is an approach commonly used for assessing organisations. To make a strong argument for the use of a best practice model it would be expected that the Dutyholder would argue for the validity of the model and obtain favourable ratings from an unbiased assessment. The use of 'leading indicators' of safety performance is another approach related to benchmarking. As catastrophes are rare, not suffering a catastrophe is not proof that safety controls are sufficient and fully effective. The idea is that 'leading' indicators allow the 'safeness' of an organisation to be assessed. These types of approaches to demonstrating safety do not fall easily within the risk assessment based model to demonstrating safety.

Most safety reports tend to describe the SMS and do not demonstrate that the SMS is designed to manage the hazards. Additionally, most safety reports are prepared by safety professionals who focus on technical issues. There is generally little content on other organisational factors.

Many safety reports strive to present a favourable picture of a site's operations. However, safety reports would be more realistic and credible if they demonstrated the adequacy of the safety control measures through defining the assumptions, limitations and potential and planned improvements.

OTHER SECTORS

Other MAH industries have slightly different approaches to SC development. Some of the differences in regulations and practices are briefly described below.

The Offshore Safety Case regime[16,20,22] uses the concept of performance standards and a verification scheme. This requires the safety requirements of a safety critical system to be defined (normally by safety assessment) and a verification scheme to be set up to ensure the safety system is suitably managed throughout its life cycle.

The railway and defence industries tend to adopt the system engineering approach to safety assessment and safety case development. In particular the specification of system requirements and apportionment of requirements allows the precise safety requirements and validation criteria to be defined. By taking a life cycle approach the safety plan describes the management arrangement responsible for delivering the safety for each phase of the life cycle[1,34,32]. The life cycle approach adopted within the rail industry stems from a British Standard[1] and industry guidance[34] not HSE published regulations[18,24] and guidance[19].

However, all the SC regulations tend to define a similar structure and content of a SC. Further, SC regulations only require limited demonstration to defined aspects not to the entire SC. Generally, the regulations only require demonstration of technical safety measures taken using risk assessment; whilst other SC sections need only present a description. For example, the Railways Safety Case assessment criteria[27] states that different terms (such as description of, particular of, particular to demonstrate) are used in the regulations to indicate the level of detail to be provided in the SC and to some extent defines the type and robustness of arguments required.

ARGUMENT WITHIN THE SAFETY CASE REGIME

A role of SC regulations should be to stimulate deep questioning by a Dutyholder of their beliefs about safety and their approach to safety, and encourage Dutyholders to look beyond their immediate horizons for safety knowledge that prompts re-evaluation and improvement. Regulations can do this by requiring Dutyholders to present a critical and comprehensive argument to justify their safety controls. They should impose the need to consider what is critical to safety and defend the arrangements; e.g. does the Dutyholder judge their organisational design to be critical and if they do then the Dutyholder should argue for why their organisation is adequate rather than just describe it. However, the regulations should not prescribe a list of the sub-arguments to be presented. It is for the Dutyholder to decide the completeness of the argument. This approach is consistent with the precautionary principle, which shifts the burden of proof in demonstrating presence of risk or degree of safety towards the hazard creator. The presumption should be that the hazard creator should provide, as a minimum, the information needed for decision-making[28].

Two tasks for the regulator are the evaluation of the validity of the argument and the veracity of its components. Validation is an intellectual task which evaluates whether the grounds, warrants, backing and rebuttals are complete and of sufficient strength to make the claim of safe operation. Verification involves checking data and confirming that what is said to occur is actually done (effectively). Additionally the regulator can take note of how the Dutyholder has gone about constructing the argument, such as how the Dutyholder has prevented vested interests from biasing the argument.

In setting out how SC's are to be evaluated, a question is whether the regulations should specify not only the safety criteria a Dutyholder is aiming for (e.g. ALARP) but whether they should set out also the standard of proof that the SC is to be judged on. Options for standards are: beyond all possible doubt, beyond a reasonable doubt, on a balance of probabilities.

Such an approach can assist in deciding when the precautionary principle should be invoked as it will bring to the surface aspects of the argument for which there is insufficient science to build confidence[28].

CONTENT OF AN ARGUMENT BASED SAFETY CASE

An argument based SC should contain a critical appraisal and justification of the Dutyholder's governance of risk. Should such a SC include a description of the SMS? Is a hazard log and a quantitative risk assessment necessary? The answer is they are if they make the argument sound and/or strong but not if they do not. It is unlikely that they are sufficient. The inclusion of a SMS description, for example, begs the question – why this SMS? The type of SC that raises concerns is one in which a SMS is described that is an adaptation of an international standard or uses a template from a respected third party organisation, and the implied argument is that this

SMS must be good because of its origin and association. The reader of a SC is less impressed by the description of the SMS than by a line of reasoning that convinces of the suitability of the SMS for the purpose of achieving ALARP (or other criteria adopted by the Dutyholder). A reader is less convinced by a claim from the Dutyholder 'we have a procedure for every hazard' than by a reasoned explanation as to why the SMS is appropriate to the Dutyholder's activities and organisation; that the principles underpinning the SMS have benefited, and will continue to benefit, from sound safety science; that the Dutyholder can assure that the principles are being put into practice in a balanced and effective manner. The implication is that the SMS is an output from an SMS governance process and the SC is a critique of this process as well as of the SMS itself.

If a SC were to be structured according to Toulmin's model it could open with a statement encompassing a *claim*, its *grounds* and *qualifier(s)*, an example being "this Dutyholder configures and manages its operation so that, using the test of risks to be 'as low as is reasonably practicable', its operations are safe to the public, employees and the environment" (Figure 2).

The SC would then contain the warrants, backing and rebuttals to underpin the argument. In regard to the rebuttal, one form of query the SC should provide defence against is 'wouldn't the XYZ model of SMS be better for this Dutyholder than the model they are using?' There are contrasting views as to how to manage safety in organisation (e.g. the conflicting viewpoint of the protagonists of high reliability organisations as opposed to Normal accident theorists[36]) and given the developing state of some areas of safety science, particularly in regard to soft issues, disputes will continue for the foreseeable future. However, one could read many, many safety cases and remain ignorant of these debates and the issues of contention. A demonstration of awareness and understanding of the limits of current safety science and of new and emerging theories and knowledge will assist greatly in bringing to light the strengths and weaknesses of arguments in a SC.

Confidence in an argument is influenced by how the argument was developed. Confidence may be undermined if the authors of the SC were at a distance from the operations, or if the SC were produced in a rush. Therefore the SC should describe how the argument has been developed, over what timescale and who has been involved.

The SC should include a plan from the Dutyholder for strengthening their argument, as well as a plan for reducing risks as is currently required.

IMPLICATIONS OF USE OF ARGUMENT

FOR REGULATORS
Proposing the use of argument for SC's appears consistent with the Regulations[12,13,16,18,20,19,20,21,22,23,24] and HSE's approach[15,17,19,25,26,27], substituting argument for demonstration provides a process for achieving the desired SC regime output.

The use of validation and verification techniques for evaluating the strength and soundness of arguments has been explored above. Use of an argument structure for the SC should aid SC assessment (and in some regimes acceptance), the validation phase; and inspection, the verification phase. The validation phase demands expertise and extensive knowledge of safety in the particular relevant industry. The verification phase involves inspection, requiring access to the organisation and a number of its staff. Verification requires an open relationship and trust between the evaluators and the parties under examination. The skills and knowledge requirements for validation and verification of a SC based on argument do not appear to be different from those currently required from the Regulator for SC assessment and site inspection activities.

Pushing safety science to the fore within Dutyholder arguments would require wider appreciation of such science within the Regulatory bodies. The counter argument to this would be that safety science is not sufficient. However bringing safety science to the fore is helpful to highlight the need for more research through questioning existing knowledge and understanding and therefore stimulating improvement.

Adoption of an argument orientated approach to SC's is likely to require the production of process orientated guidance. This guidance would not wish to provide detailed example arguments which could lead to formulaic argument based SC's and would not generate the benefits available from an argument based SC approach. Dutyholders would be likely to require guidance on how to formulate an argument and how to analyse an argument for strength and soundness. However the detailed contents and development of a Dutyholder's argument for continued safe operation would have to be unique to that MAH facility.

FOR DUTYHOLDERS
The relationship the SC has with other parts of a Dutyholder's SMS is likely to change as rather than sitting within the SMS, the argument based SC would exist at a much higher level. The SMS would form part of the argument for continued safe operation, in particular why a particular SMS model has been selected and others rejected. The detailed procedures and guidance within the SMS may be examined during the verification (inspection) phase of a SC regime but would not be contained within the SC argument. The reduction in detail which is included within the SC means there would be less requirement to update. Although detailed procedures and guidance at lower levels of the management systems will change, it is unlikely that the underlying principles will unless there are significant advances in safety science or a major change in the Dutyholder's organisational philosophy due to e.g. a change in ownership.

Due to the well appreciated fact that organisational factors have a significant influence on safety; in order to form an argument for how safe an activity is, such as operating a chemical plant, it is necessary to justify the configuration of the organisation as well as hardware. There are several routes available for doing this. This requires an approach which actively benchmarks organisational structure, policies, processes, working practices and performance against 'best practice' models. Therefore in order to form strong arguments Dutyholders are likely to need to introduce increased use of such benchmarking tools. Similarly, the use of 'leading indicators' of safety performance is likely to need to increase amongst Dutyholders so that they may base their argument on why they are safe on the selection of leading indicators used and the ratings being achieved against them. There would need to be a discussion within the Dutyholding organisation around the uncertainty as to how quickly safety management controls decay and therefore the frequency of measurement of leading indicators. These additional requirements are unlikely to be identified through the existing SC structure with the emphasis on technical risk assessment and limited organisational factor content.

For Dutyholders to increase use of benchmarking and leading indicators, many will need to increase their understanding of organisational influencing factors and improve the skill balance between technical risk and other factors. The skills required to construct arguments will need to be resourced either by developing them within the Dutyholding organisation or by outsourcing. The use of argumentation is common within the social sciences but not so common within engineering and physical science disciplines where traditionally Safety specialists emanate and the use of absolutism and 'truth' is preferred. Even where risk assessment is used as a tool, many safety specialists prefer to have a defined acceptable/unacceptable cut-off point and to assign (often arbitrary) numbers to probability and consequence to allow a simple decision on the level of risk to be made and to prescribe certain levels of control measures to the different risk levels rather than to systematically analyse whether the assessed risk is ALARP and consider further control measures which may be required. That is, often the SC is produced by following a prescriptive set of instructions which stifle true thought about whether defined activities are as safe as they need to be because it is easier to write a SC that way and then to audit the SC against the internal procedure. The use of argumentation would require a rethink on how a SC is written and a critical review of the Dutyholders activities in order to construct a sound and strong argument which can be validated and verified. It requires a much higher level assessment of the basis of operation than is generally found in current SC's.

It is almost certainly the case that Dutyholders will conclude their arguments can be strengthened by greater understanding of safety science. Consequently there will be a motivation for research and some of this will merit sector wide effort. A criticism made in several safety reviews of the rail industry was a drop in research effort following privatisation[40,37,35]. Part of the blame was the lack or weakness of facilitating organisations, but it is argued here the SC regime at the time was a contributor in that it did not drive Dutyholders into formulating and pursuing research agendas.

EXAMPLES OF USE OF ARGUMENT APPROACH TO SAFETY AND RISK
We have already explored the extent to which argument is used with existing SC's. This section provides examples of how argument has been applied to the safety and risk domain within other sectors or disciplines.

The insurance industry uses validation and verification techniques when evaluating the safety of a facility's activities. The validation part of the evaluation entails checking for the use of approved tools and techniques such as HAZOP, FMEA. The use of these tools suggests a minimum level of safety performance. The confirmed use and quality of the application of such tools is verified through site visits and inspections. Insurance companies actively seek new tools which they may add to their approved list to further improve their validation and verification capability, including tools outside the technical risk area, and may even advise clients of suitable tools to apply. The benefit of such an approach is the targeting of the insurance company's resources to allow efficient and effective, dependent on expertise of insurance company employees, evaluation of the safety of a facility's activities. The insurer's role has similarities with that of the Regulator (although there are also significant difference such as the relationship with the Public and legal compliance). However both parties aim to validate and verify the safety of a facility's activities but there are significant differences in approach, particularly to validation, which for the Regulator is how it evaluates the SC. If a SC were structured as an argument then the validation techniques used by the Regulator could incorporate the approach used by insurance companies to evaluate the claim of safe operation. For example the use of workforce involvement techniques and continuous improvement by a MAH company may be evaluated favourably at the validation phase of the SC as based on relevant predictive theories of theory. The actual extent and perceived successful application of such theories would be verified through site visits and interviews with employees.

The argument based approach to safety cases has been researched within the UK, led by the University of York, and has led to the development of Safety Argument Manager (SAM) software[29]. The research sought to develop an overall safety argument by assembling 'micro-arguments' based on the Toulmin model. The initial phase of the research found the Toulmin model too restrictive and not readily applicable to the types of argument commonly found within real safety cases. The SAM software aims to provide support for the high level argument of the safety case and for the supporting evidence, particularly safety analysis techniques. The software uses a goal based notation for structuring the high level argument of the SC and manages the interrelationships that exist between the most common safety analysis techniques e.g. between Fault Tree Analysis and Failure Modes and Effects Analysis. The author[29] considers that the safety case consists of four elements:

- Requirements – the safety objectives that must be addressed to assure safety
- Evidence – information from study, analysis and test of the system in question
- Argument – showing how the evidence indicates compliance with the requirements
- Context – identifying the basis of the argument presented.

The argument links the evidence to the requirements and all three must be valid for the defined context.

The goal structures used within SAM to present the structure of a safety argument consists of goals, strategies, solutions (roughly equivalent to claims, warrants and evidence within the Toulmin model) and context. Context may be associated with goals, strategies or solutions. The author links the Goal Structuring Notation (GSN) to the four elements of the safety case argument. Requirements are represented as top level goals. Evidence is represented as solutions. Contextual information is represented as context, assumptions, justifications and models. Argument is communicated through the structuring of goals supported by sub-goals. The GSN used within SAM extends Toulmin's form of argument representation to present a notation which the author believes applies particularly well to the safety justification domain[29]. The author[29] acknowledges that the concept of goal decomposition has been applied in areas other than argumentation, particularly in requirements engineering.

The GSN approach has been applied within the railway, aerospace and defence industries. Users are reported[29] as finding the approach helpful for understanding the scope and complexity of safety cases and providing a basis for an executive summary of the safety justification.

The 'Air Traffic Management (ATM) system criticality raises issues in balancing actors responsibility (ARIBA)' project[33] utilised the SAM tool (including GSN) to develop a safety case for an advanced ATM system; constructing the High Level Argument and identifying relevant Supporting Evidence. The ARIBA project defined the safety case as 'a consistent and coherent set of arguments used to justify the safety of a system at all stages in its lifecycle'. The project found the principal benefit of the safety case argument approach was that it provides a structure to the evidence presented to justify the system. Additionally *'that the need to structure a coherent argument from general principles through to the functions to carry out the tasks required a holistic view to be taken of the safety argument. It becomes more difficult to pre-judge the impact of changes, and to get 'locked in' to considering some narrow range of issues*[34].'

The examples included within the above references on SAM[29,33] have tended to focus on technical control measures for assuring safety and not tackled the organisational contribution. Therefore it is not clear from reviewing these sources how applicable the GSN and SAM tools would be to an overall argument based safety case incorporating organisational and technical aspects. However due to the close relationship with the Toulmin model it is foreseeable that they could be applicable.

The next example is from research seeking to build computer systems which can reason autonomously about alternative actions, informed by predictions of their possible consequences[30]. Due to difficulties in estimating and agreeing quantitative probabilities the authors have explored qualitative approaches to practical reasoning and in particular, the application of argumentation. The specific application is scientific reasoning about the possible carcinogenicity of some chemical substance. One of the benefits found by the authors of such an approach was that argumentation permits coherent reasoning about the consequences and likelihood's of alternative courses of action even when expressed in qualitative terms. Another paper by these authors[31] further develops this approach. The carcinogen risk assessment usually involves the comparison and resolution of multiple and diverse evidence, which may conflict. Use of argumentation allows the reasons for claims to be represented in association with the claims themselves and cases for and against a particular claim to be compared. This paper also states that an argumentation formulation permits the representation of quantitative and qualitative information in the reasoning process. Their argument structure is informed by Toulmin's structure.

Further research from the human health risk domain[3] explores a concept of scientific rationality which involves systematic comparison of alternative risk estimates.

The final example proposes the use of argumentation for medical decisions by artificial intelligence systems[8]. The paper asserts that argumentation has far greater representational power than traditional mathematical formalisms based on probability or other quantitative concepts, that it is more versatile and robust under conditions of lack of knowledge. The author comments that where it is possible to directly compare strict probabilistic methods and the author's argumentation based decision process, greater precision does not generally lead to better decision making. The author has found argumentation to be a very practical technique for decision making in systems because it provides a simple method for comparing the relative persuasiveness of competing claims without requiring a comprehensive body of quantitative knowledge of the world. The basic structure of the argument is drawn from Toulmin's model.

DISCUSSION

What are the likely counter-arguments to this proposal?

Firstly is it achievable? It relies on safety science. Perhaps the truth is that SC lack arguments because there is insufficient safety science to support arguments. No doubt safety science has plenty of scope for development but we do not accept there is insufficient. However, the degree of confidence in the science is inconsistent. For example; generally, a greater degree of confidence can be placed in the understanding of the science of engineering materials and structures than in the understanding of organisational behaviour. That is, the conditions under which hardware components and structures fail can be forecast with greater certainty than the performance of a team of people tackling a complex problem. However, by bringing safety science to the fore, the use of argument could drive improvement in areas where further research is required.

Secondly, it may be argued that this is a consultants agenda, and that writing a SC should involve the workforce and is a powerful motivation for improvement. This is a naïve argument for several reasons.

There is the obvious concern that the staff and workforce do not have the necessary knowledge E.g. If you were to visit a facility in an earthquake zone. Would you be convinced by the knowledge and experience of those who work there about the seismic stability of the area or by the analysis of an expert seismologist? Organisational design, management of safety and human factors are not 'common sense' but difficult issues. Many of the theories are emerging, incomplete and contradictory. Workforce involvement and continuous improvement are relevant theories of predictive safety and are likely to be evaluated favourably during the validation phase of the SC evaluation, however these approaches are a means of positively influencing safety culture and when applied successfully are consistent with an organisation's full range of activities. That is, the SC is not a particularly useful tool for developing workforce involvement and continuous improvement and there are many more effective means and in itself, involving the workforce in the SC is not going to lead to cultural change.

Additionally, writing the SC is a one-off exercise. It will be updated and reviewed, but are companies really willing to repeat the initial resource commitment every few years? This is highly unlikely. Hence if a SC initiates significant change, it is a one-off event. Use of argument within SC's would be more likely to lead to critical examination of an organisation's activities rather than a justification of the status quo. Therefore it would be more likely to initiate any required changes to improve safety.

Further, is this argument claiming that the SC is an on-line management tool? But is it not the case (usually) that procedures and other material are on-line and the SC is updated periodically to reflect the on-line material that is actually in active use? The SC is poorly suited to becoming a hands-on management tool because it is difficult to envisage how such a document could be made sufficiently dynamic to facilitate daily use in parallel to regulatory requirements. If a company claims that the SC is in the front line of their SMS it should cause concern because it suggests inflexibility and a potential lack of understanding about the function of a SMS.

It would be very beneficial for the workforce (and stakeholders) to be able to read a clear argument which explains how a MAH facility's activities lead to continuing safe operation of that facility within a SC which also forms a plan for strengthening the argument.

CONCLUSIONS

This paper seeks to show that there are weaknesses associated with the current demonstration model and proposes an alternative argument model. The use of argument for SC's appears consistent with the Regulations and HSE's approach, substituting argument for demonstration provides a process for achieving the desired SC regime output.

Some benefits of the argument approach:

- Dutyholders would be led to think about the completeness of their argument and the degree of confidence in it.
- Would encourage linking of organisational as well as hardware controls to hazards.
- Revealing soundness and strength/weakness of arguments has the potential to improve the transparency of SC's.
- Argument approach draws attention to safety science and has the potential to stimulate improvement.
- The requirement for the Dutyholder to plan to strengthen the argument (as well as reducing risks) has the potential to stimulate research and development.
- Use of an argument model would assist the Regulator during the validation and verification phases of SC assessment.

REFERENCES

1. British Standards Institute, 1999. Railway applications – The specification and demonstration of Reliability, Availability, Maintainability and Safety (RAMS). BS EN 50126:1999
2. Corax, 5th Century BC. Art of Rhetoric (reported, none of original text known to have survived)
3. Crawford-Brown, D., 2000. Scientific Models of Human Health Risk Analysis in Legal and Policy Decisions, Carolina Environment Program, University of North Carolina.
4. ENDS December 1999, Issue No. 299
5. ENDS October 2001, Issue No. 321
6. Entec UK Ltd., 2000. A Railway Safety Case Regime. Class Number EF 14. Ladbroke Grove Rail Inquiry. Part 2. Rail Safety. Core Documents. Piece Number 184
7. Fisher, A., 1993. The logic of real arguments. Cambridge University Press, Cambridge
8. Fox, J., 2000. Arguments about beliefs and actions: Decision making in the real world. Imperial Cancer Research Fund Laboratories
9. Hart, C., 1998. Doing a literature review, releasing the social science research imagination, Ch4 Argumentation Analysis. London, Sage

10. Hikins, J, 1996. Remarks on The Development of Rhetoric. Kendall-Hunt
11. Hinderer D.E, 1992. Building Arguments. Belmont, California. Wadsworth
12. HMSO, 1965. Nuclear Installations Act
13. HMSO, 1971. Atomic Energy and Radioactive Substances, Licensing and Regulation of Sites. The Nuclear Installations Regulations.
14. HSE, 1984. Control of Industrial Major Accident Hazards (CIMAH) Regulations SI 1984/1902
15. HSE, 1992. Safety assessment principles for nuclear plants
16. HSE, 1992. Offshore Installations (Safety Case) Regulations
17. HSE, 1994. Nuclear Site License Conditions, Notes for Applicants
18. HSE, 1994. Railways (Safety Case) Regulations
19. HSE Books, 1994. Railway safety cases. Railways (Safety Case) Regulations, 1994. Guidance on Regulations, L52
20. HSE, 1995. Offshore Installations (Prevention, Fire, Explosion and Emergency Response) Regulations, SI 1995/743
21. HSE, 1996. The Pipelines Safety Regulations SI1996/825
22. HSE, 1996. Offshore Installations and Wells (Design and Construction, etc.) Regulations, SI 1996/913
23. HSE, 1999. Control of Major Accident Hazards (COMAH) Regulations SI 1999/743
24. HSE, 1999. Railway Safety Regulations SI 1999/2244
25. HSE, 1999. Guide to the COMAH regulations, L111
26. HSE, 1999. Preparing Safety Report, COMAH Regulations, HSG 190
27. HSE, 2002. Railways Safety Case Assessment Manual
28. HSE, United Kingdom Interdepartmental Liaison Group on Risk Assessment (UK-ILGRA), 2002. The Precautionary Principle: Policy and Application
29. Kelly, T. P., 1998. Arguing Safety – A Systematic Approach to Managing Safety Cases, University of York
30. McBurney, P and Parsons, S., 1999. Truth or Consequences: Using argumentation to reason about risk. Presented at BPS Symposium on Practical Reasoning, London.
31. McBurney, P and Parsons, S., 2000. Dialectical Argumentation for Reasoning about Chemical Carcinogenicity, University of Liverpool
32. Ministry of Defence, 1996. 00-56 Safety Management Requirements for Defence Systems. Defence Standard
33. Pygott, C., Furze, R., Thompson, I. And Kelly, C – WP5 Final Report, Safety Case Assessment Approach for Air Traffic Management (ATM), ATM system criticality raises issues in balancing actors responsibility (ARIBA), DERA
34. Railtrack, 2000. Ensuring Safety Management Issue 3, Yellow Book 3, Volumes 1 and 2, Fundamentals and Guidance, BS EN 50126
35. Rowlands (DETR), 2000. Review of Railtrack Safety and Standards Directorate
36. Sagan, S., 1993. The limits of safety: Organisations, accidents and nuclear weapons. Princeton Paperbacks
37. Tansley, 1999. The Tansley Report-review of arrangements for standard setting and application on the main railway network; interim report
38. Toulmin, S, 1958. The Uses of Argument. Cambridge University Press, Cambridge.

39. Toulmin S, Reike R and Janik A, 1979. An Introduction to Reasoning, New York: Collier Macmillan

40. Uff, Professor J., 2000. Report into The Southall Rail Crash

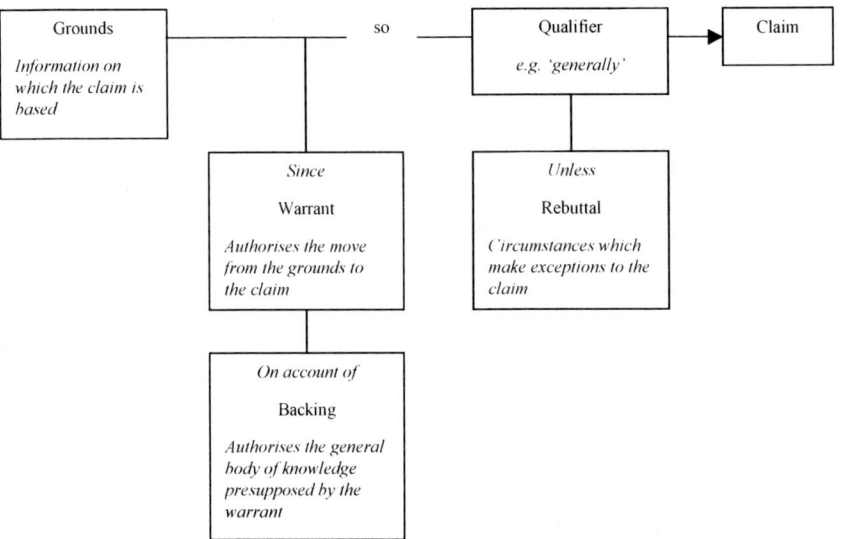

Figure 1.   Toulmin's argument structure

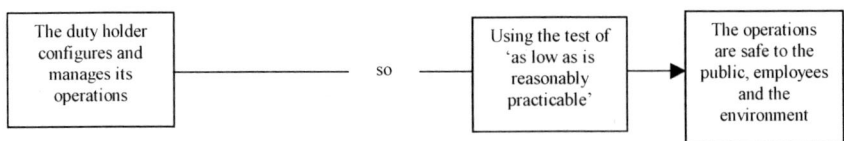

Figure 2.   An example of a dutyholders opening claim presented in Toulmin's structure

COMPETENCE ASSESSMENT AND MAJOR ACCIDENT PREVENTION

Michael Wright, Jonathan Berman and David Turner
Greenstreet Berman Ltd

Many duty holders base their safety case on the assumption that they have experienced and competent staff, particularly where the control of major accident hazards relies on people. Many major accidents have had competency deficiencies as a root cause. With the implementation of the Control of Major Accident Hazard Regulations 1999 (COMAH) on April 1st 1999, it is no longer acceptable to make untested assumptions about staff competence. The Health and Safety Executive has developed safety report assessment guidance that asks for a competence assurance system that includes the setting of appropriate competence standards, assessment and reassessment of competence. The HSE also asks that there be a specific link between identified safety critical tasks, roles and responsibilities at all levels and a targeted comprehensive management system. The report summarised in this paper provides (1) a review of current practice, (2) a view of what comprises good practice in the field of competence assessment in relation to major accident prevention, and (3) a body of advice, checklists and examples of assessment. The report has drawn together experience, standards and lessons learnt from a number of high hazard industries, particularly chemicals, offshore, nuclear and aviation. It has also given due regard to the guidance on competence assessment laid out by personnel specialists, national certification bodies and institutes. A draft version of the report was piloted and received positive feedback from Duty Holders.

Competence assessment, COMAH, Safety Case

The work described in this paper was funded by the Health and Safety Executive, managed by John Wilkinson of the Human Factors Team within the Hazardous Installations Directorate. Its contents, including any opinions and/conclusions expressed, are those of the authors alone and do not necessarily reflect HSE policy.

INTRODUCTION

WHY IS COMPETENCE ASSESSMENT CONSIDERED IMPORTANT?
Review of past major incidents indicates that the lack of certain skills or knowledge has led to errors that contributed to the incident. For example:

- The supervisors with responsibility for inspecting and maintaining the automatic train warning system of the train involved in the Southall rail crash, did not correctly understand the test procedures;
- Events during the Piper Alpha disaster demonstrated that the ability of offshore installations managers to manage major emergencies was inadequate;
- Persons responsible for developing safe systems of work have been shown, by the occurrence of major incidents such as the Hickson & Welch fire and explosion of 1992, to lack a full understanding of how processes work and how hazards may materialise.

In each case, it was assumed that a person with a certain level of experience or training would be competent and/or that the simple dissemination of a procedure would be sufficient. The individual's actual competence was not sufficiently tested.

The most recent example entails the Esso Gas Plant explosion in Longford, Australia, wherein the loss of oil circulation resulted in the plant getting colder followed by the rupture of a heat exchanger on restart due to cold metal embrittlement. The inquiry concluded that there was ample evidence that operators did not appreciate the dangers of cold metal embrittlement despite being trained about it. The inquiry reports that operators were tested during training modules and that in the event of incorrect answers further coaching was provided. However, after coaching the "re-assessment" comprised asking the operator if they now understood the matter. If the operator indicated they understood the matter it was ticked off. According to operators it took courage to say you did not understand the re-explanation. The Commission found, during questioning, that operators still did not see what was wrong with their answers. In addition, the inquiry found that operators gave answers without actually understanding them. In particular, operators knew the correct answer to a question on the action of a valve was to prevent "thermal damage" but did not know what was meant by thermal damage. They gave the answer thermal damage because that was the answer in the training manual. The assessment tested whether operators could present information received from training back to the assessor, without testing understanding.

In addition, the concern about competence is further increased by the move towards multi-skilling, delayering and downsizing. Staff are increasingly expected to take on a wider range of responsibilities with less supervision. This increases the need to check competence.

It is also one of the so-called "Ironies of automation" that training and competency become even more important as automation increases. "Perhaps the final irony is that it is the most successful automated systems with rare need for manual intervention, which may need the greatest investment in human operators training" (Bainbridge, 1987, p7).

HSE CONCERNS AND EXPECTATIONS

The Control of Major Accident Hazard Regulations 1999 (COMAH) came into force on April 1st 1999. They require that the major accident prevention policy statement (MAPPS) includes arrangements for the recruitment of competent personnel, as well as arrangements for meeting their training needs. The HSE has issued guidance that competence assessment be undertaken such that, through assessment of required skills and knowledge, it can be ensured that individuals are capable of performing their tasks safely and properly.

Against this background, it is pertinent to note that, from the experience of HSE Hazardous Installations Directorate inspectors, whilst many organisations have developed job – competence matrices, few Duty Holders have developed a set of competence assessment methods. The issues of concern include:

- It is common practice to assume a person is competent due to their completion of certain training courses or possession of qualifications and period of experience – without questioning whether their experience provides pertinent exposure to safety critical tasks, whether the training outcome is assessed or whether performance on the job in safety critical tasks was satisfactory;

- Lack of linkage between competence assessment and the COMAH major accident controls, and failure to recognise that this extends to all levels of management;
- Failure to identify tasks (and associated skills and knowledge) that could impact process safety, such as knowledge of safety critical valves, operating procedures etc;
- Adequacy of on-the-job assessment, including the competence, credibility, and independence of the assessor, what criteria are applied and the consistency in evaluation criteria;
- A presumption that achievement of general standards such as National Vocational Qualifications is enough alone to assure competence;
- A widespread view that training is a "one-off" and not part of an ongoing competency assurance process;
- Lack of systematic assessment on recruitment and little of ongoing assessment, and;
- Many sites view safety competence as "personal" safety rather than process/major accident safety.

Indeed, the HSE guide "Human Factors for COMAH Safety Report Assessors", produced for use by HSE assessment teams, states:

"The relevance of training to a safety report is that there should be establishment and assurance of competence specifically for the key responsibilities and for key safety critical and safety related tasks. Too many reports give information just about general safety training which is often more about personal safety than major accident prevention." (p6–7)

THE BENEFITS OF COMPETENCE ASSESSMENT
Competence assessment is intended to improve accident prevention in the following ways:

- Improved competency, appropriately linked to the control of major accident hazards, improves the understanding, knowledge and awareness of staff that is necessary to prevent or mitigate major accidents;
- The development of standards will provide individuals with a clear view of what competencies they need, and hence should encourage development of competencies;
- The requirement for Duty Holders to set performance standards should prompt them to provide appropriate training and, if incidentally, reveal where unrealistic performance expectations exist for staff;
- Competence assessment may reveal that sub-standard competence is occurring;
- The inclusion of competence assessment within a planned management system should reduce the likelihood of substandard staff performance being overlooked, and;
- The setting of standards and associated assessment of competence should mitigate the pressures arising from organisational change for staff to work beyond their limits and provide a baseline for change by capturing competences that need to be retained.

In addition, a demonstrable process of competence assessment should provide assurance to regulators and other stakeholders that a core aspect of the safety case, namely staff competence, is valid. This can be particularly important in the case of older plant, which can fail to meet latter-day operability standards and may lack up to date procedures, whose safety is based on the ability of staff to compensate for lower operability standards.

Furthermore, the cost of training can be significant – it is important for duty holders to be confident that finite training budgets are being deployed to best effect, and hence that they have available the means of assessing the effectiveness of their training arrangements.

HSE STUDY ON COMPETENCE ASSESSMENT

Accordingly the HSE commissioned Greenstreet Berman to undertake a study on competence assessment. The objectives were to identify good practice in assuring competence at hazardous sites, including identification of required competences, methods for their assessment, and management of the assessment/assurance process. To meet these objectives, the study included a review of competence assessment guidance in the field of Human Resources and Occupational Psychology and a survey of assessment practices in a range of safety critical industries, including chemicals, offshore oil and gas, water, aviation and nuclear.

From this work a set of practical guidelines has been developed covering;

- The link between COMAH and competence assessment, including the link between risk assessment and competence assessment;
- How you can identify appropriate competence assessment methods, and;
- Examples of appropriate competence assessment methods for tasks of varying degrees of safety criticality and complexity.

The generic guidelines are intended to assist a COMAH Duty Holder to appraise the full range of tasks onsite, to assess the safety impact of these tasks, and to identify an appropriate and practical level and approach to competence assessment.

WHAT IS COMPETENCE ASSESSMENT?

Competence is defined by the Department for Education and Employment as:

"the ability to perform the activities within an occupation or function to the standards expected in employment".

This is an outcome-based view of competence. It leads to the view that competence assessment entails the collection of sufficient evidence of workplace and/or personal performance to demonstrate that the individual can perform to the specified standard. This definition of competence is important in two respects. Firstly, it highlights the need to recognise the difference between recording a person's experience/training, and assessing their competence.

Secondly, this outcome-based view of competence assessment can be compared with the common objectives of selection and recruitment. That is, selection and recruitment processes often aim to predict whether a person has the appropriate underlying characteristics and knowledge for a job. Accordingly they test personality, aspirations, underpinning knowledge and attitudes with the expectation that they *will* be able to perform competently with the passage of time and appropriate experience and training. Standard selection processes do not necessarily require an individual to be fully competent at the time of appointment. This also highlights the difference between aptitudes and ability, wherein a person can have an aptitude but may lack the ability to apply it to good effect.

These distinctions are important in the context of COMAH, as it is insufficient to assume that a standard selection process (which focuses on aptitude rather than ability) ensures competence.

CURRENT PRACTICE

Our review of current practices within and outside the onshore hazardous industries indicates that there is wide variation in the standard of competence assessment. In some cases Duty Holders have developed systematic approaches to competence assessment and even made explicit links between the COMAH safety report risk assessment and competence assessment. In other cases reliance is placed on unstructured on-the-job training and assessment.

The most frequent method of competence assessment is that of 'observational assessment' by a supervisor or appointed trainer. In some cases the assessment of operators and maintenance staff rely on unstructured peer review, a practice considered as poor by this study. However, in some companies this judgement is guided by the use of validated task descriptions, skill and knowledge inventories, verbal test questions, and guidelines on (for example) the number of times a person needs correctly to perform a task to be deemed competent.

Three of the surveyed onshore companies have used some form of task analysis in order to identify safety critical task and define the correct way of working for use in assessing their staff. Indeed, some companies carry out a form of risk assessment to identify safety critical tasks for which they require assurance of staff competence. Assessors can then test a person's knowledge by asking them how they would carry out a particular task and probe their understanding of (say) the safety function of equipment and key safety procedures. However, two other surveyed onshore companies had no systematic approach for defining tasks.

Also, in the case of safety critical emergency roles, such as control room management of process upsets, there are examples of the application of "advanced" forms of assessment, such as the systematic use of simulator based assessment of decision making and command skills. Finally, in some cases assessor competence has been addressed by the use of trained assessors.

LESSONS FROM OTHER HIGH HAZARD SECTORS

The review of practices in other sectors highlights a number of key points. First, there are examples of competence assessment being managed as part of a comprehensive, planned and managed process. In particular, the nuclear sector has developed a process of ensuring Suitably Qualified and Experienced Personnel perform all tasks. A competence – job matrix is used to determine competence requirements for each job against which staff are assessed.

Another example can be found in the aviation sector wherein there is a set of training, experience and assessment requirements for pilot qualification covering initial appointment, progression from one grade to another, transfer between aircraft types and ongoing competence assurance. The range of tasks, experience and knowledge required is laid out in standards.

Finally, the assessment of offshore installation managers, air pilots and submarine commanders provide examples of how to assess the competence of emergency response roles. In particular, these examples entail the use of simulators and exercises based on accident scenarios taken from risk assessments, and the use of behavioural checklists by observers to guide the assessment of performance on "softer" competencies such as delegation, communication, decision making under stress, information acquisition etc.

COMPARISON OF "STANDARD" ASSESSMENT AND SAFETY RELATED COMPETENCE ASSESSMENT

In many ways the recommended approach to the assessment of competence in safety critical roles does and should mirror the approach advocated for competence assessment in general. In particular, the concept of collecting evidence of performance, the need to set performance criteria, independent credible and competent assessors and the use of standards outlining key skills and knowledge are all equally pertinent. However, it is apparent that there are some particular requirements and practices in the context of major accident prevention. These include:

- A need to ensure that the process of competence assessment is managed in a systematic and proactive manner to a standard commensurate with major accident prevention;
- High-risk industries tend to place more emphasis on certain methods due to the relative importance of certain types of tasks and the need to provide a particularly high level of competence assurance for safety critical roles. In particular, high-hazard industries tend to place more emphasis on:
  ⇒ The role of risk assessment in identifying competence needs;
  ⇒ The use of task analysis to identify the skills and knowledge entailed in complex technical tasks;
  ⇒ The development of techniques, such as the use of behavioural markers to assess "softer" skills such as communication in emergencies;
  ⇒ Licensing – again reflecting the need for a high level of assurance and very high standards of competence in certain safety critical tasks;
  ⇒ The role of simulators and exercises, due in part to the rareness of emergencies but again reflecting the importance of assessing competence for handling emergencies and rare events;
  ⇒ The need to monitor and maintain competence, in particular recognising the need for skills to be maintained to handle infrequent events and to ensure staff maintain technical skills and knowledge to operate processes and equipment.

Thus, whilst there are many commonalities between "standard" competence assessment and assessment for safety critical tasks, the characteristics of high hazard tasks and the need for a high standard of assessment does mean that specific attention must be awarded to the design of competence assessment for major accident prevention.

COMPETENCE ASSESSMENT ADVICE

In order to develop a process of competence assessment, it is necessary to answer the following points:

- What competencies (at all levels) need to be assessed to ensure that error or sub-standard performance will not contribute to a major accident?

- Are the competence expectations realistic?
- What assessment criteria and competence standards, including what level of performance evidence, are required to ensure risks are ALARP?
- What method(s) of assessment are required to acquire evidence of competence?
- What qualifications and experience do assessors need?
- How often should performance be reassessed, reflecting the level of risk and possibility of skill decay?
- What method(s) of reassessment are needed?

This entire process should be managed proactively as per any aspect of accident prevention, thereby bringing about continuous improvement. Figure 1 illustrates the process. A self-assessment checklist is provided in an annex to this paper.

IDENTIFYING SAFETY CRITICAL TASKS

It is envisaged that competence assessment commences with the identification of safety related tasks using techniques such as risk assessment and task analysis. Where necessary there should be justification for reliance on human performance, as opposed to engineered safeguards. Such an assessment should cover all forms of activity, including normal process operation, process upsets, planned and unplanned maintenance. At this stage the analysis may simply provide a task or activity inventory for which assessment is required in the context of major accident prevention.

DEFINE PERFORMANCE STANDARDS

Next, a set of performance standards and assessment criteria is defined. This entails analysing the types of competencies required, describing what comprises adequate performance and defining measurable criteria by which to judge performance. At this stage a more detailed task analysis and/or specification of competences may be required to help develop testable competence standards. The task or competence description should provide a view of:

- The correct way of doing a task (against which a person's performance can be judged), and;
- Key competences (Skills, behaviours and underpinning knowledge).

The task need only be decomposed to a level that enables the production of testable task/competence descriptions. As befits the task, competence standards tend to cover:

- Skills, such as being able demonstrate an ability to (say) interpret process instrumentation readings, diagnose faults, operate controls, enact a procedure;
- Underpinning knowledge, such as understanding the chemistry of a reaction;
- Safety behaviours and attitudes, such as safety leadership, communication, teamwork.

It is reiterated here that assessment should aim to acquire performance-based evidence that a person can carry out a task, rather than just collate evidence of underpinning knowledge. Thus, standards should denote demonstrable skills and testable definitions of what comprises "competent" performance. Examples of competence assessment criteria are given below.

- Operators involved in emergency response need successfully to carry out an emergency response procedure on three separate accident scenarios selected from the safety case;

655

- Supervisors must correctly manage an operation, such as removing a hydrocarbon pump, starting from developing the plan of work, specifying a permit to work, instructing staff, monitoring their work, checking pump integrity prior to start up … etc;
- Safety engineers must be able correctly to interpret a piping and instrumentation diagram, identify all (contrived) engineering defects and specify safety devices and engineering modifications as noted in company standards;
- A maintenance technician should be able to recollect all key safety actions required in the isolation of a hydrocarbon pump, its dismantling and restoration.

These may be augmented by "tests" of underpinning knowledge, such as:

- Minimum periods of "observed" experience – taken to be indicative of competent performance;
- Qualifications and training – used as an indication of the level of underpinning knowledge;
- Verbal or written examination of a person's knowledge and/or attitudes.

Thus, a range of criteria may be devised, each matched to the type of competence (observable skills and behaviours versus underpinning knowledge).

The competence standard may assume or require a certain level of supervision, and hence there may be a scale of competence standards for people of varying competence. This is illustrated by the Institute of Electrical Engineers competency guidelines for use with safety practitioners working on safety related Electrical, Electronic and Programmable Electronic Systems. The standards of competence are graduated for supervised practitioners, practitioners and experts.

It is common practice to use national qualifications as a means of demonstrating skills and knowledge. Whilst this is entirely reasonable it is important to:

- Ensure that the national qualifications cover the specific skills and knowledge required by the site's processes, equipment and activities, including specific safety matters;
- Recognise that NVQs by their nature are limited to assessment of on the job performance and hence may not cover infrequent safety critical activities, such as emergency response, process upsets, infrequent maintenance activities etc;
- Ensure that the form of assessment and level of performance evidence collated matches the safety criticality of the processes, equipment and activities.

It is pertinent to note that in some case studies the implementation of NVQs is guided by in-house assessment of the specific skills and knowledge associated with the site's processes, equipment and activities. In addition, it should be noted that some organisations have felt that their assessment process has been "NVQ driven" rather than driven by their range of activities. Finally, as NVQs are designed to cover all aspects of task performance, they may include tasks and activities that have relatively little bearing on major accident prevention. Hence, whilst NVQs may assist with the demonstration of safety related competences, major accident prevention may not by itself require completion of the entire NVQ syllabus.

SELECT ASSESSMENT METHOD

Once the task and type of competences are understood, an appropriate assessment method can be identified. The method of assessment should provide a valid and reliable measure of the type of competence in question, such that two different assessors would give similar results. Ideally the reliability of the assessment process would be monitored by review of actual performance, i.e. does the standard of staff performance accord with the results of competence assessment. If sub-standard performance is observed, in contrast to acceptable assessment results, the validity and reliability of the assessment process should be reviewed. In summary:

- Physical/sensory-motor competences can be demonstrated by practical "show me" assessments wherein people either complete the real task or a component of it, such as driving a road tanker to demonstrate steering skills.
- The ability to carry out a prescribed procedure of work can, usually, be demonstrated by a "show me" test wherein you attempt to complete the task.
- Cognitive skills, such as the ability to (say) assimilate process control information from a VDU and thereafter interpret it might be demonstrated by the candidate talking through the interpretation of displayed information. However, such verbalisation may interfere with some cognitive skills whilst it may not be possible to verbalise other cognitive skills, such as mental arithmetic. In these cases post-task debriefing of candidates may be appropriate.
- Whilst satisfactory completion of a task that requires the use of knowledge, such as fault diagnosis, may be indicative of underpinning knowledge, it is possible that the correct action was by luck. Accordingly, knowledge tends to be assessed through verbal or written questioning.
- Whilst psychometric personality tests may provide a prediction of interpersonal, team management and safety behaviours, observation of actual behaviour in the real or simulated work setting using behavioural observation tends to provide a more valid measure.

ASSESSOR COMPETENCE AND CREDIBILITY

Assessors should be competent in the process of competence assessment and have a certain level of knowledge and experience of the tasks being assessed. The level of expertise in assessment should be matched to the form of assessment. For example, in-house coaching on how to complete assessments may be adequate for "on the job observation" of simple operation tasks, but completion of NVQ units D32/33 may be needed for (say) assessment of more complex operational tasks. In the case of behavioural competences such as team coordination and communication assessors may need to be trained on what comprises "good performance", what are the behavioural markers and how to gauge performance against these markers. Assessors should also be credible in the eyes of the assessees. This tends to require assessors to have sufficient knowledge and experience of the operations to pose insightful questions and judge the validity of answers.

ONGOING ASSESSMENT

Finally, ongoing "competence checking" needs are determined by consideration of how competencies may decay over time and the safety criticality of the task. More frequent assessment tends to be required for higher risk tasks and tasks wherein skills may decay sooner. All persons tend to be assessed at least annually in the form of a performance appraisal based on line management observations. People involved in complex safety critical tasks, such as process managers or control room operators, may be appraised more formally every 1 to 3 years. The highest risk tasks may be assessed every six months.

There are a number of considerations regarding the form of reassessment to apply. In the case of infrequent tasks, such as emergency response or response to a process upset, normal day to day work may not provide any opportunities for performance to be demonstrated. In such a case, it may be necessary to set tasks, run simulations or exercises. On the other hand, day-to-day work may provide a valid indication of performance in the case of routine frequent task, such as road tanker driving. However, there is an additional array of sources of evidence of performance available once a person is in post. These can include:

- Standard SHE audits can cover the performance of safety critical tasks, specifically the level of adherence to safe practices and individual performance;
- The contribution of individual competence to an incident can be assessed as part of the incident analysis process;
- Many companies use behavioural observation schemes that can provide observations of safety related behaviour that can be used as the basis of one to one coaching – assuming the scheme covers behaviours that effect process safety.
- Peer review: On the job performance can be monitored and appraised by line managers.

The latter sources of evidence may compliment more formal forms of assessment.

RESPONDING TO SUB-STANDARD PERFORMANCE

It is clearly important to have a pre-planned response to the identification of sub-standard performance to enable the company to act purposefully on the results of competence assessment. The response to sub-standard performance tends to vary according to the purpose of the assessment and the safety criticality of the task.

In the case of selection, promotion and recruitment decisions the discovery of sub-standard performance tends not to pose a significant "policy" problem, in that people are simply not appointed to the position and/or required to undergo further training/experience. Once a person is in post, the discovery of sub-standard performance tends to pose a more difficult challenge. First, it is important to check whether the sub-standard performance arises from omissions in training, supervision or other factors such as inadequate procedures or equipment. If the sub-standard performance is attributed to the individual, there are at least three common responses, namely retraining staff, increasing the level of supervision and placing limits on the scope of an individual's role and responsibilities. In the case of the most safety critical roles, it is likely that a person will be required to demonstrate competence, perhaps by undergoing re-assessment, before they are re-authorised to take on their normal duties again, especially if they normally work unsupervised or are a key decision-maker such as a process supervisor.

CONCLUSION

Organisations operating high-hazard plant recognise that safety relies on the experience, commitment and competence of their staff, and this remains true. Indeed, reliance on staff competence is perhaps even greater in latter day downsized organisations and the use of automated systems for control. The COMAH regulations and the lessons learnt from major incident indicate that it is not enough to assume that exposure to training and experience assures competence. There are already examples of good practice in the development and application of competence standards and systematic assessment methods. This report provides a summary of these practices in a sufficiently general way that the diverse range of sites regulated under COMAH can apply.

REFERENCES

1.   Michael Wright, David Turner and Caroline Horbury. Competence assessment for hazardous industries. Health and Safety Executive Contract Research Report, In press 2002.
2.   Andrew Hopkins, 2000, Lessons from Longford: the Esso Gas Plant Explosion, CCH Australia Ltd, ISBN 1 86468 422 4.
3.   Human Factors for COMAH Safety Report Assessors, CIF 01/14, www.hse.gov.uk/hid/land/comah2.
4.   Bainbridge, Lisanne. 1987. Ironies of automation, in Rasmussen, J et al, New technology and human error, John Wiley & Sons, Chichester.

APPENDIX: SELF-ASSESSMENT QUESTIONS

QUESTIONS FOR PROBING THE PROCESS OF COMPETENCE ASSESSMENT

1.   Is the assessment of competence proactively managed as part of an integrated selection, training and assessment process? For example:
    - Is the need for competence assessment stated within (training) policy?
    - Have roles, responsibilities and adequate resources been allocated to the process of competence assessment?
    - Is the level of training and supervision matched to the level of individual competence?
    - Does the management of change process address issues of competence assessment (during delayering and downsizing)?
    - Are the results of competence assessment recorded?
    - Is the system of competence assessment audited and reviewed?
2.   Have all those activities and tasks that have the potential to contribute to a major accident been identified?
    Have task inventories been prepared for all those activities and tasks that have the potential to contribute to a major accident been identified?

4. Has the task analysis been cross-referenced to a risk assessment (such as a HAZOP, What if analysis or safety review) or safety case, or has an error analysis been carried out – such that the safety criticality of tasks is validated?
5. Has the safety case been reviewed so as to identify emergency scenarios (and process upsets/abnormal operating states) and safety critical tasks for which staff competence needs to be assessed?
6. Are there a set of valid and comprehensive descriptions of safety related tasks, covering operations maintenance and abnormal events, for use in assessing a person's performance?
7. Do the task analyses identify the necessary underpinning knowledge of equipment, processes, hazards and consequences?
8. In the case of supervisory and managerial roles/activities, does the task description cover specific safety tasks such as risk assessment, developing safe systems of work, effective communication, etc?
9. In the case of tasks such as emergency response and supervision, does the task analysis cover the "softer" (non-task specific) competencies of team management, communication, event recognition, delegation and so on.
10. Is competence assessment supported by a set of standards that cover:
    - A list of those tasks that staff need to be competent in;
    - A specification of the skills, knowledge, behaviours and correct working practices/procedures against which performance can be judged;
    - Guidance on what constitutes acceptable evidence of performance, such as written tests, on the job observation, simulated exercises, minimum periods of experience and qualifications;
    - Measurable criteria for judging adequacy of performance during assessment, such as correctly completing a task three times.
11. How have NVQs or any other national standards been adjusted to match site hazard/risk profile?
12. Have the standards been "graded", as appropriate to allow for different individual levels of competence and supervision?
13. Has the level of error tolerance (in the system) been taken into account in setting competence criteria? Such as:
    - Hardware or software interlocks, and automatic shut down systems;
    - Supervision.
14. Has the realism of performance standards been reviewed to ensure that competence expectations reasonable and realistic?
15. Is the scope of assessment adequate? For example, does it cover, as appropriate:
    - Process knowledge;
    - Understanding of hazards associated with the process/equipment/plant;
    - Understanding of correct operating procedures and practices;
    - Safety behaviours and attitudes;
    - Demonstration of correct performance, and;
    - Softer competencies such as safety leadership, as appropriate

16. Is there a suitable match of assessment methods to the different types of competencies? For example:
   - On and offline observation of task performance for operational skills;
   - Simulation based assessment of process upsets and emergencies;
   - Behavioural observation for "soft" competencies such as safety leadership, communication and teamwork;
   - Question/examination based assessment of knowledge.
17. Is on the job assessment guided by a set of standards and assessment criteria?
18. Is on the job assessment independent?
19. Is assessment carried out by accredited or otherwise trained trainers/assessors with pertinent process experience?
20. Have minimum "check and train" requirements and frequencies been set for staff in post?
21. Is the frequency of re-assessment matched to the frequency and safety criticality of tasks, with for example, annual simulator assessment of complex control room tasks and, annual on the job appraisals for routine plant operation?
22. Is an appropriate set of assessment methods used to check ongoing competence? Such as:
   - Simulation/exercise based assessment for infrequent tasks;
   - Verbal or written tests of retention of knowledge and/or up to datedness of knowledge;
   - On the job observation of routine tasks.
23. Is appropriate use being made of safety audit and other performance review systems to identify individual competence problems? For example:
   - Is there a system in place to report and review the cause of significant errors and refer, as necessary, people back to a "check and train" process?
   - Are the results of behavioural safety observation used to identify individuals behaving unsafely and providing them with coaching?
   - Do general SHE audits report individual competence issues?
24. Are appropriate standards used to guide the assessment of "on the job" performance of staff in post?
25. Does the company have a planned approach to the identification of sub-standard competence, such as retraining, demotion and/or increased supervision?

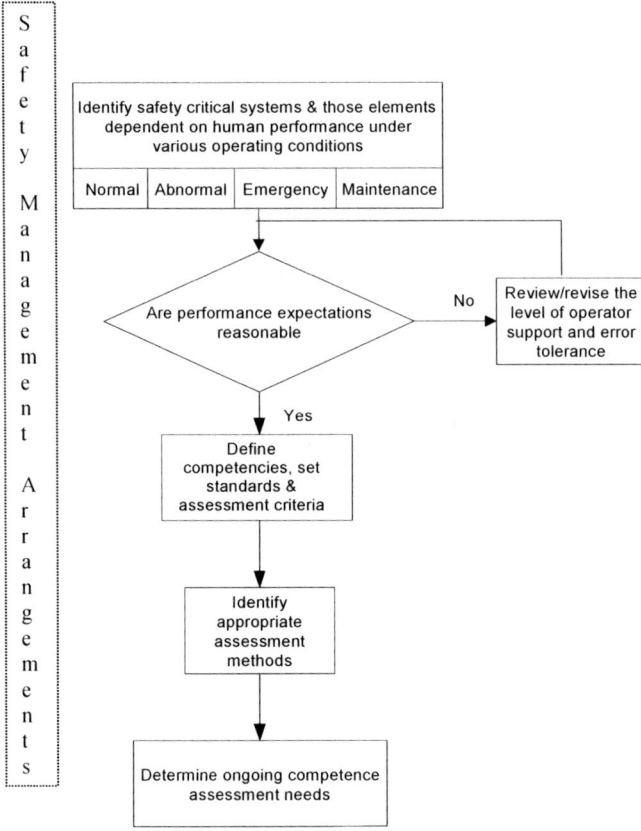

Figure 1. Overview of competence assessment process

MODELLING HIGH CONSEQUENCE, LOW PROBABILITY SCENARIOS

R.P. Cleaver, A.R. Halford and C.E. Humphreys
Advantica Ltd, Ashby Road, Loughborough, Leicestershire, LE11 3GR

By their very nature, major accidents in the gas and oil or petrochemical industries have a low frequency but potentially high consequences. However, assigning quantitative measures to the risks of such an accident is difficult, as the frequency has a high degree of uncertainty attached to it and the consequences may be sensitive to the assumptions made in representing the scenario within the analysis. It follows that the assessment of such accidents requires a robust methodology to be used. There may be cases in which a worst-case or single representative analysis may distort the analysis compared with a full evaluation. Advantica has developed a methodology that allows the different realisations of scenarios to be analysed, taking into account relevant parameters such as release and wind directions. This paper provides details about the basis of this method. It is shown that the calculations allow FN curves to be built up from the frequency and number of casualties for each realisation. This allows the contributions from different realisations to be analysed and those contributing most to be identified. It is also noted that using this method may reveal that some preconceived ideas about what is the 'worst case' are not always correct. As well as being of use when carrying out qualitative risk assessments for Safety Reports for example, it is noted that these methods are of great help in cost benefit analysis. Using worst-case assumptions can often mask the benefits of protective measures, as they appear to have little effect on the worst-case realisation. However, using the methodology allows a correct identification of suitable measures to be taken to reduce risks, as the impact of the measure on all the different realisations is assessed.

KEYWORDS: Quantified risk assessment, Major accident scenarios, Gas releases.

INTRODUCTION

Within the UK, under the Control of Major Accident Hazards Regulations (COMAH), there is a requirement for owners of certain installations to prepare Safety Reports. Sites storing an aggregate amount of certain dangerous substances fall under the remit of this legislation. Typically, major gas or oil process and storage sites exceed these threshold amounts and so require Safety Reports. Guidance is available from the Health and Safety Executive (HSE) on preparing the Safety Reports[1]. In terms of the analysis of major accident hazards, the HSE describe the three steps to be followed. These are:

- Identify all the possible major accidents and select a subset for analysis.
- Give a realistic estimate of the likelihood of each major accident hazard or an adequate summary of initiating events.
- Produce an adequate assessment of the extent and severity of the consequences for each identified major accident hazard.

In a previous paper[2], an account was given of the difficulties encountered in analysing the so-called low frequency-high consequence scenarios. In summary, it was argued in that paper that selecting a worst or single representative case may distort the analysis of these scenarios compared with a full evaluation. It was also noted that to compound the problem it is not always clear which case should be selected as 'worst' or 'representative'. A remedy was described in the form of a more realistic treatment that can be used for all such scenarios. The next section of this paper provides a more detailed account of this method. Its application to a typical high consequence scenario is described in the following Section. The paper ends with a discussion of the findings from this and other application of the methodology.

METHODOLOGY

In the first stage, information is collated on the distribution of population, both on-site and off-site. This population is then represented on a map that can be interpreted by the computer-based risk assessment package. Two methods are used for the analysis of people in buildings. The first is to represent a building or a group of buildings as a single point receiver located at the building centre. The second type of analysis is to represent the building using a grid of receivers suitably distributed to take account of its shape. The choice of method depends on the size of the building relative to the fire or explosion being considered. For larger buildings a grid of receivers is more appropriate as it is allows a greater spatial resolution in analysing the effects of incident thermal radiation or overpressure. A group of buildings that is a large distance from the release may be represented by a single receiver. An occupancy level is defined for all buildings and this is taken to vary according to the time of day. Office buildings, for example, are likely to have a higher population during the day than at night. The opposite is likely to be true for domestic buildings. The population who are outside can be represented on a similar grid of points. For on-site locations, the distribution of the points can be selected to reflect working patterns and common activities. In order to account for normal working patterns, the occupancy level at each point is taken to vary according to the time of day. For example, it is usual to consider occupancy during a normal working day (e.g. 8 to 5 Monday to Friday), at weekends and during the night. Special events, such as loading or discharge, may require particular analysis to ensure that hazards that can only arise in connection with certain activities have a population distribution appropriate to that activity.

Each high consequence scenario is analysed in detail, considering the different ways in which the scenario could occur. A representative number of different locations for the release are selected and a range of possible temperature and pressures for the process fluid are selected, as appropriate. For jetted releases, the number of different directions in which the release can point initially is considered. It may also be necessary to consider transient behaviour following the initial release, in order to account for a finite capacity of the system under consideration. The weather conditions that are assumed for each evaluation of the release are chosen to represent meteorological data for that particular location. Given that the scenario occurs, the probability with which each of these particular combinations of conditions arises is determined using a combination of the meteorological data and site information. These combinations are then simulated in the computer package.

The vulnerability at the specific grid points used in the package is evaluated for each realisation of the scenario. For flammable gases, this involves a consideration of the harm arising from any fires or explosions. For gas/vapour dispersion, the main hazard distance is taken to be when the mean in-plume concentration has decayed to the lower flammable limit (LFL). The distance to LFL is considered to represent the maximum distance within which an ignition source could ignite a release, leading to a flash fire throughout all of the source cloud or an explosion in a congested or confined region engulfed by the cloud. In principle, persons and property within this range could be affected in the event of ignition occurring, although in practice the occupants of most buildings would be afforded protection, unless gas ingress had occurred. Ignition of isolated pockets of gas beyond this LFL contour may be a possibility, but it is considered that this would not lead to general cloud ignition[3]. However, in recognition of the potential for such ignitions to cause harm, distances to half of the LFL (½ LFL) are also evaluated and used in the assessment, as noted below.

The effects of thermal radiation are determined by the dose of thermal radiation received as a function of time[4]. A typical hazard range criterion for personnel exposed to thermal radiation is taken as the distance from the fire from which persons can be expected to escape without receiving a defined dose of 1060 thermal dose units $[(kW/m^2)^{4/3}s]$, which is fatal in approximately 1% of cases. A secondary criterion is the distance from the fire from which persons can be expected to escape without sustaining injury in the form of second-degree burns (skin blistering). The dose of thermal radiation required to cause the onset of skin blistering depends on the thermal radiation flux level and the time of exposure, but for the scenarios analysed in this report it is typically of the order of 250 dose units. In calculating the "escape distance" using either criterion, a lower threshold of 1 kW/m^2 is used, to which it is assumed a person can be exposed for an indefinite time without injury. For any assumed escape speed, the "escape distance" is calculated neglecting the possibility of obtaining shelter.

Ignition of combustible material on buildings or structures can also be caused by intense thermal radiation, although this is again dependent on the thermal radiation flux level and the time of exposure. The threshold for buildings exposed to thermal radiation is taken as the flux level at which secondary fires may be started by piloted ignition of combustible materials (minimum 12 kW/m^2).

The effects of an explosion depend on the strength and duration of the overpressure wave that is generated. In the assessment of hazard distances for people inside buildings, a blast wave overpressure of 40 mbar is sometimes used. An overpressure of 40 mbar is estimated to cause 90% window glass breakage. The guidance from the Chemical Industries Association (CIA) on the safety of occupied buildings associated with major-hazard installations[5] indicates that an overpressure of 40 mbar would cause approximately 1% fatality within a population in a typical domestic building. Different levels of overpressure may be required for other buildings, such as supermarkets or sports halls. The overpressure required to cause this level of fatalities rises to 100 mbar for a typical office building. An overpressure value of 180 mbar is considered to be capable of producing fatality in 1% of the population for people who are outside.

The strategy that would normally be adopted in analysing the vulnerability for the high consequence scenarios is summarised in Table 1.

As noted in the Table, the methodology distinguishes between people who are indoors and those who are outdoors. It is assumed that the people who are outdoors at the time of the event attempt to escape to a safe distance. In the case of fires, the thermal dosage that they receive in doing this is evaluated and used in a probit relationship to infer their vulnerability. People who are indoors initially are assumed to stay inside unless the building is predicted to start to burn because of piloted ignition. A certain proportion (10%) of the people inside such buildings are then assumed to become 'trapped' fatalities. It is assumed that the remaining people try to escape from the building. The size of the fire relative to the size of the building is taken into account in this analysis and also the possibility that people could use the different exits from the building. For example, people within the piloted ignition distance within buildings are assumed to attempt escape form the nearest exit, whereas anyone outside of this distance, but still within the building, is assumed to attempt escape from the most favourable exit.

Table 1. Levels of harm used for the impact assessment

Hazard	Hazard range/dose/ metric	Effect of people within range		
		Outside	Inside "normal" building	Inside "hardened" building
Free field overpressure (mbar)	Received blast loading	Calculated based on overpressure - correlation given by Baker[3] for percentage fatalities arising for specified free field overpressure (464 mbar taken as producing 100% fatality, 300 mbar 50% fatality, 180 mbar producing 1% fatality)	Calculated based on overpressure - correlation given in CIA guidelines[5] (600 mbar taken as producing 100% fatality, 250 mbar 50% fatality, 40 mbar producing 1% fatality)	Case specific analysis based on specification for building – typically greater than 1000 mbar required to produce fatalities.
Jet fire, pool fire or fireball	Secondary fire range (based on piloted ignition of wood)	-	10% of residents of buildings completely engulfed without any fire proofing are assumed to be fatalities– the remaining 90% of residents seek to escape at the time of piloted ignition and vulnerability is calculated as for people outdoors within escape distance	

	Within escape distance (calculated for person outside attempting escape at uniform speed from event)	Percentage vulnerability calculated from Probit relationship, based on received dosage in attempting escape.	People within buildings that are located outside of the secondary fire distance are assumed to remain in buildings and to be safe.
Flash fire	Within LFL contour	Assumed fatalities 100%	Protected by building
	Between LFL contour and 0.5 LFL contour	Vulnerability reduces from 100% at edge of LFL contour to zero at edge of 0.5 LFL contour	Protected by buildings

The vulnerability information at each grid point and the frequency information for each realisation of the scenario can then be combined to determine the overall risks posed by each scenario. That is, the location specific individual risk is defined at each grid point. Further analysis of the output from the calculations enables values of the maximum number of fatalities, the average number of casualties and the societal risk (the combination of the frequency of each event with the expected number of fatalities) to be determined. This also enables the contribution to the risk from different realisations to be analysed and those contributing most to be identified.

ILLUSTRATIVE EXAMPLE
In order to demonstrate the principles of the methodology, a hypothetical LNG storage site has been created. The population distribution surrounding the site is based on one specific location in the UK, in order to give realistic numbers for subsequent analysis. The LNG Storage site and the accident scenarios have been 'superposed' for demonstration purposes and are not intended to reflect the situation at any real site.

The Site has been split into four main areas, with associated population distributions, as follows:

- Inside the administration building, with entrance building
- Inside the workshop
- Inside the control room
- Outside in the process area (within the site boundary)

It is assumed that 'night' occurs for 50% of the time and peak daytime hours and off-peak daytime hours each occur 25% of the time. A typical working day (peak hours) therefore occurs for 42 hours per week. The control and Administration building, the Workshop and the Stores were each represented using a grid of receivers. The remainder of the site was divided into 117 rectangular areas, each of which was represented by a receiver at its centre. Seven local business or commercial properties are also assumed to be situated close to the site, with a number of isolated farms of hamlets and smaller villages, and a large industrial estate nearby.

Figure 1 shows the locations of the populations assumed for this study.

It would normally be assumed that the population of domestic properties is lower during the working day, when some people are at work. However, in order to account for a number of small businesses in the settlements, it has been assumed in this analysis that the population is constant.

A typical scenario that could have onsite and offsite effects is a large spill of a flammable, volatile liquid. A release of LNG from a storage tank has been selected as an illustrative example. The failure of an LNG tank is extremely unlikely. Studies have been carried out to estimate an appropriate failure frequency for this event (see, Lees[6] for examples). For the purposes of illustration, however, it is assumed that the tank fails catastrophically leading to a release of its contents (cryogenic LNG). A number of different cases are modelled as follows:

Case 1   Total tank failure, simultaneous failure of surrounding bund – release spreads as though on flat terrain.

Case 2   Total tank failure, surrounding bund undamaged, some LNG overtops depending on spread calculation.

Case 3   Total tank failure, liquid retained in bund.

Cases of immediate and delayed ignition are considered, as appropriate. This is typical of a high consequence-low frequency event of the type considered in Safety Report for an LNG Storage site (see discussion on Glass and Johnson[7]).

For immediate ignition, a running pool fire model was used to calculate the maximum diameter of the pool fire. Radiation predictions for a steady state pool fire with this maximum diameter were used to calculate the hazard to people. The pool fire was assumed to exist until the mass of LNG burned by the fire exceeded the mass of LNG released. It is likely that after the bulk of the LNG pool had burned away, a much smaller pool fire would persist close to the release point. However, sensitivity studies showed that the additional effects of the radiation from this fire were negligible.

If no immediate ignition occurs then the dispersing clouds may ignite at a later time. There is also a possibility that local pockets of flammable gas may exist between the location of the mean in-plume LFL and half LFL contours. For the purposes of this analysis, it has been assumed that when ignition occurs, all of the gas in the cloud inside the mean in-plume LFL contour is burnt in the flash fire and that all of the local pockets of flammable gas between the LFL and half LFL are ignited. It is also assumed that the remaining LNG in the liquid pool ignites at the same time.

For delayed ignition, a running pool fire model was used to calculate the maximum diameter of a pool fire ignited in the appropriate time interval. Radiation predictions for a steady state pool fire with this maximum diameter were used to calculate the hazard to people. The pool fire was assumed to exist until the mass of LNG burned by the fire exceeded the mass of LNG released. A transient dispersion model was used to predict the maximum extent of the mean in-plume LFL and half LFL contours in the appropriate time interval.

In summary, the following scenarios were modelled:

- 5 different ignition times
- 3 pool diameters
- 11 different combinations of wind speed and atmospheric stability
- 12 different wind directions

To calculate the average fatalities, the probability of each wind speed and direction was taken from the wind rose data. Each appropriate atmospheric stability was assumed to be equally likely, that is, in a 2 m/s wind speed there was a 20% probability of each atmospheric stability between B and F, whereas a stability category of D was always used for a 10 m/s wind. The overall average number of fatalities for this scenario is calculated by assuming that immediate ignition occurs 30% of the time, delayed ignition 60% of the time and no ignition 10% of the time. It has also been assumed that case 3 happens 90% of the time, case 2 9% of the time and case 1 the remaining 1% of the time. If delayed ignition occurs, then ignition in each of the separate time intervals is assumed to be equally likely.

Figure 2 shows how the average number of fatalities varies with time of ignition. As can be seen, for this transient event, it would not be obvious in advance which set of conditions would lead to a 'worst' case. Figure 3 complements this information by showing how the relative vulnerability (or relative individual risk) varies with distance from the source of the release along a particular 'ray' emerging from the source. The vulnerabilities are shown for 5 different times after the start of the scenario. This shows that vulnerability at each distance increases and then decreases throughout the event. However, the time of maximum vulnerability is different for different distances from the source.

The FN curve for this scenario is shown in Figure 4, assuming an appropriate event frequency. This is compared with a line that can be inferred from HSE Guidance as to acceptable societal risks for any single event from a large installation[8].

DISCUSSION

An analysis of the type described above allows the calculation of individual risk at specific locations, for instance in the control room, of individual risk contours and of societal risk. The structured way in which the scenarios are handled removes a degree of subjectivity from the analysis and allows a consistent and auditable approach to be used. Using a realistic range of parameter values, when exact values cannot be know in advance, reduces the sensitivity of the analysis to the input conditions. However, like all methods of analysis, the answers that are obtained are only as good as the quality of the input data used to describe the scenario.

The results obtained can be compared with risk criteria in order to aid decision-making on the acceptability of the risks. A typical comparison is shown in Figure 4, where the FN curve is compared with a particular acceptability criterion that has been applied to determine if a scenario is totally unacceptable in all but exceptional circumstances. One of the main purposes of carrying out the assessment of major accidents is to aid the decision making process, as to whether risks are broadly acceptable or whether further safety improving measures need to be taken. The realistic analysis gives a much better representation of a major accident, as opposed to using a simple worst case or representative analysis. This in turn gives a more accurate assessment of risk, thus enabling decisions to be taken with a higher degree of confidence.

The analysis method illustrated above has been applied in practice to real sites. For example, it has been used in the preparation of COMAH reports in the UK and it is currently being used in updates to some of these reports. The method has also been used in quantitative risk assessments for sites elsewhere in the world, both for existing and new facilities. There are number of important benefits in using the methodology. It allows a better judgement to be made of the distribution of risks. For example, cases have been found where the maximum number of fatalities and the average number of fatalities peak at different times after the event initiation. Depending on the area of most concern, different mitigation measures may be proposed to tackle what is the largest contributor to the overall risk. Again, use of this approach is less subjective and gives consistency between assessments.

The analysis also allows the benefits of any specific mitigation measures to be analysed, particularly if linked to cost benefit analysis. This was illustrated elsewhere[2], where a particular liquid spill was analysed, with or without flow limiters installed to reduce the spill rate. It was shown there that the realistic analysis allowed the effects of the flow limiters to be assessed more rigorously by not only comparing the maximum number of fatalities but also the average number of fatalities. Using a simple analysis, the results of a cost benefit analysis showed little benefit in installing the suggested safety improving measures, whilst a realistic analysis gave a much better measure of the improvement. The implications of this are significant, as using a worst case analysis may mean that safety improvement measures may not be installed when they should be or, if alternatives are being assessed, a less effective option may be selected. Having automated the method it is efficient to use on a standard desktop PC to investigate such issues. Despite its complexity, changes can be assessed quickly, which is of particular use in cost benefit analysis or if the project is at a design stage. The speed of modern computers means that time taken to carry out calculations is not significant. The most time consuming stage of the process is in obtaining the appropriate input conditions to represent the scenario and discussing with site personnel the way in which the plant will respond to different types of release. Experience suggests, however, that time spent in this way is well spent.

Finally, it is noted that, in principle, the vulnerability of the environment could also be evaluated on a similar array of grid points. For example, the amount of any toxic fluid calculated to reach a sensitive area or waterway or that is calculated to percolate into the subsoil could be used in evaluating the harm to the environment in a similar way to using thermal radiation or over pressure in evaluating the harm to people. In this way the environmental consequences of major accidents could also be assessed.

REFERENCES

1. Environmental Agency, HSE and Scottish Environment Protection Agency, 1999, Preparing safety reports – control of major accident hazards regulations 1999, HSG 190, *HSE Books.*
2. Cleaver, RP., Robsinson, CG., Halford, AR., July 2002, Analysing high consequence, low frequency accidents on process and storage plants, *Chemical Engineer.*
3. Birch, AD., Brown, DR., Fairweather, M., Hargrave GK., 1989, An experimental study of a turbulent natural gas jet in a cross-flow, 1989, *Comb. Sci. & Tech,* 66; 217–232.
4. Hymes, I., 1983, The physiological and pathological effects of thermal radiation, *SRD R275.*
5. Chemical Industries Association. Guidance for the location and design of occupied buildings on chemical Manufacturing sites, February 1998, *Chemical Industries Association (CIA) publication.*
6. Lees, F., Loss prevention in process industries, 1995, 2nd Edition butterworth Heinemann
7. Glass, D., Johnson, M., Demonstrating the tolerability of risk from major accidents, *March 2003,* Submitted to Hazards XVII, Manchester March 2003.
8. HSE, Reducing risk, protecting people, 2001, *HSE Books.*

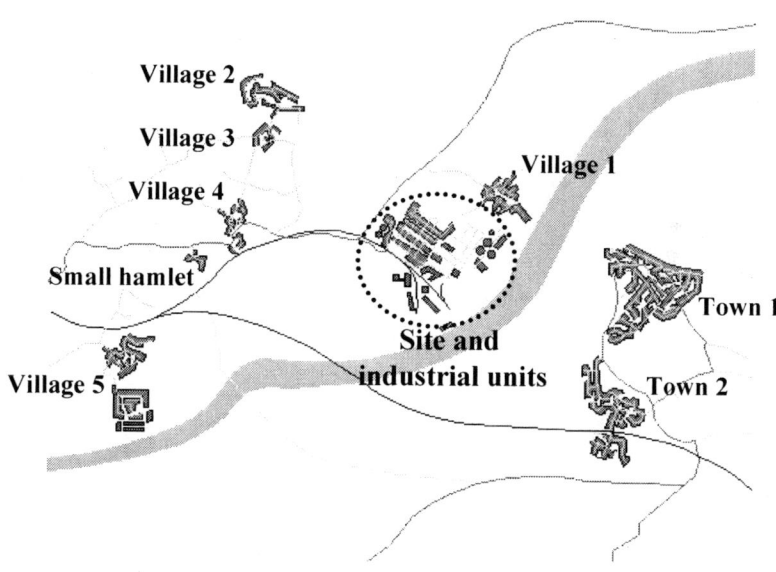

Figure 1. Location of the off-site population assumed in this hypothetical example

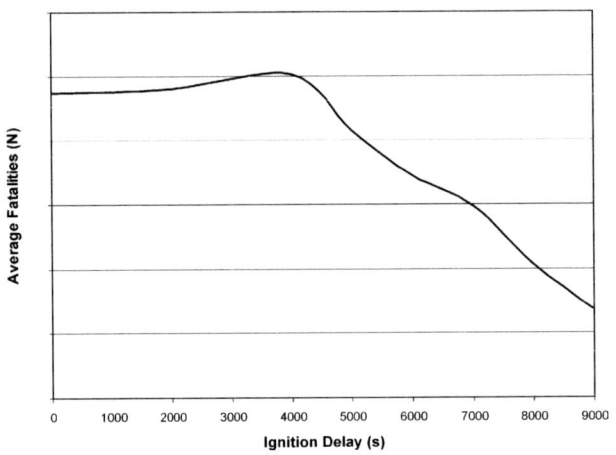

Figure 2.   Average number of fatalities for LNG tank failure

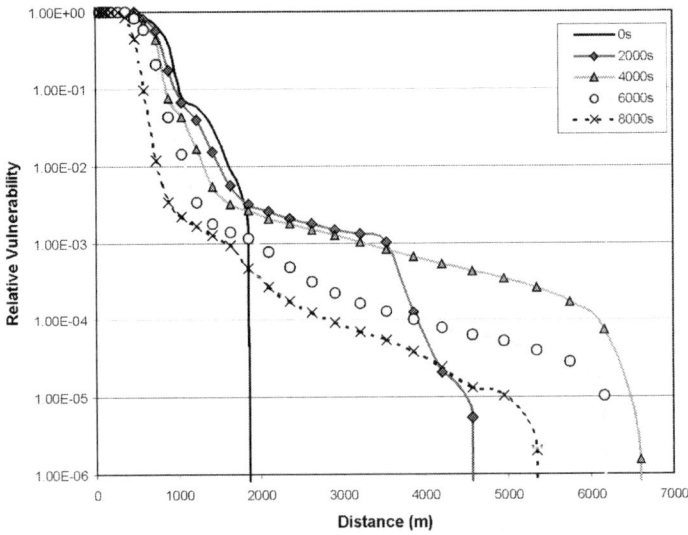

Figure 3.   Vulnerability of individuals along a line from the centre of the release for persons inside 90% and outside 10% of the time.

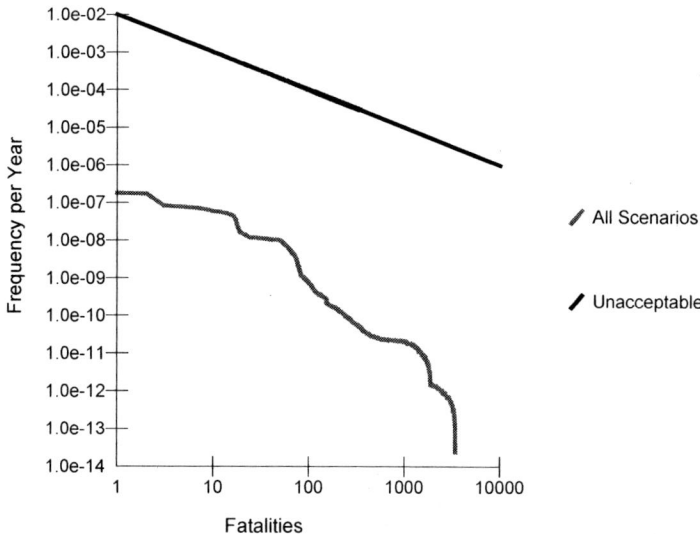

Figure 4. F-N curve for a failure of the LNG tank

EVALUATION STUDY OF RISK ASSESSMENT PROCEDURES FOR SMALL-SCALE CHEMICAL FACTORIES IN JORDAN

Dr. Mazen M. Abu-Khader
Department of Chemical Engineering, Faculty of Engineering Technology, Al-Balqa Applied University, Box: 15008, (11134) Amman, Jordan.

An evaluation study was conducted on Risk Assessment Procedures implemented in eleven small-scale chemical factories in Jordan. The assessments failed to reduce number of minor incidents occurring within the working site area in four factories. All incidents are directly or indirectly caused by human behaviour. Therefore, understanding the human factor and human behaviour is an important issue. It is quit hard to change the human behaviour over night to meet all the obligations from a simple worker with limited skills. The implementation of ISO 9001 or 9002 contributed indirectly in reducing the number of minor incidents in these chemical factories. The foreign Risk Assessment Expert did not take into account the socio-economical and human behaviour factors when conducting the Risk Assessment Procedure on these factories. It is crucial to look at the effect of the environment and human behaviour within the framework of socio-economical issue before conducting an assessment. There are also several important issues should be taken into consideration, such as: a) The level of education of the working staff, b) The Job stability and satisfaction, c) low wages, d) The character and behaviour of the daily worker. The key for a successful implementation of a Risk Assessment in a factory is to build up a two-way communication between the top management and the simple worker to fulfil his needs and provide him with sense of security and job stability.

KEYWORDS: Risk Assessment, Risk Management, Human Behaviour, Minor Incidents, ISO 9000.

INTRODUCTION

Jordan is a developing country located in the East Mediterranean Region. It has a unique history of straggling to build up a stable economy to cope with the huge increase of population due to the wars in the Middle East (1948 and 1967), the Gulf War in 1990 and political instability in the region, which created huge pressure on socio-economical levels because the country has very limited resources. Also, Jordan has a wide variety of cultures, traditions and religions among its population. The labour union law was established in 1953 and was modified several times during 1960, 1972 and 1974. That was due to the increase in the industrial growth and to cope with the new developments in labour rights issues.

Before the Risk Assessment Procedure is conducted, it is highly important to understand the nature of people that we are dealing with. It is definitely true that the result of a Risk Assessment for two identical plants in two different locations is different, (for example, one in Europe and another in the Middle East). The assessment is not just ticking off boxes in an instruction sheet or investigating the sources for human health risks.

An ideal risk assessment method which would suit all organizations does not exist, as each organization possesses its own unique characteristics[1]. However, the requirements for an ideal risk assessment method have been studied by several researcher[2,3,4]. A method should be complete in its coverage of all necessary components, and to perform a risk analysis and risk management at the same time.

Risk Assessment is not new and in simple words it is an action we take as individuals, almost in every working day. As individuals, we spontaneously analyse, assess and decide upon risky situations without thinking of the uncertainty of the outcomes. The technical development of Risk Assessment for human health and occupational standards started as early as 1930s. The US National Research Council (NCR) has established in 1983 all the work related to the systematic and quantitative approach to Risk Assessment. The NRC regarded the process of risk assessment as an activity conducted by the application of objective science and scientific principles, while risk management was viewed as a decision making process that 'entail (ed) considerations of political, social, economic and engineering information with risk-related information ... to develop, analyse, compare and select the appropriate regulatory response ... The selection necessarily requires the use of value judgments on such issues as the acceptability of risk and the reasonableness of the costs of control'[5].

The main typical elements of Operational Risk Management (ORM) process can be summarized by the following steps[6]:

1.   Identify hazards:
     a) Outline the major steps in the operation (operational analysis).
     b) A preliminary hazard analysis is conducted by listing all of the hazards associated with each step in the operational analysis along with possible causes for those hazards.

2.   Assess hazards:
     For each hazard identified, the associated degree of risk in terms of probability and severity is determined.

3.   Make risk decisions:
     Developing risk control options. Start with the most serious risk and selection of controls that will reduce the risk to a minimum level. The communication with higher authority is necessary when assistance is needed to implement controls.

4.   Implement Controls:
     Engineering and administrative controls are to be implemented considered. These would include modification of equipment, standard operating procedures, work rotations, personal protective equipment, etc., that are also standard controls for chemical and physical hazards encountered in occupational settings.

5.   Supervise:
     Conduct follow-up evaluations of the controls to ensure they remain in place and have the desired effect. Monitor for changes that may require further operational risk management, and take corrective action when necessary.

From the previous five elements of the ORM, it is clear that the human factor has not been considered as a source of health risk. On the other hand, the major safety issue in the development of technological societies is the consideration of the human element as the source of, and contributor to, accidents[7], and that all accidents in artificial systems are directly or indirectly caused by human behaviour. For those of us interested in genuine safety, and not just legally prompt cosmetic action, understanding the human factor and human behaviour is unavoidable[8]. There is a need to provide officials, professional safety managers and workers with a better understanding of the different categories of human behaviours and the performance-shaping factors behind these categories. A simulated model[9] on human behaviour and human interactions was described based on three structured frameworks:

1. *Human error event frameworks* aim at representing human errors within the socio-technical and organizational context.
2. *Human behaviour frameworks* describe how individual human behaviour is modelled, with respect to the data and the boundary conditions due to the environmental and the organizational context, as represented in the previous framework.
3. *Human–machine interaction frameworks* account for the interaction between models of operator and plant and produce an overall simulation of a range of accident dynamics

From the presented frameworks, it is clear that human behaviour is the key element in constructing the simulated model.

Workers with limited skills represent high percentage of the working force in most of the small-scale chemical factories. Therefore, it is quit difficult to fulfil all the technical requirements to meet the entire obligation from a simple worker. It is necessary to use simplified procedures and construct an easy scheme to implement. For an Example, a UK scheme to help small firms control health risks from chemicals in the workplace has been developed by a working party set up by the UK Health and Safety Commissions Advisory Committee on Toxic Substances (ACTS). Two important criteria which were applied to the development of the toxicological basis for the scheme were firstly, the approach taken must be simple and transparent, such that it can be readily understood and consistently applied by small and medium sized enterprises and secondly, the best use should be made of any available hazard information[10]. It is acknowledged that safety, health and environmental protection measures have become important and shall be initiated by the highest level of management of the company, and not be left to staff experts, but must be felt and appreciated by the company as a whole[11]. To achieve this is as much a technical as an organizational task and it requires good communication. The human factor is a major element in safety.

From all of the above, it is crucial to look at the characteristic behaviour of the working labours with the frame of present environment and socio-economical issues before conducting an assessment. This evaluation study is trying to prove that human factor is an essential part of any successful assessment, and studying all characteristic behaviours which may lead to a better understanding of what we can call unique cultural behaviour inside the factory.

EVALUATION STUDY
This industrial evaluation study was conducted on a set of eleven factories working in the field ranging from chemical production to chemical mixing and packaging. The working force ranges from 80 up to 120. The factories are located in different industrial areas in the Greater Amman Municipality. All the selected factories have obtained either ISO 9001 or 9002 for more than one year before carrying out the Risk Assessment. The Risk Assessment Procedures were done through foreign agencies. Most of these factories have primitive and simple safety scheme implemented inside the factory's premises.

From the results of the study, it was noticed that the Risk Assessment Procedures for four out of eleven factories failed to reduce the number of accidents which can be classified as minor incidents occurring in the working location. Table 1 illustrates the number of incidents for the eleven factories before and after conducting the Risk Assessment. It shows that there was a clear drop in number of both minor and major incidents after obtaining either ISO 9001 or 9002. This may be due to restrictions implemented on workers, and limiting their access to certain departments which minimize contact and friction with other employees. Also, the creation of Job Description helped to identify the activities of each worker within a group of workers. Also, Table 1 shows that after conducting the Risk Assessment the drop in number of minor incidents what was not up to expectation.

Table 1. Evaluation results of the factories before and after conducting risk assessment procedures.

Implementation process	Minor incidents	Major incidents
Before obtaining the ISO 9001/9002	105	7
After obtaining the ISO 9001/9002 and before conducting the risk assessment	59	4
After conducting the risk assessment	33	1

Figure 1 shows clearly without any doubt that increase in number of minor incidents is directly related to increase in the percentage of workers with limited skills within the working force of the factory. Also, Figure 2 compares the change in the number of minor incidents after implementing the ISO 9000 and the Risk assessments methods. Figure 3 concludes that there are four factories failed to reduce minor incidents more than 50% of the its total number of incidents recorded.

DISCUSSION OF THE EVALUATION OUTCOMES
Based on the collected information on the eleven factories and by looking at the managerial scheme inside each factory, it can easily be noticed that these factories can be divided into two different types of managerial systems: Seven of these factories implement modern management systems where as Four factories apply classical management systems. The modern management systems open direct lines among all different working levels of the working force in the factories through controlled communication channels. An evaluation scheme is implemented for each worker to monitor his personal activity, performance and productivity. This concludes that before conducting any Risk Assessment, it is important to have a full picture of the management system implemented. This will help in building up faith in the assessment procedure and that it is not just a extra pile of paper work for the staff.

The four factories were the assessment failed to reduced the minor incidents use classical management systems. It is noticed that the factories employees more than 35% of workers with limited skills on daily paid basis which create the sense of job instability for the workers. Also they lack training opportunities to improve their working capabilities. Also most of these workers have low wages which reflects heavily on their performance and work commitment. Several recommendations were given to these factories to promote the concept of safety culture. The implementation of these recommendations has further reduced minor incidents in the first six months of implementation. These recommendations were:

1) Set-up of several controlled short breaks during the working day for the workers to do their private activities such as; doing prayers, having snacks, smoking, and drinking coffee and tea.

2) Install facilities to listen for traditional music in the working area.

3) The use of a simplified coding scheme to identify hazard chemicals, chemicals handling and storage procedures.

4) The use of clear signs for dangerous locations.

5) Minimize the paperwork used for documentations by labours with limited skills.

6) Set up social activities one every month between the management and the working staff.

7) Promote the idea of salary bonus for best worker of the month for those of low wages, especially for workers with limited skills. This is to encourage work commitment and job stability.

FUTURE TRENDS OF RISK ASSESSMENTS

There are several future trends of Risk Assessment should be taken into consideration. These trends be can summarized in the following points:

a) The concept of damage as primarily being limited to the risk to human and occasionally ecological health needs to be broadened to include a range of economic and social considerations.

b) It important to understand the cultural behaviour of workers inside the factory because the worker behaviour is a reflection of this safety culture. Also, to examine the regular activities for the simple worker during a normal working day.

c) The implementation of a simple coding system and installing signs in dangerous locations. Then making easy access for information when needed, and minimizing the paperwork among large percentage of workers with limited skills.

d) The assessments must allow for multiple exposure pathways, inter-media transfers of pollutants together with secondary environmental effects, as well as assessments that draw into the decision making framework technical, health, economic, social and other issues. Environmental quality standards for substances in air, water, soil, biota and foods will have to be developed in a manner that ensures internal consistency and coherence[12].

e) Risk Assessment needs to take greater attention on what may cause minor risks that could lead indirectly to major incidents. In most occasions major incidents occur from minor ones.

f) Involving the top management into the Risk Assessment process will definitely help in verifying what kind of risk the worker is exposed to. Also, it is important to open channels with all the parties involved in the Risk Assessment specially between the top management, the operational personnel in charge, and the Risk Assessment Expert.

g) The challenge in the coming years will be to embed the use of science in risk assessment and risk management within a socio-political framework, and to subsume within the decision making process the nature of things and the nature of man. This is not to diminish the status or role of scientists and of experts in environmental decision making[13].

CONCLUSIONS

The chemical industries in Jordan have limited experience in the issues of security and labour safety. Therefore implementing procedures and instructions as listed in the manuals is not always a successful way in conducting Health Risk Assessment. It is important to take into consideration the management system implemented inside the selected factory for the assessment. It is highly crucial to build up confidence and faith from both sides: the management and the simple worker so that the proposed safety system and modifications will be definitely fruitful, and not just an extra pile of paper work.

The foreign Risk Assessment Expert usually misses out such a vital matter when carrying out human health assessment. It is quit hard to change the human behaviour over night to meet all the obligations from a simple worker. The assessor should always keep in his mind that conducting an assessment will not be necessarily the same for the same factory in a different location. Complicated and long list of procedures are not necessary the best way to implement Risk Management. A simple scheme of procedure may prove more fruitful and easy to implement by simple workers inside the factory. Not every program or scheme implemented in Developed Countries can be easily implemented in Developing Countries. Verifying all the obstacles is an important step before implementing an assessment procedure. We teach the baby to crawl first, stand up and then walk.

Human behaviour is the main issue to be included in any assessment conducted not only cultural behaviour, but also behaviour based on the sense of job instability and security. Also it is important to implement Operational Risk Management parallel with risk assessment methods when dealing with minor incidents in small factories with high percentage of workers with limited capabilities and where the availability of training programs is limited and expensive.

REFERENCES
1. Lichtenstein, S., 1996, Factors in the selection of a risk assessment method, *Information Management & Computer Security*. 4/4, pp. 20–25
2. Anderson, A.M., 1991, Comparing risk analysis methodologies, *Proceedings of the IFIP TC11 Seventh International Conference on Information Security*, North Holland, New York, NY, Amsterdam, pp. 301–11.
3. Clark, R.,1989, Risk management – a new approach, *Proceedings of the Fourth IFIP TCII International Conference on Computer Security*, North Holland, New York, NY, Amsterdam.
4. Moses, R.,1993, A European standard for risk analysis, *Proceedings of COMPSEC International*, Elsevier, Oxford, 1993.

5.   National Research Council, 1983. Risk assessment in the Federal Government: managing the process. Washington DC: National Academy Press, 1983.
6.   CNO, April 3,1997, (Chief of Naval Operations).OPNAV Instruction 3500.39. "Operational Risk Management", Department of the Navy, Washington, DC.
7.   Sheridan, T.B., 1985, Forty-five years of Human–Machine systems: history and trends. Keynote Address. *Proceedings of 2nd IFAC Conf. on Analysis, Design and Evaluation of Human– Machine Systems*. Varese, Italy, Pergamon, Oxford.
8.   Sundstrom-Frisk C., 1999, Understanding human behaviour: A necessity in improving safety and health performance, *Journal of Occupational Health and Safety*, vol.15, issue 1, pp. 37–45.
9.   Cacciabue, P.C.,1998, Modelling And Simulation Of Human Behaviour For Safety Analysis And Control Of Complex Systems, *Safety Science*, vol.28, No.2, pp. 97–110.
10. Brooke, M. I., 1998, A UK Scheme to Help Small Firms Control Health Risk from Chemicals: Toxicological Considerations, *Ann. Occup. Hug.*, vol. 42, no.6, pp. 377–390.
11. Pasman, H.J, 2000, Risk informed resource allocation policy: safety can save costs, *Journal of Hazardous Materials*, vol.71, pp. 375–394
12. Van de Meent, D. and de Bruijn, JHM, 1995, A modelling procedure to evaluate the coherence of independently derived environ-mental quality objectives for air, water and soil, *Environ Toxicol Chem*, (14), pp.177–186.
13. Eduljee, G.H., 2000, Trends in risk assessment and risk management, *The Science of the Total Environment*, (249), pp.13–23.

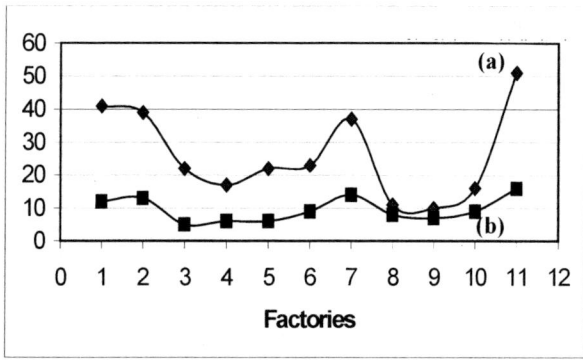

Figure 1.   The relationship between the percentage of labour with limited skills **(a)** and the number of minor incidents **(b)**

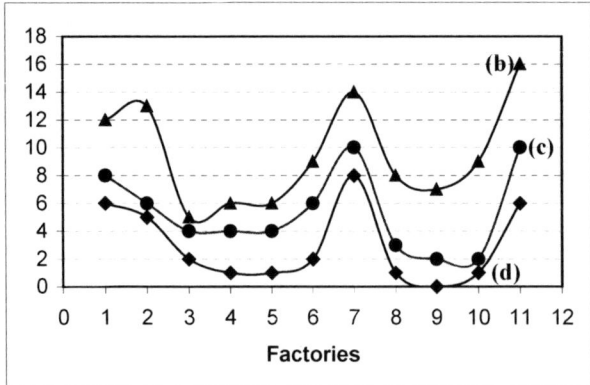

Figure 2. The change in number of minor incidents **(b)**, after implementing ISO 9000 **(c)**, and after conducting risk assessment method **(d)**

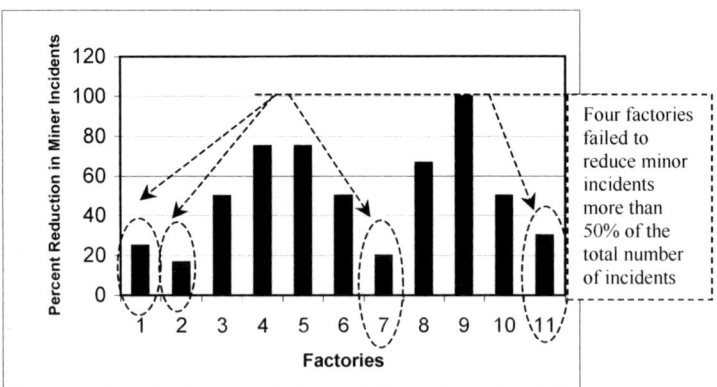

Figure 3. Percentage of reduction in minor incidents in each chemical factory

QRA STUDY OF AN ACTIVATED CARBON FILTER SAFEGUARD SYSTEM

M. Sam Mannan, Yanjun Wang, and Harry H. West
Mary Kay O'Connor Process Safety Center, Chemical Engineering Department, Texas A&M University System, College Station, Texas 77843-3122, USA (mannan@tamu.edu)

The Mary Kay O'Connor Process Safety Center (MKOPSC) at Texas A&M University conducted a Quantitative Risk Analysis (QRA) for subsystems of the VX neutralization process. The process is conducted in a negative pressure containment system. The very large air handling system also acts to direct any fugitive emissions or small leaks to the carbon adsorption filter systems, the final safeguard in preventing highly toxic chemicals from escaping to the atmosphere.

The activated carbon filter system concentrates the low level fugitive emissions, thereby creating a potential for a more significant catastrophic release if the carbon filter system fails. The risk trade off between capturing low-level fugitive emissions and the potential for a large-scale toxic chemical release must be compared. The fault trees for the air handling system and the carbon filter systems are highlighted. In particular, the uncertainty in the reliability data and their respective impact on the overall failure rate and system availability are emphasized.

The importance of this study was to point out a specific failure mode that was not adequately addressed with safeguards in the original process design, thereby creating an unacceptable risk and requiring additional safeguards. The study also verified that the safeguards in the original design for all other identified failure modes reduced the risks to generally acceptable levels.

KEYWORDS: Process Safety, Quantitative Risk Assessment, Loss Prevention, Fault Tree Analysis, Failure Modes

INTRODUCTION

O-ethyl S-(2-diisopropylaminoethyl) methylphosphonothiolate ($C_{11}H_{26}NO_2PS$) is commonly referred to as VX (Chemical Abstract Service Number 50782-69-9). It is an organophosphorous ester and is a lethal nerve agent. It enters into the body by respiration or skin absorption. It is very persistent and does not evaporate under the normal temperature. VX is generally stored in ton containers. During the VX neutralization process, the ton containers are punched, washed, decontaminated and cut apart in a negative air pressure containment building. The VX agent is collected and then sent to two subsequent reactors. In these reactors, the VX agent reacts with aqueous sodium hydroxide (NaOH) to produce hydrolysate. After the target demilitarization level (330 ppb) is achieved, the hydrolysate product is then sent to a supercritical water oxidation systems for post-treatment to remove the organics.

The US Army System Safety Program Plan defines a risk assessment code (RAC)(Table 3) and three concerned hazard scenarios: VX agent release, personnel injury/illness, and system loss. In this study, the Fault Tree Analysis (FTA) methodology was employed to estimate the risk level of these three major events in the VX neutralization subsystem and the associated support system. Tables 1 and 2 present the hazard severity level definition and frequency level definition for each incident. The RAC matrix (Table 3)

683

is then defined based on the hazard severity level and the hazard frequency level. The defined risk levels range from 1 to 4. Each has a specific acceptability criteria and resolution authority, which is shown below the RAC matrix.

The VX neutralization process is designed to operate in a negative pressure containment system. The very large air handling system is vital for orderly and safe operation. The carbon adsorption filter systems are the final safeguards in preventing highly toxic chemicals from escaping to the atmosphere. The aim of this study is to evaluate the activated carbon filter system from a safety point of view.

METHODOLOGY

In this QRA project, FTA methodology was applied to the VX neutralization processes to obtain the frequency of the major hazard incidents. The hazard frequency level definition (Table 2) was then applied to acquire the corresponding RAC for each scenario. If the calculated RAC code was unacceptable, possible safety mitigations were recommended and compared by revising and reevaluating the fault trees to bring the system to an acceptable risk level.

Consequence analysis was first employed to determine a hazard severity level of II – Critical. FTA was then used to calculate the occurrence frequency of potential incidents. The first step of FTA is to identify the undesired top events. As mentioned, three major hazards were defined: agent release in-plant, personnel injury, and system loss. In the second step, the backward-reasoning FTA technique is applied to each scenario until external or primary basic events, whose failure rate data are available, are reached. It is essential to include necessary and sufficient events that can contribute to the top event. Using the process information, failure rate data, and human error probability; an estimation of the probability of the identified hazardous incident can be accomplished.

Average failure rate data for process equipment are available in literature in many databases. Data exists on probability of failure on demand for safeguards as well as human errors. In this study, the American Institute of Chemical Engineers (AIChE) Center for Chemical Process Safety (CCPS) publication, "Guidelines for Process Equipment Reliability Data with Data Tables," [1] was the primary source of equipment failure data. Other sources include the offshore reliability data (OREDA) [2], Mechanical Reliability [3], and specific technical articles in the MKOPSC Library that provide special process equipment or process instrumentation failure data. Other equipment failure data can be obtained from the Government Industry Data Exchange Program (GIDEP) [4], IEEE Standard 500 1984 [5], and European Industry Reliability Data Handbook (EIREDA) [6]. Additional equipment reliability and failure rate databases are available for purchase or by a combination of financial support and equipment reliability data sharing. Hartford Steam Boiler and the AIChE special interest group on reliability are examples of groups that have additional data. However, the use of these special-purchase failure rate databases was not appropriate for this study.

In some instances, failure data can be represented by statistical distribution functions, thereby permitting Monte Carlo analysis techniques to predict a distribution of failure estimates. However, the dearth of applicable failure probability distribution data for this new VX neutralization technology does not justify such complex modeling methods.

HVAC SYSTEM DESCRIPTION

The VX neutralization facility was designed to provide agent containment through the use of effective physical separation between the toxic and nontoxic areas and a ventilation system design using progressively negative differential pressures. Areas are categorized according to the potential contamination level. Areas with the higher potential of contamination must be maintained at lower (negative) pressure. Thus air is controlled to flow progressively from the areas of the least probable contamination to the areas of the highest probable contamination. The building's design also facilitates the HVAC design goals with construction to provide appropriate airlocks to minimize air leakage. The process vent streams cannot be discharged directly to the outside because they contain trace amounts of VX agent and volatile organic compounds (VOCs). These vent streams are filtered at the main activated carbon filter units to remove VOCs and agent before transferring into the building's exhaust air streams.

The VX neutralization facility includes a cascade ventilation system that is vital for orderly and safe operations because it provides controlled air temperature, air pressure, and flow to confine the contaminants within special areas. The functions of the cascade HVAC system are:

1. Protect the equipment/building areas and the site's environment.
2. Provide sufficient air volume to remove agent and VOCs from the contaminated areas.
3. Maintain negative room pressure to prevent diffusion or leakage of possible contaminants to the outside.
4. Control and minimize the spread of contamination by maintaining a certain direction of airflow.
5. Receive and filter the process vent stream to minimize discharge to the atmosphere.

The cascade ventilation system was designed to have a total of four supply air handling units and eight exhaust activated carbon filter units. During normal operation, three supply air handling units and six exhaust activated carbon filter units are online. The air handling units supply 100% outside air. The air is then transferred to areas of successively more contamination potential. The pressure in each room is monitored and alarmed if an out-of-range pressure condition occurs. If the pressure between Category C and A/B rooms equalizes, isolation dampers are closed to confine the possible contamination in specific areas. Air entering the supply units passes across an air tempering hot water coil with face and bypass dampers. The air then passes through media particulate filters rated at 30% and 85%. Next, the air is heated by a hot water coil or cooled by chilled water to the desired temperature. Constant-flow, variable-speed centrifugal fans are used to overcome the pressure drops through the coils, filters, and ductwork to move the air to the various Category C areas. Differential pressure gauges are installed to monitor the filter loading and alarm on high-pressure differential.

The supply air-handling system is started manually at the main HVAC control panel. Interlock logic prevents more than the prescribed number of units in the system from being operated at a time. In the event of a loss of airflow in an operating supply unit, a low-limit flow transmitter signals an alarm at the main HVAC control panel and resets the interlock logic so that the standby fan can be automatically started.

HIGH-HUMIDITY EXHAUST SCENARIO
Based upon preliminary failure mode analysis studies, a number of scenarios were evaluated. This study describes a particular scenario that has the potential for compromising the carbon filter systems. Other scenarios, such as fire within the carbon bed systems, had adequate safeguards defined in the original design.

The performance of the activated carbon bed filters is critical to successful containment of any agent that might be present in the exhausts. The performance of any activated carbon bed filter is known to degrade when the relative humidity of the exhaust air exceeds 70%. However, experimental measurements or calculations are not available for the efficiency of carbon bed filters under high-humidity conditions. In this case, each exhaust filter contains six carbon beds in series. Agent monitors are located after the first and second beds. However, the conservative assumption of a common mode failure is that VX agent breakthrough on one carbon bed due to high relative humidity will subsequently bypass all remaining carbon beds.

The cascade HVAC system depends on operating system heat loads to increase the air temperature and thereby reduce the relative humidity of the air before it reaches the exhaust activated carbon filters. Incoming air is cooled to 55°F; therefore, the exhaust air would normally be about 55% relative humidity at 72°F in the summer. Lower relative humidity can be expected in the winter.

Furthermore, the HVAC system is designed to remain operational during an emergency, such as a VX leak. However, once a VX leak is known, all other operations will cease, thereby reducing the process heat load. Preliminary analysis of the heat loads suggests that non-process heat loads add about 10°F of sensible heat to the HVAC exhaust. Hence, humidity control procedure systems are required to ensure activated carbon bed performance. This can be further compounded upon the loss of utility power because only essential power will be online, and sensible heat to the HVAC exhaust will be minimal. No mitigation credit can be taken for process system shutdown upon agent breakthrough on the first carbon bed, because in this scenario, it is already assumed that the process system has been previously shut down because of a large leak event. Depending on the post-release procedures to bring the HVAC system back to operational status, in addition to a potential agent release scenario, the high humidity failure event would also be an HVAC system loss category failure.

A high humidity condition in the exhaust air could also be the result of excess steam during the steam cleaning of the VX containers after the contents have been transferred to the neutralization reactors.

RESULTS & RECOMMENDATIONS
In the original HVAC design, redundant moisture sensor-activated exhaust control system was not included to ensure that the relative humidity of the exhaust manifold upstream of the carbon filter systems is below 70%. RAC hazard Severity Level II (agent release) was assigned to this event because the HVAC carbon filters would be inoperable, hence, any simultaneous VX leak would be outside engineering controls. "Agent release through the HVAC system" events were found to be at RAC 2 (undesirable) prior to implementing recommended changes (as shown in the attached fault tree in Figure 1).

Note that one can calculate the impact of the humidity control interlock system by setting the probability of "supply air heater fails to operate on high humidity" under Gate

158 to 1. Hence the impact of recommended control features on the RAC matrix can be easily evaluated.

An exhaust air humidity control interlock system with redundant humidity sensors is recommended to bring this event to RAC 3 (acceptable with control). In the high humidity scenario, the supply air heaters will be turned on and the supply air chillers turned off until the exhaust air relative humidity is controlled to an acceptable limit. This system will ensure that the relative humidity contacting the activated carbon filters will remain under the design limit of 70% relative humidity. The humidity sensors should be inspected on a 4- to 6-month preventive maintenance schedule. In particular, a 2 out of 3 voting logic control system may be considered to add even higher reliability to the relative humidity safeguard system.

The heat loads created by cleanup or repair crews during a maintenance period are not known. Hence, the complete reliance on space temperature in neutralization cubicles to activate a high relative humidity safeguard system was not considered appropriate in this study. This space temperature automated system is somewhat similar to the dual exhaust air humidity sensors systems recommended in the base case hereunder, but relies on an indirect measurement of the relative humidity at the exhaust air carbon filters.

Another alternative strategy of using operating procedures for the operator to respond to a high humidity alarm was considered and rejected because the human error failure rates are not sufficiently reliable enough to achieve failure frequencies consistent with RAC 3.

CONCLUSIONS
The HVAC system maintains an environment of progressively negative differential pressures to control and minimize possible contamination. The performance of the activated carbon filter systems is critical to capturing the agent and VOCs that might be present in the process exhaust. High humidity is a common cause failure of the series activated carbon filters. The original design without the exhaust moisture control system was found to be unacceptable according to RAC matrix. The results of this study indicate the humidity safeguards of the activated carbon filters will bring it to "Acceptable with controls". The study also verified that many other identified failure modes, such as carbon filter fires and errors during unloading/reloading carbon were adequately protected by safeguards.

REFERENCES
1. Guidelines for Process Equipment Reliability Data with Data Tables, Center for Chemical Process Safety AICHE, *American Institute for Chemical Engineers*, 1989.
2. Offshore Reliability Data Handbook, OREDA-84, P.O. Box 370, N-1322, HOVIK, Norway, 1984. Distributed by *Pennwell Publishing Company*, Tulsa, OK.
3. Moss, T.R. (editor), "Mechanical Reliability", IPC Science & Technology, Guiford, U.K., 1980.
4. The Government Industry Data Exchange Program (GIDEP), http://www.gidep.corona.navy.mil/
5. IEEE Standard 500-1984 Guide to the Collection and Presentation of Electrical, Electronic, Sensing Component, and Mechanical Equipment Reliability Data for

Nuclear-Power Generating Stations, *The Institute of Electrical and Electronics Engineers*, New York, NY, 1983.
6.   European Industry Reliability Data Handbook (EIREDA), EUORSTAT, Paris, 1991.

ACKNOWLEDGEMENTS
This research was funded by the Mary Kay O'Connor Process Safety Center of the Department of Chemical Engineering at Texas A&M University.

Table 1.   Hazard severity level definition

Severity level	Agent release in-plant	Personnel injury/illness	System loss
I—Catastrophic	> IDLH outside engineering controls	Illness, death, or injury involving permanent total disability	> 25% and/or >1 month to repair
II—Critical	≥ AEL outside of engineering controls	Injury involving permanent partial disability	10% to 25% and/or 1 week to 1 month to repair
III—Marginal	≥ AEL inside nonagent areas	Injury involving temporary total disability	< 10% and/or 1 day to 1 week to repair
IV—Negligible	< AEL nonagent areas	Injury involving only first aid or minor supportive treatment	No system loss downtime, or repairs completed within 1 day

AEL = airborne exposure level ($VX = 0.00001$ mg/m^3)
IDLH = immediately dangerous to life and health ($VX = 0.02$ mg/m^3)

Table 2.   Hazard frequency level definition (event/year)

Qualitative frequency	Agent release in-plant	Personnel injury/illness	System loss
A — frequent	A ≥ 1E–01	A ≥ 10.0	A ≥ 1.0
B — probable	1E–01 > B ≥ 1E–02	10.0 > B ≥ 1.0	1.0 > B ≥ 1E–01
C — occasional	1E–02 > C ≥ 1E–03	1.0 > C ≥ 1E–02	1E–01 > C ≥ 1E–02
D — remote	1E–03 > D ≥ 1E–04	1E–02 > D ≥ 1E–04	1E–02 > D ≥ 1E–03
E — improbable	1E–04 > E ≥ 1E–06	1E–04 > E ≥ 1E–06	1E–03 > E ≥ 1E–06
F — rare	1E–06 > F	1E–06 > F	1E–06 > F

Table 3. RAC matrix

Qualitative frequency	Severity level			
	I (catastrophic)	II (critical)	III (marginal)	IV (negligible)
A — frequent	1	1	1	3
B — probable	1	1	2	3
C — occasional	1	2	3	4
D — remote	2	2	3	4
E — improbable	3	3	3	4
F — rare	4	4	4	4

Acceptability criteria:

RAC	Description	Resolution authority
1	Unacceptable	Assistant secretary of the army
2	Undesirable	Product manager
3	Acceptable with controls	Program manager-safety
4	Acceptable	

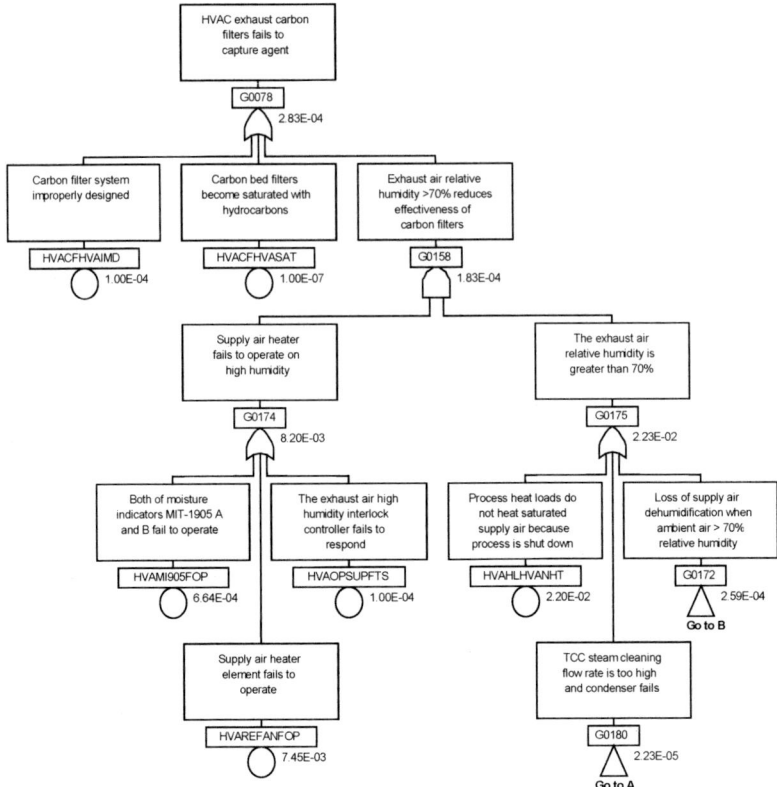

Figure 1. Fault tree for an activated carbon filter safeguard system

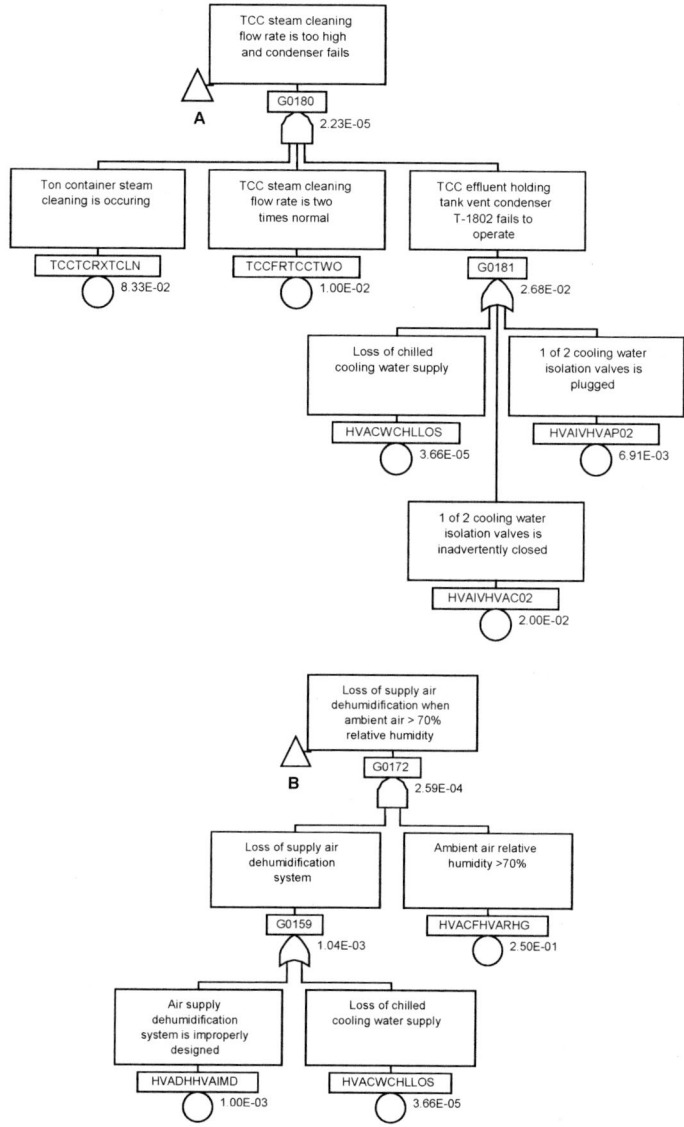

Note: The probability shown is based on 16-month.

Figure 1 (continued).   Fault tree for an activated carbon filter safeguard system

A WALK IN THE CHEMICAL PARK – PROCESS SAFETY PERSPECTIVES

Dieter Dambmann & Lee Allford
Clariant & The European Process Safety Centre

Chemical parks play host to many chemical manufacturers within Europe and offer many business advantages to both the site owners and guest companies. Clearly high density hazardous operations have a major implication for process safety management but such physical arrangements in Europe are hardly novel. Increasingly the challenge for management of major accident hazards within chemical parks is to attain a seamless working cooperation between separate legal enterprises so that overall safety integrity is preserved and then improved.

KEYWORDS: Chemical park, process safety

INTRODUCTION

There are many terms in popular use to describe a chemical park. Industrial park or multi-occupier site conveys the same meaning however chemical park appears to be in widespread usage across Europe and is the term which is used throughout this paper

A chemical park can be defined as a chemical manufacturing complex which possesses controlled entrance and exit points and accommodates several separately owned chemical manufacturing companies. The perimeter normally is secured by fencing and gives rise to the terms inside the fence and outside the fence which respectively delineate hazardous and non hazardous operations. There are no public roads as such inside the fence.

CHEMICAL PARK ORGANISATION

Typically a chemical park has evolved from a single legal entity whose safety organisation has controlled all activities inside the fence and has therefore offered a single point of contact to the outside world as illustrated in Figure 1. It is the loss of a single point of contact that can often lead to concern amongst local authorities charged with regulating hazardous operations on chemical parks[1].

In order to save costs many organisations have invited guest companies onto site either to operate existing facilities or build new facilities with a view to providing the guests with site services and infrastructure. The advantage for the host (often referred to as the major user) is that the centralised overhead is shared with the guest companies. The advantage for the guests is that they have on site an already existing and practised site service which allows them to concentrate on their core business. In some countries it is state policy to encourage the ongoing viability of large sites or at least prevent their disuse not least due to the possible change of land use and consequent land remediation issues.

Another option to the major user concept is the separate site operating company where the site services and infrastructure are operated by a company separate to its users and clients as illustrated in Figure 2.

Such an example in Germany is InfraServ on former Hoechst sites[6]. Another example in the UK is ETOL who supply services to a chemical park in Teeside[5]. The advantage to users is that they receive a customer focussed service which may operate on a lower cost basis compared to services operated by a major user or indeed single occupier.

IMPLICATIONS

So why should chemical parks in themselves have concerns for process safety?

The first is that the growing issue of chemical parks has gone hand in hand with the European chemical giants restructuring their resources in order to compete in the global market. Large company safety departments have fragmented. Safety professionals may find themselves working for different companies on the same chemical park whereas before they had all worked for the same organisation on the same site. Fortunately in this case the former company relationship can sustain the exchange and sharing of information between individuals which is pre-requisite for safe operation between neighbouring hazardous installations[1].

In the medium to long term for sustainable networking a more formal arrangement needs to be in place – out of which naturally flows informal sharing of valuable information. An example in Germany of such a network is the Industrial Practices Interest Group, IGR,[2] which arose from the break up of Hoechst AG and provides technical information on process safety to its subscribers from several guest companies on the same chemical park. Clearly it was felt important on the break up of Hoechst that its process safety knowledge base was preserved as far as possible.

The second concern is again symptomatic of large company fragmentation. Increasingly process safety managers find themselves in slimmed down departments and devoting much of their time to promoting process safety within their own organisations. They are therefore more inward looking than before. This can militate against exchange of information between guest companies on a chemical park even between seasoned professionals who have worked together in the same company.

In effect the process safety concerns raised about chemical parks is mainly one due to parallel developments – i.e. large company restructuring and cost cutting etc rather than the physical reality of several hazardous operations working in close proximity to each other. Such an arrangement has existed for almost as long as the European chemical industry itself.

However independent legal entities co-existing on a chemical park do pose extra challenges in several aspects and some are mentioned below.

One issue is emergency planning. In the case of the major user and the site operating company the threshold between normal operation and emergency will need careful consideration and dissemination within the chemical park. The authority of the site emergency services during an emergency will take precedence although undoubtedly cooperation from the guest companies will be vital in addressing the threat.

Another issue is waste treatment. How shall the relationship between the users of the chemical park and the chemical park operator be configured as far as legal responsibility for waste is concerned? In Germany each individual chemical park user is responsible for the waste which his operation(s) produce.

This responsibility includes

1. classification of the waste according to the Waste Code Regulations and according to the relevant Water, Chemicals and Hazardous Substances laws
2. making the waste available (for the steps outlined below)
3. allowing appropriate parties to take over the waste (transfer)
4. gathering the waste together
5. transporting the waste
6. storing the waste
7. pre-treatment of waste so that it can go on to be recycled or disposed of
8. recycling and disposal

Each chemical park user must take care of these activities. The scope of his involvement in each one depends on how the disposal of waste is organised in that particular chemical park. The user can undertake disposal himself if he has his own waste treatment plant or he can outsource this task to a third party outside the chemical park or to the operator of the chemical park. In the last two cases, "disposal by a user company's own means" (*Eigenentsorgung*) becomes "disposal by a party external to the user company via a third party" (*Fremdentsorgung durch einen Dritten).*

The third party can gather together, transport and store and also recycle or dispose of waste insofar as such waste is **not** normal household or business type waste which would usually be transferred to the official public waste disposal authority (i.e. the local council rubbish collection service).

Whether a chemical park user's industrial waste can be transferred to the operator of the chemical park for disposal depends on the agreement of the local authority and local town council. If such an agreement has been made, chemical park users must transfer such waste to the chemical park operator. The chemical park operator can dispose of such waste if he has suitable processing plant. If not, he must transfer the waste to the legal public authority. It may also be possible that such waste could be transferred to a "third party" commercial waste disposal company which is itself situated on the chemical park and thus a user of the chemical park.

Finally an important issue affected by chemical parks is insurance. This is a factor which can be at the forefront of minds when considering either opening up a chemical park to guests or becoming a guest on a park. The proliferation of third party hazardous operations clearly has an influence on decision making although the overall risk (the number & size of hazardous operations) compared to a single occupier site may remain unchanged or even reduced[3].

Almost all of the above aspects can be addressed in advance through site contracts and uniform rules[4]. How is adherence to these rules monitored? Clearly a balance needs to be struck between on the one hand excessive interference by the service provider into the business of a user and on the other hand a laissez faire and irresponsible approach. There is evidence to suggest that the contractual or rule based relationships which exist on a chemical park has resulted in a greater preparedness and response to the threat of major accident hazards.

Neighbouring companies on the same chemical park must manage productively their relationships with each other. Many companies on chemical parks own shareholdings in each other so it is unlikely that any disputes will remain unresolved for long. Chemical park democracy or the ability for each company on a chemical park to influence important decisions affecting potentially their own business is an important issue as far as process safety is concerned. The marketing and commercial competition between chemical parks suggests this aspect will become ever more transparent.

An illustration of the competition is offered by Degussa AG who in 1997 launched an initiative to bring investors to the main production site of Marl in the Ruhr Valley and transform the site into a chemical park. Now under the auspices of ChemSite, the park has seen 10 new companies start up since 1997 and still up to 100% of the various sites available for new arrivals as Table 1 details.

Table 1. Land available for chemical park development; ChemSite Initiative, Ruhr[7]

Location	Area available for investors in ha (acres)	Total area in ha (acres)
Marl chemical park	60 (148)	650 (1,605)
Gelsenkirchen-Scholven	4 (10)	250 (617)
Gelsenkirchen-Horst	40 (99)	165 (408)
Castrop-Rauxel	15 (37)	106 (262)
Bottrop	12 (30)	12 (30)
Intermunicipal industrial park Dorsten/Marl	67 (166)	88 (217)

FUTURE WORK

In conclusion the legacy of a chemical park on a brownfield site will often play a significant part in how it is perceived by employees, the authorities and the public. It would be interesting to contrast the experiences of chemical parks within Europe to that of chemical parks on greenfield sites in developing countries outside of Europe. Does starting from a clean sheet of paper result in vastly different approaches to that in Europe. Can we also learn from examples where the chemical park concept has not worked particularly well from a process safety perspective? Finally can we identify the major chemical parks in Europe and arrange exchanges of knowledge and experience specific to process safety so that we can distill out some general guidance and best practice on the management of chemical parks? As far as the last question is concerned EPSC is in discussion with our US counterpart, CCPS, about future collaboration in such guidance.

REFERENCES

Papers 1 to 6 were given at an EPSC members-only workshop held in February 2002
1. Roper, W, Ciba, Change in EHS management of industrial parks
2. Westphal, F, Siemens Axiva, A new approach to managing knowledge in industrial parks

3. Bartholome, C, Solvay, Risk management in industrial parks
4. Soeder, JM, Bayer, Bayer's chemical parks
5. Lewis, A, ETOL, Asset protection and emergency planning, the view of the operating company
6. Dambmann, D, Clariant, Clariant's experience on management of safety of multi-occupier sites
7. www.chemsite.de

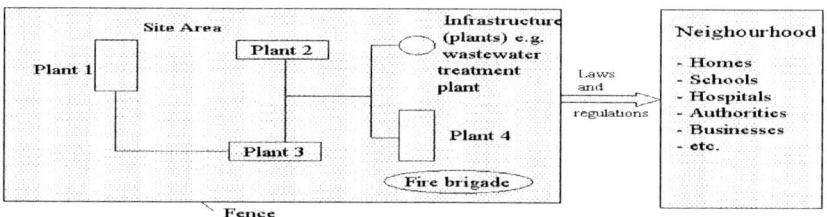

Figure 1.   Conventional chemical site (several plants, one legal entity)

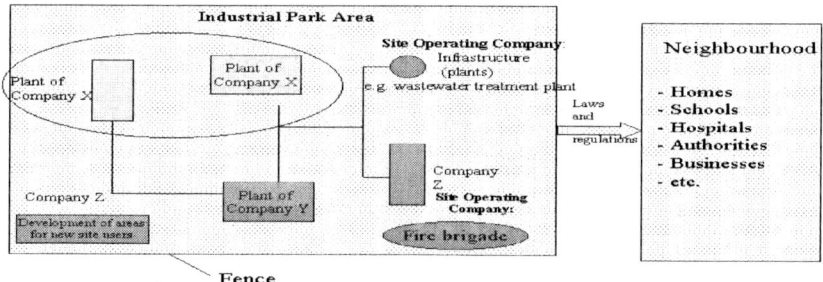

Figure 2.   Chemical park (several plants, several legal entities)

"HOW DOES SAFETY PERFORMANCE AFFECT CORPORATE VALUE"

M. Hobbs and G.C. Stevens
Arthur D Little, Science Park Cambridge CB4 4DW

It is conventional to measure corporate safety performance using measures of accident frequency such as lost time injury rate. However, it has long been recognised that safety losses include many effects in addition to injury to staff such as Property Loss, Third Party Liability, Business Interruption, Environmental remediation and Loss of Corporate Reputation.

This paper discusses the use of loss profiles to evaluate property loss and provides a practical illustration of the extension of the approach to third party liability. The probabilistic treatment of business interruption and environmental remediation costs is also discussed.

Finally the effect of major accidents on corporate value is reviewed. Two different patterns of response are outlined, one in which effective management response to an emergency leads to recovery of initial share value declines and the other in which market recovery does not occur and may in some cases lead to the financial failure of the firm. The relationship between corporate reputation and loss of corporate value following a major accident is examined and illustrated with case histories.

Corporate Reputation; Risk Management; Loss Profiles

INTRODUCTION

Major accidents often lead to initiatives to improve safety performance through improved regulation. These initiatives may be internal to the company, for example Union Carbide's development of its Episodic Risk Management program following the Bhopal accident, or they may be the spur to national or international regulation, for example the European Union directives adopted in the aftermath of the Seveso catastrophe. Even where there is a comprehensive safety management framework developed in response to historical disasters, for example in the UK rail industry, a succession of accidents may still occur which in the case of Railtrack contributed to a weakening of the company position.

This paper examines the methods used to assess potential losses resulting from major accidents and extends the thinking to include corporate reputation. Case histories are used to indicate two paths following such an event, one leading to recovery and continued operation and the other to erosion of the company's position, and eventually to its bankruptcy.

ASSESSMENT OF SAFETY PERFORMANCE

Quantified risk analysis (QRA) has become conventional in the assessment of safety performance over the past 20 years. When applied to a selection of petrochemical facilities in the Rijnmond area it was a cutting edge technique[1] but has now become a routine part of facility assessment. This does not mean that the methods are applied consistently. Recent reviews of QRA work carried out in fulfillment of the Seveso II directive[2] shows just how much disparity can be found in results from different practitioners.

Typical output from QRA studies is provided as individual risk of fatality contours showing the risk exposure from the facility to a 'theoretical' individual with full time exposure.

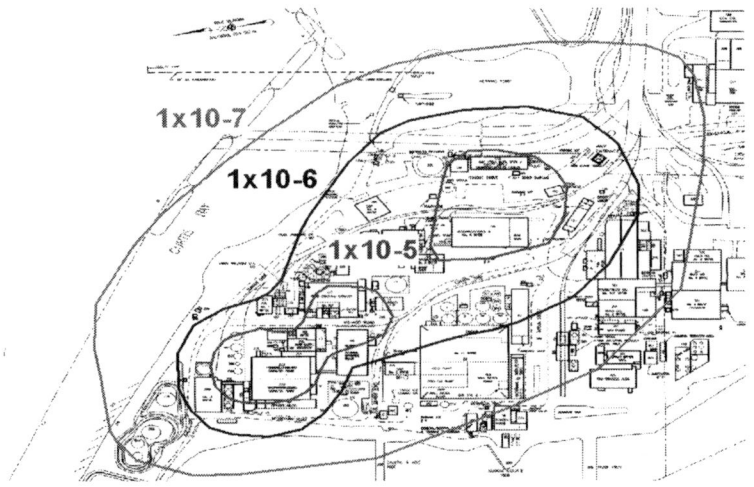

Figure 1. Individual risk of fatality Contours superimposed on a facility

The alternative approach uses Societal Risk expressed as a so called FN curve which plots the number of fatalities N against the Frequency of N or more fatalities occurring per year.

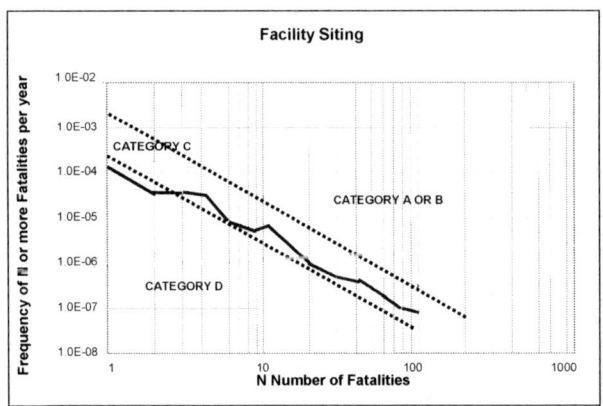

Figure 2. FN curve used for prioritisation
Category A, B, C and D refer to risk classifications used
for the assessment of the siting of the facility

The Individual Risk of fatality contours can be applied to those working on the site as well as to neighbours who are potentially third parties affected by any major accident. Risk acceptability criteria have been promulgated by regulatory authorities using both individual risk of fatality criteria[3] and societal risk criteria[4] for guidance on risk acceptability.

These well established techniques provide a numerical basis for the assessment of safety performance.

RISKS AFFECTING BUSINESS CONTINUITY

Information from accident histories and insurance sources provides a basis for assessing property loss expected after an incident. The information can be presented in the form of Loss Profiles showing the relationship between size of annual loss and probability of occurrence. Methods have been described for the derivation of these curves by matrix ranking after hazard identification by a technique such as HAZOP analysis[5,6]. The results allow the benchmarking of the plant under study against industry wide loss expectations.

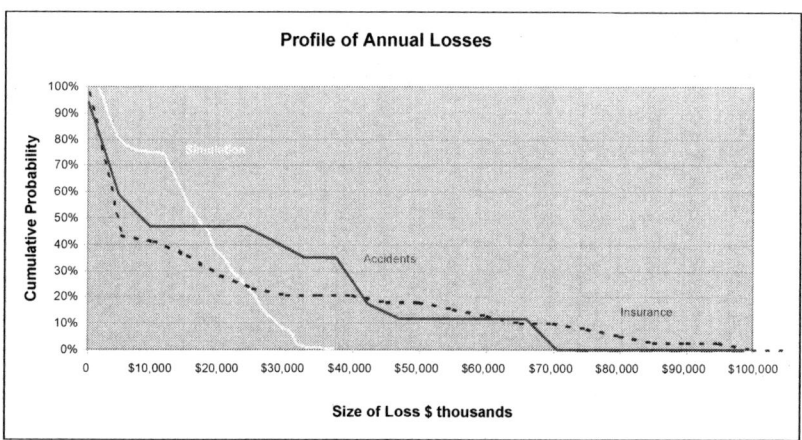

Figure 3. Loss expectation for specific plant (simulation) and industry sources for accidents and insured losses. The curve 'Insurance' shows the cumulative profile of claims, 'Accidents' the profile using losses reported by owners and 'Simulation' the curve generated by Monte Carlo simulation based on HAZOP team assessment

A similar approach has been useful for the assessment of environmental liabilities. Following site survey during which the potential for groundwater contamination has been evaluated, the simulation approach can be used to provide a profile of potential liabilities showing the size of liability against the probability that claims of that size will develop.

The use of Loss Profiles provides a quantitative approach to assessing the property loss expected after an incident.

The techniques used to simulate property losses can also be applied to business interruption where the consequences are reflected in terms of the discounted cash flow impact of complete or partial shutdown following an incident. The following illustration shows the technique applied to the revenue generated by an offshore development. The cumulative profile of discounted cash flow Net Present Value is shown for two cases:

1. The expected case which allows a small probability for failure to agree contract terms and an allowance for reservoir uncertainty
2. A 'downside case' which includes an assessment of technical failures of production or transport system and contract default without legal redress

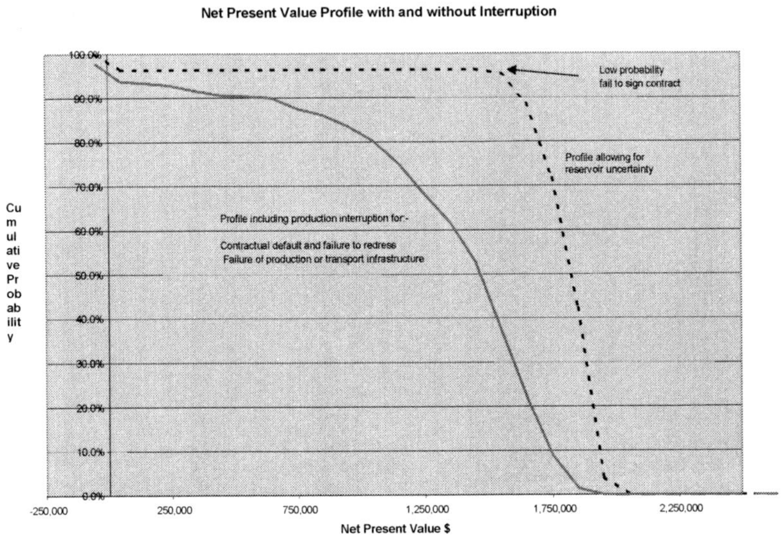

Net Present Value Profile with and without Interruption

Figure 4. Simulation shows the impact of technical and legal risks on the expected project net present value

The cumulative net present benefit curves were generated by applying Monte Carlo simulation to a discounted cash flow model of the offshore project. A risk identification exercise was conducted with the project team to recognise the major sources of risk and to assess their impact. The assessment included Political, Economic, Legal and Social risks as well as Technical failures affecting construction cost, schedule or service factor after commissioning.

This technique is capable, in principle, of dealing with all of the risks affecting the contribution the project is expected to make to the firm's capital. A similar approach could be applied to each of the major assets owned or under development within the firm (or in which the firm participates for example through joint venture companies). An aggregated profile for the contribution to capital of the firm could be built up in this way.

The approach outlined above provides a method to numerically assess most of the loss expected after an incident except impact on corporate reputation. As case studies show, companies may suffer considerable erosion in value, in extreme cases leading to bankruptcy, not simply through failure of their tangible assets but also as a result of loss of Corporate Reputation. The loss of reputation may arise for a number of reasons, for example financial or professional impropriety. However, for a number of businesses where safety is of paramount importance, major accidents can threaten Corporate Reputation. Where the firm is relatively large and the losses comparatively small, such an accident may not cripple the entire company but can have severe local consequences for the facility involved. In other cases a succession of major losses or even one major accident can initiate a process leading to collapse of the company value.

PUBLIC PERCEPTION OF RISK

Case studies were collected for two industries where safety is an important concern; refining/petrochemicals and rail/air transportation. Past reporting of public concern can have a profound effect on energy company operations following major ecological issues (Exxon Valdez, Shell Brent Spar) and also at a local level when incidents have been reported in terms of local worries about a plant safety record (BP Grangemouth). In transportation concern is expressed by the travelling public who expect a higher standard of safety for example on Rail or Air services than on the road. Reasons for this perception of risk have been discussed in terms of the extent to which the exposure to risk is voluntary and the activity is seen to be of benefit. The investment that different industries need to make to prevent a fatality varies as a reflection of this perception of risk[7].

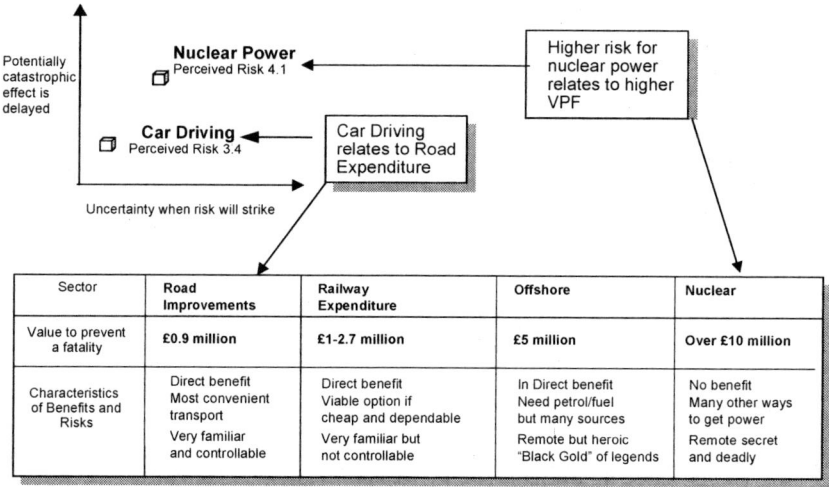

Figure 5. Relationship between perception of risk and value to prevent a fatality (VPF)

CASE STUDIES OF LOSS FOLLOWING ACCIDENTS

A SERIES OF ACCIDENTS AT A REFINERY/PETROCHEMICAL COMPLEX
In the first case study we examined the published public and regulatory response to two accidents which occurred at BP Grangemouth refinery within a few days of each other in June 2000. In the first a steam pipe burst after a steam valve was turned off without the pressure being throttled back. When the pipe ruptured, a member of the public passing suffered three cracked ribs (her dog was also hurt). In the following days there was a serious fire which was contained within the complex but which also caused public concern. The events were the subject of a court hearing in January 2001 resulting in punitive fines totalling £ 1 million levied against BP companies. In addition to loss of local confidence, the local Member of Parliament raised concerns about the management of safety at the plant. The reporting of the accidents in June 2000 was accompanied by lists of previous incidents at Grangemouth (including a serious explosion in 1987 causing two deaths, power failure in July 1999 and again in May 2000 and failures on Catalytic Cracker and a gas compressor in November 1999).

The effect of these events (the accident and then the court hearing) can be judged by the following graph that plots BP share price against a group of oil majors over the same period.

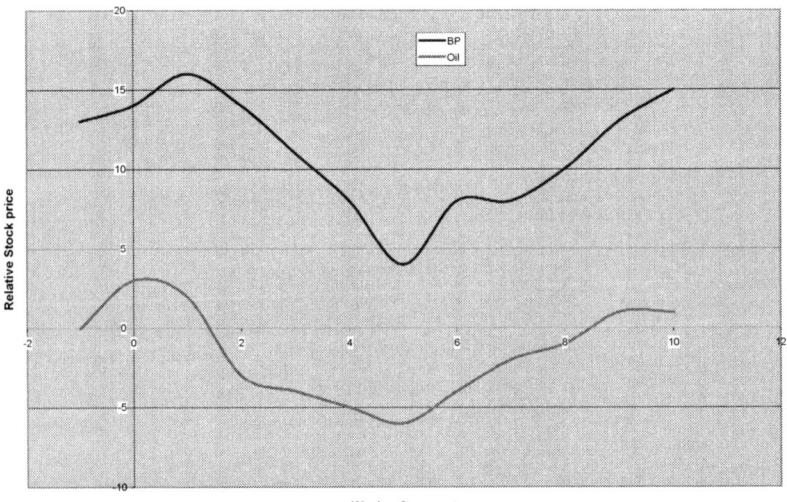

Figure 6. Relative stock prices of BP and oil group during June 2000 when 2 accidents occurred at Grangemouth

At the time of the incidents in June 2000 (which attracted national coverage) there was a decline in value of BP stock but that was matched by similar falls in other oil stocks. The fall and subsequent recovery in oil industry stock matched that experienced by BP. If the difference between the relative stock prices is compared to the start and end difference for a period of 10 weeks around the accident no effect can be seen.

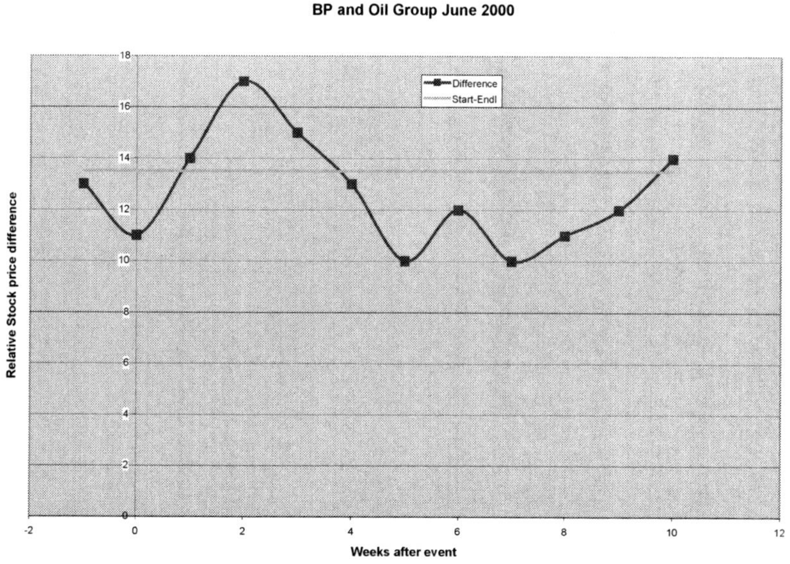

BP and Oil Group June 2000

Figure 7. Range of difference between BP and oil group relative share price during June 2000

At the time of the court hearing, and the imposition of a heavy fine, there is a downward dip of about 5% relative value in BP stock some 2 to 4 weeks after the verdict of the court.

The BP price recovered within 6 weeks of trading and thereafter matched the industry average for a period of several months. The pattern suggests

1.1 For this very large company, the two accidents themselves had a negligible impact on company valuation.

1.2 The imposition of a heavy penalty may have produced a downward movement in value but this was recovered over the following weeks

It is clear however from accompanying press coverage, that negative local feeling could affect company operations for example in public reaction to planning applications or permitting for expansion or plant modification.

BP and Oil Group around January 2001

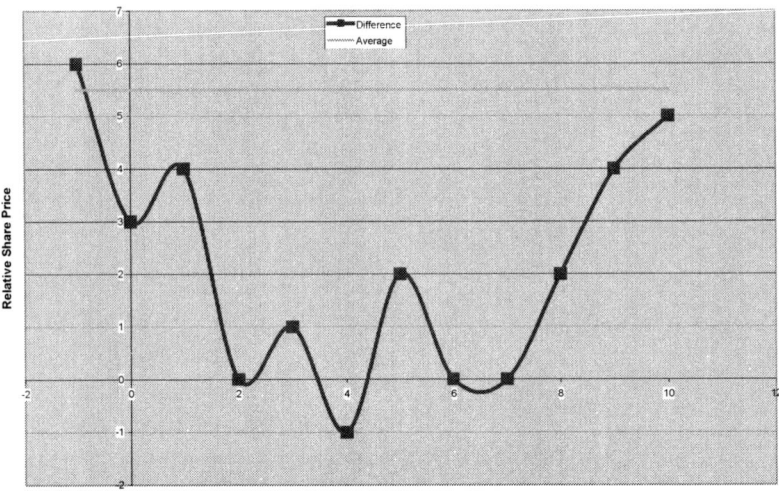

Figure 8. Range of difference between BP and oil group during January 2001 after court fined BP £1 million

TANKER GROUNDING AND POLLUTION INCIDENT
Accidents generating ecological impact can also affect company valuations. The following graph shows the pattern of annual earnings reported by Exxon and reveals the loss associated with the Exxon Valdez incident.

The anticipated loss of operating income also produced a dip in company stock price. Knight and Pretty[9] show a much greater fall lasting longer than was experienced following the BP incidents. An initial fall of about 10% was exacerbated by a further fall to 20% loss of value 3 months after the incident possibly as a reflection of perceptions at the time of the company response. However, as the situation came under control, stock values rose and over the long term the impact of the Exxon Valdez incident is hard to detect.

The long-term performance of the Exxon stock has been summarised in the following terms: 'Exxon's focus on asset stewardship has driven exceptional operational and financial performance'.

This strong affirmation of the reputation of the company finds a quantitative expression in the Reputational Quotient developed by Frombrun et al[8]. Three Reputation Quotient surveys have been published, one of which included a number of major oil companies. In general as a group these companies scored in the middle range of corporate reputation quotients, lead by Chevron (66.2%) followed by Exxon (65.2%) and BP and Texaco both at 65%.

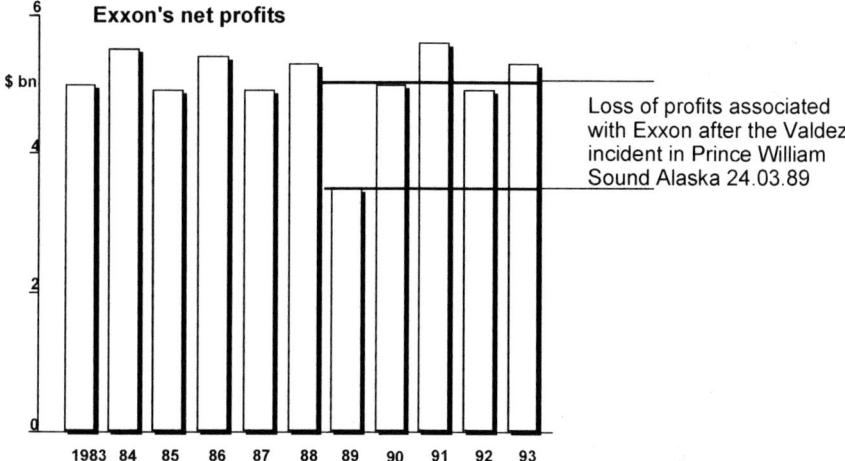

Figure 9. Impact of Exxon Valdez incident on Exxon annual net profit

Knight and Pretty[9] found from their case studies, two groups of companies: 'recoverers' and 'non-recoverers':

- Recovering companies who suffered a fall in value between 5 and 10% immediately following a major incident and then recovered over a period of a month or two
- Failing companies who suffered somewhat more than 10% loss of value but who's management was not able to effect recovery

AIRLINE CRASHES

The examples looked at above show strong companies (like Exxon and BP) which recovered from major accidents. This is not always the case however. The airline industry provides a number of instances that enable the impact of major air disasters on company value to be examined.

British Airways has a good corporate reputation (using Reputational Quotient it scored 72.5% in 2000)[8]. The charts below shows the impact of the Concorde crash and the attack on the World Trade Centre in September 2001.

Even though the accident on 25th July involved another airline Air France, the first crash of Concorde in its operating history lead to a drop of about 25% in the relative BA stock value. The position was quickly recovered in the next 15 weeks trading. The response to the World Trade Centre attack was entirely different. Even though its aircraft were not directly involved, BA along with the rest of the airline industry suffered an immediate loss in value of about 30%. The difference chart shows this change did not affect the relative position of BA to the rest of the industry. Recovery begins about 8 weeks after the initial event but still after 10 weeks the air group remains below its pre disaster level.

British Airways and Air group around July 2000

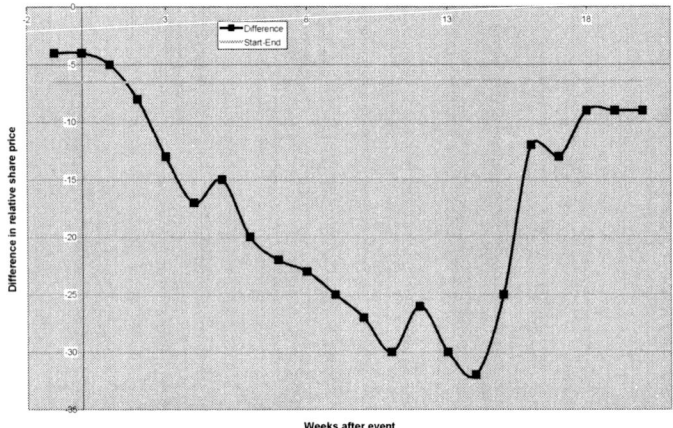

Figure 10. Impact of Air France concorde crash on British Airways

British Airways and Air Group around September 2001

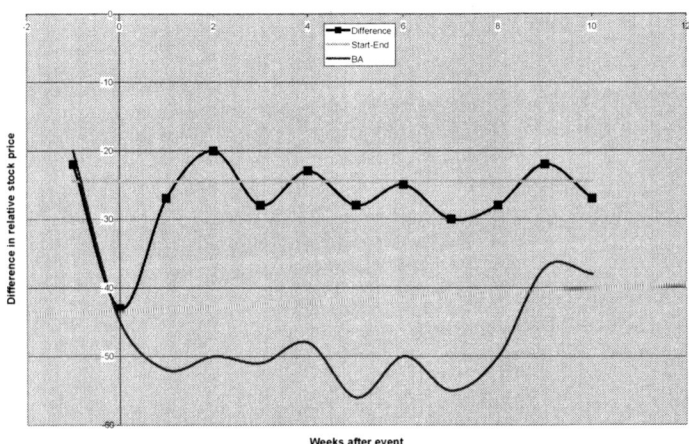

Figure 11. Impact of WTC attach (9/11) on British Airways

The pattern of events suggests some additional criteria to those proposed by Knight and Pretty namely

3.1 Company value can be affected by a major accident suffered by a company carrying out similar operations. The immediate market assumption appears to be 'if it can happen to one company why not to another'. When stable and reliable operations are demonstrated to continue, confidence returns and value is quickly recovered

3.2 A few very rare disasters have the ability to affect valuation of the entire market sector. The depth of the fall and the duration of recovery can be protracted.

One of the cases considered by Knight and Pretty[9] was the crash of ValuJet on 11th May 1996 which suffered a 25% loss of value after the accident and did not recover. Figure 12 illustrates stock price movements and citations of media reporting. The stock had been rising but the company had attracted criticism that it was 'cutting corners' on maintenance. After the crash there was an abrupt fall in stock price and a huge increase in media speculation about the company.

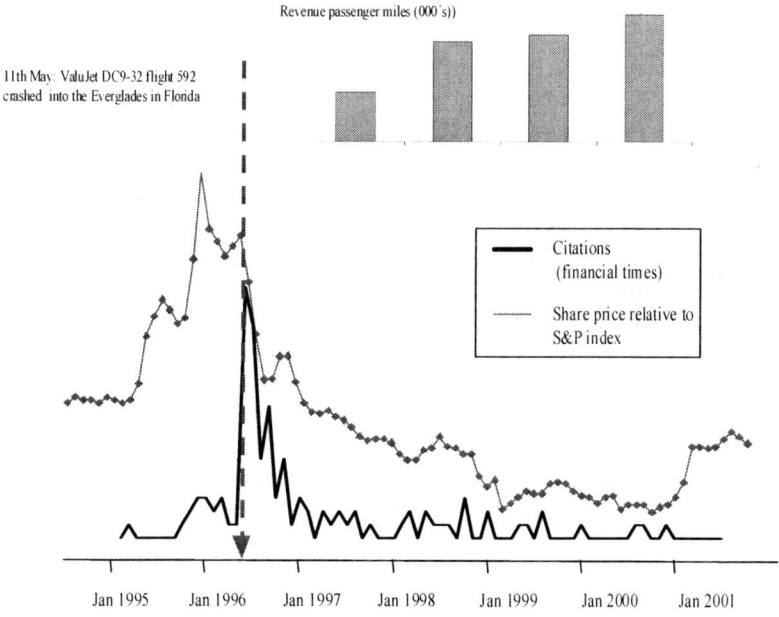

Figure 12.   ValuJet stock price movements

There was no recovery in company value and the concern went into bankruptcy proceedings, from which it emerged under a new name. Despite a steady growth in revenue passenger miles the downward trend in company value was not reversed until 5 years after the initiating accident.

It remains with the market to decide if 'Cut Price' airlines remain vulnerable to this pattern of events in the future. At the time of writing, media continue to pay attention to the safety standards in such operations.

A SUCCESSION OF SERIOUS RAILWAY ACCIDENTS

The final case study draws on a succession of accidents that have occurred in the UK rail industry since 1997. The following graph shows the pattern of share price movement for Railtrack and a selection of rail group stock. The sequence of accidents involving multiple passenger fatalities was as follows

- Southall 19th September 1997
- Ladbroke Grove 5th October 1999
- Hatfield 17th October 2000
- Potters Bar 10th May 2002

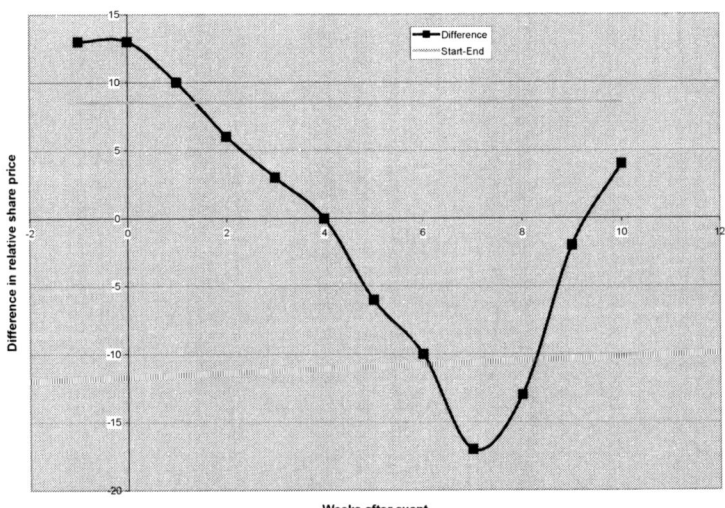

Figure 13.   Railtrack stock price movements following the Ladbroke Grove accident

At the time of the Southall accident (the first accident involving large numbers of fatalities since Clapham in 1988) Railtrack stock was rising relative to the market and the price continued in this fashion until late 1998. The graph below shows the reaction of the Railtrack stock to the Ladbroke Grove incident which fell by 25% against the trend of the rail industry stocks. The fall built up over a period of 7 weeks as Railtrack became implicated through negative interpretation of its handling of signal siting issues around the crash site.

The Hatfield incident a year later precipitated a further fall in relative stock price, initially 25% of value in the 8 weeks following the accident. It became clear after this third major event that Railtrack needed to radically improve its track renewals activities and in the short term this produced great service disruption. Figure 14 shows that on this occasion there was no evidence of recovery of value, unlike the curve after Ladbroke Grove. Two months later the stock fell a further 50% in value and the company entered administration in October 2001. The case extends the ideas of Knight and Pretty[9] to suggest that a company can, through a succession of accidents, shift from being a 'recoverer' to becoming a 'non-recoverer'. Specifically the case suggests:

4.1   If a company is well regarded and manages the immediate aftermath of a bad accident in an effective manner, it can recover any initial losses in value

4.2   If a succession of accidents occurs confidence in the company falters. At first this affects the strength and duration of recovery but in the end can contribute to collapse of company reputation

4.3   Media play a part in this process, focussing increasing coverage on the company implicated at each successive accident

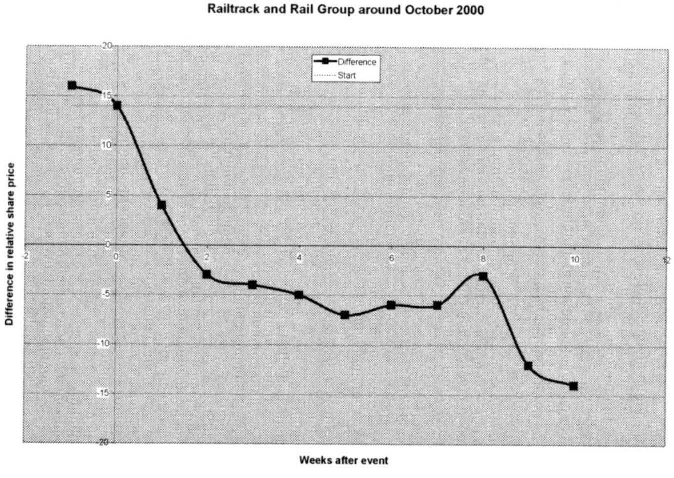

Railtrack and Rail Group around October 2000

Figure 14.   Railtrack stock price movements following the Hatfield accident

The sequence of events with Railtrack is not typical of the rail sector as a whole. The following data shows a similar trend for the Services Company, Jarvis which has a significant presence as an infrastructure maintenance contractor working for Railtrack.

Jarvis and Rail Group around October 1999 and 2000

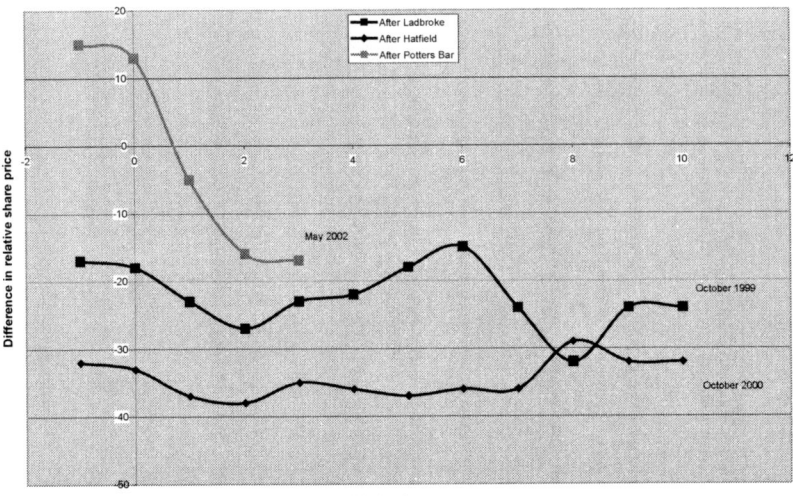

Weeks after events

Figure 15.   Jarvis stock price movement after Ladbroke Grove, Hatfield and Potters bar

The price pattern against the rail industry sector is quite different to Railtrack. The price suffered no adverse fall at the time of Ladbroke Grove or Hatfield. The situation is entirely different for Potters Bar, an accident where questions have been asked about the standard of maintenance at the points involved in the derailment. Following the accident there was an immediate drop of 30% in relative share price. At the time of writing it is unclear if Jarvis price will recover. Again the market, influenced by media interpretation, will be the ultimate judge.

To summarise, the case studies have suggested the following, which extend the original analysis by Knight and Pretty

- Accidents which cause little interruption to overall operations have relatively little impact on corporate value. Corporate reputation may be negatively affected at the local level and this may impact plans for site expansion or regulatory permitting.
- Major incidents cause an immediate drop in market value of the company. The extent of the drop varies from case to case in the range –5% to –30% in the examples reported here and recovery is shown to occur after initial falls as high as 25%.

- Loss of company value can occur even if the accident occurs to another company with similar operations.
- Critical to success is achieving recovery following the initial loss. Some companies, typically those with strong corporate reputation based on sound management recover their value within a few weeks of trading.
- A succession of incidents, particularly associated with negative media reporting, weakens corporate reputation. Recovery of value following an accident and an initial loss becomes progressively harder. In extreme cases company value may not recover and this can be a factor contributing to bankruptcy.

A hypothesis is presented to account for these observations in the following section.

HYPOTHESIS FOR CORPORATE RECOVERY FOLLOWING A MAJOR ACCIDENT

The crucial element under the control of management is not the measures it has in place to avoid major accidents. Of course such measures are required and need to be backed by a thorough safety management system but the case histories show that company value, as reflected by the share price can be negatively affected by a major accident experienced by another company with similar operations. What is crucial is the ability of management to regain control of the situation, be seen to have matters in hand and in this way rebuild corporate reputation. The following figure illustrates the hypothesis by considering 'leading edge' and trailing edge' companies.

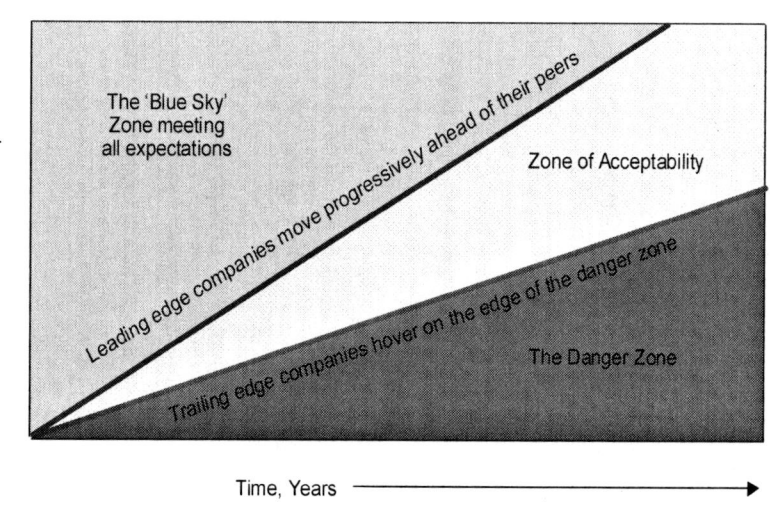

Figure 16.   Leading and trailing edge companies

The hypothesis proposes that the stakeholders in a company have expectations about company performance. Leading edge companies move progressively ahead of their peers seeking to meet all expectations. This is not a static situation but evolves as patterns of risk change in response to emerging and latent issues.

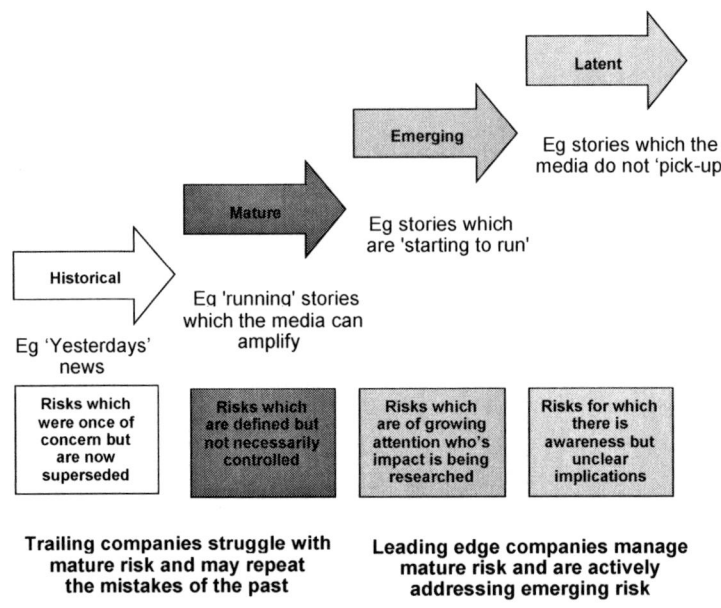

Figure 17. Development of stakeholder perceptions

It is significant that by choosing to cut costs some companies coincidentally shift towards the 'trailing edge'. Their operation seeks to be acceptable but no more, to comply with regulations with just sufficient resources to make the grade. Where safety is a key element in the service (for example in air or rail transportation) a trailing edge position which may lead for example to criticism of 'cutting corners' on maintenance can make the company vulnerable in the case of a major accident.

Immediately following a major accident (here described as a trigger event) the case studies indicate that management handling of the situation is critical to quick recovery of the initial loss of value. This recovery process can be pictured in terms of interaction with the various stakeholders related to company operations. The process is illustrated as an event tree showing how 6 stakeholders, media, financial markets, regulators, customers, staff and public action groups may all interact to help value recovery or, in adverse cases, maintain or even deepen the fall in value.

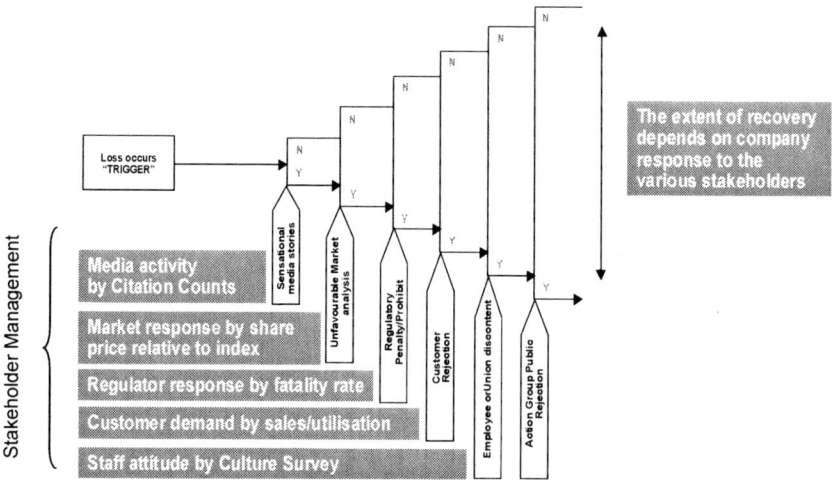

Figure 18. Stakeholder interactions following a loss event

As well as managing stakeholders, a leading edge company will be expected to have well established programmes to manage risks to the business. The following diagram shows how these risks and their controls interact to contribute to the trigger event.

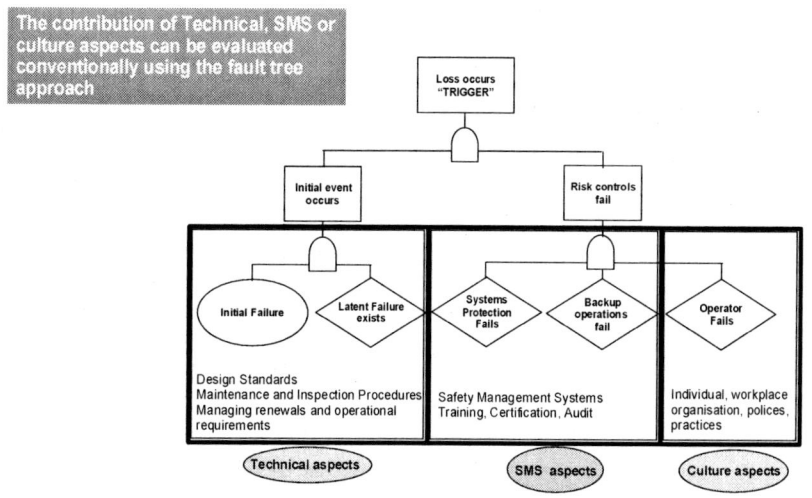

Figure 19. Fault tree for trigger event

The fault tree leading to the trigger event and the event tree leading to a recovery of value or a spiral of loss can be combined into an overall model.

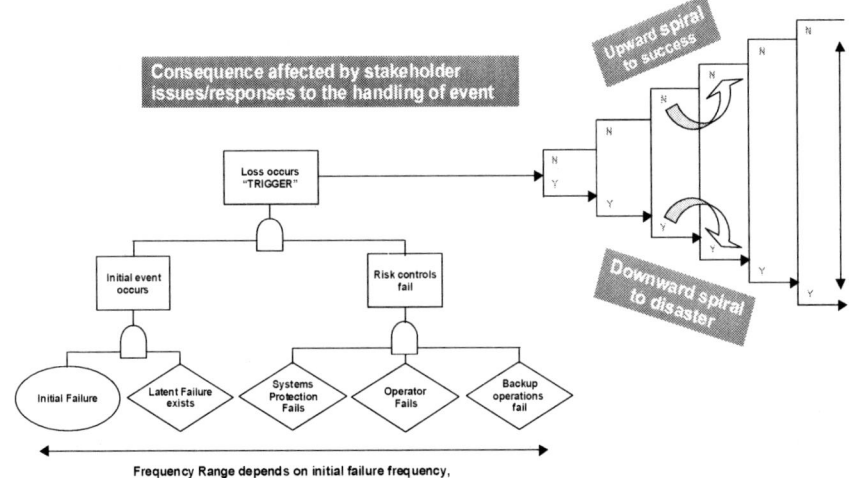

Figure 20. Trigger events and the spiral of recovery or loss

The hypothesis summarised in the preceding diagrams needs to be systematically tested. One idea would be to develop scales to measure the extent to which a company was 'leading edge or trailing edge'. The reputation index goes some way in this respect but is not available for all the companies considered in the case studies. A second step would be to develop measures of the extent of stakeholder interaction to try to relate specific successes or failures to the value recovery experienced by a company after the initial loss associated with the trigger event.

SUMMARY

The cases we have examined confirm and extend the effect of major accidents on corporate value described by Knight and Pretty[9]. They identified two different patterns of response, one in which effective management after the emergency leads to recovery of initial losses and the other in which market recovery does not occur and may in some cases lead to the financial failure of the firm. These earlier findings can be qualified as follows

- The initial accident need not be experienced by the company suffering the loss. A major accident by a competitor carrying out similar operations can trigger a loss in company value which can be recovered if reliable operations are demonstrated.
- A few very rare disasters have the ability to affect valuation of an entire market sector. The depth of the fall and the duration of recovery can be protracted.

- Accidents causing little interruption to overall operations have relatively little impact on corporate value. Corporate reputation may however, be negatively affected at the local level and this may impact plans for site expansion or regulatory permitting.
- Major incidents cause an immediate drop in market value of the company. The extent of the drop varies from case to case in the range –5% to –30% in the examples reported here and recovery is shown to occur after initial falls as high as 25%.
- Critical to success is achieving recovery following the initial loss. Some companies, typically those with strong corporate reputation based on sound management recover their value within a few weeks of trading.
- A succession of incidents, particularly where the media focuses increasing coverage on the company implicated at each successive accident can weaken corporate reputation. Recovery of value following an accident and an initial loss becomes progressively harder. In extreme cases company value may not recover and this can be a factor contributing to bankruptcy.

A hypothesis is presented which explains these findings in terms of 'leading edge' and 'trailing' companies.

REFERENCES

1. Risk Analysis Report to the Rijnmond Public Authority, D.Reidel Publishing Co., 1981 ISBN 90-277-1393-6
2. F. Markert et al. *Sources and Magnitudes of Uncertainties in Risk Analysis of Chemical Establishments. First insights from an European Benchmark Study* 10th International Symposium on Loss Prevention and Safety Promotion in the Process Industries (June 2001).
3. HSE: *Risk Criteria for Land Use Planning in the Vicinity of Major Industrial Hazards*, HMSO, 1989
4. Dutch National Environmental Policy Plan: *Premises for Risk Management*, Second Chamber of States General, 1988-89 session, 21137, nos.1-2)
5. G.C Stevens: *Prioritisation of Safety related Plant Modifications* 7th International Symposium on Loss Prevention (May 1992)
6. G.C. Stevens and M Marchi: *A Benefit/Cost approach for prioritising expenditure during plant turnaround* 10th International Symposium on Loss Prevention and Safety Promotion in the Process Industries (June 2001).
7. New Scientist 28 Sept 1996 p37
8. C J Frombrun and Naomi Gardberg *Who's tops in corporate reputation?* Corporate Reputation Review March 2000
9. R F Knight and D J Pretty *The impact of catastrophes on Shareholder Value* Oxford Executives Research Briefings 1997

FACTORS INFLUENCING THE SAFE MANAGEMENT OF CONTRACTORS ON MAJOR HAZARD INSTALLATIONS

Christopher J. Beale (MIChemE)
Ciba Specialty Chemicals, Water & Paper Treatments,
PO Box 38, Bradford, West Yorkshire, BD12 0JZ, UK.

Contractors are increasingly being used on major hazard installations and now perform many critical roles which can directly cause and prevent hazards. Operating companies have to manage the interfaces with these contractors carefully as experience has shown that fragmented systems and inadequate control of contractors have contributed to a number of recent accidents. This paper summarises the reasons for using contractors and the typical roles performed by different types of contractor (such as design, installation, maintenance, outsourcing of non-core activities, specialist EHS (Environment, Health and Safety) support). Recent incidents in a number of industries which have been exacerbated by poor contractor management are then reviewed and learning points from these incidents are identified. The types of problem which can be caused by contractor interfaces are then summarised and good practices/risk controls are identified for minimising these problems. These contractor management issues are critical within the COMAH (Control of Major Accident Hazard) Safety Case regime for chemical manufacturing companies.

KEYWORDS: COMAH, contractors.

WHY COMPANIES USE CONTRACTORS

All organisations use contractors to some extent and must decide on the appropriate boundaries of the firm. Economic theory would suggest that contractors will be used when work can be completed more cheaply by contractors than by in-house staff.

Table 1 illustrates different ways in which organisations can be structured regarding their use of contractors. Companies can thrive with all of these three structures. It is, however, clear that some organisations are more effective at managing within their structures than others.

Table 1. Different organisational structures

Type	Philosophy	Examples
Virtual Organisation	Strategy, marketing, R&D, product development completed in-house. All other activities outsourced.	High technology companies in Silicon Valley, USA.
Focused on core competencies	Devote resources to critical activities. Non-core activities are outsourced.	Many UK manufacturing companies.
Highly vertically integrated	Maintain full control over whole supply chain to minimise reliance on third parties.	Steel manufacturers purchase iron ore mines.

Positive reasons for using contractors include:

1. Gaining access to specialist skills and staff.
2. Allowing managers to focus their efforts on core company activities without being distracted by peripheral activities.
3. Meeting fluctuating patterns of workload.
4. Internal costs are too high.

Unfortunately, many organisations use contractors for poorly conceived reasons such as:

1. Everybody else is using contractors, it must be a good idea.
2. It's corporate policy.

TYPES OF CONTRACTOR

A typical chemical company will use the wide range of contractors illustrated in Figure 1.

Many of these contractors will therefore play a critical role in the prevention and management of major hazards on the site. Errors made by contractors or caused by poor contractor management would therefore have been expected to have contributed to a number of accidents on chemical sites. In reality, there are relatively few accidents which have been attributed to contractor errors.

This is probably because until relatively recently, many accident investigations did not identify underlying causes such as safety management system and organisational failures. For example, a chemical leak from a flange may have been attributed to human error by a fitter without considering that the fitter was a contractor and contractor supervision had been inadequate.

Table 2 summarises a number of accidents which have occurred in industries with major hazard potential in which contractor management had been identified as one of the causes of the accident.

SPECIFIC PITFALLS

CHASE TO THE LOWEST STANDARDS

There is a real danger that the skill and commitment of workers will drop when the use of contractors is driven excessively by cost reduction pressures. Railway industry Trade Unions have identified the following contractorisation mechanism which causes particular concern:

- Work is contracted out.
- The contractor employs subcontractors.
- The subcontractors use temporary agency workers.
- The agency workers have low pay, no job security and no career development opportunities.

In these cases, committed, well trained and motivated staff have been replaced with poor quality staff who have no commitment to the overall goals of the operating company. This should be a cause for concern in major hazard industries.

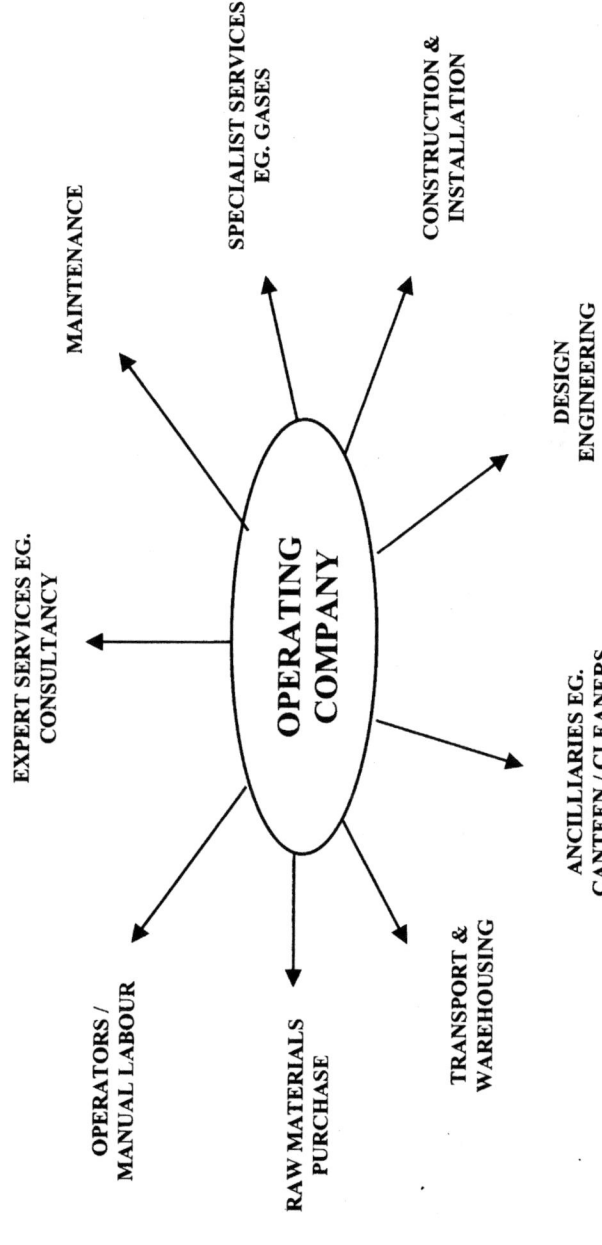

Figure 1. Types of contractor used by chemical companies

Table 2. Accidents involving contractor management

Ref	Company and date	Industry	Causes	Consequences	Contractor issues	Reference
A	Albright & Wilson. Avonmouth. 3rd October 1996.	Chemical	Incorrect chemical delivered to storage tank by haulage company.	Explosions and fire creating a 100 m black plume of smoke.	Paperwork for two chemical tanker loads was mixed up by haulage company, causing wrong chemical to be sent to site.	(TCE, 1996)
B	AEA. Dounreay. Scotland. 7th May 1998. *(HSE Audit Report Following Incident)*	Nuclear	Mechanical digger damaged power cables supplying part of site.	Loss of power to part of site for significant period.	Contractorisation so weakened technical and management base that company could not manage it's technical operations. Over delegation of control to contractors. Unable to act as 'intelligent customer'.	(HSE, 1998)
C	Kaiser Aluminium. Louisiana. USA. 5th July 1999.	Mineral Processing	Explosion in milling facility.	Multiple serious injuries.	Striking permanent staff replaced with 400 temporary workers.	(CHI, 2000)

D	EVC, Merseyside, 8th March 2000.	Chemical	Bellows leak in section of 12" diameter pipework.	Release of 500 kg of hydrogen chloride gas to air.	Plant under control of contractor during commissioning.	(HSE, 2001a)
E	Railtrack, Hatfield, 17th October 2000.	Rail	Derailment, mechanical failure of rail.	Multi-fatality rail accident.	Management and maintenance of rail infrastructure by contractors. Delays in responding to known problems with bureaucratic contractor management system.	(HSE, 2001b)
F	Railtrack, Potters Bar, 10th May 2002.	Rail	Derailment, loose bolts on points.	Multi-fatality rail accident.	Points maintained by contractor but many were loose. Different standards of work across UK rail network.	(HSE, 2002)

PROVISION OF INADEQUATE INTERNAL RESOURCES

The economics of contracting often appear attractive because the operating company has not made adequate budgetary provisions for the internal costs of contract management. The most common error is to allow insufficient resources for supervising and checking contractor performance. This tends to cause a gradual erosion in the standard of work, which accelerates as the contractor realises that the supervision activity has been cut back.

Figure 2 illustrates the contractor management lifecycle.

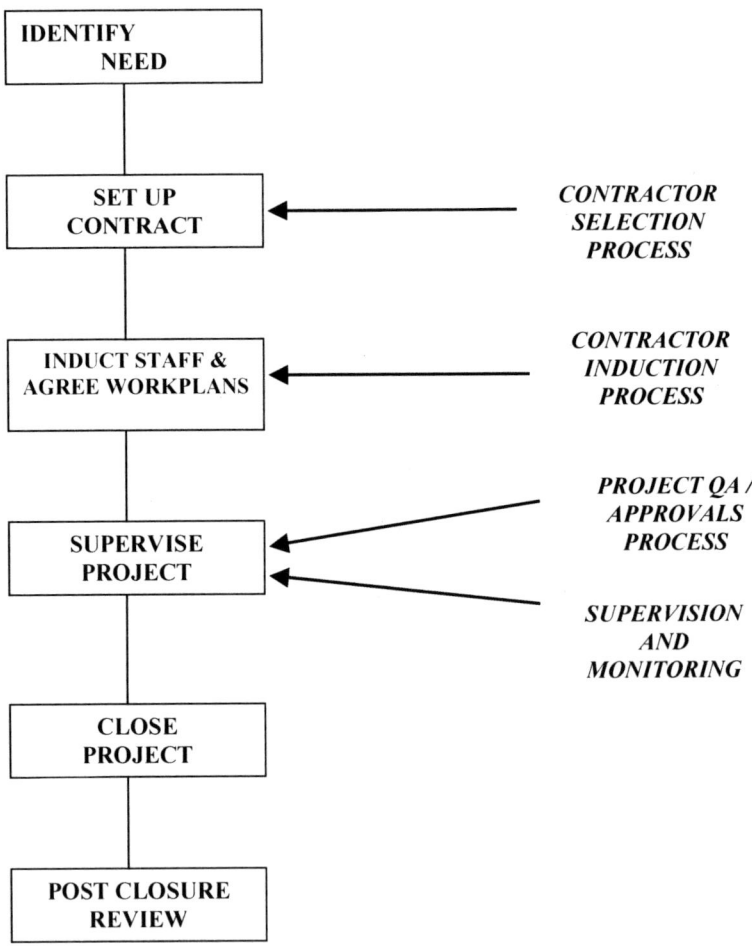

Figure 2.   Contractor management lifecycle

SYSTEM INCOMPATABILITIES

Operating companies tend to use complicated safety and business management systems. Contractors will have their own systems and will have to climb a learning curve to properly understand the operating company's systems.

There is a risk that work may be completed deliberately or accidentally without complying with corporate systems because:

- Critical standards and systems were not agreed before the contract was signed.
- Some critical standards were not defined and agreed in the original contract.
- The contractor fundamentally fails to understand one or more critical corporate standards.
- The contractor works to a different standard because it is felt that this is more suitable than the corporate standard.

If contract supervision is poor, these problems may not be identified until a late stage in the project, exacerbating their impacts.

These problems often affect the human factors aspects of projects as the impact of system incompatabilities is often:

- Poor or no training.
- Inadequate project documentation.
- EHS (Environment, health and safety) studies which are difficult to use or different from the standard format.

UNMANAGEABLE SYSTEMS

From a theoretical viewpoint, it may be argued that any organisational structure involving contractors can be made to work effectively. In reality, the more complicated the structure, the higher the risk becomes that significant problems will occur. Problems normally occur because:

- There are inadequate internal resources to adequately supervise and manage the operation.
- Bureaucracy slows down the fluent operation of the overall system, delaying critical operations.
- Communications between all of the operations are imperfect, driven by the complexity of the arrangements or the fact that different contractors have different and incompatible business objectives.

One of the best examples of how systems can become unmanageable can be found it the post privatisation UK railway industry.

- An operating company (COMPANY 1) will operate the front end service to the customer using rolling stock which is leased from a separate company (COMPANY 2) and track which is managed by another separate company (COMPANY 3).
- The track is inspected by a specialist company (COMPANY 4) and repairs are completed by another company (COMPANY 5) on approval of COMPANY 3.
- COMPANY 1 may operate a number of franchises in different regions of the UK. In each region, it may be dealing with different companies performing the roles COMPANY 1 to 5.

Acceptable safety standards can only be achieved if the interfaces between all of these companies are carefully managed. On a practical level, this has not been achieved.

Similar problems may well exist on some UK chemical sites with:

- An operating company running the site.
- Engineering design activities outsourced to a number of specialist design companies.
- Construction activities outsourced to separate companies.
- Maintenance activities outsourced to other separate companies.

These activities all need to be carefully co-ordinated to minimise risks. The operating company also needs to ensure that it's overall contracting strategy is mutually consistent and workable.

CONTRACTORS WHO DON'T COMMUNICATE

Contractors will be under financial pressure to complete agreed scopes of work as quickly as possible. This pressure will often discourage them from fully communicating with other contractors and the operating company. Particular problems occur when errors are made by a design company and are not detected by a separate construction company. These problems can be minimised using control methods such as regular team meetings and formal design approval systems or by structuring contracts so that one company is responsible for both design and construction activities.

CONTRACTORS WHO ARE GOOD AT BUILDING SHOPPING CENTERS AND DANGEROUS WHEN WORKING ON CHEMICAL SITES

General fitting and maintenance work is often performed by specialist contracting companies who operate in a wide range of industries. Some of these companies do not have adequate experience of working on high hazard sites. Their safety culture often does not match that of the operating company because of factors such as:

- Poor awareness of the hazards around them.
- Lack of familiarity with site safe systems of work.
- Temptation to use working practices deployed in other less hazardous industries.

Experience must therefore be carefully assessed before the contractor is selected.

REMOVAL OF CORE TECHNICAL KNOWLEDGE

Operators of high hazard installations must retain a sufficient depth of technical knowledge such that they can safely run the facility. Care has to be taken when staff with key technical knowledge are replaced with contractors. When this happens, the internal knowledge that has been built up over many years is often dissipated causing problems when:

- Unexpected events occur.
- People require unwritten knowledge which is held in people's heads.
- Organisations forget about past errors and repeat these errors causing easily avoidable accidents.

Examples of each of these problems from the chemical industry are:

- A chemical reactor exceeds it's safe working envelope. It is suspected that this has been caused by a hardware or design fault. Experienced engineers will be required at short notice who have a good working knowledge of the plant.
- A certain type of valve performs poorly with the chemicals handled on the site and should not be used.
- Reliance on one temperature probe at the base of the reactor is inadequate as experience has shown that it often becomes coated with a viscous layer of polymer.

SPECIFIC RISKS

OPERATORS
Risks tend to be increased when temporary staff are used on short term contracts. The staff then have a:

- Poor awareness of plant hazards.
- Poor understanding of the implications of doing the job wrongly.
- Lack of experience in the event that unexpected conditions occur on plant.
- High risk of causing human errors.
- Lack of recognition of unsafe acts.

These risks can be effectively controlled by adequate induction training and supervision.

RAW MATERIALS, TRANSPORT AND WAREHOUSING
Most companies rely on external suppliers for the provision of some or all of these services. The most dangerous errors are likely to involve:

- Delivery of the wrong chemical to a plant or storage tank.
- Incorrect labelling or paperwork associated with a chemical delivery.

    Better quality suppliers will often be able to provide additional safety features such as:

- Dedicated tanker sizes and designs, minimising offloading leak risks.
- Deliveries in tankers fitted with dry link couplings to minimise operator chemical contact during the offloading process.
- Facilities for reprocessing offspec material.

ANCILLIARIES
In general, this group of contractors has a relatively minor impact on major hazard risk levels. The two main problems tend to be:

- Blocking escape routes or critical access points for emergency services by parking vehicles in inappropriate locations.
- Lack of awareness of emergency plans.

These problems can easily be prevented by induction training and site supervision.

One particular group of ancilliary contractors does often, however, play a critical role in managing major hazards: security staff. In an emergency situation, they will have to liase with the emergency services and media, respond to alarms, control access to and from the site and man the emergency control center. The critical role of security staff must therefore be emphasised and agreed before a contract is signed for this type of service. Additional emergency management training will probably be required for these staff.

DESIGN
Design activities can be performed successfully by contractors as long as operating company staff are fully involved in the design process and effectively approve all designs.
   Specific problems which can be encountered include:

–   Lack of corporate knowledge and site experience.
–   Specification of equipment which is incompatible with the rest of the site, which then causes problems with spares and maintenance.
–   Use of inappropriate design or EHS standards.

CONSTRUCTION AND INSTALLATION
These activities often occur under severe time pressure. Contractors often select the fastest installation method rather than the best pipe layout, causing future operability problems. Manual valves may be poorly located and excessive pipework or leak sources may be installed.
   Revealed or unrevealed damage may occur to adjacent equipment either because of human error or because the contractor was not aware of the importance of this equipment. Typical damage would range from bent pipes because fitters have stood on them, removal of pipes which were not shown on installation drawings but performed an important role and accidental activation of safety systems. This type of damage should be identified and corrected during the project commissioning phase but the more of these errors that occur, the less the likelihood that they will all be identified and corrected.
   A crucial part of installation is the handover phase. Some of this involves work which improves the human factors aspects of plant operation such as plant manuals, documentation, drawings and operator training. If the handover phase is not completed properly, the operators will find it difficult to operate the plant.
   Arrangements also have to be put in place for dealing with unforeseen post commissioning problems. The maintenance staff will be climbing a learning curve and critical plant knowledge will still rest with the design/installation contractors.

MAINTENANCE
This is high risk activity as the contractor may (i) be working next to live plants and (ii) any errors could cause future accidents when the plant is switched on and runs live. There are numerous examples of accidents caused by poor maintenance and these have been highlighted by recent experience in the UK rail industry.
   Fundamentally, the contractor must have the skill and discipline to be able to work safely under a Permit To Work system. Permits must be agreed, understood and followed.

Short cuts must not be taken. Contractors may be tempted to take short cuts when they are 'custom and practice' in other industries.

Problems often occur when work is inadequately supervised. If the bolts on the points at Potters Bar had not been adequately tightened (HSE, 2002), one person must have made an error in tightening the bolts in the first place, but another person or organisation must also have made an error in not detecting the problem. Operating companies using contractors for maintenance activities need to think carefully about defining the responsibilities for supervision and inspection activities. Some degree of independent checking will be required.

SPECIALIST SERVICES

Two particular problems tend to undermine the quality of work completed by specialist external consultants:

- Inadequate involvement of operating company staff. The work may then be based on incorrect assumptions and base data.
- Lack of clarity in the report. Staff fail to understand the logic and analysis in the report and do not incorporate it's recommendations correctly.

AVOIDING PROBLEMS WHEN USING CONTRACTORS

A number of factors contribute to the successful management of contractors. These include:

1. Carefully select contractors and check that they have experience in high hazard industries.
2. Develop a good working relationship based on trust and underpinned by a sound legal contract.
3. Provide clear scopes for the contractor and agree any required changes to these scopes as the contract progresses.
4. Clearly specify all responsibilities and interfaces between parties affected by the contract.
5. Define the EHS systems and standards which are to be used in the contract.
6. Train and induct contractor staff before they start working at the site.
7. Allocate adequate internal staff resources so that the operating company can manage the contract effectively.
8. Supervise the work completed by the contractor carefully and provide feedback quickly so that small issues do not escalate into major problems.
9. Formally close the contract when the work has been completed satisfactorily and identify any areas where improvements could be made in the future with the contractor.

Table 3 lists some of the key controls that operating companies use for managing contractors.

Table 3.   Key controls for managing contractors

Control	Objective
EHS questionnaires	Review EHS performance and safety management systems for potential new suppliers.
Approved supplier lists	Only allow suppliers who meet the site EHS standards to work on the site.
Contractor inductions	Train contractors about site risks and systems.
Specialist training	Specialist training for high risk activities like roof work, vessel entry and permit-to-work.
'Passport' scheme	Only allow suppliers who meet the site EHS standards to work on the site.
Site supervision	Independently identify problems quickly.
Design/drawing approval system	Check by operating company that design/project is fit for purpose, including EHS issues.
Contractor audits	Ensure that contractor performance is being maintained.
Workplace inspections	Spot check of compliance with site systems.

CONCLUSIONS

Every organisation uses contractors in some part of it's operations. Some companies are heavily reliant on contractors and operate as virtual organisations. Others limit their use of contractors to specific non-core activities. All of these contractor structures can be made to work but some may require large amounts of effort and internal resource for contractor supervision.

Historically, contractor accident rates may have been underestimated with many accidents involving contractors being attributed to other causes such as 'equipment failure' or 'maintenance error'. Recent events in the railway industry have focused attention on the critical importance of effective contractor management. Contractors on major hazard sites therefore play a critical role in the prevention and control of major accidents on these sites. This paper has highlighted some of the problems which can occur with contractors and some of the techniques which can be used to control these problems.

REFERENCES

(CHI, 2000)   'USWA: Kaiser Aluminium again loses on legal manoeuvre to avoid responsibility for Gramercy explosion', Chemical Hazards in Industry, page 4, September 2000.

(HSE, 1998)   'Safety audit of Dounreay, 1998', HSE Books, 1998.

(HSE, 2001a)   'COMAH major accidents notified to the European Commission, England, Scotland and Wales, 1999–2000. Report of the Competent Authority', Health and Safety Executive, 2001.

(HSE, 2001b)   'Train derailment at Hatfield, 17 October 2000. Second HSE interim report', UK Health and Safety Executive, published on internet, http:///www.hse.gov.uk/railway/hatfield/interim2.htm, 23 January 2001.

(HSE, 2002)   'Train derailment at Potters Bar, 10th May 2002. A progress report by the HSE Investigation Board to the end of June 2002', HSE website, July 2002.

(TCE, 1996)   'Wrong paperwork caused Albright & Wilson blast', The Chemical Engineer, page 7, 24th October 1996.

TOP MANAGEMENT BEHAVIOURS – THE DETERMINING ROLE IN CHANGING SAFETY CULTURE

Mr Roderick Prior
Managing Director, SHExcellence CC, Johannesburg, South Africa

Abstract
Behavioural safety programmes have traditionally been focussed on employees engaged in, and close to the operating environment. Two significant South African chemical companies (African Explosives and Sasol Polymers) have undergone significant successful safety culture change programmes starting in the late 1990s. In both cases the role, involvement, and behavioural change of the senior management team was highly significant in achieving the change. The lack of safety involvement of top management in both a visible sense and in more hidden activities was highlighted as an issue in the HAZARDS VI SYMPOSIUM. Although the motivation for changing culture was very different for the two companies both approaches lead to substantial safety improvement. Practical examples of senior management change are given particularly in prioritising resource allocation and demonstrating commitment in public.

Safety, Behaviour, Management involvement, Culture

INTRODUCTION

Over the past 8 years two substantial South African chemical companies have undergone significant changes to their safety cultures and have demonstrated large improvements in safety performance.

AECI Explosives Ltd (now known as AFRICAN EXPLOSIVES LIMITED) has been manufacturing and selling commercial explosives for 106 years. The products manufactured in the period discussed in the paper comprised both traditional and modern explosives. These included products based on nitro-glycerine, ammonium nitrate, and emulsions. All explosive products have considerable hazards in manufacture and use. Expertise in making and using explosives requires a blend of experience and knowledge of chemical/physical/ engineering principles. African Explosives Limited (AEL) has its major manufacturing site at Modderfontein close to Johannesburg. The total employment was about 4500 people. The author held the positions of Production and Business Director for this company over a period of some 8 years.

POLIFIN LIMITED (now known as SASOL POLYMERS) is based on chlor-alkali technology and produces a variety of monomers and polymers, chlor-alkali products and mining reagents. It has four major production sites and employs about 4400 people. At the time of the initiation of a safety culture change POLIFIN was a joint venture between AECI Ltd and Sasol Ltd and was a separate listed company with considerable independence. The author of the paper is closely familiar with the change processes in POLIFIN and the role played by top management.

A framework for analysing and highlighting the role of senior management is outlined in the next section. In the description of the case studies many examples of visible and hidden behaviours of senior mangers will be described and it is hoped that these will be of use to other managers.

FRAMEWORK FOR ANALYSING SAFETY AND HEALTH BEHAVIOURS

Safety management needs to be applied in three major areas for any operation i.e. plant (hardware), safety management systems, and people. In the people category a variety of safety and health behaviours involving all employees, including the most senior managers, is essential in improving safety performance and establishing a world-class safety culture. Ronny Lardner of the Keil Centre, Edinburgh has analysed these behaviours and pointed out that conventional behavioural safety programmes have focussed on the general safety behaviours of frontline personnel. Figure 1 describes the categories of behaviours.

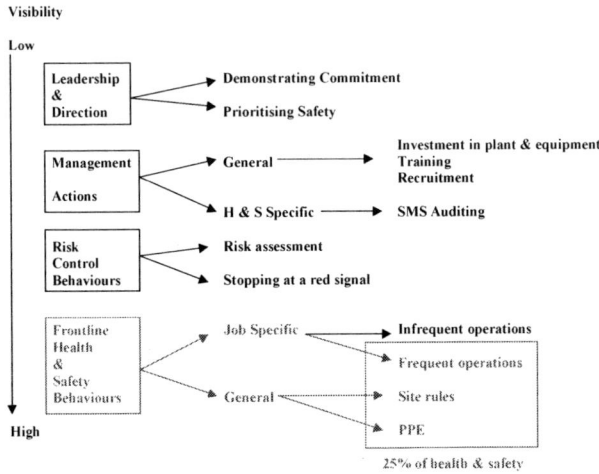

Figure 1.   Categories of health and safety behaviours

In general there is no focus on top management behaviours. These can be visible and hidden. In the two South African case studies described in this paper, probably 75% of the employee population has been subjected to safety behavioural programmes. This is the reverse of the UK analysis as seen in the above figure. However the impact of top management behaviour, both positive and negative, has a multiplier effect of considerable magnitude. The above framework has been used to analyse the two companies from a top management safety behaviour perspective. Both companies feel that top management behaviour has been the determining role in changing safety culture and improving safety performance.

DRIVERS FOR CHANGE IN AEL

The parent company AECI and AEL, its explosives arm, have always been leaders in safety management in South Africa. Features of safety management in the 1980's and early 1990's were the reliance on the experience of many employees where service of 20/30 years was common. A lot of the good practice and standards stemmed from learning from accidents and incidents. Worldwide sharing of safety experience is also common via various mechanisms. Training and retraining has also been very strong in AEL. Line management was held responsible for safety performance. Management involvement included setting targets, drawing up plans and doing occasional inspections.

Prior to 1993 the last serious accident had occurred in 1985. A steady improvement in accident rates was recorded over this period. In 1993 two serious accidents occurred involving 3 fatalities. These were investigated and plant specific improvements made but with no review of the entire safety programme.

A number of shortcomings were detected via audits in 1994 and management decided that a radical change was needed. To that end, visits to DuPont in the USA and ICI in the UK were undertaken to look at new approaches. Before changes could be implemented a major event involving the death of 8 employees in a nitro-glycerine plant changed the entire approach to safety. This watershed event had numerous serious consequences apart from the tragic deaths. These were:

3 Week strike
Government inquiry
Internal inquiry with Union representation
Major impact on morale
Company image and loss of business

Whilst the Government inquiry failed to establish a specific cause of the accident the impact of the event was such as to drive a massive improvement effort. It was felt that the basics of safety had to be re-established. Based on the investigation, ICI/AEL audits, and DuPont/ICI principles and practices, a comprehensive improvement plan was drawn up. This covered hardware, management systems and people aspects.

AEL SAFETY IMPROVEMENT PLAN – THE DUPONT INFLUENCE

The plan was detailed covering all aspects of safety improvement. Only those areas where top management involvement was significant will be covered in this paper. The DuPont experience was perhaps the most influential aspect of the thinking around change. Whilst much can be learnt from reading about their approach to safety, spending significant time in their operations enabled the author to have a clear vision of what 'world class' safety looked like. The amazing consistency and apparent simplicity of the safety culture left an indelible impression. Time was spent at levels of the organisation to understand the DuPont approach.

DuPont had well designed plant and sophisticated safety management systems but the real difference in their safety performance stemmed from the people element. Safety was driven from the top and was seen as the prime business goal. A strong belief was developed that a similar culture could be developed in AEL over time.

INVOLVEMENT OF TOP MANAGEMENT IN SAFETY IMPROVEMENT IN AEL

A number of critical actions were agreed. These were:

1) Safety to be the dominant objective before production, costs etc. This was put in a policy document outlined in flow chart form. All employees were asked to hold

management accountable against this document. The document was put on all notice boards.

2) Line management, including directors, to be held fully responsible and to be involved in key activities
3) Safety auditing to be introduced – to be lead by senior line management
4) Full Union and employee involvement
5) Performance Management System to reflect at least 30% of the total score for safety performance
6) DuPont's STOP programme to be introduced

BEHAVIOURAL SAFETY CHANGES

It was made crystal clear to the entire company via briefs, discussions and illustrated in decision making that safety (SHE) was the primary objective of the company. SHE was the first item on all meeting agendas including Board meetings. Over time a change in safety orientation was seen to evolve at the highest level. At the Board level any proposal had to clear SHE hurdles before any further debate was allowed. Capital projects were closely scrutinised to ensure that SHE requirements were met and exceeded.

A system of safety auditing called active monitoring was put in place. This approach was based on quality management principles and had been successfully used in the Ardeer Factory of ICI Explosives. By auditing the inputs to the safety process like training, emergency procedures, PPE management etc safety effort can be proactive as opposed to the reactive approach of only focussing on incidents and accidents. Senior mangers and specialists were trained in auditing principles and a 3-year plan put in place to audit all systems with the high priority systems tackled first.

Each senior manager, including Personnel, Finance etc lead monthly audits. By including all managers very strong team ownership of safety become apparent. Emphasis was placed on quick corrective action and close out of actions. A vast amount of work was created at the start of the audits but resources were deployed to complete the work. Often simple inexpensive measures were used. Credibility of the process was seen as of paramount importance. In leading the audits, management behaviour could be seen to emphasise that safety was a critical objective. Interaction with employees improved communication immensely. Coaching and influencing opportunities were found during the audits. A very large gap was found between what supervisors thought was happening and reality. The auditing performance of senior managers was reviewed monthly and results displayed publically. A sample is shown below in Figure 2.

Performance management at all levels and across all functions had safety as the dominant performance measure for all functions. In 1997 the Board of Directors refused a performance bonus because of a dip in safety performance.

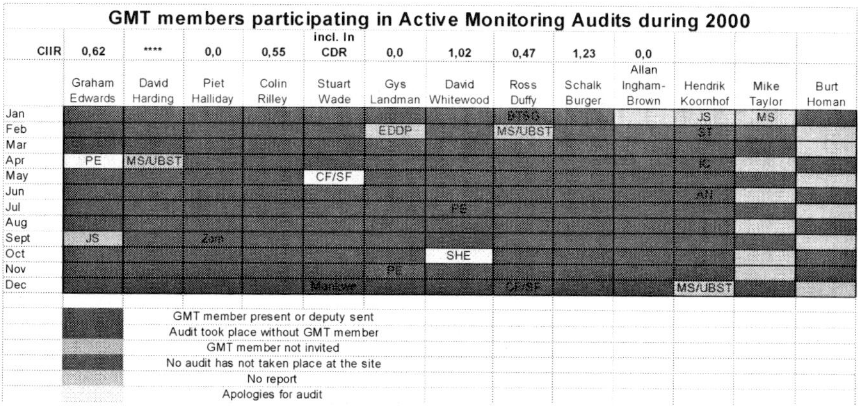

GMT members participating in Active Monitoring Audits during 2000

CIIR	0,62	****	0,0	0,55	incl. In CDR	0,0	1,02	0,47	1,23	0,0			
	Graham Edwards	David Harding	Piet Halliday	Colin Rilley	Stuart Wade	Gys Landman	David Whitewood	Ross Duffy	Schalk Burger	Allan Ingham-Brown	Hendrik Koornhof	Mike Taylor	Burt Homan
Jan								BTSG			JS	MS	
Feb					EDDP		MS/UBST				ST		
Mar													
Apr	PE	MS/UBST									RC		
May					CF/SF								
Jun											AH		
Jul							PE						
Aug													
Sept	JS		Zorn										
Oct							SHE						
Nov						PE							
Dec			Manlowe				CF/SF				MS/UBST		

GMT member present or deputy sent
Audit took place without GMT member
GMT member not invited
No audit has not taken place at the site
No report
Apologies for audit

Figure 2. Sample AEL active monitoring summary

It was necessary to show that managers were capable of changing old habits to provide a moral and visible basis for others to change. The parking conventions were changed, starting at Head Office, to the safer parking rule of reverse parking. This caused a level of discomfort amongst senior managers. It was important to show that managers would comply with unpopular rules as well. All mangers were required to pass defensive driving courses and their vehicles were subject to inspection of safety standards. The DuPont rule for having one hand on the stair handrail at all times was also implemented. A set of safety guidelines for senior managers was issued with examples of appropriate behaviour. This can be seen as Appendix 1. The main point of these activities was to show that everybody was capable of changing their safety behaviours.

Union/management workshops were held to understand safety issues and possible solutions. These were co-chaired by senior representatives from both parties. This resulted in a Safety Charter being negotiated. The Charter contained important principles that had been agreed.

The DuPont 'STOP' programme was implemented progressively. DuPont's Safety Training Observation Program, known as STOP, is a positive behaviour modification programme and has been used in DuPont and other companies for some years. Employees and supervisors are trained to observe, correct, prevent and report unsafe acts systematically. The programme is based on the well-known DuPont safety principles. In addition to direct intervention observation cards are filled in to create a data-base for detecting safety trends. The programme is aimed at the workforce and first line supervision. It focuses on specific areas like the use of PPE. The programme is straightforward and easy to use. It fits in well with other safety initiatives.

STOP had a very slow start but gathered pace over the years and is currently a strong element of AEL's safety programme. 500 Observation cards are being received monthly.

The top management behaviours in AEL did not form part of a formal safety behavioural programme. They evolved from good practice in world-class companies, ideas generated internally and common sense. Fortunately there was a lot of emphasis on the 'optics' which meant that all employees could see that top management 'was walking the talk'.

AEL PROGRESS TO WORLD CLASS
Figure 3 illustrates performance over the past years.

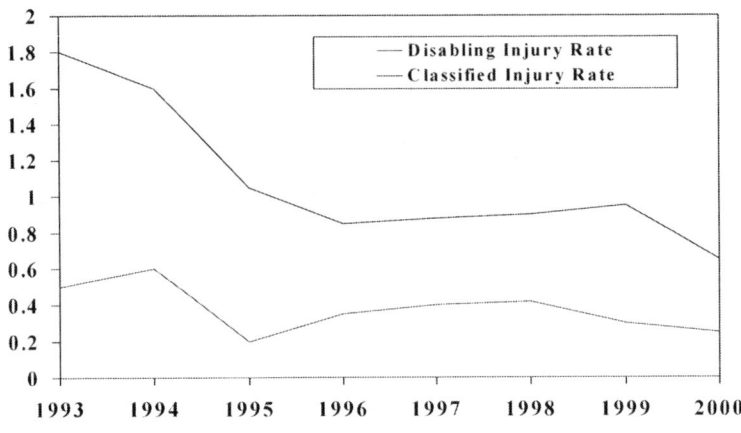

Figure 3.   AEL safety performance

The integrated improved plan was implemented in late 1994 and continued over succeeding years. Old technologies were exited but the emphasis on people and particularly senior manager behaviour continued. It was evident to all that the safety culture of the company had changed significantly. An improvement in safety performance was seen almost immediately

Some impressive achievements were noted. In the period 1997–1999 10 million man hours were recorded without a single Reportable Incident(3 day absence). This was achieved by the 3000 workers at the AEL Modderfontein Factory. This is comparable to some DuPont sites. In the period 1995–1997 Modderfontein was adjudged the best performing site in ICI Explosives(from about 20 sites worldwide).

POLIFIN: DRIVERS FOR CHANGE
In 1996 the top management team at POLIFIN decided to adopt a values driven approach to changing the company culture. This was seen as a necessary approach to meeting the needs of competing in the global markets and establishing a strong independent identity separate from that of its owners, AECI Ltd, SASOL Ltd. A vision of world class was adopted. Relevant shared values were needed to guide change and behaviours required to support the values. Following a consultative process with 500 employees 5 values were distilled. These were grouped under the headings of:

CUSTOMERS
PEOPLE (three values)
ENVIRONMENT

Sets of behaviours were developed to underpin the values and employees at all levels were required to interpret these for their own use. Under the PEOPLE banner one of the values pertaining directly to safety stated '*we will work safely and provide safe working conditions*'. One of the supporting behaviours is to take personal responsibility for safety.

One of the major areas for improvement was POLIFIN's safety performance and the creation of a new safety culture was needed. In 1996 POLIFIN's Recordable Injury Rate was 3.5 and world class was judged to 0.6.

The values were seen as the vehicle to deliver a number of important business goals including safety and customer service. The values were seen to work together and a failure in one would impact on the others.

POLIFIN COMMUNICATIONS – TOP MANAGEMENT INVOLVEMENT

The roll out of the values involved a number of communication channels .In addition to the conventional approaches such as videos, newsletters, extensive use was made of industrial theatre. The videos and newsletters involved the Managing Director and other senior mangers prominently as actors and contributors. This conveyed a strong sense of ownership by top management.

Industrial theatre is a well-known technique used extensively in South Africa by large companies to convey important concepts and messages to semi-literate and illiterate people. A play is acted out by professional actors. Humour, developed by negative role-playing, is the device used to make points and get messages across. POLIFIN's play "Tomatoes for Africa" ran to 54 performances across all the sites in 1997 and each one was hosted and facilitated by the relevant General Manager. The play highlighted, in a humorous fashion, safety issues such as training, right to refuse to work under unsafe conditions, looking out for colleagues' safety, taking short cuts and many others. The General Manager played a critical role in linking episodes in the play to the safety value.

LEOPARD PROCESS

In 1997 POLIFIN began the implementation of a behaviour based safety programme named LEOPARD using the BST^{TM} process. This was the first application in Africa. LEOPARD stands for '*Let Employees Observe Peers And Observe Deviations*'. Management at the highest level has driven the programme and total buy-in was obtained from the workforce. Identified behaviours are used to conduct peer-based observations on a no-name no-blame basis. This programme is still running and has made a significant impact on safety in POLIFIN.

POLIFIN: SPECIFIC SAFETY BEHAVIOURS OF SENIOR MANAGEMENT

A variety of behaviours have characterised management behaviour change in POLIFIN:

- General Manager sits in on all Lost Workday Case incident investigations
- Lead item at all business meetings
- SHE to be discussed at informal gatherings e.g. tea session at start of day
- Refusal to run plants with temporary labour if safety not guaranteed. (strikes)
- Demonstrate interest in people rather than the statistics. Focus on the person who got hurt.

- Talk about injuries rather than accidents. Want unhurt and healthy people to go home.
- Translate statistics into real life in a credible way e.g. What does a Recordable Injury Rate of 1 mean and is it acceptable?
- General Managers trained as observers in LEOPARD and do regular observations
- General Managers meets directly with safety representatives
- All plant safety meeting minutes signed off by GM
- Shut down of hazardous old units like Carbide Plant
- Senior Manager ejected from plant for not wearing PPE and voluntarily participating in a disciplinary enquiry. One rule for all.
- 3 Monthly management inspections of other areas.
- Sponsoring home safety seminars.
- Show caring for safety of employees in transit to work. R2 million bridge over road.
- R25 million spent on Sasolburg town infrastructure to improve traffic and pedestrian safety, Cows are a hazard.

The devastating impact of negative behaviour was also noted. A new manager with a different safety background put his public quote 'safety is costing us too much' into effect and the Departmental safety record deteriorated markedly with his employee's safety behaviours reverting to previous low standards.

POLIFIN: INCORPORATION OF SAFETY INTO THE BUSINESS MODEL
Some managers have incorporated safety into a business model so as to ensure it is not seen as an add-on. Figure 4 illustrates the model for a services department.

The values are linked to strategic focus areas. In turn the focus areas are developed into Key Performance Areas and Key Performance Indicators for individuals. A set of performance contracts exists from the Managing Director to first line supervision. This links the MD to the shop floor. Any significant decision in any of the strategic focus areas has to be tested against the requirements (and inherent values) of the other areas. Therefore SHE is considered in each decision.

RECOGNITION AND REWARDS
At the senior level up to 4% is available on safety performance. This comes out of a total of 15% incentive pay that is available. The award is based on Classified Injury Incidence Rate.

An innovative set of recognition awards is available via the POLIFIN *Bateleur* ceremony. Based on the *Oscar* awards the Managing Director awards the most deserving groups prizes in various categories relating to the 5 values. In addition to the team safety award other business team awards include safety as one of the criteria.

IMPROVEMENT IN POLIFIN SAFETY PERFORMANCE
The accident record for the past 8 years is shown in the figure below.

A sharp improvement is evident from 1996 to 1998 and following a deterioration in 2000 the trend is again down. The early improvement is attributed to the values programme and the later improvement due to applying the BST^{TM} process more rigorously as well as the developing safety culture.

740

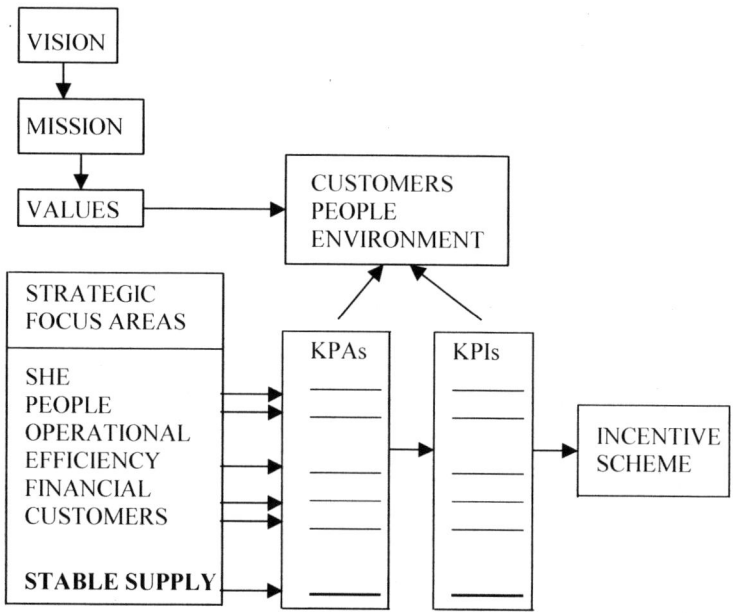

Figure 4.   POLIFIN business model incorporating safety

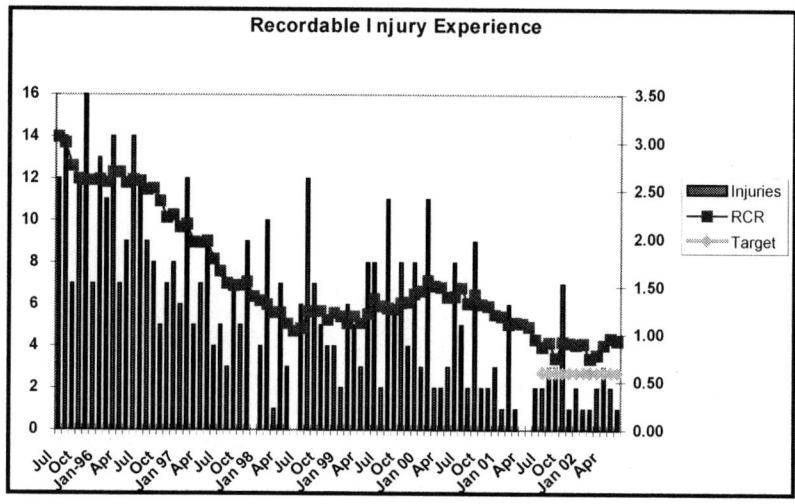

Figure 5.   Poli fin recordable experience 1996–2002

CONCLUSIONS

All employee behaviours are important in developing and maintaining an effective safety culture. It is clear that the safety behaviour of senior management is critical in improving and attaining world-class safety standards. Both positive and negative behaviours have a '*multiplier*' effect because of the role model status of persons having authority and the ability to direct resources to different strategic areas. Negative behaviour can have a particularly devastating effect on safety culture and performance.

In the case study of the two South African chemical companies senior management played a determining role in initiating and sustaining safety changes. Their role was not part of a conventional safety behavioural programme. The role evolved out of best practices seen in world-class companies and the application of common sense. Many examples have been quoted in the paper both of hidden and visible behaviours. It is hoped that these will assist other senior managers to demonstrate their commitment in a concrete fashion.

In changing a safety culture in a company, attention needs to be paid to all levels of employees including those at the very highest level. With creative thought and lessons learnt from others meaningful change can involve everybody. Top management, in demonstrating that it can change behaviour, creates a powerful base for corporate behavioural change.

ACKNOWLEDGEMENTS

This paper was made possible with the support of African Explosives Limited and Sasol Polymers in the interests of sharing safety information. In particular Dr G. Edwards and Mr T. Munday were particularly helpful.

Mr R. Lardner of the Keil Centre kindly gave permission to use the diagram illustrating Categories of Critical Health and Safety Behaviours.

APPENDIX 1

AECI Explosives Limited–Excellence in SHE Behaviours for Managers and Supervision

KEY AREA 1 - SETTING THE STANDARD

Sets unambiguous high standards and follows them without compromise. Standard has been defined, relevant documents can be produced. Manager can describe supporting actions.

MANAGEMENT/SUPERVISOR BEHAVIOURS

VEHICLES
- Always wears seatbelt and checks others wear theirs
- Drives in a safe manner demonstrating defensive driving skills
- Observes speed limits and other regulations

OFFICE
- Never walks past a trip hazard
- Office materials in a safe and stable position

- Proper use of office furniture
- Holds onto handrail when going up or down stairs
- Stops others carrying out unsafe or unhealthy acts
- Talks about SHE and communicates SHE news with high priority
- Takes a positive interest in checking office safety e.g. SHE policy posted, fire wardens, safety meetings

ON SITE
- Discusses SHE issues with employees
- Wears PPE and asks for if not offered
- Asks about emergency procedures if not told
- Reviews the safety reporting systems and asks about progress
- Compliments good work
- Does not walk past deviations and checks action is being taken
- Gives immediate feedback to local management on observations

KEY AREA 2 - COMMUNICATION
The manager/supervisor will be accessible to all employees on SHE issues and will ensure that the team maintains a high level of accessibility to employees. He will check on a regular basis that there is an open and multilateral communication process in operation. Employees will be able to tell you want the GM expects.

MANAGEMENT/SUPERVISOR BEHAVIOURS
- Employees can tell you who the manager/supervisor is
- Employees can describe the current business performance is, including SHE goals
- Employees can correctly outline the priorities
- The manager/supervisor can describe the current employee concerns
- The local manager can demonstrate that there is an effective process to report to them all reportable SHE issues on the same day
- The local manager personally reports all such incidents to his superior
- Uses attitude surveys
- Is there active fostering of employee involvement at all levels and encourage full empowerment of all employees

KEY AREA 3 – RESPONSIBILITY AND ACCOUNTABILITY
The manager/supervisor will ensure that there are effective processes to enable each person to understand the scope of their responsibility and that the person is competent to fulfil the job requirements.

MANAGEMENT/SUPERVISOR BEHAVIOURS

- SHE targets are included in personal objectives and are subject to review through the performance management system
- The incumbent always follows the reward and disciplinary system
- Employees feel able to stop activities for SHE reasons irrespective of short-term business needs
- Is it apparent that SHE management competency is taken into account when selecting employees for new positions ?
- The manager/supervisor will always ensure that the SHE requirements are fully addressed in the development, evaluation and approval of all changes
- The manager/supervisor ensures that equipment is maintained fit for purpose

KEY AREA 4 – AUDITING/COMPLIANCE

The manager/supervisor ensures that audit processes are in place to give assurance that requirements are being met.

MANAGEMENT/SUPERVISOR BEHAVIOURS

- The manager/supervisor can demonstrate understanding of SHE policies/standards
- The manager/supervisor can describe the audit process for the unit
- The manager/supervisor can describe the review process for maintaining the improvement plans and taking of corrective action
- Has easy access to SHE data
- Can describe examples of where deviations found
- SHE items are the first item on business meeting agendas

IMPROVING SAFETY PERFORMANCE AND CULTURE WHILST UNDERTAKING BUSINESS RE-ENGINEERING

Geoff Entwistle
Octel Corporation

Octel Corporation manufactures and markets transport fuel additives and specialty chemicals. The Company is growing mainly by acquisition in these markets. This activity is being funded in part by earnings from the Company's traditional Lead Additive business but this is a declining market. In view of this Company background, the Lead Additive business at the Cheshire Manufacturing Park site has been significantly re-engineered since early 2000. It has become a Business Unit in its own right. Its management team is now stewarding more site assets, has lost experienced operators through voluntary severance schemes, has taken operators from other (redundant) plant areas and has changed the way its operating teams are organised. The Business Unit handles and stores hazardous materials requiring it to produce a Safety Report under the COMAH Regulations.

All the above organisational change activity has been carried out with the express objective of requiring an improved safety performance and safety culture. Since early 2000, the lost time accident frequency rate for the Operations team of the Lead Additive Business Unit at the Cheshire Manufacturing Park has reduced from 1.29 to zero. Similarly, the minor accident frequency rate for this work group has reduced from 17.4 to 2.6. These figures are for an operational workforce reduced from about 150 to about 100 employees. Other Safety, Health and Environment (SHE) performance indicators are also showing positive trends. This paper will give a brief summary of Octel's recent business history to provide a context for the rest of the paper. It will then describe the Business Unit's organisational change from a SHE culture and performance improvement viewpoint. The rebuilding of the Lead Additive Business Unit's Safety Management System (SMS) will then be described.

KEYWORDS: operational change, management standards, performance indicators

BUSINESS AND ORGANISATIONAL BACKGROUND

Octel Corporation was created as an independent company on the New York Stock Exchange in mid 1998. At the time of the "spin" from its parent company, Octel was a major world supplier of transport fuel additives and specialty chemicals. As part of the "spin" Octel acquired a significant debt burden of about $350 m. The Corporation's vision was to create a world class company by organic growth and acquisition in the petroleum additives and specialty chemicals market areas. The company's Lead Additive business is a major contributor to the company's earnings and yet is in a world market declining at a rate of about 15% per annum. The Company objective was to ensure that as the world demand for Lead Additive reduced Octel would be the last manufacturer of this product. In order to achieve this the production cost base needed to be minimised to overcome competition.

Octel's Ellesmere Port site has been a manufacturing site for Lead Additives for transport fuels for over 50 years. The site was an integrated chemical site with plants producing products which were the raw materials for the Lead Additive manufacturing plant. In addition the Lead Additive plant operated the Site Effluent treatment facilities. Other plants were built on the site over the 50 years to make other petroleum fuel additives, such as detergents, and specialty chemical additives, such as a biodegradable chelator. After the "spin" a review of all site facilities was undertaken to seek minimum cost operations. It was decided that plant assets should be regrouped into Business Units, with each Unit having its own operational organisation focussed on that business' needs. As a result certain plants were shutdown and their products bought in from lower cost producers whilst in other cases operational assets were retained but transferred to more appropriate Business or Service Units.

The closure of some operational assets created land, offices, laboratories and warehousing that would no longer be required by Octel. The 85 acre Octel Ellesmere Port site was launched as Cheshire Manufacturing Park in June 2000 to attract independent companies to rent, lease or build on the site. It was anticipated that in order to attract tenants the site would need to have a good safety record. The prime objective of the site reorganisation was therefore to create minimum cost operations with a "safe operation" culture. Whilst similarly significant organisational changes have been made in the various Octel Business and Service Units on Cheshire Manufacturing Park, this paper focuses on the changes made by the formation and subsequent operation of the Lead Additive Business Unit as they have impacted on this Business Unit's Safety, Health and Environment (SHE) performance.

LEAD ADDITIVE BUSINESS RE-ENGINEERING
The re-engineering of the Lead Additive Business Unit will be described in two parts, asset re-organisation and organisational change.

ASSET RE-ORGANISATION
The review of operational assets following the floatation of Octel in 1998 identified that two raw materials used by the Lead Additive plants and manufactured on site could be provided by lower cost suppliers. As a result the plant manufacturing one of the raw materials was closed in 1999 and road tanker offloading bays were constructed for imports onto site. The plant manufacturing the other raw material was closed in 2000 and road offloading facilities constructed in addition to the already available ship offloading facility. Both closed manufacturing facilities had storage tankage which was to be retained and these plus the new assets were to be transferred to the ownership of the operational team of the Lead Additive Business Unit which was formed in early 2000. The redundant assets were not previously operated by the Lead Additive team. The controlrooms for the closed assets were to be de-manned so that consideration needed to be given to the retention of instrumentation required for the retained storage assets and this instrumentation relocated to existing control points where the new operational teams were based.

Both of the above asset changes were treated as "off-line" projects using the company's normal projects procedures. These include hazard studies appropriate to the stage of development of the project. Both projects involved plant personnel on the project teams. The formal hand-over of the projects included the outcome of training and competency assessments made on personnel who were to operate the facilities.

The Lead Additive Business Unit (LABU) was created at the same time as the other Business and Service Units on the site. The Lead Additive business operational team had previously managed the Site Effluent plant but clearly this was not an asset solely for the benefit of this Business Unit. This asset, and its operational personnel, were transferred out of the Lead Additive Business Unit as one entity.

ORGANISATIONAL CHANGE

Some organisational change had occurred in Octel across the 1990s as the company strove to maintain profitability in the Lead Additive business as the market for this product declined. Until the mid 1990s little of this change was formally assessed for its impact on safety. With the need for more radical asset and organisational management becoming evident in the mid 1990s a more involving management style was encouraged. Higher operational standards with regular open communications with all employees were demanded. Emphasis was to be placed on teamworking and flexibility with greater individual responsibility. A joint management union group (Review Group) was set up in 1995 and new terms and conditions agreements were drawn up with employee representation. In 1996 Ellesmere Port site moved from an 8 hour to a 12 hour shift working pattern. Again in 1996, with plant closures looking more likely, the Review Group reached agreement that enforced redundancies would be avoided wherever practicable. A voluntary severance scheme was opened that year and throughout the restructuring to date voluntary severance has been maintained. Lessons have been learned moving from one severance scheme to the next including that leavers needed to clearly understand that the business would come first and their departure, if agreed, would depend upon fully trained replacements being available.

The Review Group was involved during the formation of LABU in 2000. That group assisted in the formulation of the re-organisation arrangements and arranged the regular briefing of the changes to employees. Union, management and employee representation worked together to arrive at an organisational structure which would allow reduced manning by using four operating teams led by Leading Operators with all teams reporting to a Shift Team Leader. The number of teams was determined by process plant area boundaries.

Octel had developed its own risk assessment methodology for organisational changes. All the proposed changes relating to the Lead Additive Business Unit were then assessed using this methodology. The assessment process centres on a meeting of those persons affected by the change and those individuals work through a set of change related questions. In most cases, Union representation attended such meetings. The assessment starts by defining the objective and scope of the proposal and then considers its impact on relevant individuals' workload. The criticality of the change with respect to safety procedures and good practice is then considered. The second issue covered by the assessment is whether there will be an effect on a critical activity. Typically these activities are defined as key roles within company procedures (e.g. emergency procedure role, permit to work role, etc). The assessment then questions whether the proposal will impact on a company procedure. Finally the assessment seeks to question whether the person or persons who are to take up the new duties are competent to do so. The conclusion of the assessment is the acceptance or rejection of the proposal by the senior manager of the business unit along with any action plan arising from the assessment. The involvement of those affected by the proposed changes has the benefit of "buy in" to the changes with any issues aired and the ability to influence the outcome.

The current LABU organisational structure is given below. The re-engineering reduced the LABU headcount from approximately 150 to 100. The voluntary severance scheme meant that some more experienced personnel took advantage of the access to an early retirement pension. Also, some of the personnel who had worked in plant areas which were to be closed down decided to stay with the company and some became LABU employees. The organisational change action plans took account of individual's additional training needs but the general re-engineering process was provided with a separate "off line" training support project. The re-engineering process looked at the balance of skills across the shift teams and some individuals were moved across shifts in order to redress imbalances caused by those who wished to leave the company via the severance scheme and those brought into the Business Unit from closed plants.

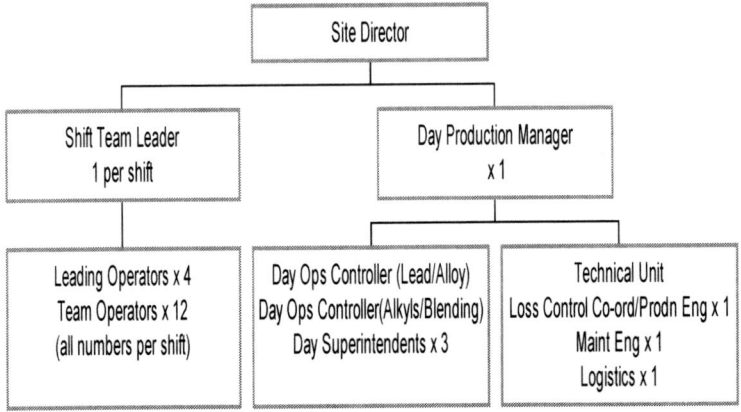

Figure 1.   Current LABU operations organisational structure

My personal experience of this organisational change assessment process has been good. The original proposal for the organisation was that there would not have been a mechanical engineering support person in the Technical Unit. This would have meant that the Business Unit would have had to bring in such a person from the central maintenance team when required. The identification as to when such a person would be needed was a concern. The likelihood that such an individual would not be sufficiently familiar with the detail of the Business Unit to contribute effectively was also a concern. Both concerns contributed to the inclusion of the Maintenance Engineer skill in this Unit. I have also been able to influence my role as Loss Control Co-ordinator/Production Engineer in that I have justified the need within our Business Unit for me to carry out Loss Control work for 50% of my time and Production Engineering work for the other 50%. This allows me to demonstrate the implementation of the SHE standards I am promoting using leadership by example. I also deputise for the Day Production Manager, which allows us to share values and gives a similar opportunity. My role is seen as part of the day to day asset management team.

RE-ENGINEERING TRAINING

The approach to training during LABU re-engineering followed work that had previously been tested out on another plant area on site by using the NVQ standard. This was seen to have the benefits of using a Nationally recognised standard which would fit with our operations and for our employees would give them transferable skills if they should leave the company at a later date. Training for the shift operating teams started from a team matrix of operational tasks and who needed to do what. This led to individual training plans. Training was then arranged to suit team and individual needs. All LABU operating personnel, whether they had been in LABU for some time or had transferred into LABU as part of re-engineering, were given general awareness safety training courses covering the more important SHE issues (accident/incident reporting, permit system, change control, risk assessment, COSHH assessments and IPC). Personnel who had recently transferred into LABU were given an Induction Course to familiarise them with LABU activities and then made top priority for assessment of their training needs for their particular roles within LABU. Specific training then followed using plant information and task books presented at the Induction Course. This approach allowed formal assessment of each individual on plant orientation, emergency response, safety equipment function and application along with the knowledge imparted from the general awareness course. Training to highlight how common plant and equipment used in LABU works, common problems with it and the safe way to operate it was delivered by craftsmen. Personnel who had worked in LABU for some time went through the same process after "new" personnel to LABU had been trained.

"Leading Operator" was a new role in the LABU organisation. Individuals placed in these roles were experienced LABU operators. This new position is in effect a first line supervisor role as they operate plant, lead their team and carry out company safety system roles such as permit to work process authoriser and emergency response forward control point leader. The previous LABU organisation included "supervisors"; these roles were removed in the re-engineering and their activities split between the "Leading Operators" and "Shift Team Leaders". "Leading Operators" have more responsibilities than operators and so were given more extensive training in the key site safety and operational systems for their role. This included hands on emergency simulation training. Their initial competency in what they had been taught was then checked by use of practice examples for each activity in which they had been trained. Their performance was observed and recorded and if satisfactory the individual was then "passed out" in that particular activity. "Shift Team Leader" is also a new role, though it is very similar to the previous Shift Manager role. "Shift Team Leaders" have experienced a training plan similar to that followed by the "Leading Operators" though their assessments are based on their particular responsibilities and duties (e.g. "Shift Team Leader" is the Site Controller in an emergency).

Formal revalidation of initial competency is now under review.

LABU SHE PERFORMANCE AND SAFETY MANAGEMENT SYSTEM STATUS AT TIME OF RE-ENGINEERING

At the time re-engineering started in early 2000, the LABU Operations lost time accident frequency rate was 1.29 per hundred thousand hours worked. The LABU minor accident frequency rate was 17.35 on the same basis. LABU operations had a head count of over 150 people. Octel had started to use propriety audit systems in the early 1990s and had established formal safety procedures to comply with the chosen system. The central safety function at that time had generated many safety procedures to formalise the way the site safety systems were to operate. Works areas had then translated those centrally generated procedures into local procedures and instructions. These, for the plant area which is now called LABU, had been assembled in the early to mid 1990s but since that time there had been a re-deployment of resources, particularly at management level, without updating the safety management system procedures. In effect whilst the safety systems had not changed the local procedures had not been updated to be consistent with the personnel in post, and, their needs to operate to the procedures. An audit of the status of the application of safety systems within LABU was therefore carried out in early 2000.

The audit of LABU safety systems looked at what we considered should be in place in order for us to be convinced that our people were competent for the activities they were to perform and that our plant, equipment and working environment were "fit for purpose". In relation to plant and equipment there was a further subdivision by considering equipment that was deemed "critical" separately to "non critical" equipment. "Critical" equipment was that which had been identified by previous plant studies as critical to prevent a "major accident hazard". The audit indicated that as far as the "people" element was concerned the re-engineering training project would deliver initial competence in a way which would have an audit trail. The need for ongoing training needs analysis, once training responsibility was handed over to line management, was not adequately clear at the time of the audit. This element also indicated that there was limited supervisory observation of plant conditions and little structured contact between supervisors and operators on a day to day basis and had led to a lack of adequate leadership. The "plant and equipment" element of the audit indicated that there were inadequate routine checks of certain equipment and overall that there was informal ownership of key documents for the safe operation of the plant. The latter had led to certain documents (e.g. safety management system) not being driven so as to be kept up to date. Similarly, the company as a whole has a SHE Policy and at the time of the audit all business units were working to that Policy. By its nature the Policy could not detail responsibilities of individuals within each Business Unit nor the local arrangements for implementation of the Policy. There was no formal ownership of local implementation of the SHE Policy.

The results of the audit were presented to LABU shift and day management personnel and circulated for all LABU operating personnel to see.

STRATEGY TO REBUILD LABU SAFETY MANAGEMENT SYSTEM

Corporately, Octel was starting to produce instructions on what was expected to be included in a Business Unit's safety management system. These supported the company SHE Policy. In the re-engineering, Octel had retained a small central operational safety function which reported into the Site Director. That group started to produce safety procedures which all Business Units on the site were expected to implement. Locally, LABU's audit had indicated what was needed for LABU. The key elements of good health and safety management have been well publicised (e.g. HSG 65). The previous safety management system had obviously struggled to be implemented and its effectiveness, as measured by the lost accident frequency rate, was not considered adequate for a re-engineered LABU. The previous safety management system procedures were extensive and it was considered that this probably contributed to the difficulty of implementation.

We decided to produce what we thought we needed primarily from observations in LABU in a format that made responsibilities and ownership very clear and which was flexible enough for easy of future modification. The system would need to provide a clear audit trail and would as far as possible be computer based. It would meet the needs of Octel SHE Policy and in due course would incorporate Corporate and Site safety procedure demands. It would follow the key elements for good health and safety management, and any relevant guidance notes as produced by the HSE. The development of the system would be led by the Loss Control Co-ordinator but all LABU personnel would get the opportunity to input into what was being developed and in due course the continuing development would pass into line management. It was seen to be key to the activity that LABU employees would have involvement in the improvement activities consistent with the day to day role of each individual. Documentation was viewed differently to the way it had been seen in the past. The system was the seen as the "activity process" and documentation was seen as aiding clarification of the process and providing evidence of the activity. Documentation would be created providing it added value to the process and not to "fill a shelf". The arrangement chosen is depicted in Figure 2 below.

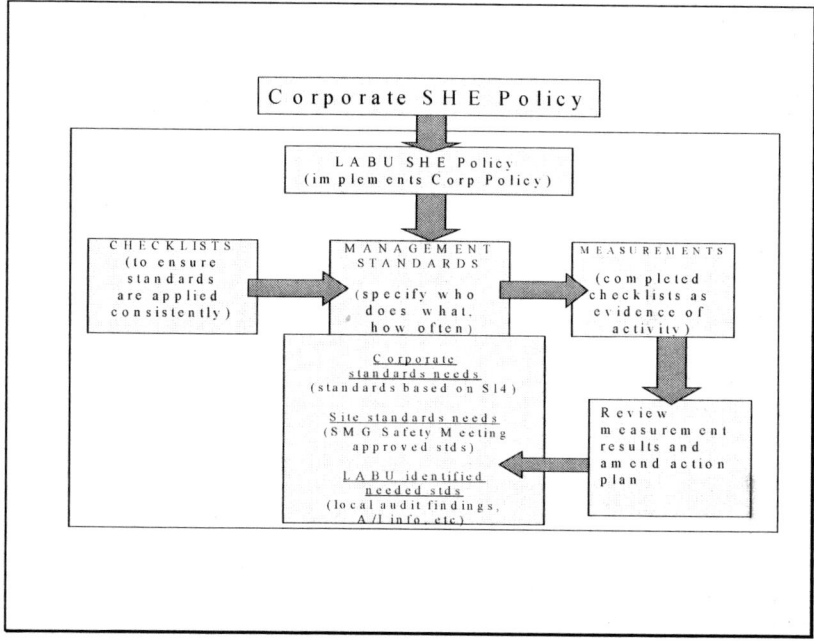

Figure 2.   LABU safety management system structure

The first action was to set up a "safety improvement meeting". This meeting was initially set up informally chaired by the Director responsible for Operations with representation from Day Operations personnel, the Shift Team Leader and a safety representative from the shift that was on duty that day. This meeting has now been set up for a monthly frequency and its timing is scheduled such that each of our five shifts can attend the meeting twice per year. Each shift has set up its own monthly meeting called "safety forum". The Shift Team Leader, Leading Operators and safety representatives of that shift attend these meetings. The minutes of the "safety forum" are circulated to all shifts and to day line management. The "safety improvement meeting" minutes are circulated to Day and Shift line managers and all safety representatives. This first action was to allow a communication path to be set up prior to development of any policies, standards or other safety improvement activity.

The next action was to draw up a LABU SHE Policy. This states that LABU is committed to implementing the Corporate SHE Policy. It then shows the LABU organisational structure and lists the SHE responsibilities of all persons within LABU commensurate with their role. The arrangement for implementation of the Policy (as shown in Figure 2 above) and a brief description of how this works is provided in the Policy. In summary ; a management standard is written for each SHE system activity we have in place (e.g. loss control inspections). The standard lists the tasks to be done, who is to do the tasks and when they are to be done. Where a number of persons in the Business Unit may be required to carry out actions listed in the standard, the activity may be open to interpretation and so a checklist is supplied to ensure consistency of application. Completed checklists act as evidence of activity and are a basis for review of what has been found for any future improvements. The LABU Policy requires that LABU has an annual action plan for continuous improvement. The site Director responsible for Operations signs the Policy.

Our site is required to produce a Safety Report under the COMAH Regulations and the site has a Major Accident Prevention Policy (M.A.P.P.). The next task was to produce a LABU M.A.P.P. to give a similar level of detail as in the LABU SHE Policy as to how LABU would implement the site M.A.P.P. The structure of the LABU M.A.P.P. is similar to the LABU SHE Policy.

Both of the above Policies were briefed out to all LABU Operations personnel by the relevant line supervisor, and each person was given a summary card of the Policies to keep in their personal copy of the company handbook. Copies of the Policies are on LABU notice boards and retained by Shift Team Leaders and safety representatives. The Policies are updated annually (or potentially at shorter intervals in the event of significant organisational change). Having produced these Policies, work then started on assembling an "annual action plan" ; one part focussing on "general safety improvements" and the other part focussing on "major hazard safety improvements". These plans support, and are the immediate focus of, the SHE Policy and M.A.P.P. respectively.

Initially the communication of the new safety management system approach gave rise to concerns that this would be a new initiative that would create more work for individuals and then fade away as just another initiative. There was essentially a lack of trust between operators and line management that what was being said would become reality. This showed itself as lack of interest in the development and implementation of the system. This was not an easy issue to overcome and has required considerable attention over last three years. Over this time we have built relationships between operators and line managers and Day support personnel in order to develop trust and earn respect. We deliberately encouraged a "hands on" approach initially in order to develop relationships and earn respect. This has led to unsolicited comments of management approachability. My role in the Business Unit (50% split between Production Engineering and Loss Control Co-ordination) has allowed me to build relationships across the Business Unit in carrying out my day to day tasks so that safety management is seen as being grown from within the Business Unit and is an integral part of Business Unit operations. There is a realisation by operators that not all problems will be solved immediately, and an appreciation of the need to tackle the highest risks first.

THE ANNUAL SAFETY IMPROVEMENT ACTION PLAN

There is an annual action plan which lists the activities for continuous safety improvement. This plan is generated towards the end of each year and is discussed with and approved by the Site Director for implementation. The Site Director in effect commits that resources will be made available to carry through the plan. Activities within the plan are included after consideration of recommendations by external auditors and internal review of accident, incident and near miss basic causes (i.e. lessons learned over previous year of a systematic nature rather than related to an individual event).

The LABU annual action plan for year 2000 was constructed with a mix of activities to comply with some of the demands of corporate and site safety procedures (see previous section) but was primarily based on issues identified in the audit of LABU activities. In the main these were actions to write "SHE management standards". Principal topics covered were around communications, document control and leadership activities. The Director responsible for Operations approved the plan for implementation. Developments in the Safety Management System are an agenda item on the monthly "safety improvement meeting". Recognising that not all the corporate demands could be met in year 2000, draft plans for years 2001 and 2002 were made. These subsequent annual plans have included actions from audits of LABU, and the site, by others during year 2001. I was the actionee for all items on the year 2000 plan but, with the explicit intention of promoting safety as a line management responsibility, subsequent annual plans increased the activities of line managers as the plans' actionees. In the year 2002 plan line managers have approximately 50% of the actions assigned to them, the 2003 plan is intended to raise this value to 80%.

SHE SYSTEM MANAGEMENT STANDARDS

The previous safety management system was based upon a number of safety procedures which were extensive in nature and more of a reference style than something which gave the flow process of a system. Our new SHE management standards aim to be short, typically two to three pages long, and take the reader through the system flow process. The first page describes the intent of that standard. The subsequent page(s) take the reader through the activities to be carried out to achieve the intent. The person(s) responsible for carrying out each activity is identified in the tabular layout of the standard. The draft version of such a standard is circulated via the monthly "safety improvement meeting" for comments from Day and Shift personnel and after a suitable comment time is approved by the person who has control of the resources to be used in the standard.

Where more than one person may be assigned to carry out a task in a standard (e.g. inspect plant areas in our Loss Control Inspection standard) we have developed simple checklists so that each individual who inspects a plant area looks at the same process or safety critical items. The checklists have been drawn up for plant areas which are the responsibility of specific operator(s) rather than operator teams. The standard requires that the person conducting the inspection has a discussion with that area operator(s). The means by which the implementation of the standard is to be monitored is also identified as an activity within the standard along with the person who is to carry out that task. Completed checklists are retained as evidence that the activity has been performed and so that quality checks on the activity can be made.

The underlying causes of accidents were one element of the original internal audit in year 2000. The predominant basic causes of the accidents were improper motivation, lack of knowledge/skill and use of inadequate work standards. This was consistent with other findings from the audit which showed the need to improve leadership, motivation and ownership of plant activities and key process documentation. The rebuilding of the safety management system focussed on devising management standards (with the purpose of simply and clearly defining whose responsibility it is to carry out specific tasks) to overcome these issues. The management standards devised and implemented in LABU to date include;

- Monthly Safety Improvement Meeting. This meeting bring together shift teams' safety representatives and Day and Shift line managers. It is used for communication of LABUs previous month safety Performance Indicators and Site wide Performance Indicators, to receive and discuss Safety Representatives feedback on safety issues arising from shift teams activities, and to discuss activity progress on the safety management system development.

- Loss Control Inspections (line managers view specific aspects of plant areas and discuss findings with area operators). This encourages line managers to view the state of plant equipment and gives opportunity to motivate area operators towards work standard improvement.

- Management standard for defining who should carry out what tasks to implement key Site Safety Procedures. Topics covered include Basis of Safety studies, Fork Lift Truck operations, health surveillance/Lead hygiene Rules, control of ignition sources, minimum personal protective equipment to be worn on non specific tasks, shift hand-over documentation, safety monitoring activity, control of plant changes, accident/incident and near miss reporting, employee selection and placement)

It is my belief that the clarification of who is expected to do what to implement these standards, the communication of this information and the monitoring that such implementation is taking place has had the effect of changing attitudes of, and the Safety Culture within, the LABU team membership. We aim to positively promote safety communications, near miss reporting, operator involvement in topics relating to their normal work activities (e.g. general risk assessments and alarm priority reviews) and line management appreciation of plant condition and people behaviours. We have management commitment. We are aware of the importance of involving **all** business unit personnel and have identified their personal responsibilities. All business unit personnel are asked to participate in SHE improvement activity relevant to their individual jobs. The result has been greater motivation of LABU personnel towards identifying SHE issues and assisting in their solution.

As well as a monitoring exercise to check that a standard is operating as intended, certain standards also have a pro-active Performance Indicator assigned to them. For example, the Loss Control Inspection standard it states who is to carry out an inspection and at a what frequency. LABU as a Business Unit has a pro-active Performance Indicator that we will carry out a minimum number of such inspections per month. A wall chart is provided in the Shift Team Leaders' office to aid checking who has inspected which plant area over the month. Our Performance Indicators in general are discussed in the next section.

An example LABU management standard is given above in Figure 3. The LABU standard is based on a Site Safety Procedure and identifies who in LABU is to do what to comply with the site procedure. We extend the site requirements by compiling a summary of all accidents, incidents and near misses within LABU over the previous week. The summary is to include the root causes of each event. The briefing in LABU includes this detail.

PERFORMANCE INDICATORS

LABU works to a suite of Performance Indicators each year. Three years ago the Indicators were predominantly "reactive" type indicators (e.g. lost time accident frequency rates, the number of minor accidents, very serious and serious incidents, quantities of volatile organic carbons emitted, etc). We have since aimed to change the balance of "reactive" and "pro-active" Indicators such that by the start of 2002 we were operating with 35% of our Indicators as "reactive" type and 65% "pro-active". The Indicators are assembled and communicated monthly both to the Octel Board and to all LABU Operations personnel – the latter through the monthly "safety improvement meeting". Our "reactive" Indicators cover safety, health and environment issues (e.g. accident frequency rates, numbers of incidents, medical surveillance results, organic carbon emissions). Our "pro-active" Indicators include items such as number of near-miss reports, number of overdue actions from serious and very serious incidents, number of Loss Control Inspections and permit-to-work checks completed, number of safety procedures/standards briefed, number of environmental improvement actions completed, etc.

LABU SHE MANAGEMENT STANDARD

Subject: **Weekly Safety Review**

Procedure: 1	Issue: **REV 1**

PURPOSE AND METHOD

A meeting is held weekly to review key incidents and accidents across the Cheshire Manufacturing Park site. The meeting is chaired by the Site Director and attended by Business and Service Unit management and Site Operational Safety personnel (who act as secretariat). The purpose of the meeting is to discuss the key incidents from the last seven days of site operations to identify site learning points from those incidents. A Weekly Review Sheet is produced by Site Operational Safety personnel to include site accident statistics. Each Business/Service Unit then adds its own Unit statistics and the completed Sheet is briefed to all Unit personnel as soon as practicable (normally next working day/next shift cycle)

LEAD BUSINESS UNIT PERSONNEL RESPONSIBILITIES

R = Primary Responsibility r = Secondary Responsibility	PRODUCTION MANAGER	LOSS CONTROL CO-ORDINATOR	SHIFT TEAM LEADERS
1) Shift Team Leaders are to compile a LABU A/I investigation status sheet every week for use at the Weekly Safety Review meeting and for the subsequent Review briefings.			R
2) Production Manager and Shift Team Leader attend Weekly Safety Review meeting (normally Friday 0800 hrs) and contribute to learning point discussion	R		R
3) Site Safety Manager issues the Weekly Safety Review sheet and Shift Team Leader adds LABU A/I investigation information same day as meeting.			R
4) Production Manager cascade briefs the Review sheet information to Day personnel at the Friday staff meeting or at another scheduled meeting time. A record is made of those briefed each week (using toolbox talk record sheet). This is retained in Production Manager's office.	R		
5) Shift Team Leaders brief the Review sheet to their operator teams and keep a record of those briefed using the toolbox talk record sheet. This is retained by the Shift Team Leader.			R
6) Understanding of briefs will be checked by 3 monthly personnel monitoring surveys for each day group and shift team that is briefed.		R	

Figure 3.   Example LABU safety management system standard

ACTIVITY ACTION TRACKING

The annual "safety improvement" and "major hazard" action plans and a number of the "pro-active" Performance Indicators generate activity which obviously needs to be measured in order that we can monitor our performance towards managing the improvements which the activity is intended to bring. Initially our activity was progressed by minutes of various meetings which were stewarding the activity. We found that once we had sparked interest in particular actions and activity through our "safety improvement" meeting, etc, there was a great desire from the operational teams for feedback on the outcome of the activity. We needed a better way of checking action progress. In 2002 we collected actions together from the various stewarding type meetings and created a simple Microsoft Access® database for "action tracking". The database links with our e-mail system and allows us to run reports of outstanding actions – from the various sources – which can then be e-mailed to actionees and their managers. The database is updated with completed actions (as advised by the actionees) and re-issued monthly. Actionee responses are archived to create an audit trail for closed actions. Actions are categorised by source of action (for easy of stewarding through actionee groups) and can be assigned priorities (e.g. to distinguish very serious incident actions from minor incident actions).

CURRENT SHE IMPROVEMENT STATUS IN LEAD ADDITIVE BUSINESS UNIT

The LABU lost time accident frequency rate was 1.29 in January 2000. By July 2001 this had dropped to zero. That situation has continued to the time of writing this paper (August 2002). The minor accident frequency rate for LABU has also shown a consistent downward trend over the last two and a half years. Figure 4 below gives a plot of the minor accident frequency rate since January 2000 to the time of writing.

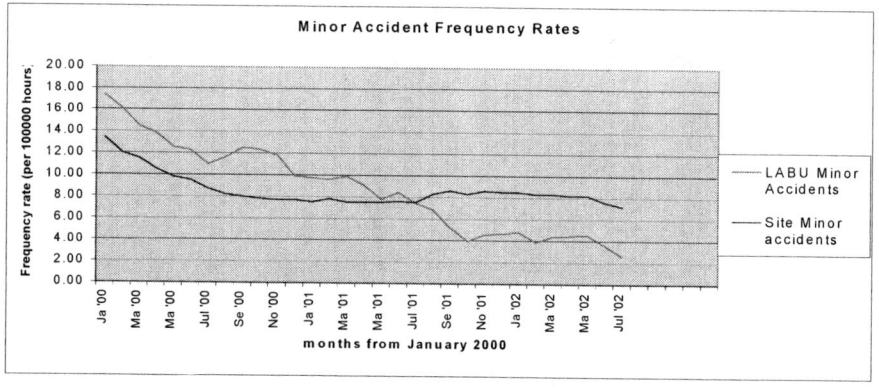

Figure 4.   Minor accident rates for LABU and Cheshire manufacturing park

The number of minor accidents attributed to the business unit has fallen over the last three years from 58 in 1999, to 29 in year 2000, to 10 in year 2001 and is expected to be no more than 6 in year 2002 (80% reduction in three years). Minor incident reporting has fallen slightly from 276 in year 2000 to an anticipated 200 by end of year 2002 (28% reduction in same three years). Additionally near miss reporting has increased by 60% over the last three years. We positively promote incident reporting as we use accident/incident (A/I) basic cause analysis at year end (looking for systematic causes) to develop activities for the subsequent year improvement plan. It is therefore considered that the changes have not discouraged reporting of minor incidents and that the benefits seen are genuine.

To date we have created management standards with the specific aim of promoting communications of SHE issues, improved leadership, clear ownership of key documentation, control of change and clarity on a range of site safety procedures. The format of the management standards has allowed individual responsibilities to be listed across all the management standards to clarify what is expected of each role. Below is a brief summary of those activities we have implemented or are implementing;

- Competency; general awareness training for all LABU employees completed, specific training for Leading Operators, Shift Team Leaders completed, now embarking on safety system training for these groups, "basis of safety" training completed for operators working on "major hazard" plant, starting similar training for maintenance personnel

- Control; Policies and standards approved by owner of the resources named in these documents, management standard for loss control tours implemented over last 18 months, lists for individual role responsibilities being generated from SHE Policy, M.A.P.P. and the range of management standards on various SHE topics, monthly stewarding meeting to implement SHE and M.A.P.P. action plans, pro-active safety Performance Indicators measured and monitored monthly, operators reviewing previously completed general risk assessments

- Co-operation; LABU employees involved in the business re-engineering, human factor assessment completed with operators working on LABU "major accident hazard" potential plant, Director chairs monthly Safety Improvement meeting, all employees can see the monthly Performance Indicator results, new Contractor Induction course notes produced

- Communication; monthly Shift (Safety Forum) and Day (Safety Improvement) meetings held involving supervisory, operating personnel and safety representatives, weekly review meetings with same personnel for previous week's accident and incident reports

The introduction of the formal Safety Improvement Meeting revamped a "Safety Committee" activity for Safety Representatives to meet face to face with Day and Shift team line managers and Operations Director to air any safety concerns with formal recording of those concerns. In addition this meeting brought focus to communicating SHE performance information and SHE safety management system development. The establishment of the "safety forum" (the shift safety meeting) and accident/incident weekly briefing meetings on each shift gave a similar communication link within each shift. The Loss Control Inspection standard requires line managers to spend a defined length of time on plant looking at the state of the plant, equipment and local environment with the need then to discuss the activity with the relevant plant operator. All of our management standards drive home the ownership and responsibilities for SHE for each individual member of the Business Unit appropriate to their normal role in the Unit.

We have come a long way in a relatively short time period but have done so with belief and support in what we were doing throughout from our Director responsible for Operations. We have clarified roles and responsibilities and made access to such information as simple as possible with regular meetings and briefings to communicate the various facets of activity. We started from a position of giving everyone a chance to see the overview of where we were heading and reinforced this as we progressed. We have aimed throughout to try to ensure that all LABU personnel have been encouraged to get involved with the SHE management system or elements within its development – and how it impacts on each persons individual role - as it has progressed. Safety representatives now come forward and not only make suggestions for improvements but are actively encouraged to chase actions and actionees to ensure suggestions become reality.

We have obviously improved our safety performance over the last two and a half years and there is evidence of improved safety culture within LABU as described earlier. For me this has been achieved by a combination of issues, none of which are particularly new or unique. We have had belief and support from Director level, a reasonable resource allocation, a simple and concise approach to clarify who is responsible for what, an efficient means of chasing actions, feedback on what we are doing and involvement of all relative to each individual's day to day job.

COMAH SAFETY REPORT REGIME – EVALUATING THE IMPACT ON "NEW ENTRANT" ESTABLISHMENTS

Richard Thomas
The Health and Safety Executive, Hazardous Installations Directorate (HID), Central Division, St Anne's House University Road, Bootle, L20 3RA

In April 1999 the COMAH[1] regulations superseded the previous onshore major hazard regulations - CIMAH -. Although the CIMAH[2] regulations included safety report requirements for the high hazard "top tier" (TT) establishments the COMAH regulations introduced a more rigorous regime because of the need to demonstrate that all necessary measures have been taken to control major accident hazard risks. The COMAH regime brought into scope of top tier requirements a significant number of establishments not previously covered by TT requirements. – COMAH TT "new entrant" establishments. Using these "new entrant establishments" as a study group HID is undertaking a project to identify the impact of being under the COMAH safety report regime has on health and safety performance. The first stage of the project (to be completed by January 03) is to obtain views from industry and inspectors along with baseline data on incident performance. This paper presents the methodology used in this first stage of the project and briefly considers future stages of the project.

KEYWORDS: COMAH, impact evaluation

INTRODUCTION

HSE is currently undertaking an impact evaluation programme looking at its key activities to gain an understanding of whether these activities are meeting their objectives and that resources are being used in the most effective way. As part of this impact evaluation programme HSE has identified a need to evaluate safety case type regimes to inform the debate[3] on the value of these regimes for regulating high hazard industries. To contribute to this HID is conducting a project to identify the impact that the COMAH safety report regime has on the management of major hazard risks. Although the COMAH regulations are enforced in the UK by the "Joint Competent Authority" (HSE working with EA & SEPA), this study is to investigate the impact of HSE's activities on the major accident risks to people. This is envisaged to be a long-term project looking at the effect of the regime over the safety report 5-year cycle and beyond. This paper sets out the methodology the first stage of the project is using.

The specific question the project is designed to explore is **"What is the impact of the COMAH Safety report regime on Health and Safety Performance in respect of Major Accident Hazards?"**

BACKGROUND

In April 1999 the COMAH regulations superseded the previous onshore major hazard regulations - CIMAH -. Although CIMAH included safety report requirements for the high hazard "top tier" (TT) establishments the COMAH regulations introduced a more rigorous regime. COMAH also brought into scope of top tier requirements a significant number (approximately 120) of establishments not previously covered by CIMAH TT requirements. These "COMAH TT new entrant" establishments provide an opportunity to capture data on their performance as they go through the safety report process. This project uses them as a study group and will compare their performance with those establishments that were TT under CIMAH and those that are LT under COMAH.

At the end of 2000, The Health and Safety Executive, in response to the Revitalising Health and Safety initiative and the setting of targets for the Health and Safety Commission for the first time since its inception, commissioned a review of the evidence of the impact of their work[4]. One of the outcomes of the research has been the development by HSE of a more comprehensive evaluation framework. A key part of this has been the development of guidance on impact evaluation methodology for HSE[5] the project follows this guidance which draws on experience elsewhere in evaluation of government policy.

PROJECT METHODOLOGY

OVERVIEW
The complex nature of COMAH sites along with the full range of Health and Safety duties on establishments means that there is no single performance measure that can be effectively used to indicate the impact of the COMAH safety report regime on an establishment. For example the key objectives of COMAH are to reduce COMAH major accidents and mitigate successfully those that do occur. So we could just measure the numbers of COMAH major accidents that occur, however COMAH Major Accidents are few in number annually and on their own provide a poor measure of performance. Additional information could be gained from looking at the wider picture of reportable incidents (injuries and dangerous occurrence) under the Reporting of Injuries, Diseases and Dangerous Occurrences Regulations 1995[6] (RIDDOR) from these establishments but a significant number of these incidents may have little or no link to performance in respect of Major Accident Hazards. (eg Slips, trips and falls).

Consequently to establish a pragmatic picture of the performance of establishments and the impact the COMAH safety report regime has, a range of data will need to be gathered. This project sets out to obtain both quantitative data i.e. details of accidents, DO's and enforcement action that does have some relationship to major hazards and qualitative data i.e. views of the establishment's operator and HSE inspectors - on the establishment's performance. To this end the methodology is divided into three distinct parts as listed below:

- Survey of establishment operators
- Survey of regulatory inspectors
- Provision of COMAH focussed quantitative data

To account for counterfactual issues (i.e. the performance these "new entrant" establishments would have returned had they not been brought into the COMAH TT regime) the project will obtain data from the three broad groupings of COMAH establishments listed below:

- TT establishments (excluding new entrant establishments)
- LT establishments
- TT new entrant establishments

The quantitative and qualitative data obtained from these groups will be analysed on a comparative basis to provide information on the relative performance of the groups.

With the new entrant establishments having to submit their safety reports in February 2002 the aim is to capture their views whilst the safety report process is still current in their minds. By reviewing the information gathered the aim is to identify changes in performance of the new entrant establishments during their first years under the COMAH safety report regime.

OPERATOR SURVEY

To provide objectivity and independence from HSE this part of the project has been contracted out to ENTEC UK Ltd. This ensures that:

- Operators feel able to freely express their views (both positive and negative) on the value of the COMAH regime, (data provided to HSE from the survey will ensure anonymity).
- Appropriate expertise and resources are applied to obtain quality data within the timescale of the project.

The aim, objectives and survey method used are set out below.

Aim

To obtain qualitative data, from operators of identified new entrant COMAH establishments and counterfactual establishments, that shows whether changes in management and control of major hazard risks have occurred as a result of being within the COMAH safety report regime.

Objectives

Data to be gathered (from "new entrant" and counterfactual group operators) regarding:

- Views/perceptions of duty holders of the effect of being under the COMAH safety report regime including any negative effects.
- What those in the safety case regime do differently (better or worse) compared with
  o before they were in the regime
  o those not in the regime.
- Specific arrangements (either hardware or systems) in place or being put in place as a result of the COMAH safety report regime and the writing of the safety report.
- Specific key COMAH requirements and their effect. Including views on whether they would have been addressed without COMAH.
- What they learned about how they control their major accident hazards from being under the safety report regime.

Method
Overview of approach
The survey is being conducted in the following stages:

Stage 1 – define the key issues to be addressed to answer the question "What is the impact of the COMAH Safety report regime on Health and Safety Performance in respect of Major Accident Hazards?" and agree the "qualitative data" to be collected during the survey

Stage 2 – develop and carryout postal questionnaires for target and counterfactual groups;

Stage 3 – analyse questionnaire findings and establish objectives for follow-up interviews;

Stage 4 – conduct face-to-face and telephone interviews at a selection of establishments;

Stage 5 – analyse, interpret and report the survey findings.

The method is described in detail below.

Stage 1: define key issues to answer and agree qualitative data to be collected
To identify the data to be collected through the survey a workshop with HSE and Entec was held where the question being asked was explored. The range of potential impacts of the COMAH requirements were discussed informed by evidence from HSE inspectors, technical press and Entec. Key issues to deal with and areas for questions were identified. Issues included:

- Difficulty in collecting statistically significant safety data for major hazards due to the low frequency of events
- Demonstrating that improvements have occurred because of regulations
- The qualitative nature of the data to be collected

The key question areas identified were the identification and measurement of changes (improvements) in:

- Safety management systems
- Procedures or methods of working (including emergency arrangements)
- Changes to hardware, control systems, etc.
- Overall health and safety activity (i.e. What effect has the focus on Major Hazard aspects had on occupational health and safety issues.)
- Resource expenditure on Health and safety issues.

This stage of the project resulted in an improved understanding of the question being asked and the data needed to answer it.

Stage 2: develop and carryout postal questionnaire

Questionnaire design.   The postal survey is being carried out using a questionnaire based on the output from stage 1. The questionnaire is designed to maximise the data collected whilst minimising the impact on the organisations being surveyed and achieve the following goals:

- address all of the project objectives, without making the questionnaire too large,
- the sequence and wording of questions be easy to understand and minimise risk of biased responses;

- questions about sensitive issues such as approach to health and safety management, must be asked in a non-controversial and answerable manner
- target and counterfactual groups must be able to answer the questions in a consistent fashion that allows comparison
- questions must identify whether safety performance has changed due to COMAH regulations or other driving factors.

The final structure and content of the questionnaire is designed to:

- elicit the changing nature of safety management and the contribution made to this change by the safety report regime;
- identify differences between the target and counterfactual groups;
- estimate the costs (both direct and indirect) of compliance with HSE regulations, especially the additional costs of becoming a top-tier COMAH site;
- identify measures for long term (linked to the 5 year COMAH safety report cycle) impact evaluation.

With the postal survey involving approximately 300 questionnaires it was important to test the questionnaire at an early stage to ensure it's ease of use and that the questions had the appropriate breadth and depth required to address the issues posed by this project. To this end the questionnaire was piloted at a small number of representative establishments and revised in line with the feedback received.

Sample structure and size. All 120 of the "new entrant" target sites received a questionnaire. Due to the relatively small data set it is essential that as many responses are received as possible. Each target site not replying in an appropriate time will be followed up by telephone. However, a 100% response is highly unlikely and a goal of 60% is thought to be more realistic.

To achieve the project objective of identifying specific changes attributable to the new COMAH regime, it is essential to involve the counterfactual groups to allow the impact of becoming a top-tier site to be assessed. Initially the questionnaire will be sent to approximately the same number of establishments from each counterfactual group, as to the target group. It is recognised, however, that the counterfactual data will not be as useful as that collected from the target establishment and so they will not be followed up with the same rigour. The goal is to achieve a representative sample of counterfactual establishments of similar profile to the target group. The focus will be on sites that are only just below the top-tier cut-off point and those that are just above it as the issues at these establishments will be the most similar to the target sites.

Stage 3: Analyse and interpret questionnaire responses. The purpose of analysis at this stage in the project is to prepare for the follow-up interviews. This includes identifying establishments to be included in the follow-up and the topics for discussion.

Along with a summary of data collected from the postal survey, stage 3 of this project will also result in a pro-forma question sheet to be used during the follow-up interviews. The question will be focussed on areas where it proved difficult to collect data in a postal survey, and where interesting trends have emerged. The use of a pro-forma will ensure consistency of data collected in the interviews.

Stage 4: Conduct follow-up interviews

Sample structure and size. The purpose of the follow-up interviews (these will be carried out by a combination of face to face and telephone interviews) is to validate and enhance the results from the postal survey. It is not intended to interview every participant. Instead establishments will be selected giving a similar profile, but numbering approximately 15–20% of the original. The structure of the sample will be selected according to the following two factors:

i)   Representation of all sizes of organisation
The individual responses to the postal survey will be from establishments of different sizes. Therefore it is proposed that a selection of small, medium and large sites are selected for the follow-up interviews.

ii)   Representation of all sectors
The selection of establishments for follow up interviews will also strive to ensure a representative sample of all the sector types. It is possible that the distribution of sector types would be related to the size of the establishments.
The focus for follow-up interviews will be on the target establishments. A small number of interviews with counterfactual establishments to act as control will also be conducted.

Stage 5: Reporting
The report from this survey will aim to answer the key question "What is the impact of the COMAH Safety report regime on Health and Safety Performance in respect of Major Accident Hazards?" from the operator's perspective. In particular the report will:

- Identify the changes in management of major hazard risks that occur as a result of the COMAH safety report regime.
- Identify the effect on wider health and safety issues.
- Make recommendations on longer-term data needs for ongoing evaluation of the safety report regime giving consideration to the 5 year COMAH safety report cycle.

INSPECTOR SURVEY
The level of independence required for the Inspector survey is different form the operator survey. This survey is to be carried out in house by HID's quasi independent audit and benchmarking team.

Aims
- To obtain qualitative data, from Regulatory Inspectors (RI's) of identified "new entrant" COMAH establishments and counterfactual establishments, to show the impact on management of major hazard risks as a result of being within the COMAH safety report regime.
- To identify benefits from the safety report process for the regulation of health and safety. Specifically with regard to the efficiency and effectiveness of regulatory interventions at the establishments.

Objectives

To gather data from "new entrant" and counterfactual group RI's regarding:

- Views/perceptions of RI's of the effect of being under the COMAH safety report regime including any negative effects. Need to consider
  - Does COMAH SR regime provide extra regulatory leverage for achieving risk reduction – possible comparison here with LT MAPP regime
  - Does the information gained from the SR process lead to better targeted interventions and planning
  - Use of the SR as a reference document for investigation of incidents and complaints.
- What those in the safety report regime do differently (better or worse) compared with
  - before they were in the regime
  - those not in the regime (Comparison here will be between MAPP and SR regime)
- Specific health and safety arrangements (either hardware or systems) in place or being put in place as a result of the COMAH safety report regime and the writing of the safety report.

Method

Overview of approach

The survey is being conducted in the following stages:

Stage 1 – Define the key issues to explore with RI's
Stage 2 – Develop and carry out e mail questionnaire
Stage 3 – Analyse the findings and identify any data that requires clarification
Stage 4 – Report the findings of the survey

The method is described in more detail below.

Stage 1: Defining the key issues to explore

Based on the in-depth consideration of issues for the operator survey HID considered further the information to be gathered from inspectors to provide their perspective. It was identified that data needed to be compatible with the operator survey to enable a balanced view of the impact of COMAH to be developed. From this two key areas were identified to explore

- Information from RI's on the changes they had seen at COMAH sites regarding the management and control of major accident hazards. This would provide a set of data to cross reference/validate information gained from the operator survey.
- Views from RI's on the positive and negative impact of COMAH for the regulation of their sites.

Stage 2: Develop and carry out the questionnaire survey

For the inspector survey a different level of anonymity was required with individual responses being treated as confidential between the audit and benchmarking team and the inspector involved. It was decided that the most efficient to carryout the survey was via email, as this would make for easier tracking of the survey and provide a quick straightforward route to follow up any issues where clarification of data was required. The questionnaire was designed to achieve similar goals to the operator survey the key ones being:

- address the project objectives, without making the questionnaire too long,
- the questionnaire to be easily understood, minimise risk of biased responses and be consistent with the operator survey;
- elicit the changing nature of safety management and the contribution made to this change by the safety report regime;
- identify differences between the target and counterfactual groups;
- elicit views from RI's on both the positive and negative impact COMAH has had on health and safety and the way interventions with operators are conducted.

To ensure the questionnaire was user friendly, provided the required information and was manageable via the email route the questionnaire was piloted with a small number of RI's covering a range of COMAH establishments. Following the pilot the questionnaire was revised in line with the feedback before being issued for completion.

Sample structure and size. The operator survey targeted approximately 300 establishments, a mixture of top tier, new entrant top tier and lower tier establishment groups. The aim with inspector survey was to cover the same types of establishments but minimise demands on RI's. Because RI's regulate a large number of establishments it was possible to keep the target population to approximately 45 RI's (all safety report assessment managers) that were identified as having involvement with several establishments across the three groups. This target group was designed to give coverage of establishments similar in number, make up and geographic location to the operator survey.

Stage 3: Analyse the questionnaire findings and identify data requiring clarification
The responses will be analysed to identify differences and similarities between the three groups with particular reference to the new entrant group. The outcome for COMAH sites regarding the management and control of major accident hazards will be compared/validated against the findings from the operator survey. With regard to the impact on how RI's carryout interventions at site the analysis will look for both consistent and conflicting views from RI's with a view to identifying any common themes or concerns.
   If required RI's will be contacted to provide clarification on their responses.

Stage 4 Report the findings of the survey
The report from this survey will aim to answer the key question "What is the impact of the COMAH Safety report regime on Health and Safety Performance in respect of Major Accident Hazards?" from the RI's perspective. In particular the report will:

- Identify the changes in management of major hazard risks that occur as a result of the COMAH safety report regime.
- Identify the positive and negative impact COMAH has on the way RI's interventions with operators are conducted.

QUANTITATIVE DATA

Quantitative data overview
The overall objective of COMAH is to reduce COMAH major accidents and mitigate successfully those that do occur. So as a start point in looking at quantitative data the numbers of COMAH major accidents can be measured and tracked over time.

Fortunately COMAH Major Accidents are few in number annually but consequently provide a poor outcome measure. Also as new entrants to COMAH the key target group of this project will have no Major Accident history. For these two reasons other quantitative data measures have been identified to provide data for this evaluation. The quantitative data types that provide useful measures are reported incidents (COMAH Major accidents, accidents & DO's) and enforcement data (notices and prosecutions). Enforcement data when considered with intervention activity will give some indication of the level of compliance at groups of establishments.

Data Sources
HID information systems will provide quantitative data on an ongoing basis and back to 1996 on the following:

- Accidents
- Dangerous Occurrences
- Notices issued
- Prosecutions taken

Target groups
To obtain comparative data from which trends/differences between new entrant establishments and control/counterfactual groups can be identified data from the following groups will be obtained.

- All TT establishments
- All LT establishments
- TT new entrant establishments

Data requirements
To obtain information on both overall performance and COMAH related performance the following data will be obtained for the above target groups:

- COMAH Major accidents reported to Europe
- Total reported accidents broken down by type (fatal, major, over 3 day)
- Total reported dangerous occurrences
- COMAH related reported accidents (See below)
- COMAH related reported dangerous occurrences (See below)
- Total enforcement notices
- Total prosecutions
- COMAH related notices (See below)
- COMAH related prosecutions. (See below)

COMAH related data
Incident data
Although COMAH major accidents are uncommon there are incidents that occur at establishments, which if they hadn't been controlled could have carried on to become a major accident. An analysis of the RIDDOR reporting requirements identifies types of Dangerous Occurrences and Accidents that could be considered as potential major accident precursors. Listed below is the common agreed set of DOs & accidents that HID uses to monitor major accident precursors.

Dangerous occurrence/accident	RIDDOR category
Dangerous occurrence	Pressure systems
Dangerous occurrence	Electrical short circuit
Dangerous occurrence	Explosives
Dangerous occurrence	Explosion or fire
Dangerous occurrence	Escape of flammable substances
Dangerous occurrence	Escape of substances
Accident	Loss of consciousness by asphyxia or exposure to harmful substance
Accident	Chemical or hot metal burn to the eye.

Enforcement data
Some legal provisions have strong links with Major Accident issues when considered in the context of COMAH establishments. Those provisions listed below (when used at COMAH establishments) are considered to provide an indication of formal enforcement used to correct deficiencies linked to the control of major accident hazards. This can be tentatively used to give an indication of major accident compliance levels at COMAH establishments when considered against intervention activity at the establishment.

- The Management of Health and Safety at Work Regulations 1992/1999
- The Control of Industrial Major Accident Hazards Regulations (CIMAH) 1984
- The Control of Major Accident Hazards Regulations (COMAH) 1999.
- The Pressure Systems Safety Regulations 2000
- The Pressure Systems and Transportable Gas Containers Regulations 1989
- The Fire Certificate (Special Premises) Regulations 1976
- The Highly Flammable Liquids and Liquefied Petroleum Gases Regulations 1972
- The Explosives Act 1875/1923
- The Control of Explosives Regulations 1991

Time periods for data collection
Data will be obtained for 12 month periods based on HID work planning years (i.e. 1st April to 31st March) from 1st April 1996 to 31st March 2002.

Normalising the data
To take account of variations in

i.   numbers of establishments subject to COMAH
ii.   inspection activity at establishments (re enforcement data).

The aim is to try and normalise the data obtained in some way, possibly as set out below.

For incident data
By number of incidents per 100 establishments

For enforcement data
By time spent by inspectors on COMAH work at establishments.

Data regarding the time spent on COMAH work by inspectors will come from COMAH charging records. This data is only been available from 1999. Prior to this obtaining comparable data may not be possible but options are being explored.

Data analysis
The incident and enforcement data obtained for the identified target COMAH groups will be analysed on a yearly basis looking for trends over years and significant differences between establishment groups.

RESULTS

The operator and inspector surveys are planned to return findings by October 2002, at this point quantitative data for years up to April 2002 will have been obtained. The data from these three parts of the project will then be brought together reviewed and analysed to identify what information they provide regarding the impact the COMAH safety report regime has had on the Health and Safety Performance, in respect of Major Accident Hazards, of the COMAH "new entrant" establishments. It is planned that the use of the three parts of the project will enable triangulation of the data providing an element of assurance to the project findings.

Note – As an HSE research project the findings of the Operator Survey will be made available in an HSE Contract Research Report. The dissemination arrangements for the findings of the complete project have yet to be agreed.

USE OF THE PROJECT FINDINGS

HID's aim from the project is to learn from both the positive and negative findings of the project and use them to inform:

i.   HID policy on the COMAH safety report regime and in particular the five-year update process.
ii.   HSE's overall policy on permissioning regimes.
iii. HSE's thinking on regulating high hazard industries.

FUTURE STAGES OF THE PROJECT

As set out in the introduction to this paper this project is the first stage of a much longer plan to evaluate the impact of the COMAH safety report regime. It is envisaged that the process will cover at least the five-year period of the COMAH safety report cycle and probably beyond. It is hoped that the findings and experience gained from this project will provide a strong steer to the key long-term measures that should be used to provide data on the impact of the COMAH safety report regime.

The measures currently being considered for long-term use are:

- Annual analysis of the quantitative data as set out in this paper
- Further questionnaire surveys of operators, inspectors and other stakeholders at key stages in the COMAH cycle, specifically at the 5 year update and at the midpoint during inspection informed by the results of SR assessment.
- Monitoring of other new entrant groups as they come within scope of COMAH, for example as a result of CHIP 3 or amendments to Seveso II

REFERENCES

1. A Guide to the Control of Major Accident Regulations 1999, L111, HSE Books PO Box 1999 Sudbury Suffolk CO10 2WA ISBN 0-7176-1604-5
2. Control of Industrial Major Accident Hazards Regulations 1984 Statutory Instrument Number 1902. The Stationary Office http://www.tso.co.uk/site.asp
3. HSE Discussion Document DDE15 Regulating higher hazards: exploring the issues, HSE website, http://www.hse.gov.uk/disdocs http://www.hse.gov.uk/disdocs/closed/dde15.htm
4. The impact of the HSC/E: A review Prepared by The Institute for Employment Studies for the Health and Safety Executive. Contract Research Report 385/2001, HSE website http://www.hse.gov.uk/research/crr_pdf/2001/crr01385.pdf
5. HSE General Administrative Procedure (GAP) 6 Policy and Project Impact Evaluation. HSE Information Services Caerphilly Business Park, Caerphilly, United Kingdom. CF83 3GG
6. A guide to the Reporting of Injuries, Diseases and Dangerous Occurrences Regulations 1995, L73, HSE Books PO Box 1999 Sudbury Suffolk CO10 2WA ISBN 0717624315

KEY ELEMENTS OF RISK DECISIONS IN THE CONTROL OF MAJOR ACCIDENTS HAZARDS

Andrew G. Rushton
Methodology and Standards Development Unit, Hazardous Installations Directorate, HSE

COMAH Safety Reports must show that all measures necessary are in place to prevent control or mitigate relevant major accidents.

Many Operators consider that they have in place all reasonably practicable risk reduction measures, but have difficulty presenting their position to the Competent Authority, particularly in relation to "predictive aspects". Usually, most fundamentally, this is because the Operator does not set out how their process of risk management provides the necessary assurance that the risk is not intolerable and all necessary measures are in place. Often the description of necessary steps leading to risk decisions is incomplete.

This paper clarifies the flow of information which needs to exist in a risk control system and which, therefore, underlies the regulators' view of how predictive aspects are handled in safety reports. The link to a specific goal-based standard (BS IEC 61508) is described.

The Operator should be able to show that the relevant decision makers were, where necessary, aware of the risk criteria, the "risk picture" and the costs of practicable risk reduction measures.

Where initial attempts at the demonstrations required by COMAH have been ineffective, this is likely to be remedied by setting out the thought process underlying risk decisions more clearly, rather than just by providing more detail.

KEYWORDS: COMAH, risk, information, predictive, criteria

INTRODUCTION

The "Seveso II" directive aimed to improve control of major accident hazards from hazardous substances. It is implemented in the United Kingdom, in part, by the Control of Major Accident Hazards Regulations (COMAH)[1].

"Operators" of "establishment" where COMAH Regulation 7 applies ("top tier COMAH sites") are required to prepare a Safety Report and present it to the Competent Authority (CA), comprised of the Health and Safety Executive (HSE) and the Environment Agency or the Scottish Environmental Protection Agency. A requirement of the Safety Report is that it includes demonstrations which indicate acceptable control of major accident hazards.

Operators of COMAH establishments have found the preparation and presentation of demonstrations in Safety Reports challenging, particularly in relation to "predictive aspects"[2].

This paper aims to set out the background against which these predictive aspects are viewed by the regulators (the "Competent Authority"). It should help Operators to understand the CA's perspective and help Operators to avoid common faults in the preparation and presentation of predictive aspects of their reports.

Key steps necessary to support risk decisions are identified. The relationship between case-specific risks and risk criteria is described. The implications for reporting a risk analysis within a Safety Report are identified. Attention to these implications can help lead to successful demonstration that all necessary measures may have been taken.

KEY STEPS IN SUPPORT OF RISK DECISIONS

The essence of making the necessary demonstrations in a Safety Report is to show that all necessary measures to prevent control and mitigate major accidents have been taken. In relation to safety, this generally involves the following four key steps in support of risk decisions:

i)   identification of potential major accidents or "major accident scenarios";

ii) assessment of the approximate consequences and approximate expected frequency of the identified major accidents;

iii) classification of the overall risk position (with respect to individuals and groups) as either broadly acceptable or tolerable if risks are reduced as low as reasonably practicable (ALARP), *in other words a decision that the overall position is not intolerable*;

iv) consideration of what further measures could be taken and justification for not taking these measures, *in other words a decision that further measures are not reasonably practicable*.

Sometimes reference to and use of industry-wide standards may be considered to cover the later steps. This will be relatively rare for top-tier COMAH sites and even then will usually only apply to individual risk to people at the establishment. More commonly - for high and multiple hazards, for complex and novel plant, and for assessment of risk with respect to groups, or *societal risk* - case-specific attention to all four steps is necessary.

In the process of classifying an overall risk position, or "risk picture", it may be found that the position is intolerable. In such cases the position will need to be altered (by suspending the activity or introducing further risk reduction measures). An intolerable position should not, therefore, be an outcome of risk analysis that is reported. For this reason this outcome has been excluded from the above steps.

In the course of considering further measures, the justification for not taking a measure may be found to be too weak. In such cases the measure is necessary and will need to be taken. A position of not having taken a necessary measure should not be an outcome of risk analysis that is reported. For this reason this outcome has been excluded from the above steps.

In the sense that further measures may become available or may become more cost-effective, or improved science may lead to a revised estimate of probability or consequence of an accident, maintaining risk "ALARP" is a process of continuous review and improvement. This does not mean that, where reasonably practicable measures have been identified, then it is satisfactory to describe a forward plan which will, if and when implemented, achieve the demonstrations required.

EXPERIENCE IN REVIEWING REPORTED RISK ANALYSIS

Many Operators consider that they have in place all reasonably practicable risk reduction measures, but have difficulty presenting their position to the Competent Authority. Usually, most fundamentally, this is because the Operator does not set out how their process of risk management achieves the steps set out above. Often the steps are incomplete or the later steps are completely omitted (or merely implied). Sometimes the flow of information from one step to the next is unclear, so that the robustness of the overall approach is not evident.

When they are challenged about how their report achieves the steps set out above, many Operators perceive that the CA cannot be satisfied without much more quantification and/or much more detailed information. More quantification or detailed information may indeed be appropriate, but this is not usually the key to improving the Report. No amount of additional quantification or information may compensate for incompleteness in reporting the necessary steps taken in support of risk decisions.

Sometimes the problem will lie in the description of the risk control process in the Safety Report, rather than in the operating practices at the establishment. However, the presumption should be that where hazards are high there might be unusual risk reduction measures (i.e. measures not prescribed by codes or standards) which are necessary in the particular circumstances. The Operator needs to set out the thought process – considering what can go wrong with what result, assessing the risk picture, considering what defences are in place and why more cannot be done – which will ensure that such unusual, but necessary, measures are implemented.

In order to conduct and report risk analysis and risk control successfully, an Operator must understand the relationship of risk criteria to case-specific risk. The Operator must employ suitable risk criteria when considering both the risk posed by their operations and the opportunities for further risk reduction measures.

RISK CRITERIA AND THEIR RELATIONSHIP TO RISKS FROM MAJOR ACCIDENT SCENARIOS

A criterion which, for example, discriminates between a position where the risk to an individual at work is intolerable and a position where the risk is tolerable if risks are rendered ALARP may be termed a "risk criterion".

The classification of the overall risk positions (in relation to individuals and groups) requires an Operator to have in mind risk criteria.

The COMAH Regulations are "goal-setting", in recognition of the scale, complexity and evolving nature of the regulated activities. It is generally agreed that such activities are not suited to prescriptive controls.

The Regulations are not prescriptive, so there is no recipe to follow in making the required demonstrations. There is, on the other hand, a philosophy of risk control, underlying COMAH and other legislation, which provides a framework for adequate treatment of "predictive" aspects in Safety Reports.

HSE has set out indicative criteria for risks to people[3] which will be used in support of regulatory decisions. Clearly the relationship of any Operator-specified criteria to those indicated by HSE will be considered when HSE makes regulatory decisions as part of the COMAH CA.

A significant feature of risk criteria is that they relate to the relationships of a risk controller to the person or group at risk. These relationships are not restricted to COMAH regulated hazards, nor further restricted to individual major accident hazards. Therefore there is strictly no mapping between risk criteria and particular hazards or scenarios. For this reason methods which imply such a mapping, such as frequency-consequence matrices must be used with caution[4]. Individual risk to the more active people at work in the process industries from hazards other than those regulated by COMAH will rarely be negligible. Whilst much of the risk analysis in a Safety Report will be focussed on major accident scenarios as defined in the COMAH regulations, the risk picture which the Operator reports (and reflects on in reaching decisions on risk reduction measures) should be informed by a broader perspective.

SUPPORT TOOLS FOR RISK ANALYSIS

Sophisticated tools supporting risk analysis (e.g. for gas dispersion modelling) may be appropriate aids to judgement. This does not imply that such tools are always required. Whether or not such tools are used, it will usually be necessary to show that all the steps in risk analysis necessary to prepare for risk decisions are understood and followed when managing major hazards at the establishment.

In other words, sophisticated tools may be a necessary support to risk decisions but are not a substitute for risk decisions.

With or without a sophisticated approach, the Operator needs to be making decisions about what can go wrong, what the consequences would be, what measures are in place and why no more should be done.

DECISIONS ON WHAT ARE THE NECESSARY MEASURES

Figure 1 sets out, in outline, the flow of information in major hazard risk analysis and decision-making. This flow of information is essential to ensuring that a proper conclusion is reached on what are the necessary risk control measures.

In principle, Operators will decide that the necessary measures are complete (in the sense that they are all the measures necessary). This will allow them to set out a *justification for not adopting any further risk reduction measure* (final box in the Figure). To reach this conclusion, the Operator generally needs to make two critical decisions. Firstly, having in mind the risk to groups and individuals (the *risk picture*), a decision can be made that the risk is not intolerable. Secondly, and where risks are only tolerable if rendered ALARP, having in mind the available *further practicable risk reduction measures* and their costs, a decision can be made whether or not to deploy each measure.

Knowledge of the *risk picture* requires knowledge of the *risk criteria*, the *COMAH risks* and other relevant risks (*other risks*). Knowledge of the *COMAH risks* requires knowledge of the *consequences* of major accident scenarios and the predicted *frequency* of major accident scenarios. The *consequence assessment* and *frequency assessment* generally requires many inputs some of which are indicated in Figure 1.

Knowledge of the available *further practicable risk reduction measures* (and their costs) in turn requires a mechanism for identifying (and costing) such measures.

The process of risk analysis is complete when the decision maker is able to conclude either that the risks are broadly acceptable or that no further risk reduction measures are reasonably practicable (because the cost is grossly disproportionate to the benefit[4]).

From a regulator's perspective, the most satisfactory way of demonstrating that the risk analysis and decision making may have been concluded in a satisfactory way is to show that the decision makers had all the information that they needed to follow the process outlined above (and have followed this or a comparable process).

QUANTIFICATION OF ELEMENTS OF THE RISK ANALYSIS

Some approximate quantification of consequences of major accident scenarios, frequencies of major accident scenarios and costs of risk reduction measures will generally be necessary in order to achieve the purposes of risk analysis to support risk decisions reliably and transparently. Using judgement to allocate case-specific variables (such as predicted frequency) to bands of values, within quantified limits, will usually be both appropriate (because of the lack of accuracy in available estimation methods) and sufficient. The correct allocation is crucial to discovering whether the necessary measures are complete. Where it is proportionate (because the consequences are high), or where the uncertainty in the judgement is high, then the decision will be sensitive to the depth and strength of the analysis. In such cases it will be reasonable to support the judgement with further quantitative or qualitative analysis (more detailed analysis and/or "sensitivity" analysis) or to upgrade the allocation to a more pessimistic band.

Some would regard the process of allocation to (quantified) bands as qualitative, others may regard it as unnecessarily quantitative. The case for this minimal level of quantification is clearly seen wherever the decision making process involves collaboration.

For example, one person may estimate the frequency of a scenario as "unlikely". If another person attaches a different meaning to the word "unlikely" when setting out the risk criteria, or if a third person has another meaning in mind when judging the benefit of eliminating the scenario, then it is difficult to see how the overall decision making process can be relied upon to deliver all the necessary measures.

It has to be accepted that estimates of each contributory factor underlying a decision will be uncertain, but this is not a justification for not making the necessary estimates.

Methods for proceeding with risk analysis have been discussed by Carter et al[5], who also describe an approach to estimating the overall position in relation to societal risk. Gadd et al[6] have recently reviewed pitfalls in risk assessment, many of which are particularly pertinent to the support of risk decisions in the control of major accident hazards.

DEMONSTRATION

Whatever support tools are used, the Operator usually and ultimately will rely on an expert group or person making decisions (about whether to deploy a risk reducing measure).

It may well be that "expert judgement" is used, and is appropriate, for some or all contributory decisions. It will generally be practicable for an expert to express their reasoning in a (minimally) quantitative way. Where there is not minimal quantification, then auditing the decision making process or revisiting a decision in the light of changing circumstances will be difficult or impossible.

For decisions on whether the risks are tolerable (if ALARP) or broadly acceptable, the Operator should be able to show that the decision making person or group was aware of the relevant criteria and the risk picture and show that the decision was informed by a reasonable (but not necessarily detailed) estimate of the consequences and likelihood of the hazard in question. For a justification of not implementing a further risk reduction measure, the operator should additionally be able to show that the person or group had some awareness of the approximate cost of the proposed measure.

Where these conditions for decision making cannot be shown, then it is hard to see how the Operator can be confident that appropriate decisions are being made and how the decisions arising can be revisited as circumstances change (for example in response to increased activity or reduced cost of a risk reduction measure). In particular it is hard to see what check there is against a drift into an intolerable position.

The Safety Report needs therefore to provide answers to these questions:

- How does the Operator perform hazard identification (HAZID) and what satisfies the Operator that this approach to HAZID is suitable and sufficient in all the circumstances of their business?
- How does the Operator estimate the frequency and consequences of potential major accidents?
- How does the Operator judge the overall risk position to be tolerable (or more rarely broadly acceptable)?
- How does the operator ensure that opportunities to take further measures are identified and reviewed appropriately?
- What justification does the operator have for not taking any practicable further measures and what satisfies the operator that this approach to justifying taking any further measures is suitable and sufficient in all the circumstances of their business?

A Report is unlikely to be satisfactory in relation to predictive aspects if it does not clearly and preferably explicitly deal with these questions.

This does not mean that every Report must have a uniform and high level of quantification and detail. It does mean that a narrative style of report which makes clear the flow of information leading to decisions on risk tolerability and deployment (or not) of measures is helpful. Where links between details of fact and decisions of risk control are not presented, then an increase in detail is unlikely to remove concerns about predictive aspects.

GOAL SETTING FOR IMPLEMENTING INSTRUMENTED PROTECTIVE SYSTEMS

BS IEC 61508[7] supports the specification and implementation (etc.) of safety-related systems utilising electrical, electronic or programmable electronic technologies. Publication of a related process-industries sector-specific standard (BS IEC 61511) is imminent.

The standard links the "safety requirements specification" of the safety-related systems to risk criteria through a "Safety Integrity Level" (SIL). The "necessary risk reduction" is the reduction in risk that has to be achieved to meet the "tolerable" risk for a specific situation. In other words, the standard supposes that, for a particular application, a target of risk reduction by the protective system can be identified and this will lead to specification of the necessary SIL. Figure 2 shows the outline of how risk reduction can be allocated.

In the context of COMAH, the target risk reduction may be relatively complex. For example the aim may be to reduce an intolerable risk to a tolerable one but then to reduce the risk further if reasonably practicable.

To apply BS IEC 61508, therefore, an Operator of a top tier COMAH site will generally need to set out the overall approach to risk control in order, in this case, to generate the target risk reduction for a protective system.

Often Safety Reports describe the SIL achieved in the implemented protective system without reference to any process for allocating risk reduction to that system. This does not provide any justification for the selection or acceptance of the system implemented. Sometimes it is stated or implied that, on the basis of the SIL achieved and the consequent reduction of risk from a particular hazard or collection of hazards, the Operator can conclude that they have directly met an appropriate risk criterion. This approach, which supposes risk criteria apply to the relationship between risks from equipment or hazards to individuals or groups, is not generally compatible with HSE's view that risk criteria are applicable to the relationship between all the risks in the control of one *dutyholder* (the Operator in the context of COMAH) and the individual or group at risk. The annex outlines how these two views differ and why they are not compatible (in general and in the context of COMAH in particular).

In some cases the Operator has treated the electrical/electronic/programmable electronic system (for which a SIL has been established) in isolation without reference to other (e.g. mechanical) risk reduction measures. Clearly this can make the task of showing that the risks are acceptably low more difficult.

DISCUSSION

In the COMAH regime, it is not generally sufficient for an Operator to employ competent people, on whom they rely to make decisions about – for example – the appropriate SIL of an instrumented protective system. One purpose of the Safety Report is to make these decisions and their basis explicit (however much or little they rely on sophisticated decision aids). If the Operator cannot write down how it makes these decisions, then the charitable conclusion is that it does not know how it makes these decisions but it has employed staff who do know; the less charitable conclusion is that the necessary decisions have not been made.

Of course the CA is more likely to challenge Operators' risk decisions where they do not appear to be proportionately sophisticated, but the CA will be most concerned to see evidence of the application of a complete approach to risk decision making which, in principle, *can* deliver appropriate decisions, however unsophisticated.

CONCLUSIONS

Operators of COMAH establishments have found the preparation and presentation of demonstrations in Safety Reports challenging, particularly in relation to "predictive aspects".

There is a philosophy of risk control, underlying COMAH and other legislation, which provides a framework (but no prescription) for adequate treatment of "predictive" aspects in Safety Reports.

The Operator should be able to show that the relevant decision makers were, where necessary, aware of risk criteria, the "risk picture" and the costs of practicable risk reduction measures. A model for the essential flow of information, needed to support risk decisions in the control of major accident hazards, has been presented (Figure 1).

Where initial attempts at the demonstrations required by COMAH are ineffective, this is likely to be remedied by setting out more clearly the thought process underlying risk decisions, rather than just by providing more detail.

The regulator is, of course, more likely to challenge Operators' decisions where they do not appear to be proportionately sophisticated, but the regulator will be most concerned to see evidence of a complete approach to risk decision which can deliver appropriate results.

DISCLAIMER
The views expressed in this paper are those of the author alone and are not a statement of HSE policy.

1   Health & Safety Executive, 1999, 'A guide to the Control of Major Accident Hazards regulations 1999, L111, HSE Books.
2   Health & Safety Executive, 2002, COMAH Safety Report Assessment Manual (revised), available at http://www.hse.gov.uk
3   Health & Safety Executive, 2001, 'Reducing Risks, Protecting People, HSE's decision-making process', HSE Books, see also http://www.hse.gov.uk/dst/alarp1.htm etc.
4   Middleton M & Franks A, September 2001, 'Using Risk Matrices', The Chemical Engineer.
5   Carter DA, Hirst IL, Maddison TE and Porter SR, 2003, 'Appropriate risk assessment methods for major accident establishments', Trans I Chem E (B) Proc Safety and Envtl Prot, accepted for publication.
6   Gadd S, Keeley D and Balmforth H (2002) 'Good practice and pitfalls in risk assessment' HSL research report, in preparation.
7   BS IEC 61508, 1998, Functional safety of electrical/electronic/programmable electronic safety related systems.

KEY ELEMENTS OF RISK DECISIONS IN THE CONTROL OF MAJOR ACCIDENTS HAZARDS

ANNEX: RISK IN THE CONTROL OF A DUTYHOLDER AND RISK FROM A HAZARD

INTRODUCTION

The question is often asked: "What do I have to do to make the risk from my plant (or equipment) acceptable". There is no simple answer. This annex describes some features of the type of risk approach described by HSE which lead to the answer being context dependent. Two extreme positions are described which, hopefully, shed light on why there is no simple answer.

FEATURES OF HSE'S RISK APPROACH WHICH COMPLICATE ITS APPLICATION TO SINGLE HAZARDS

The risk criteria set out in "R2P2"[3] suggest that, in general, a dutyholder must find that risks are not intolerable and (where risks are not broadly acceptable) must find that further risk reduction measures are not reasonably practicable.

These criteria, however, relate to all the risks arising in the relationship of a risk controller (the dutyholder) with the person or group at risk. These relationships are not restricted to risks arising from COMAH regulated hazards, nor further restricted to any individual major accident hazard. Therefore there is strictly no mapping between risk criteria and risk from a singular hazard (perhaps associated with one plant or one piece of equipment).

Application of the "ALARP" principle requires the costs and benefits of further risk reduction measures to be considered, but the costs and (particularly) the benefits are affected by the operating context of the hazard. Most typically, in the context of COMAH, the benefit to be gained will increase with the number of people likely to be affected by a particular scenario.

Societal risk criteria are different in quality to individual risk criteria. Generally, satisfying individual risk criteria does not guarantee that any societal risk criteria have been satisfied.

Three needs arise, therefore, which can affect the risk picture and/or the reasonable practicability of further risk reduction measures:

i)   the need to consider all relevant risk under control of the dutyholder (not just the risk from one plant/equipment);

ii) the need to distinguish between ostensibly identical hazards where one instance of the hazard has potential for harming more than one person (or, more generally, where the cost and benefit of a risk reduction measure in one application can be distinguished from the cost and benefit of the same measure in a different application);

iii) the need to consider societal risk.

In principle a collection of risks, each one "tolerable" if it existed in isolation, can produce an "intolerable" position.

The benefit of deploying a further risk reduction measure will generally increase if more than one person could be harmed by the hazard that is to be controlled.

A single measure may reduce the risks from several hazards, so that in the particular application of that measure the benefit is increased.

When the societal risk picture is evaluated it may be that further measures are required (which would not have been required on the basis of individual risk assessment).

IMPLICATIONS FOR RISK MANAGEMENT

The problem is that individuals in a risk-controlling organisation will often only have direct responsibility for one hazard or a small sub-set of the hazards controlled by the dutyholder. It is tempting to try to re-cast the overall requirement as a need to find that risks *from my plant* are not intolerable (etc.). In the context of COMAH this temptation should usually be resisted and there is a need for the bigger picture to be considered.

At one extreme there may be no significant societal risks and, with the exception of a single hazard, there are no significant risks of death to an individual. In this case application of a general approach (for risk of death) *as if* it applied to the single hazard will not lead to serious error. At the other extreme, there may be significant societal risks and several hazards each of which has potential (in a single incident) of causing harm to many people.

CONCLUSION (OF ANNEX)

A risk control approach intended to encompass all the risks arising in the relationship of a risk controller to a person or group cannot generally be applied to single hazards where other hazards are significant.

Whether or not risk reduction measures are reasonably practicable can depend on the operating context of a hazard. In the context of COMAH, where the emphasis is on sites presenting many hazards and on scenarios in which many people could be harmed, a strong dependence of the reasonable practicability of measures on the particular operating context can be expected.

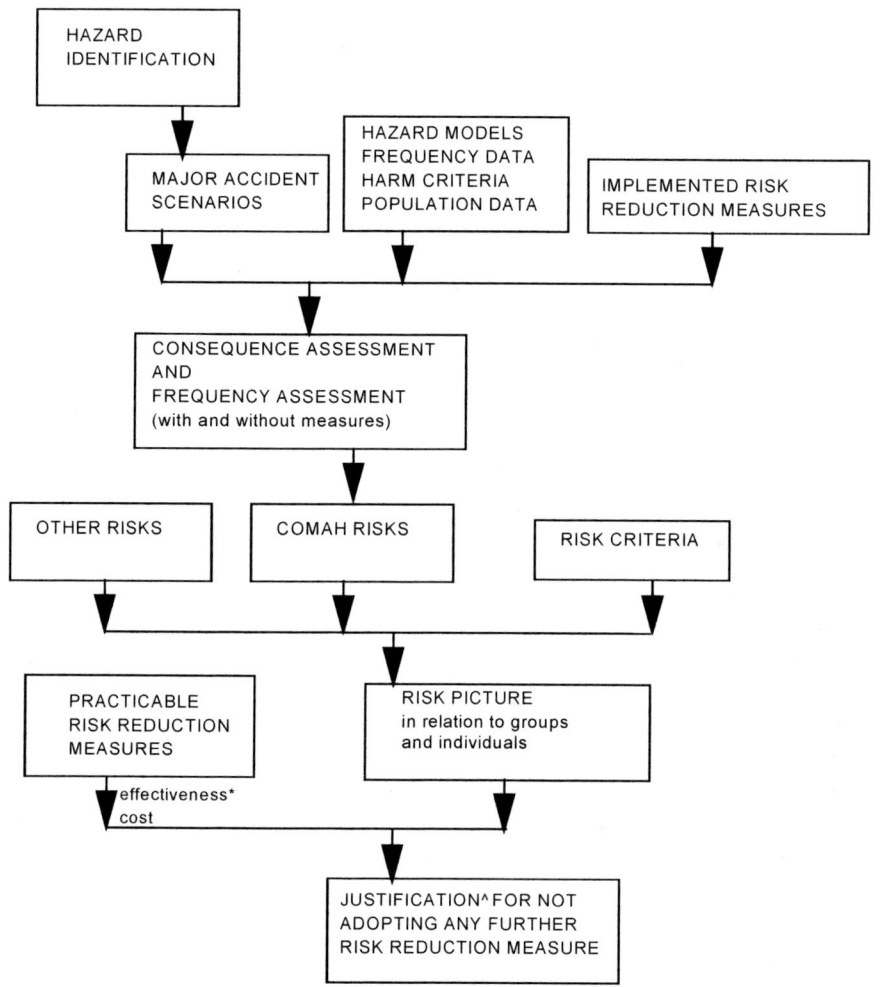

* which scenarios are affected, how much they are relieved and how reliably, will determine the benefit

^ justification will be that risk is broadly acceptable or that further measures are not reasonably practicable

Figure 1. The flow of information in major hazard risk analysis
Note that in the process of establishing an acceptable position there will be iteration
(not shown in the figure) if any element of the "risk picture" is intolerable or any further risk
reduction measure is found to be reasonably practicable.

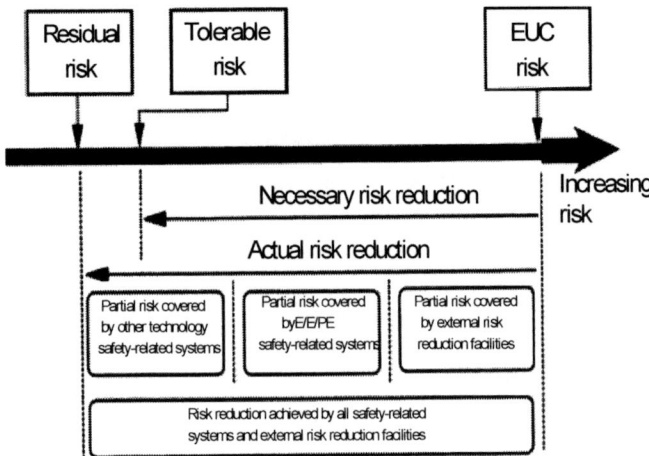

Figure 2. "Risk Reduction General Concepts", reproduced from BS IEC 61508 - 5, Figure A.1, courtesy of the British Standards Institution.
EUC   = equipment under control
E/E/PE = electrical/electronic/programmable electronic

HUMAN FACTORS AND COMAH: A REGULATOR'S PERSPECTIVE

Martin Anderson MErgS, EurErg, MIOSH
HM Specialist Inspector of Health & Safety (Human Factors)
UK Health And Safety Executive

This paper outlines recent experience of the HSE's Human Factors Team in assessing human factors issues on major hazard installations. It presents the 'Top Ten' human factors issues that have arisen out of inspection and audit of major hazard sites and from assessment of COMAH safety reports. Our involvement on major hazard sites reveals that most duty holders do not adequately address human factors. Primary weaknesses include an imbalance between hardware and human issues, and focussing on the human contribution to personal safety, rather than to the initiation and control of major accident hazards.

However, following targeted inspection and awareness-raising in the industry, the profile of human factors and effective consideration of these issues is steadily increasing. We are beginning to see the results of these efforts reflected in contact with sites. We will continue to work with major accident sites and industry bodies to develop and share emerging best practice.

KEYWORDS: human factors, major accident hazards, safety management, COMAH

INTRODUCTION

Human failures are implicated in the majority of serious accidents in hazardous industries. Some recent examples include Bhopal, Texaco Milford Haven, Chernobyl, Piper Alpha and Flixborough to name a few. As technical safety measures improve, we can expect the significance of human factors in major accidents to increase.

To help address these issues, the UK Health & Safety Executive (HSE) set up a new Human Factors Team in the Hazardous Installations Directorate in 1999. This Team provides site inspectors with specialist advice and support during inspections, investigations and enforcement; as well as preparing industry guidance on human factors issues. The Team is composed of a balanced mix of experienced field inspectors, psychologists and ergonomists.

The focus of the Team's activities is on those sites that fall within the scope of the Control of Major Accident Hazards Regulations 1999 (COMAH), although we are also active in the railway industry. Over the last three years, the team has become involved at numerous chemical sites across the whole of the UK, including most oil refineries.

In addition to our inspection and assessment activities, we develop guidance and standards, train field inspectors, set policies for the field, manage applied research and promote human factors to industry (either directly or through intermediaries such as the Institute of Petroleum, Institution of Chemical Engineers and the Chemical Industries Association). We are also involved in the European Commission PRISM network co-ordinated by the European Process Safety Centre (EPSC), the aims of which are to develop and disseminate best practice guidance on human factor topics.

WHAT WE MEAN BY 'HUMAN FACTORS'

The HSE document HS(G)48[1] presents a simple introduction to generic industry guidance on human factors. This guidance provides a useful definition:

'Human factors refer to environmental, organisational and job factors, and human and individual characteristics, which influence behaviour at work in a way which can affect health and safety'

This definition includes three interrelated aspects that must be considered – the job, the individual and the organisation. In other words, human factors is concerned with what people are being asked to do, who is doing it and where they are working. Human factors interventions will not be effective if they consider these aspects in isolation. The scope of our interventions thus includes organisational systems and is considerably broader than traditional views of human factors/ergonomics.

It is deficiencies in either of these three areas, or in the interactions between them, that lead to human performance problems. There are three types of human failures that may lead to major accidents:

- **Errors** are physical actions that were not as intended;
- **Mistakes** are also errors, but errors of judgement or decision-making;
- **Violations** differ from the above in that they are intentional (but usually well-meaning) failures, such as taking a short-cut or non-compliance with procedures.

The likelihood of these human failures is determined by the condition of a finite number of 'performing influencing factors', such as time pressure, workload, competence, morale, noise levels and communication systems. Given that these factors influencing human performance can be identified, assessed and managed; potential human failures can also be predicted and managed. In short, human failures are not random events.

THE ROLE OF THE HUMAN FACTORS SPECIALIST IN COMAH

Our involvement is usually instigated at the request of the local site field inspector where they have a concern or where there has been an accident with a human factors aspect. Furthermore, our contribution to the assessment of COMAH Safety Reports may lead us to visit the site to follow-up issues identified in the assessment.

Our objectives range from obtaining an overview of the site's approach to human factors to undertaking an in-depth investigation of a particular human factors issue. Generally, our involvement on site varies from one to five days, depending on the nature of the intervention, although in certain cases this may be extended. When on site, we undertake several activities, including:

- Interviewing a cross-section of personnel, from Directors/senior management to front-line operators and their representatives;
- Reviewing documentation not obtained prior to the visit;
- Verification inspection (i.e. comparing what we have been told with what we observe);
- Initial feedback to site personnel.

786

Where necessary, we support formal enforcement action by the site field inspector, i.e. improvement and prohibition notices, or prosecution. Following our visit, we will prepare an inspection/audit report for the field inspector and possibly revisit the site to present our findings/recommendations and agree a way forward. We will monitor progress through contact with the field inspector and may revisit the site for further inspection as necessary.

Human factors is often seen as a rather nebulous concept and so it is convenient to break the subject down into a series of discrete topics. As a result of our site visits and assessment of COMAH safety reports, a small group of topics has emerged and we promote these as our 'top ten':

1. Organisational change and transition management
2. Demanning and staffing levels
3. Training and competence
4. Safety culture
5. Alarm handling
6. Fatigue from shiftwork and overtime
7. Compliance with safety critical procedures
8. Safety critical communications (e.g. shift handover)
9. Ergonomic design of interfaces
10. Maintenance error.

Although general HSE guidance on human factors is available [HS(G)48], the Human Factors Team have a programme of producing guidance specific to the above topics. For example, a free information sheet is available on alarm management. Forthcoming guidance will include organisational change, competency, fatigue and human factors in design. We have also assisted with the production of guidance published on the Institute of Petroleum website[2].

EXPERIENCE OF REGULATING COMAH SITES - POSITIVE ISSUES

The Human Factors Team has a distinct advantage of having visited a broad sample of Major Accident Hazard (MAH) sites over the past three years, enabling us to construct a picture of best practice in human factors in the process industry. We are therefore able to facilitate the sharing of what works and what doesn't across the industry, through published guidance, seminars and individual site contact.

Clearly, the efforts of the Team in promoting these issues are beginning to reveal themselves in our contact with MAH sites. For example, some sites have addressed issues that we have raised at regional one-day events held in conjunction with industry bodies and associations (including consideration of the top-ten topics listed above). This has been reflected in the structure and content of their COMAH safety report submissions and in information available on site inspections.

As our capabilities are increasingly recognised within HSE, we are now finding that we are becoming involved at an earlier stage of the design lifecycle. For example, we are being consulted prior to site modifications (including proposals for organisational change) and are also involved in specifications for human factors in the design of new process plants. This opportunity to be involved at such an early stage will increase the impact of our involvement.

Although our approach has been new to many sites, we have received positive feedback following our interventions. For example, some sites apply the lessons learnt from an intervention to other installations in the company. Other sites have commented that we have provided them with a different perspective on their organisation, not obtained from previous 'independent' audits.

Although our interest is in safety improvements, several companies have experienced significant quality and productivity gains following human factors interventions. For example, at one site, the analysis of procedures and task design led to reduced start-up times.

Once their awareness has been raised, some sites have clearly embraced the issues and are developing their in-house capability in human factors. We are seeing an increasing number of companies having a human factors champion on site, who acts as an 'intelligent customer' in dealings with the competent authority and external consultants. This person should be highly visible, have influence, a link to various project teams and access to human factors technical advice and support where necessary.

Over the past couple of years or so, we have recognised that the major hazards industry is readdressing the balance between hardware and human factors. Given that the regulation of these issues has developed rapidly since the formation of a dedicated team of specialists, we expect that 'emerging best practice' will continue to develop across the industry.

EXPERIENCE OF REGULATING COMAH SITES - NEGATIVE ISSUES
Although many MAH sites are managed by multi-national, blue-chip companies, the experience of the Team is that their consideration of human factors issues could be significantly improved. The main failings apparent in relation to human factors are discussed in detail below. These weaknesses have all been observed at numerous installations and are common threads rather than isolated occurrences.

FOCUS ON ENGINEERING ISSUES
Despite the growing awareness of the significance of human factors in safety, particularly major accident safety, many sites do not address these issues in any detail. Their focus is almost exclusively on engineering and hardware aspects, at the expense of 'people' issues. From reading many safety reports it would appear that these sites are unmanned, such is the lack of reference to human performance aspects.

For example, a site may describe alarm systems as being safety-critical and describe the assurance of their electro-mechanical reliability, but fail to address the reliability of the operator in the control room who must respond to the alarm. If the operator does not respond in a timely and effective manner then this safety critical system will fail and therefore it is essential that the site addresses and manages this operator performance.

Due to the 'ironies of automation'[3], it is not possible to engineer-out human performance issues. All automated systems are still designed, built and maintained by human beings. For example, an increased reliance on automation may reduce day-to-day human involvement, but increases maintenance, where performance problems have been shown to be a significant contributor to major accidents[4].

Furthermore, where the operator moves from direct involvement to a monitoring and supervisory role in a complex process control system, they will be less prepared to take timely and correct action in the event of a process abnormality. In these infrequent events the operator, often under stress, may not have 'situational awareness' or an accurate mental model of the system state and the actions required.

FOCUS ON OCCUPATIONAL SAFETY

The majority of MAH sites tend to focus on occupational safety rather than on process safety. Those sites that consider human factors issues rarely focus on those aspects that are relevant to the control of major hazards. For example, sites consider the personal safety of those carrying out maintenance, rather than how human errors in maintenance operations could be an initiator of major accidents. This imbalance runs throughout the safety management system, as displayed in priorities, goals, the allocation of resources and safety indicators.

For example, 'safety' is measured by Lost-Time Injuries, or LTIs. The causes of personal injuries and ill-health are not the same as the precursors to major accidents. Therefore, measures such as LTIs are not an accurate predictor of major accident hazards and sites may thus be unduly complacent in this respect. Notably, several sites that have recently suffered major accidents demonstrated good management of personal safety, based on measures such as LTIs. Therefore, the management of human factors issues in major accidents is quite different to traditional safety management.

In his analysis of the explosion at the Esso Longford gas plant, Hopkins (2000)[5] makes this point very clearly:

'Reliance on lost-time injury data in major hazard industries is itself a major hazard.'

and,

'An airline would not make the mistake of measuring air safety by looking at the number of routine injuries occurring to its staff'.

Clearly, a safety management system that is not managing the right aspects is as effective in controlling major accidents as no system at all.

Performance indicators more closely related to major accidents may include the movement of a critical operating parameter out of the normal operating envelope. The definition of a parameter could be quite wide and include process parameters, manning levels or the availability of control/mitigation systems. Many performance indicators will be site specific and further examples are given below:

- Number of accidental leakages of hazardous substances;
- Environmental releases;
- Time taken to detect and respond to releases;
- Activation of protective devices;

- Process disturbances;
- Response times for process alarms;
- Process component malfunctions;
- Maintenance delays (hours);
- Number of outstanding maintenance activities;
- Frequency of checks of critical components;
- Number of inspections/audits;
- Emergency drills;
- Procedures reviews;
- Compliance with safety critical procedures;
- Staffing levels falling below minimum targets;
- Non-compliance with company policy on working hours.

It is critical that the performance indicators should relate to the control measures outlined by the site risk assessment and/or detailed in the COMAH safety report. Furthermore, they should measure not only the performance of the control measures, but also how well the management system is monitoring and managing them.

FOCUS ON THE SHORT-TERM

Where sites do consider human factors aspects in relation to major hazards this is usually in response to a major incident, an inspection by the HSE or both. In these cases, companies tend to view the initiative as having a short-term benefit, such as satisfying the requirements of the regulator, rather than as improving the long-term safety performance of the site. For example, human factors issues may be addressed in the COMAH safety report in order to facilitate acceptance of the report by the competent authority, rather than to make a real difference in major hazard safety. Where sites focus on one area of plant, we encourage the roll-out of a human factors programme to other areas of the site.

LACK OF OWNERSHIP

This issue is related to the short-term outlook discussed above. Sites often consider human factors issues in relation to an immediate need to address a discrete topic. External consultants may be engaged to facilitate the intervention and too frequently, the expertise remains outside of the company reducing ownership of these issues by the site. In these cases, we propose that a senior member of site management adopts the role of a human factors champion.

LACK OF REALISM

We are often informed by a site that operators are well-trained and experienced, partly as justification for relying on human actions. However, it cannot be stressed enough that highly skilled, motivated and experienced people do make errors, whether unintentional or not. It is human nature to take short cuts or break rules, for example when the pressure or inconvenience is high enough. It is also the case that unintentional errors occur, for example when workload is high, a task is complicated, or the situation is abnormal.

MAH installations frequently assume that an operator will perform certain actions in the event of a process upset. However, this assumption often fails to take account of the fact that human behaviour in an emergency situation is different to that in normal operations. Where there is reliance on operator actions in controlling or mitigating a MAH, this should be demonstrated to be realistic. In many cases, manual interaction or intervention could be replaced with reasonably practical physical measures. MAH control should not rely on the heroic actions of operators (Lucas[6] termed this the 'superman approach' to risk control).

On occasion, quantitative data is quoted by the operator without justification, for example: 'the probability of the operator failing to respond to the alarm is 0.0001' (the 'magic number' approach, Lucas). Such data is to be treated with caution, and we will require the site to make the assumptions explicit and demonstrate that the data is specific to the site.

FAILURE TO IDENTIFY SAFETY CRITICAL ASPECTS

Our experience is that MAH sites fail to produce an inventory of safety-critical tasks, roles, responsibilities and procedures. Without the identification of areas where human intervention is safety-critical, any consideration of human factors will be unfocussed. Human factors analyses, although productive, can be resource-intensive and in order for resources to have maximum impact, they should be targeted where their impact will be the greatest. Again, reference should be made to the MAH risk assessment; and the role of human intervention in the initiation and control of MAH scenarios reviewed.

THE WAY FORWARD

At its inception, the Human Factors Team outlined a strategy which can be summarised as follows:

- Increase awareness of the importance of human factors among UK chemical sites;
- Improve the integration of human factors in design, risk assessment and COMAH safety reports;
- Encourage continuous improvement and sharing of good practice;
- Codify knowledge in a useful way for HSE and transfer to field inspectors.

Over the past three years we have made significant progress towards achieving these objectives, including interventions on a large number of major hazard installations across the UK.

However, there remain considerable weaknesses in the approaches taken to human factors on many of the most hazardous sites in the country, operated by some of the world's largest chemical companies. If further major accidents are to be prevented, duty holders are urged to examine whether any of the failings discussed in this paper apply to their organisation.

REFERENCES

1. Reducing error and influencing behaviour (HSG48), HSE Books 1999, ISBN 0 7176 2452 8
2. Institute of Petroleum, www.petroleum.co.uk

3. Bainbridge, L. (1987). Ironies of automation. In New Technology and Human Error. Edited by Rasmussen, J., Duncan, K. & Leplat, J. John Wiley and Sons Ltd.

4. Improving maintenance - a guide to reducing human error HSE Books 2000, ISBN 0 7176 1818 8

5. Hopkins, A. (2000). Lessons from Longford: The Esso Gas Plant Explosion. CCH Australia Ltd. ISBN 1 86468 422 4

6. Lucas, D. (2002). We are all human: How human factors impact health and safety performance. IOSH Annual Conference, Manchester, 15–16 April 2002.

DEVELOPING BEST PRACTICE SAFETY PROCEDURES THROUGH IT SYSTEMS

Prof. Dr. ing. Darabont Alexandru,Ph.D., Ştefan Kovacs, Apostol George
INCDPM Bucharest, Romania

The paper presents the joint research of Romanian and foreign specialists in the design of a structure for developing and disseminating optimal best practice safety procedures (BPSP's) using the existing safety experience from Europe and Romania. This step will be a great leap forward towards Romania's safety integration into the European Union. The BPSP's are developed gradually, starting with a primary frame model (PFM) and a expert structure that helps towards the integration of existing European BPSP's and also towards the development of Specific Frame Models (SFM) that will develop into optimally BPSP's in time, by adding safety knowledge layers. At the end of the development of frame structure, a national IT system will be developed around a BPSP server that will store the PFM, SFM's and the BPSP's developed around the time and will disseminate them to all the interested parties.

Best practice safety procedures, IT structures, expert systems, frame structures, knowledge layers

GENERAL ASPECTS

"Knowledge is worth thousand pounds; especially when you need it and when you don't have it" says an old Romanian proverb. Safety knowledge is worthless; considering the possibility to save lives and to eliminate/reduce occupational accidents and incidents, everyone must be interested in obtaining and using safety knowledge.

However, there is a slight paradox regarding the useful safety knowledge, in the right place and at the right time – if apparently, safety knowledge is available for everyone, everywhere, in large quantities, in reality the phenomenon of "knowledge disappearing" is met very often when safety problems must be solved quickly and efficiently. Figure 1 illustrates this aspect.

Safety knowledge is needed:

- At the right place;
- In the right moment;
- In a right format.

So we could speak about useful safety knowledge (USK) as the safety knowledge that is fulfilling all these requirements.

Using cognitive psychology studies[1] we could connect our USK with two basic knowledge processes used to perform tasks:

- External representations-used to be a memory aid;
- Mental images-match external representations in the extraction of information about perceptual relations

Using the system approach we could see that mental images are extensively used by workers in order to keep and improve safety as perceptual relations are defining the workplace.

In this respect we could also speak about safety functional information[2].

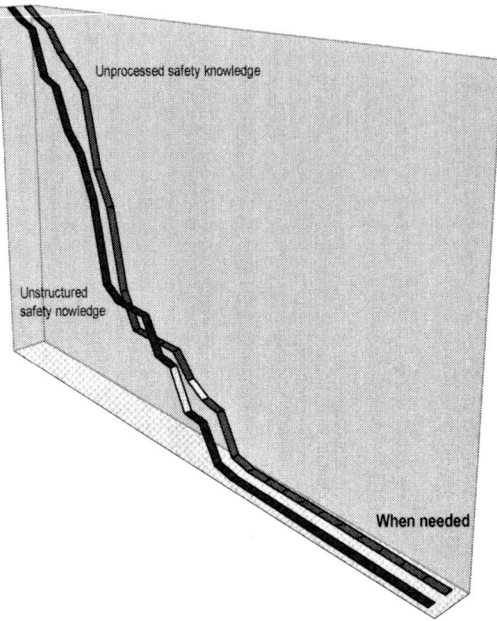

Figure 1. The loss of unstructured safety knowledge and unprocessed safety knowledge in time

Why does the big store of existing knowledge lie fallow? One reason is that relevant findings are not easily accessible at the time decisions need to be made. The existing knowledge is difficult to tap into. It is dispersed in inaccessible reports, obscure journal papers, remote libraries, and in unknown people's heads. The existing knowledge, therefore, is of limited practical use to people who need to make safety-related decisions here and now. The question is how do we put safety knowledge into the hands of the decision makers so that safety is explicitly and quantitatively considered?

While much of what we know about safety lies unused, that must not deter us from continuing to increase our knowledge. The development of safety knowledge is often a long process filled with imperfect data and analysis tools. It is only through continual reexamination of the existing knowledge base and the conduct of new research that we are able to improve on our knowledge.

At the state level, information systems and knowledge-based decision making have traditionally suffered from fragmentation and overlap. Many groups with different objectives have been collecting safety-related information for decades. Unfortunately this information is owned by groups who

- are part of a variety of state or local bureaucracies;
- are often unwilling or unable to coordinate and share information for the purpose of making better safety decisions in general and specifically better safety decisions, or
- have been rewarded for individual accomplishments rather than for statewide programs.

Since the mid-1960s, technology has provided the means to share information by establishing common reference points and system platforms. However, the owners of data systems have been reluctant to share their information because of a perceived loss of control as well as the inability to see the benefits of knowledge-based decisions. The costs associated with bringing divergent systems together always provided the necessary rationale for business as usual and the many database owners did not present a united front when approaching top management for the funds to achieve the desired goals. This situation was compounded by the perception that sharing information represented loss of control and, therefore, a diminishing ability to reach the goals for which the data were originally established. The concept of knowledge-based decisions was either ignored or not comprehended.

As technology has expanded and resources have diminished, there has been some movement toward sharing data systems and establishing knowledge-based information systems. However, significant institutional barriers to the full use of information decisions still exist.

SAFETY KNOWLEDGE AS A NEED

The ability to anticipate the safety consequences of an action could be defined as *safety knowledge*. The richer the body of safety knowledge, the larger the scope of rational safety management.

It would be difficult to make the case against knowledge-based safety management. But the most of safety activities cannot (yet) be called knowledge based because there are two types of impediments.

The first type has to do with inadequacies of knowledge. To serve day-to-day decision making

- Knowledge must exist,
- Knowledge must be practically available, and
- Professionals must be trained to be safety knowledgeable and able to apply that knowledge to decision making.

The second type involves the reluctance to use explicit safety knowledge even when available. Organizational self-interest and the inertia of habit or ingrained professional practice are sometimes barriers.

These two types of impediments are interrelated. It is difficult to ask an organization or profession to use explicit safety knowledge if the knowledge does not exist or is not easily available, or if trained people can not be hired. Conversely, safety knowledge will not come into being, nor will professionals be trained in safety, if organizations make no use of safety knowledge and if professions do not insist on it.

FACTUAL SAFETY KNOWLEDGE AND THE SAFETY MANAGEMENT PROCESS

Factual knowledge is the most used by professionals in their daily activity. Regarding safety, it could be presumed[3] that the most of the existing and immediately usable knowledge is factual.

Many attempt to influence safety by premeditated actions and programs. The set of all these premeditated prevention actions and programs could be considered as the *active* part of safety management. Then, there is the set of activities and programs that influence safety but which are not premeditated as far as their safety consequences go. The set of all their actions constitutes the *passive*[4] part of safety management. The actions by both the active and the passive parts of this amorphous safety management system jointly determine the safety future of a country: how many will be killed, how many injured, how much property destroyed.

Two prototype styles of safety management can be taken into account, marking two ends of a scale.

The *pragmatic style* stems from the confluence of two main sources. It rests on widespread popular beliefs about safety, and on the self-interest of organizations. The popular beliefs may pertain to the safety effect of police enforcement, of passing laws, of firmer punishment of better safety education etc. The self-interest of organizations may pertain to the need to show concern and initiative, and to maintain a budget, influence, manpower or income. Actors and organizations adhering to the pragmatic style appear to do what is widely believed to be right. It is good public relations.

The *rational style*, in contrast, is rooted in the idealistic desire to reduce the harm of accidents efficiently. One wishes to foresee the consequences of decisions and actions, to balance costs and gains, and to improve the management of safety in the light of what can be learned from experience. Figure 2 presents the two styles.

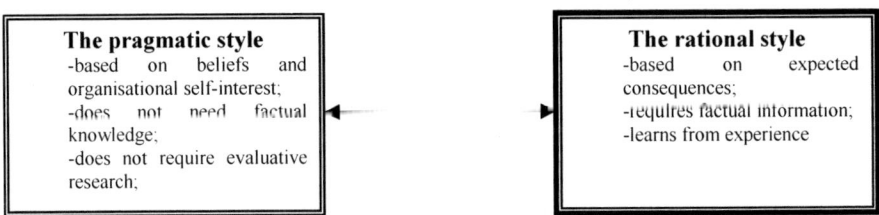

Figure 2. Two safety management styles

It is obvious that the most efficient[5] and computer friendly management style is the rational one.

Whether a real actor or organization is close to the pragmatic end of the scale can be ascertained by asking a few questions:

- Does the actor or organization require that extant factual knowledge about the safety consequences of decisions be ascertained?
- Does it employ or buy advice from people who have been trained in and have acquired factual knowledge about safety?
- Does it do evaluative research to learn about the success or failure of its actions?

If the answer to these questions is "NO", the style of the actor or organization is close to pragmatic. If the answer is "YES", there is still no assurance that the style is close to rational. In the amorphous and multifaceted safety management system there is one common trait - most decisions are made by the managerial and political echelons, few are made by the professionals who are the carriers of factual knowledge. In the delivery of health it is accepted that decisions about diagnosis and treatment rest with professionals trained in the matter. In safety delivery decisions unfortunately, the decisions are finally taken by the accountants. Therefore, even if factual knowledge is available to those making decisions, it's influence on the decision making is impossible to ascertain.

SAFETY KNOWLEDGE SPECIFIC TASKS

For the purposes of this paper we can identify four top level safety knowledge subtasks:

- *Propose*–the Propose subtask is responsible[6] for proposing prevention solutions; typically the proposals are partial, attempting to satisfy some of the safety goals; in this respect a safety goal hierarchy will be developed, considering the possible effects on the human operator, starting with death and ending with slight discomfort; during the process, there will be many instances of the propose task;
- *Verify*–takes partial or complete safety specifications and checks to see if the main safety goals are satisfied;
- *Critique*–if a Verify action identifies some safety goals not satisfied the Critique subtask attempts to determine why-to determine which of the safety commitments is responsible for the failure;
- *Modify*[7]–makes use of the results of Critique and makes changes in the general safety design

Safety problem solving takes place in the context of an external world that we perceive and act on multiple modalities and image and reason in multiple modalities as well[8].

KNOWLEDGE DIRECTED INFORMATION PASSING

Suppose a piece of unstructured knowledge that states "If history of noxious exposure consider changement of workplace" But what if there is no mentioning of noxious exposure in the worker record, even if the workplace analysis indicated chemical risks ? We could expect that a competent safety specialist could infer possible noxious exposure, considering workplace analysis and indicate the need of changing workplaces. Mittal[9] noted that inference involved in such exercises is not classificatory but involves a form of generic task named knowledge directed information passing.

The knowledge about each safety concept –the general safety domain knowledge could be stored in a frame structure –the well known safety concepts could be organised as a frame hierarchy.

This reasoning model involves accessing a frame that stores the desired datum or information on how the datum value could be obtained, including possible default values.

Best practice safety procedures are a perfect application of knowledge directed information passing.

BEST PRACTICE SAFETY PROCEDURES

Best practice safety procedures could be described by their ability to stimulate the mental images and external representations of the workers in order to assure, preserve and improve safety at their workplace. In order to do this BPSP must:

- Be the most efficient and safe way to perform an activity;
- Be as explicit as possible;
- Specific;
- Be training friendly[10]- a procedure that can not be easily explained and understood is not worth to be included as BPSP;
- Be traceable considering the logic flow of the work process;
- Be formulated in an understandable language;
- Take into account all the significant risks at the workplace and offer efficient methods in order to eliminate/prevent/reduce them;

Of course that there are many other attributes of the BPSP. The above mentioned attributes were considered significant in connection with the IT system[11] for developing BPSP that will be sketched below.

BPSP's are developed taking into account three elements:

- The development and dissemination at the enterprise level of a primary frame model[12] (PFM) for BPSP; the PFM will direct the development of BPSP's considering all the required elements;
- The development of extensive checklists in order to help the BPSP builders to collect all the necessary data for their procedures, taking into account all the elements of man-machine system:
  o The human operator;
  o The task;
  o The machine;
  o The work environment;
- The development of an expert system[13] that will use as an input the developed checklists and will generate, on the PFM basis, activity specific frame models (SFM)- the base for the future BPSP's; one of the specific subcomponent of the above

mentioned expert system will allow the import of foreign specific BPSP; the way from SFM to BPSP involves adding safety knowledge layers to SFM by all the safety specialists(willing to help) involved in a specific activity; in order to stimulate our safety specialists, the BPSP's could be used freely by the participants at their development.

Figure 3 shows the stages of development of the IT system.

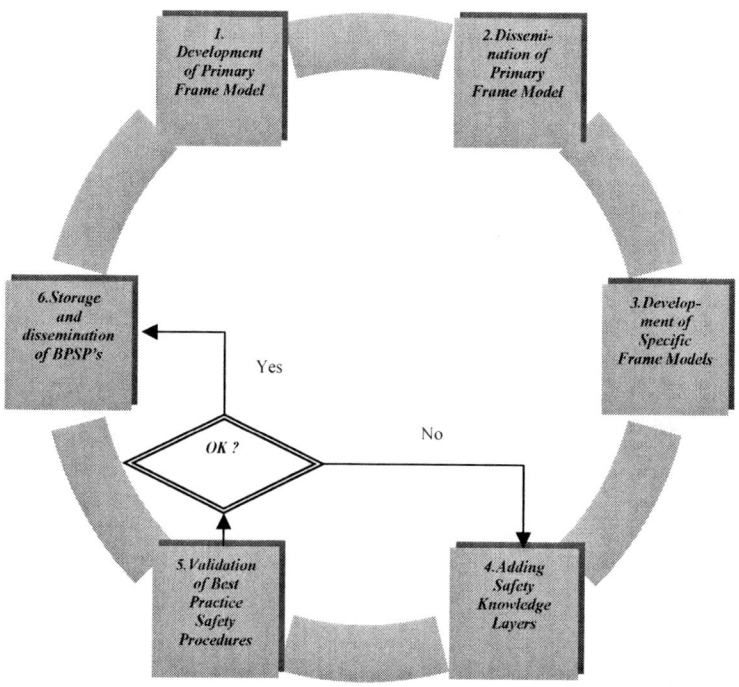

Figure 3. Steps in the development of BPSP's

It is easy to see that the process is tuned so that the developed procedures will be optimal . Step 5 is required as a supplemental check-up[14] in order to store and disseminate just the best procedures developed .If the procedure is sufficiently evolved it will be stored in the BPSP server; if not, it will be presented to more specialists in order to add supplemental safety knowledge layers.

Figure 4 shows the Primary Frame Model. The model is built as a general support structure[15] for the design and development of Specific Frame Models, so as that the developers will be sure that all the necessary aspects are included in their procedures.

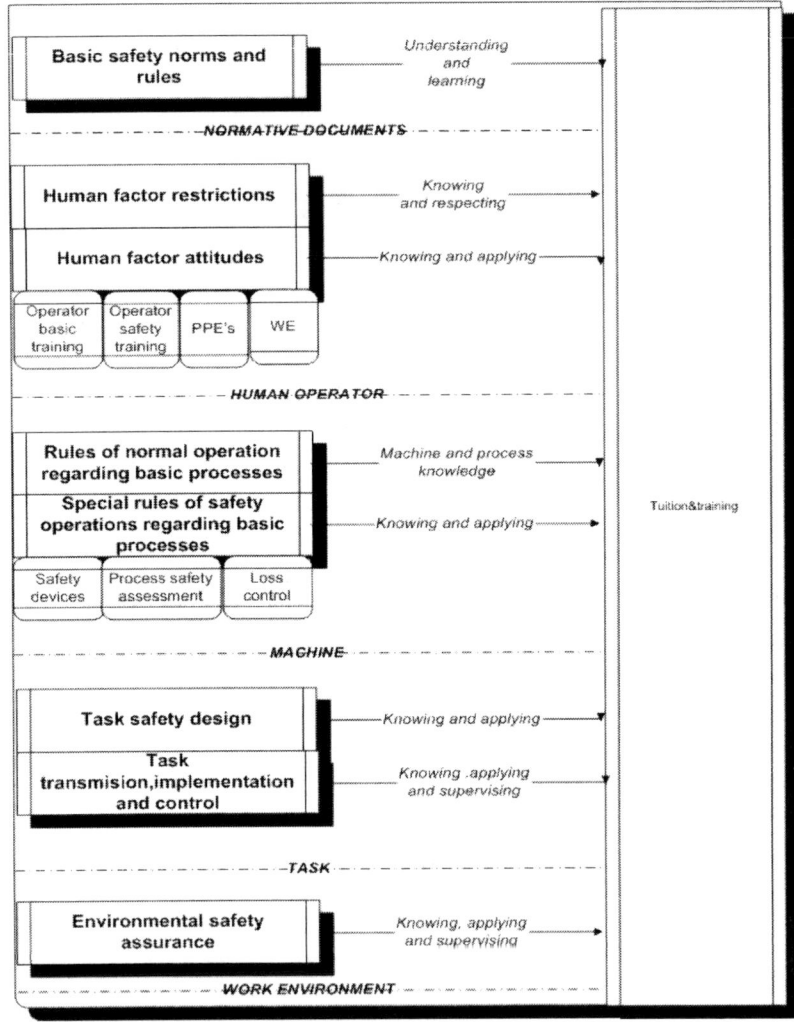

Figure 4.   The primary frame model

Figure 5 shows the process of development of a best practice safety procedure, using all the elements mentioned above.

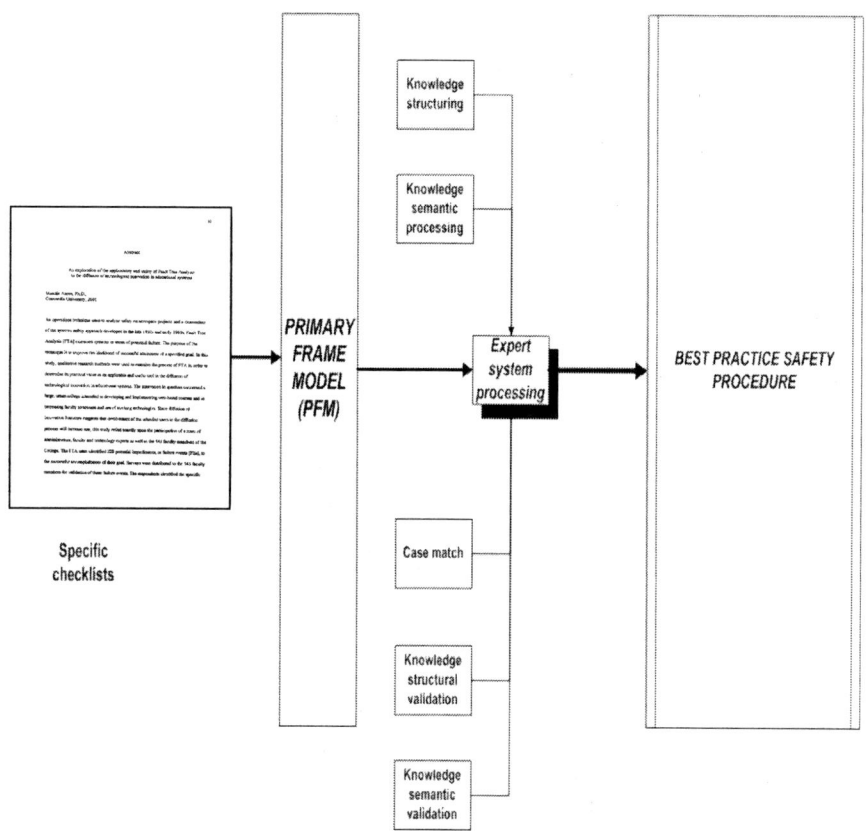

Figure 5.   Development of a BPSP

Figure 6 shows in detail a checklist screen. The user (and further SFM and BPSP developer) is supposed to have all the necessary safety knowledge in his domain, at his workplace. This knowledge is collected and processed in a natural way[16].

Figure 7 presents the preliminary result of a system run-up, concerning the human component.

This result is further processed[17] taking into account PFM and also the other components of man-machine system-resulting specific frame models and at the end of the way the best practice safety procedure. Expert systems are the best tool for acquiring and preserving knowledge. The expert system used logic blocs as shown in figure 8.

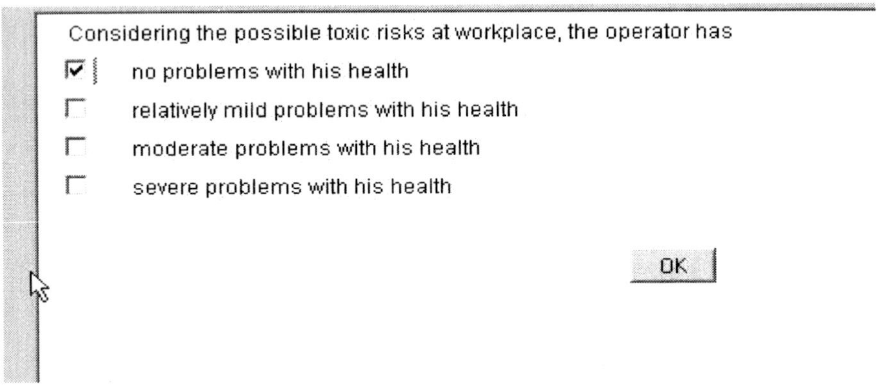

Considering the possible toxic risks at workplace, the operator has

☑   no problems with his health

☐   relatively mild problems with his health

☐   moderate problems with his health

☐   severe problems with his health

OK

Figure 6. A safety checklist screen

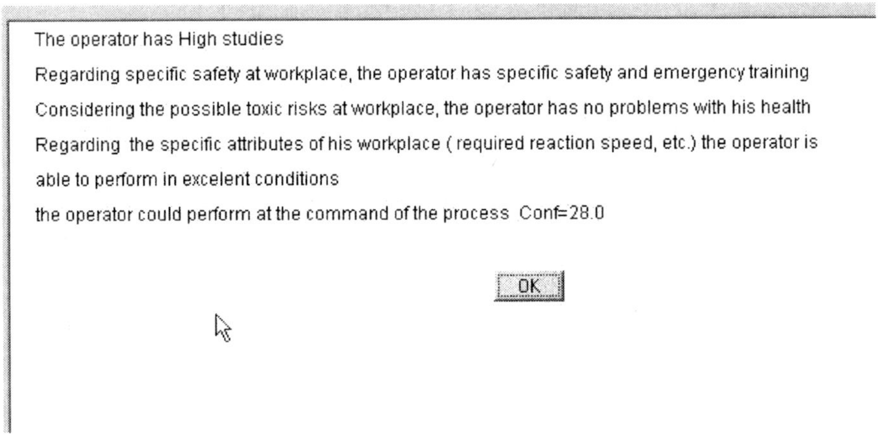

The operator has High studies

Regarding specific safety at workplace, the operator has specific safety and emergency training

Considering the possible toxic risks at workplace, the operator has no problems with his health

Regarding the specific attributes of his workplace (required reaction speed, etc.) the operator is able to perform in excelent conditions

the operator could perform at the command of the process Conf=28.0

OK

Figure 7. A preliminary result

There are two key components to improving the safety knowledge base of the system

•   the development of improved analysis methods;
•   the collection of better data.

Over the past decade, the safety analysis field has seen clear advances in the use of more appropriate methods.

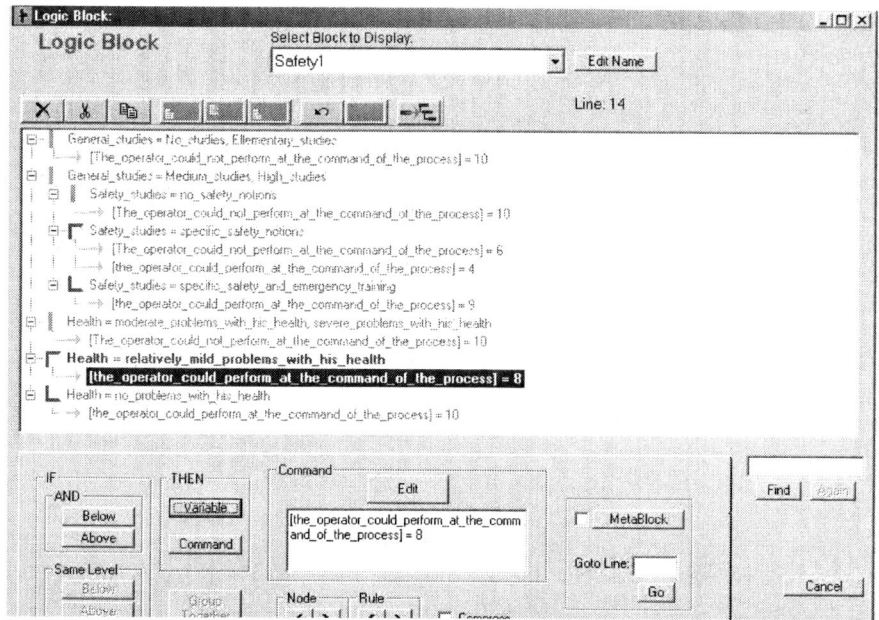

Figure 8.   Logic block

There is now nearstandard use of more appropriate Poisson and negative-binomial models in research studies, the introduction of Bayesian procedures, and innovative use of other statistical methods such as metanalysis.

There continue to be needs for further improvements, primarily the development of a similar reference concerning safety modeling and the development of improved problem identification methodologies.

Unfortunately, it does not appear that the second component of improved safety research—safety data—is enjoying the same pace of advancement.

CONCLUSIONS

The research regarding best practice safety procedures is in a middle stage-involving the fine tuning of the primary frame model and also of the expert system. The research was supported by the Romanian National Research Institute for Occupational Safety and by our research partners from Germany and Austria.

The final development of this best practice system will contribute to the improvement of safety and health in our country, considering best practice safety procedures as:

• The safe way to do things at the workplace;
• A specific safety document, more accurate and precise as the normative one;

- An efficient training tool;
- A dynamic check-up instrument in order to control safety at workplace;
- A quality booster-best practice procedures being included usually in the Total Quality Management Systems.

Also under development, some tools that allow the Internet knowledge mining after significant safety knowledge that could be included into our BPSP's will be ready probably in 2004.

The development of our fully functional prototype to a national scale implies some important aspects as:

- The acceptance of best practice safety procedures as a modern and efficient safety instrument;
- The usage of such instrument at the floor level of the enterprise for:
  o   Training;
  o   Performing;
  o   Control.
- The existance of necessary resources.

The development and implementation of this tool at a national level needs an significant investment, considering the size of the approach, the resources needed and the human contribution. The development of such system could be an international joint project, considering the possible safety benefits.

Of course that there are many problems regarding the development and implementation process and more. For example, one big issue is the property of best practice safety procedures. In this case, a mixed approach, starting with a free access for the developers of procedures and including then usage taxes on various levels will be the most appropriate in our opinion. Safety is invaluable and the avoiding of incidents and accidents will bring important benefices. The final ideea will be the development of a safety knowledge sphere, based on BPSP, with regional storage servers and BPSP development packages in every significant enterprise in Romania.

A connexion with existing or future safety networks (SafetyNet,S2S, HarshNet, Prism) will be invaluable to bring foreign experience into our BPSP's.

REFERENCES

1.   Chandrasekaran B., 1999, Multimodal perceptual representations and design problem solving, in *Visual and spatial reasoning in design:computational and cognitive approaches*, MIT,Cambridge, USA
2.   Suwa M., Tversky B., 1997, what do architects and students perceive in their design sketches? A protocol analysis, in *Design studies*, 18(4)
3.   Hauer, E. A., 1997, Observational Before–After Studies in Road Safety. Elsevier Science, Inc
4.   Bebge E.J.,1990,By-products of safety evaluation, in *Personnel Journal*, nr.29, pg.94–99
5.   Hauer E.,1998,Knowledge and the management of safety,in *Traffic Safety Summit Proceedings*, Kananaskis,Alberta,Canada

6. Gomez F.,Chandrasekaran B.,1981,Knowledge organization and distribution for medical diagnosis, in *IEEE Trans.Systems, Man and Cybernetics*, vol.11,No 1, pg. 34–42
7. Steels L ,1996, Self-organising vocabularies. In: Langton CG, Shimohara K (eds) *Proceedings of the ALIFE V*. MIT Press, Cambridge MA, pp 179–184
8. Goel V.,1995,Sketches of thought, MIT Press, Cambridge, USA
9. Mittal S.,1980, Design of a distributed medical diagnosis and database system, doctoral dissertation, Dept. of Computer and Information Science, Ohio State University
10. Oliphant M, Batali J ,1997, Learning and the emergence of coordinated communication Centre for Research in *Language Newsletter*, 11(1)
11. Tijsseling A, Harnad S ,1997 Warping similarity space in category learning by backprop nets. In: Ramscar M, Hahn U, Cambouropolos E, Pain H (eds) *Proceedings of SimCat 1997: Interdisciplinary Workshop on Similarity and Categorization*. Department of Artificial Intelligence, Edinburgh University, pp 263–269
12. Nakisa RC, Plunkett K ,1998 Evolution of a rapidly learned representation for speech in *Language and Cognitive Processes*, 13: 105–127
13. Rumelhart DE, McClelland JL, the PDP Research Group (eds) ,1986, Parallel distributed processing: Explorations in the microstructure of cognition (Vol 1: Foundations). MIT Press, Cambridge MA
14. Searle JR ,1982, The Chinese room revisited in *Behavioral and Brain Sciences*, 5: 345–348
15. Parisi D ,1997, An Artificial Life approach to language in *Mind and Language*, 59, 121–146
16. Steels L, Kaplan F ,1999 Collective learning and semiotic dynamics. In: Floreano D, Nicoud JD, Mondada F (eds*) Proceedings of ECAL99 the Fifth European Conference on Artificial Life (Lecture Notes in Artificial Intelligence)*. Springer-Verlag, Berlin, pp 679–688
17. Pylyshyn ZW ,1984 Computation and cognition. MIT Press, Bradford Books, Cambridge MA

THE USE OF A SAFETY CASE APPROACH TO SUPPORT DECISION MAKING IN DESIGN

W.A.T. Alder and J. Perkins
Binnie Black and Veatch, Redhill, UK

In many of the high hazard industries the safety case and safety report approach is required under major hazards legislation. The benefit of the approach is the demonstration that the risks of an operation or a facility are reduced to as low as reasonably practicable and that continued operation is justified on health and safety grounds. The principles behind the development of a safety case are the development of an argument, the presentation of information to support the argument and a permanent record of that information for future use. These features are in themselves useful to organisations to assist risk management and possibly to meet wider requirements of the Health and Safety at Work etc. Act. The Safety Case approach has been adopted during in major infrastructure projects specifically to assist decision-making. The approach has been used to assess levels of safety management and residual risks with certain design options. In areas where there has been concern over health and safety aspects of the options, the safety case has been used as the forum for analysing and presenting the arguments for or against the options. A process of consultation with the stakeholders using the safety case has been carried out to gain agreement on which option to select.

Safety Case, infrastructure projects, CDM regulations.

INTRODUCTION

Safety cases or safety reports are used widely in the high hazard industries for a variety of purposes, but principally to demonstrate the adequacy of safety of facilities or operations. Their production is required under a variety of Health and Safety regulations and other legal requirements.

The need for such a demonstration of adequacy is obviously due to the potentially disastrous consequences in the event of a major accident, and hence the need to be stringent in analysis of the potential for the accident to occur and the control of its likelihood and mitigation of severity.

However the concept of demonstration of the adequacy of safety has potential for application beyond the high hazard industries. In particular where approval or acceptance from a range of stakeholders is required, demonstration in the form of a safety case presents a useful way forward.

The authors have developed and used a safety case approach as a means of demonstrating the adequacy of safety of options during the design phase of projects, and as a vehicle for stakeholder consultation and acceptance. The particular applications in which it has been used has been in major infrastructure projects in the United Kingdom which are not subject to any of the UK major hazard legislation but are subject to the CDM (Construction Design and Management Regulations, 1990) and other safety regulations.

This paper discusses the broad issues of safety cases and the approach used by the authors and discusses the benefits and some of the problems of the approach.

AN OVERVIEW OF SAFETY CASES AND SAFETY REPORTS

The high hazard industries are the principal users of safety cases or safety reports in the UK where legislation requires their production. The Control of Major Accident Hazards Regulations, 1999[1] (COMAH Regulations) require companies to produce safety reports if they operate or are intending to operate plant or processes that have inventories of hazardous materials exceeding certain threshold values. The Offshore Installations (Safety Case) Regulations, 1992[2] require a safety case for fixed and mobile installations for oil and gas extractions in UK waters, while the Railway Safety Case Regulations[3] require safety cases from a range of operators on the UK Rail networks. In the nuclear industry, safety cases are required to be produced under Condition 14 of an operators Nuclear Site Licence issued under the Nuclear Installations Act, 1965[4].

The major requirement for all of these safety cases and reports is to provide a *demonstration* of the adequacy of safety with control measures in place to conform to the ALARP principle[8]. Safety cases have been defined as a documented body of evidence that provides a convincing and valid argument that a system is adequately safe for a given application in a given environment[9].

For example the COMAH regulations call for the safety case to demonstrate that major accidents are identified and necessary measures are taken. Demonstration in this case is intended to be "make the case/argument" rather than "prove beyond doubt"[10]. Views on the composition of a Safety Case under the earlier CIMAH regulations were that is should consist of facts about the site, reasoned arguments about the risk from the site and conclusions[5]. The offshore installations safety case regulations require that a demonstration that measures will reduce risks to as low as reasonably practicable. Nuclear site licence conditions phrase the requirement slightly differently in that a safety case consists of "documentation to justify safety" which amounts to a demonstration.

The role and purpose of the safety case in the Railway industry was considered in detail in the Ladbroke Grove Rail Inquiry. Many of the conclusions are applicable to safety cases used in any of the high hazard industries. In particular Lord Cullen, in his report following the inquiry[6], notes that a safety case is seen as providing an appropriate means of managing safety and providing an adequate assurance of safety by independent reviewers. He also notes that a safety case should show how the duty holder has reduced the risks associated with its operation to as low as reasonably practicable. During the inquiry evidence, from expert witness Peter Waite, was given that in producing the safety case, an argument must be constructed to give confidence that the operators had considered all the risks and that its principal purpose was to a tool, a route map and a record of commitments for management to set out how they organise their operation to work safely[6].

In discussing the principles of permissioning regimes, the HSE note in a recent discussion document that the duty-holder must carry through the assertions and assumptions in the safety case to practices on the ground and monitor and evaluate their implementation[7].

Overall the content of safety cases is common throughout the major hazard industries and can be summarised as:

i)   A descriptive section which provides plant and operations details, including design standards. This provides a demonstration that the facility was designed and built appropriately;

ii) A description of safety management arrangements. This provides a demonstration that the facility is operated and maintained appropriately;
iii) An assessment of the residual risks. This shows that the residual risks are as low as reasonably practicable;
iv) Conclusions as to the adequacy of the safety based on i) to iii) above.

A report with these contents is likely to meet the requirements and suggestions discussed above.

SAFETY CHALLENGE FACED IN MAJOR INFRASTRUCTURE PROJECTS

The construction of major infrastructure, such as transport links, utilities, airports or major building works and its subsequent operation and maintenance presents a range of major health and safety issues, which generally affect workers but can also impact on members of the public and the environment. Where the major hazard safety regulations don't apply – a wide range of health and safety legislation is in place from the Health and Safety at Work etc. Act through general safety regulations such as the Management of Health at Work Regulations, 1999 to specific design and construction requirements such as the Construction (Design and Management) Regulations 1994 and many others.

In addition to health and safety issues there are concerns over the practicality of options for the design such as can it be maintained, will it work, what emergency or standby provisions need to be in place, how much does it cost, how much will it cost to operate and what are the non-safety related risks with it such as business interruption and asset loss or damage in the event of plant or equipment failure.

In any major infrastructure project there are many parties involved or interested in the decision-making including:

- the final owners, who are likely to have business objectives to meet,
- operations teams, who will have to deal with any of the problems of operating the facility successfully,
- the designers/project managers, who are responsible for ensuring the right design for the right price and to programme.
- emergency services, who may be called in to provide fire and rescue activities,
- Maintainers, who may be third party maintenance contractors who need suitable and safe access and facilities,
- Other stakeholders which may include the public, government (national and local), environmental groups, etc.

The immediate requirements of these stakeholders may not necessarily be compatible with all of the requirements under the health and safety legislation and attempting achieve the right balance can be a source of conflict and ultimately cause delay, cost increases and in the worst case sub-standard levels of safety and environmental protection. Therefore the challenge is identifying, demonstrating and agreeing the best way forward in contentious areas where safety is an issue. The safety case was proposed as a way of meeting this challenge.

SAFETY CASE PROCESS ADOPTED FOR INFRASTRUCTURE PROJECTS
A number of projects have been trialled by the authors using the safety case as a tool for aiding decision-making. The team creating the safety case included health and safety specialists experienced in producing safety cases in the high hazard industries as well as those with experience in CDM and the specific health and safety issues related to the construction industry. The team also included members of the design team who were looking at the various options involved.

PRODUCTION OF AN OPTIONS SAFETY CASE REPORT
The first stage in our process was to identify where a safety case was required. Unlike the high hazard industries where a safety case is provided for an entire facility, the safety case approach was used to deal with specific issues raised for the facility. So for example, safe maintenance and access may have been identified as a concern with a number of possible options to deal with the problem. If there is no consensus on which option to use, the safety case can be specifically produced to deal with this issue.

Typically these issues were raised from risk workshops where the risk register was being reviewed or updated – or, where projects use them, from HAZOP studies.

The next stage in our process was to carry out a more detailed review of the hazards from the identified options. This typically involved face-to-face discussions with the interested parties, including operators and maintainers, the design team and many of the third parties. These meetings were intended to include staff at all levels – from the operatives who would carry out activities in the facility to Senior managers with business objectives to meet. In addition to these meetings, a series of interviews with experts in the particular area of concern were carried out. For example if the issue was related to confined space access, then suppliers of safety equipment and confined space rescue organisations were consulted.

From these meetings the hazards could be further developed and understood and the practicalities and effectiveness of using available safeguards, both hardware and managerial could be understood. With a better understanding of these, the options could be reviewed and amended.

The following stage required assessment of the options. This involved 3 elements:

- A review against codes and standards and the opinions of experts as to best practice. This included the hardware safeguards in place to control risks,
- A review of the management systems to see if the systems were in place to deal with operating and maintaining the options – or whether they could be put in place. This included the emergency arrangements that may be required,
- A review of the residual risks. This included documenting the hazards identified, some qualitative risk assessment and some quantitative risk assessment including cost benefit analysis and determination of whether risks were reduced to as low as reasonably practicable.

These 3 elements relate to the items i) to iii) noted earlier as the first 3 elements of the safety case. In carrying out the elements the review was intended to confirm or dispute that an option was justifiable on health and safety grounds, i.e. that it met codes and best practice, that it could be managed safely and that risks were reduced to as low as reasonably practicable.

The results of these studies lend weight to a particular option being preferential. The next step was to return to the stakeholders and discuss the findings. From this, consensus could be gained as to the viable option.

Having gained agreement in principle from the interested parties on the way forward for the design, the Safety Case Report could now be constructed. Each of the 3 elements described above was documented. Additionally, detailed descriptions of the facility, its method of operation and maintenance and any emergency response required for hazardous situations. This additional information was included in order to define the "safe working envelope", i.e. the limits of operation which were assessed and therefore to which the conclusions of the report apply. This is common to safety cases produced in the high hazard industries. A series of appendices were attached to the safety case including:

- the detailed reviews against codes, guidelines and regulations,
- proposals for 3rd party activities, such as maintenance programmes or emergency standby services (such as in the case of confined space emergency teams),
- minutes of meetings with all of the interested parties.

This last item is particularly important. The meetings with interested parties contained discussions of their agreements and the conditions by which they gave agreement. For example a maintenance team leader was happy to see a reduction in access points to a particular area, but only when it was explained that stairs would be provided at the access points rather than ladders.

The final element to the safety case was the conclusions – the justification of adequacy of safety. This included the conclusions that:

- the design option met relevant codes and standards, legislatory requirements were met,
- the facility could be operated in the way described and that the arrangements under the management system in place by the operator was or would be sufficient to allow safe operation,
- the risks had been reduced to as low as reasonably practicable. The approach to this differed from project to project and included cost benefit analysis using the "value of lives saved" approach, an approach based on placing the risks in context of other risks faced by the project, or by comparison to other industrial activities. The method of selection of the risk criteria was based on discussion with the parties as to what they felt most useful in understanding risk.

Having produced a final report this had then to be represented to all of the interested parties to ensure they agreed with the scope, their concerns were covered and they agreed with the conclusion. Each party was then asked to sign off against the document and the design option could go ahead. The finalised report would then allow close-out of HAZOP actions or seen as a control measure in the risk register for the project.

FURTHER CHANGES TO THE DESIGN

In any major infrastructure project, as the design progresses so changes are proposed from agreed options. The safety case was used as a means of controlling the changes which may have an impact on health and safety. Each change was submitted to the safety case team who reviewed the assessments in the report and commented on the impact. At this stage, the minutes of meetings with parties became very important. The reasons for acceptance of the option could be reviewed and any impact noted. A recommendation was made on any option, either that it did not invalidate the conclusions of the study, in which case the report was revised in line with changes and circulated to interested parties for their sign-off, or that it did invalidate the conclusions, with an explanation of why, and that the option should not go ahead.

In keeping with the premise that a safety case is a route map, the additional analyses, even for options that are not recommended, were added as appendices to the report. This provides a record of subsequent decisions and provides evidence of the designer's duties to identify, assess and control hazards.

BENEFITS OF THE APPROACH

Where our safety case approach was adopted it was felt to be of significant benefit to the project. The benefits identified include:

- A documented compliance with appropriate standards, codes and regulatory requirements.
- The safety case report became a central repository for all documentation relating to the health and safety issues, and can provide a useful reference later in the design or during operation, maintenance or decommissioning.
- The safety case process provided a route to achieve consensus between the interested parties. As the arguments in the safety case were developed logically and systematically with conclusions drawn on the adequacy of safety, and as the parties helped to shape and develop the arguments, there was a much better management of conflict and delays were limited.
- The process of holding meetings with all interested parties at all levels allowed a degree of workforce involvement in the design and in health and safety. This was beneficial to the workforce as their participation in the discussions provided them with additional understanding of the hazards of the new facility, and was beneficial to the design as the views and opinions of those who would have to make the facility work safely were heard and acted upon.
- The Safety Case Report became a living document and a tool which could provide a baseline for safety against which changes to the design could be measured. When further changes were suggested later on (particularly cost-saving measures) a rapid evaluation of the impact on safety could be made and the change justified or prevented.
- The safety case process provided demonstration that the client and the designers had identified hazards, assessed risks and provide suitable controls and safeguards, and that risks had been reduced to as low as reasonably practicable.

CONCLUSIONS

The majority of safety cases produced are in response to a requirement of specific major hazards health and safety regulations. However an approach which has been based on the already established principles of safety case production has been adapted for major infrastructure projects, particularly in areas where contentious decisions had to be made. The approach was used as it was anticipated that there would be programme and cost benefits as the systematic approach would provide reasoned arguments as to the most appropriate option.

Additional benefits included those which safety cases are known to provide such as providing a route map to the safety of the design, and a repository of information, but also included benefits such as consensus building and workforce involvement as a result of particular aspects of the approach used.

REFERENCES

1. Control of Major Accident Regulations, 1999, *Statutory Instrument 1999/743*
2. Offshore Installations (Safety Case) Regulations, 1992, *Statutory Instrument 1992/2885*
3. Railway (Safety Case) Regulations, 1994, *Statutory Instrument 1994/237*
4. HSE, 2002, Nuclear Site Licence Conditions, *www.hse.gov.uk/nsd.*
5. K.Cassidy, 1989, Chapter 13 of Safety cases within the Control of Industrial Major Accident Hazards (CIMAH) Regulations, 1984. Edited by FP Lees and ML Ang. *Butterworth & Co. (Publishers) Ltd.*
6. Ladbroke Grove Rail Inquiry, Part 2 Report; The Rt. Hon. Lord Cullen PC. *HSE Books.* 2001.
7. HSE, 2000, Regulating Higher Hazards: Exploring the issues. *DDE 15.*
8. HSE, 2000, Reducing Risks, Protecting People. *DDE 11.*
9. P Bishop, R Bloomfield, 1998, A Methodology for Safety Case Development. *Safety Critical Systems Symposium, Birmingham, UK.*
10. TJ Britton, 1999, Examining Safety Reports and Evaluating Safety Management Systems. *European Conference - Seveso 2000, Athens, Greece*

DEMONSTRATING THE TOLERABILITY OF RISK FROM MAJOR ACCIDENTS

David Glass and Mark Johnson
Transco LNG Storage, Solihull, UK

A risk assessment framework for major accidents that meets the requirements of the Control of Major Accident Hazards (COMAH) Regulations is presented. The framework has salient features that link directly to the requirements of the regulations and the interpretation of law by the Competent Authority for COMAH. A scheme for categorising risk levels based upon a realistic approach to consequence analysis was developed within the framework, with methods for demonstrating that risks are tolerable. The risk assessment framework and the supporting methods are applicable to new developments as well as existing establishments, and can be used during conceptual design to assess the relevance or adequacy of proposed risk reduction measures. The framework has been applied to an LNG production and storage facility in the UK, and has allowed the completion of the assessment of the facility's safety report by the Competent Authority for COMAH.

KEYWORDS: COMAH, risk, tolerability, ALARP

INTRODUCTION

Transco LNG Storage owns and operates five LNG production and storage facilities in the UK, storing between 21,000 and 84,000 tonnes of LNG each. In addition, the facilities have inventories of LPG (liquid ethylene, propane and butane) and high-pressure natural gas. Safety reports for each of the facilities have been produced previously to meet the requirements of the former CIMAH Regulations in the UK. The introduction of the Control of Major Accident Hazards (COMAH) regulations meant that the existing safety report format needed substantial revision to meet more stringent requirements.

Three safety reports for LNG storage facilities were submitted in February 2000, and the remaining two were submitted in February 2001. The reports passed initial assessment by the Competent Authority for COMAH (the Health and Safety Executive, the Environment Agency and the Scottish Environmental Protection Agency SEPA). However the risk assessment methods employed were subsequently found to require further development.

As a result of development work and discussions with the Competent Authority, a risk assessment framework was developed to meet the requirements of UK regulation and the Competent Authority. The framework has allowed the successful completion of the assessment of the exemplar safety report from Transco LNG Storage.

PRINCIPLES ADOPTED FOR RISK ASSESSMENT

One of the key regulations in COMAH[1] states that operators of establishments must demonstrate that they have taken all measures necessary to prevent, or reduce, the consequences of major accidents. This demonstration is an important part of the safety reports for establishments such as the LNG storage facilities, and a very strong emphasis is placed by the Competent Authority on the assessment of this demonstration. It is essential that the demonstrations made reflect UK Government policy and its interpretation by the Competent Authority when this is available via guidance documents or other publications. Where no guidance is available, company policy may be used provided it is supported by a justification.

The Competent Authority for COMAH has published guidance documents which cover the preparation of safety reports[2] and the tolerability and control of risk[3]. Two principles are contained within these documents that can be used to derive a framework for risk assessment for major accidents:

- Demonstration that all necessary measures have been taken to reduce risk can be:
  o Utility-based (meeting the minimum levels of protection expected)
  o Equity-based (showing that further improvements are not beneficial in terms of cost)
  o Technology-based (showing that state-of-the-art technology is applied)
- Demonstration must be proportional to the magnitude of the hazard or risk

The revised basis of demonstration for the LNG storage sites was essentially equity-based. Technology-based demonstrations are usually hard to achieve, as the best practice available is often well in excess of requirements for acceptable working in a specific situation (e.g. the use of clean-room technology for painting); Even so, demonstration of the use of appropriate codes and standards for the design, operation and modification of the facilities was still seen as essential. The use of utility-based demonstrations was felt to lack depth based on the initial risk assessments in the original safety reports. These showed that none of the individual risks from major accidents at the facility were intolerable.

An additional concept that arises from the guidance documents[2,3] is that of "broadly acceptable" risk. This represents levels of risk that would be of no concern to workers at the facilities or the general public. If a major accident scenario has a broadly acceptable level of risk assigned to it, then no further demonstration is needed that all measures necessary have been taken to reduce risk. Nevertheless the facility is still obliged to adopt risk reduction measures that are simple and cheap to implement. This is consistent with "going the extra mile" for safety[3].

The guidance for preparing safety reports[2] indicates that for major accidents, societal risk is usually more relevant than individual risk. To assess the tolerability of risks from major accidents at the LNG storage installations, the suggested upper limit for societal risk (a frequency of not more than 1 in 5000 years for any major accident producing 50 or more fatalities) from HSE guidance[3] was therefore adopted. Figure 1 shows the resultant societal risk plot for a typical scenario; a line with a slope of –1 on the logarithmic scale has been constructed through the HSE recommended point, in line with published information[4]. As there was little published guidance available regarding the lower boundary for tolerability, below which societal risks were regarded as broadly acceptable, a decision was made to set a limit four orders of magnitude below this, based on suggested published values[5]. Risks that fall below this value are regarded as broadly acceptable. Risks that are up to two orders of magnitude below the intolerable level are regarded as tolerable if "As Low As Reasonably Practicable" (ALARP), and are subject to an equity-based demonstration to establish that this is the case. As an interim measure, risks that are in the two orders of magnitude below this are regarded as though they are in a border region between broadly acceptable and tolerable if ALARP. Subsequent discussion documents published by HSE suggest that these risks are more likely to be regarded as broadly acceptable.

Proportionality in equity-based demonstrations needs to apply to both the identification and the assessment of risk reduction measures. This was achieved by assigning stringent methods (e.g. detailed cost-benefit analysis) to the major accidents with the greatest risk in the tolerable if ALARP region, and assigning progressively simpler and more qualitative methods to accidents with lower risks.

The use of cost-benefit analysis for demonstration needs to follow UK Government practice to be acceptable to the Competent Authority. An acceptable method is defined in the HM Treasury "Green Book"[6]. An additional requirement is the demonstration of "gross disproportion" between costs and benefits, with greater disproportion for more severe consequences of accidents (e.g. greater offsite effects, greater numbers of fatalities).

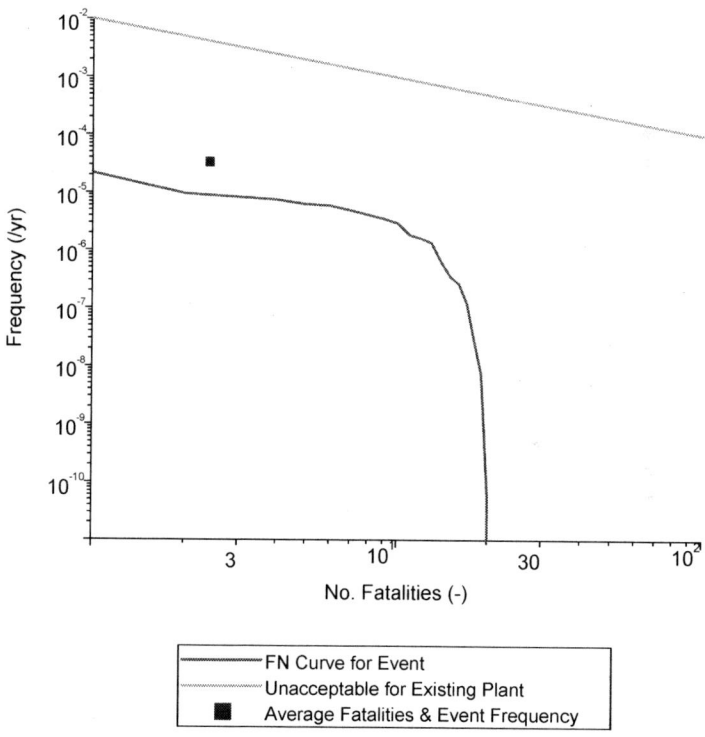

Figure 1. Typical F-N plot for a major accident scenario

RISK ASSESSMENT FRAMEWORK
Given the principles outlined above, together with specific requirements from the COMAH regulations and guidance documents[1,2,3], a framework for demonstrating that all necessary measures have been taken to control risks from major accidents was developed. Useful feedback was obtained from the Competent Authority for COMAH during its development.

The key steps in the framework are as follows:

1. Identify all major accidents, and select a representative set of major accident scenarios for detailed analysis.
2. Determine the consequences and frequency of occurrence of the major accident scenarios, for a range of realizations.
3. Assign an appropriate "Risk Category" to the scenario, based on the magnitude of the risk. (This is expressed numerically through the "potential loss of life per year" i.e. the product of the frequency of the event and the expected number of fatalities it produces.)

Then for each scenario:

4. Identify risk reduction measures using a method appropriate for the risk ranking, and rank the measures qualitatively in order of perceived cost-benefit.
5. Examine the benefits of top risk reduction measure using a method appropriate for the risk ranking.
   a) If the measure is not of benefit, then it is assumed that all necessary measures have been applied.
   b) If the measure is of benefit, adopt the measure, re-evaluate the risk ranking and examine the next risk reduction measure using a method appropriate for the new risk ranking. Continue until no further measures are found to be worthwhile.

The outcome of this work will either be a demonstration that no additional risk reduction measures are worthy of application, or a list of risk reduction measures to be implemented to an appropriate timescale. Once any worthwhile measures are implemented, all necessary measures will have been taken to prevent, or reduce the consequences of, major accidents as the risks from the scenario are demonstrated as ALARP.

APPLICATION TO AN LNG STORAGE FACILITY
The framework described above was applied to the Transco LNG Storage facility at Partington, near Manchester. This facility has four LNG storage tanks, each with a capacity of 21,000 te, and storage for liquid ethylene, propane and butane for use as refrigerants in its two liquefaction plants. The facility also stores gas oil (Diesel fuel) to drive firewater pumps. In the past, it used to store odorant to odorise exported natural gas from vaporised LNG, although this has now been removed.

Details on each of the steps and how they were applied within Transco LNG Storage are given below. The process is shown as a diagram in Figure 2.

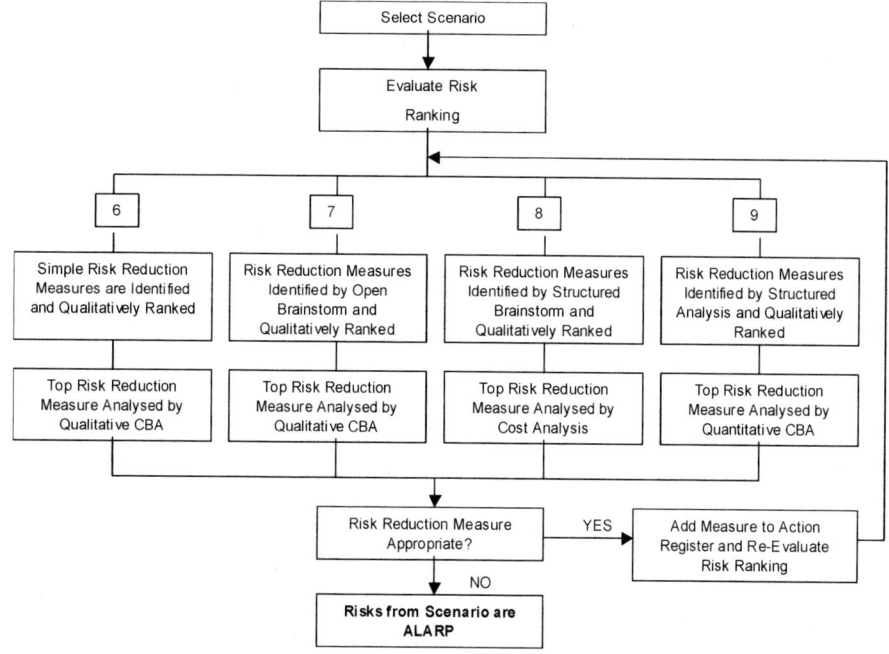

Figure 2.   Identification and analysis of risk reduction measures

MAJOR ACCIDENT IDENTIFICATION

The major accident identification had already been completed as part of the development of the original safety reports.

The identification was achieved using a panel of experts, covering the following expertise:

- Chairman (experienced in similar study leadership e.g. HAZOP)
- Engineering (process, mechanical, control & instrumentation, electrical)
- Process safety (major hazards)
- Operations (representative from shift operations team and/or the facility Safety Representative)
- Management (representative from facility management team)

The facility was divided into functional areas based on a combination of activity, dangerous substance stored and location, and keywords representing initiating events for major accidents were applied to each area. The consequences of the initiating event being realised (the major accidents) were agreed, and a summary of the control and mitigating measures for the initiating event was recorded to assist in the development of the safety report. In conducting the panel discussions, best practice for HAZOP studies was found to work well.

After all areas were analysed, scenarios were selected that represented releases from the largest sources of dangerous substance or specific events occurring in sensitive locations. In the case of the LNG storage installation at Partington these covered:

- The LNG storage tanks
- The major LNG pipelines (fill and export)
- The LPG storage tanks and pipelines
- The largest natural gas pipeline
- The odorant storage tank
- The largest pressure relief valves for each dangerous substance
- The gas oil storage tank
- The most significant congested or confined volumes

In this particular case, the resultant set of major accident scenarios had already been assessed by the competent authority as acceptable.

CONSEQUENCES AND FREQUENCY OF MAJOR ACCIDENTS
Consequence analysis of the major accident scenarios was carried out using models developed by Advantica Technology. These models are generally based on experimental data, often obtained at large or full scale using LNG, LPG and high-pressure natural gas[7]. Major accidents with high consequences were analysed over a range of realisations using methods discussed elsewhere[8]; this ensured that the worst-case realisation was identified. The output from these analyses was an estimate of the expectation value for fatalities and a corresponding frequency of occurrence for each scenario, obtained from the integration of a societal risk plot for the more severe major accidents. An example of such a plot is shown in Figure 1.

Although none of the major accident scenarios for the LNG storage facilities could lead to a Major Accident to the Environment (MATTE), a method was developed to obtain a severity term for such events. This involved comparing the likely response of the media to an environmental accident with the response for fatalities. Table 1 shows the anticipated responses of the media for fatalities, derived from published information[9]. The likely response of the media to an environmental accident would be based on local knowledge and recent articles in the press and this would be used to obtain an equivalent number of fatalities using Table 1.

Table 1. Guide to media response in the event of a major accident

Acute effects	Typical media attention
>5 Fatalities on-site	International news. Outcry threatens to close operation.
Fatalities off site	
1–5 Fatalities, injuries off-site	Headline National news. Continuing local attention.
Low Probability of 1 Fatality	Considerable local, some National attention. Local outcry.

RISK CATEGORY DETERMINATION
Given a slope of −1 for the upper bound on tolerable risk (Figure 1), the product of the number of fatalities and frequency is constant for events lying on this line. Given also that the lower limit of tolerability has been set four orders of magnitude below the upper bound, categories of risk emerge based on the product of the number of fatalities and the frequency of occurrence as shown in Table 2. Risks at Category 10 are deemed to be unacceptable, and if ever encountered an action plan would be created immediately to reduce risk. Risks that fall below Category 5 are deemed to be broadly acceptable and the scenario is not subject to analysis for risk reduction measures (unless a measure becomes known that is simple and cheap to implement).

As a result of this work, the number of scenarios falling into each category was as follows:

Category 9:   0
Category 8:   2
Category 7:   3
Category 6:   4

Risks for the remaining scenarios were judged to be broadly acceptable.

IDENTIFICATION AND ANALYSIS OF RISK REDUCTION MEASURES
To determine appropriate risk reduction measures for each scenario, an expert panel was set up similar to the one used for the identification of major accidents described above. To ensure proportionality in determining measures, four methods were used corresponding to the four relevant risk categories in Table 2:

- *Structured Analysis*: the initiating events for the scenario were examined in turn to highlight additional controls or mitigation measures that could be adopted.
- *Structured Brainstorm*: this was a less structured discussion than the analysis above, focussing on prevention, protection and mitigation measures in turn.
- *Open Brainstorm*: The panel was free to propose measures in any order
- *Open Brainstorm (Operations)*: This was similar to the brainstorm above, but only operations staff were involved. The brainstorm focussed on operational improvements.

Table 2.   Risk categories for major accidents

Product, P (in units of fatalities per year)	Risk ranking number
$<10^{-6}$	4 or 5
$10^{-6}-10^{-5}$	6
$10^{-5}-10^{-4}$	7
$10^{-4}-10^{-3}$	8
$10^{-3}-10^{-2}$	9
$>10^{-2}$	10

To ensure proportionality in examining the cost-benefit of risk reduction measures, four methods were developed corresponding to the four relevant risk categories in Table 1:

- *Cost-Benefit Analysis*:   A method was developed by Advantica Technology based on principles in the HM Treasury "Green Book"[6]. A key requirement of the cost-benefit method developed was to ensure that appropriate "gross disproportionality" was achieved when comparing costs against benefits. The method increased the magnitude of the quantified benefits with increasing total risk of fatalities, increasing number of offsite fatalities and increasing difference between expected and maximum fatalities over the range of realisations.
- *Cost Analysis*: The panel decided on a sum that would be appropriate for the risk reduction measure, and an assessment was then made of the actual cost of implementation either by knowledge of similar work or from an independent assessment by a quantity surveyor. If the assessed sum was less than the sum decided by the panel, then the measure would be implemented.
- *Qualitative analysis*: The relative costs and benefits of the risk reduction measures were discussed qualitatively by the panel.
- *Qualitative Analysis (Operations)*: As above, but using operations staff only.

The panel discussions to determine risk reduction measures were generally successful, once the panel had the "feel" for what was required. It became apparent that proportionately more time was spent on scenarios with higher risk, as intended to achieve proportionality in identifying measures.

Determining costs for risk reduction measures where necessary was also straightforward. An idea that was proposed was to obtain quotations from contractors for the work. However, this was rejected on the grounds that contractors would lose patience with the facility if too many requests for quotations were made without business necessarily being generated. A quantity surveyor was employed to provide an independent, realistic assessment of costs in the absence of other information such as the costs of previous similar projects.

RESULTS OF APPLYING THE FRAMEWORK

Once the methods described above had been applied, certain risk reduction measures dealing with inspection issues were found to be worthy of adoption, and are currently being included in maintenance regimes. A modification to an enclosure to reduce the risk of explosions was shown by consequence modelling to have negligible benefit compared to the cost of the modification. As a result, risks from major accidents at the Partington LNG storage facility have been shown to be ALARP.

The results of the work were submitted to the Competent Authority in December 2001, and in March 2002 the Competent Authority stated that the safety report for the Partington LNG storage installation meets the minimum requirements of the COMAH regulations.

CURRENT AND FUTURE WORK

At present (2002), work is underway to apply the framework to the remaining LNG storage facilities belonging to Transco LNG Storage. The automated models for consequence analysis are allowing the work to proceed at a faster pace.

In addition, plans are being developed to convert the Isle of Grain LNG storage facility to allow importation of LNG from ships, via a new jetty and a 3.5 km LNG pipeline. The framework presented above has been used to assess the risks from major accidents involving the new jetty and the pipeline, and to assess the cost-benefits of risk reduction measures proposed as part of the conceptual design. So far, risk levels from the jetty and pipeline have been shown to be relatively low, and a number of worthwhile risk reduction features have been examined and assessed. As a result, the project is being developed in the knowledge that an appropriate level of expenditure will be allocated to ensure that the risks are as low as reasonably practicable.

ACKNOWLEDGEMENTS

The authors are grateful to the Management Team of Transco LNG Storage for permission to publish this paper.

The innovation and input from staff at Advantica Technology Ltd, Loughborough, in developing and applying the risk assessment framework is acknowledged.

REFERENCES

1. Health and Safety Executive, 1999, A Guide to the Control of Major Accident Hazards 1999, L111, HSE Books.
2. Health and Safety Executive, 1999, Preparing Safety Reports: Control of Major Accident Hazards 1999, HSG190, HSE Books.
3. Health and Safety Executive, 2001, Reducing Risk, Protecting People, HSE Books.
4. Health and Safety Executive, 1989, Quantified Risk Assessment: Its Input to Decision Making, HMSO.
5. Greenwood, B., Seeley, L., Spouge, J., Risk Criteria for Use in Quantitative Risk Analysis, International Conference and Workshop on Risk Analysis in Process Safety, 21 – 24/10/1997, Atlanta, USA.
6. HM Treasury, 1997 Appraisal and Evaluation in Central Government ("The Green Book"), 2ed, HMSO.
7. Cleaver, R.P., Dodson, M.G., 1998, Safety Assessments for LNG Storage Sites. IGRC 98, San Diego, USA.
8. Cleaver, R.P., Humphries, C.E., "Modelling, High Consequence, Low Probability Scenarios", Hazards XVII, March 2003, Manchester, UK.
9. Middleton, M., Franks, A., Using Risk Matrices, *The Chemical Engineer*, September 2001, 34–37.

USING THE MEASUREMENTS

John Saxton, B Sc. Tech., C.Eng. F.I.Chem.E The Saxton Consultancy
Ray Black, Safety Software Limited

Most organisations report accidents/incidents in some form, even if only to meet legal requirements. More enlightened ones include "Near Misses" in this process. Such reporting is often seen as a chore rather than an essential part of the loss control programme. This is particularly true at the level of the "reporters" – typically first line supervision. This paper outlines how an effective reporting system can turn this negative into a positive to become an essential tool for performance improvement. It concludes with a demonstration of a powerful web-based product suited to this application.

Accident reporting; Performance measurement, Accident Analysis Software.

TRADITIONAL REPORTING SYSTEMS

We are all familiar with the routine task of reporting unwanted events within our work area. At the earliest, basic level, this was filling in the "Accident Book" – which is still required of us, with serious, specified events having to be reported to the authorities. Injuries were entered in the book, together with the briefest detail of what happened and the names of anyone involved. Developments from this simple process followed two key routes, one was to collate the information to look for trends. The other was the realisation that not all losses were injuries, so "incidents" or hardware damage came to be included. From this grew the concept of "near misses" or, more accurately, potential losses, being a valuable learning tool to prevent further unwanted events. In parallel with this, the increasing compensation claim culture has driven organisations to record more and more detail related to the occurrence.

Clearly, early systems were paper based, allowing limited use of the data. This tended to be:

A monthly statistical report to management, typically including; number of Lost Time Accidents (LTA), Lost Time Accident Frequency (LTAF), graphs showing how the frequency was moving and time since the last LTA. A short description of the more serious events might be included. There may have been analysis by someone "responsible for safety" to look for trends in accident types, parts of the body injured, areas of the organisation having the accidents, etc.

The emphasis was very much on recording the circumstances surrounding an unwanted event, with minimum investigation, except for major accidents, and at best, identification of immediate causes. In well run organisations of current average size, too, they become meaningless, since there may well be less than 1 LTA/annum.

LATER DEVELOPMENTS

In the last quarter of the 20th century, two developments have come together to give us much more power to use unwanted event reporting to improve loss control performance. The first was the work done by the International Loss Control Institute and others on accident causation and hence the way we can use sound investigation to establish basic causes and thus prevent a recurrence.[1] The second has been the rapid changes in computer technology which allows gathering more data more easily, manipulating it more effectively and disseminating the results more quickly and widely.[2]

USING THE DATA

Any information gathered, however detailed and comprehensive is of no value if it is not used to improve loss control performance and hence that of the business. The objective of any reporting system should be to gather data and process it in a way that can strengthen the improvement programme.[3] Typically, it can be used to help in:

- Measuring performance
- Forming a basis for effective investigation and establishment of basic causes and hence areas of deteriorating control
- Highlighting trends
- Benchmarking
- Defining priorities
- Identifying risks

MEASURING PERFORMANCE

To achieve the objectives set out above, information needs to be translated into more creative performance indicators than the traditional ones, although these will still be necessary for some purposes. Some examples (which we have coined the expression "KUEPI's" (Key Unwanted Event Performance Indicators) are:

UNWANTED EVENT REPORTING FREQUENCY

It is now widely recognised that the difference between an injury, damage to equipment or a "near miss" is luck, hence the need to record all unwanted events. This is the list that will give us the maximum learning opportunity. If we then take the number of such reports and turn it into a frequency by dividing by a number of worked hours, we have a performance indicator that tells us something about the loss control culture of the organisation or given sections of it. A low number can demonstrate a lack of awareness by employees, a reluctance to report – perhaps a fear culture in place or a "so what" attitude to safety. We know that the unwanted events are happening, to learn from them we need to know what they are. A good starting target is 2 unwanted events reports/employee/annum.

AUDIT/INSPECTION REPORTS

There are many occasions in the working year when we examine the way we do things in the loss control area. This can range from the formal, external audit through safety team monthly inspections to the supervisor's daily job walk, yet we seldom turn these into proactive indicators of performance. The formal audit may give us a score to compare with the last one but beyond that, there is often little "measurement". There are many opportunities here:

- How many breaches of safety rules were observed/hour – just a tick for each one in a note book during the tour.

- How many unsafe conditions were observed/hour.
- How many examples of good practice were observed.

PRODUCTION LOSS OR DOWNTIME DUE TO UNWANTED EVENTS

Readily understood measures of success by everyone within an organisation are "How much have we produced?" or "How much downtime have we had?". To demonstrate how these values have been affected by unwanted events, therefore, can be a powerful tool. Again, generation of the indicator can be quite simple – an estimate of lost production or time on the report form which is then translated into the percentage of available output or on-stream time.

COST OF UNWANTED EVENTS

This is more difficult to establish than the production loss measure due to the range of costs which might be involved. There could be output loss, as above, repair costs, additional labour, investigation costs, compensation payments, increased insurance premiums and many more. Even if only estimates are made, rather than the actual costs, it is well worth the effort to demonstrate the impact on the business of failing to control our operations. Conversely, it is valuable to be able to show that investment in safety measures such as training have prevented or reduced a large loss. It is illuminating to show these figures as a percentage of turnover, operating costs or profit.

BASIS FOR INVESTIGATION

It has long been recognised that the best way of preventing recurrence of accidents is to establish the basic causes by effective investigation, then correct the deficiencies. It follows from this that **ALL** unwanted events should be investigated, the depth of examination reflecting the potential harm which could arise. A correctly designed reporting system is a vital first step to carrying out investigations in a way that not only gets to the right answer but keeps the process manageable. In addition to the typical description of conditions, witnesses names etc., therefore, questions should be included such as:

- Is there an up to date Risk Assessment for the task being carried out?
- Was a Permit To Work issued?
- Was the person trained/licensed to carry out the task?

The use of a **potential** severity rating system, typically a matrix one, is valuable here to guide on the depth of investigation required for each event.

HIGHLIGHTING TRENDS

We have already mentioned that some trends have been followed traditionally from the limited information available. The more powerful tools we now have allow us quicker analysis and hence more options to spot changes where correction can further improve performance. Examples are:

MOVEMENT IN "KUEPI'S"

Any of the chosen performance indicators can be turned into trend graphs, preferably on a rolling average basis to highlight improvement or deterioration in a particular measure

BASIC CAUSES

This is the most powerful driver for reporting and from the investigations described above, we can draw up lists of deficiencies within the management system that require correction.

MOTIVATION SHIFTS

This can be identified from movements in reporting frequencies, unwanted events occurring where the correct work method has been defined and training given etc.,

CHANGES IN EVENT SERIOUSNESS

Analysis of the potential severity ratings can quickly show whether current events are drifting to potentially more serious losses or improving as a result of the application of the loss control programme. This can be done by the simple recording of how many reports in a given category or applying a weighting process.

BENCHMARKING

It is a natural reaction not just to know whether your own unit is improving or deteriorating but if the right standards are being achieved and how you compare with others, both within and without the organisation. One of the difficulties in doing this traditionally has been the disparity between operating unit size and activity type. The KUEPI's described above go some way to overcoming this problem.

One of the additional benefits of Benchmarking is that if you can agree common performance indicators with others in your industry, trade association or geographical area, not only can you measure where you stand with regard to the outside world, you can trade experiences mutually to improve performance.

DEFINING PRIORITIES

It is self-evident that not all corrective actions can be put in place immediately and with increasing cost pressures, the need to prioritise these actions becomes ever more important. Meaningful performance measurements allow this when combined with a potential severity rating system for the harm which could arise from an unwanted event. Clearly, we could all deduce that preventing an unwanted event that could harm many people would take preference over one which might cause minor disruption to the operation but in most cases the options are not so obvious. It is important, therefore to have a disciplined, objective method of establishing priorities. Any method needs to extend beyond unwanted event reporting, since corrective actions arise for other sources such as inspections and task observations.

RISK ASSESSMENT

Past experience is a key part of identifying hazards and evaluating risks thus analysis of unwanted event reports should always form part of the risk assessment process. Hazards can be identified by the presence in a report of one not anticipated at the time of the initial assessment, such as noise from a nearby unit. Items cropping up repeatedly can indicate a greater incidence of risk than anticipated, a change in circumstances, people taking short cuts because procedures are inadequate or lack of understanding by those carrying out the task. All indicate a need for a re-assessment and give additional data to make that exercise more effective.

REQUIREMENTS OF A REPORTING AND MEASURING SYSTEM

The objective of any reporting or measuring system must be to gather data from unwanted events and subsequently manipulate that data to make a contribution to the performance improvement programme. This requires careful design and any system needs to be:

- A live system – out of date information not only reduces effectiveness, it reduces credibility
- Simple to use – those entering data and using it will be busy people with a whole range of other tasks to perform.
- Readily understood by everyone using it – people will not grapple with the incomprehensible!
- Transparent – inputs and outputs clear and available to all – no opportunity for being influenced by those concerned.
- Seen to be of value to everyone using it – if there is something in it for me and those with whom I work, I will use it!

EXAMPLE OF AN EFFECTIVE, FLEXIBLE SYSTEM

A number of high quality reporting and evaluation systems are in use, either being designed in-house or bought as commercial products. Each has its own particular strengths. The one with which we are closely associated is "AIRSWEB" and we will now demonstrate some of its features.

DESCRIPTION

The key features of this system are:

It is web-based - there is minimal need for hardware at the point of making the report – just a web browser to the intranet or internet.

All the screens have a banner at the top giving everyone with access to the system a summary of the Company's current position – constantly updated.

All the input screens have the same basic format – see Fig 1.

The data can be manipulated to give trend information, graphs etc., for routine reports – see Fig 2

Defined actions from the accident investigation are automatically built in to an "Action Tracking" module, which will also incorporate actions arising from other sources such as audits. – see Fig 3

A cost tracking facility is included which allows not just an estimation of the cost of an unwanted event but one of the savings which have resulted from a loss control related investment

The ability to report monthly on LTA's etc., is incorporated, as is a severity rating matrix, which can be used to give objective priority setting procedures.

There is the capability of attaching electronic files with support data, e.g. photographs, witness statements, etc.

REFERENCES

1. Bird, F.E., and Germain, G.L., 1986, Practical Loss Control Leadership, International Loss Control Institute.
2. Kibblewhite, J.,1998, Using personal Computers to Manage Safety, Industrial Relations Services
3. H.S.E., 1997, Successful Health and Safety Management (HSG 65) HSE Books

SAMPLES OF SCREENS

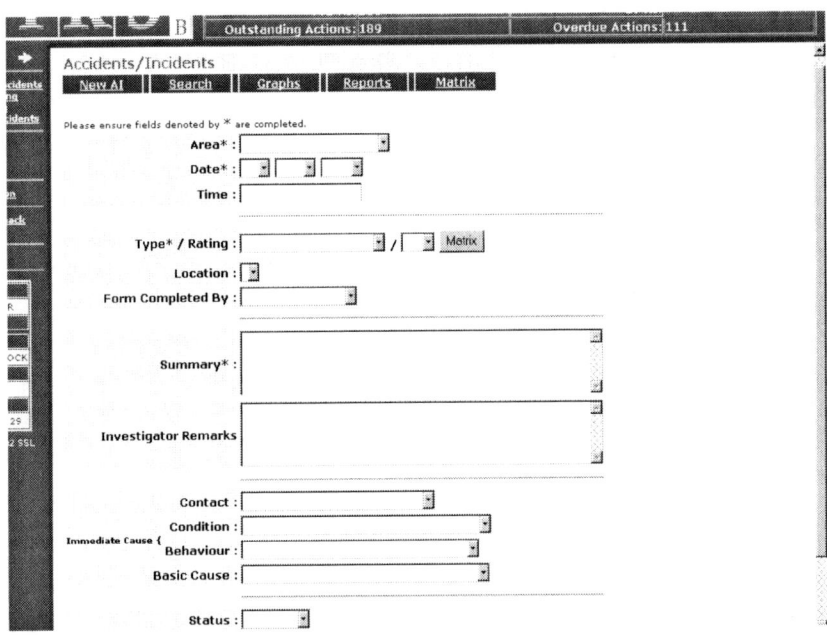

Figure 1. Unwanted event input screen

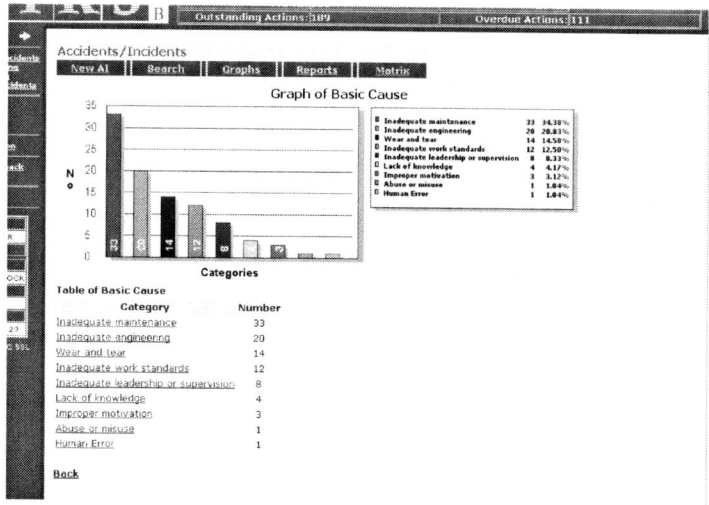

Figure 2. Basic cause analysis graph

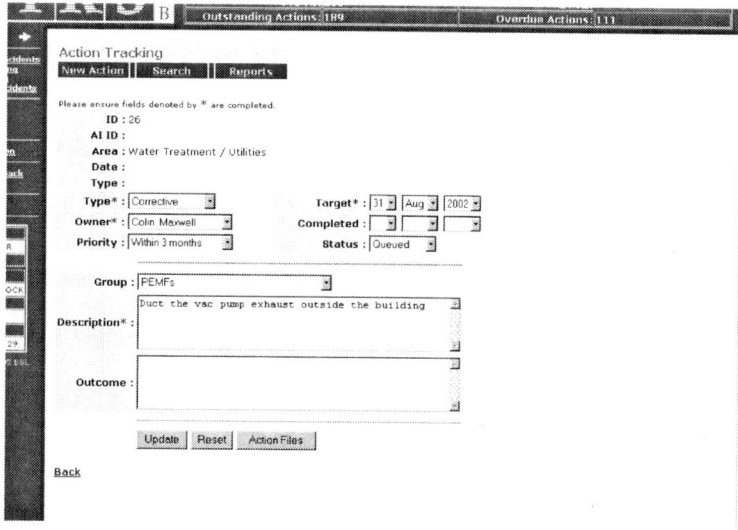

Figure 3. Typical action tracking screen

Index

S

V

W

Z